# Mathematical Modeling with Maple

**William P. Fox**

*Naval Postgraduate School*

BROOKS/COLE
CENGAGE Learning™

Australia • Brazil • Japan • Korea • Mexico • Singapore • Spain • United Kingdom • United States

**Mathematical Modeling with Maple**
William P. Fox

Editor in Chief: Michelle Julet

Publisher: Richard Stratton

Senior Sponsoring Editor: Molly Taylor

Assistant Editor: Shaylin Walsh

Editorial Assistant: Alexander Gontar

Associate Media Editor: Andrew Coppola

Senior Marketing Manager:
    Jennifer Pursley Jones

Marketing Coordinator: Michael Ledesma

Marketing Communications Manager:
    Mary Anne Payumo

Content Project Manager: Jill Clark

Art Director: Jill Ort

Senior Manufacturing Buyer: Diane Gibbons

Rights Acquisition Specialist, Image:
    Mandy Groszko

Production Service: MPS Limited,
    a Macmillan Company

Cover Designer: Harold Burch and Wing
    Ngan

Cover Image: Maplesoft

Compositor: MPS Limited, a Macmillan
    Company





For product information and technology assistance, contact us at
**Cengage Learning Customer & Sales Support, 1-800-354-9706**

For permission to use material from this text or product,
submit all requests online at **www.cengage.com/permissions**
Further permissions questions can be emailed to
**permissionrequest@cengage.com**

Library of Congress Control Number: 2010943534

ISBN-13: 978-0-495-10941-9

ISBN-10: 0-495-10941-X

**Brooks/Cole**
20 Channel Center Street
Boston, MA 02210
USA

Cengage Learning is a leading provider of customized learning solutions
with office locations around the globe, including Singapore, the United
Kingdom, Australia, Mexico, Brazil and Japan. Locate your local office at:
**international.cengage.com/region**

Cengage Learning products are represented in Canada by
Nelson Education, Ltd.

For your course and learning solutions, visit **www.cengage.com**

Purchase any of our products at your local college store or at our
preferred online store **www.cengagebrain.com**

**Instructors:** Please visit **login.cengage.com** and log in to access
instructor-specific resources.

Printed in the United States of America
1 2 3 4 5 6 7 15 14 13 12 11

# CONTENTS

## Chapter 4   Model Fitting   78



## Chapter 5   Modeling with Proportionality and Geometric Similarity   91



## Chapter 6   Empirical Model Construction   120

## Chapter 9    Models Using Unconstrained Optimization: Maximization and Minimization with Several Variables    237

# Chapter 10    Modeling Optimization with Constraints    275



# Chapter 11    Modeling with Linear Systems of Equations Using Linear Algebra Techniques    314



# Chapter 12    Modeling First-Order Ordinary Differential Equations (ODEs)    344

## Chapter 13  Modeling with Systems of Differential Equations   385

# PREFACE

The study of mathematical modeling is essential for anyone who desires to use applied mathematics to solve real-world problems. I present mathematical modeling topics using Maple as the computer algebra system for solving mathematical equations as well as obtaining plots that help us in our visualization and interpretation of the results for better analysis. I present cogent applications of applied mathematics, demonstrate an effective use of a computational tool (in this case, Maple) to assist in doing the mathematics, provide discussions of the results obtained using Maple, and stimulate thought and analysis of additional applications.

This book is as an introductory course to start to prepare students to apply the mathematical modeling process by formulating, building, solving, analyzing, and criticizing mathematical models. It is intended for a first course at the sophomore or junior level for applied mathematics or operations research majors that introduces these students to mathematical topics that they will revisit within their major. This book can also be used as a beginning graduate-level text for mathematics education majors because modeling has a bigger influence in secondary education with the newest National Council of Teachers of Mathematics (NCTM) standards. This text also serves to introduce many additional mathematics topics that students may study more in depth later in their majors.

Although calculus (either engineering or business) is the prerequisite material, many sections and chapters require multivariable calculus. In addition, the use of linear algebra is required in some chapters. For students without the necessary background, these chapters can be omitted when you design your specific course. I realize that here are more chapters in this text than could ever be covered within one semester. The increased number of topics and chapters provide flexibility for the coverage based on the background of your students.

## Goals and Orientation

This course bridges the study of mathematics topics and the applications of mathematics to various fields of mathematics, science, and engineering. This text affords the student an early opportunity to see how assumptions drive the models as well as an opportunity to put the mathematical modeling process together. The student investigates real-world problems from a variety of disciplines such as mathematics, operations research, engineering, computer science, business, management, biology, physics, and chemistry.

This text provides introductory material to the entire modeling process. Students will find themselves applying the process and enhancing their problem-solving capabilities to become competent, confident problem solvers for the 21st century.

Students are introduced to the following facets of modeling:

- *Creative and Empirical Modeling*. Students learn the modeling process by identifying the problem to be solved, making assumptions and collecting data, proposing a model (or building a model), testing their assumptions, refining the model as necessary, fitting the model to the data if appropriate, and analyzing the mathematical structure of the model to appraise the sensitivity of the results when the assumptions are not strictly met.

- *Model Analysis*. Given a model, students will learn to work backward to uncover the assumptions, access how well those assumptions fit the scenario, and estimate the sensitivity of the results when the assumptions are not strictly met.

- *Model Research*. The students investigate a specific area to gain understanding of behavior and to learn how to apply what has been already created or developed to other scenarios.

## Student Background and Course Content

I introduce modeling early. Chapters 1 through 7 can be taught without too much calculus background, but care should be taken in Chapter 4 not to discuss the development of the normal equations for least squares because they require multivariable calculus. Single-variable calculus (differentiation and integration) is a prerequisite for Chapters 8 and 12. Multivariable calculus is a prerequisite for Chapters 9 and 10. Linear algebra or matrix algebra (solving systems of equations with matrices) is a prerequisite for Chapters 11 and 13. Chapters 14 through 16 introduce probability and stochastic models, which requires some knowledge of basic probability and statistic concepts. Chapter 17 is an introduction to Game Theory based on linear programming.

## Organization of Text

Chapter 1 introduces Maple and its basic command structure. Because the book uses Maple as the tool in mathematical modeling, the chapter provides the foundation or cornerstone of using technology in the modeling process. Chapter 2 introduces modeling as a process. Scenarios are developed within the scope of the modeling process. Student thought is required. Chapter 3 begins modeling with discrete dynamical systems and includes analysis of the models and their long-term behavior from affine dynamical systems through systems of discrete dynamical systems. Chapter 4 introduces model fitting and emphasizes the use of the least-squares approach in model fitting. In addition, two other techniques are described. Chapter 5 provides more on our simplifying assumptions using proportionality arguments and the use of a powerful assumption, geometric similarity, in constructing proportionality arguments. Chapter 6 introduces empirical model building using various techniques such as simple one-term models, divided difference tables to explore low-order polynomials fits, and cubic spline.

Chapters 7 through 10 deal with optimization techniques. Chapter 7 covers linear programming formulations through sensitivity analysis. Chapter 8 reviews single-variable unconstrained optimization and applied models, and it covers numerical search methods. Chapter 9 covers multivariable unconstrained techniques, applied models, and numerical search methods. Chapter 10 covers constrained optimization methods: equality constrained with Lagrange multipliers and inequality constraints with Kuhn–Tucker conditions. Chapter 11 covers the use of linear algebra techniques to model and solve models such as the Leontief model. Chapters 12 and 13 address modeling with differential equations. Here a few techniques for solving closed-form differential equations are covered. Chapters 14 through 16 cover stochastic models. Chapter 14 introduces discrete probability and their models. Chapter 15 introduces continuous probability distributions and their model. Chapter 16 introduces Monte Carlo simulation. Chapter 17 introduces Game Theory, Nash equilibrium and Nash arbitration.

# Student Projects

The backbone of this course is the student project. In each project, students get to apply the mathematical modeling process and the mathematical tools they have learned. Each chapter and many sections have their own student projects. I have seen student growth in the project submission from their first project to the final submissions. Student projects take time to apply the modeling process, so I typically do not assign more than one project to a student. Most of these projects are designed to be student group projects (three or fewer student working together), although they can be done as individual projects.

# Technology

Technology is fundamental to serious mathematical modeling. Maple was chosen as our technology for this text, but any technology could be used—from graphing and symbolic calculators through spreadsheets.

### Emphasis on Numerical Approximations

Numerical solutions techniques are used in the dynamical systems, explicative modeling with some numerical analysis approaches, and in optimization search procedures. These are the methods most easily employed in iterative and recursive formulas. Early on, the student is exposed to numerical techniques. These numerical procedures are algorithmic and iterative.

### Focus on Algorithms

All algorithms are provided with step-by-step formats to aid students in learning to do mathematical modeling with these methods and Maple. Examples follow the summary to illustrate its use and application. Throughout I emphasize the process and interpretation, not the rote use of formula.

### Modeling and Applications

Each chapter includes examples of models, real-world applications, and problems and projects. Problems are modeled, formulated, and solved within the scenario of the application. These models and applications play an important role in student growth in working in today's complex world.

### Exercises

Many exercises are provided at the end of each section and chapter so the student can practice the solution techniques and work with the mathematical concepts discussed. Review problems are given at the end of each chapter, some of which combine elements from several chapter sections. Projects are also provided at the end of each chapter to enhance understanding of the concepts and their application to real-world-type problems.

### Computer Usage

Maple is the exclusive computer algebra system used for this text. Tutorial labs are developed for exposure to Maple's syntax and command structure. Emphasis is on providing the student with the ability to use Maple to assist with in mathematical modeling. We illustrate graphing in both two and three dimensions. Each modeling venue has an accompanying Maple worksheet.

### Additional Resources

*Solution Builder*
www.cengage.com/solutionbuilder
Available to adopters by signing up at the web address listed above, Solution Builder allows instructors to create customized, secure PDF printouts of solutions matched exactly to the exercises assigned for class.

*Companion Website*
www.cengage.com/math/fox
Additional online material for this text includes data sets and Maple worksheets to accompany examples and problems in the text.

# Acknowledgments

I need to begin by thanking Frank R. Giordano for being my mentor and choosing to involve me with mathematical modeling. I am indebted to my colleagues who taught with me over the years and who were involved with problem and project development: Jeff Appleget, David Cameron, Chris Fowler, Paul Grimm, Dan Hogan, Steve Horton, Mike Huber, Mike Jaye, Rickey Kolb, Gary Krahn, Steve Maddox, Jack Piccuito, Jack Pollin and Rich West. A special thank to William Hank Richardson, my colleague at Francis Marion University, who assisted in developing and refining many of our Maple programs used within this text.

I would also like to thank Roger Lipsett for his careful accuracy review of this text. I am also appreciative of the editorial and production professionals at Brooks/Cole Cengage. In particular, I'd like to thank Richard Stratton, Molly Taylor, Shaylin Walsh, Jill Clark, and Edward Dionne.

I want to especially acknowledge my wife, Diane Hamilton Dix Fox, whose encouragement, patience, motivation, and faith in my abilities inspired me to complete this text.

William P. Fox

1

# Introduction to Maple

Maple is a symbolic computation system or computer algebra system (CAS) that manipulates information in a symbolic or algebraic manner. You can use these symbolic capabilities to obtain exact, analytical solutions to many mathematical problems. It also provides numerical estimates when desired, or where exact solutions do not exist.

Maple 10 and higher is different from previous versions of Maple. With Maple 10, you can create professional-quality documents, presentations, and custom computational tools. You can access the power of the Maple computational engine through a variety of interfaces: standard worksheet, classical worksheet, command line version, graphing calculator, and Maplet applications. Although the commands are entered in a similar manner as in previous versions, Maple appears in a *document-type format* on the screen.

**Standard Worksheet** This is a full-featured graphical user interface that offers features that help create documents that show all assumptions, the calculations, and any margin of error in your results. You can even hide the computations to focus on problem setup and final results.

**Classic Worksheet** The basic worksheet environment works best for older computers with limited memory and is not available on the Mac OS X system.

**Command-Line Version** The command-line interface, without graphical-user-interface features, is used for solving large, complex problems or batch processing.

**Maplesoft Graphing Calculator** The graphical interface to the Maple computational engine allows you to perform simple computations and create customizable, zoomable graphs.

**Maplet Applications** The graphical user interface contains windows, textbox regions, and other visual features that give you point-and-click access to the power of Maple. It allows you to perform calculations and plot functions without using the worksheet or command-line interfaces.

Maple's extensive mathematical functionality is most easily accessed through all these interfaces. Previous older versions relied on its advanced worksheet-based graphical interface. A worksheet is a flexible document for exploring mathematical ideas or mathematical alternatives and even creating technical reports.

Experimental mathematical modeling, a natural stepping stone to statistical analysis, has an obvious coupling with computers, which can quickly solve equations and plot and display data to assist in model test and evaluation. The computer software algebra system Maple is a powerful tool to assist in this process. When dealing with real-world problems, data requirements can be immense. When evaluating immigration trends, for example, and the political, social, and economic effects of these trends, thousands of data points are used—in some cases, millions of data points. Such problems cannot be analyzed by hand, effectively or efficiently. The manipulation required to plot, curve fit, and statistically analyze with goodness-of-fit techniques cannot feasibly be done without the assistance of a computer software system.

The Maple system is easy to learn and can be applied in many mathematical applications. Maple is also an extremely powerful software package, with more than 5,000 built-in definitions and mathematical functions that cover every mathematical interest: calculus, differential equations, linear algebra, statistics, and group theory, to mention only a few. The statistical package reduces many standard time-consuming statistical questions into one-step solutions, including mean, median, percentile, kurtosis, moments, variance, standard deviation, and so forth. There are many references for Maple, and a short list would include:

- Maple Quick Reference,
- Maple Flight Manual,
- Maple Language reference manual,
- Maple Library reference manual,
- Maple Release 13 and higher release Notes, and
- Maple Release 13 and higher Getting Started.

This chapter presents a quick review of some basic Maple commands. It is not intended to be a self-contained tutorial, but it will provide a quick review of the basics before more sophisticated commands are discussed in the modeling chapters. There are many good references for those new to Maple. In addition, there is a self-contained tutorial on the companion Web site.

 ## THE STRUCTURE OF MAPLE

As noted, Maple is an example of a CAS. It is composed of thousands of commands to execute operations in algebra, calculus, differential equations, discrete mathematics, geometry, linear algebra, numerical analysis, linear programming, statistics, and graphing. It has been logically designed to minimize storage allocation while remaining user friendly. Maple allows a user to solve and evaluate complicated equations and calculations, analytically or numerically, such as optimization problems, least-square solutions to equations, and equations that involve special functions.

### Notation and Conventions

Throughout this book, different types of fonts and styles are used to distinguish between Maple commands, Maple output, and other information. Maple commands are copied directly from Maple 10, and the output will immediately follow the Maple commands. In the first example in Table 1.1, the variable $a$ has been assigned the value $2x^3 + \frac{5}{6}x$. Notice that the symbol $:=$ is used to indicate this assignment. In the second example, the value of $a$, or $2x^3 + \frac{5}{6}x$, is differentiated with respect to $x$.

If you are in the Worksheet or document mode and type the command as in $a:=2*x\textasciicircum3+5*x/6$, you will type it this way, but the screen version appears as follows:

$a:=2*x^3+\dfrac{5}{6}*x;$

$$a := 2x^3 + \frac{5}{6}x$$

**Table 1.1** Maple Command examples in the classical worksheet

| Command | Output |
| --- | --- |
| $a:=2*x\textasciicircum3+5*x/6;$ | $a: = 2x^3 + 5/6x$ |
| $diff(a,x);$ | $6x^2 + 5/6$ |

Again, be aware that the assignment of the variable $a$ is typed the same in both cases; they just appear differently.

This chapter will describe Maple's user-interface features. A brief review of standard Windows terms will be conducted before a description of the Maple software. "Clicking" refers to a single click of the button on the computer mouse when the cursor is on the item of interest on the computer screen. A "double-click" simply means clicking twice quickly on top of the item of interest.

## 1.2   A GENERAL INTRODUCTION TO MAPLE

Maple has five types of windows: Worksheet, Help, 2-D, 3-D, and Animation. In this book, the Worksheet, Help, and 2-D windows are used most often. We provide a brief explanation of each.

The Worksheet is where all interaction between Maple and the user occurs. Within a Worksheet, commands (input) and text (remarks for clarification) are entered by the user, and numerical or symbolic results are produced (e.g., output and graphics). The user may manipulate these interactions to create a flowing document that can be saved by clicking File and then Save As followed by the document name. The name of the document can be any word or group of combined letters or numbers. Once a document has initially been saved, it can be retrieved by clicking File and Open and entering the document name. Then it can be resaved after modification by clicking File and then Save.

The input and text regions of the worksheet can be modified to change a document, but the graphic and output regions cannot be modified once Maple inserts them into a document. The input region is identified by the > prompt, which proceeds all command entries in Maple. (*Note:* Maple only recognizes the Maple-generated > prompt. If the > symbol is typed by the user, Maple does not respond to it as an input prompt but as the "greater than" relation.) The input commands, output characters, and text regions are all of different font sizes and colors to assist the user in distinguishing between them. Text regions assist in documentation and explanation of the input/output regions of the document, and they may be placed anywhere in a document. Graphic regions, once they are generated by input commands, can be copied and pasted into a worksheet or into another document. Once the graphic is pasted into the worksheet, it can no longer be edited or manipulated. The output regions are generated by the user's input commands and cannot be manipulated once they appear in a document, although the user is allowed to delete these results.

The Maple menu bar is located at the top of the screen and immediately below the Maple title. The menu bar provides easy access and *collocation* (i.e., a sequence of words or terms that occur together more often than would be expected by chance) of many commonly used options. The menu bar has File, Edit, View, Insert, Format, Options, Window, and Help choices much like any Windows software. Immediately below the Maple menu bar is the Maple toolbar (see Figure 1.1). The toolbar provides accelerated access to the most commonly used options.

Maple syntax requires either a colon or a semicolon after every statement. Maple will not process any command without one of these punctuation symbols. Multiple statements may be on the same line, but each must have its own colon or semicolon. The colon suppresses output of the command, whereas the semicolon signals that the results are to be printed to the screen immediately after the Enter key for the command is pressed. A large set of commands is either readily available in Maple memory or stored separately in Maple packages, which assist in more efficient memory storage. Standard commands such as addition and multiplication are not contained in packages. When using commands stored in packages (such as graph-plotting commands), the command *with(package name):* is issued before any of the commands is used in that package. The command is required only once. Some of the Maple packages will be used in this manuscript. Specifically packages such as *with(plots):* will be used for all plotting commands, *with(linalg):* or *with(LinearAlgebra):* for all linear algebra commands and *with(stats):* for all linear regression commands.

FIGURE 1.1
Maple toolbar

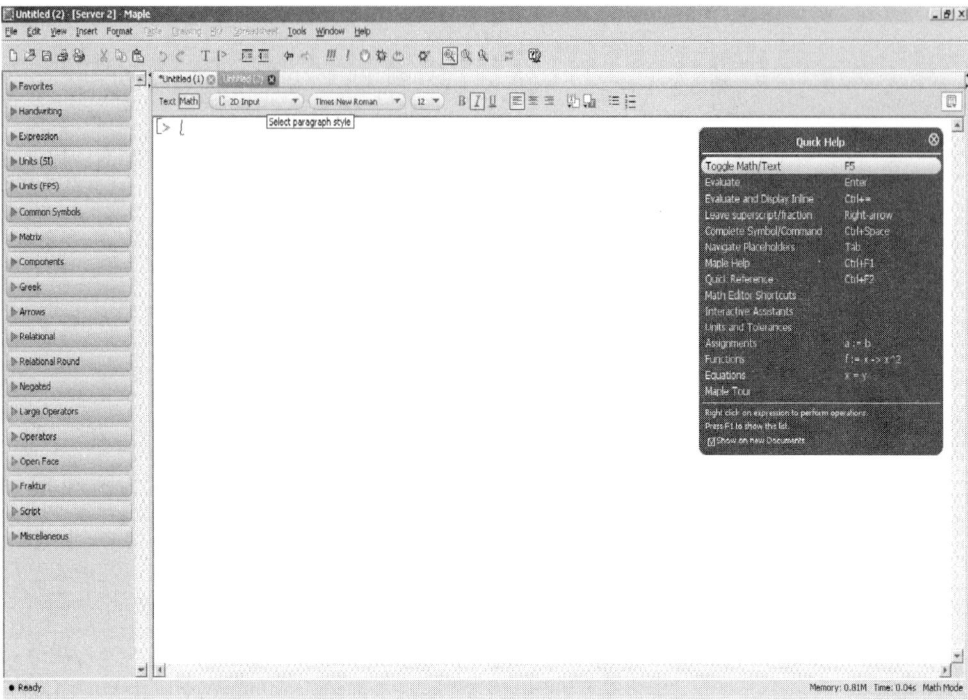

## The Help Command

Maple's Help database can provide all information found in the Maple Library Reference Manual. However, Help can be found immediately to assist the user in solving problems without leaving the document: by clicking on Help in the menu bar, typing "help" at the > prompt, or typing "?" at the > prompt (see Figure 1.2). When using "?" or "help" at the > prompt, the

FIGURE 1.2
Maple Help bar

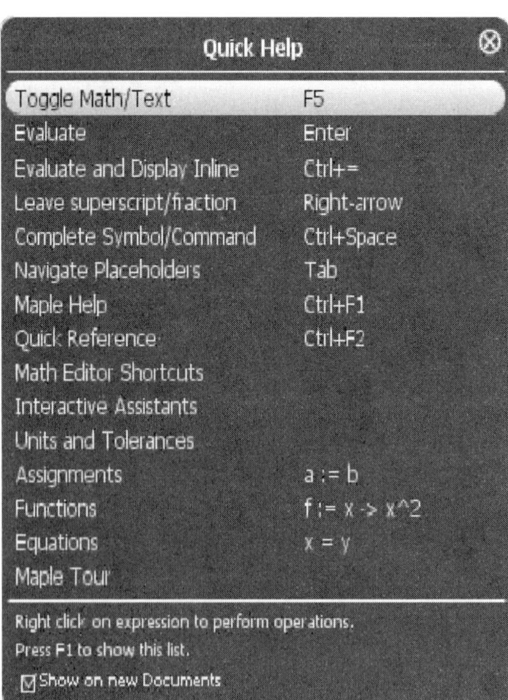

**Table 1.2** Entering a data list

| Command | Output |
| --- | --- |
| *ages:=[18,20,25,19,19];* | ages := [18,20,25,19,19] |

user must type the keyword for the Help search. If a specific syntax command is in question (for example, the user desires to learn Maple's syntax for differentiation), the user need only type "?differentiate" at the > prompt. This procedure is perhaps the most convenient one for help on syntax. The Help on the menu bar gives the user more options.

## Data Entry

Commonly, a set of data will describe a process. Entering the data is the first step to analyzing the data. Maple provides a convenient method for manually entering data via a list. A list is a group of data to which Maple's many operations are applied. Suppose that five college students' ages are known: 18, 20, 25, 19, and 19. Table 1.2 demonstrates the command required to enter the data into Maple.

*ages:=[18,20,25,19,19];*

$$ages := [18, 20, 25, 19, 19]$$

The use of brackets in the command indicates a list. The brackets will maintain an initial ordering of the data and allow for duplicate values, and they can be manipulated by the methods described later in this book. Commas are required between elements, and blanks, periods, or any other separators may not be substituted for commas. Any alphanumeric may be included in a list, including integers and rational numbers. If any datum should be missing, a placeholder can be entered in its place, such as the letter x. For example, if there was a sixth student in the group but the age was unknown, the symbol could be used to represent the sixth student's age. Table 1.3 presents the required command.

## Recovering Data from a File

Typically, data collection is done by an outside agency, for examination by an analyst. When a survey is conducted on the state or national level, the number of data points is typically in the thousands, which would lead to tremendous time loss if manual input were required. To alleviate this problem, Maple has two separate commands to recover data from a file: *importdata* and *readdata*.

*Importdata* is part of Maple's statistical package and requires the *with(stats)*: command. To use *importdata*, the file name and the number of columns or data sets are required; the default is one data set. Missing data in the file will be converted to the keyword *missing*. Given a data file named "data1" that contains two sets of data {2, 4, 6} and {3, 5} with the third element in the second set missing, Table 1.4 presents the Maple command and output.

*Readdata* is a more flexible command. The file name, number of columns, and type of data are required to use the command. The type of data indicates either integer or float, with float being the default. *Readdata* requires a separate command prior to use (similar to the *with( ):* command); the command is *readlib(readdata):*, which is required only once. An example of this

**Table 1.3** Entering data with a missing data point

| Command | Output |
| --- | --- |
| *ages:=[18,20,25,19,19,x];* | ages := [18,20,25,19,19,x] |

**Table 1.4** Reading data from a file using *importdata*

| Command | Output |
| --- | --- |
| *newdata:=importdata ('data1',2);* | newdata := [2.0,4.0,6.0], [3.0,5.0,'missing'] |

**Table 1.5** Reading data from a file using *readdata*

| Command | Action |
| --- | --- |
| *readdata(data2,3);* | read 3 columns of floats for the file "data" |
| *readdata(data2,integer,3);* | read 3 columns of integers |
| *readdata(data2,integer);* | read first column of integers |
| *readdata(data2);* | read first column of floats |

command follows with a data file named "data2" containing 12 data points in three columns. Table 1.5 presents four separate commands that describe the possible importation of the data file using the *readdata* command. The file "data2" is composed of three columns of data, each including four data points each, described in three sets: {1, 11, 15, 17}, {3, 22, 5, 78}, and {5, 55, 55, 70}, as shown in Figure 1.3.

### Data Entry and Verification

Figure 1.3 illustrates entering, verifying, and naming data for the length and weight of bass caught during a fishing derby. In subsequent chapters, several models for predicting the weight of a bass as a function of some readily measurable dimension will be suggested. In Figure 1.3, the data is entered, named, assigned to an array, and printed to verify correct entry.

### Correcting Erroneous Entries

In the above illustration, Maple displayed the bass data immediately after its input. This was done to help the user verify that all elements were entered correctly. Reentering each command is typically the best method for correcting an erroneous entry with many errors. However, other commands can expedite smaller-scale error correction, including the use of the following commands: *op* and *subsop*. This section demonstrates that by using these two commands on a defined data set, data can be inserted, deleted, and replaced with ease. Once the data set has been entered, a data point may have been left out or forgotten. Appending an element to the *wt* data set from Figure 1.3 is done by the command, *[op(wt),X]*, where *X* is the new element.

**FIGURE 1.3**
Data entry and verification

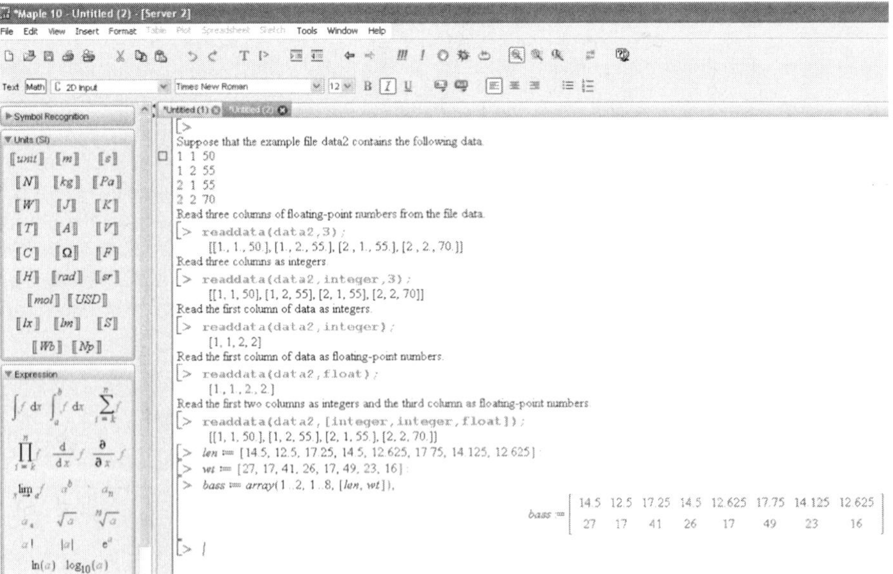

**Table 1.6** Selecting a range from a data set

| Command | Output |
| --- | --- |
| *newweight:=op(2..5,wt);* | newweight := 17, 41, 26, 17 |
| *newweight:=wt[2..5];* | newweight := [ 17, 41, 26, 17 ] |

**Table 1.7** Replacing data in a data set

| Command | Output |
| --- | --- |
| *newwt:=subsop (2=15,wt);* | newwt := [27, 15, 41, 26, 17, 49, 23, 16] |
| *newwt:=subsop (7=NULL,wt);* | newwt := [27, 17, 41, 26, 17, 49, 16] |

There is a difference between the two commands: *op( )* produces a list separated by commas, and *wt[a..b]* produces a list from element #a to element #b. Used in a slightly different fashion, *op* can also select a portion of a data set as a list. This is done by *op(j..k,wt)*, in which *wt* is the data set and *j* and *k* are row values in the data set $j < k$. As a result, the command selects a range from the *wt* data set. Table 1.6 presents these two methods of selecting the range.

*Subsop* has two main purposes: (1) to replace a data point with a new value and (2) to delete an element from a data set. Table 1.7 demonstrates an example of each using the *wt* data set.

**Transformation and Functions**

As described, the symbol := is used to assign the value on the right-hand side of the statement to the name on the left-hand side of the statement. An example, *x:=3*; assigns the value 3 to *x*; to unassign *x*, use the command *x:='x'*. Functions can be defined using the mapping arrow symbol →. The function can be evaluated numerically or symbolically; Table 1.8 provides a short example.

A mathematical model suggests an existing relationship among the selected variables. In the two-dimensional case, the model suggests a functional relationship between a dependent and an independent variable. Given values of an independent variable, a function can transform the given data to yield predicted values for the dependent variable. Transforming data requires the understanding of algebraic operations and functions that are used in Maple. Table 1.9 presents the regular algebraic operations that Maple recognizes.

Many other functions are also recognized by Maple. Table 1.10 is an abbreviated listing.

**Table 1.8** The mapping command

| Command | Output |
| --- | --- |
| *f:=x->x^2+2;* | f := x—>x$^2$ + 2 |
| *f (5);* | 27 |
| *f (x-1);* | (x − 1)$^2$ + 2 |

**Table 1.9** Maple algebraic functions

| Symbol | Operation |
| --- | --- |
| + | Addition |
| - | Subtraction |
| * | Multiplication |
| / | Division |
| ^ | Exponents |

**Table 1.10** Other Maple functions

| Command | Operation |
| --- | --- |
| *abs* | Absolute value of real or complex argument |
| *argument* | Argument of a complex number |
| *ceil(x)* | Smallest integer $\geq x$ |
| *conjugate* | Conjugate of a complex number |
| *exp* | The exponential function: exp(x) = sum(x^i/i!,i-0..∞) |
| *factorial* | The factorial function, factorial(*n*) = *n*! |
| *floor* | Floor(*x*) = greatest integer $\leq x$ |
| *ln* | Natural logarithm (with base E = 2.718 . . .) |
| *log* | Logarithm to arbitrary base |
| *log10* | Log to base 10 |
| *max,min* | Maximum and minimum of a list of real values |
| *RootOf* | Function for expressing roots of algebraic expression |

Some other functions that are available will be discussed in a later chapter when considering statistical operations that may be performed on columns of data: sums, mean, standard deviation, and so forth. Table 1.11 presents a few examples of data transformations with the correlating Maple commands.

### Column Operations

This section presents the operations that can be performed directly on worksheet columns. Among the available operations are *matadd*, *evalm*, and *scalarmul*. These operations all require the *with( LinearAlgebra)*: command prior to use. Table 1.12 demonstrates these commands with six examples using two columns of data: c1:=[1,2,3] and c2:=[4,5,6].

### Arrays and Matrices

Arrays and matrices are structured devices used to store and manipulate data. An array is a specialization of a table; a matrix is a two-dimensional array. Both *array* and *matrix* are part of the linear algebra package and require the *with( LinearAlgebra)*: command prior to use. Table 1.13 presents a few examples of the use of the *array* and *matrix* commands.

To observe multiple columns of data in a chart an *array* or *matrix* command may be used. Table 1.14 presents an example with output.

### Saving and Printing a Worksheet

To start Maple from Windows, enter Maple by double-clicking on the Maple icon. Once Maple has been started, it will automatically open an empty worksheet, with a flashing cursor to the right of a character prompt >. To save the worksheet, click on File and then on Save As and

**Table 1.11** Examples of Maple commands in the classic worksheet

| Expression | Maple Command |
| --- | --- |
| $x^2$ | x^2 |
| $2x^2 + 2.5x + 9$ | 2*x^2+2.5*x+9 |
| $(x^2 + 2)^{0.5}$ | (x^2+2) ^ (.5) |

**Table 1.12** Column operations

| Operation | Command under *with linalg* | Command under *with* (*Linear Algebra*)<br>with c1:=<<1>\|<2>\|<3>><br>and c2:=<<3>\|<4>\|<5>> | Output |
| --- | --- | --- | --- |
| Summing columns: c1 + c2 | *c3:= matadd( c1,c2,1,1 );* | c3:=MatrixAdd(c1,c2); | c3:=[ 5  7  9 ] |
| Summing a constant into a column | *c4:=evalm( 3+c2 );* | c4:=MatrixAdd(Constant (1,1,3),c2,3); | c4:=[ 7  8  9 ] |
| Multiplying a constant into a column | *c5:=scalarmul( c1,3 );* | c5:=Multiply(c1,3,inplace); | c5:=[ 3  6  9 ] |
| Adding multiples of columns | *c6:=matadd( c1,c2,4,2 );* | c6:=Add(c1,c2,4,2); | c6:=[ 12  18  24 ] |
| Using a function on a column | *c7:=map( x→ln( x ),c1 ):evalf(' );* | c7:=evalf(map(x→ln(x),c1)); | c7:=[ 0 .693 1.099 ] |
| Taking each value in a column to a power | *c8:=map( x→x^2,c1 );* | c8:=evalf(map(x→x^2,c1)); | c8:=[ 1. 4.  9. ] |

**Table 1.13** Array and matrix commands

| Command | Output |
| --- | --- |
| *with( LinearAlgebra):*<br>*a:=array( [1,2,2,3,4] );* | a := [ 1 2 2 3 4 ] |
| *b:=array( [[1,2],[3,5]] );* | $b := \begin{bmatrix} 1 & 2 \\ 3 & 5 \end{bmatrix}$ |
| *c:=array( 1..5,[9,3,1,8,3] );* | c := [ 9 3 1 8 3 ] |
| *d:=<<7,2>\|<3,3>>; or d:=matrix( [[7,3],[2,3]] );* | $d := \begin{bmatrix} 7 & 3 \\ 2 & 3 \end{bmatrix}$ |

**Table 1.14** Charts using matrices

| Command | Output |
| --- | --- |
| *a:=array([1,2,2,3,4]):* | no output |
| *c:=array(1..5,[9,3,1,8,3]):* | no output |
| *with(LinearAlgebra):* | no output |
| *achart:=matrix(2,5,[a,c]);* | $achart := \begin{bmatrix} 1 & 2 & 2 & 3 & 4 \\ 9 & 3 & 1 & 8 & 3 \end{bmatrix}$ |

specify a name for the worksheet. Once the worksheet has been named and saved, click on File and then on Save to resave the worksheet. To open a previously saved worksheet, click on File and then on Open and specify the name of the worksheet.

After reopening a document, the document will contain only the commands, not the initial results. This is because of Maple's kernel, or internal state. After a command is executed, the result is stored in memory (the kernel). If the document is closed and then reopened, Maple recovers the commands but does not recall the results, unless the kernel itself is saved. Saving the kernel is done by selecting "Save Kernel State" under the Options menu followed by saving the document. This saves time if you are reusing the same document time and time again, but it does take up a lot of memory of the computer or disk space.

After completion of our work in Maple, the document can be saved and then printed by clicking File and Print (which calls up the print window) and then clicking again on Print.

To exit Maple, click on File and the click on Exit, or type "quit," "done," or "stop" at the Maple prompt. *Note:* There is no opportunity to save your worksheet using one of these latter three commands, but saving your worksheet file is automatic by simply clicking on Exit.

## 1.3   MAPLE QUICK REVIEW

### Assignments and Basic Mathematics

For example, let's compute $(1.2^{4.1} + 4.3(9.8))/34$.

> *(1.2^4.1+4.3*(9.8))/34;*

$$1.301522146$$

Again, note that every Maple command ends in a semicolon. All Maple statements are entered after the > prompt. Maple output is centered in the page.

To assign a label to a number or an expression, we use :=. For example, let's assign two Maple statements: *a* to 11.5 and *b* to 9.

> *a:=11.5;*
>

$$a := 11.5$$

> *b:=9;*

$$b := 9$$

> *a:=11.5;b:=9;*

$$a := 11.5$$
$$b := 9$$

We can use different command lines or the same command line to enter multiple commands.

If we separated the two commands with a colon (:) instead of the semicolon, only the second command would be displayed in the output, even though both commands were executed.

> *a:=11.5:b:=9;*

$$b := 9$$

Once the assignments have been made, we can perform arithmetic operations. For example, let's compute $a^2 + b^3$, $a^2b^3$, and $\sqrt{a^2 + b^3}$.

> *a^2+b^3;a^2*b^3;sqrt(a^2+b^3);*

$$861.25$$
$$96410.25$$
$$29.34706118$$

A very useful command is *evalf*. This command produces the decimal equivalent of a given expression.

> *c:=evalf(a^2/b^3);*

$$c := 0.1814128944$$

For functions, we assign *f* using the same assignment sequence.

> *f:=x^2+21.6*x-1;*

$$f := x^2 + 21.6x - 1$$

To evaluate this type of functional expression, we use the *sub* command (substitution). For example, we want to substitute $x=3$ for *x*.

> *subs(x=3,f);*

$$72.8$$

To use functional notation, such as f(3), we must start with a different assignment. To create a function assignment $f(x)$, we use the operator arrow as follows:

> *f:=x->x^2+21.6*x-1;*

$$f := x \to x^2 + 21.6x - 1$$

To obtain $f(3)$ we type $f(3);$.

> *f(3);*

$$72.8$$

We can even substitute variables such as $(x+h)$ for *x*:

> *f(x+h);*

$$(x + h)^2 + 21.6\,x + 21.6h - 1$$

These two forms of expressions are important in both programming and plotting functions as we shall see later.

Changing between these forms is easy using the *unapply* command.

> *f:=x^3+3*cos(x)-4;f:=unapply(f,x);f(x);*

$$f := x^3 + 3\cos(x) - 4$$
$$f := x \to x^3 + 3\cos(x) - 4$$
$$x^3 + 3\cos(x) - 4$$

Maple can easily handle functions of more than one variable. For example, consider a surface area defined by $\pi(x^2 y^3 + 3)$; we want to evaluate the surface area at the point (2,5).

> s:=Pi*(x^2*y^3+3);

$$s := \pi(x^2 y^3 + 3)$$

> s1:=(x,y)->Pi*(x^2*y^3+3);

$$s1 := (x,y) \rightarrow \pi(x^2 y^3 + 3)$$

> s1(2,5);

$$503\,\pi$$

> evalf(%);

$$1580.221105$$

The expression evalf(%) contains the symbol %, which means "insert the last expression typed."

## Algebra and Calculus

Let's return to our expression for >f:=x^2+21.6*x-1;

$$f := x^2 + 21.6\,x - 1$$

Let's factor $f$. We can use the factor command or the solve command. The solve command has more utility.

> factor(f);

$$(x + 21.64619749)\,(x - 0.04619749036)$$

> solve(f,x);

$$0.04619749036, -21.64619749$$

Let's consider the function $ax^2 + bx + c$. We use the solve command and obtain the result:

> restart;
> f:=a*x^2+b*x+c;

$$f := ax^2 + bx + c$$

> solve(f,x);

$$-\frac{b - \sqrt{b^2 - 4ac}}{2a}, \; -\frac{b + \sqrt{b^2 - 4ac}}{2a}$$

Note that the results are the quadratic formula. We also used the command "restart." Restart told Maple to forget all previous assignments to $f$, $a$, $b$, and $c$.

In calculus, we can differentiate and integrate in one and many variables. The commands for differentiation and integration are:

**diff or Diff**—differentiation or partial differentiation

and

**int**—definite and indefinite integration

For example, let's differentiate the expression $f = 2x^2 + 24.1\,x - 1$ and then find the area under the curve (in general and from $x = 1$ to $x = 4$).

> $f := 2*x^2+24.1*x-1;$

$$f := 2x^2 + 24.1x - 1$$

> $diff(f,x);$

$$4x + 24.1$$

> $Diff(f,x);$

$$\frac{d}{dx}(2x^2 + 24.1x - 1)$$

> $Int(f,x);$

$$\int 2x^2 + 24.1x - 1\ dx$$

> $int(f,x);$

$$0.6666666667x^3 + 12.05000000x^2 - 1.x$$

> $int(f,x=1..4);$

$$219.7500000$$

Often we want to find critical points of the first derivative [where $f'(x) = 0$]. We can use the solve command as follows:

> $solve(diff(f,x)=0,x);$

$$-6.025000000$$

### Plotting and Graphs

Maple has an extremely detailed and developed plot command, which provides plots in both two and three dimensions. For these modeling purposes, 2-D plots will typically be used. The plot function is called by $plot(f, hr, vr, options)$, where $f$ is the function to be plotted, $hr$ is the horizontal range and $vr$ is the vertical range. In addition, many other options can be added after the vertical range to control a variety of options. Table 1.15 presents a short list of these options (defaults are in bold font).

### Continuous Plots

The plot command, $plot(f):$, will generate a 2-D plot of $f$ with a default range of $-10$ to $10$ on the $x$-axis. No other command information is required; however, the axes ranges and options may

**Table 1.15** Plot command options

| Option | Description |
| --- | --- |
| scaling = | constrained or **unconstrained** |
| style = | point, **line**, patch or patchnogrid |
| title = 'title' | **no title** |
| thickness = | **0**, 1, 2, or 3 |
| axes = | framed, boxed, **normal** or none |
| view = | [xmin..xmax,ymin..ymax, **entire curve**] |

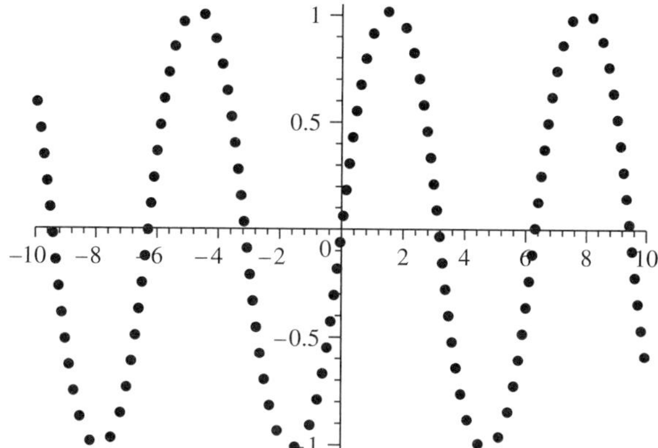



**FIGURE 1.4**

A sine plot with the default range

be specified to generate a specific plot. The default range can be specified as a finite range or an infinite range. By constraining the scale, equal units occur in both the $x$ and $y$ direction. However, a plot is generally easier to see when the scale is constrained, although it would be distorted (e.g., a circle would appear as an ellipse). Maple automatically scales the axes to spread the data over as large a space as possible, but this procedure does not imply that the area of interest will be plotted most effectively. As a result, the view option must be employed to ensure that the correct portion of the plot is best displayed. To demonstrate the plot command with a variety of options, a few examples are provided in Figures 1.4, 1.5, and 1.6. Note the distortion in both Figures 1.4 and 1.5.

**Scatterplots**

The previous plots demonstrate functions with continuous x-values; either defaulted to $-10$ to 10 or selected by the user. However, Maple can also plot discontinuous sets of data. Using the data provided, Figure 1.7 presents an example of plotting the ages of five people versus respective weights.

| Age (years) | 1 | 5 | 13 | 17 | 24 |
|---|---|---|---|---|---|
| Weight (lbs) | 15 | 40 | 90 | 160 | 180 |

Age–Weight Data

```
> age:=[1,5,13,17,24]:
> wt:=[15,40,90,160,180]:
> agwt:={seq([age[I],wt[I]],I=1..5)}:
plot(agewt, style=point, symbol, diamond, thickness=2);
```

**FIGURE 1.5**

A sine plot with an infinite range

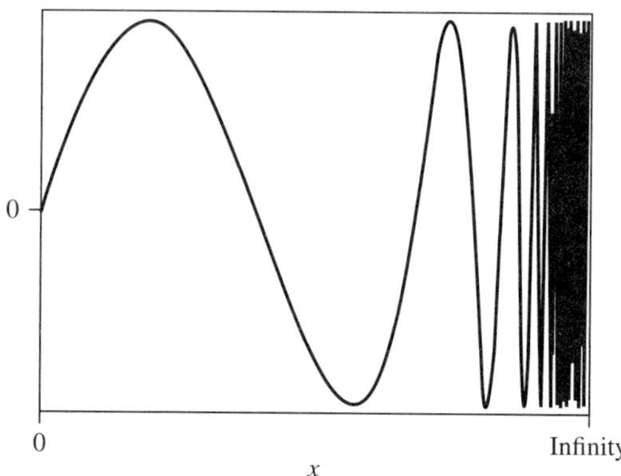

**FIGURE 1.6**

A sine plot with a constrained scale

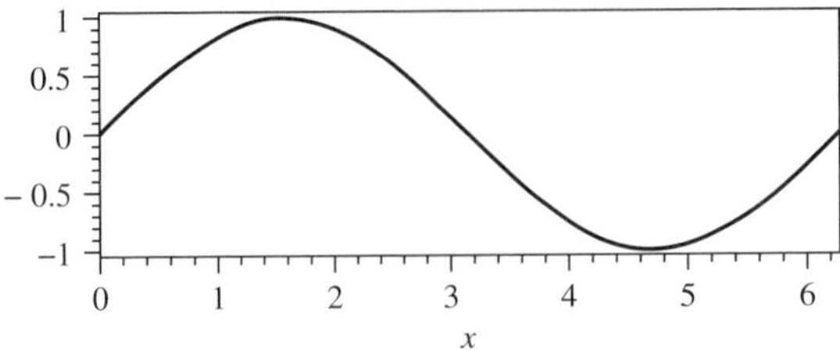

**FIGURE 1.7**

Age versus weight data plot

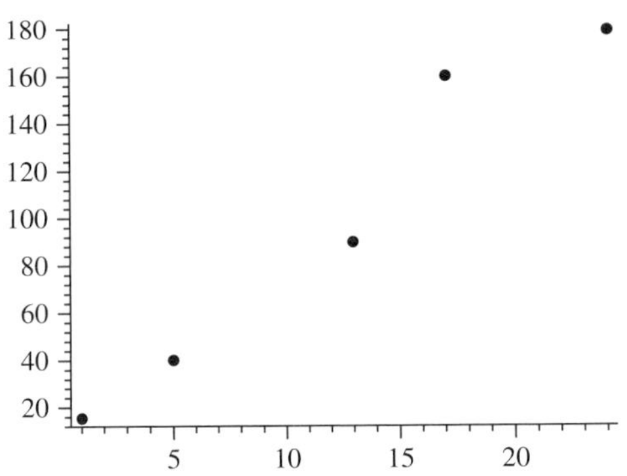

```
> with(plots):
> age:=[1,5,13,17,24]:
> wt:=[15,40,90,160,180]:
> agewt:={seq([age[i],wt[i]],i=1..5)};
```

$$agewt := \{[1,15], [5,40], [13,90], [17,160], [24,180]\}$$

```
> pointplot({seq([age[i],wt[i]],i=1..5)},
style=point,symbol=diamond);
```

### Multiple Plots

Multiple functions can also be plotted on one set of axes. One method requires that both functions have the same domain as presented in Figure 1.8. The second method uses the display command that requires that the *with(plots):* command be called once. This method does not restrict the domain (see Figure 1.9).

**FIGURE 1.8**

Multiple functions on one plot

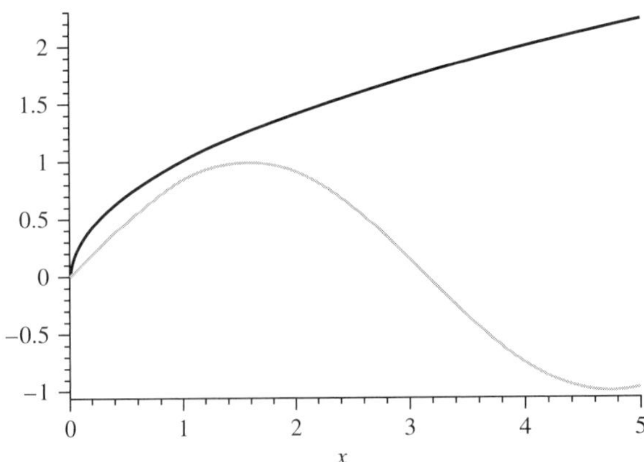

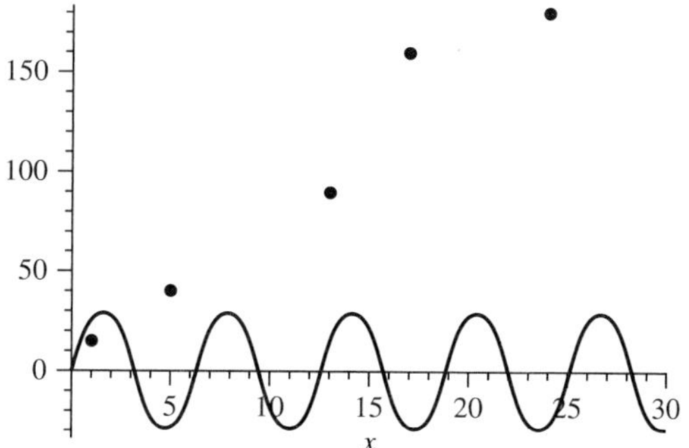



**FIGURE 1.9**
Multiple functions on one plot with different domains

> *plot({sin(x),x^(−.5)},x=0..5,scaling=constrained,axes=framed,thickness=2);*
> *with(plots):*
> *curve:=plot(30*sin(x),x=0..30):*
> *points:=pointplot({seq([age[i],wt[i]],i=1..5)},style=point,symbol=diamond):*
> *display({curve,points});*

## 1.4  MAPLE TRAINING

A Maple training of basic features is provided with this text. This training is a self-contained tutorial. These features include the following:

- numeric computations
- symbolic computations
- programming basic Maple procedures
- visualization
- animation
- calculus
- linear algebra
- student package
- other Maple packages

This training serves as a basic refresher of Maple commands.

## 1.1  EXERCISES

Perform the following operations in Maple.

1. $\sqrt{2.3^3\,(4.5)}$

2. $11.3^3 + 5.1^2$

3. $\sqrt[3]{21.6}$

4. $a = 8,\ b = 7$
   $(a^2 - b^2)$

5. $2(11.5) + 6.2^2\,(.7)$

Enter the following functions. Obtain a graph. Find roots and solve for the intercepts.

**6.** $f(x) = -x^2 + 3x + 3$

**7.** $f(x) = x^2 - 3x - 1$

**8.** $f(x) = -0.1213x^4 + 3.462x^3 - 29.22x^2 + 64.68x + 97.69$

**9.** $g(x) = x^3 - 2x^2 - 5x + 6$

**10.** $s(x) = 2x^3 - 3x^2 - 11x + 7$

Enter the following date sets into Maple and obtain a scatter plot:

**11.**

| $x$ | 1 | 3 | 8 | 10 |
|---|---|---|---|---|
| $y$ | .7 | 5 | 15.2 | 36 |

**12.**

| $t$ | 7 | 14 | 21 | 28 | 35 | 42 |
|---|---|---|---|---|---|---|
| $P$ | 8 | 41 | 133 | 250 | 280 | 297 |

**13.**

| $x$ | 29 | 48 | 72.7 | 92 | 118 | 140 | 165 | 199 |
|---|---|---|---|---|---|---|---|---|
| $y$ | .49 | .82 | 1.23 | 1.54 | 1.97 | 2.34 | 2.74 | 3.30 |

Perform the required function in Maple.

**14.** $\dfrac{d}{dx}(1.104x - 0.542x^2)$

**15.** $\displaystyle\int (1.104x - 0.542x^2)$

**16.** $\displaystyle\int_{x=1}^{x=5} (1.104x - 0.542x^2)$

**2**

# Introduction, Overview, and the Process of Mathematical Modeling

## Introduction

Two observation posts 5.43 miles apart pick up a brief radio signal. The sensing devices were oriented at 110° and 119°, respectively, when a signal was detected. The devices are accurate to within 2° (that is, ±2° of their respective angle of orientation). According to intelligence, the reading of the signal came from a region of active terrorist exchange, and it is inferred that a boat is waiting to pick up terrorists. It is dusk, the weather is calm, and there are no currents. A small helicopter leaves a pad from post 1 and is able to fly accurately along the 110°-angle direction. This helicopter has only one detection device: a searchlight. At 200 ft, it can just illuminate a circular region with a radius of 25 ft. The helicopter can fly 225 miles in support of this mission because of its fuel capacity. Where do we search for the boat? How many search helicopters would give a "good" chance of finding the target?

Consider the poor air quality in the Los Angeles basin from traffic pollution. Photochemical smog permeates the Los Angeles basin most days of the year. Although this problem is not unique to the Los Angeles area, conditions in the basin are well suited to this phenomenon. As early as 1542, explorer Juan Rodriquez Cabrillo named the San Pedro Bay the "Bay of Smokes" because of heavy haze from the native fires that covered the area. The surrounding mountains and frequent inversion layers create the stagnant air that give rise to these conditions. Can we build a model to examine this?

In the sport of bridge jumping, a willing participant attaches one end of a bungee cord to him- or herself, attaches the other end to a bridge railing, and then drops off a bridge. Is this a safe sport? Can we describe the typical motion?

You are a new city manager in California. You are worried about how well your city's water tower would survive an earthquake. You need to analyze the effects of an earthquake and see if design improvements are needed to prevent catastrophic failure.

Maybe you have flown lately. Most airplanes are full these days. As a matter of fact, most of the time an announcement is made that the plane is overbooked and the airline is looking for volunteers to take a later flight. Why do airlines overbook? Should they overbook? How does this affect passengers?

These are all events that we can model using mathematics. This textbook will help you become a confident problem solver using the techniques of mathematical modeling.

## 2.1   THE MODELING PROCESS

In Chapter 1, we examined Maple, a computational tool that you'll find useful as you learn the art of mathematical modeling. In this chapter, we turn our attention to the process of modeling and examine many different scenarios in which mathematical modeling can play a role.

Mathematical modeling is more of an *art* than a science. Modelers must be creative—that is, they must be willing to be more artistic or original in their approach to solving a problem. They must be inquisitive by questioning their assumptions, variables, and hypothesized relationships. Science is very important, and understanding science enables one to be more creative in viewing and modeling a problem. Creativity is extremely advantageous.

To gain insight, we should consider one framework that will enable a modeler to address the largest number of problems. The key is that *something is changing for which we want to know the effects*. We call this the **system** under analysis. The real-world system can be very complicated or very simplistic. This requires a process that allows for both types of real-world systems to be modeled within the same process.

Consider a ball being dropped from the top of a 20-foot-tall building. Our first inclination is to use the equations about distance and velocity that we used in high school mathematics class. These equations are simplistic and ignore many factors that could affect the fall of the ball such as wind speed, air resistance, the ball's mass, and other factors. As we add more factors, we can improve the precision of the model. Adding these factors makes the model more realistic and more complicated to produce. Understanding this model might be a first start in building a model for a bungee jumper as described earlier. The systems are similar for part of the model: The free-fall portion has similar characteristics.

Figures 2.1 and 2.2 provide closed-loop processes for modeling. Suppose we want to understand some real-world behavior or phenomenon such as the ball dropped from the 20-foot-tall building as mentioned above. Perhaps we want to know something about the force with which the ball hits the ground, or how high the ball will bounce after its initial hit of the surface below. Thus, we can apply the ideas in Figure 2.1 through considering the possible models for free-fall and considering which, if any, apply to the real-world observation that we want to model. Perhaps we decide that we need to consider friction and decide to construct our own model. Perhaps we decide to experiment and using Figure 2.2 as a guide, we collect data in order to formulate a mathematical model. This mathematical model can be one we derive or select from a collection of already built mathematical models. Then we analyze the model that we used and reach mathematical conclusions about it. Next, we interpret the model so that we can either make predictions about what has occurred or offer explanations as to why something has occurred. Finally, we test our conclusion about the real-world system with new data. We may refine or improve the model to improve its ability to predict or explain the phenomena. We might even reformulate a new mathematical model.

### Mathematical Modeling

We will **build** some mathematical models describing change in the real world. We will solve these models and **analyze** how good our resulting mathematical explanations and predictions

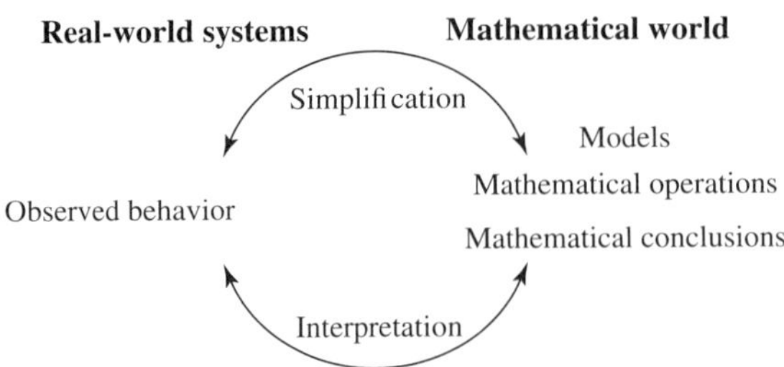


**FIGURE 2.1**
Modeling real-world systems
with mathematics

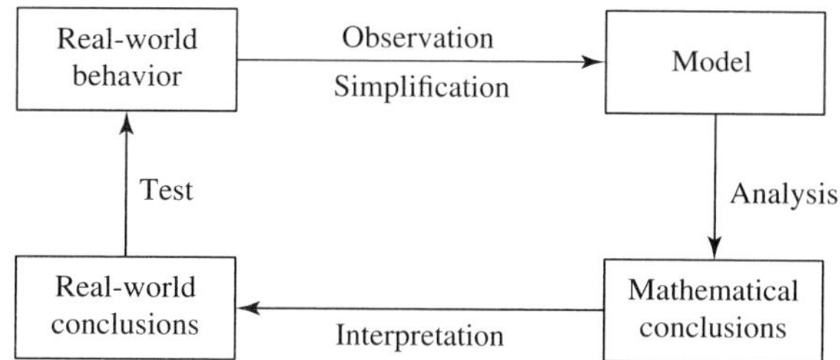


**FIGURE 2.2**

In reaching conclusions about a real-world behavior, the modeling process is a closed system.

are. The solution techniques that we employ in subsequent chapters take advantage of certain characteristics that the various models enjoy. Consequently, after building the models, we will **classify** the models based on their mathematical structure.

When we observe change, we are often interested in understanding why change occurs the way it does, perhaps to analyze the effects of different conditions, or perhaps to predict what will happen in the future. Often, a mathematical model can help us understand a behavior better, while allowing us to experiment mathematically with different conditions. For our purposes, we will consider a mathematical model to be a mathematical construct designed to study a particular real-world system or behavior. The model allows us to use mathematical operations to reach mathematical conclusions about the model as illustrated in Figure 2.1.

**Models and Real-World Systems**  A system is an assemblage of objects joined by some regular interaction or interdependence. The solar system, the United States economy, a fish population living in a lake, a satellite orbiting Earth, delivering mail, locations of service facilities—all are examples of systems. The person modeling is interested in understanding:

- how a particular system works,
- what causes change in the system, and
- the sensitivity of the system to certain changes.

The person modeling is also interested in predicting:

- what changes will occur in the system, and
- when change will occur.

A basic technique used in constructing a mathematical model of some system is a combined mathematical–physical analysis. In this approach, we start with some known physical principles or reasonable assumptions about the system. Then we reason logically to obtain conclusions.

Figure 2.2 suggests how we can obtain real-world conclusions from a mathematical model. First, observations identify the factors that seem to be involved in the behavior of interest. Often we cannot consider, or even identify, all the relevant factors, so we make simplifying assumptions excluding some of them. Next, we conjecture tentative relationships among the identified factors we have retained, thereby creating a rough "model" of the behavior. We then apply mathematical reasoning that leads to conclusions about the model. These conclusions apply only to the model, and may or may not apply to the actual real-world system in question. Simplifications were made in constructing the model, and the observations on which the model is based invariably contain errors and limitations. Thus, we must carefully account for these anomalies and test the conclusions of the model against real-world observations. If the model is reasonably valid, then we can then draw inferences about the real-world behavior from the conclusions drawn from the model. In summary, we have the following procedure for investigating real-world behavior:

1. Through observation, we identify the primary factors involved in the real-world behavior, possibly making simplifications.
2. We conjecture about tentative relationships among the factors.
3. We apply mathematical reasoning to the resultant "model."
4. We interpret the mathematical conclusions in terms of the real-world system.
5. We test the model conclusions against real-world observations.

There are various kinds of models, such as the mathematical model given by difference equations. As an example, we will use the modeling process to construct difference equations as models. These models take the form of an equation or system of simultaneous equations involving the change of a function of one or more variables. One of our tasks is to build a library of models and to recognize various real-world situations to which they apply. Another task is to formulate and analyze new models. Still another task is to learn to solve an equation or system in order to find more revealing or useful expressions relating to the variables. And we often study the models graphically to gain insight into the behavior under investigation. Through these activities, we hope to develop a strong sense of the mathematical aspects of the problem, its physical underpinnings, and the powerful interplay between them.

Most models simplify reality. Generally, models can only approximate real-world behavior. One very powerful simplifying relationship that we will use is **proportionality**.

**Definition**  Two variables—$y$ and $x$—are **proportional** to each other if one is always a constant multiple of the other. In symbols,

$$y = k\,x$$

for some nonzero constant $k$. We write $y \propto x$.

This definition implies that the graph of $y$ versus $x$ lies along a straight line through the origin. The graphical test is a useful method to determine if a data collection reasonably can be assumed to be a straight line through the origin and whether the graph supports the simplifying proportionality assumption. Here is an example.

**Testing the Simplifying Assumption of Proportionality**  Consider a spring–mass system such as the one in Figure 2.3.

We conduct an experiment to measure the stretch of the spring as a function of mass placed on the spring. We measure only how far the spring stretched from its original position, and these data are displayed in Table 2.1.

The plot (Figure 2.4) looks reasonably like a straight line through the origin. Our next step is to calculate the slope. We will use the following data pairs to find the slope: (50, .1) and (550, .875).

$$slope = \frac{.875 - .1}{550 - 50} = 0.00155$$

We then examine how close our model fits the data by plotting the line with our data.



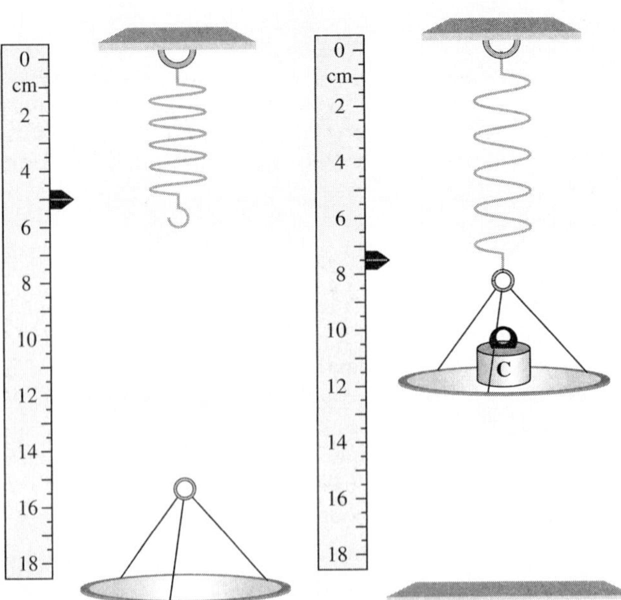

**Table 2.1** Spring–Mass System

| Mass (grams) | Stretch (m) |
| --- | --- |
| 50 | .1 |
| 100 | .1875 |
| 150 | .275 |
| 200 | .325 |
| 250 | .4375 |
| 300 | .4875 |
| 350 | .5675 |
| 400 | .65 |
| 450 | .725 |
| 500 | .80 |
| 550 | .875 |

**FIGURE 2.4**
Plot of spring–mass data

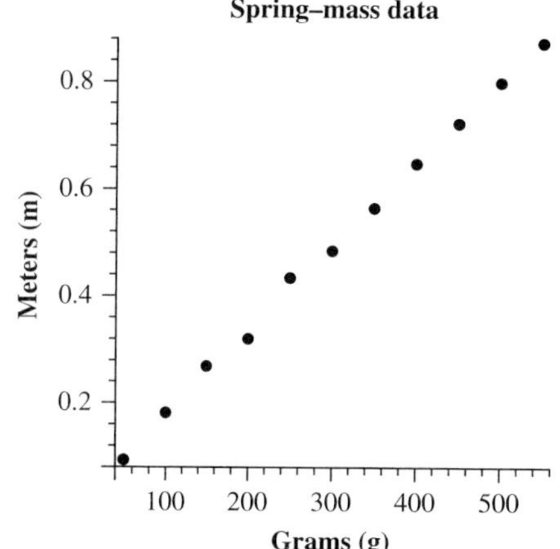


**FIGURE 2.5**
Data from spring–mass system with proportionality line

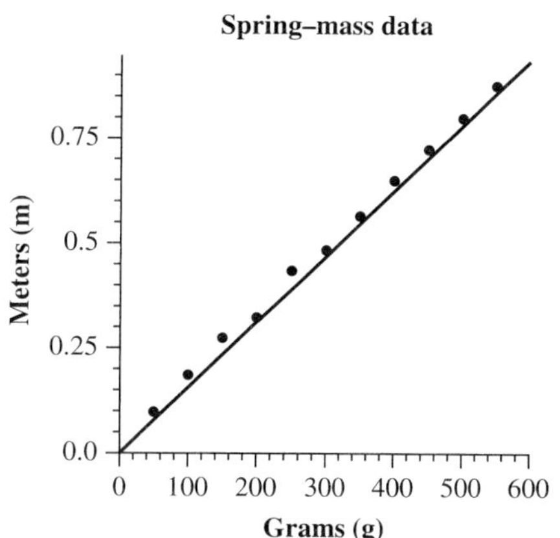


The graph (Figure 2.5) reveals that the simplifying proportionality assumption for the spring–mass model is reasonable. Therefore, we might use this model for spring–mass systems.

Next, let's summarize a *process* for formulating a model.

## Model Construction

Let's focus our attention on the process of model construction. An outline is presented as a procedure to help construct mathematical models. In the next section, we will illustrate this procedure with a few examples.

These eight steps are summarized in Figure 2.6. These steps act as a guide for thinking about the problem and getting started in the modeling process.

Let's discuss each step in more depth.

**FIGURE 2.6**
Mathematical modeling process

| | |
| --- | --- |
| **Step 1** | Understand the problem or the question asked. |
| **Step 2** | Make simplifying assumptions. |
| **Step 3** | Define all variables. |
| **Step 4** | Construct a model. |
| **Step 5** | Solve and interpret the model. |
| **Step 6** | Verify the model. |
| **Step 7** | Identify the strengths and weaknesses of the model. |
| **Step 8** | Implement and maintain the model for future use. |

**Step 1   Understand the Problem or the Question Asked** Identifying the problem to study is usually difficult. In real life, no one walks up to us and hands us an equation to be solved. Usually, the problem is implied in a comment such as "We need to make more money," or "We need to improve our efficiency." We need to be precise in our formulation of the mathematics to describe the situation.

**Step 2   Make Simplifying Assumptions** We start by brainstorming the situation. We make a list of as many factors, or variables, as we can. Realize that we usually cannot capture all the factors that influence a problem. The task is simplified by reducing the number of factors under consideration. We do this by making simplifying assumptions about the factors, such as holding certain factors as constants. We might then examine to see if relationships exist between the remaining factors (or variables). Assuming simple relationships might reduce the complexity of the problem. Once we have a shorter list of variables, classify them as independent variables, dependent variables, or neither.

**Step 3   Define All Variables** It is critical to define all our variables and provide the mathematical notation for each.

**Step 4   Construct the Model** Using the tools in this text and our own creativity, we build a model that describes the situation and whose solution helps us to answer important questions.

**Step 5   Solve and Interpret the Model** We take the model we constructed in steps 1–4 and solve it. Often this model might be too complex or unwieldy, so we cannot solve it or interpret it. If this happens, we return to steps 2–4 and simplify the model further.

**Step 6   Verify the Model** Before we use the model, we should test it out. There are several questions we must ask. Does the model directly answer the question, or does the model allow for the answer to the question to be found? Is the model usable in a practical sense (can we obtain data to use the model)? Does the model pass the common-sense test?

   We like to say we *collaborate* the reasonableness of our model.

**Step 7   Strengths and Weaknesses** No model is complete without self-reflection in the modeling process. We need to consider not only what we did right but also what we did that might be suspect and what we could do better. This reflection also helps in refining models.

**Step 8   Implement and Maintain the Model** A model is pointless if we do not use it. The more user friendly the model, the more it will be used. Sometimes the ease of obtaining data for the model can dictate its success or failure. The model must also remain current. Often this entails updating the parameters used in the model.

## 2.1 | EXERCISES

Using steps 1−3 of the modeling process above, identify a problem from one of the following 10 scenarios that you could study. There are no right or wrong answers, just measures of difficulty.

1.  The population of deer in your community.

2.  A new outdoor shopping mall is being constructed. How should you design the illumination?

3.  A farmer wants a successful season with his crops. He thinks that this is accomplished by growing a lot of anything as long as he uses all his land.

4.  Ford Motor Company bought Volvo. Are Volvos still "top quality" cars?

5.  A minor Forbes 500 company wants to go mobile with Internet access and computer upgrades, but cost might be a problem.

6.  Starbucks has many varieties of coffee available. How can Starbucks make more money?

7.  A student does not like math or math-related courses. How can a student maximize her chances for a good grade in a math class to improve her overall GPA?

8.  Freshmen think that their first semester courses should be pass–fail for credit or no-credit only.

9. Alumni are clamoring to fire the college's head football coach.

10. Stocking a fish pond with bass and trout.

## 2.2   ILLUSTRATED EXAMPLES

We now demonstrate the modeling process that was presented in the previous section. Emphasis is placed on problem identification and choosing appropriate (usable) variables. We do not build the models because these modeling example are repeated later in the book and the models are completed and discussed there.

### Example 1      The Size of Prehistoric Creatures

## Scenario

*Titanis walleri*—also known as the "terror bird." No, we didn't make this up! *Titanis* really lived about 2 million to 3 million years ago on the oak and grass savannas of what is now Florida. Dr. Robert Chandler, of Georgia College and State University and one of the world's experts on fossil birds, has dredged up fossilized bits and pieces of this bizarre predatory bird from the Santa Fe River near Gainesville, Florida.

Here are some facts about the terror bird:

- It was giant and flightless.
- It lived in South America roughly 3 million years ago.
- It is suspected to have been a fierce hunter that would lie in wait and then ambush its prey by attacking from tall grasslands.
- It probably pinned down its prey with its beak and used its 4- to 5-inch inner toe claws and beak to shred its prey.
- It had arms, not wings, that were more powerful than the arms of a *Velociraptor* (Chandler, 1994).

We realize that bones give a good indication of size (height) but not reliable body weight (weight is not easily fossilized). We want to model *Titanis*'s weight based on the information that can be determined from the fossils.

## Understanding the Problem

What is the relationship between the bones found and the original "grown" size of the creature? This might be too general, so let's restrict our problem. The fossil is a femur bone that is intact. We want to know if we can find a relationship between the weight of an animal and the size of its femur (measured as circumference). If we can find a relationship, then we can use the model to predict the size (in weight measures) of the terror bird as a function of the circumference of the femur.

## Assumptions

We assume that the thickness (measured by circumference) of the femur is important to the size of the animal. We might conjecture that thicker bones support more weight. We know that the terror bird is an ancestor of modern birds, but should birds be used to help build the model? We might assume that bird data are sufficient and use only those data to build a model. Or should we rely on species that lived about when the terror bird roamed Earth like the dinosaurs and use their data to help build the model? We could assume that dinosaur data are more appropriate to use when building a model. We might want to try both approaches and compare results. If bird data are available, we might address the appropriateness of using modern bird data to build a model for a bird that lived 2 million years ago and did not even fly. If more terror bird fossils are discovered, then we might attempt to model their size and get a sense of differing terror bird sizes.

---

**Example 2**    ## Prescribed Drug Dosage

## Scenario

Consider a patient who needs to take a newly marketed prescribed drug. To prescribe a safe and effective regimen for treating the disease, we must maintain a blood concentration above some effective level and below an unsafe level.

## Understanding the Problem

Our goal is a mathematical model that relates dosage and time between dosages to the level of the drug in the bloodstream. *What is the relationship between the amount of drug taken and the amount in the blood after time* t? By answering this question, we are empowered to examine other facets of the problem of taking a prescribed drug.

## Assumptions

We should know the disease in question and the type (name) of the drug that is to be taken. We will assume the drug is digoxin, a drug taken for heart disease. We need to know or to find decaying rate of digoxin in the bloodstream. This might be found from data that have been previously collected. We need to find the safe and unsafe levels of digoxin based on the drug's effects within the body. This will serve as the bounds for our model. Initially, we might assume that patient size and weight have no effect on the drug's decay rate. We might assume that all patients are about the same size and weight. All are in good health, and no one takes other drugs that affect the prescribed drug. We assume all internal organs function properly. We might assume that we can model this using a discrete time period even though the absorption rate is a continuous function. These assumptions help simplify the model.

---

**Example 3**    ## Determining Heart Weight

Let's assume we are interested in building a model that models heart weight as a function of the size of the heart. We might have access to data that relate, for the following seven mammals, their heart weight in grams and their diameter of the left ventricle of the heart measured in mm.

**Table 2.2** Data for Mammals' Heart Sizes

| Animal | Heart Weight (grams) | Left Ventricle Diameter (mm) |
|---|---|---|
| Mouse | 0.13 | 0.55 |
| Rat | 0.64 | 1.0 |
| Rabbit | 5.8 | 2.2 |
| Dog | 102 | 4.0 |
| Sheep | 210 | 6.5 |
| Ox | 2030 | 12.0 |
| Horse | 3900 | 16.0 |

## Problem Identification

Find a relationship between heart weight and the diameter of the left ventricle of the heart.

## Assumptions

We will assume that the heart weights listed are typical for healthy animals. We further assume that the diameters are all measured the same way. We might assume that all mammals are scale models of other mammals and that all mammals are in good health. We assume that the data were collected the same way for each mammal.

---

**Example 4**        ## The Bridge Too Far

Consider an engineering design for a truss bridge as shown in Figure 2.7. Trusses are lightweight structures capable of carrying heavy loads. In civil-engineering bridge design, the individual members of the truss are connected with rotatable pin joints that permit forces to be transferred from one member of the truss to another. The figure shows a truss that (1) is held stationary at lower-left endpoint 1, (2) is permitted to move horizontally at the lower-right endpoint 4, and (3) has pin joints at 1, 2, 3, and 4. A load of 10 kilonewtons (kN) is placed at joint 3, and the forces on the members of the truss have magnitudes given by $f_1, f_2, f_3, f_4$, and $f_5$ as shown. The stationary support member has both a horizontal force $F_1$ and a vertical force $F_2$, but the movable support member has only the vertical force $F_3$.

  If the truss is in static equilibrium, then the forces at each joint must add to the zero vector, so the sum of the horizontal and vertical components of the forces at each joint must be zero. This produces a system of linear equations.

**FIGURE 2.7**
Bridge truss

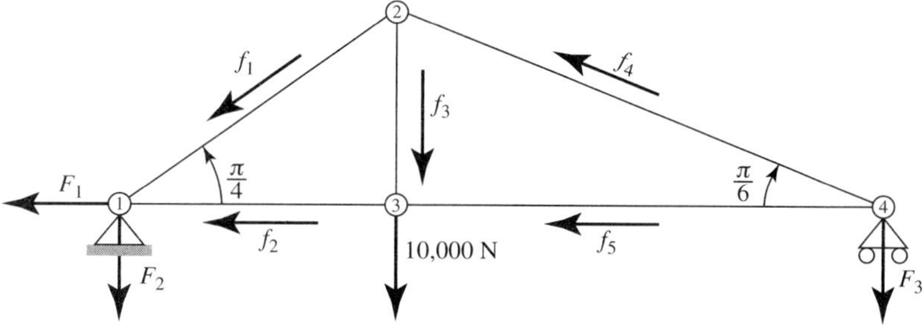


## Problem Identification

Determine the forces on each joint as a function of the angles between joints.

## Assumptions

We assume the bridge is sturdy and will not falter or collapse. We assume the motion of the bridge caused by normal motion is negligible. We assume a design such as the one shown in the figure.

---

**Example 5**        ## Oil-Rig Placement

Consider an oil-drilling rig that is 8.5 miles offshore. The drilling rig is to be connected by underwater pipe to a pumping station. The pumping station is connected by land-based pipe to

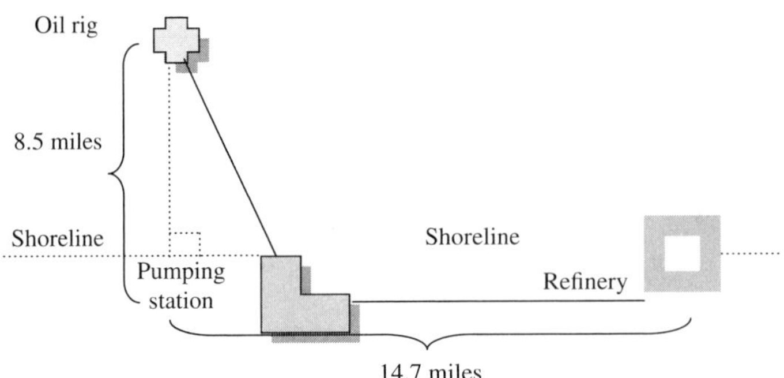




a refinery 14.7 miles down the shoreline from the drilling rig (see Figure 2.8). Underwater pipe costs $31,575 per mile, and land-based pipe costs $13,342 per mile. You are to determine where to place the pumping station to minimize the cost of piping.

## Problem Identification

Build a model to minimize the cost of a pipe from the oil rig to the refinery.

## Assumptions

We assume that the costs are based on solid estimates and will not fluctuate during the building of the pipeline. We assume no weather delays or any other natural or unnatural delays such as disasters or storms to building. All workers are competent in their jobs. The materials are unflawed.

In subsequent chapters, we will build some of these models and discuss their results.

---

**Example 6**    Military to the Rescue

Recall the chapter-opening scenario with the two listening posts detecting suspicious activity.

## Problem Identification

Build a mathematical model of the region that is under investigation and then build a model for an efficient method to search the region.

## Assumptions

Two observation posts are 5.43 miles apart. The sensing devices were oriented at 110° and 119°, respectively, when a signal was detected. The devices are accurate to within 2° (that is, ±2° of their respective angle of orientation). According to intelligence, the reading of the signal came from a region of active terrorist exchange, and it is inferred that a boat is waiting to pick up terrorists. It is dusk, the weather is calm, and there are no currents. A small helicopter leaves a pad from post 1 and is able to fly accurately along the 110°-angle direction. This helicopter has only one detection device: a searchlight. At 200 ft, it can just illuminate a circular region with a radius of 25 ft. The helicopter can fly 225 miles in support of this mission because of its fuel capacity.

## 2.2 | EXERCISES

1. How would you approach a problem concerning a drug dosage? Do you always assume the prescribing physician is right?

2. In modeling the size of any prehistoric creature, what information would you like to be able to obtain? What additional assumptions might be required?

3. In the oil-rig problem, what other factors might be critical in obtaining a "good" model that predicts reasonably well? What variables could be important that were not considered?

4. For the model in Example 3, are the assumptions about the data reasonable? How would you collect data to build the model? What other variables would you consider?

5. In the bridge example, what is the impact of the location of the bridge?

6. For Example 6, review your algebra and trigonometry and construct the region to be searched. Describe the shape and then find the overall area to be searched.

## 2.2 | PROJECTS

1. Is Michael Jordan the greatest basketball player of the 20th century? What variables and factors need to be considered?

2. What kind of car should you buy when you graduate from college? What factors should be in your decision? Are car companies modeling your needs?

3. Consider domestic decaffeinated coffee brewing. Suggest some objectives that could be used if you wanted to market your new brew. What variables and data would be useful?

4. Replacing a coaching legend at a school is a difficult task. How would you model this? What factors and data would you consider? Would you equally weigh all factors?

5. How would you go about building a model for the "best pro football player of all time"?

6. Rumors abound in major league baseball about steroid use. How would you go about creating a model that could imply the use of steroids?

7. The American League won 13 straight all-star games from 1997 to 2009, the longest winning streak ever recorded by either the American or the National League. Help the National League prepare to win by designing a model for players or a lineup that could help change the all-star game outcome.

# 3

# Discrete Dynamical Models

## Introduction to Modeling with Dynamical Systems and Difference Equations

Consider financing a new car. Once you have looked at the makes and models to determine what type of car you like, it is time to consider the "costs" and the finance packages that lure potential buyers into the car dealerships. This process can be modeled as a dynamical system. Payments are made typically at the end of each month. The amount owed is predetermined, as we will see later in the chapter.

Let's begin with some basic definitions that we will use in this chapter.

A **sequence** is a *function* whose domain is the set of all nonnegative integers, and whose range is a subset of the real numbers. A **dynamical system** is a relationship among the terms in a sequence. A **numerical solution** is a table of values that satisfy the dynamical system.

Let's start with a brief review of the concept of a function because it is part of the previous definition. A fundamental concept in mathematical modeling is a function. We start with the concept of a relation. A *relation* is a set of inputs and outputs, often written as ordered pairs (input, output). As an ordered pair, a relation tells how one thing depends on another thing. A **function** is a special type of relation and is defined more formally as, "For each element in the domain, there is exactly one element in the range." This definition introduces two more concepts: *domain* and *range*. The domain of a function is the set of all of the *independent variables* (allowed inputs), and the range is the set of all possible *dependent values* (outputs).

This concept of function, domain, and range can be seen more clearly in a dynamical system model. Let's define a recurrence relation as an equation of the form

$$a(n + 1) = f(a(n), a(n - 1), \ldots, n)$$

where $f$ is a function.

Recurrence relations, also known as *recursive formulas*, occur in many branches of applied mathematics. A discrete dynamical system (or *difference equation*) is a sequence defined by a recurrence relation. Suppose we have a function, $y = f(x)$. A *first-order* discrete dynamical

""

system is a sequence of numbers $a(n)$ for $n = 0, 1, \ldots$ such that each number after the first is related to the previous number by the relation

$$a(n + 1) = f(a(n))$$

**Remark:** Many books and articles refer to sequences of numbers using the relationship

$$a(n + 1) - a(n) = g(a(n))$$

This relationship is called a *first-order difference equation*.

**Remark:** If one were to take the smallest argument (or variable of the function) appearing in a dynamical system and subtract it from the largest, the resulting number would be the order of the dynamical system. For example, $A(n + 2) = 0.5\, A(n + 1) + 2\, A(n)$ is a second order system because $n + 2$ is the largest and $n$ is the smallest, so $n + 2 - n = 2$.

Let's define the dynamical system $A(n)$ to be the amount of antibiotic in our bloodstream after $n$ time periods. The domain is nonnegative integers representing the time periods from 0, 1, 2, ... that will be the inputs to the function. Because the domain is discrete, our function is a discrete function. The range is the values of $A(n)$ determined for each value of the domain. Thus, $A(n)$ also represents the dependent variable. For each input value of the domain from 0, 1, 2, ... , the result is one and only one $A(n)$—thus, $A(n)$ is a function.

Very often in mathematics you see a table of information. For example, consider the following table for population in the United States. This is a table of discrete values. The years are the discrete inputs (as time periods), and the outputs are the size of the population in that year.

| Year | U.S. Population ($\times 10^6$) |
|---|---|
| 1960 | 179.3 |
| 1970 | 203.3 |
| 1980 | 226.5 |
| 1987 | 243.4 |
| 1990 | 250.0 |
| 1999 | 279.0 |
| 2002 | 287.7 |
| 2005 | 295.6 |
| 2008 | 304.1 |

Let's state that there are three components to dynamical systems: (1) an equation for sequences representing $A(n)$, (2) a well-defined time period $n$, and (3) at least one starting value. This starting value is called an **initial condition**. For example, if we start with no antibiotic in our system, then $A(0) = 0$ mg is our initial condition, but if we started after we took an initial 200-mg tablet, then $A(0) = 200$ mg would be our initial condition. An example of a discrete dynamical system with its initial condition would be:

$$A(n + 1) = 0.5\, A(n),\ A(0) = 500$$

# Maple Commands for Discrete Dynamical Systems

We will use some commands and libraries from Maple such as *with(plots)*, and we will add some new commands to our Maple toolbox:

> **rsolve** - recurrence equation solver
>
> Calling Sequence
>
> > **rsolve(eqns, fcns)**
> >
> > **rsolve(eqns, fcns, 'genfunc'(z))**
> >
> > **rsolve(eqns, fcns, 'makeproc')**
>
> Parameters
>
> > eqns - single equation or a set of equations
> >
> > fcns - function name or set of function names
> >
> > z - name, the generating function variable

> **seq** - create a sequence
>
> Calling Sequence
>
> > **seq(f, i = m..$n$)**
> >
> > **seq(f, i = x)**
>
> Parameters
>
> > f - any expression
> >
> > i - name
> >
> > m, $n$ - numerical values
> >
> > x - expression

Many of the models that we will solve have closed-form solutions so that we can use the command **rsolve** to obtain the closed solution and then use the sequence command (**seq**) to obtain the numerical values in the solution.

**Remark:** Many dynamical systems do not have closed-form analytical solutions, so we cannot use the **rsolve** and **seq** commands to obtain solutions. When this occurs, we will write a small program using Proc to obtain the numerical solutions. Proc is a valid expression of a sequence of commands that can be assigned to a name. We will use the Proc command later to create programs in Maple. To plot the solution to the dynamical systems, we will use plot commands to plot sequential data pairs. We will illustrate all these commands in our examples.

## 3.1    MODELING DISCRETE CHANGE

We are interested in modeling discrete *change*. Modeling with discrete dynamical systems employs a method to explain certain discrete behaviors or make long-term predictions. A powerful paradigm that we use to model with discrete dynamical systems is:

$$\text{Future value} = \text{present value} + \text{change}$$

The dynamical systems that we will study with this paradigm will differ in appearance and composition, but we will be able to solve a large class of these seemingly different dynamical systems with similar methods. In this chapter, we will use iteration and graphical methods to answer questions about the discrete dynamical systems.

We will use flow diagrams to help us see how the dependent variable changes. These flow diagrams help show the paradigm and put it into mathematical terms. Let's consider financing a new Ford Mustang. The cost is $25,000. You can put down $2,000, so you need to finance $23,000. The dealership offers you 2 percent financing over 72 months. Consider the following flow diagram for financing the car below that depicts this situation.

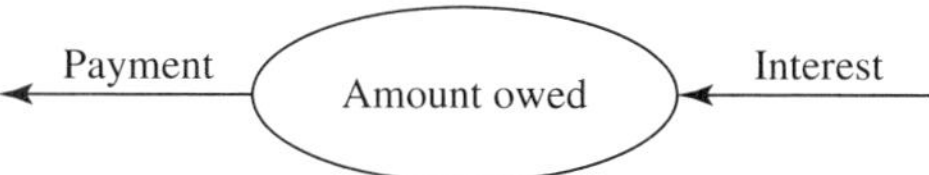


We use this flow diagram to help build the discrete dynamical model. Let $A(n) =$ the amount owed after $n$ months. Notice that the arrow pointing into the circle is the interest to the unpaid balance. This increases your debt. The arrow pointing out of the circle is the monthly payment that decreases your debt.

$$A(n + 1) = \text{the amount owed in the future}$$
$$A(n) = \text{amount currently owed}$$
$$A(n + 1) = A(n) + i\,A(n) - P$$

where $i$ is the monthly interest rate, and $P$ is the monthly payment.

We will model dynamical systems that have only **constant coefficients**. A third-order discrete dynamical system with constant coefficients may be written in the form

$$a(n + 3) = b_2\,a(n + 2) + b_1\,a(n + 1) + b_0\,a(n)$$

where $b_0$, $b_1$, and $b_2$ are arbitrary constants.

A **discrete dynamical system** is a "changing system," where system change at each discrete iteration depends on (is related to) its previous state (or states) of the system. For the Tower of Hanoi example we will soon illustrate, the number of discrete moves $a(n)$ that it takes to move $n$ disks, depends on the number of discrete moves it took to move $n-1$ disks. For the drug dosage problem we will also model, the amount of drug in the bloodstream after $n$ hours depends on the amount of drug in the bloodstream after $n - 1$ hours. For financial matters, such as a mortgage balance or credit-card balance, the amount you still owe after $n$ months depends on the amount you owed after $n - 1$ months. You will also find this process concerning states useful when we discuss Markov chains as an example of discrete dynamical systems.

A **discrete dynamical system** is a system in which the state of the system after $n$ iterations (a sequence of numbers), $a(n)$ for $n = 0, 1, \ldots$ , such that each iteration after the first iteration is related only to the previous iteration by the relation

$$a(n + 1) = f(a(n))$$

where $f(a(n))$ is some function of $a(n)$—that is, $a(n + 1) = 2\,a(n)$. If the function of $a(n)$ is just a constant multiple of $a(n)$, a constant coefficient, we will say that the discrete dynamical system is **linear**. If the function of $a(n)$ involves powers of $a(n)$ [such as $a^2(n)$] or a functional relationship [such as $a(n)/a(n - 1)$], we will say that the discrete dynamical system is **nonlinear**. A **sequence** is a function whose domain is the set of nonnegative integers ($n = 0, 1, 2, \ldots$).

Here is a summary of some key ideas about discrete dynamical systems that we will use in this chapter:

- A *sequence* is a function whose domain is the set of all nonnegative integers and whose range is a subset of the real numbers.
- A *dynamical system* is a relationship among terms in a sequence.
- A *numerical solution* is a table of values satisfying the dynamical system.
- A *recurrence relation* is an equation of the form:

$$a(n + 1) = f(a(n), a(n - 1), \ldots) \text{ where } f \text{ is a discrete function.}$$

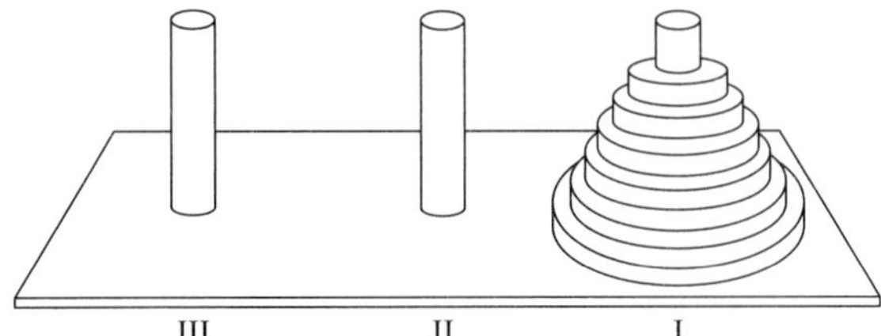



**FIGURE 3.1**
Tower of Hanoi

- A *discrete dynamical system* or *difference equation* is a sequence defined by a recurrence relation.
- *Iteration* is the process of obtaining a number in a sequence from previous numbers.

### Iteration

A game called the Tower of Hanoi, invented by French mathematician Edouard Lucas (1842−1891), consists of a board with three upright pegs and seven disks with different outside diameters. The seven disks begin all on one peg, with the largest disk on the bottom and the smallest disk on the top (Figure 3.1). The object of the game is to transfer all the disks, one at a time, in the smallest possible number of moves, to the third peg to form an identical pyramid. During the transfers, the player is not permitted to place a larger disk on top of a smaller disk, and this is why the second peg is needed. Can you do this problem? If so, how many moves does it take you to complete the transfer?

Let us start with the simplest problem, the problem in which there is only one disk on the first peg. This problem is, perhaps, too trivial. We simply move the disk from the first peg to the third peg—one move. Suppose we have two disks. Then we move the smaller disk from the first peg to the second peg, the larger disk from the first peg to the third peg, and finally the smaller disk from the second peg to the third peg—a total of three moves.

If we have three disks, we begin to see a recursion process. We know from the previous paragraph that we can move the two smaller disks from the first peg to the second peg in three moves. Next, we move the largest disk from the first peg to the third peg. Finally, we again know from the previous paragraph that we can move the two smaller disks from the second peg to the third peg in three moves. So we can move three disks from the first peg to the third peg in $3 + 1 + 3 = 7$ moves. Can you guess how many moves seven disks require? Draw a picture and try it on your own, or use a quarter, nickel, and dime. Here is a plausible flow diagram for the Tower of Hanoi.

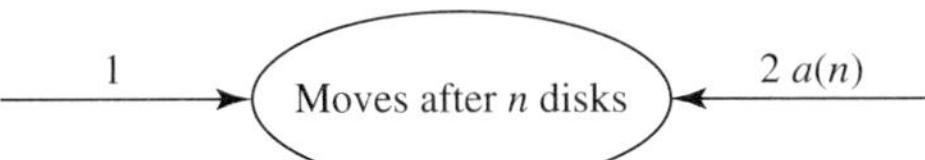



As Lucas told his story, his game with the seven disks was just a model to illustrate the legendary Tower of Hanoi, which had 64 disks. This tower was attended by a mystic order of monks who moved one disk a minute according to the long-established ritual that no larger disk could be placed on a smaller disk. The monks believed, so Lucas's story went, that as soon as all sixty-four disks were transferred, Earth would collapse in a cloud of dust. This legend seems somewhat alarming, so let us use induction to find a formula for the number of moves required to transfer $n$ disks from the first peg to the third peg. Suppose that we have $n + 1$ disks. Then we can move $n$ smaller disks from the first peg to the second peg in $a(n)$ moves, where $a$ is the number of moves for $n$ disks. Then we move the largest disk from the first peg to the third peg. Finally we move the $n$ smaller disks from the second peg to the third peg in $a(n)$ moves. Therefore $a(n + 1) = a(n) + 1 + a(n) = 2\,a(n) + 1$. So the recursion relation that gives the number of moves required is $a(n + 1) = 2\,a(n) + 1$, and we have the initial value $a(1) = 1$, meaning that we start with the case where we have one disk to move.

In this particular case, it turns out there is a formula for the number of moves required to move $n$ disks from the first peg to the third peg. We will consider two methods of solving this recursion relation to find this formula. The first method is *iteration*. Later we will learn a better method by explicitly solving difference equations.

Let us construct a table and iterate this recursion relation, written as

$$a(n + 1) = 2\,a(n) + 1$$

| Number of Disks | Recursion Relationship | Number of Moves |
|---|---|---|
| 1 | $a(1) = 1$ | 1 |
| 2 | $a(2) = 2(1) + 1$ | 3 |
| 3 | $a(3) = 2(3) + 1$ | 7 |
| 4 | $a(4) = 2(7) + 1$ | 15 |
| 5 | $a(5) = 2(15) + 1$ | 31 |
| 6 | $a(6) = 2(31) + 1$ | 63 |
| 7 | $a(7) = 2(63) + 1$ | 127 |
| $N$ | $a(n + 1) = 2\,a(n) + 1$ | $2\,a(n) + 1$ |

In this method, *iteration*, we repeatedly used the given recursion relation, each time for a different value of $n$. To find out how many minutes it would take to move 64 disks, we would have to *iterate* the given recursion relation 64 times (see Table 3.1). We will learn to use technology to help us with this.

Table 3.1 shows that $1.84467 \times 10^{19}$ minutes are needed. There are 1,440 minutes in a day and 525,600 minutes in a year. Our value is equivalent to $3.51345129 \times 10^{13}$ years. Imagine how long that amount of time really is.

The power of discrete dynamical systems is that they can always be solved by iteration. Between the table of iteration values and a graph of those values, we are able to analyze difficult modeling problems.

In Maple, to obtain the iteration we can either write a procedure, ***Proc***, or use the **rsolve** and **seq** commands to see the values. We illustrate both below to obtain the values seen in Table 3.1.

**Table 3.1** Output from Tower of Hanoi

| Number of Disks | Number of Moves |
|---|---|
| 1 | 1 |
| 2 | 3 |
| 3 | 7 |
| 4 | 15 |
| 5 | 31 |
| 6 | 63 |
| 7 | 127 |
| 8 | 255 |
| 9 | 511 |
| 10 | 1023 |
| 11 | 2047 |
| 12 | 4095 |
| . | . |
| . | . |
| . | . |
| 62 | $4.61169 \cdot 10^{18}$ |
| 63 | $9.22337 \cdot 10^{18}$ |
| 64 | $1.84467 \cdot 10^{19}$ |

---

| **Example 1** | ## Tower of Hanoi |

> *rsolve( A(n+1)=2*A(n)+1,A(k) );*

$$A(0)\, 2^k + 2^k - 1$$

*rsolve( {A(n+1)=2*A(n)+1,A(1)=1},A(k) );*

$$2^k - 1$$

Thus, we have a closed-form analytical solution for the Tower of Hanoi:

$$2^k - 1$$

We start with $k = 1$ and obtain the sequence $a(1) = (2^1) - 1 = 1$, $a(2) = (2^2) - 1 = 3$, $a(3) = (2^3) - 1 = 7, \ldots$.

We can obtain the sequence from 0 to 64 moves with the sequence command as follows:

> *seq( $2^i$−1,i=1..64 );*

1, 3, 7, 15, 31, 63, 127, 255, 511, 1023, 2047, 4095, 8191,
16383, 32767, 65535, 131071, 262143, 524287, 1048575,
2097151, 4194303, 8388607, 16777215, 33554431,
67108863, 134217727, 268435455, 536870911, 1073741823,
2147483647, 4294967295, 8589934591, 17179869183,
34359738367, 68719476735, 137438953471, 274877906943,
549755813887, 1099511627775, 2199023255551,
4398046511103, 8796093022207, 17592186044415,
35184372088831, 70368744177663, 140737488355327,
281474976710655, 562949953421311, 1125899906842623,
2251799813685247, 4503599627370495, 9007199254740991
18014398509481983, 36028797018963967,
72057594037927935, 144115188075855871,
288230376151711743, 576460752303423487,
1152921504606846975, 2305843009213693951,
4611686018427387903, 9223372036854775807,
18446744073709551615

When the dynamical systems does not have a closed-form analytical solution, the **rsolve** command returns nothing, so we cannot use the sequence (**seq**) command. We recommend using the Proc command to create a program to iterate the discrete dynamical system. We illustrate the use of a simple Proc below. We call our program *tower* and use inputs $N$, $M$, and $S$, where $N$ is the first nonnegative integer in our domain, $M$ is our first range value [consider $(N, M)$ the initial condition], and $S$ is the final nonnegative integer value from our domain that we need.

> *tower:=proc(N,M,S) option remember;*
> *T(N):=M;*
> *for i from N to S do*
> *T(i+1):=2*T(i)+1;*
> *print(i,T(i));*
> *end do;*
> *end;*

$$\begin{aligned}
&tower := \text{proc } (N,\, M,\, S)\\
&\quad \text{local } i;\\
&\quad \text{option } remember;
\end{aligned}$$

$$\text{T}(N) := M;$$
$$\text{for } i \text{ from } N \text{ to } S \text{ do } \text{T}(i + 1) := 2 \times \text{T}(i) + 1; \text{ print } (i, \text{T}(i + 1)) \text{ end}$$
$$\text{do} \quad \text{end proc}$$

The following command will produce all the iterations from $T(1)$ to $T(64)$. We illustrate only the first and last output pairs below.

$> tower(1,1,64);$

$$1,1$$
$$\ldots$$
$$64,18446744073709551615$$

---

| | |
|---|---|
| **Example 2** | ## Drug Dosage Problem |

Suppose that a doctor prescribes that her patient take a pill containing 100 mg of a certain drug every hour. Assume that the drug is immediately ingested into the bloodstream once taken. Also, assume that every hour the patient's body eliminates 25 percent of the drug that is in his/her bloodstream. Suppose that the patient had 0 mg of the drug in his bloodstream prior to taking the first pill. How much of the drug will be in his bloodstream after 72 hours?

## Problem Statement

Determine the relationship between the amount of drug in the bloodstream and time.

## Assumptions

The system can be modeled by a discrete dynamical system. The patient is of normal size and health. There are no other drugs being taken that will affect the prescribed drug. There are no internal or external factors that will affect the drug absorption rate. The patient always takes the prescribed dosage at the correct time.

## Variables

Define $a(n)$ to be the amount of drug in the bloodstream after period $n$, $n = 0, 1, 2, \ldots$ hours.

## Flow Diagram

The kidneys remove 25 percent of the 100-mg dose (see Figure 3.2).

## Model Construction

Let's define the following variables:

$$a(n + 1) = \text{amount of drug in the system in the future}$$
$$a(n) = \text{amount currently in system}$$

**FIGURE 3.2**
Flow diagram for drugs in system

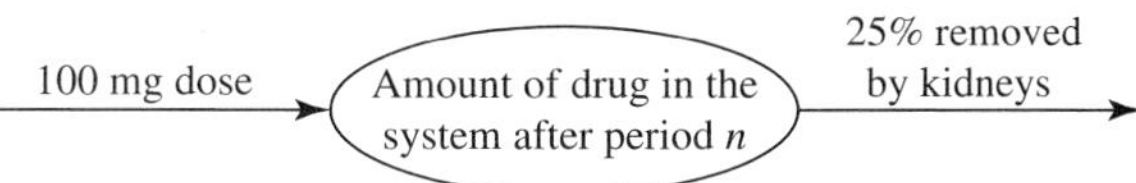

We define change as follows: Change = dose − loss in system

$$\text{Change} = 100 - 0.25\, a(n)$$

so, Future = Present + Change is

$$a(n + 1) = a(n) - 0.25\, a(n) + 100$$

or

$$a(n + 1) = 0.75\, a(n) + 100$$

Because the body loses 25 percent of the amount of drug in the bloodstream every hour, there would be 75 percent of the amount of drug in the bloodstream remaining every hour. After one hour, the body has 75 percent of the initial amount, 0 mg, to the 100 mg that is added every hour. So the body has 100 mg of drug in the bloodstream after one hour. After two hours, the body has 75 percent of the amount of drug that was in the bloodstream after one hour (100 mg), plus an additional 100 mg of drug added to the bloodstream. So there would be 175 mg of drug in the bloodstream after two hours. After three hours, the body has 75 percent of the amount of drug that was in the bloodstream after two hours (175 mg), plus an additional 100 mg of drug added to the bloodstream. So there would be 231.25 mg of drug in the bloodstream after three hours.

Let's write down the relationships that we just developed. If we let $a(n)$ (we say "$a$ at $n$" or "$a$ of $n$") be the number of milligrams of drug in the bloodstream after $n$ hours, and the initial amount in the bloodstream is 0 mg, then

$$a(0) = 0$$
$$a(1) = 0.75(0) + 100 = 100$$
$$a(2) = 0.75(100) + 100 = 175$$
$$a(3) = 0.75(175) + 100 = 231.25 \text{ mg}$$

We could write these equations where we do not substitute the numerical values that we calculated:

$$a(0) = 0$$
$$a(1) = 0.75\, a(0) + 100$$
$$a(2) = 0.75\, a(1) + 100$$
$$a(3) = 0.75\, a(2) + 100$$

Here we see that the amount of drug in the bloodstream is related to the amount of drug in the bloodstream after the previous time period. Specifically, the amount of drug in the bloodstream after any of the first three time periods is 0.75 times the amount of drug in the bloodstream after the previous time period plus an additional 100 mg that is injected every hour.

We see that a pattern has developed that describes the amount of drug in the bloodstream. We are now prepared to conjecture (make an educated guess) about the amount of drug in the bloodstream after any hour. Mathematically, we say that the amount of drug in the bloodstream after $n$ hours is

$$a(n + 1) = 0.75\, a(n) + 100$$

The relationship above describes the amount of drug in the bloodstream after $n$ hours. With this, we can see the change that occurs every hour within this "system" (amount of drug in the bloodstream), and that the state of the system, after any hour, depends on the state of the system after the previous hour. This is a *discrete dynamical system* (DDS).

We want to find the value of $a(72)$. Let's illustrate the iterative technique for analyzing a DDS in Maple (see Figure 3.3).

```
> restart;
> drug:=rsolve({a(n+1)=.75*a(n)+100,a(0)=0},a(n));
```

$$drug := 400 - 400 \left(\frac{3}{4}\right)^{n}$$

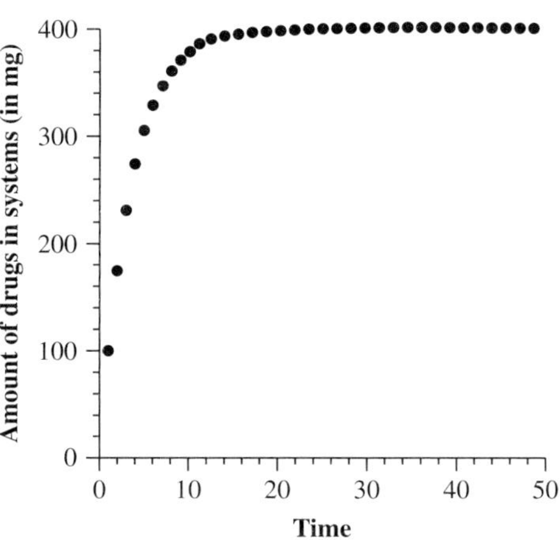

**FIGURE 3.3**
Behavior of drugs in our systems

> *L:=limit(drug,n=infinity);*

$$L := 400$$

> *with(plots):*
> *pointplot({seq([i,−400*(3/4)^i+400],i=0..48)});*
> *drug_table:=seq(evalf(−400*(3/4)^i+400),i=0..48);*

> *drug_table* := 0., 100.000, 175.00000, 231.2500000, 273.4375000, 305.0781250, 328.8085938, 346.6064453, 359.9548340, 369.9661255, 377.4745941, 383.1059456, 387.3294592, 390.4970944, 392.8728208, 394.6546156, 395.9909617, 396.9932213, 397.7449160, 398.3086870, 398.7315152, 399.0486364, 399.2864773, 399.4648580, 399.5986435, 399.6989826, 399.7742370, 399.8306777, 399.8730083, 399.9047562, 399.9285672, 399.9464254, 399.9598190, 399.9698643, 399.9773982, 399.9830487, 399.9872865, 399.9904649, 399.9928487, 399.9946365, 399.9959774, 399.9969830, 399.9977373, 399.9983030, 399.9987272, 399.9990454, 399.9992841, 399.9994630, 399.9995973

## Interpretation of Results

The DDS shows that the drug reaches a value where change stops and eventually the concentration in the bloodstream levels at 400 mg. If 400 mg is both a safe and effective dosage level, then this dosage schedule is acceptable. We discuss this concept of change stopping (equilibrium) later in this chapter. We introduced a new Maple command, **limit** in the above example.

Calling Sequences

    limit(**f**, **x** = **a**, **dir**)

    limit(**f**, **x** = **a**, **dir**)

Parameters

    f - algebraic expression

    x - name

    a - algebraic expression; limit point, possibly **infinity**, or **-infinity**

    dir - (optional) symbol; direction chosen from: **left**, **right**, **real**, or **complex**

<table>
<tr><td>

## Example 3

</td><td>

# Time Value of Money

You want to purchase a $1,000 savings certificate that pays 12 percent interest a year compounded monthly at 1 percent per month.

## Problem Statement

Find a relationship between the amount of money invested and the time over which it is invested.

## Assumptions

The interest rate is constant over the entire time period. No additional money is added or withdrawn other than by interest.

    The following represents the worth of the certificate with interest accumulated each month.

## Variables

Let $a(n)$ = the amount of money in the certificate after month $n$ where $n = 1, 2, 3, \ldots$ (see Figure 3.4).

## Model Construction

</td></tr>
</table>

$$a(n + 1) \text{ is the future}$$
$$a(n) = \text{present}$$
$$a(n + 1) = a(n) + 0.12/12 \; a(n) \text{ or } a(n + 1) = a(n) + 0.01 \; a(n)$$

or

$$a(n + 1) = 1.01 \; a(n)$$

We know that we initially had $1,000, so

$$a(0) = 1,000$$

**Remark:** We always divide the annual interest rate by the compounding period in order to recompute the actual interest rate being used in the problem. Here the annual rate is 12 percent or 0.12, and we pay monthly (12 months per year). So, we use $0.12/12 = 0.01$ as the interest rate.

    We then use the discrete time periods (1, 2, 3, ...) to iterate as follows:

$$a(1) = 1,010 = 1,000 + 0.01 \; (1,000) = a(0) + 0.01 \; a(0)$$
$$a(2) = 1,020.10 = a(1) + 0.01 \; a(1) = a(0) + 0.01 \; a(0) + 0.01 \; [a(0) + 0.01 \; a(0)]$$
$$= a(0) + 0.02 \; a(0) + 0.01^2 \; a(0) = (1 + 0.01)^2 \; a(0)$$

Let's call $(1+i)=r=1.01$:

$$a(2) = a(0) \; r^2$$
$$a(3) = 1,030.30 = a(2) + 0.01 \; a(2) = a(0) \; r^3$$
$$\ldots$$
$$a(n) = a(0) \; r^n$$

<table>
<tr><td>



</td><td>

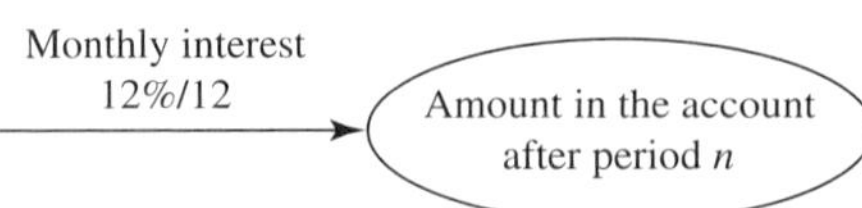


</td></tr>
</table>

Let's check.
Iterate to $a(12)$

$$a(12) = 1.01^{12}\, a(0) = (1.01)^{12}\, 1.000 = 1126.83$$

so

$$a(k) = (1.01)^k\, a(0) \text{ for } k = 0, 1, 2, \dots$$

In Maple/we use the following commands:

> *restart;*
> *money:=rsolve( {m(n+1)=1.01*m(n),m(0)=1000},m(n) );*

$$money := 1000\left(\frac{101}{100}\right)^n$$

> *L:=limit(money,n=infinity);*

$$L := \infty$$

> *with(plots):*
> *pointplot( {seq([i,1,000. * (101./100. )^i],i = 0..48)} );*

See Figure 3.5.

> *money_table:=seq(evalf(10008(101/100)^i),i=0..48);*

> $money\_table :=$ 1000., 1010.000000, 1020.100000, 1030.301000, 1040.604010,
> 1051.010050, 1061.520151, 1072.135352, 1082.856706, 1093.685273, 1104.622125,
> 1115.668347, 1126.825030, 1138.093280, 1149.474213, 1160.968955, 1172.578645,
> 1184.304431, 1196.147476, 1208.108950, 1220.190040, 1232.391940, 1244.715860,
> 1257.163018, 1269.734649, 1282.431995, 1295.256315, 1308.208878, 1321.290967,
>
> 1334.503877, 1347.848915, 1361.327404, 1374.940679, 1388.690085, 1402.576986,
> 1416.602756, 1430.768784, 1445.076471, 1459.527236, 1474.122509, 1488.863734,
> 1503.752371, 1518.789895, 1533.977794, 1549.317571, 1564.810747, 1580.458855,
> 1596.263443, 1612.226078

Can we ever reach \$10,000 in our account? We can iterate using technology to see that it will take about 19.33 years to have \$10,000 in the account. The money continues to grow as long as the money is left in the account. You will have \$1,000,000 in about 58 years.

> *solve(evalf(10000=1000*(101/100)^n),n);*

$$231.4078926$$

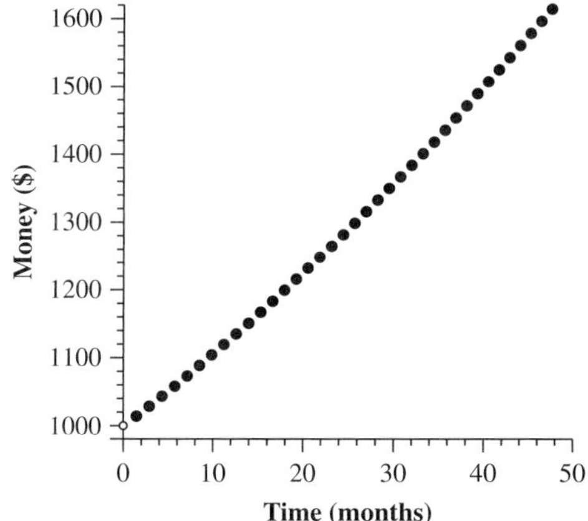


**FIGURE 3.5**

Plot of DDS for money in a certificate that grows over time

> *years:=231.407/12;*

$$years := 19.28391667$$

> *solve(evalf(100000=1000*(101/100)^n),n);*

$$694.2236777$$

> *years:=694.223/12;*

$$years := 57.85191667$$

---

| **Example 4** | ## Simple Mortgage |

Five years ago, your parents purchased a home by financing \$80,000 for 20 years, paying monthly payments of \$880.87 with a monthly interest of 1 percent. They have made 60 payments and wish to know what they actually owe on the house at this time. They can use this information to decide whether they should refinance their house at a lower interest rate for the next 15 or 20 years. The change in the amount owed each period increases by the amount of the interest and decreases by the amount of the payment.

## Problem Identification

Build a model that relates the time with the amount owed on a mortgage for a home.

## Assumptions

Initial interest was 12 percent. Payments are made on time each month. The current rate for refinancing is 6.25 percent for 15 years and 6.50 percent for 20 years.

## Variables

Let $b(n)$ = amount owed on the home after $n$ months. (Consult the flow diagram in Figure 3.6.)

## Model Construction

$$b(n + 1) = b(n) + 0.12/12 \, b(n) - 880.87, \; b(0) = 80{,}000$$
$$b(n + 1) = 1.01 \, b(n) - 880.87, \; b(0) = 80{,}000$$

## Model Solution

Graphical: First, we plot the DDS over the entire 20 years (240 months). Note the graph (Figure 3.7) shows that it reaches 0 in 240 months.

**FIGURE 3.6**
Flow diagram for mortgage example

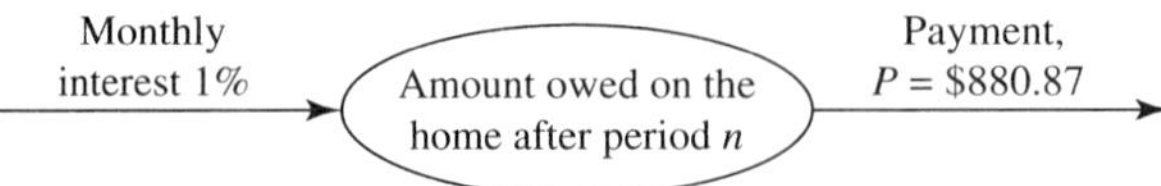

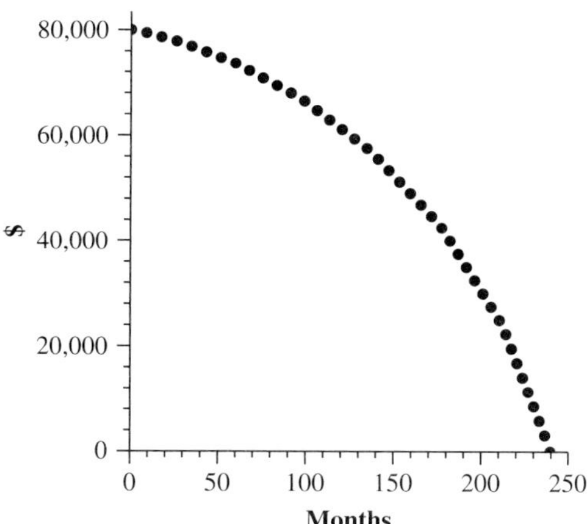

**FIGURE 3.7**
Mortgage payments
over time

```
> restart;
> mortgage:=rsolve( {m(n+1)=1.01*m(n)−880.87,m(0)=80000.},m(n));
```

$$mortgage := -8087 \left(\frac{101}{100}\right)^{n} + 88087$$

```
> with(plots):
> pointplot( {seq( [i,−8087*(101/100)^i+88087],i=0..240)});
> money_table:=seq(evalf(88087−(8087*(101/100)^i,i=0..240));
```

$money_table := $ 80000., 79919.13000, 79837.45130, 79754.95581, 79671.63537,
79587.48173, 79502.48654, 79416.64141, 79329.93782, 79242.36720, 79153.92088,
79064.59008, 78974.36598, 78883.23964, 78791.20204, 78698.24406, 78604.35650,
78509.53007, 78413.75536, 78317.02292, 78219.32315, 78120.64638, 78020.98284,
77920.32267, 77818.65589, 77715.97246, 77612.26218, 77507.51480, 77401.71995,
77294.86715, 77186.94582, 77077.94528, 76967.85473, 76856.66328, 76744.35991,
76630.93351, 76516.37284, 76400.66658, 76283.80324, 76165.77127, 76046.55898,
75926.15458, 75804.54612, 75681.72158, 75557.66880, 75432.37549, 75305.82924,
75178.01754, 75048.92771, 74918.54699, 74786.86246, 74653.86108, 74519.52970,
74383.85499, 74246.82354, 74108.42178, 73968.63599, 73827.45236, 73684.85687,
73540.83544, 73395.37380, 73248.45753, 73100.07211, 72950.20284, 72798.83486,
72645.95321, 72491.54274, 72335.58817, 72178.07405, 72018.98479, 71858.30464,
71696.01769, 71532.10786, 71366.55894, 71199.35453, 71030.47808, 70859.91286,
70687.64198, 70513.64841, 70337.91489, 70160.42404, 69981.15828, 69800.09986,
69617.23086, 69432.53317, 69245.98850, 69057.57839, 68867.28417, 68675.08701,
68480.96788, 68284.90756, 68086.88664, 67886.88551, 67684.88436, 67480.86320,
67274.80183, 67066.67985, 66856.47665, 66644.17142, 66429.74313, 66213.17056,
65994.43226, 65773.50659, 65550.37166, 65325.00537, 65097.38543, 64867.48928,
64635.29418, 64400.77711, 64163.91488, 63924.68404, 63683.06088, 63439.02148,
63192.54170, 62943.59711, 62692.16309, 62438.21472, 62181.72687, 61922.67414,
61661.03087, 61396.77118, 61129.86889, 60860.29759, 60588.03056, 60313.04087,
60035.30128, 59754.78429, 59471.46213, 59185.30675, 58896.28982, 58604.38272,
58309.55654, 58011.78211, 57711.02994, 57407.27023, 57100.47293, 56790.60767,
56477.64374, 56161.55017, 55842.29568, 55519.84863, 55194.17712, 54865.24889,
54533.03138, 54197.49169, 53858.59661, 53516.31258, 53170.60570, 52821.44176,
52468.78618, 52112.60404, 51752.86008, 51389.51868, 51022.54387, 50651.89930,
50277.54830, 49899.45379, 49517.57832, 49131.88411, 48742.33295, 48348.88628,
47951.50513, 47550.15019, 47144.78169, 46735.35950, 46321.84310, 45904.19153,
45482.36345, 45056.31708, 44626.01025, 44191.40036, 43752.44436, 43309.09880,
42861.31979, 42409.06298, 41952.28362, 41490.93645, 41024.97582, 40554.35558,

40079.02914, 39598.94943, 39114.06892, 38624.33961, 38129.71300, 37630.14013, 37125.57153, 36615.95725, 36101.24682, 35581.38929, 35056.33318, 34526.02651, 33990.41678, 33449.45095, 32903.07546, 32351.23622, 31793.87858, 31230.94736, 30662.38684, 30088.14070, 29508.15211, 28922.36363, 28330.71727, 27733.15444, 27129.61598, 26520.04214, 25904.37257, 25282.54629, 24654.50175, 24020.17678, 23379.50854, 22732.43363, 22078.88796, 21418.80684, 20752.12491, 20078.77616, 19398.69392, 18711.81086, 18018.05897, 17317.36956, 16609.67325, 15894.89999, 15172.97899, 14443.83877, 13707.40716, 12963.61123, 12212.37735, 11453.63112, 10687.29743, 9913.30041, 9131.56341, 8342.00905, 7544.55914, 6739.13472, 5925.65603, 5104.04263, 4274.21308, 3436.08521, 2589.57606, 1734.60176, 871.07781, −1.08137

After paying for 60 months, your parents still owe $73,395 out of the original $80,000.

$>mortgage60:=subs(n=60,(evalf(-8087*(101/100)^n+88087)));$

$$mortgage\ 60 := 73395.37380$$

They have paid in $52,852.20, and only $6,605 went toward the principal payment of the home. The rest of the money went toward paying only the interest. If the family continues with this loan, then it will make 240 payments of $880.87, or $211,400.80 total in payments. This is $133,400.80 in interest. They have already paid $46,647.20 in interest. They would pay an additional $86,753.60 in interest over the next 15 years. What should they do?

## Alternative 1

Refinance the home at 6.50 percent for 15 years. Assume no closing costs for this refinancing. Only refinance what they owe: $73,395.

### Model Construction

$$B(n + 1) = B(n) + 0.065/12\ B(n) - \text{payment}, \quad B(0) = 73,395$$

The monthly payment is now $639.34. We will show later how this was calculated.

The total payment over the lifetime of the refinance is $115,081.20. We have previously paid a total of $52,852.20 for 5 years. The new total of payments is $167,933.40 as compared to $211,400.80. Refinancing is the smarter move because not only are the monthly payments $241.53 a month less ($639.34 versus $880.87) but also the total package saves the family a total of $43,467.40 over the lifetime of the mortgage.

---

## Example 5     The Spotted Owl

There is a general fear that perhaps one-quarter of Earth's current plant and animal species will be extinct by the end of the century, primarily as a result of human activity. At the center of a current controversy in the Pacific Northwest of the United States is the northern subspecies of the spotted owl.

The National Forest Management Act of 1976 legislates that timber harvesting on the National Forest lands must be managed so that viable populations of native vertebrate species are maintained over a wide domain. Home-range territory for a single nesting pair of spotted owls consists of a huge tract of land, ranging from 2,000 to 4,200 acres of forest. The Forest Service management plan calls for 500 to 600 spotted owl habitat areas, each containing 1,000 to 3,000 acres of suitable habitat per nesting pair. This management plan is at the center of a heated controversy between the timber industry and environmental organizations.

The timber industry claims that management is too costly. Based on current values, the land set aside for the spotted owls would yield about \$2.2 billion in profit for the timber industry. In general, the value of timber set aside to accommodate a single nesting pair is about \$8 million. To emphasize the magnitude of the economics, about 44 percent of Oregon's economy and 28 percent of Washington's depend on national forest resources.

Environmentalists claim that continued logging of the forests will eventually drive the owl to extinction. Many in the timber industry believe that the spotted owl is expendable by virtue of its impact on jobs and local economies.

Our goal is to model the population of female spotted owls and predict what will happen to the species. You will need the following biological information about the spotted owl.

1. The owl is a pulse breeder, with a short breeding season in late April to early May. Accordingly, we adopt a discrete model to project the population from year to year.

2. There are three distinct stages in the life of an owl:

   a. juvenile (first year),

   b. subadult (second year), and

   c. adult (third year and beyond).

## Resulting Model

See Figure 3.8. Let

$J(n)$ = number of juvenile owls population after $n$ years
$S(n)$ = number of subadult owl population after $n$ years
$A(n)$ = number of adult owl population after $n$ years
 $j$ = proportion of juveniles alive at $n$ years who survive to become subadults at time $n + 1$ years
 $s$ = proportion of subadults alive at $n$ years who survive to become adults at time $n + 1$ years
 $a$ = proportion of adults alive at $n$ years who survive to become adults at time $n + 1$ years
 $b$ = average number of juveniles produced by an adult female

The resulting difference equations are:

$$J(n) = b\,A(n)$$
$$S(n + 1) = j\,J(n)$$
$$A(n + 1) = s\,S(n) + a\,A(n)$$

We want to develop a second-order difference equation that models the number of adult female owls, $A(n)$, after $n$ years. Rather than treat this as a system of DDS, we will do this by substitution.

$$A(n + 2) = a * A(n + 1) + b * j * s * A(n), \quad n = 0, 1, 2, 3 \ldots \text{years}$$

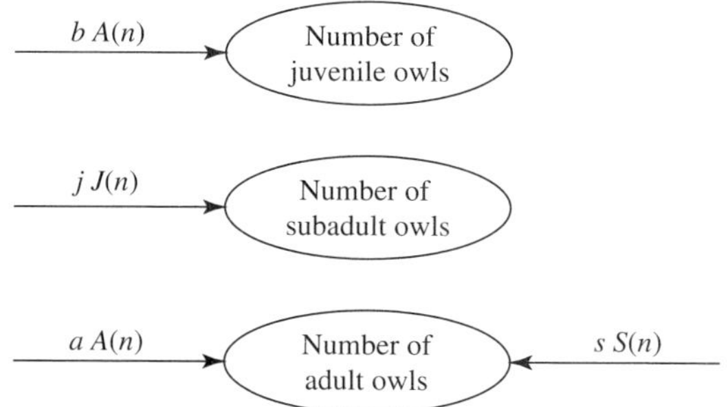


**FIGURE 3.8**
Flow diagram for spotted owl example

Given that the adult spotted owl population was 1,500 in 1989 and 1,400 in 1990, we write down initial conditions for this problem as:

$$A(0) = 1500, \; A(1) = 1400$$

We found that several groups have studied this situation and have estimated the values for the survival and birth parameters.

| Parameter | U.S. Forest Service | Lande |
|---|---|---|
| $a$ | .97 | .94 |
| $j$ | .34 | .30 |
| $s$ | .97 | .50 |
| $b$ | .24 | .26 |

We will look at only the Lande provided parameters in our illustration and iterate to a result.

## Lande Model

$$A(n + 2) = 0.94 \, A(n + 1) + 0.0390 \, A(n)$$

We can determine the long-term behavior from both numerical and graphical techniques. Remember that mathematics does not distinguish that these numbers represent live owls, so mathematically the solution will go below zero. It is up to the modeler to recognize that this cannot really happen and terminate the solution at zero.

| Lande | |
|---|---|
| $n$ | $b(n)$ |
| 0 | 1,500 |
| 1 | 1,400 |
| 2 | 1,344.116 |
| 3 | 1,289.711 |
| 4 | 1,237.522 |
| 5 | 1,187.445 |
| 6 | 1,139.395 |
| 7 | 1,093.288 |
| 8 | 1,049.048 |
| 9 | 1,006.598 |
| 10 | 965.8651 |
| 11 | 926.7808 |
| 12 | 889.2782 |
| 13 | 853.293 |
| 14 | 818.7641 |
| 15 | 785.6324 |
| 16 | 753.8413 |
| 17 | 723.3368 |
| 18 | 694.0665 |
| 19 | 665.9808 |

The plot of the values in the numerical table using these parameters shows that the model tends to zero (see Figure 3.9). This implies that if these parameters are correct that the spotted owl will become extinct unless some environmental group acts to prevent it.

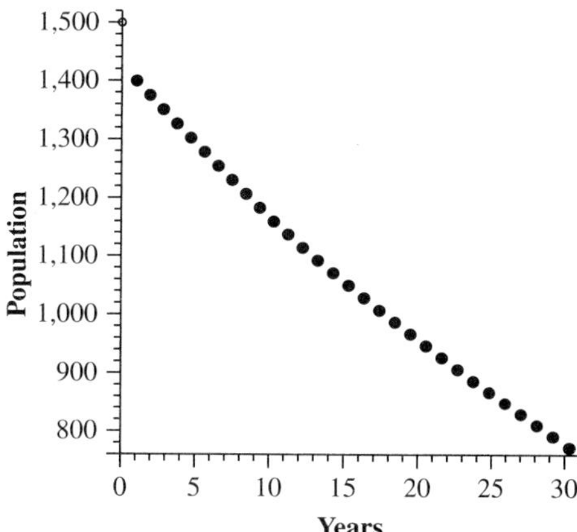

**FIGURE 3.9**
DDS graphical solution to spotted owls over time



## 3.1 | EXERCISES

Iterate and graph the following DDS. Explain their long-term behavior. Try to find realistic scenarios that these DDS might explain.

1. $a(n + 1) = 0.5\,a(n) + 0.1$, $a(0) = 0.1$

2. $a(n + 1) = 0.5\,a(n) + 0.1$, $a(0) = 0.2$

3. $a(n + 1) = 0.5\,a(n) + 0.1$, $a(0) = 0.3$

4. $a(n + 1) = 1.01\,a(n) - 1{,}000$, $a(0) = 90{,}000$

5. $a(n + 1) = 1.01\,a(n) - 1{,}000$, $a(0) = 100{,}000$

6. $a(n + 1) = 1.01\,a(n) - 1{,}000$, $a(0) = 110{,}000$

7. $a(n + 1) = -1.3\,a(n) + 20$, $a(0) = 9$

8. $a(n + 1) = -4\,a(n) + 50$, $a(0) = 9$

9. $a(n + 1) = 0.1\,a(n) + 1{,}000$, $a(0) = 9$

10. $a(n + 1) = 0.9987\,a(n) + 0.335$, $a(0) = 72$

## 3.1 | PROJECTS

1. **Sewage Plant**. A sewage-treatment plant processes raw sewage to produce usable fertilizer and clean water by removing all contaminants. The process is such that each hour 12 percent of the remaining contaminants in the process tank are removed. What percentage is left after one day? How long would it take to lower the amount of sewage by about half? How long until the level of sewage is down to 10 percent of its original level?

   Problem Identification (PID)
   *Assumptions:*
   *Variables:*
   *Model:*

2. **Cooling a Hot Soda**. A can of soda has been sitting out at room temperature (~72°) for several days. If that same soda is placed in a refrigerator that is set at 40°, how long will it take the soda to reach 50°? 45°? 40°? Assume a cooling rate of 0.0081 degrees per minute (obtained experimentally).

   *PID:*
   *Assumptions:*
   *Variables:*
   *Model:*

3. **Buying a New Car**. Part 1. You wish to buy a new car soon. You initially narrow your choices to a Saturn, a Ford Focus, and a Toyota Corolla. Each dealership offers you its prime deal:

| | | | |
|---|---|---|---|
| Saturn | $11,900 | $1,000 down | 3.5% interest for up to 60 months |
| Focus | $11,500 | $1,500 down | 4.5% interest for up to 60 months |
| Corolla | $10,900 | $500 down | 6.5% interest for up to 48 months |

You have allocated $475 a month at most on a car payment.

Use a dynamical system to compare the alternatives, choose a car, and establish your exact monthly payment.

Part 2. Nissan is offering a first-time-buyer's special. Nissan is offering 6.9 percent for 24 months or $500 cash back. If the regular interest rate is 9.00 percent, what is the amount financed at 6.9 percent so that these two options are equivalent for a new car buyer?

Check out your local paper or dealership. What kind of car could you get from Nissan with the equivalent option?

4. You are also a bit worried about the comfort of a smaller car and its ride. You are in luck: *Consumer Reports* did a study and compared the rides of four small cars in terms of their suspension systems. Consider the following alternative models for suspension systems available for a small car. Analyze each alternative, describe the behavior of each suspension system over time, and choose an alternative suspension system. Each car sits at 0.2 initially. It is determined that under 0.6 maintains a smooth ride.

Alternative 1: $S(n + 1) = 2\,S(n)\,(1 - S(n))$
Alternative 2: $S(n + 1) = 3\,S(n)\,(1 - S(n))$
Alternative 3: $S(n + 1) = 3.6\,S(n)\,(1 - S(n))$
Alternative 4: $S(n + 1) = 3.7\,S(n)\,(1 - S(n))$

5. **D. B. Cooper**. On November 24, 1971, a cold and rainy evening just before Thanksgiving, a middle-aged man giving the name Dan B. Cooper purchased a plane ticket on Northwest Airlines flight 305 from Portland, Oregon, to Seattle, Washington. He boarded the plane and flew into history.

After Cooper was seated, he demanded that the flight attendant bring him a drink and $200,000 in ransom. His note read, "Miss, I have a bomb in my suitcase, and I want you to sit beside me." When she hesitated, Cooper pulled her into the seat next to him and opened his case, revealing several sticks of dynamite connected to a battery. He then ordered her to have the captain relay Cooper's ransom demand for the money and four parachutes: two front emergency chutes and two main back parachutes. The FBI delivered the parachutes and ransom to Cooper when the 727 landed in Seattle to be refueled. Cooper, in turn, allowed the 32 other passengers he had held captive to go free.

When the 727 was again airborne, Cooper instructed the pilot to fly at an altitude of 10,000 feet on a course destined for Reno, Nevada. He then forced the flight attendant to enter the cockpit with the remaining three crew members and told her to stay there until landing. Alone in the rear cabin of the plane, he lowered the aft stairs beneath the tail. Then, in the dark of the night, somewhere over Oregon, D. B. Cooper stepped with the money into history. When the plane landed in Reno, the only thing the FBI found in the cabin was one of the back parachutes. Despite the massive efforts of manhunts conducted by the FBI and scores of local and state police groups, no evidence was ever found of D. B. Cooper, the first person to hijack a plane for ransom and parachute from it. One package of marked bills from the ransom was found along the Columbia River near Portland in 1980. Some people surmise that Cooper might have been killed in his jump to fame and part of the ransom washed downstream.

Consider the following items dealing with Flight 305 and its famous passenger. Using Excel to assist you, complete each one as directed.

    **a.** Suppose Cooper invested the ransom over a period of 10 years, in increments of $20,000 per year, in a small local bank in rural Utah at a fixed rate of 6 percent, compounded annually, starting on January 1, 1972. Further, if he left the money to accumulate, what would be the value of the account on December 31, 2004?

    **b.** Suppose that instead of playing it safe, Cooper gambled half of the money away in Reno the night after the hijacking and then invested the other half with Night Wings Federal in Reno on January 1, 1972. If the account paid 6 percent interest compounded quarterly, what would this money be worth at the end of this year?

    **c.** Another possibility for Cooper would have been to invest 75 percent of the total with a "loco" investor, Smiling Pete's Federal Credit Union, at a rate of 3 percent compounded monthly. What would this investment be worth at the end of this year?

    **d.** Compare these three methods of investing. Contrast their different patterns of growth over time.

**6. The Spotted Owl Problem Revisited.** Example 4 analyzed the spotted owl population using the Lande survival and birth parameters. Several other groups have studied this situation and have estimated the values for the survival and birth parameters.

| Parameter | U.S. Forest Service | Lande |
|:---:|:---:|:---:|
| $a$ | .97 | .94 |
| $j$ | .34 | .11 |
| $s$ | .97 | .71 |
| $b$ | .24 | .24 |

    **a.** Using the U.S. Forest Service's data and the initial conditions from Example 4, iterate the spotted owl populations for 85 years.

    **b.** What is the long-term behavior of the system for each of the data sets? Do these results surprise you? What does this say about the survival of the spotted owl population?

    **c.** An independent contractor studied the problem and came up with the following data: $a = 0.7, j = 0.75, s = 0.8, b = 0.5$

According to these new data, will the spotted owl population survive?

## 3.2   EQUILIBRIUM VALUES AND LONG-TERM BEHAVIOR

### Equilibrium Values

Let's go back to our original paradigm,

$$\text{Future} = \text{present} + \text{change}$$

When change stops, the change equals zero and future equals the present. The value for which this happens, if any, is the equilibrium value. This gives us a context for the concept of the equilibrium value.

We will define the *equilibrium value* (or fixed point) as the value where change stops. The value $a(n)$ is an equilibrium value for the DDS, $a(n + 1) = f(a(n), a(n - 1),...)$ if for a value $a(n) = A_{ev}$ all future values equal $A_{ev}$.

Formally, we define the equilibrium value as follows:

> The number *ev* is called an **equilibrium value** or **fixed point** for a discrete dynamical system if $a(k) = ev$ for all values of $k$ when the initial value $a(0)$ is set at *ev*; that is, $a(k) = ev$ is a constant solution to the recurrence relation for the dynamical system.

Another way of characterizing such values is to note that a number $a$ is an equilibrium value for a dynamical system $a(n + 1) = f(a(n), a(n - 1), ..., n)$ if and only if $a$ satisfies the equation $ev = f(ev, ev, ..., ev)$.

Using this definition, we can show that a linear homogeneous (a DDS is homogeneous if the only terms that appear in the recurrence relationship are ones that involve the $a(n)$ sequence variable) dynamical system of order 1 only has the value 0 as an equilibrium value.

In general, dynamical systems may have no equilibrium values, a single equilibrium value, or multiple equilibrium values. Linear systems have unique equilibrium values. The more nonlinear a dynamical system is, the more equilibrium values it may have.

Not all DDSs have equilibrium values, and many DDSs have equilibrium values that the system will never achieve. However, we already know that for a first-order equation, if $a(0) = ev$, and every subsequent iteration value of the DDS is equal to $ev$, i.e., $a(k) = ev$ for all value of $k$, then $ev$ is an equilibrium value. For example, the DDS $a(n + 1) = 2\,a(n) + 1$ (Tower of Hanoi) has an equilibrium value of $ev = -1$. Although $a(n)$ has a mathematical $ev$, the value $-1$ makes no sense in context of the Tower problem. If we begin with $a(0) = -1$ and iterate, we get

$$a(1) = 2\,a(0) + 1 = 2\,(-1) + 1 = -2 + 1 = -1, \text{ so } a(1) = -1$$
$$a(2) = 2\,a(1) + 1 = 2\,(-1) + 1 = -2 + 1 = -1, \text{ so } a(2) = -1$$
$$a(3) = 2\,a(2) + 1 = 2\,(-1) + 1 = -2 + 1 = -1, \text{ so } a(3) = -1$$

and so on.

So we say that when $a(0) = -1$, $a(k) = -1$ for all values of $k$, or in general, when $a(0) = ev$, then $a(k) = ev$ for all values of $k$.

We can use this observation to find equilibrium values and to find out whether or not a DDS has an equilibrium value. Let's look at the DDS $a(n + 1) = 2\,a(n) + 1$, when $a(0) = ev$, $a(k) = ev$ for all value of $k$, so that $a(1) = ev$, $a(2) = ev$, $a(3) = ev$, ... , $a(n) = ev$, $a(n + 1) = ev$. Substituting $a(n) = ev$ and $a(n + 1) = ev$ into our DDS yields

$$a(n + 1) = 2\,a(n) + 1$$
$$ev = 2\,ev + 1$$
$$-ev = 1$$
$$ev = -1$$

So, the equilibrium value for the DDS is $-1$.

Now, let's consider the DDS $a(n + 1) = a(n) + 1$. Using our definition of equilibrium values, we write

$$a(n + 1) = a(n) + 1$$
$$ev = ev + 1$$
$$ev - ev = 1$$
$$0 = 1$$

The statement $0 = 1$ is not true, so the DDS **does not have an equilibrium value**.

### Examples

Consider the following DDS and find their equilibrium value, if they exist.

    **a.** $a(n + 1) = 0.3\,a(n) - 10$
    **b.** $a(n + 1) = 1.3\,a(n) + 20$
    **c.** $a(n + 1) = 0.5\,a(n)$
    **d.** $a(n + 1) = -0.1\,a(n) + 11$

### Solutions

    **a.** $ev = 0.3\,ev - 10$
      $0.7\,ev = -10$
      $ev = -10/0.7 = -14.29$
    **b.** $ev = 1.3\,ev + 20$
      $-0.3\,ev = 20$
      $ev = -20/.3 = -66.66667$
    **c.** $ev = 0.5\,ev,\ ev = 0$
    **d.** $ev = -0.1\,ev + 11$
      $1.1\,ev = 11\ ev = 10$

Above we said that DDSs that have equilibrium values **may not ever attain** their equilibrium value, given some initial condition $a(0)$. For the Tower of Hanoi, the equilibrium value was $-1$. Suppose that we begin iterating the DDS with an initial value $a(0) = 0$:

$$a(1) = 2(0) + 1 = 3$$
$$a(2) = 2(3) + 1 = 7$$
$$a(3) = 2(7) + 1 = 15$$
$$a(4) = 2(15) + 1 = 31$$

and so on.

The values continue to get larger and will never reach the value of $-1$. Because $a(n)$ represents the number of moves of the disks, the value of $-1$ makes no real sense in the context of the number of disks to move.

We will study equilibrium values in many of the applications of discrete dynamical systems. In general, a linear, nonhomogeneous discrete dynamical system, where the nonhomogeneous part is a constant, will have an equilibrium value. (Can you find any exceptions?) Linear, homogeneous discrete dynamical systems will have an equilibrium value of zero. (Why?)

### A Graphical Approach to Equilibrium Values

We can examine the plot of the iterations using technology. If the values reach a specific value and remain constant, then that value is an equilibrium value (change has stopped).

As seen, dynamical systems often represent real-world behavior that we are trying to understand. At times, we want to predict future behavior and gain deeper insights into how to influence or alter the behavior. Thus, we have great interest in the predictions of the model. How does it change in the future?

**Models of the Form: $a(n + 1) = r\,a(n)$, $r$ Is a Constant**   Let's revisit our savings account problem where we invest \$1,000 at 12 percent a year compounded monthly (see Figure 3.10).

This sequence, with $r > 1$, grows without bound. The graph seems to suggest that there is no equilibrium value (where the graph levels out to a constant value). Analytically, $ev = 1.1\ ev + 1,000$, $ev = -909.09$. There is an equilibrium value of $-909.09$, but that value will never be reached in our savings account problem.

Now, what happens if $r$ is less than 0? We replace $r = 1.01$ with $r = -1.01$ in the previous example. First, we can analytically solve for the equilibrium value.

$$ev = -1.01\ ev + 1,000$$
$$2.01\ ev = 1,000$$
$$ev = 497.5124378$$

The definition implies that if we start at $a(0) = 497.5124378$, we stay there forever. If we plot the solution, we note the oscillations between positive and negative numbers, each growing

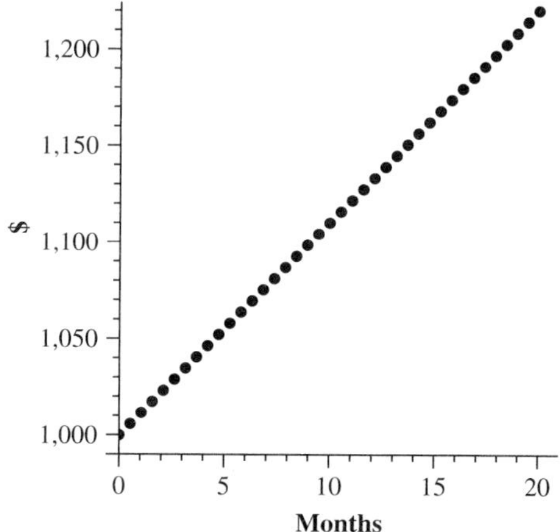



**FIGURE 3.10**
Savings account revisited

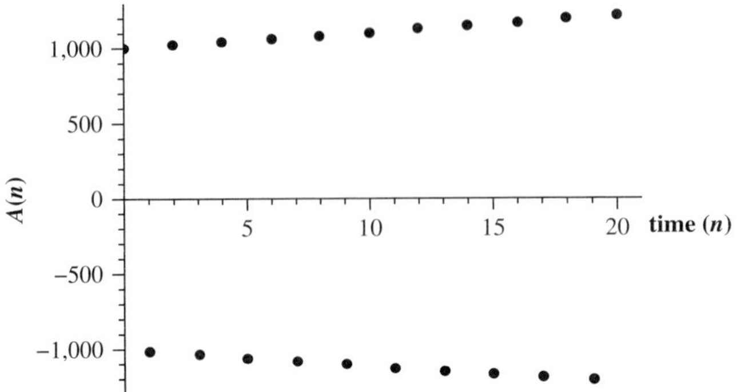

without bound as the oscillations fan out (see Figure 3.11). Although there is an equilibrium value, the solution to our example does not tend toward this equilibrium value.

Let's examine values of $r$, $0 < r < 1$. Let's take a look at a drug dosage model, $a(n + 1) = 0.5\, a(n)$ with $a(0) = 20$, where half of what is in the system is discarded each time period.

The equilibrium value is zero.

**Models of the Form $a(n + 1) = r\, a(n) + b$, Where $r$ and $b$ Are Constants**  Let's return to our drug dosage problem and consider adding the constant dosage each time period (time periods might be 4 hours). Our model is $a(n + 1) = 0.5\, a(n) + 16$ mg. We will also assume that there is an initial dosage applied prior to beginning the regime (see Figure 3.12). We will let these initial values be as follows:



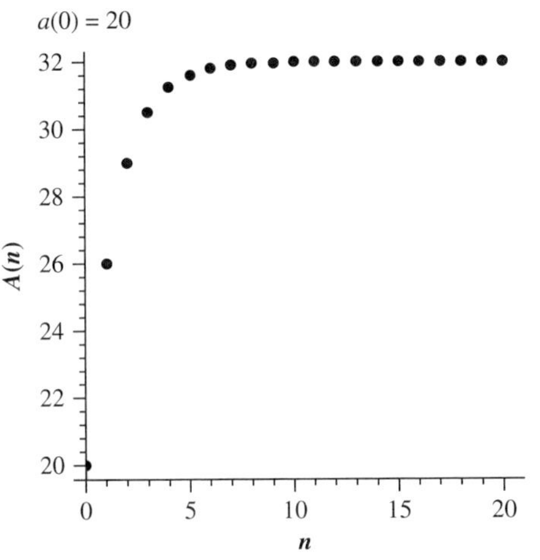

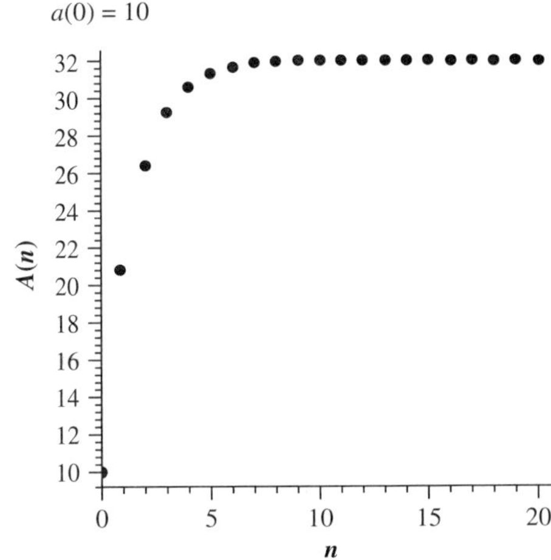

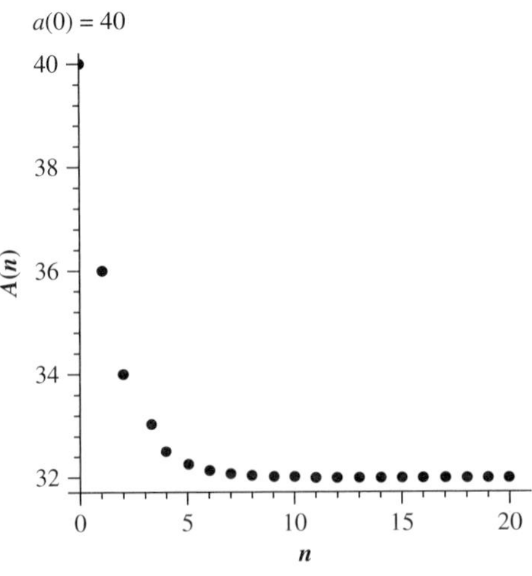

Regardless of the starting value, the future terms of $a(n)$ approach 32. Thus, 32 is the equilibrium value. We could have solved for this algebraically as well.

$$a(n + 1) = 0.5\,a(n) + 16$$
$$ev = 0.5\,ev + 16$$
$$0.5\,ev = 16$$
$$ev = 32$$

Another method of finding the equilibrium values involves solving the equation $a = ra + b$, and solving for $A$ (where $a$ is $ev$) we find:

$$a = \frac{b}{1 - r}, \text{ if } r \neq 1$$

Using this formula in our previous example, the equilibrium value is

$$a = 16/(1 - 0.5) = 32$$

## Stability and Long-Term Behavior

For a dynamical system, $a(n + 1)$ with a specific initial condition, $a(0) = a_0$, we have shown that we can compute $a(1)$, $a(2)$, and so forth. Often these particular values are not as important as the long-term behavior. By long-term behavior, we refer to what will eventually happen to $a(n)$ for larger values of $n$. There are many types of long-term behavior that can occur with DDS; we will only discuss a few here.

If the $a(n)$ values for a DDS eventually get close to the equilibrium value, $ev$, no matter the initial condition, then the equilibrium value is called a **stable equilibrium value** or **an attracting fixed point**.

**Example: Consider the DDS**

$$A(n + 1) = 0.5\,A(n) + 64$$

With initial conditions $A(0) = 0$ or $A(0) = 150$, the $ev$ is 128. The $ev$ is stable (see Figure 3.13).

```
> p1:=pointplot({seq([i,–128*(1/2)^i+128],i=0..20)}):
> p2:=pointplot({seq([i,22*(1/2)^i+128],i=0..20)}):
> display(p1,p2);
> table:=seq([i,128–128*(.5)^i,22*(.5)^i+128],i=0..20);
```

> table:= [ 0, 0., 150.], [1, 64.0, 139.0 ], [2, 96.00, 133.50 ], [3, 112.000, 130.750 ],
> [4, 120.0000, 129.3750 ], [5, 124.00000, 128.68750 ], [6, 126.000000, 128.343750 ],
> [7, 127.0000000, 128.1718750 ], [8, 127.5000000, 128.0859375 ],
> [9, 127.7500000, 128.0429688 ], [10, 127.8750000, 128.0214844 ],
> [11, 127.9375000, 128.0107422 ], [12, 127.9687500, 128.0053711 ],

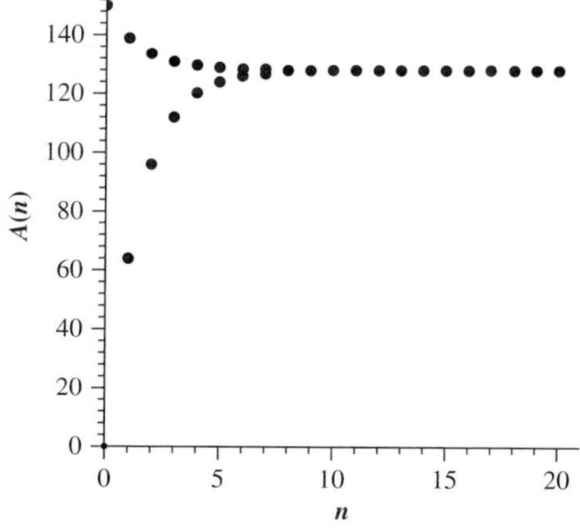


**FIGURE 3.13**

Stable equilibrium value

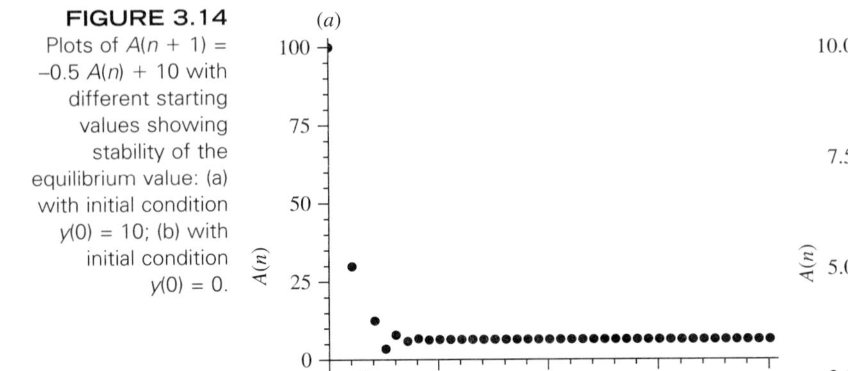

**FIGURE 3.14**
Plots of $A(n + 1) =$ $-0.5\,A(n) + 10$ with different starting values showing stability of the equilibrium value: (a) with initial condition $y(0) = 10$; (b) with initial condition $y(0) = 0$.

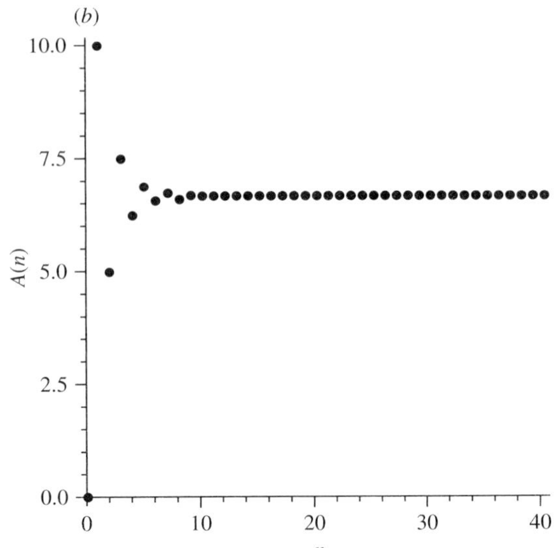

[13, 127.9843750, 128.0026855], [14, 127.9921875, 128.0013428 ],
[15, 127.9960938, 28.0006714], [16, 127.9980469, 128.0003357 ],
[17, 127.9990234, 128.0001678], [18, 127.9995117, 128.0000839 ],
[ 19, 127.9997559, 128.0000420 ], [20, 127.9998779, 128.0000210 ]

Notice that both sequences are converging on 128 as the attracting fixed point.

**Example: Consider the DDS, $A(n + 1) = -0.5\,A(n) + 10$ with $ev = 6.66667$**
With $A(0) = 100$, Figure 3.14a is the plot of behavior.
With $A(0) = 0$, Figure 3.14b is the plot of behavior.
The table of values more clearly shows that these tend to the equilibrium value.

$> rsolve(\{a(n+1)=-.5 \cdot a(n)+10, a(0)=100\}, \{a\});$

$$\left\{ a(n) = \frac{280}{3}\left(-\frac{1}{2}\right)^n + \frac{20}{3} \right\}$$

$> evalf\left( seq\left( \frac{280}{3}\left(-\frac{1}{2}\right)^n + \frac{20}{3},\ n=0..25 \right) \right);$

100., $-40.$, 30., $-5.$, 12.50000000, 3.750000000, 8.125000000,
5.937500000, 7.031250000, 6.484375000, 6.757812500,
6.621093750, 6.689453125, 6.655273438, 6.672363281,
6.663818359, 6.668090820, 6.665954590, 6.667022705,
6.666488647, 6.666755676, 6.666622162, 6.666688919,
6.666655540, 6.666672230, 6.666663885

**Example: Consider the DDS for the Financial Model, $A(n + 1) = 1.1\,A(n) + 100$, $A(0) = 100$**
The $ev$ value is $-1,000$. If the DDS ever achieves an input of $-1,000$, then the systems stays at $-1,000$ forever. But when we start at typical values like \$100 to begin the process, we find the values tend to move away from the $ev$. When this occurs, we say the $ev$ is unstable or a repelling fixed point (see Figure 3.15).

$> rsolve(\{B(n+1)=1.1 \cdot B(n)+100, B(0)=100\}, B(k));$

$$1100\left(\frac{11}{10}\right)^k - 1000$$

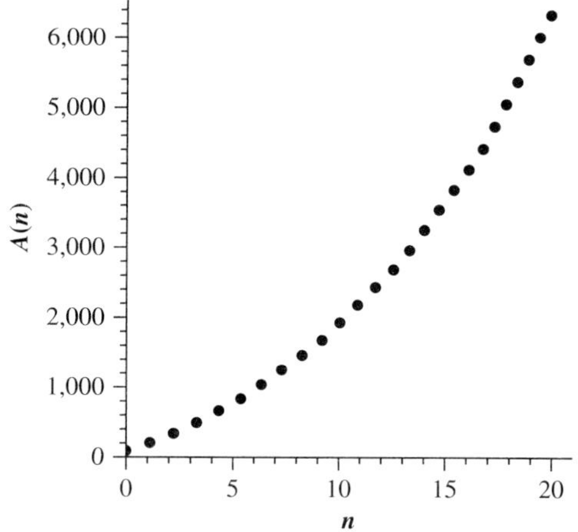



**FIGURE 3.15**

Unstable equilibrium value as the solution tends to move away from the equilibrium value

$$> evalf\left(seq\left(1100\left(\frac{11}{10}\right)^{k}-1000,\ k=0..20\right)\right);$$

> 100., 210., 331., 464.1000000, 610.5100000, 771.5610000,
> 948.7171000, 1143.588810, 1357.947691, 1593.742460,
> 1853.116706, 2138.428377, 2452.271214, 2797.498336,
> 3177.248169, 3594.972986, 4054.470285, 4559.917313,
> 5115.909045, 5727.499949, 6400.249944

$$> with\ (plots):$$

$$> pointplot\left(\left[seq\left(\left[i,\ 1100\left(\frac{11}{10}\right)^{i}-1000\right],\ i=0..20\right)\right]\right);$$

The values tend to increase over time and never move toward $-1,000$. Therefore, the *ev* is unstable.

Often, we characterize the long-term behavior of the system in terms of its stability. If a DDS has an equilibrium value and if the DDS tends to the equilibrium value from starting values near the equilibrium value, then the DDS is said to be stable.

Thus, for the dynamical system $a(n + 1) = r\,a(n) + b$, where $b \neq 0$:

| Value of $r$ | DDS Form | Equilibrium | Stability of Solution | Long-Term Behavior |
|---|---|---|---|---|
| $r = 0$ | $a(n + 1) = b$ | $b$ | Stable | Stable equilibrium |
| $r = 1$ | $a(n + 1) = a(n) + b$ | None | Unstable | |
| $r < 0$ | $a(n + 1) = r * a(n) + b$ | $b/(1 - r)$ | Depends on $\|r\|$ | Oscillations |
| $\|r\| < 1$ | $a(n + 1) = r * a(n) + b$ | $b/(1 - r)$ | Stable | Approaches $b/(1 - r)$ |
| $\|r\| > 1$ | $a(n + 1) = r * a(n) + b$ | $b/(1 - r)$ | Unstable | Unbounded |

If $r \neq 1$, an equilibrium exists at $a = b/(1 - r)$.
If $r = 1$, no equilibrium value exists.

### Relationship to Analytical Solutions

If a discrete dynamical system has an *ev* value, we can use the *ev* value to find the analytical solution.

Recall the mortgage example from Section 3.1,

$$B(n + 1) = 1.00541667\,B(n) - 639.34,\ B(0) = 73,395$$

The *ev* value is found as 118031.9274.

The analytical solution may be found using the following form:

$$B(k) = (1.00541667^k)C + D \text{ where } D \text{ is the } ev.$$
$$B(k) = (1.00541667^k)C + 11{,}8031.9274, \ B(0) = 73.395$$

Because $B(0) = 73{,}395 = 1.00541667^0 \, (C) + 118{,}031.9274$

$$C = -44{,}636.92736$$

Thus,

$$B(k) = -44{,}636.92737 \, (1.00541667^k) + 118{,}031.9274$$

Let's assume we did not know the payment was \$639.34 month. We could use the analytical solutions to help find the payment.

$$B(k) = (1.00541667^k)C + D$$

We build a system of two equations and two unknowns.

$$B(K) = (1.00541667^k)C + D$$
$$B(0) = 73{,}395 = C + D$$
$$B(180) = 0 = 1.00541667^{180} \, C + D$$
$$C = -44{,}638.70, \ D = 118{,}033.7$$
$$B(K) = -44{,}638.70(1.00541667)^k + 118{,}033.7$$

$D$ represents the equilibrium value, and we accepted some round-off error. From our model form

$$B(n + 1) = 1.00541667 \, B(n) - P$$

we can find $P$.

Solving analytically for the equilibrium value,

$$X - 1.00541667X = -P$$
$$X = P/.00541667$$

$X$ is 11,8033.70, so

$$11{,}8033.70 = P/.00541667$$
$$P = 639.34$$

### The Limit

As we study the behavior of equilibrium values, we need to understand the concept of a *limit*. When you study calculus, you use this definition or a related definition for the limit. For now, we need a simple definition so that we all think of the same thing when we say *limit*.

We are interested in what happens to $a(k)$ as $k$ gets large without bound (In other words, when we iterate many times.) It may happen that for large values of $k$, $a(k)$ is close to or equal to some number we'll call **L**. If increasing $k$ (taking more iterations) causes $a(k)$ to get even closer to **L** until $a(k)$ is nearly equal to or equal to **L** for very large values of $k$, we call **L** the limit of $a(k)$. We write this as $\lim_{k \to \infty} a(k) = \mathbf{L}$. If $a(k)$ has a limit, then we also say $a(k)$ converges to **L** or $a(k)$ converges. It is important to understand that if $a(k)$ *converges* to **L**, then further increases in $k$ will never cause $a(k)$ to move too far away from **L**. Also remember that we are looking at large values of $k$. How large is large may depend on the problem. You will have to develop a feel for how large $k$ must be.

If increasing $k$ causes $a(k)$ to increase and $a(k)$ does not start to converge on a number **L**, we say $a(k)$ is diverging and the limit does not exist.

For example, consider the difference equation $a(n + 1) = 0.9a(n) + 2$ with $a(0) = 1$. Iterating to an accuracy of six decimal places produces the following:

$$a(150) = 19.999997$$
$$a(151) = 19.999998$$
$$.\ .$$
$$.\ .$$
$$.\ .$$
$$a(156) = 19.999999$$
$$.\ .$$
$$.\ .$$
$$.\ .$$
$$a(166) = 20.000000$$

and for $k > 166$, $a(k) = 20$.

For this difference equation, the limit of $a(k)$ is 20.

As another example, consider the difference equation $a(n + 1) = -0.1a(n) + 11$ with $a(0) = 1$. To an accuracy of six decimal places, we see the following:

$$a(4) = 9.999100$$
$$a(5) = 10.000090$$
$$a(6) = 9.999991$$
$$a(7) = 10.000001$$
$$a(8) = 10.000000$$

and for $k > 8$, the limit of $a(k)$ is 10.

Even though $a(4)$ is less than 10 and $a(5)$ is greater than 10, $a(5)$ is closer to 10 than $a(4)$. Increasing $k$ caused $a(k)$ to get closer to 10. Thus, the limit as $k \to \infty$ of $a(k)$ is 10.

If $a(k)$ does not converge to **L**, then it *diverges*. There are several ways in which it may diverge. If $a(k)$ gets infinitely large as $k$ gets infinitely large, then $a(k)$ diverges. If $a(k)$ gets infinitely large in the negative direction as $k$ gets infinitely large, then $a(k)$ also diverges. Also, $a(k)$ may oscillate between large positive and large negative values, always getting further from 0 as $k$ gets infinitely large. That is also an example of $a(k)$ diverging. If $a(k)$ oscillates in a pattern between two or more fixed values as $k$ gets infinitely large, then $a(k)$ diverges. If $a(k)$ shows absolutely no pattern of behavior as $k$ gets infinitely large, then $a(k)$ diverges.

An example of a divergent difference equation is

$$a(n + 1) = 4a(n) + 2 \text{ with } a(0) = 1$$

Iteration of this difference equation produces these results:

$$a(10) = 1,747,626$$
$$a(11) = 6,990,506$$
$$a(12) = 27,962,026$$
$$a(13) = 111,848,106$$

Further increase in $k$ causes an increase in $a(k)$. Thus, $a(k)$ diverges. Another divergent difference equation is

$$a(n + 1) = -4a(n) + 2 \text{ with } a(0) = 1$$

In this case, iteration produces:

$$a(10) = 629,146$$
$$a(11) = -2,516,582$$
$$a(12) = 1,006,630$$
$$a(13) = -40,265,318$$

As $k$ gets larger, $a(k)$ gets further from 0, always oscillating between positive and negative values.

If the dynamical system has a limit, then that limit is a stable equilibrium value (fixed-point attractor).

<table><tr><td>**3.2**</td><td># EXERCISES</td></tr></table>

1. For the following DDS, find the equilibrium value if one exists. Classify the DDS as stable or unstable.

   **a.** $a(n + 1) = 1.23\, a(n)$
   **b.** $a(n + 1) = 0.99\, a(n)$
   **c.** $a(n + 1) = -0.8\, a(n)$
   **d.** $a(n + 1) = a(n) + 1$
   **e.** $a(n + 1) = 0.75\, a(n) + 21$
   **f.** $a(n + 1) = 0.80\, a(n) + 100$
   **g.** $a(n + 1) = 0.80\, a(n) = 100$
   **h.** $a(n + 1) = -0.80\, a(n) + 100$

2. Build a numerical table for the following initial value DDS problems. Observe the patterns and provide information on equilibrium values and stability.

   **a.** $a(n + 1) = 1.1\, a(n) + 50,\ a(0) = 1010$
   **b.** $a(n + 1) = 0.85\, a(n) + 100,\ a(0) = 10$
   **c.** $a(n + 1) = 0.75\, a(n) - 100,\ a(0) = -25$
   **d.** $a(n + 1) = a(n) + 100,\ a(0) = 500$

<table><tr><td>**3.3**</td><td># MODELING NONLINEAR DISCRETE DYNAMICAL SYSTEMS</td></tr></table>

**Introduction**

In this section, we build nonlinear discrete dynamical systems to describe the change in behavior of the quantities we study. We also will study systems of DDS to describe the changes in various systems that act together in some way or ways. We define a nonlinear DDS: If the function of $a(n)$ involves powers of $a(n)$ [such as $a^2(n)$] or a functional relationship [such as $a(n)/a(n - 1)$], we will say that the discrete dynamical system is **nonlinear**. A **sequence** is a function whose domain is the set of nonnegative integers ($n = 0, 1, 2, \ldots$). We will restrict our model solution to the numerical and graphical solutions. Analytical solutions may be studied in more advanced mathematics courses.

<table><tr><td>**Example 1**</td><td>## Growth of a Yeast Culture</td></tr></table>

We often model population growth by assuming that the change in population is directly proportional to the current size of the given population. This produces a simple, first-order DDS similar to those seen earlier. It might appear reasonable at first examination, but the long-term behavior of growth without bound is disturbing. Why would growth without bound of a yeast culture in a jar (or controlled space) be alarming?

Certain factors affect population growth. Things include resources (food, oxygen, space, etc.) These resources can support some maximum population. As this number is approached, the change (or growth rate) should decrease, and the population should never exceed its resource supported amount.

## Problem Identification

Predict the growth of yeast in a controlled environment as a function of the resources available and the current population.

## Assumptions and Variables

We assume that the population size is best described by the weight of the biomass of the culture. We define $y(n)$ as the population size of the yeast culture after period $n$. There exists a maximum carrying capacity, $M$, that is sustainable by the resources available. The yeast culture is growing under the conditions established.

## Model

$$y(n + 1) = y(n) + k\, y(n)\, (M - y(n))$$

where

$y(n)$ is the population size after period $n$
$n$ is the time period measured in hours
$k$ is the constant of proportionality
$M$ is the carrying capacity of our system

In our experiment, we first plot $y(n)$ versus $n$ and find a stable $ev$ of approximately 665. Next, we plot $y(n + 1) - y(n)$ versus $y(n)$ $(665 - y(n))$ to find the slope, $k$, is approximately 0.00082, With $k = 0.00082$ and the carrying capacity in biomass is 665. This model is

$$y(n + 1) = y(n) + 0.00082\, y(n)\, (665 - y(n))$$

Again, this is nonlinear because of the $y^2(n)$ term. The following is the solution iterated in Maple (there is no closed-form analytical solution for this equation) from an initial condition, biomass, of 9.6:

```
> biomass:=proc(N,M,S) option remember;
> B(N):=M;
> for i from N to S do
> B(i+1):=B(i)+.00082*B(i)*(665−B(i));
> end do;
> end;
```

$$biomass := \text{proc}\ (N, M, S)$$
$$\text{local } i;$$
$$\text{option } remember;$$
$$B(N) := M;$$
$$\text{for } i \text{ from } N \text{ to } S \text{ do } B(i + 1) := B(i) + 0.00082 \times B(i) \times (665 - B(i)) \text{ end}$$
$$\text{do end proc}$$

```
> biomass(0,9.6,30);
```

$$664.9999559$$

```
>pointplot( {seq( [i,B(i)],i=0..30)});
```

The model shows stability in that the population (biomass) of the yeast culture approaches 665 as $n$ gets large (see Figure 3.16). Thus, the population is eventually stable at approximately 665 units.

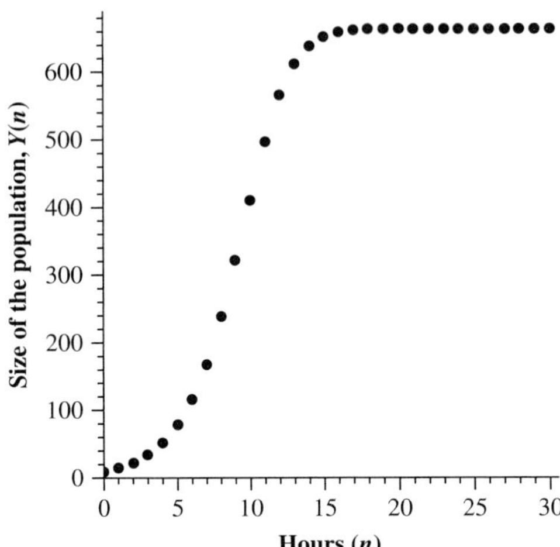

**FIGURE 3.16**

Plot of DDS from growth of a yeast culture

---



## Spread of a Contagious Disease

There are 1,000 students in a college dormitory, and some students have been diagnosed with meningitis, a highly contagious disease. The health center wants to build a model to determine how fast the disease will spread.

## Problem Identification

Predict the number of students affected with meningitis as a function of time.

## Assumptions and Variables

Let m($n$) be the number of students affected with meningitis after $n$ days. We assume all students are susceptible to the disease. The possible interactions of infected and susceptible students are proportional to their product (as an interaction term).

The model is

$$m(n + 1) - m(n) = k\, m(n)\, (1{,}000 - m(n))$$

or

$$m(n + 1) = m(n) + k\, m(n)\, (1{,}000 - m(n))$$

Two students returned from spring break with meningitis. The rate of spreading per day is characterized by $k = 0.0090$. It is assumed that a vaccine can be in place and students vaccinated within 1 to 2 weeks.

$$m(n + 1) = m(n) + 0.00090\, m(n)\, (1{,}000 - m(n))$$

```
> spread:=proc(N,M,S) option remember;
> for i from N to S do
> Sp(i+1):=Sp(i)+.0009*Sp(i)*(1000−Sp(i));
> end do;
> end;
```

```
spread := proc( N, M, S)
option remember;
local N1, M1, S1, i;
N1 := N;
M1 := M;
S1 := S;
Sp(N1) := M1;
for i from N1 to S1 do
Sp(i+ 1) := Sp(i) + 0.0009*Sp(i)*(1000−Sp(i))
end do
end proc
```

> *spread(0,2,30);*

$$1000.000000$$

> *seq( Sp(i),i=0..30);*

2, 3.7964, 7.200188612, 13.63369992, 25.73673985,
48.30366391, 89.67704188, 163.1486049, 286.0266287,
469.8204854, 694.0007626, 885.1280963, 976.6368108,
997.1724263, 999.7100470, 999.9709290, 999.9970921,
999.9997092, 999.9999709, 999.9999971, 999.9999997,
1000.000000, 1000.000000, 1000.000000, 1000.000000,
1000.000000, 1000.000000, 1000.000000, 1000.000000,
1000.000000, 1000.000000

> *with(plots):*
> *pointplot( {seq( [i,Sp(i)],i = 0..15)});*

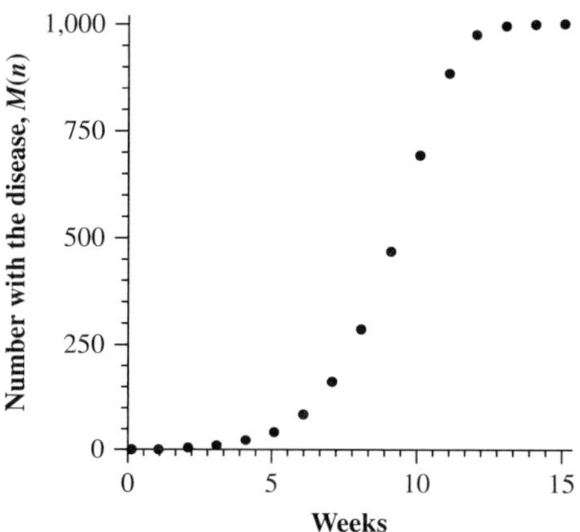

**FIGURE 3.17**
Plot of DDS for the spread of a disease

## Interpretation

The results show that most students will be affected within 2 weeks (see Figure 3.17). Because only about 10 percent will be affected within one week, every effort must be made to get the vaccination at the school and get the students vaccinated within one week.

## 3.3   EXERCISES

Consider the model $a(n + 1) = r\, a(n)\, (1 − a(n))$. Let $a(0) = 0.2$. Determine the numerical and graphical solution for the following values of $r$. Find the pattern in the solution.

1. $r = 2$
2. $r = 3$
3. $r = 3.6$
4. $r = 3.7$

For problems 5–8, find the equilibrium value by iteration and determine if it is stable or unstable.

5. $a(n + 1) = 1.7\,a(n) - 0.14\,a(n)^2$
6. $a(n + 1) = 0.8\,a(n) + 0.\,1\,a(n)^2$
7. $a(n + 1) = 0.2\,a(n) - 0.2\,a(n)^3$
8. $a(n + 1) = 0.1\,a(n)^2 + 0.9\,a(n) - 0.2$

9. Consider a spreading rumor through a company of 1,000 employees, all working in the same building. We assume it is similar to the spread of a contagious disease in that the number of people hearing the rumor each day is proportional to the product of the number hearing the rumor and the number who have not heard the rumor. This is given by $r(n + 1) = r(n) + 1{,}000\,k\,r(n) - k\,r(n)^2$, where $k$ is the parameter that depends on how fast the rumor spreads. Assume $k = 0.001$ and further assume that four people initially know the rumor. How soon will everyone know the rumor?

# 3.3 | PROJECTS

1. Consider the contagious disease as the Ebola virus. Look it up on the Internet and find out how deadly this virus actually is. Now consider an animal research laboratory in Reston, Virginia, a suburb of Washington, D.C., with a population of 856,900 people. A monkey with the Ebola virus has escaped captivity and infected one employee (unknown at the time) during its escape. This employee reports to university hospital later with Ebola hemorrhagic fever symptoms. The Infectious Disease Center (IDC) in Atlanta gets a call and begins to model the spread of the disease. Build a model for the IDC with the following growth rates to determine the number of humans infected after two weeks:

   a. $k = 0.00025$
   b. $k = 0.000025$
   c. $k = 0.00005$
   d. $k = 0.000009$
   e. List some ways of controlling the spread of the virus.

2. Consider the spread of a rumor concerning termination among 1,000 employees of a major company. Assume that the spreading of a rumor is similar to the spread of contagious disease in that the number hearing the rumor each day is proportional to the product of those who have heard the rumor and those who have not heard the rumor. Build a model for the company with the following rumor growth rates to determine the number of employees who have heard the rumor after one week:

   a. $k = 0.25$
   b. $k = 0.025$
   c. $k = 0.0025$
   d. $k = 0.00025$
   e. List some ways of controlling the spread of the rumor.

# 3.4  SYSTEMS OF DISCRETE DYNAMICAL SYSTEMS

In this section, we examine models of systems of difference equations (DDS). For selected initial conditions, we build numerical solutions to get a sense of long-term behavior for the system. For the systems we study, we will find their equilibrium values. We then explore starting values near the equilibrium values to see whether, by starting close to an equilibrium value, the system will:

1. remain close,
2. approach the equilibrium value, or
3. not remain close.

What happens near these values gives great insight concerning the long-term behavior of the system. We can study the resulting pattern of the numerical solutions.

---

**Example 1**     Merchants Located Downtown and in Malls

Let's consider the attempt to revitalize the downtown section of a small city with merchants. There are merchants downtown and others in the large mall. Suppose historical records determined that 60 percent of the downtown merchants remain downtown while 40 percent move to the mall. We find the 70 percent of the mall merchants want to remain in the mall but 30 percent want to move to downtown. Build a model to determine the long-term behavior of these merchants based upon these historical data.

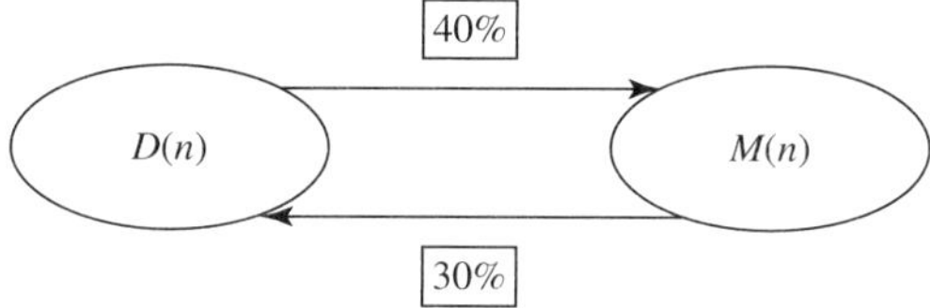


# Problem Identification

Determine the relationship of the merchants over time.

# Assumptions and Variables

Let $n$ represent the number of business months. We define the following:

$D(n)$ = the number of merchants operating downtown at the end of $n$ months

$M(n)$ = the number of merchants operating at the mall at the end of $n$ months

We assume that no other incentives are given to the merchants for either staying or moving.

# The Model

The number of merchants downtown in any time period is equal to the number of downtown merchants who stay downtown plus the number of mall merchants who relocate downtown.

The same is true for the number of mall merchants: In any time period, it is equal to the number who remain in the mall plus the number of downtown merchants who move to the mall. Mathematically, this is written as:

$$D(n + 1) = 0.60\, D(n) + 0.30\, M(n)$$
$$M(n + 1) = 0.4\, D(n) + 0.7\, M(n)$$

There are initially 150 merchants in the mall and 100 merchants downtown. We seek to find the long-term behavior of this system.

We rewrite the model as a system of DDS:

$$D(n + 1) = 0.6\, D(n) + 0.3\, M(n)$$
$$M(n + 1) = 0.4\, D(n) + 0.7\, M(n)$$
$$D(0) = 150 \text{ and } M(0) = 100 \text{ merchants, respectively.}$$

```
> u:=n–>if n=0 then 150 else .6*u(n–1)+.3*v(n–1) fi ;
> v:=n–>if n=0 then 100 else .4*u(n–1)+.7*v(n–1)fi ;
> u(0);
```

$$150$$

```
> v(0);
```

$$100$$

```
> merchants:=seq(u(n),n=0..10);
```

$$merchants := 150, 120.0, 111.00, 108.300, 107.4900, 107.24700, 107.174100,$$
$$107.1522300, 107.1456690, 107.1437007, 107.1431102$$

```
> downtown:=seq(v(n),n=0..10);
```

$$downtown := 100, 130.0, 139.00, 141.700, 142.5100, 142.75300, 142.825900,$$
$$142.8477700, 142.8543310, 142.8562993, 142.8568898$$

```
> with(plots):
> n1:=pointplot( {seq( [n,u(n) ],n=0..10)},title=`Retail`):
> n2:=pointplot( {seq( [n,v(n) ],n=0..10)}):
> display( {n1,n2});
```

Analytically, we can solve for the equilibrium values (see Figure 3.18). We let $X = D(n)$ and $Y = M(n)$. From the DDS, we obtain the equations:

$$X = 0.6X + 0.3\,Y$$
$$Y = 0.4\,X + 0.7\,Y$$

and both equations reduce to $X = \frac{3}{4}\,Y$. There are two unknowns, so we need a second equation. From the initial conditions, we know that $X + Y = 250$. We can use the equations $X + Y = 250$ and $X = \frac{3}{4}\,Y$ to find the equilibrium values:

$$X = 107.1428571 \qquad \text{and} \qquad Y = 142.85714329$$

We iterate from near those equilibrium values, and we find the sequences tend toward those values. We conclude the system has *stable* equilibrium values.

Go back and change the initial conditions and see what behavior follows.

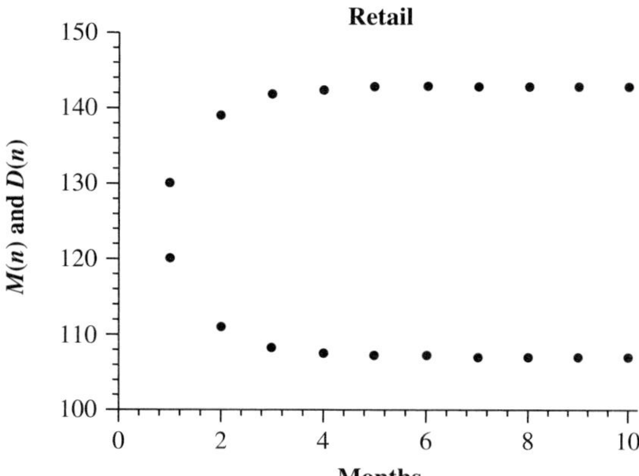


**FIGURE 3.18**
Plot of DDS for merchant relocation example

## Interpretation

The long-term behavior shows that eventually (without other influences) 107 of the 250 merchants will be downtown, and about 143 will be in the mall. We might want to try to attract new businesses to the community and add incentives for operating the business downtown.

## Example 2

## Competitive-Hunter Models

Competitive-hunter models involve species vying for the same resources (such as food or living space) in the habitat. The effect of the presence of a second species diminishes the growth rate of the first species. We now consider a specific example concerning trout and bass in a small pond.

Hugh Ketum owns a small pond that he uses to stock fish and eventually allows fishing. He has decided to stock both bass and trout. The fish and game warden tells Hugh that after inspecting his pond for environmental conditions he has a solid pond for growth of his fish. In isolation, bass grow at a rate of 20 percent and trout at a rate of 30 percent. The warden tells Hugh that because both species compete for the same resources, the interactions affect trout more than bass. We measure interaction by $\alpha \cdot x \cdot y$ where $\alpha$ represents the rate of competition. They estimate the interaction affecting bass is 0.0010 bass * trout and for trout is 0.0020 bass * trout. Assume no other changes in the habitat occur.

## Model

Let's define the following variables:

$B(n)$ = the number of bass in the pond after period $n$.
$T(n)$ = the number of trout in the pond after period $n$.
$B(n) * T(n)$ = interaction of the two species.
$B(n + 1) = 1.20 \, B(n) - 0.0010 \, B(n) * T(n)$
$T(n + 1) = 1.30 \, T(n) - 0.0020 \, B(n) * T(n)$

The equilibrium values can be found by allowing $X = B(n)$ and $Y = T(n)$ and solving for $X$ and $Y$.

$$X = 1.2 \, X - 0.001 \, X * Y \tag{3.1}$$
$$Y = 1.3 \, Y - 0.0020 \, X * Y \tag{3.2}$$

We rewrite these equations as

$$0.2\ X - 0.001\ X * Y = 0 \tag{3.3}$$
$$0.3\ Y - 0.002\ X * Y = 0 \tag{3.4}$$

We factor $X$ out of (3.3) and $Y$ out of (3.4) to obtain

$$X\ (0.2 - 0.001\ Y) = 0$$
$$Y(0.3 - 0.002\ X) = 0$$

Solving, we find $X = 0$ or $Y = 200$ and $Y = 0$ or $X = 150$

We want to know the long-term behavior of the system and the stability of the equilibrium points.

Hugh initially considers 151 bass and 199 trout for his pond. The solution is left to the student as an exercise. From Hugh's initial conditions, bass will grow without bound, and trout will eventually die out. This is certainly not what Hugh had in mind.

---

<table>
<tr><td>**Example 3**</td><td></td></tr>
</table>

## Fast-Food Tendencies

Consider that your student union center desires to have three fast-food chains available to students, serving: burgers, tacos, and pizza. These chains run a survey of students and find the following information concerning lunch: 75 percent who ate burgers will eat burgers again at the next lunch, 5 percent will eat tacos next, and 20 percent will eat pizza next. Of those who ate tacos last, 20 percent will eat burgers next, 60 percent will stay with tacos, and 35 percent will eat pizza next. Of those who ate pizza, 40 percent will eat burgers next, 20 percent tacos, and 40 percent pizza again.

We formulate the problem as follows. Let $n$ represent the $n$th days lunch and define:

$$B(n) = \text{the number of burger eaters in the } n\text{th lunch.}$$
$$T(n) = \text{the number of taco eaters in the } n\text{th lunch.}$$
$$P(n) = \text{the number of pizza eaters in the } n\text{th lunch.}$$

Formulating the system, we have the following dynamical system:

$$B(n + 1) = 0.75\ B(n) + 0.20\ T(n) + 0.40\ P(n)$$
$$T(n + 1) = 0.05\ B(n) + 0.60\ T(n) + 0.20\ P(n)$$
$$P(n + 1) = 0.20\ B(n) + 0.20\ T(n) + 0.40\ P(n)$$

Analytically, we let $X = B(n)$, $Y = T(n)$, and $Z = P(n)$ so that

$$X = 0.75\ X + 0.2\ Y + 0.4\ Z$$
$$Y = 0.05\ X + 0.6\ Y + 0.2\ Z$$
$$Z = 0.2\ X + 0.2\ Y + 0.4\ Z$$

These equations reduce to

$$X = 20/9\ Z$$
$$Y = 7/9\ Z$$
$$Z = Z$$

Because we have 14,000 students, then $X + Y + Z = 14000$

We substitute and solve for $Z$ first.

$$4\ Z = 14{,}000$$
$$Z = 3{,}500$$
$$X = 20/9\ Z = 20/9\ (3500) = 7{,}777.77$$
$$Y = 7/9\ Z = 7/9\ (3{,}500) = 2{,}722.222$$

Suppose the campus has 14,000 students who eat one of the three fast-food lunches. The graphical results also show that an equilibrium value is reached at a value of about 7,778 burger eaters, 2,722 taco eaters, and 3,500 pizza eaters. This allows the fast-food establishments to plan for a projected future. By varying the initial conditions for 14,000 students, we find that these values are stable equilibrium values.

## 3.4 | EXERCISES

1. What happens to the merchant problem if 200 merchants were initially in the mall and 50 were in the downtown portion?

2. Determine the equilibrium values of the bass and trout. Can these levels ever be achieved and maintained? Explain.

3. Test the fast-food models with different starting conditions summing to 14,000 students. What happens? Obtain a graphical output and analyze the graph in terms of long-term behavior.

## 3.4 | PROJECTS

1. Owls and Hawks

   PID: Predict the number of owls and hawks in the same environment as a function of time.

   Assumptions

   The variables:
   $O(n)$ = number of owls at then end of period $n$
   $H(n)$ = number of hawks at the end of period $n$

   Model:
   $$O(n + 1) = 1.2\, O(n) - 0.001\, O(n)\, H(n)$$
   $$H(n + 1) = 1.3\, H(n) - 0.002\, H(n)\, O(n)$$

   a. Find the equilibrium values of the system.
   b. Iterate the system from the following initial conditions and determine what happens to the hawks and owls in the long term.

   | Owls | Hawks |
   | --- | --- |
   | 150 | 200 |
   | 151 | 199 |
   | 149 | 210 |
   | 10 | 10 |
   | 100 | 100 |

2. George and Gracie play racquetball very often and are very competitive. Their racquetball matches consist of two games. When George wins the first game, he wins the second game 65 percent of the time. When Gracie wins the first game, she wins the second only 45 percent of the time. Model this as a DDS and determine the long-term percentages of their racquetball games. What assumptions are necessary?

3. It is getting close to election day. The influence of the new Independent Party is of concern to the Republicans and Democrats. Assume that in the next election that 75 percent of those who vote Republican vote Republican again, 5 percent vote Democratic, and 20 percent vote Independent. Of those that voted Democratic before, 20 percent vote

Republican, 60 percent vote Democratic, and 20 percent vote Independent. Of those that voted Independent, 40 percent vote Republican, 20 percent vote Democratic, and 40 percent vote Independent.

**a.** Formulate and write the system of discrete dynamical systems that models this situation.

**b.** Assume that there are 399,998 voters initially in the system. How many will vote Republican, Democratic, and Independent in the long run? (*Hint:* You can break down the 399,998 voters in any manner that you desire as initial conditions.)

**c.** New scenario: In addition to the above, the community is growing (18-year-olds + new people − deaths − losses to the community, etc.). Republicans predict a gain of 2,000 voters between elections. Democrats estimate a gain of 2,000 voters between elections. The independents estimate a gain of 1,000 voters between elections. If this rate of growth continues, what will be the long-term distribution of the voters?

## 3.4 | FURTHER READING

Sandefur, James. *Elementary Mathematical Model: A Dynamic Approach*. Belmont, CA: Brooks-Cole Publishers, 2002.

## 3.5 | MODELING OF PREDATOR–PREY MODEL, SIR MODEL, AND MILITARY MODELS

### Example 1 — A Predator–Prey Model: Foxes and Rabbits

In the study of the dynamics of a single population, we typically take into consideration such factors as the *natural* growth rate and the *carrying capacity* of the environment. Mathematical ecology requires the study of populations that interact, thereby affecting each other's growth rates. In this module, we study a very special interaction in which there are exactly two species, one in which predators eat the prey. Such pairs exist throughout nature—for example,

- lions and gazelles,
- birds and insects,
- pandas and eucalyptus trees, and
- Venus fly traps and flies.

To keep our model simple, we will make some assumptions that would be unrealistic in most of these predator–prey situations. Specifically, we will assume that

- the predator species is totally dependent on a single prey species as its only food supply,
- the prey species has an unlimited food supply, and
- there are no other threats to the prey other than the specific predator.

## The Discrete Lotka–Volterra Model for Predator–Prey Models

Vito Volterra (1860–1940) was a famous Italian mathematician who retired from a distinguished career in pure mathematics in the early 1920s. His son-in-law, Humberto D'Ancona, was a biologist who studied the populations of various species of fish in the Adriatic Sea. In 1926, D'Ancona completed a statistical study of the numbers of each species sold on the fish markets of three ports: Fiume, Trieste, and Venice. The percentages of predator species (sharks, skates, rays, etc.) in the Fiume catch are shown in the following table.

Percentages of Predators in the Fiume Fish Catch

| 1914 | 1915 | 1916 | 1917 | 1918 | 1919 | 1920 | 1921 | 1922 | 1923 |
|------|------|------|------|------|------|------|------|------|------|
| 12   | 21   | 22   | 21   | 36   | 27   | 16   | 16   | 15   | 11   |

We may assume that proportions within the harvested population reflect those in the total population. D'Ancona observed that the highest percentages of predators occurred during and just after World War I, when fishing was drastically curtailed. He concluded that the predator–prey balance was at its natural state during the war and that intense fishing before and after the war disturbed this natural balance—to the detriment of predators. Having no biological or ecological explanation for this phenomenon, D'Ancona asked Volterra if he could come up with a mathematical model that might explain what was going on. In a matter of months, Volterra developed a series of models for interactions of two or more species. The first and simplest of these models is the model that we will use for this scenario.

Alfred J. Lotka (1880–1949) was an American mathematical biologist (and later actuary) who formulated many of the same models as Volterra, independently and at about the same time. His primary example of a predator–prey system comprised a plant population and a herbivorous animal that was dependent on that plant for food.

We repeat our two key assumptions:

- The predator species is totally dependent on the prey species as its only food supply.
- The prey species has an unlimited food supply, and there are no threats to its growth other than the specific predator.

If there were no predators, the second assumption would imply that the prey species grows exponentially without bound, that is, if $x = x(n)$ is the size of the prey population after a discrete time period $n$, then we would have $x(n + 1) = a\,x(n)$.

But there *are* predators, which must account for a negative component in the prey growth rate. Suppose we write $y = y(n)$ for the size of the predator population at time $t$. Here are the crucial assumptions for completing the model:

- The rate at which predators encounter prey is jointly proportional to the sizes of the two populations.
- A fixed proportion of encounters lead to the death of the prey.

These assumptions lead to the conclusion that the negative component of the prey growth rate is proportional to the product $xy$ of the population sizes—that is,

$$x(n + 1) = x(n) + ax(n) - bx(n)\,y(n)$$

Now we consider the predator population. If there were no food supply, the population would die out at a rate proportional to its size; that is, we would find $y(n + 1) = -cy(n)$.

We assume that is the simple case that the natural growth rate is a composite of birthrates and death rates, both presumably proportional to population size. In the absence of food, there is no energy supply to support the birthrate. But there is a food supply: the prey. And what's bad for hares is good for foxes. In other words, the energy to support growth of the predator population is proportional to deaths of prey, so

$$y(n + 1) = y(n) - cy(n) + px(n)\,y(n)$$

This discussion leads to the discrete version of the Lotka−Volterra predator−prey model:

$$\begin{aligned}
x(n + 1) &= (1 + a)\,x(n) - bx(n)\,y(n) \\
y(n + 1) &= (1 - c)\,y(n) + px(n)\,y(n) \\
n &= 0,1,2,\dots
\end{aligned} \tag{3.5}$$

where $a$, $b$, $c$, and $p$ are positive constants.

The Lotka–Volterra model, Equation 3.5, consists of a system of linked differential equations that cannot be separated from each other and that cannot be solved in closed form. Nevertheless, they can be solved numerically and graphed in order to obtain insights about the scenario being studied.

In our foxes and rabbits scenario, let's assume this discrete model explained above. Further, data investigation yields the following estimates for the parameters $\{a,b,c,p\}$ = $\{0.039,0.0003,0.12,0.0001\}$.

We use Maple to investigate the model's results.

```
> predprey:=proc(fox,rabbit,a,b,c,d,n)
> option remember;
> f(0):=fox;
> r(0):=rabbit;
> for i from 1 to n do
> f(i):=evalf((f(i–1)–(c)*f(i–1)+d*f(i–1)*r(i–1)));
> r(i):=evalf(r(i–1)+(a)*r(i–1)–b*f(i–1)*r(i–1));
> end do;
> end;
```

$$predprey := \mathrm{proc}\textit{(fox, rabbit, a, b, c, d, n)}$$

local $i$;

option *remember*;

$\quad$ f(0) := $fox$;

$\quad$ r(0) := $rabbit$;

for $i$ to $n$ do

$\quad$ f$(i)$ := evalf(f$(i-1) - c \times$ f$(i-1) + d \times$ f$(i-1) \times$ r$(i-1)$);

$\quad$ r$(i)$ := evalf(r$(i-1) + a \times$ r$(i-1) - b \times$ f$(i-1) \times$ r$(i-1)$))

end do

end proc

```
> predprey(110,900,.039,0.0003,.12,0.0001,100);
```

$$898.3459060$$

```
> seq([i,r(i),f(i)],i=1..100);
```

[1, 905.4000, 106.7000 ], [2, 911.7287460, 103.5566180 ],
[3, 918.9615035, 100.5713784 ], [4, 927.0746346, 97.74493550 ],
[5, 936.0454903, 95.07722828 ], [6, 945.8522812, 92.56762197 ],
[7, 956.4739313, 90.21503697 ], [8, 967.8899153, 88.01806564 ],
[9, 980.0800826, 85.97507757 ], [10, 993.0244677, 84.08431437 ],
[11, 1006.703088, 82.34397480 ], [12, 1021.095728, 80.75229119 ],
[13, 1036.181715, 79.30759821 ], [14, 1051.939677, 78.00839473 ],
[15, 1068.347286, 76.85339992 ], [16, 1085.380994, 75.84160405 ],
[17, 1103.015742, 74.97231512 ], [18, 1121.224663, 74.24520169 ],
[19, 1139.978760, 73.66033261 ], [20, 1159.246568, 73.21821416 ],
[21, 1178.993795, 72.91982481 ], [22, 1199.182947, 72.76664793 ],
[23, 1219.772925, 72.76070251 ], [24, 1240.718609, 72.90457170 ],
[25, 1261.970417, 73.20142898 ], [26, 1283.473852, 73.65506128 ],
[27, 1305.169028, 74.26988845 ], [28, 1326.990193, 75.05097765 ],
[29, 1348.865238, 76.00405146 ], [30, 1370.715215, 77.13548758 ],
[31, 1392.453872, 78.45230771 ], [32, 1413.987207, 79.96215274 ],
[33, 1435.213070, 81.67324051 ], [34, 1456.020829, 83.59430187 ],
[35, 1476.291128, 85.73449012 ], [36, 1495.895762, 88.10325802 ],
[37, 1514.697710, 90.71019609 ], [38, 1532.551363, 93.56482519 ],

[39, 1549.302996, 96.67633621 ], [40, 1564.791532, 100.0532696 ],
[41, 1578.849649, 103.7031282 ], [42, 1591.305291, 107.6319176 ],
[43, 1601.983625, 111.8436115 ], [44, 1610.709496, 116.3395415 ],
[45, 1617.310405, 121.1177169 ], [46, 1621.620028, 126.1720853 ],
[47, 1623.482255, 131.4917532 ], [48, 1622.755705, 137.0601956 ],
[49, 1619.318614, 142.8544935 ], [50, 1613.073958, 148.8446483 ],
[51, 1603.954614, 154.9930331 ], [52, 1591.928307, 161.2540482 ],
[53, 1577.002046, 167.5740508 ], [54, 1559.225740, 173.8916268 ],
[55, 1538.694654, 180.1382616 ], [56, 1515.550412, 186.2394482 ],
[57, 1489.980296, 192.1162416 ], [58, 1462.214704, 197.6872340 ],
[59, 1432.522723, 202.8708839 ], [60, 1401.205964, 207.5880929 ],
[61, 1368.590895, 211.7648892 ], [62, 1335.020090, 215.3350524 ],
[63, 1300.842888, 218.2425082 ], [64, 1266.405997, 220.4433287 ],
[65, 1232.044605, 221.9072046 ], [66, 1198.074473, 222.6182974 ],
[67, 1164.785387, 222.5754316 ], [68, 1132.436234, 221.7916408 ],
[69, 1101.251780, 220.2931329 ], [70, 1071.421138, 218.1177775 ],
[71, 1043.097763, 215.3132439 ], [72, 1016.400747, 211.9349309 ],
[73, 991.4171294, 208.0438214 ], [74, 968.2049345, 203.7043836 ],
[75, 946.7966502, 198.9826165 ], [76, 927.2028972, 193.9443100 ],
[77, 909.4160924, 188.6535654 ], [78, 893.4139435, 183.1715964 ],
[79, 879.1626698, 177.5558106 ], [80, 866.6198818, 171.8591574 ],
[81, 855.7370884, 166.1297148 ], [82, 846.4618273, 160.4104848 ],
[83, 838.7394330, 154.7393618 ], [84, 832.5144695, 149.1492389 ],
[85, 827.7318640, 143.6682202 ], [86, 824.3377776, 138.3199102 ],
[87, 822.2802527, 133.1237537 ], [88, 821.5096724, 128.0954067 ],
[89, 821.9790649, 123.2471195 ], [90, 823.6442828, 118.5881204 ],
[91, 826.4640816, 114.1249887 ], [92, 830.4001196, 109.8620105 ],
[93, 835.4168963, 105.8015119 ], [94, 841.4816441, 101.9441675 ],
[95, 848.5641845, 98.28928197 ], [96, 856.6367584, 94.83504457 ],
[97, 865.6738364, 91.57875774 ], [98, 875.6519156, 88.51704026 ],
[99, 886.5493055, 85.64600702 ], [100, 898.3459060, 82.96142698 ]

```
> with(plots):
> pointplot( [seq( [i,r(i) ],i = 0..100) ], title = `Rabbits`);
```

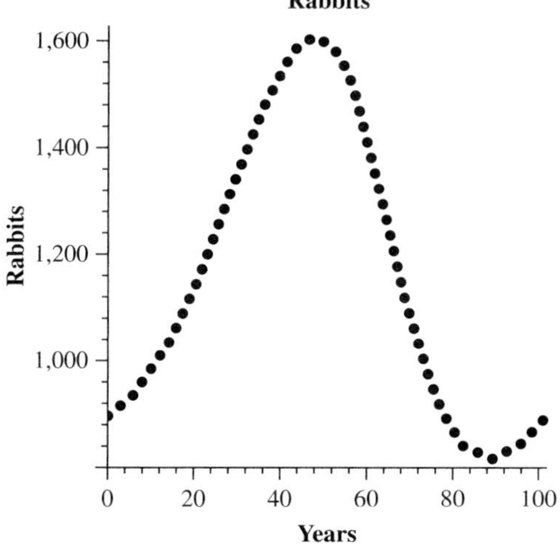

> *pointplot( [seq( [i,f(i)],i = 0..100)],title = `Foxes`);*

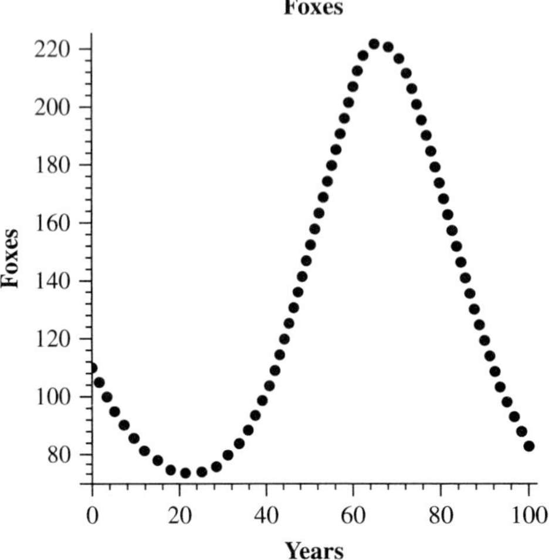


> *pointplot( [seq( [r(i),f(i)],i = 0..100)],title = `Predator Prey`);*

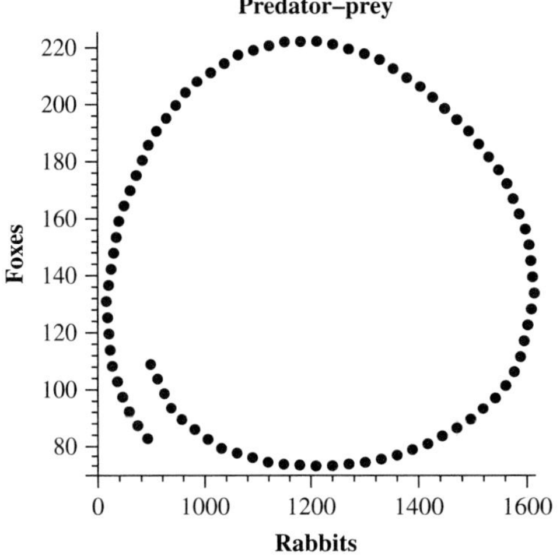


If we ran this model for more iterations, we would find the plot of foxes versus rabbits spirals in a similar fashion as before. We conclude that the model appears reasonable. We could find the equilibrium values for the system. There are two sets of equilibrium points for rabbits and foxes: (0,0) and (1200,130). The orbits of the spiral indicate that the system is moving away from (1200,130), so we conclude the system is not stable. In the exercise set, you will be asked to do more explorations.

## Example 2    Discrete SIR Models of Epidemics

Consider a disease that is spreading throughout the United States, such as the new flu. The Centers for Disease Control and Prevention is interested in knowing and experimenting with a model for this new disease before it actually becomes a real epidemic. Let us consider the population being divided into three categories: susceptible, infected, and removed. We make the following assumptions for our model:

- No one enters or leaves the community, and there is no contact outside the community.
- Each person is susceptible, $S$ (able to catch this new flu); infected, $I$ (currently has the flu and can spread the flu); or removed, $R$ (already had the flu and will not get it again— which includes death).
- Initially every person is either $S$ or $I$.
- Once someone gets the flu this year, they cannot get again.
- The average length of the disease is two weeks, over which time the person is deemed infected and can spread the disease.
- Our time period for the model will be per week.

The model we will consider is the SIR model (Allman, 2004).
Let's assume the following definition for our variables.

$S(n)$ = number in the population susceptible after period $n$.

$I(n)$ = number infected after period $n$.

$R(n)$ = number removed after period $n$.

Let's start our modeling process with $R(n)$. Our assumption for the length of time someone has the flu is two weeks. Thus, half the infected people will be removed each week:

$$R(n + 1) = R(n) + 0.5\ I(n)$$

The value, 0.5, is called the *removal rate per week*. It represents the proportion of the infected persons who are removed from infection each week. If real data are available, then we could do data analysis to obtain the removal rate.

$I(n)$ will have terms that both increase and decrease its amount over time. It is decreased by the number that are removed each week: $0.5 * I(n)$. It is increased by the numbers of susceptibles who come into contact with infected people and catch the disease: $aS(n)I(n)$. We define $a$ as the rate in which the disease is spread, or the transmission coefficient. We realize this is a probabilistic coefficient. We will assume, initially, that this rate is a constant value that can be found from initial conditions.

Let's illustrate as follows. Assume we have a population of 1,000 students in the dorms. Further assume that the school nurse has reported three students with flu-like symptoms coming to the infirmary at the same time. The next week, 5 students came in to the infirmary with flu-like symptoms. $I(0) = 3,\ S(0) = 997$. In week 1, the number of newly infected is 30.

$$5 = a\ I(n)\ S(n) = a(3) * (995)$$
$$a = 0.00167$$

Let's consider $S(n)$. This number is decreased only by the number that becomes infected. We may use the same rate, $a$, as before to obtain the model:

$$S(n + 1) = S(n) - aS(n)I(n)$$

Our coupled SIR model is

$$\begin{aligned}
R(n + 1) &= R(n) + 0.5I(n) \\
I(n + 1) &= I(n) - 0.5I(n) + 0.00167I(n)S(n) \\
S(n + 1) &= S(n) - 0.00167S(n)I(n) \\
I(0) &= 3,\ S(0) = 997,\ R(0) = 0
\end{aligned} \tag{3.6}$$

The SIR model, Equation 3.6, can be solved iteratively and viewed graphically. Let's iterate the solution and obtain the graph to observe the behavior, in order to obtain some insights.

```
> with(plots):
> SIRModel:=proc(x,y,z,g,a,n)
```

```
> option remember;
> R(0):=x;
> Inft(0):=y;
> S(0):=z;
> for i from 1 to n do
> R(i):=evalf(R(i–1)+g*Inft(i–1));
> Inft(i):=evalf(Inft(i–1)–g * Inft(i–1)+a*S(i–1)*Inft(i–1));
> S(i):=evalf(S(i–1)–a*S(i–1)*Inft(i–1));
> end do;
> end;
```

$SIRModel := \text{proc}(x, y, z, g, a, n)$
local $i$;
option remember;
   $R(0) := x$;
   $Inft(0) := y$:
   $S(0) := z$;
for $i$ to $n$ do
   $R(i) := \text{evalf}(R(i-1) + g \times Inft(i-1))$;
   $Inft(i) := \text{evalf}(Inft(i-1) - g \times Inft(i-1) + a \times S(i-1) \times Inft(i-1))$; end do
   $S(i) := \text{evalf}(S(i-1) - a \times S(i-1) \times Inft(i-1))$ end proc

```
> SIRModel(0,3,997,0.5,0.00167,25);
```

$$9.647351413$$

```
> seq([i,R(i),Inft(i),S(i)],i = 0..25);
```

[0, 0, 3, 997 ], [1, 1.5, 6.49497, 992.00503 ],
[2, 4.747485, 14.00736666, 981.2451483 ],
[3, 11.75116833, 29.95726650, 958.2915651 ],
[4, 26.72980158, 62.92065225, 910.3495461 ],
[5, 58.19012770, 127.1175708, 814.6923014 ],
[6, 121.7489131, 236.5068349, 641.7442519 ],
[7,240.0023305,371.7208436,388.2768258],
[8,425.8627523,426.8925058,147.2447418],
[9,639.3090052,318.4185731,42.2724216],
[10,798.5182918,181.6880279,19.79368024],
[11,889.3623058,96.84979274,13.78790144],
[12,937.7872022,50.65493988,11.55785793],
[13,963.1146721,26.30519248,10.58013539],
[14,976.2672683,13.61737811,10.11535352],
[15,983.0759574,7.038722526,9.885320049],
[16,986.5953187,3.635559905,9.769121407],
[17, 988.4130987, 1.877092051, 9.709809309],
[18, 989.3516447, 0.9689837893, 9.679371545],
[19, 989.8361366, 0.5001550821, 9.663708358],
[20, 990.0862141, 0.2581492404, 9.655636659],
[21, 990.2152887, 0.1332372543, 9.651474025],
[22, 990.2819073, 0.06876614010, 9.649326512],
[23, 990.3162904, 0.03549119344, 9.648218389],
[24, 990.3340360, 0.01831744945, 9.647646536],
[25, 990.3431947, 0.009453847589, 9.647351413]

> *pointplot( [seq( [i,R(i)],i = 0..25)]);*

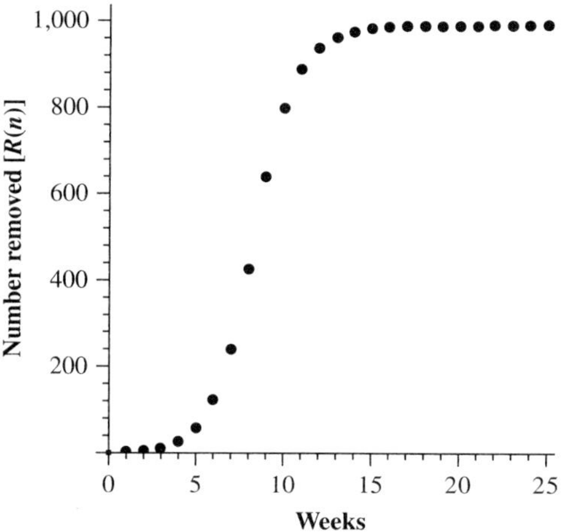

> *pointplot( [seq( [i,Inft(i)],i = 0..25)]);*

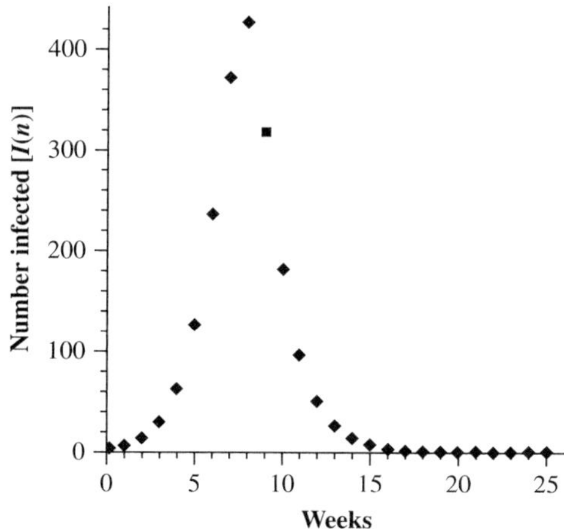

> *pointplot( [seq( [i,S(i)],i = 0..25)]);*

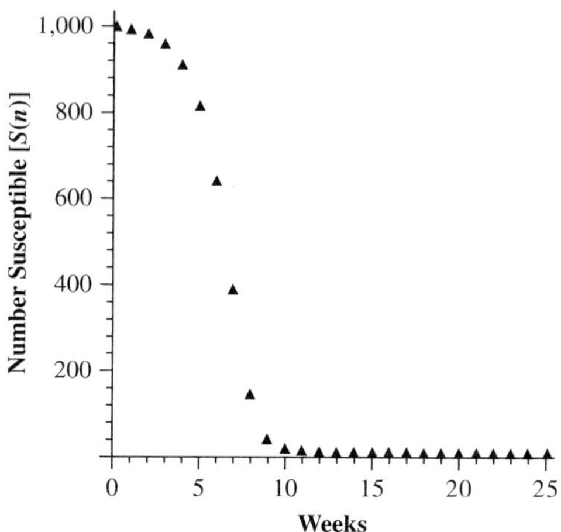

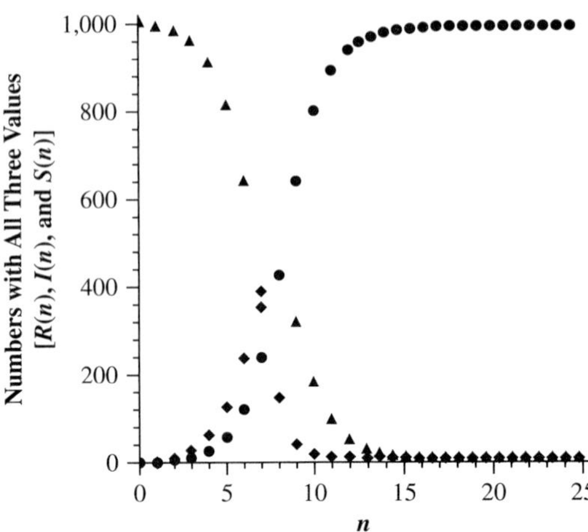



The peak of the flu epidemic occurs around week 8, at the maximum of the infected graph. The maximum number is slightly larger than 400; from the table, it is 427. After 25 weeks, slightly more than nine people never get the flu. You will be asked to check for sensitivity to the coefficient in the exercise set.

---

## Example 3 — Modeling Military Insurgencies

Insurgent forces have a strong foothold in the city of Urbania, a major metropolis in the center of the country of Ibestan. Intelligence estimates that the insurgents currently have a force of about 1,000 fighters. The local police force has approximately 1,300 officers, many of whom have no formal training in law-enforcement methods or modern tactics for addressing insurgent activity. Based on data collected over the past year, approximately 8 percent of insurgents switch sides and join the police each week, whereas about 11 percent of police switch sides and join the insurgents. Intelligence also estimates that around 120 new insurgents arrive from the neighboring country of Moronka each week. Recruiting efforts in Ibestan yield about 85 new police recruits each week as well. In armed conflict with insurgent forces, the local police are able to capture or kill approximately 10 percent of the insurgent force each week on average, while losing about 3 percent of the police forces.

## Problem Statement

Build a mathematical model of this insurgency. Determine the equilibrium state (if it exists) for this DDS.

We define the variables:

$P(n)$ = the number of police in the system after time period $n$.

$I(n)$ = the number of insurgents in the system after time period $n$.

$n = 0, 1, 2, 3, \ldots$ weeks

## Model

$$P(n + 1) = P(n) - 0.03\, P(n) - 0.11\, P(n) + 0.08\, I(n) + 85, \quad P(0) = 1{,}300$$
$$I(n + 1) = I(n) + 0.11\, P(n) - 0.08\, I(n) - 0.1\, I(n) + 120, \quad I(0) = 1{,}000$$

```
> INSURGENTModel:=proc(x,y,a,b,c,d,e,f,g,h,n)
> option remember;
> govt(0):=x;
> Insurg(0):=y;
> for i from 1 to n do
> govt(i):=evalf((1−a−b)*govt(i−1)+c*Insurg(i−1)+d);
> Insurg(i):=evalf((1−f−g)*Insurg(i−1)+e*govt(i−1)+h);
> end do;
> end;
```

$$INSURGENTModel := \mathrm{proc}(x,\, y,\, a,\, b,\, c,\, d,\, e,f,\, g,\, h,\, n)$$
$$\mathrm{local}\ i;$$
$$\mathrm{option}\ remember;$$
$$\mathrm{govt}(0) := x;$$
$$\mathrm{Insurg}(0) := y;$$
$$\mathrm{for}\ i\ \mathrm{to}\ n\ \mathrm{do}$$
$$\mathrm{govt}(i) := \mathrm{evalf}((\,1 - a - b)\times \mathrm{govt}(i-1) + c \times \mathrm{Insurg}(i-1) + d);$$
$$\mathrm{Insurg}(i) := \mathrm{evalf}((\,1 - f - g)\times \mathrm{Insurg}(i-1) + e \times \mathrm{govt}(i-1) + h)$$
$$\mathrm{end\ do}$$
$$\mathrm{end\ proc}$$

```
> INSURGENTModel(1300,1,000,.03,.11,.08,85,.11,.08,.1,120,52);
```

$$1582.999558$$

```
> seq([i,govt(i),Insurg(i)],i = 0..52);
```

[0, 1300, 1000 ], [1, 1283.00, 1083.00 ], [2, 1275.0200, 1149.1900 ],
 [3, 1273.452400, 1202.588000 ], [4, 1276.376104, 1246.201924 ],
 [5, 1282.379603, 1282.286949 ], [6, 1290.429415, 1312.537054 ],
 [7, 1299.772261, 1338.227620 ], [8, 1309.862354, 1360.321597 ],
 [9, 1320.307352, 1379.548569 ], [10, 1330.828208, 1396.463636 ],
[11, 1341.229350, 1411.491285 ], [12, 1351.376544, 1424.958082 ],
[13, 1361.180475, 1437.117047 ], [14, 1370.584572, 1448.165831 ],
[15, 1379.555998, 1458.260284 ], [16, 1388.078981, 1467.524593 ],
[17, 1396.149891, 1476.058854 ], [18, 1403.773614, 1483.944748 ],
[19, 1410.960888, 1491.249790 ], [20, 1417.726347, 1498.030526 ],
[21, 1424.087100, 1504.334929 ], [22, 1430.061700, 1510.204223 ],
[23, 1435.669400, 1515.674250 ], [24, 1440.929624, 1520.776519 ],
[25, 1445.861598, 1525.539005 ], [26, 1450.484094, 1529.986760 ],
[27, 1454.815262, 1534.142393 ], [28, 1458.872516, 1538.026441 ],
[29, 1462.672479, 1541.657659 ], [30, 1466.230945, 1545.053253 ],
[31, 1469.562873, 1548.229071 ], [32, 1472.682397, 1551.199754 ],
[33, 1475.602841, 1553.978862 ], [34, 1478.336752, 1556.578980 ],
[35, 1480.895925, 1559.011807 ], [36, 1483.291441, 1561.288234 ],
[37, 1485.533698, 1563.418410 ], [38, 1487.632453, 1565.411803 ],
[39, 1489.596854, 1567.277248 ], [40, 1491.435474, 1569.022997 ],
[41, 1493.156348, 1570.656760 ], [42, 1494.767000, 1572.185741 ],
[43, 1496.274479, 1573.616678 ], [44, 1497.685386, 1574.955869 ],
[45, 1499.005902, 1576.209206 ], [46, 1500.241812, 1577.382198 ],
[47, 1501.398534, 1578.480001 ], [48, 1502.481139, 1579.507440 ],
[49, 1503.494375, 1580.469026 ], [50, 1504.442684, 1581.368982 ],
[51, 1505.330227, 1582.211260 ], [52, 1506.160896, 1582.999558 ],

*> pointplot( [seq( [i,govt(i) ],i = 0..52 ) ] );*

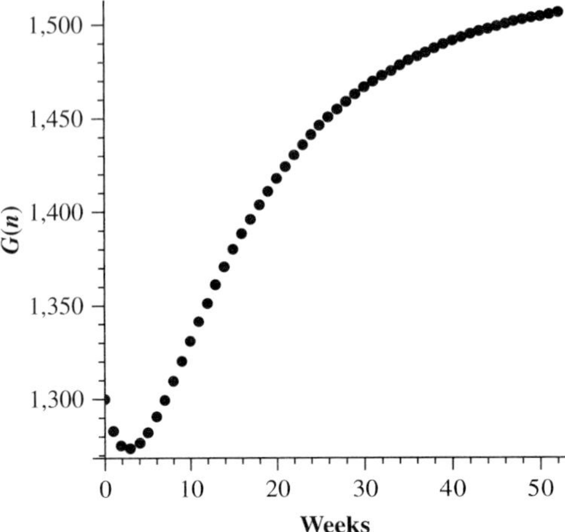

*> pointplot( [seq( [i,Insurg(i) ],i = 0..52 ) ] );*

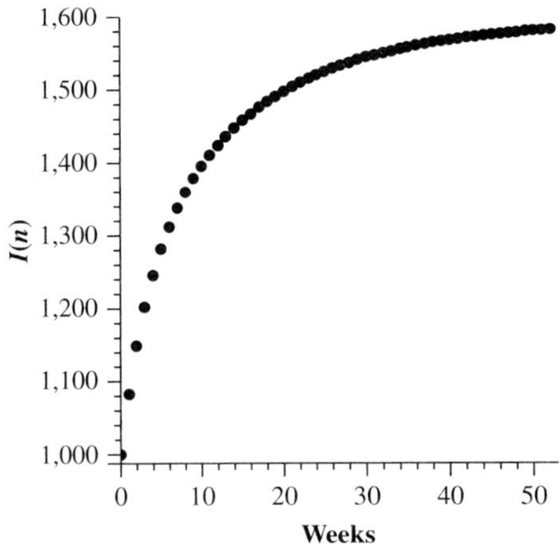

*> with( plots ):*
*> p1:=pointplot( [seq( [i,govt(i) ],i=0..52 ) ],symbol=diamond ):*
*> p2:=pointplot( [seq( [i,Insurg(i) ],i=0..52 ) ],symbol=circle ):*
*> display( {p1,p2} );*

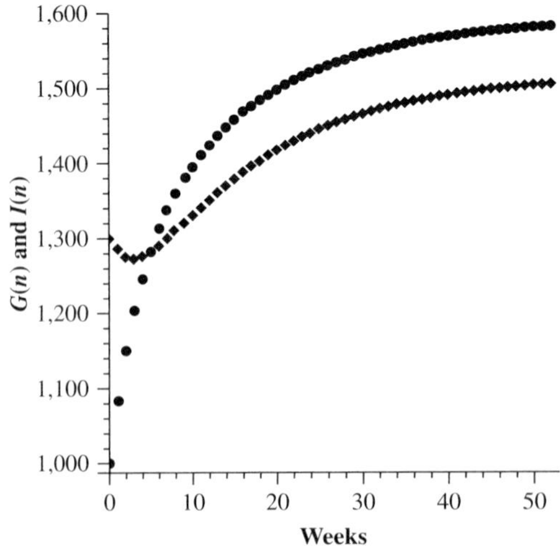

In this insurgency model, we see that, under the current conditions, the insurgency overtakes the government within five to 10 weeks. If this is unacceptable, then we must modify conditions that affect the parameters in such a way as to obtain a governmental victory. You will be asked to experiment with these parameters in the exercises.

## 3.5 | EXERCISES

1. Find the equilibrium values for the predator–prey model presented.

2. Find the equilibrium values for the SIR model presented.

3. Find the equilibrium values for the insurgency model presented.

4. In the predator–prey model, determine the outcomes with the following sets of parameters.

   **a.** Initial foxes are 200, and initial rabbits are 400.
   **b.** Initial foxes are 2,000, and initial rabbits are 10,000.
   **c.** Birthrate of rabbits increases to 0.1.

5. In the SIR model, determine the outcome with the following parameters changed.

   **a.** Initially, five are sick; 10 are ill the next week.
   **b.** The flu lasts one week.
   **c.** The flu lasts four weeks.
   **d.** There are 4,000 students in the dorm; five are initially infected, and 30 more are infected the next week.

6. In the insurgency model, determine what values enable a reversal of the outcome as predicted in our model under the current conditions.

# 4

# Model Fitting

## Introduction

Consider a spring mass system such as the one in Figure 4.1 (we saw this problem in Chapter 2).

We conducted an experiment to measure the stretch of the spring as a function of mass placed on the spring. We measure only how far the spring stretched from its original position; these data are displayed in Table 4.1.

The data plot, seen in Figure 4.2, looks reasonably like a straight line through the origin. Our next step was to calculate the slope. We found the model as $F = 0.00155\ S$. We now want a more exact fit of our line to the data. Model fitting, especially with least squares, will be how we obtain a better fit.

In previous chapters, we have demonstrated Maple's capability to perform various transformations on a data set and to plot the resulting transformed data to assist in determining a proposed proportionality model graphically. In particular, we discussed methods to enter data, transform data, obtain a scatterplot, test a proportionality relationship, and estimate the parameters of a model. This chapter describes how to determine the parameters of a model analytically, according to some criterion of "best fit," and to test the adequacy of the model.

Suppose it is proposed that a parabolic model might best explain a behavior being studied, and you are interested in selecting that member of the parabolic family, say, $y = kx^2$. Recall that if $y$ is proportional to $x^2$ then we may only want to fit the model, $y = kx^2$, instead of $y = ax^2 + bx + c$. Using Maple, $y$ versus $x^2$ can be plotted, and we could use the graph to assist in selecting two points to obtain an estimate of the slope of the line, as demonstrated in Chapter 2 with proportionality as a simplifying assumption. In this chapter, we want to find the "best fit" line using a specific criterion.

This chapter focuses on the analytical methods to arrive at a model for a given data set using a prescribed criterion. Again, from the family $y = kx^2$, the parameter $k$ can be determined analytically by using a curve-fitting criterion, such as least-squares, Chebyshev's, or minimizing the sum of the absolute error, and then solving the resulting optimization problem. Although we briefly present these other criterion, we concentrate on the presentation of least squares in this chapter. We present the Maple commands that solve the least-squares optimization problem with analysis of the "goodness of the fit" of the resulting model.

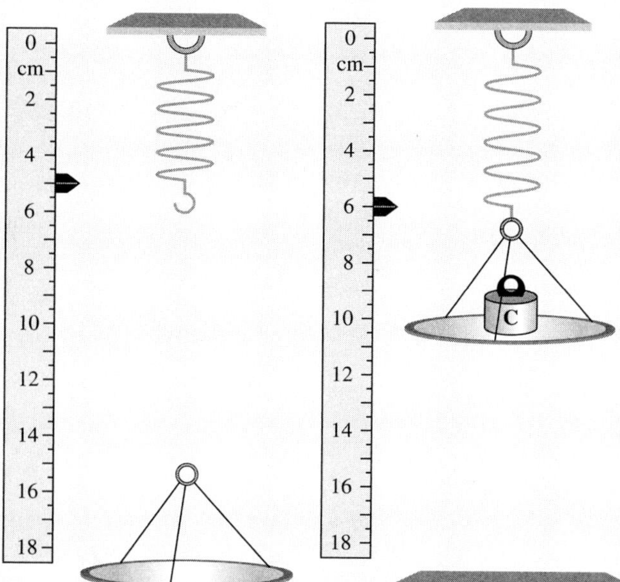

**FIGURE 4.1**
Spring–mass systems

**Table 4.1** Spring–Mass System

| Mass (grams) | Stretch (m) |
|---|---|
| 50 | .1 |
| 100 | .1875 |
| 150 | .275 |
| 200 | .325 |
| 250 | .4375 |
| 300 | .4875 |
| 350 | .5675 |
| 400 | .65 |
| 450 | .725 |
| 500 | .80 |
| 550 | .875 |

**FIGURE 4.2**
Plot of spring–mass data

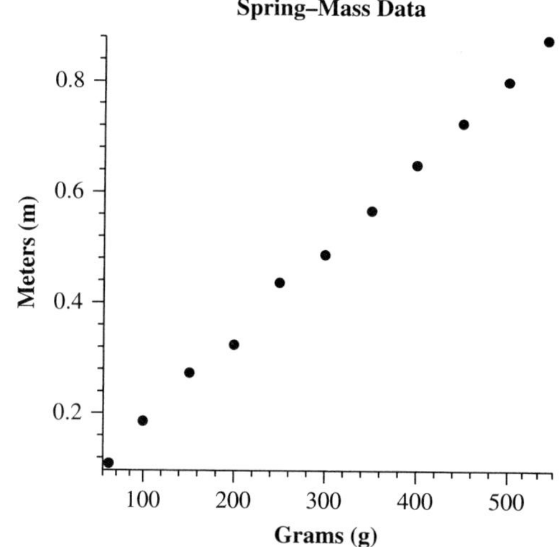


## 4.1 THE DIFFERENT CURVE-FITTING CRITERION

We will briefly describe three curve-fitting criterion: least squares, Chebyshev's, and minimizing the sum of the absolute error. We start with least squares.

**Criterion 1**  **Least Squares**

The method of least-squares curve fitting, also known as **ordinary least squares** and **linear regression**, is simply the solution to a model that minimizes the sum of the squares of the deviations between the observations and predictions. In Maple, we can fit a curve to a set of data points using the least-squares method. We might use the fit command within the statistical package. We will illustrate several commands to obtain least-squares. Least squares will find the parameters of the function, $f(x)$, that will minimize the sum of squared differences between the real data and the proposed model as shown in Equation 4.1.

**Table 4.2** Least-Squares Data Points

| X | 0.5 | 1.0 | 1.5 | 2.0 | 2.5 |
|---|-----|-----|-----|-----|-----|
| Y | 0.7 | 3.4 | 7.2 | 12.4 | 20.1 |

$$\text{Minimize } S = \sum_{j=1}^{m} [y_1 - f(x_j)]^2 \tag{4.1}$$

For example, to fit a proposed proportionality model $y = kx^2$ to a set of data, the least-squares criterion requires the minimization of Equation 4.2.

$$\text{Minimize } S = \sum_{j=1}^{5} [y_i - kx_j^2]^2 \tag{4.2}$$

Minimizing Equation 4.2 is achieved using the first derivative, setting it equal to zero, and solving for the unknown parameter, $k$:

$$\frac{ds}{dk} = -2\sum x_j^2(y_j - kx_j^2) = 0. \text{ Solving for } k\text{: } k = \left(\sum x_j^2\, y_j\right)\bigg/\left(\sum x_j^4\right) \tag{4.3}$$

Given the data set in Table 4.2, we will find the least squares fit to the model, $y = kx^2$. Solving for $k$:

$$k = \left(\sum x_j^2\, y_j\right)\bigg/\left(\sum x_j^4\right) = (195.0)/(61.1875) = 3.1869$$

and the model $y = kx^2$ becomes $y = 3.1869\, x^2$.

In Chapter 9, we discuss more fully the optimization process; and in Chapter 11 (Section 11.2), we discuss solving the normal equations as a system of equations. In this chapter, we discuss the internal Maple commands.

In Maple, the fit command fits the model type $Y = b + b_1X_1 + b_2X_2 + \ldots + b_kX_k$ to the data set given in the $k$ specified columns. To fit the quadratic model $y = A_0 + A_1x + A_2x^2$ to a data set of $x$ values, $xv$, and of $y$ values, $yv$, the Maple fit command required to solve a full quadratic equation is *quadraticfit:=[leastsquare[x,y],y=a*x^2 + b*x + c,{a,b,c}]] ([xv,yv]);* where $x$ and $y$ are command calls which will use the two data sets named in the command (in the case xv and yv), while a, $b$, and c are the coefficient variables for which a least-squares solution will be fit. To fit our example, using only, $y = Ax^2$, we use the following command:

*quadraticfit:=[leastsquare[x,y],y=A*x^2 ,{A}]] ([xv,yv])*

Because fit is part of the statistics package, the *with(stats)*: or *with(Statistics)*: command must be entered once prior to using the fit command. First, we illustrate the procedure with the model, $y = ax^2 + bx + c$ (see the visual fit in Figure 4.3):

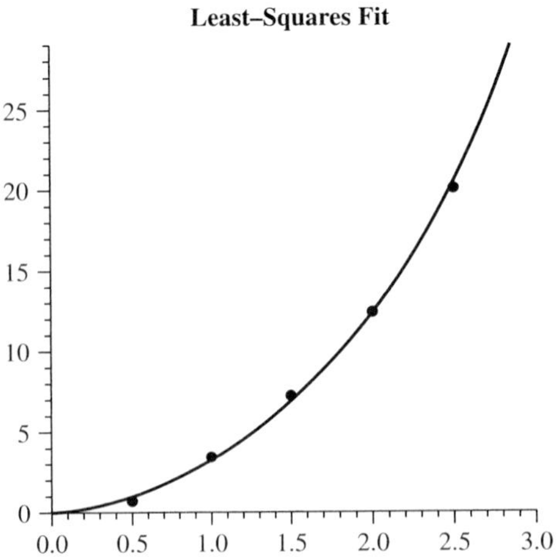


**FIGURE 4.3**

Least-squares fit plotted with the original data

> *with* (*stats*) :
> *with* (*plots*):*Digits=5:*
> *xv:=[.5, 1, 1.5, 2, 2.5];*

$$xv := [0.5, 1, 1.5, 2, 2.5]$$

> *yv:=[.7, 3.4, 7.2, 12.4, 20.1];*

$$yv := [0.7, 3.4, 7.2, 12.4, 20.1]$$

> *eqfit1:=fit[leastsquare[[x,y], y=a*x^2+b*x+c, {a,b,c}]]([xv,yv]);*

$$eqfit1 := y = 3.257142857\ x^2 - 0.2114285714\ x + 0.12000000$$

> *f :=unapply(rhs(eqfit1), x);*

$$f := x \rightarrow 3.257142857\ x^2 - 0.2114285714\ x + 0.12000000$$

> *xy:= {seq([xv[i],yv[i]],i=1..5)}:*
> *plot1:=plot(xy, style=point, symbol=diamond):*
> *plot2:=plot(f(x),x=0..3):*
> *display( {plot1,plot2},title='Least Squares Fit');*

We now illustrate the newer commands in Maple.

> *with(Statistics):*
> *Fit(a*x^2+b*x+c,xv,yv,x);*

$$3.25714287143\ x^2 - 0.21142857143\ x + 0.12000000000$$

Next, we illustrate the fit command applied to the proportionality model $y = kx^2$ for the data set for our example. As obtained previously, the least-squares model is $y = 3.1870\ x^2$ (rounded to three decimal places). Because the two data sets used in this example, *xv* and *yv*, have been entered previously, they can be called without reentering the data. Having previously invoked both with(plots): and with(stats): commands, they need not be repeated in this continuation example either. (See also Figure 4.4.)

> *eqfit2:=fit[leastsquare[[x,y],y=k*x^2, {k}]]([xv,yv]);*

$$eqfit2 := y = 3.1870\ x^2$$

> *f:=unapply(rhs(eqfit2),x);*

$$f := x \rightarrow 3.1870\ x^2$$

**FIGURE 4.4**

The least-squares $kx^2$ fit plotted with the data

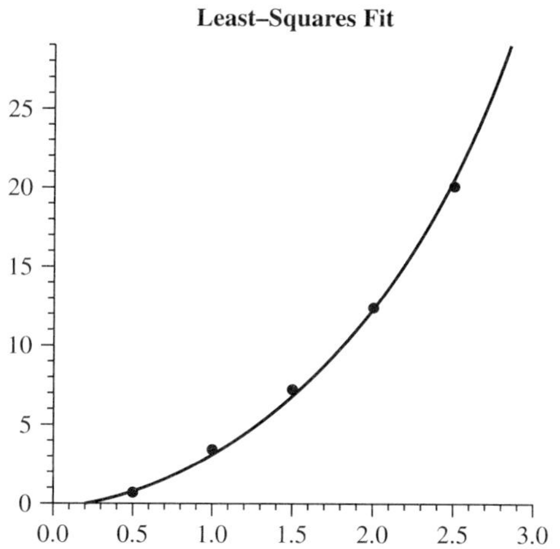

> $xy:=\{seq([xv[i],yv[i]],i=1..5)\}:$
> $plot1:=plot(xy,style=point,symbol=diamond):$
> $plot2:=plot(f(x),x=0..3):$
> $display(\{plot1, plot2\},title=\text{'}Least\ Squares\ Fit\text{'});$

The least-squares techniques will be applied to the proportionality models in Chapter 5, as well as to many future models that we will be discussing.

---

| **Example 1** | # A Least-Squares Fit of Bass Fishing Derby Data |

Let's consider a fishing derby where we want to construct a simple model to predict the weight of a fish that we catch as function of its length. Assume we have developed a proportionality model, $W = kL^3$. The Maple commands have been previously shown. Let's demonstrate how to determine a constant of proportionality, using the least-squares criterion with the bass fishing data. This example illustrates the fit command analytically fitting the model $W = k\ len^3$ to the same data set.

> $len:=[14.5, 12.5, 17.3, 14.5, 12.6, 17.8, 14.1, 12.6];$

$$len := [14.5, 12.5, 17.3, 14.5, 12.6, 17.8, 14.1, 12.6]$$

> $wt:=[27, 17, 41, 26, 17, 49, 23, 16];$

$$wt := [27, 17, 41, 26, 17, 49, 23, 16]$$

> $lenwt:=\{seq([len[i],wt[i]],i=1..8)\}:$
> $lenwtfit:=Fit(ax^2+bx+c,xv,yv,x);$

$$lenwtfit := 0.0084011\ x^3$$

> $plot4:=plot(lenwt,style=point,symbol=diamond,color=black):$
> $plot5:=plot(f(x),x=0\ ..18,\ title=\text{'}Fitting\ the\ bass\ fish\text{'}):display(\{plot4, plot5\});$

The least-squares estimate of the proportionality constant in this model is $k = 0.0084365$. Thus, the model is $W = 0.0084365\ L^3$. The graph of the least-squares fit with the original data above ( see Figure 4.5) shows that the model does capture the trend of the data.

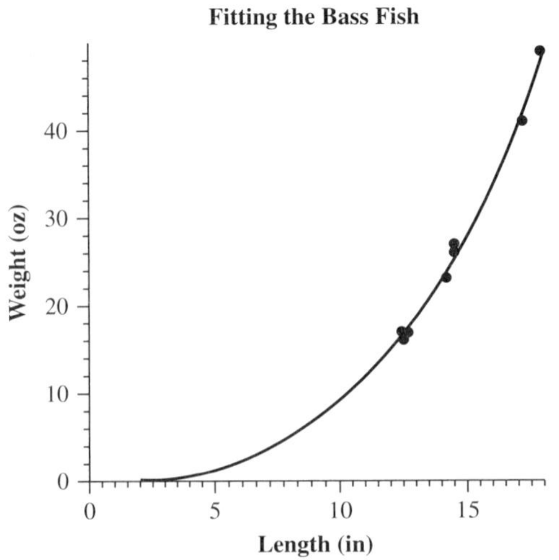



**FIGURE 4.5**
Plot of the model $y = 0.0084011\ x^3$ with the data

**Criterion 2**    **Minimize the Sum of the Absolute Deviations**

The model is to find the value of the parameters in $f(x)$ that

$$\text{Minimize} \sum_{i=1}^{n} |y_i - f(x_i)| \tag{4.4}$$

Assume we have the following data and want to use this criterion in Equation 4.4 to fit the model $W = kL^3$.

Data Available

| Length (in) | 12.5 | 12.625 | 12.625 | 14.125 | 14.5 | 14.5 | 17.27 | 17.75 |
|---|---|---|---|---|---|---|---|---|
| Weight (oz) | 17 | 16 | 17 | 23 | 26 | 27 | 43 | 49 |

**Minimize the Sum of the Absolute Deviations Using $W = KL^3$**

We formulate the model as:

*Minimize $S = |17-k*12.5\wedge3|+|16-k*12.625\wedge3|+...+|49-k*17.75\wedge3|$*

The method that we will eventually use to solve this problem for the value of $k$ that minimizes this sum is a numerical search technique called *golden section search*. We will learn about using golden section search in Chapter 9 on single-variable optimization.

**Criterion 3**    **Chebyshev's Criterion or Minimize the Largest Error**

We define $W$ for weight and $l$ for the length, then the largest error as $R = \text{Max}\,(W_i - kl_i^3)$. Our goal is to minimize this largest error. We use the following generalized formulation:

$$\text{Minimize } R \tag{4.5}$$
$$\text{Subject to:}$$
$$R \pm R_i \geq 0$$

This formulation, Equation 4.5, with our data, would look like:

$$\text{Minimize } R$$
$$\text{Subject to:}$$
$$R + (17 - k*12.5\wedge3) \geq 0$$
$$R - (17 - k*12.5\wedge3) \geq 0$$
$$\vdots$$
$$R + (49 - k*17.75\,\wedge3) \geq 0$$
$$R - (49 - k*17.75\wedge3) \geq 0$$
$$R \geq 0$$
$$k \geq 0$$

This is a linear programming problem that finds the value of $k$ that minimizes the largest $R$. We will study linear programming models in Chapter 7 on linear programming.

---

## **4.1** | EXERCISES

Fit the following data with the models using least squares.

**1.**

| $x$ | 1 | 2 | 3 | 4 | 5 |
|---|---|---|---|---|---|
| $y$ | 1 | 1 | 2 | 2 | 4 |

    **a.** $y = b + ax$
    **b.** $y = ax^2$

2. Data for the stretch of a spring:

| $x\,(\times 10^{-3})$ | 5 | 10 | 20 | 30 | 40 | 50 | 60 | 70 | 80 | 90 | 100 |
|---|---|---|---|---|---|---|---|---|---|---|---|
| $y\,(\times 10^{5})$ | 0 | 19 | 57 | 94 | 134 | 173 | 216 | 256 | 297 | 343 | 390 |

    **a.** $y = ax$
    **b.** $y = b + ax$
    **c.** $y = ax^2$

3. Data for the ponderosa pine.

| $x$ | 17 | 19 | 20 | 22 | 23 | 25 | 28 | 31 | 32 | 33 | 36 | 37 | 39 | 42 |
|---|---|---|---|---|---|---|---|---|---|---|---|---|---|---|
| $y$ | 19 | 25 | 32 | 51 | 57 | 71 | 113 | 140 | 153 | 187 | 192 | 205 | 250 | 260 |

    **a.** $y = ax + b$
    **b.** $y = ax^2$
    **c.** $y = ax^3$
    **d.** $y = ax^3 + bx^2 + c$

4. Given Kepler's data in the following table, fit the model $y = ax^{3/2}$.

| Body | Period (s) | Distance from Sun (m) |
|---|---|---|
| Mercury | $7.60 \times 10^{6}$ | $5.79 \times 10^{10}$ |
| Venus | $1.94 \times 10^{7}$ | $1.08 \times 10^{11}$ |
| Earth | $3.16 \times 10^{7}$ | $1.5 \times 10^{11}$ |
| Mars | $5.94 \times 10^{7}$ | $2.28 \times 10^{11}$ |
| Jupiter | $3.74 \times 10^{8}$ | $7.79 \times 10^{11}$ |
| Saturn | $9.35 \times 10^{8}$ | $1.43 \times 10^{12}$ |
| Uranus | $2.64 \times 10^{9}$ | $2.87 \times 10^{12}$ |
| Neptune | $5.22 \times 10^{9}$ | $4.5 \times 10^{12}$ |
| Pluto | $7.82 \times 10^{9}$ | $5.91 \times 10^{12}$ |

## 4.1 | PROJECT

1. Write a Maple program to find the least squares for the general proportionality model: $y$ is proportional to $x^n$.

## 4.1 | FURTHER READING


Burden, R. L., & J. D. Faires. *Numerical Analysis.* 8th ed. Belmont, CA: Brooks-Cole Publishers, 2005.

Cheney, E. W., & D. Kincaid. *Numerical Mathematics and Computing.* Monterey, CA: Brooks-Cole Publishers, 1984.

Neter, J., W. Wasserman, and M. H. Kutner. *Applied Linear Statistical Models,* 4th ed. Boston: McGraw-Hill Publishing, 1996.

## 4.2  PLOTTING THE RESIDUALS FOR A LEAST-SQUARES FIT

In the previous section, you learned how to obtain a least-squares fit of a model, and you plotted the model's predictions on the same graph as the observed data points in order to get a visual indication of how well the model matches the trend of the data. A powerful technique for quickly determining if the model is adequate or inadequate is to plot the actual deviations or residuals between the observed and predicted values as a function of the independent variable or the model.

We plot the residuals in the $y$-axis and the model or the independent variable on the $x$-axis. The deviations should be randomly distributed and contained in a reasonably small band that is commensurate with the accuracy required by the model. Any excessively large residual warrants further investigation of the data point in question to discover the cause of the large deviation. A pattern or trend in the residuals indicates that a predictable effect remains to be modeled, and the nature of the pattern gives clues on how to refine the model, if a refinement is warranted. We illustrate the possible patterns for residuals in Figures 4.6 (a)–(e). Our intent is to provide modelers with knowledge about the adequacy of the models they have found. We will leave further investigations into correcting the patterns to follow on courses in statistical regression.

**FIGURE 4.6**

Patterns for residuals: (a) no pattern, (b) curved pattern, (c) fanning pattern, (d) outliers, (e) linear trend

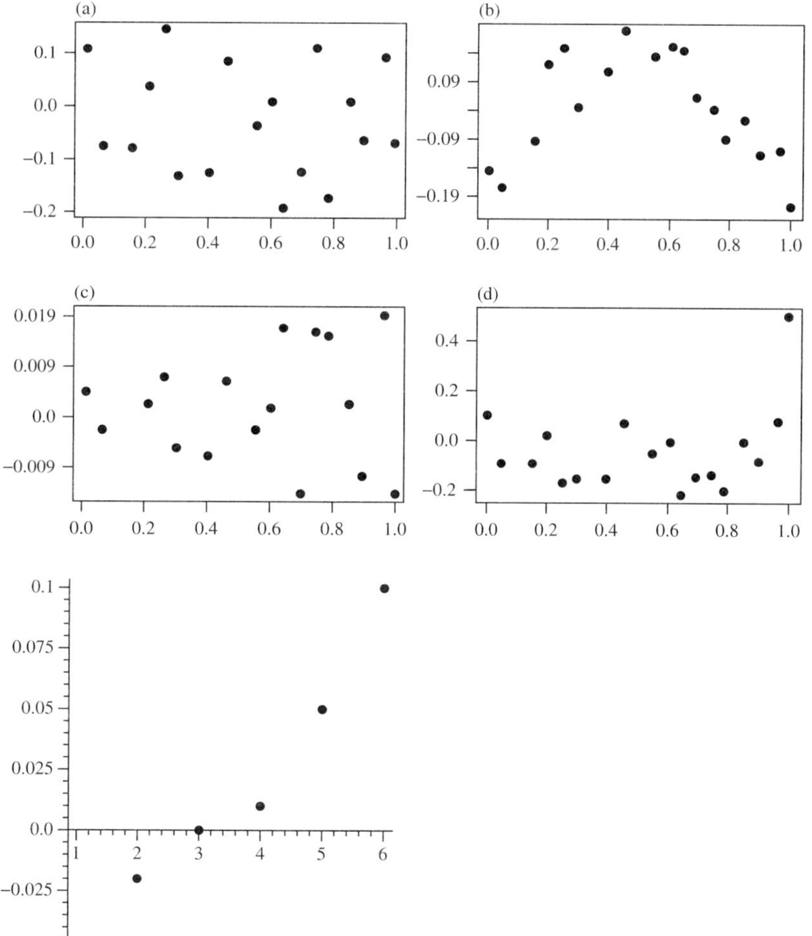

This section explains how to use Maple to compute residuals. After fitting a specified model, the difference between the observed and predicted values can be calculated by using the array manipulations.

---

**Example 1**        ## Bass Fish

We return to our bass fish example from the previous section. In this example, the relationship between the weight of a bass and its length could be reasonably modeled by the following expression:

$$W = 0.0084011 \, \text{len}^3$$

The residuals are the differences between the predicted and observed values. Mathematically, these are called *errors*, *deviations*, or *residuals*. They are found by

$$y_i - f(x_i)$$

In our example, the errors are found using

$$W_i - 0.0084011 * L_i^3$$

Again, we will use Maple to collect and plot this information. The observed values for the weight of the bass fish are stored in the array that we labeled *wt*, whereas the predicted values must be calculated using the formula $W = 0.0084365 \, \text{len}^3$. Then the residuals are calculated by subtracting each observed value from each predicted value. To analyze the results, the residuals are plotted using graphical methods and the modeler's interpretation of the provided plot (see Figure 4.7). We need the linalg library [with(linalg)], which allows us to easily do the subtraction to get the residuals.

```
> with (linalg):with(LinearAlgebra):
> predict:=map(x→f(x),len);
```

$$predict := [25.612, 16.408, 43.498, 25.612, 16.806, 47.381, 23.550, 16.806]$$

```
> resid:=matadd(wt,predict,1,–1);
```

$$resid := [1.388, 0.592, -2.498, 0.388, 0.194, 1.619, -.550, -.806]$$

```
> wtpred:={seq([len[i],resid[i]],i=1..8)}:
> plot(wtpred, style=point, symbol=circle,color=black,thickness=3,
title='Residual Plot');
```

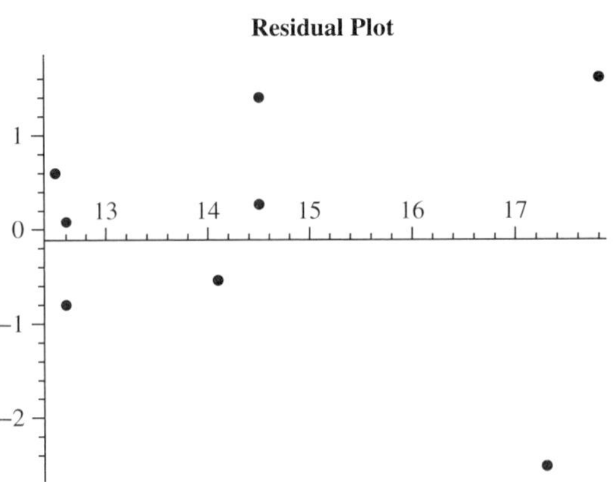

We illustrate with the new *with(LinearAlgebra):* command sequence.

> *c1:=matrix(1,8,predict): c2:=matrix(1,8,wt):*
> *resids:=MatrixAdd(c1,c2,1,−1);*

$$reisds := [1.388, 0.592, -2.498, 0.388, 0.194, 1.619, -.550, -.806]$$

However, we cannot obtain the residual plot easily.
Cut and paste the the output of resids into the [ ] of

$$errors := [];$$
$$errors := [1.388, 0.592, -2.498, 0.388, 0.194, 1.619, -.550, -.806];$$

> *wtpred:={seq([xv[i],errors[i]],i=1..8)};*
> *plot(wtpred,style=points,symbol=circle,color=black,thickness=3,title='Residual Plot');*

Note that the residuals are randomly distributed and contained in a relatively small band about zero. There are no outliers, or unusually large residuals, and there appears to be no pattern in the residuals. Based on these aspects of the plot of the residuals, the model approximates the data.

---

**Example 2**

## Population Example

We decide to fit a cubic equation to the following data, where $T$ represents years and $P$ represents the size of the population.

| $T$ | 7 | 14 | 21 | 28 | 35 | 42 |
|---|---|---|---|---|---|---|
| $P$ | 125 | 275 | 800 | 1,200 | 1,700 | 1,650 |

> *xdata:=[7, 14, 21, 28, 35, 42];*

$$xdata := [7, 14, 21, 28, 35, 42]$$

> *ydata:=[125, 275, 800, 1200, 1700, 1650];*

$$ydata := [125, 275, 800, 1200, 1700, 1650]$$

> *pointplot({seq([xdata[i],ydata[i]],i=1..6)},title='Scatterplot of Raw Data')*

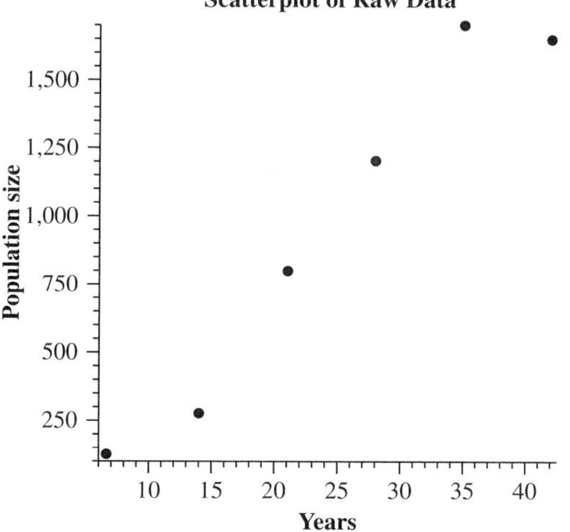

There appears to be a change in concavity in the data, so we decided to try a cubic equation.

```
> with(stats):with(plots):with(linalg):with(Linear Algebra):with(Statistics):
> eqfit:=Fit(ax^3+bx^2+cx+d,xdata,ydata,x);
```

$$eqfit := -0.1066299536\ x^3 + 7.436426952\ x^2 - 95.78136810\ x + 466.6666667$$

```
> eq_function:=x->eqfit
```

We use the *map( )* command to obtain the predicted values that we call *Yvaluespredicted*.

```
> Yvaluespredicted:=map(x->xdata,eqfit);
```

$$Yvaluespredicted := [124.0079365, 290.6746037, 747.2222227,$$
$$1274.206349, 1652.182538, 1661.706345]$$

Find the residuals:

```
> Residuals:=seq(ydata[i]–Yvaluespredicted[i],i=1..6);
```

$$Residuals := [0.9920635, -15.6746037, 52.7777773, -74.206349$$
$$47.817462, -11.706345]$$

```
> pointplot( {seq( [xdata[i], Residuals[i] ], i = 1 ..6)}, title = 'Residual Plot');
```

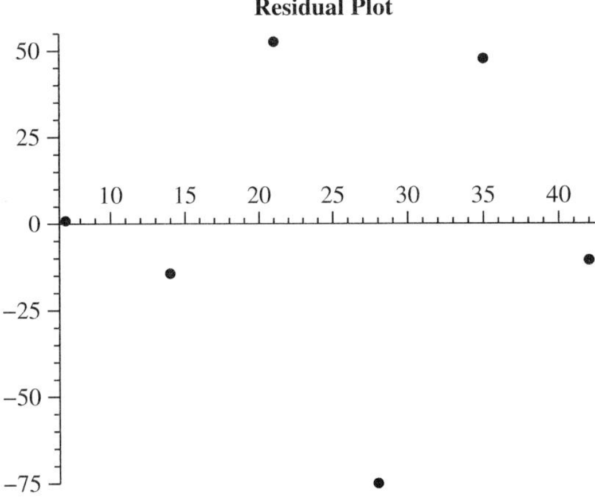


There does not appear to be any pattern to the residual plot, so we conclude the model is adequate.

---

## 4.3   BOUNDING THE CHEBYSHEV'S CRITERION

As discussed earlier, the fit command determines the parameters of a model, such that the sum of the squared residuals is minimized. Other curve-fitting criteria exist, although these lead to different optimization problems. For example, the Chebyshev criterion determines the parameters of a model such that the largest absolute deviation is minimized. In other words, the Chebyshev criterion determines the parameters of the function type $y = f(x)$ that minimizes the number.

$$\text{Maximize } |y_j - f(x_j)|\ , j = 1,2,...,m \tag{4.5}$$

The formulation of a specific problem yields a mathematical program, which is either linear or nonlinear. Denote the largest of the residuals that results from a Chebyshev fit by $c_{max}$. The value of $c_{max}$ will be as small as possible if the Chebyshev criterion is used to determine the model parameters. Therefore, we might be interested in knowing whether we should pursue the Chebyshev's model or just stay with a least-squares fit. We can find a bound on $c_{max}$. This bound may be obtained using the results of a least-squares fit as follows:

$$D \le c_{max} \le d_{max} \qquad (4.6)$$

where $D = \left(\left(\Sigma(d_j)^2\right)/m\right)^{0.5}$ and $d_{max}$ is the largest of the residuals $d_j$ resulting when the least-squares criterion is used. If the difference between $D$ and $d_{max}$ is large, and minimizing the largest absolute deviation is important in a particular application, then we may wish to investigate the Chebyshev criterion further. We illustrate the commands and outputs for the bounds for Equations 4.5 and 4.6 using our bass fishing data and model. Maple's *sum* command requires a *set* and will not operate on a list, which requires the residuals to be reentered as a *set*.

```
> residsum:=sum('(resid[k])^2', 'k = 1..8');
```

$$residsum := 12.28352$$

```
> bigd:=((residsum)/8)^(.5);
```

$$bigd := 1.53544$$

```
> resids:=map(x \rightarrow abs(x),resid) :
> Resid:=(resids[1], resids[2].resids[3], resids[4],
resids[5], resids[6], resids[7], resids[8]);
```

$$Resid := 1.388, 0.5915, 2.4988, 0.3879, 0.1945, 1.6195, 0.55032, 0.80548$$

```
> dmax:=max(Resid);
```

$$dmax := 2.4988$$

The bounds on $c_{max}$ are $1.53544 \le c_{max} \le 2.4988$. We will use as a rule of thumb that if the interval is large then we might want to fit the coefficient using the Chebyshev's criterion (we will cover the Chebyshev's criterion in the chapter on linear programming). We consider this interval small and would not find the Chebyshev's fit.

## 4.3  EXERCISES

Bound the Chebyshev's criterion for the following problems and state whether or not you would find the Chebyshev's fit.

**1.** Use the following data and models.

| $x$ | 1 | 2 | 3 | 4 | 5 |
|---|---|---|---|---|---|
| $y$ | 1 | 1 | 2 | 2 | 4 |

**a.** $y = ax^2$

**2.** Data for the stretch of a spring data and the following models.

| $x\ (\times 10^{-3})$ | 5 | 10 | 20 | 30 | 40 | 50 | 60 | 70 | 80 | 90 | 100 |
|---|---|---|---|---|---|---|---|---|---|---|---|
| $y\ (\times 10^{5})$ | 0 | 19 | 57 | 94 | 134 | 173 | 216 | 256 | 297 | 343 | 390 |

   **a.** $y = ax$
   **b.** $y = ax^2$

3. Data for the ponderosa pine and the following models.

| $x$ | 17 | 19 | 20 | 22 | 23 | 25 | 28 | 31 | 32 | 33 | 36 | 37 | 39 | 42 |
|---|---|---|---|---|---|---|---|---|---|---|---|---|---|---|
| $y$ | 19 | 25 | 32 | 51 | 57 | 71 | 113 | 140 | 153 | 187 | 192 | 205 | 250 | 260 |

   **a.** $y = ax^2$
   **b.** $y = ax^3$

4. Use Kepler's data and the model $y = ax^{3/2}$.

| Body | Period ($s$) | Distance from Sun (m) |
|---|---|---|
| Mercury | $7.60 \times 10^6$ | $5.79 \times 10^{10}$ |
| Venus | $1.94 \times 10^7$ | $1.08 \times 10^{11}$ |
| Earth | $3.16 \times 10^7$ | $1.5 \times 10^{11}$ |
| Mars | $5.94 \times 10^7$ | $2.28 \times 10^{11}$ |
| Jupiter | $3.74 \times 10^8$ | $7.79 \times 10^{11}$ |
| Saturn | $9.35 \times 10^8$ | $1.43 \times 10^{12}$ |
| Uranus | $2.64 \times 10^9$ | $2.87 \times 10^{12}$ |
| Neptune | $5.22 \times 10^9$ | $4.5 \times 10^{12}$ |
| Pluto | $7.82 \times 10^9$ | $5.91 \times 10^{12}$ |

5

# Modeling with Proportionality and Geometric Similarity

You are with a famous archaeologist on a dig in Southern California. The archaeologist finds a femur bone of what appears to be a somewhat prehistoric birdlike creature. Of initial concern to the archaeologist is determining the size of this creature based on the bone. She turns to you as the mathematical scientist on this expedition. You tell her that there are modeling tools that will help determine the size of this creature. One of these tools is using proportionality with geometric similarity.

In our previous modeling work, we examined the concept of proportionality as a simplifying assumption. In this chapter, we formalize the concept of proportionality and use it to test relationships among variables chosen in our modeling process. Then we will present another simplifying argument called *geometric similarity*.

## 5.1 PROPORTIONALITY

Let's begin with a formal definition of proportionality. Two positive quantities $x$ and $y$ are said to be proportional (to each other) if one quantity is a constant multiple of the other. This definition implies that $y = kx$ for some constant $k$. We will write $y \propto x$ to indicate that the quantity $y$ is proportional to the quantity $x$. Thus,

$$y \propto x \text{ if and only if } y = kx \text{ for some nonzero constant } k$$

In this chapter, we will restrict ourselves to models in which $k$ is *positive* ($k > 0$). It makes sense that if $y \propto x$, then $x \propto y$ because the positive constants of proportionality are reciprocals of each other. Thus, if $y = kx$, then $x = (1/k) y$.

Let's visualize the geometric interpretation of proportionality. In Figure 5.1, we plot the geometric interpretation of $y \propto x$. The constant $k$ represents the slope of the straight line that intersects the origin. The slope, $k$, is defined to be the $\dfrac{rise}{run} = \dfrac{\Delta y}{\Delta x}$.

Examples of proportionalities include:

$$y \propto x^2 \text{ if and only if } y = k_1 x^2 \text{ for } k_1 > 0$$
$$y \propto e^x \text{ if and only if } y = k_2 e^x \text{ for } k_2 > 0$$

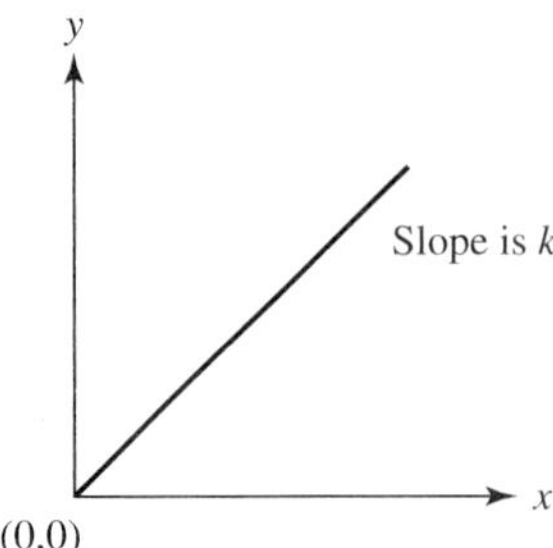

**FIGURE 5.1**

Graphical interpretation of proportionality



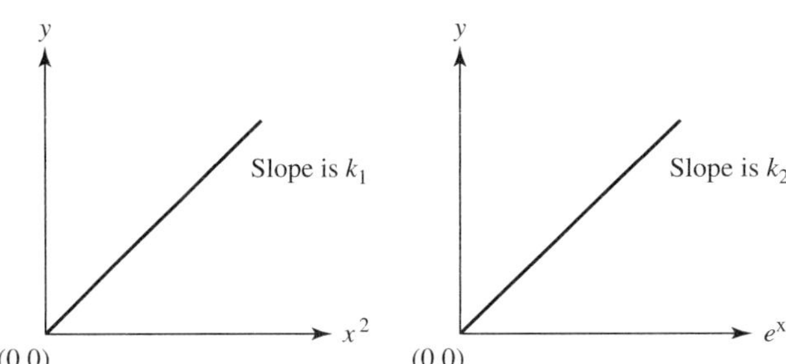

**FIGURE 5.2**

A geometric interpretation of proportionality



Graphically, we can present the proportionality by altering the $x$ axes to the quantity representing $x$ in the proportionality argument. These are illustrated in Figure 5.2.

Because in model $y = kx^2$, $k > 0$ then $x \propto y^{1/2}$ because $x = \left(\frac{1}{\sqrt{k}}\right)y^{\frac{1}{2}}$. This leads us to consider a *transitive property for proportionality*:

$$\text{If } y \propto x \text{ and } x \propto z, \text{ then } y \propto z.$$

Thus, any variables proportional to the same variable are proportional to one another.

There have been many scientific discoveries that are proportionality arguments. We list a few in Table 5.1.

### Testing and Estimating Proportionality Arguments

The simplest method of constructing a model is by applying proportionality arguments to a set of data. By plotting the transformed data set for the proposed proportionality model, we can obtain an initial estimation of the proportionality constant directly from the graph by finding the slope of the line that makes that line pass through the origin. The procedure for determining a proportionality constant for a model is presented in Table 5.2.

In this chapter, we demonstrate the Maple commands necessary to perform this method of *proportionality* model construction. First, we present the graphical plotting methods to check

**Table 5.1** Famous proportionality arguments

| | |
|---|---|
| Hooke's law, $F = kS$ | where $F$ is the restoring force in a spring stretched or compressed a distance $S$. |
| Newton's second law, $F = ma$ or $a = 1/m\,F$ | where $a$ is acceleration of a mass $m$ subjected to a net external force $F$. |
| Ohm's law, $E = iR$ | where $E$ is voltage, $i$ represents the induced current, and $R$ is the resistance. |
| Boyle's law, $V = k/P$ | where under a constant temperature $k$ the volume $V$ is inversely proportional to the pressure, $P$. |
| Kepler's third law, $T = cR^{3/2}$ | where $T$ is the period in any measured dimension and $R$ is mean distance to the sun. |

**Table 5.2** Procedure to Estimate a Proportionality Argument

1. Enter the data observed for the dependent and independent variables.
2. Plot the raw data points to check for trends and to identify potential data outliers.
3. Perform the transformations that support a hypothesized proportionality.
4. Plot the transformed data to test the hypothesized proportionality.
5. Estimate the constant of proportionality—that is, find the slope using least squares.

the proposed proportionality, and then we present the commands for estimating the constants of proportionality. Next, several modeling applications are investigated using proportionality.

This description of proportionality graphically depicts a line through the origin. Thus, a quick test of a proportionality model is the plot of the data that will approximate a straight line projected through the origin. Other types of proportionality models involve transformations such as $y \propto x^2$ or $y \propto e^x$. Maple can quickly transform the original data, $x$, into many such functions—here, $x^2$ and $e^x$—as we have previously seen in our introduction to Maple. A sub-model involving a proportionality argument must reasonably approximate a line and it *must* pass through the origin.

Once a proportionality argument has been successfully identified, the next step is determining the slope of the line that "best" fits the data graphically. Initially, a large plot is advantageous to determine if the data approximates a line. Quite often, this plot will not include the origin, which necessitates a second plot to verify that the best lines does pass through the origin. Consequently, two plots may be required to accurately plot the transformed data. This procedure is illustrated by several examples. Because the line passes through the origin, we recommend using (0,0) as one of these two points.

| Example 1 | Kepler's Law as a Proportionality Model |
|---|---|

Let's consider Kepler's third law of planetary motion and use the data typically available.

| Body | Period (s) | Distance from Sun (m) |
|---|---|---|
| Mercury | $7.60 \times 10^6$ | $5.79 \times 10^{10}$ |
| Venus | $1.94 \times 10^7$ | $1.08 \times 10^{11}$ |
| Earth | $3.16 \times 10^7$ | $1.5 \times 10^{11}$ |
| Mars | $5.94 \times 10^7$ | $2.28 \times 10^{11}$ |
| Jupiter | $3.74 \times 10^8$ | $7.79 \times 10^{11}$ |
| Saturn | $9.35 \times 10^8$ | $1.43 \times 10^{12}$ |
| Uranus | $2.64 \times 10^9$ | $2.87 \times 10^{12}$ |
| Neptune | $5.22 \times 10^9$ | $4.5 \times 10^{12}$ |
| Pluto | $7.82 \times 10^9$ | $5.91 \times 10^{12}$ |

Because of the data's magnitudes, we decide to compute the period in days (we divide the period by the product of 60 sec/min * 60 min/hr * 24 hrs/ day) and to scale the distance (divide by 1,000,000,000).

New data that we used are created by the map command.

**Step 1**    Enter the data.

```
> period:=[7.60*10^6,1.94*10^7,3.16*10^7,5.94*10^7,3.74*10^8,9.35*10^8,2.64*10^9,
   5.22 *10^9,7.82*10^9];
```

$$period := [0.760000000 \ 10^7, \ 0.1940000000 \ 10^8, \ 0.3160000000 \ 10^8, \ 0.5940000000 \ 10^8,$$
$$0.3740000000 \ 10^9, \ 0.9350000000 \ 10^9, \ 0.2640000000 \ 10^{10}, \ 0.5220000000 \ 10^{10},$$
$$0.7820000000 \ 10^{10}]$$

> *period_days:=map( x->x/(60*60*24),period );*

$$period\_days := [87.96296296, \ 224.5370370, \ 365.7407407, \ 687.5000000, \ 4328.703704,$$
$$10821.75926, \ 30555.55556, \ 60416.66667, \ 90509.25926 \ ]$$

*distance := [5.79*10^10, 1.08*10^11, 1.5*10^11,*
*2.28*10^11, 7.79*10^11, 1.43*10^12, 2.87* 10^12,*
*4.5* 10^12, 5.91* 10^12 ];*

> $distance := \ [5.790000000 \ 10^{10}, \ 1.080000000 \ 10^{11}, \ 1.500000000 \ 10^{11},$
$2.280000000 \ 10^{11}, \ 7.790000000 \ 10^{11}, \ 1.430000000 \ 10^{12},$
$2.870000000 \ 10^{12}, \ 4.500000000 \ 10^{12}, \ 5.910000000 \ 10^{12}]$

> *distance1:=map( x → x/( 1000000000 ),distance );*

$$distance1 := [57.90000000, \ 108.0000000, \ 150.0000000, \ 228.0000000,$$
$$779.0000000, \ 1430.000000, \ 2870.000000, \ 4500.000000,$$
$$5910.000000]$$

> *data:=seq( [period_days[i],distance1[i]],i=1..9);*

$$data := [87.96296296, \ 57.90000000], \ [224.5370370,$$
$$108.0000000], \ [365.7407407, \ 150.0000000], \ [687.500000,$$
$$228.0000000], \ [4328.703704, \ 779.0000000],$$
$$[10821.75926, \ 1430.000000], \ [30555.55556, \ 2870.000000]$$
$$[60416.66667, \ 4500.000000], \ [90509.25926, \ 5910.000000]$$

**Step 2**    Plot the original data.

> *pointplot( {seq( [period_days[i],distance[i]],i=1..9)},style=point,title='Original_data');*

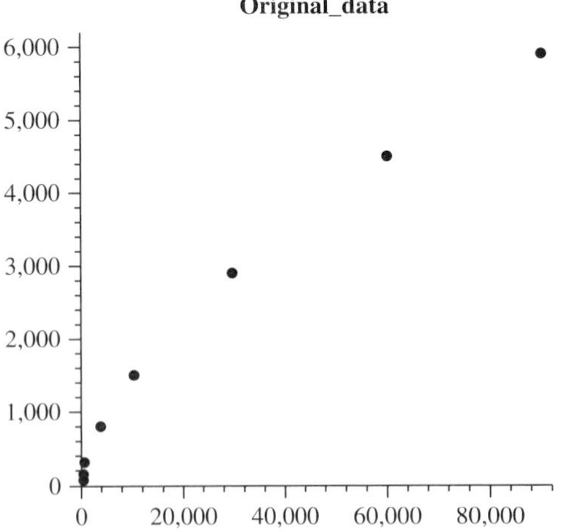


We see that the data trends are increasing and concave down.

**Step 3**    We use Kepler's third law ($T \propto R^{3/2}$) and transform the data. Period remains the same, but distance is transformed to distance raised to the 3/2 power. In Maple, we will use the command *d1:=map( x->x^3/2,distance1 ).*

**Step 4**     We plot the transformed data to see if it is a reasonable line through the origin.

> $d1:=map(x \rightarrow x^{\wedge}(3/2), distance1);$

$$d1 := [440.5729667, 1122.368925, 1837.117305, 3442.724502,$$
$$21742.33517, 54075.93734, 1.537527333 \; 10^5,$$
$$3.018691769 \; 10^5, 4.543402591 \; 10^5]$$

> $pointplot(\{seq([period\_days[i],d1[i]],i=1..9)\});$

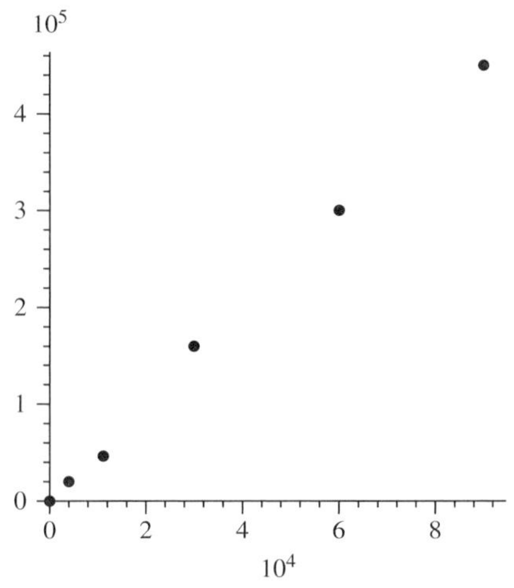

> $pointplot(\{seq([period\_days[i],d1[i]],i=1..9)\},style=point, title='Proportionality\_Test');$

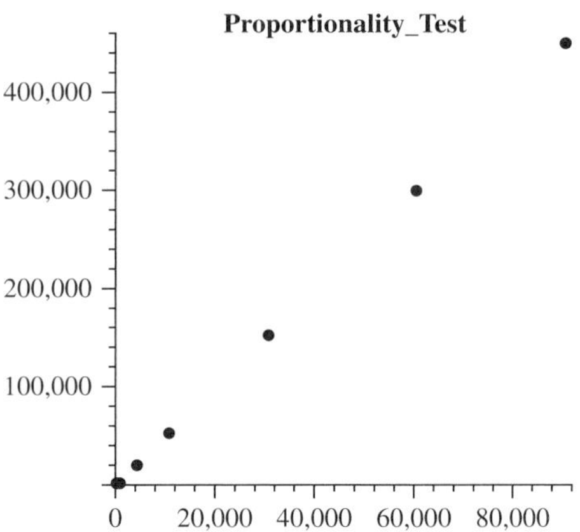


The plot does look like it is a line through the origin. We will return to this plot later.

**Step 5**     Estimate the slope. We determine that the ninth point falls on the line we desire, so we find the slope by period_days[9]/d1[9] (since the data are in numerical order) or we could use least squares.

Slope by use of points:

> $slope:=period\_days[9]/d1[9];$

$$slope := 0.1992102999$$

The model for our data with period in days and distance is

$$Period = .199210299 * Distance^{3/2}$$

We use Maple to obtain a plot of the original data and a plot of our model overlaid:

> *pt:=pointplot( {seq( [distance1[i],period_days[i]],i=1..9) },style=point):*
> *curve:=plot(.199210299*x^(3/2),x=0..4000):*
> *display(pt,curve);*

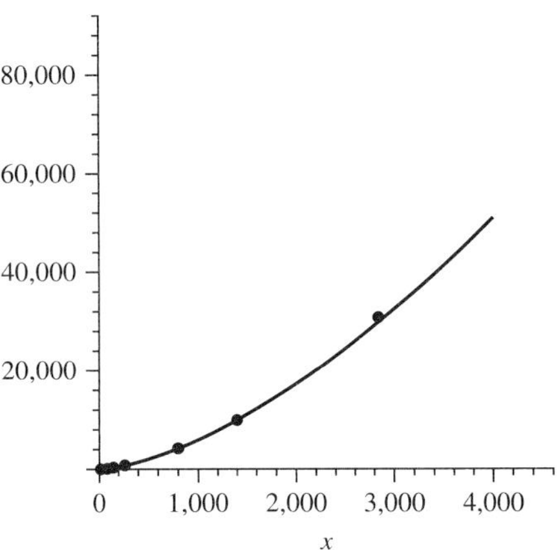

We interpret this as a good visual fit. The points appear to fall on the curve that we have found. Slope by use of least squares:

> *modfit:=Fit(a*x^(2/3),distance,period_days,x);*

$$modfit := 0.19944\ x^{3/2}$$

> *p1:=pointplot( {seq( [distance1[i], period_days[i]], i=1..9) })*
> *p2:=plot$\left(.19944 \cdot x^{\left(\frac{3}{2}\right)}, x=0..6000,\ title='Least\ squares\ Fit\right)$:*
> *display ( {p1,p2} );*

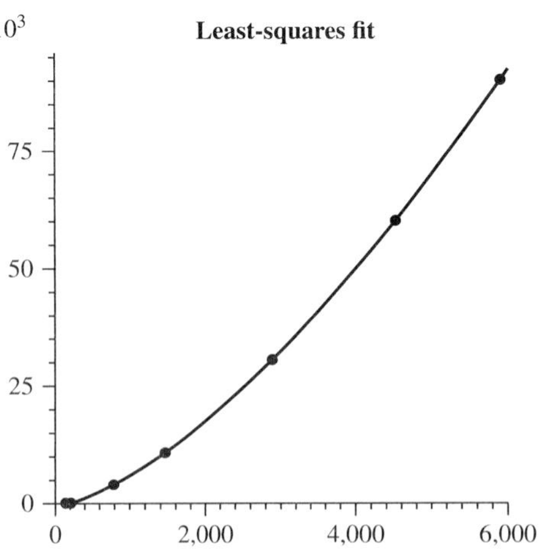

We interpret this as a good visual fit. The points appear to fall on the curve that we have found. We examine the residuals from our model.

> *f:= unapply(rhs(modfit),x);*

$$f := x \rightarrow 0.19944\ x^{3/2}$$

> *with (linalg):with(LinearAlgebra):*
> *predict:=map(x → f(x), distance1);*

$$predict := [87.867, 223.83, 366.35, 686.67, 4336.4, 10785.,$$
$$30664., 60205., 90616.]$$

> *resid:=matadd (period_days, predict,1,−1);*

$$resid := [0.096\ \ 0.71\ \ -.61\ \ 0.83\ \ -7.7\ \ 37.\ \ -108.\ \ 212.\ \ -107.]$$

If we use *with(LinearAlgebra)* then the command are as follows:

> *c1:=matrix(1,9,predict):c2:=matrix(1, 9, period_days):*
> *resid:=MatrixAdd (c1,c2,−1,1);*
> *wtpred := {seq ([distance1[i], resid[i]], i= 1 ..9)} :*
> *plot(wtpred,style=point,symbol = diamond,color = black,thickness = 3, title = 'Residual Plot');*

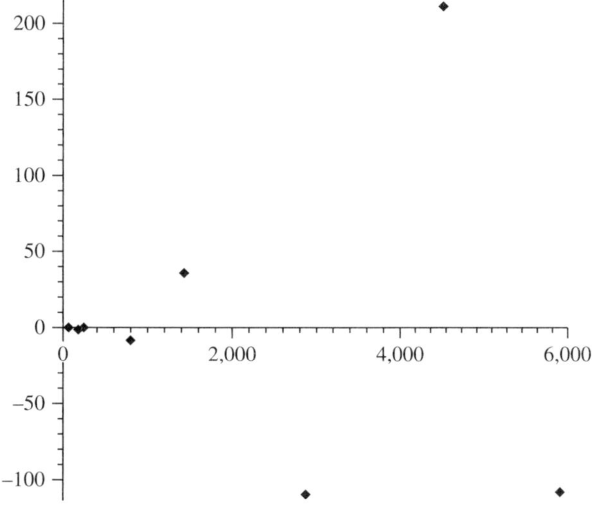


There appears to be a slight fanning out trend to our residuals and, therefore, we conclude that our model is not adequate. We would suggest going back and refining the model.

| Example 2 | Bass Fish Derby |
| --- | --- |

The bass fishing derby is analyzed to determine which fishing club member caught the most fish by weight. In this example, an analyst wishes to test the model volume $\propto$ length$^3$. The data for the fish length (len) and its volume (vol) is provided.

Length and Volume Data Points

| Length | 0.6 | 1.0 | 2 | 4 | 7 | 20 |
| --- | --- | --- | --- | --- | --- | --- |
| Volume | 0.1 | 0.7 | 6 | 100 | 210 | 4,000 |

**Step 1**   We enter the data into Maple and verify that it is correct.

> *len:=[0.6, 1, 2, 4, 7, 20];*

$$len := [0.6, 1, 2, 4, 7, 20]$$

> *vol:=[0.1, 0.7, 6, 100, 210, 4000];*

$$vol := [0.1, 0.7, 6, 100, 210, 4000]$$

**Step 2**   We plot the data, checking for trends, and identify any potential outliers found.

> *datalv:=seq([len[i],vol[i]],i=1 ..6);*

$$datalv := [0.6, 0.1], [1, 0.7], [2, 6], [4, 100], [7, 210], [20, 4000]$$

> *pointplot( {seq([len[i],vol[i]],i=1..6)},title='Raw Data Plot');*

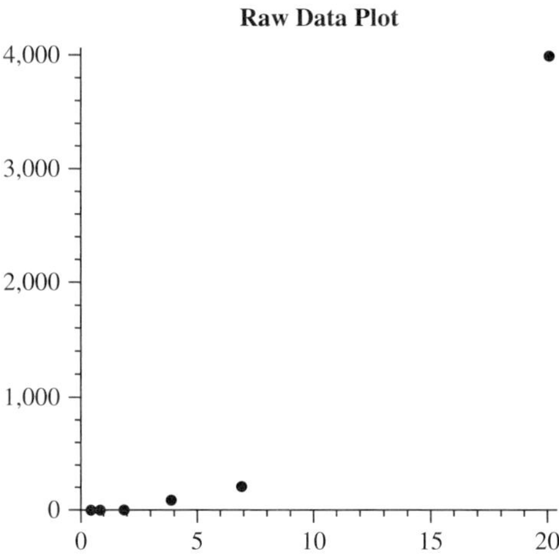


We examine the trends and state that the data appear to be concave up and increasing.

**Step 3**   Through proper proportionality arguments, we find that a possible model is vol = $k$ len$^3$. We use Maple to transform the data for length to length raised to the third power.

> *len3:=map(x→x^3,len);*

$$len3 := [0.216, 1, 8, 64, 343, 8000]$$

> *len3vol:=array(1..2,1..6,[len3, vol]);*

$$len3vol := \begin{bmatrix} 0.216 & 1 & 8 & 64 & 343 & 8000 \\ 0.1 & 0.7 & 6 & 100 & 210 & 4000 \end{bmatrix}$$

We illustrate the use of array here; previously we used seq. Either command will work for viewing the transformed data.

**Step 4**   **a.** We plot the transformed data to test for a straight line.

> *data3:=seq([len3[i],vol[i]],i=1..6);*

$$data3 := [0.216, 0.1], [1, 0.7], [8, 6], [64, 100], [343, 210], [8000, 4000]$$

*pointplot( {seq([len3[i],vol[i]],i=1..6)},title='Proportionality Test Plot');*

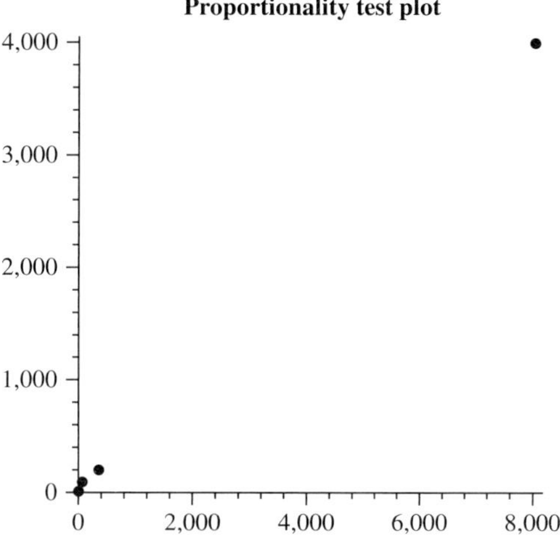

The data appear to be a straight line through the origin.

**b.** Use slope formula. To evaluate if the data project through the origin, we find the slope using either the slope formula or least-squares method. The line is drawn through the data points on the graph.

We next find the proportionality constant. Use the line that was drawn by hand onto the graphs of the data in step 4; then select two points $(x_1, y_1)$ and $(x_2, y_2)$ on this line. The line segment between the two points approximates the hand drawn best line (see Figure 4.6). The proportionality constant is the slope of the line segment between the two chosen points, slope $= \frac{y_2 - y_1}{x_2 - x_1}$. We present a solution for the slope using two points from the best line drawn onto our plot.

> *slope:=evalf((4000–100)/(8000–64));*

$$slope := 0.4914314516$$

> *c1:=pointplot({seq([len3[i],vol[i]],i=1..6)},*
> *title='Proportionality Test Plot') :*
> *c2:=plot(slope*x,x=0..8000, thickness=1):*
> *display({c1,c2});*

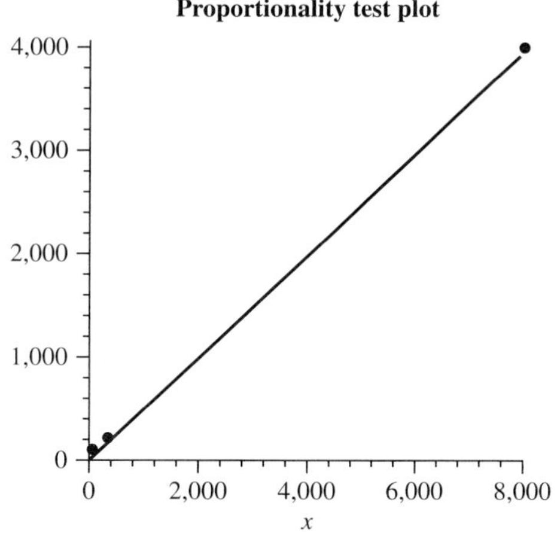

**c.** Use least squares.

> *modfit:=Fit(a*x^3,len,vol,x);*

$$modfit := 0.5002740674\ x^3$$

> *plot1:=plot(0.5002740674*x^3,x=0..20,title='Least Squares Fit', thickness=2):*
> *plot2:=pointplot( {seq( [len[i],vol[i]],i=1..6)} ):*
> *display( {plot1,plot2} );*

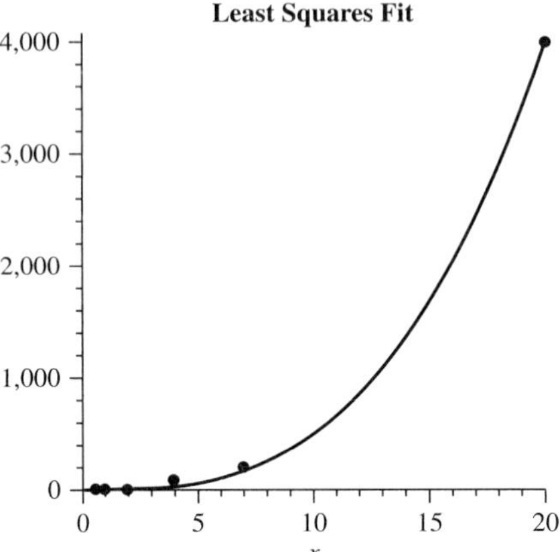


This graph gives visual proof that the proportionality model is reasonable.

**Step 5**     Examine the residuals.

> *f:=map(x–>distance,modfit);*

$$f := x \rightarrow 0.5002740674\ x^3$$

> *with(linalg):with(LinearAlgebra):*
> *predict:=map(x→distance,modfit);*

$predict := [0.1080591986, 0.5002740674, 4.002192539, 32.01754031, 171.5940051, 4002.192539]$

> *c1:=matrix(1,6,predict):c2:=matrix(1,6,vol):*
> *resid:=MatrixAdd(c1,c2,−1,1);*

$resid := [−0.0080591986, 0.1997259326, 1.99780746, 67.98245969, 38.4059949, −2.192539]$

> *wtpred:={seq( [len[i],resid[i]],i=1..6)}:*
> *plot(wtpred,style=point,symbol=circle,color=black,*
> *thickness=3,title='Residual Plot');*

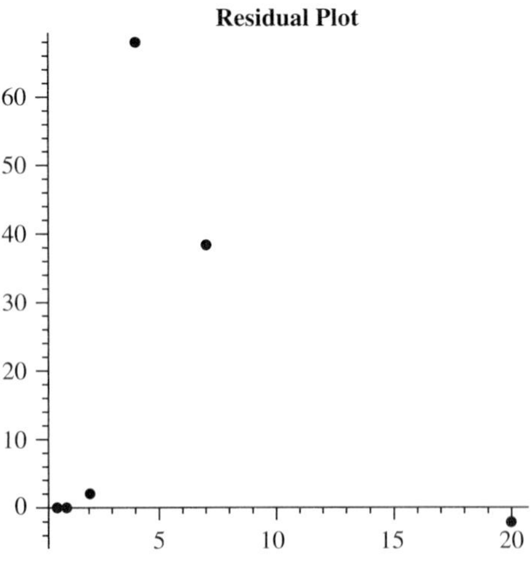


There might appear to be a fanning out of the residuals.

**Example 3**      Scaled Engineering Design

A design engineer for Ford Motor Company desires to build a scale model of a new sport utility vehicle to demonstrate to the production manager. The new SUV is planned to be 16.5 feet long. If the designer wants to build his model so that 1 foot = 1/2 inch, how long will the model be in inches?

$$y = kx$$
$$1 \text{ ft} = k((1/24 \text{ ft})$$
$$k = 24$$
$$16.5 \text{ ft} = 24\, x$$
$$x = 16.5/24 = 0.6875 \text{ feet or } (0.6875 \times 12 = 8.25 \text{ inches}).$$

## 5.1   EXERCISES

1. Explain the meaning of the following proportionality argument:

$$w \propto F/u$$

2. A road map has a scale of 1 inch = 6 miles. You measure the distance from home to the ski resort you plan to visit as 11.75 inches. How many miles will you be traveling? What assumptions are you making?

3. Determine if any of the following data sets follow the stated proportionality argument.

   **a.** $F \propto S$

| $S$ | 19 | 57 | 94 | 134 | 173 | 216 | 256 | 297 | 343 |
|---|---|---|---|---|---|---|---|---|---|
| $F$ | 10 | 20 | 30 | 40 | 50 | 60 | 70 | 80 | 90 |

   **b.** $y \propto x^3$

| $x$ | 17 | 19 | 20 | 22 | 23 | 25 | 28 | 31 | 32 | 33 | 36 | 37 | 38 | 39 | 41 |
|---|---|---|---|---|---|---|---|---|---|---|---|---|---|---|---|
| $y$ | 19 | 25 | 32 | 51 | 57 | 71 | 113 | 141 | 123 | 187 | 192 | 205 | 252 | 259 | 294 |

   **c.** $d \propto v^2$

| $v$ | 20 | 25 | 30 | 35 | 40 | 45 | 50 | 55 | 60 | 65 | 70 |
|---|---|---|---|---|---|---|---|---|---|---|---|
| $d$ | 22 | 28 | 33 | 39 | 44 | 50 | 55 | 61 | 66 | 72 | 77 |

   **d.** Model $Y \propto X^2$.

| $X$ | $Y$ |
|---|---|
| 1 | 4 |
| 2 | 11 |
| 3 | 22 |
| 4 | 35 |
| 5 | 56 |
| 6 | 80 |
| 7 | 107 |
| 8 | 140 |
| 9 | 175 |
| 10 | 215 |

**e.** Model $Y \propto X^3$.

| X | Y |
|---|---|
| 1 | 0 |
| 2 | 1 |
| 3 | 2 |
| 4 | 6 |
| 5 | 14 |
| 6 | 24 |
| 7 | 37 |
| 8 | 58 |
| 9 | 82 |
| 10 | 114 |

**f.** Model $Y \propto e^x$.

| X | Y |
|---|---|
| 1 | 6 |
| 2 | 15 |
| 3 | 42 |
| 4 | 114 |
| 5 | 311 |
| 6 | 845 |
| 7 | 2,300 |
| 8 | 6,250 |
| 9 | 17,000 |
| 10 | 46,255 |

## 5.2 GEOMETRIC SIMILARITY

Geometric similarity is a concept related to proportionality and is very useful in simplifying this portion of the mathematical modeling process. The definition is as follows.

Two objects are called *geometrically similar* if there is a one-to-one correspondence between points of the object such that the ratio of distances between corresponding points is constant for all possible pairs of points. You can think of geometrically similar objects as scale models of one another.

Although we want to consider realism, we will begin with only two-dimensional objects. Consider the rectangles in Figure 5.3. Let $l$ denote the distance from A to B in rectangle 1 and let $l'$ denote the distance from A' to B' in rectangle 2. All corresponding points in the figures and their associated distances are also marked. For rectangles that are scale models, it must be true that

$$\frac{l}{l'} = \frac{w}{w'} = k \text{ for a constant } k > 0$$

**FIGURE 5.3**

Similar rectangles

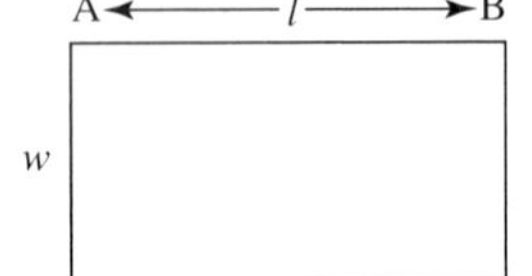
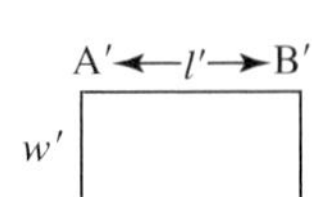

**FIGURE 5.4**

(a) Geometrically similar circles; (b) geometrically similar triangles

*(a)*

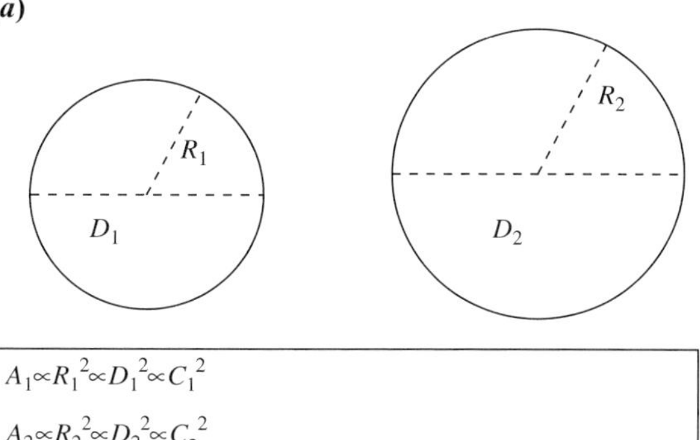

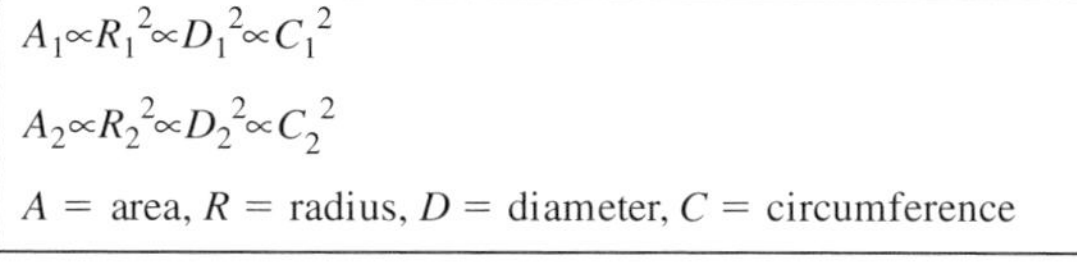

$$A_1 \propto R_1^2 \propto D_1^2 \propto C_1^2$$

$$A_2 \propto R_2^2 \propto D_2^2 \propto C_2^2$$

$A$ = area, $R$ = radius, $D$ = diameter, $C$ = circumference

*(b)*

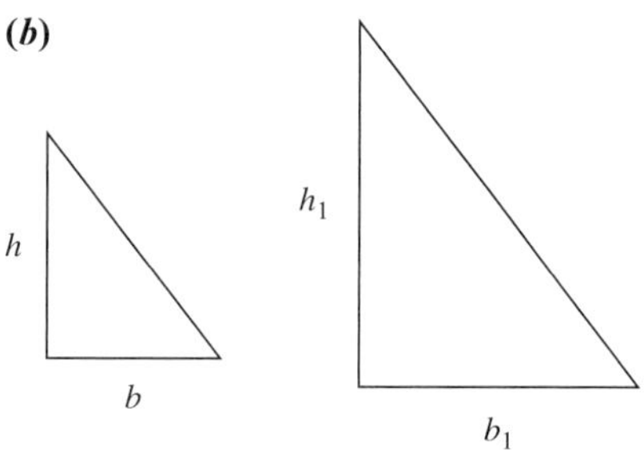

One major advantage is in simplification for certain computations. Let's consider the area of the similar rectangles $A = lw$ and $A' = l'w'$. The ratio of $A/A' = (lw)/(l'w') = k^2$. These ratios not only are related but also reduce to the square of the constant of proportionality in the ratio of corresponding points. Thus, we can readily use any **characteristic dimension** from the rectangles in a proportionality argument, for example, $A \propto l^2$ or $A \propto w'^2$. Because the constant works for all geometrically similar objects, it makes no difference what dimension we ultimately use.

This works for circles and triangles as well. These are depicted in Figures 5.4 (a) and (b). In fact, by geometric similarity, the following hold for similar circles and their characteristic dimensions:

$$A = \pi R^2, \text{ so } A \propto R^2 \tag{5.1}$$

$$R = D/2,\ A = \pi(D/2)^2, \text{ so } A \propto D^2 \tag{5.2}$$

$$C = 2\,\pi R,\ \pi(C/2\pi)^2, \text{ so } A \propto C^2 \tag{5.3}$$

By geometric similarity, the following hold for similar triangles and their characteristic dimensions:

$$A = \tfrac{1}{2}\,bh$$

$$A \propto b^2 \tag{5.4}$$

$$A \propto h^2 \tag{5.5}$$

Now we can move from two-dimensional to three-dimensional objects. In three dimensions we can relate surface area $S$ and volume $V$ by this same approach. Consider the three-dimensional box in Figure 5.5. The volume of box 1 is $V = lwh$; for box 2, it is $V' = l'w'h'$. Forming the ratio, $V/V' = lwh/l'w'h' = k^3$. Thus, the volume is proportional to any dimension cubed when dealing with geometrically similar objects.

How critical is the object's shape? It is not critical at all to the argument. Consider two spheres that are geometrically similar; we want to model information about the volume of the smaller sphere. The volume of a sphere is $V = 4/3\,\pi r^3$. Thus, the ratio of $V/V' = (4/3\,\pi r^3)/(4/3\,\pi r'^3) = k^3$.

So,

$$V \propto (\text{any dimension})^3 \tag{5.6}$$

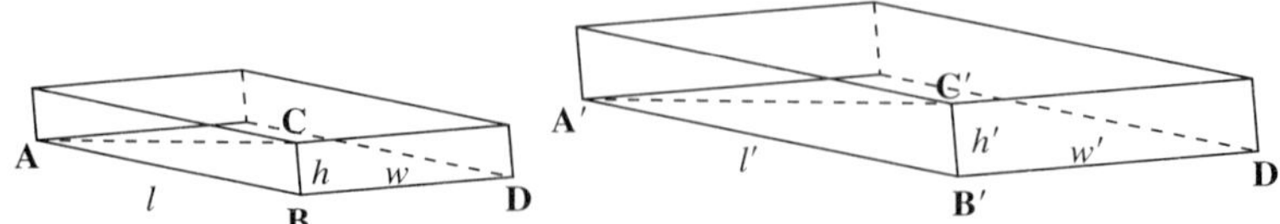


For example, consider building a model to relate heart weight to some dimension of the heart. Under constant density, the amount of volume displaced is equal to the weight of the object. Thus, $W \propto V$ (assuming constant density). If we assume that all hearts are geometrically similar and close to spherical shape, then we move directly to Equation 5.6: $V \propto$ (any dimension)$^3$. We don't need to model the exact shape of the heart at all. Let's assume that our known dimension is the diameter of the heart:

$$W \propto V \propto d^3$$

Because weight is proportional to volume and volume is proportional to $d^3$, we can conclude that $W \propto d^3$.

Now that we have discovered a possible relationship between heart weight and some characteristic dimension, we must test its validity. We do this by gathering data for heart weights and diameters, plotting weight versus diameter$^3$ and determining the reasonableness of a straight line through the origin fitting the data.

## Example 1   Heart Weight

Let's examine the following seven mammals, their heart weights in grams, and the assumed diameter of the left ventricle measured in mm.

| Animal | Heart Weight (g) | Left Ventricle Diameter (mm) |
| --- | --- | --- |
| Mouse | 0.13 | 0.55 |
| Rat | 0.64 | 1.0 |
| Rabbit | 5.8 | 2.2 |
| Dog | 102 | 4.0 |
| Sheep | 210 | 6.5 |
| Ox | 2,030 | 12.0 |
| Horse | 3,900 | 16.0 |

## Problem Identification

Find a relationship between heart weight and the diameter of the left ventricle of the heart.

## Assumptions

Mammals are geometrically similar, so we use Equation 5.6: $V \propto$ dimension$^3$. Hearts have constant weight density ($V \propto W$). We could also evoke Archimedes' law that states that an upward force acting on a body wholly or partly submerged in a fluid is equal to the weight of the fluid displaced and acts through the center of gravity of the fluid displaced or through the center of buoyancy. In common terms, one displaces an equal volume of water to its weight. Thus, under certain conditions we might assume that volume is proportional to weight.

# Model

$$W \propto V \propto D^3$$

Let's test the "reasonableness" of this explicative model relating heart weight to the diameter of the left ventricle.

Using Maple commands, we obtain the following analysis:

> *weight:=[.13,.64,5.8,102,210,2030,3900];*

$$weight := [\,0.13, 0.64, 5.8, 102, 210, 2030, 3900\,]$$

> *diameter:=[.55,1,2.2,4,6.5,12,16];*

$$diameter := [\,0.55, 1, 2.2, 4, 6.5, 12, 16\,]$$

> *pointplot( {seq( [diameter[i],weight[i]],i=1..7)},style=point, title='Original_data');*

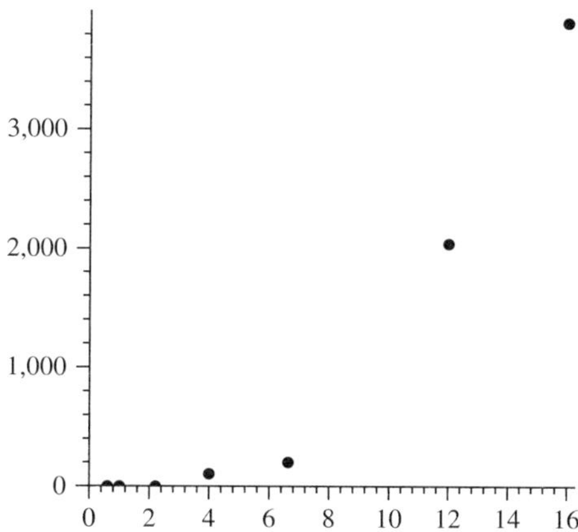


The trend is increasing and concave up.

> *d3:=map(x->x^3,diameter);*

$$d3 := [\,0.166375, 1, 10.648, 64, 274.625, 1728, 4096\,]$$

> *pointplot( {seq( [d3[i],weight[i]],i=1..7)},style=point, title='Proportionality_test');*

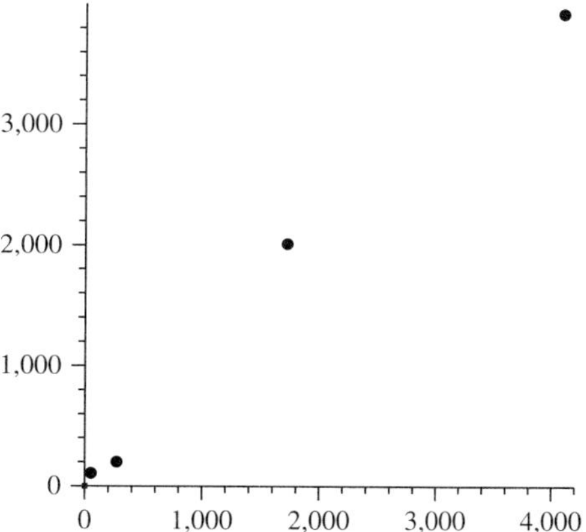

The data look like a reasonable line through the origin.

> *slope:=weight[7]/d3[7];*

$$slope := \frac{975}{1024}$$

> *evalf(%);*

$$0.9521484375$$

> *model:=slope*x^3;*

$$model := \frac{975\, x^3}{1024}$$

> *pt:=pointplot( {seq([diameter[i],weight[i]],i=1..7)},style=point):*
> *curve:=plot(slope*diam^3, diam=0..20):*
> *display(pt,curve);*

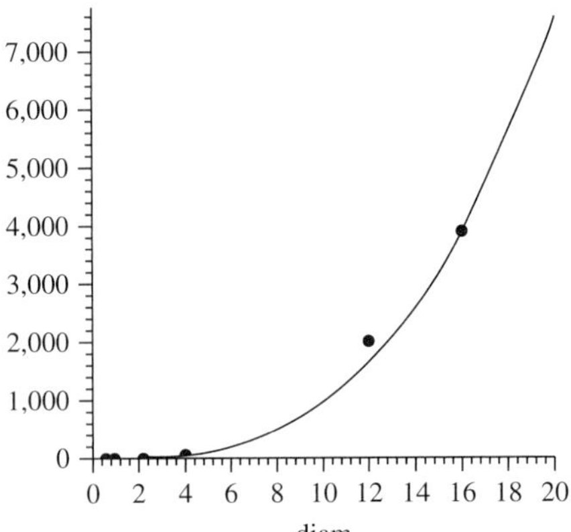


The curve overlays nicely with the data. It appears to be a reasonable fit. Using the least-squares fit, we obtain

> *with(LinearAlgebra):with(linalg):*
> *modfit:=Fit(a*x^3,diameter,weight,x);*

$$modift := y = 0.9850662612\, x^3$$

> *predict:=map(x→distance,modfit);*

$$predict := [0.16389, 0.98506, 10.489, 63.044, 270.52, 1702.2, 4034.8]$$

> *c1:=matrix(1,7,weight):c2:matrix(1,7,predict):*
> *resid:=MatrixAdd(c1,c2,1,−1);*

$$resid := [−0.03389, −.34506, −4.689, 38.956, −60.52, 327.8, −134.8]$$

> *wtpred:={seq([diameter[i],resid[i]],i=1..6)}:*
> *plot(wtpred,style=point,symbol=diamond,color=black,*
> *thickness=3,title='ResidualPlot');*

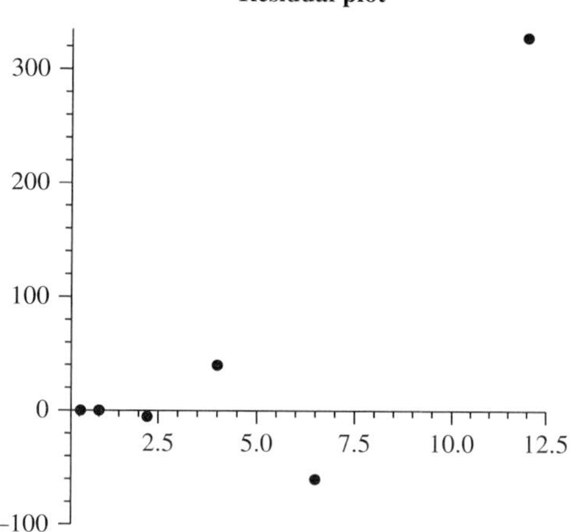



There appears to be a fanning out of the residuals, which indicates the model may not be adequate.

<table>
<tr><td>**Example 2**</td><td>

## Crew Races

</td></tr>
</table>

Crew is a popular sport along the East Coast of the United States and, of course, in England. Let's consider building a model to help a crew coach choose a team by finding a relationship between racing times and number of crew members in the racing shell. If you have been to a rowing regatta, then you might have observed that the more rowers there are in a boat, the faster the boat travels. It is therefore reasonable to investigate whether there is a mathematical relationship between the speed of a boat and the number of crew members.

## Assumptions (Partial List)

1. The total force exerted by the crew is constant for a particular crew throughout the race.
2. The drag force, $F_d$, experienced by the boat as it moves through the water is proportional to the square of the velocity times the wetted surface area of the hull.
3. Work is defined as the product of force and distance. Power is defined as work per unit time.

The data were obtained from a 1998 Occoquan Crew Racing (George Mason Crew Club) (times obtained from http://www.gmu.edu/org/crew/98Chase.htm).

| Number in Boat | Distance | Race 1 | Race 2 | Race 3 | Race 4 | Race 5 | Race 6 |
|---|---|---|---|---|---|---|---|
| 1 | 2,500 m | 20:53 | 22:21 | 22:49 | 26:52 | | |
| 2 | 2,500 m | 19:11 | 19:17 | 20:02 | | | |
| 4 | 2,500 m | 16:05 | 16:42 | 16:43 | 16:47 | 16:51 | 17:25 |
| 8 | 2,500 m | 9:19 | 9:29 | 9:49 | 9:51 | 10:21 | 10:33 |

*Hint:* When additional rowers are added to a shell, it is not obvious whether the amount of force is proportional to the number in the crew or the amount of power is proportional to the number in the crew. Which assumption appears the most reasonable? Which yields a more accurate model?

## Problem Identification

Build a mathematical model that relates the speed of the crew shell to the number of rowers in the crew shell.

## Assumptions

1. The total force exerted by the crew is constant for a particular crew throughout the race.
2. The drag force, $F_d$, experienced by the boat as it moves through the water is proportional to the square of the velocity times the wetted surface area of the hull.
3. Work is defined as the product of force and distance. Power is defined as work per unit time.
4. Racing shells are geometrically similar.
5. Crew persons are geometrically similar and provide equal power.
6. Rowers achieved a constant velocity instantaneously and maintained it until the end of the race.

## Model Construction

By Newton's second law of motion:

$$\sum F = ma = m \cdot \frac{dv}{dt} = 0$$

(By assumption 6, constant velocity.)
Therefore,

$$F_p - F_d = 0, \text{ or } F_p \propto F_d$$

The total force exerted by the crew is constant for a particular crew throughout the race. The drag force experienced by the boat as it moves through the water is proportional to the square of the velocity times the wetted surface area of the hull. $F_d \propto Sv^2$, where $S$ is the wetted surface area and $v$ is the velocity. So, $F_p \propto Sv^2$. Work is defined as force times a distance. Power is defined as work per unit time: $W = Fd$ and $P = \dfrac{W}{t}$.

The force used to overcome drag is the rowing power generated by the crew. The power is then modeled as

$$P = F_d v \propto F_p v$$

Substituting, we get a model relating velocity to both power and surface area:.

$$v \propto (P/S)^{1/3}$$

## Model Choices

1. $F \propto n$, where $n$ is the number of crew members, or
2. $P \propto n$, where $n$ is number of crew members.

With model 2 and substitution, we obtain

$$v \propto (n/S)^{1/3}$$

We now use the assumption that the shells are geometrically similar. This implies that $V \propto l^3$. $V$ also $\propto n$ because their weight determines the displacement. Because $S \propto l^2$ ( by definition), we obtain that

$$S \propto n^{2/3}$$

Substituting into the previous equation yields

$$v \propto (n/S)^{(1/3)} \propto (n/n^{(2/3)})^{(1/3)} \propto n^{(1/9)}$$

Velocity is the change in distance over the change in time. Since distance is constant velocity $\propto 1/$time. Thus, because our data contain times and number of crew, our final model becomes

$$T \propto n^{(-1/9)}$$

Now let's return to our data and see how well the model works. We will average the racing time for convenience.

Using Maple, we obtain the following results:

```
> t1:=[1253,22*60+21,22*60+49,26*60+52];
```

$$t1 := [\, 1253, 1341, 1369, 1612\,]$$

```
> t2:=[19*60+11,19*60+17,20*60+2];
```

$$t2 := [1151, 1157, 1202\,]$$

```
> t3:=[16*60+5,16*60+42,16*60+43,16*60+47,16*60+51,17*60+25];
```

$$t3 := [\, 965, 1002, 1003, 1007, 1011, 1045\,]$$

```
> t4:=[9*60+19,9*60+29,9*60+49,9*60+51,10*60+21,10*60+33];
```

$$t4 := [\, 559, 569, 589, 591, 621, 633\,]$$

```
> with(Statistics):
> Mean(t1);
```

$$1393.750000$$

```
> Mean(t2);
```

$$1170.$$

```
> Mean(t3);
```

$$1005.500000$$

```
> Mean(t4);
```

$$593.6666667$$

```
> avgtime:=[1393.75,1170,1005.5,593.6667];
```

$$avgtime := [\, 1393.75, 1170, 1005.5, 593.6667\,]$$

```
> crew:=[1,2,4,8];
```

$$crew := [1, 2, 4, 8]$$

> *pointplot( {seq( [crew[i],avgtime[i] ],i=1..4 )},style=point, title=`Original_data`);*

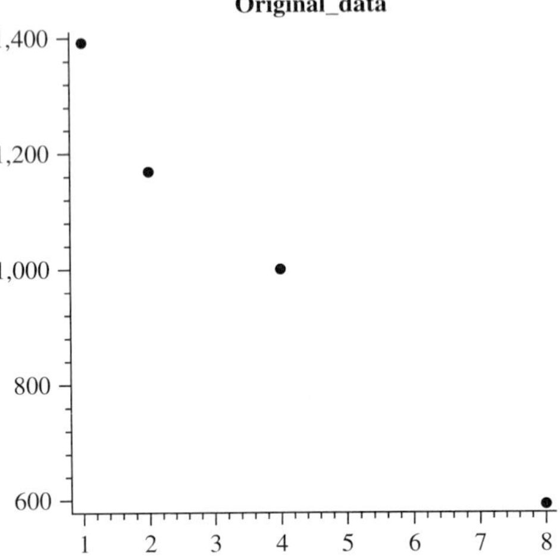


The trend is increasing and concave up.

> *ncrew:=evalf(map(x->x^(-1/9),crew));*

$$ncrew := [1., 0.9258747125, 0.8572439828, 0.7937005260]$$

> *pointplot( {seq( [ (ncrew[i]-.3),avgtime[i] ],i=1..4 )},style=point, title=`Proportionality_Test`, view=[0..1,0..1500 ]);*

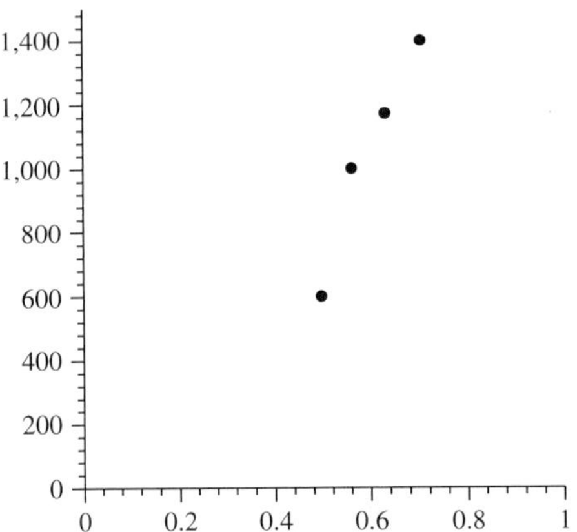


The data look like a somewhat reasonable line through the origin. Note we subtracted .3 from each data point to move the data for a line through the origin. The original proposed proportionality was a line, but it did not pass through the origin.

> *slope:=avgtime[4]/(ncrew[4]-.3);*

$$slope := 1202.483426$$

> *evalf(%);*

$$1202.483426$$

> *model:=slope*x^(-1/9);*

$$model := \frac{1202.483426}{x^{(1/9)}}$$

```
> pt:=pointplot( {seq([crew[i],avgtime[i]],i=1..4)},style=point):
> curve:=plot(slope*n^(-1/9), n=0..10):
> display(pt,curve);
```

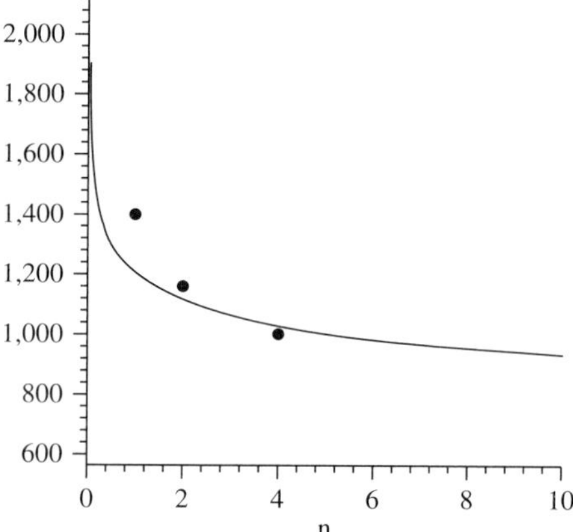

The curve does only a fair job in fitting the data.

---

**Example 3**

## Modeling the Terror Bird

Refer back to Example 1, The Size of Prehistoric Creatures, in Chapter 2 (pages 23–24). According to Dr. Robert Chandler, a paleontologist from Georgia College and State University, *Titanis walleri*, the so-called terror birds, "were the most dangerous birds to have ever lived. They were probably heirs to the dinosaurs in that they were able to kill as efficiently as some of the larger meat-eating dinosaurs." Scientists want to model the size of this prehistoric bird.

## Problem Identification

We want to predict the size of the terror bird (in weight measure) as a function of the fossil bone. This needs some refinement, so we narrow our fossil to the femur and choose its circumference as the measure to use. Predict the weight of the terror bird as a function of the circumference of its femur.

## Assumptions and Variables

The terror bird was a large, flightless bird. "Like the ostrich, the bird sacrificed the ability to fly to become a more efficient killer," said Chandler. The characteristics of the bird 2 million years ago is similar to the characteristics of birds today. We will assume that the terror birds were geometrically similar to other large birds of today. With the assumption of geometrical similarity, we can state that the volume of the bird is proportional to any characteristic dimension cubed and that the area of the bird is proportional to any characteristic dimension squared.

$$V \propto l^3 \text{ and } S \propto l^2$$

If we assume a constant weight density, then a volume displaces an amount equal to its weight, $V \propto W$. Thus, we can use Equation 5.6:

$$V \propto W \propto l^3$$

We will let the characteristic dimension be the circumference of the femur. The femur was chosen because it supports the body weight. Thus, $W = kl^3$, $k > 0$.

## Testing the Model

We gather data on various birds. The data are shown in the following table.

Bird Data

| Femur Circumference (cm) | Weight (kg) |
| --- | --- |
| 0.7943 | 0.0832 |
| 0.7079 | 0.0912 |
| 1.000 | 0.1413 |
| 1.122 | 0.1479 |
| 1.6982 | 0.2455 |
| 1.2023 | 0.2818 |
| 1.9953 | 0.7943 |
| 2.2387 | 2.5119 |
| 2.5119 | 1.4125 |
| 2.5119 | 0.8913 |
| 3.1623 | 1.9953 |
| 3.5481 | 4.2658 |
| 4.4668 | 6.3096 |
| 5.8884 | 11.2202 |
| 6.7608 | 19.9500 |
| 15.136 | 141.25 |
| 15.850 | 158.4893 |

```
> restart;
> len:=[.7943,.7079,1,1.122,1.6982,1.2023,1.9953,2.2387,2.5119,2.5119,3.1623,3.5481,4.4668,
5.8884,6.7608,15.136,15.850]:
> wt:=[.0832,.0912,.1413,.1479,.2455,.2818,.7943,2.5119,1.4125,.8913,1.9953,4.2658,6.3096,
11.2202,19.95,141.25,158.4893]:
> lenwt:=array(1..2,1..17,[len,wt]);
> with(plots):
pointplot( {seq([len[i],wt[i]],i=1..17)},style=point,symbol=box,view=[0..16,0..160]);
> len3:=map(x->x^3,len):
> len3wt:=array(1..2,1..17,[len3,wt]);
> pointplot( {seq([len3[i],wt[i]],i=1..17)}, style=point,symbol=box);
> fx2:=(wt[17]/len3[17])*x;
> with(plots):
> points:=pointplot( {seq([len3[i],wt[i]],i=1..17)}, style=point,symbol=box):
> curve:=plot(fx2,x=0..8000):
> display( {points,curve});
> with(plots):
> slope:=evalf((wt[17]-0)/(len3[17]-0));
> curve:=plot(slope*r^3,r=0..20,thickness=3):
> points:=pointplot( {seq([len[i],wt[i]],i=1..17)}, style=point,symbol=box):
> t1:=textplot([10,100,'The model is plotted as a solid line;'],align=ABOVE):
```

> *t2:=textplot( [20,40,'the length/volume are data points.'],align=ABOVE):*
> *display( {points,curve,t1,t2} );*

$$lenwt := [\ 0.7943\ ,\ 0.7079,\ 1,\ 1.122,\ 1.6982,\ 1.2023,\ 1.9953,\ 2.2387,\ 2.5119,$$
$$2.5119,\ 3.1623,\ 3.5481,\ 4.4668,\ 5.8884,\ 6.7608,\ 15.136,\ 15.850\ ]$$
$$[\ 0.082,\ 0.0912,\ 0.1413,\ 0.1479,\ 0.2455,\ 0.2818,\ 0.7943,\ 2.5119,\ 1.4125,$$
$$0.8913,\ 1.9953,\ 4.2658,\ 6.3096,\ 11.2202,\ 19.95,\ 141.25,\ 158.4893\ ]$$

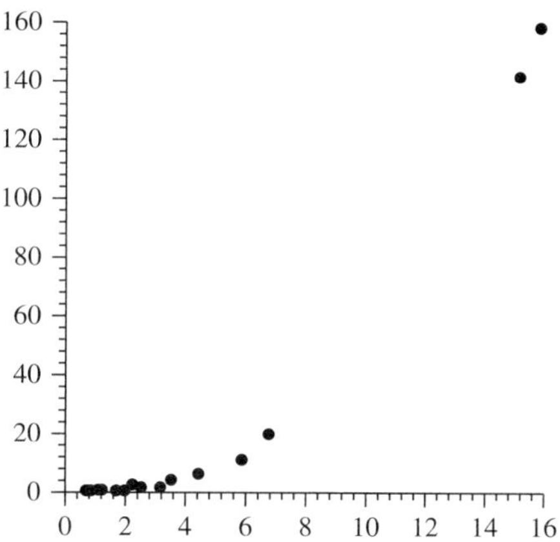

The data are concave up and increasing.

$$len3wt := [0.5011337908\ ,\ 0.3547445540\ ,\ 1\ ,\ 1.412467848\ ,\ 4.897410518\ ,$$
$$1.737955056\ ,\ 7.943732436\ ,\ 11.21986671\ ,\ 15.84918876\ ,\ 15.84918876\ ,$$
$$31.62344680\ ,\ 44.66707919\ ,\ 89.12294365\ ,\ 204.1699922\ ,\ 309.0254632\ ,$$
$$3467.634835\ ,\ 3981.876625\ ]$$

$$[\ 0.082\ ,\ 0.0912\ ,\ 0.1413\ ,\ 0.1479\ ,\ 0.2455\ ,\ 0.2818\ ,\ 0.7943\ ,\ 2.5119\ ,\ 1.4125\ ,$$
$$0.8913\ ,\ 1.9953\ ,\ 4.2658\ ,\ 6.3096\ ,\ 11.2202\ ,\ 19.95\ ,\ 141.25\ ,\ 158.4893\ ]$$

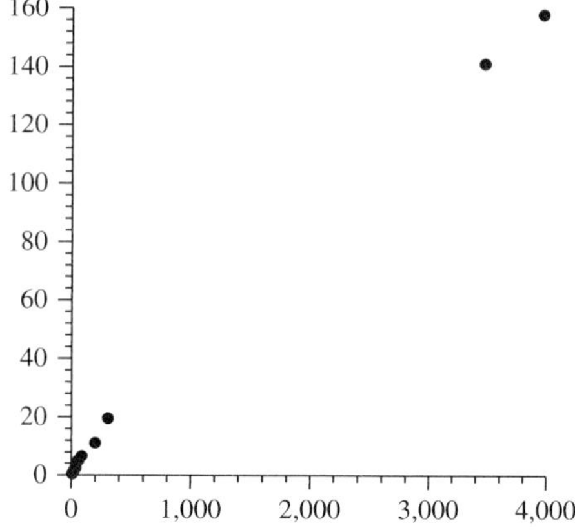

$$fx2 := 0.03980266465\ x$$

Appears to be a line through (0,0).

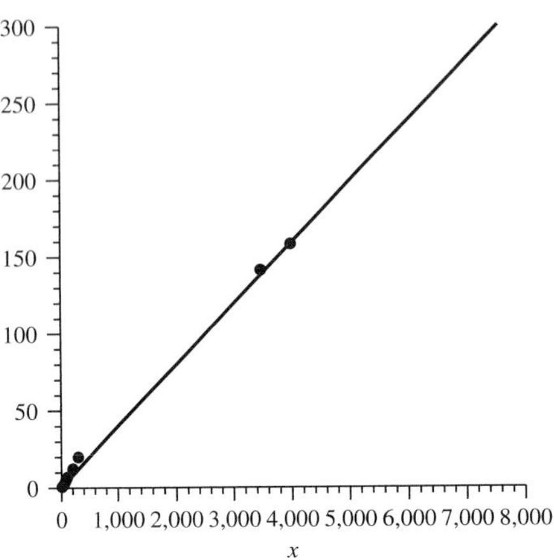

$$slope := 0.03980266465$$

With a sloped line, we clearly see the line is through (0,0).

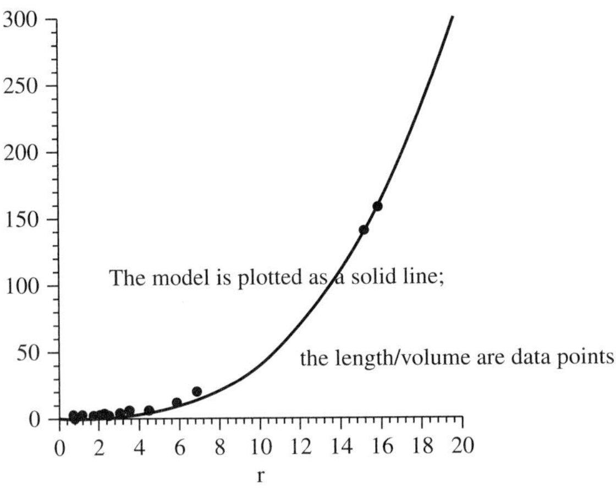


Visually, the model and data follow the same trend. We may use the model to predict the size of the terror bird:

$$Weight = .03980266465*21^3 = 368.61 \text{ kg}$$

## 5.2 | EXERCISES

1. Draw the graphical interpretation of the following.
   a. Force is proportional to acceleration: $F \propto A$.
   b. Interest is proportional to the product of the principle and time: $I \propto PT$.
   c. Distance is proportional to time: $D \propto T$.
   d. Force is inversely proportional to the square of the distance: $F \propto 1/d^2$.

2. Are the temperature scales, Celsius and Fahrenheit, proportional? Gather some data and test your conclusion.

3. In a recent year, 85 out of every 100 Americans had some form of health-care coverage. The population at that time was about 246 million. How many had health-care coverage?

4. The model and prototype are scale models at a scale of 1" = 25'. If the wing span of the model is 5.5", then what is the length of the real wingspan?

5. An architect is designing a new parking garage. The model for the garage is 36" tall, and the scale is 1" is 3.3 feet. How tall will the real garage be?

## 5.2 | PROJECTS

For each project, build a proportionality model that relates the given data. Then validate the model with graphical techniques.

1. **Terror Bird, Part II: The Dinosaur Data** From Chapter 2, recall these facts about *Titanis walleri*, the so-called terror bird:
   - It was giant and flightless.
   - It lived in South America 2 million to 3 million years ago.
   - A fierce hunter, it would lie in wait and ambush its prey and attack from the tall grasslands.
   - It probably pinned down its prey with its beak and used its 4- to 5-inch inner toe claws and beak to shred its prey.
   - It had arms, not wings, that were more powerful than the arms of the *Velociraptor* (Chandler, 1994)—that is, about the size of an NFL linebacker.
   - Bones give a good indication of size (height) but not reliable body weight (weight is not easily fossilized)
   - We want to infer its weight.

   Bones give a good indication of size (height) but not of reliable body weight (weight is not easily fossilized), which is what we want to infer. Model the size of the terror bird and use the supporting data to test your modeling assumptions.

   Dinosaur Data

   | Name | Femur Circumference (mm) | Weight (kg) |
   | --- | --- | --- |
   | Hypsilophodontidae | 103 | 55 |
   | Ornithomimdae | 136 | 115 |
   | Thescelosauridae | 201 | 311 |
   | Ceratosauridae | 267 | 640 |
   | Allosauridae | 348 | 1,230 |
   | Hadrosauridae–1 | 400 | 1,818 |
   | Hadrosauridae–2 | 504 | 3,300 |
   | Hadrosauridae–3 | 512 | 3,500 |
   | Tyrannosauridae | 534 | 4,000 |

   a. Compare and contrast the result from the dinosaur and bird data sets. Are the answers different?
   b. Discuss the issues of (1) domain and range and (2) interpolation versus extrapolation for the data.
   c. Check the proportionality arguments you built with both of these data sets. Comment on which one you think yields a better prediction model.

2. **Weightlifting** In an old TV show called *Superstars*, top athletes from various sports competed against one another in a variety of events. Of course, the athletes varied considerably in height and weight. To compensate for this in the weight-lifting competition, the athlete's body weight is subtracted from his lift. What kind of relationship does this suggest? Use the data to show this relationship.

The following data display the winning lifts at the 1996, 2000, and 2008 Olympic Games (weights are in lbs).

Men's Olympic Games

| Weight Class | Max Weight (lbs) in Class | Total Lifted (lbs) |
|---|---|---|
| 1996 | 119 | 633 |
| | 130 | 678 |
| | 141 | 787 |
| | 167.5 | 809 |
| | 183 | 633 |
| | 200.5 | 886 |
| | 218 | 926 |
| | 238 | 948 |
| | 238+ | 1,008 |
| 2000 | 123 | 674 |
| | 136 | 718 |
| | 152 | 790 |
| | 169 | 812 |
| | 187 | 862 |
| | 207 | 895 |
| | 231 | 939 |
| | 231+ | 1041 |
| 2008 | 123 | 644 |
| | 136 | 703 |
| | 152 | 767 |
| | 169 | 806 |
| | 187 | 868 |
| | 207 | 895 |
| | 231+ | 1016 |

Physiological arguments have been proposed that suggest that the strength of a muscle is proportional to its cross-sectional area. Using this submodel for strength, construct a model that relates lifting ability and body weight. List all assumptions. Do you have to assume that all weight lifters are geometrically similar? Test your model with the data provided.

Now consider a refinement to the above model. Suppose there is a certain amount of body weight that is independent of size in adults. Suggest a model that incorporates this refinement and test it against the data provided.

Criticize the use of the above data. What data would you really like in order to handicap the weight lifters? Who is the best weight lifter according to your models? Suggest a rule of thumb for the *Superstars* show to handicap the weight-lifters.

We also are provided weightlifting data for women in the 2008 Olympics. How does your above model work for their data?

Women's 2008 Olympic Games

| Max Weight in Class (lbs) | Total Pounds |
|---|---|
| 105 | 467 |
| 116 | 487 |
| 127 | 537 |
| 138 | 531 |
| 152 | 630 |
| 165 | 621 |
| > 165 | 718 |

3. **Birds** Warm-blooded animals use large quantities of energy to maintain body temperature because of the heat loss through the body surface. In fact, biologists believe that the primary energy drain on a resting warm-blooded animal is maintenance of body temperature.

   **a.** Construct a model that relates blood flow through the heart to body weight. Assume that the amount of energy available is proportional to the blood flow through the lungs, which is the source of oxygen. Assuming the least amount of blood needed to circulate, the amount of available energy will equal the amount of energy used to maintain the body temperature.

   **b.** The following data relate weights of some type of birds to their heart rates measured in beats per minute. Construct a model that relates heart rate to body weight. Discuss the assumptions of your model. Use the data provided to check your model.

| Bird | Body Weight (g) | Heart Rate (Beats/min) |
|---|---|---|
| Canary | 20 | 1,000 |
| Pigeon | 300 | 185 |
| Crow | 341 | 378 |
| Buzzard | 658 | 300 |
| Duck | 1,100 | 190 |
| Hen | 2,000 | 312 |
| Goose | 2,300 | 240 |
| Turkey | 8,750 | 193 |
| Ostrich | 71,000 | 60–70 |

Data from A. J. Clark, *Comparative Physiology of the Heart*, p. 99. New York: Macmillan, 1977.

4. **Heart Rate: Mammals** Warm-blooded animals use large quantities of energy to maintain body temperature because of the heat loss through the body surface. In fact, biologists believe that the primary energy drain on a resting warm-blooded animal is maintenance of body temperature.

   **a.** Construct a model that relates blood flow through the heart to body weight. Assume that the amount of energy available is proportional to the blood flow through the lungs, which is the source of oxygen. Assuming the least amount of blood needed to circulate, the amount of available energy will equal the amount of energy used to maintain the body temperature.

   **b.** The following data relate the weights of some mammals to their heart rates in beats per minute. Construct a model that relates heart rate to body weight. Discuss the assumptions of your model. Use the data provided to check your model.

| Mammal | Body Weight (g) | Pulse Rate (beats/min) |
|---|---|---|
| *Vespergo pipistrellus* | 4 | 660 |
| Mouse | 25 | 670 |
| Rat | 200 | 420 |
| Guinea pig | 300 | 300 |
| Rabbit | 2,000 | 205 |
| Small dog | 5,000 | 120 |
| Large dog | 30,000 | 85 |
| Sheep | 50,000 | 70 |
| Human | 70,000 | 72 |
| Horse | 450,000 | 38 |
| Ox | 500,000 | 40 |
| Elephant | 3,000,000 | 48 |

Data from A. J. Clark, *Comparative Physiology of the Heart*, p. 99. New York: Macmillan, 1977.

5. **Lumber Cutters** Lumber cutters wish to use readily available measurements to estimate the number of board feet of lumber in a tree. Assume they measure the diameter of the tree in inches at waist height. Develop a model that predicts board feet as a function of diameter in inches.

   The following data are provided for your test:

| $x$ | 17 | 19 | 20 | 23 | 25 | 28 | 32 | 38 | 39 | 41 |
|---|---|---|---|---|---|---|---|---|---|---|
| $y$ | 19 | 25 | 32 | 57 | 71 | 113 | 123 | 252 | 259 | 294 |

   The variable $x$ is the diameter of a ponderosa pine in inches, and $y$ is the number of board feet divided by 10.

   a. Consider the following two assumptions, allowing each to lead to a separate model. Completely analyze each model.

   (1) Assume all trees are right circular cylinders and all trees are about the same height.

   (2) Assume all trees are right circular cylinders and the height of a tree is proportional to its diameter.

   b. Which model appears better and why? Be sure to justify your reasoning.

6. **Crew Racing Shells** If you have been to a rowing regatta, then you might have observed that the more rowers there are in a boat, the faster the boat travels. It is therefore reasonable to investigate whether there is a mathematical relationship between the speed of a boat and the number of crew members.

   **Assumptions (Partial List)**

   a. The total force exerted by the crew is constant for a particular crew throughout the race.

   b. The drag force experienced by the boat as it moves through the water is proportional to the square of the velocity times the wetted surface area of the hull.

   c. Work is defined as the product of force and distance. Power is defined as work per unit time.

|  | Time (sec) | |
|---|---|---|
| **Number in Crew** | **Race 1** | **Race 2** |
| 1 | 429.6 | 430.2 |
| 2 | 412.2 | 406.2 |
| 4 | 379.8 | 367.8 |
| 8 | 346.8 | 343.8 |

   *Hint:* When additional rowers are added to a shell, it is not obvious whether the amount of force is proportional to the number in the crew or whether the amount of power is proportional to the number in the crew. Which assumption appears the more reasonable? Which yields a more accurate model?

7. **Vehicular Braking Distance** A popular rule of thumb often given to students in driver education classes is the *two-second rule* to prescribe a safe following distance. The rule states that if you stay two seconds behind the car in front of you, you have the correct distance no matter what your speed. Because the amount of time is constant (2 seconds), the rule suggests a proportionality between stopping distance and speed, $v$. To test this rule, we pose the following problem: *Predict a vehicle's total stopping distance as a function of its speed.*

In the development of this model, total stopping distance is calculated as the sum of the reaction distance, $d_r$, and the braking distance, $d_b$. The following submodels are hypothesized in that development:

- $d_r \propto v$
- $d_b \propto v^2$

These submodels yield that the total stopping distance, $d$, is represented by an equation of the form $d = k_1 v + k_s v^2$. Test these submodels using the following data.

Vehicular Braking Distance Data Points

| $v$ | 20 | 25 | 30 | 35 | 40 | 45 | 50 | 55 | 60 | 65 | 70 | 75 | 80 |
|---|---|---|---|---|---|---|---|---|---|---|---|---|---|
| $d_r$ | 22 | 28 | 33 | 39 | 44 | 50 | 55 | 61 | 66 | 72 | 77 | 83 | 88 |
| $d_b$ | 20 | 28 | 40.5 | 52.5 | 72 | 92.5 | 118 | 148.5 | 182 | 220.5 | 266 | 318 | 376 |
| $d$ | 42 | 56 | 73.5 | 91.5 | 116 | 142.5 | 173 | 209.5 | 248 | 292.5 | 343 | 401 | 464 |

# Empirical Model Construction

Consider the following problem: California's State Water Commission is requiring data from La Mesa housing on the rate of water use, in gallons per hour, and the total amount of water used each day. The Department of Engineering and Housing (DEH) in Monterey does not have the sophisticated equipment that measures the flow of water in or out of the main water tank. Instead, DEH can measure only the level of water in the tank, within 0.5 percent accuracy, every hour. More importantly, whenever the level in the tank drops below some minimum level $L$, a pump fills the tank up to the maximum level, $H$, but there is no measurement of the pump flow at these times, either. Thus, one cannot readily relate the level in the tank to the amount of water used while the pump is working, which occurs once or twice per day, for a couple of hours each time. The following table contains the time, in seconds, since the first measurement, and the level of water in the tank in hundredths of a foot. For example, after 3,316 seconds, the depth of the water in the tank reached 31.10 feet. The tank is a vertical circular cylinder with a height of 40 feet and a diameter of 57 feet. Usually, the pump starts filling the tank when the level drops to about 27.00 feet and the pump stops when the level rises back to about 35.50 feet. The data are displayed in Table 6.1.

The information provided by the data and the variables they represent do not lend themselves to relating the variables in order to test a model. We might find it easier to plot the data, look at the patterns in the data, and use those patterns to allow us to build a model to fit the data. We can use the model for analysis.

In Chapter 5, we assumed the modeler has developed a model relating the variables under consideration with the purpose of explaining, in some sense, the observed behavior. If collected data then corroborates the reasonableness of the assumptions underlying the hypothesized relationships, the parameters of the model can be chosen that "best fit" the model type to the collected data according to some criterion (such as least squares), as we saw in Chapter 4.

In many practical cases, the modeler is unable to construct a tractable model form that satisfactorily explains the behavior. Nevertheless, if it is necessary to *predict* the behavior, the modeler may conduct experiments to investigate the behavior of the dependent variable(s) for selected value(s) of the independent variable(s) within some range of interest. In essence, the modeler desires to construct an *empirical model* based on the collected data. In this situation, the modeler is strongly influenced by the data that have been carefully collected and analyzed, so the modeler seeks a curve that captures *the trend of the data*, in order to either interpolate between the collected data points or predict outside their range. This is where we say, "the data speak."

**Table 6.1** Water Flow Data

| Time (sec) | Level (0.01 ft) |
| --- | --- |
| 0 | 3,175 |
| 3,316 | 3,110 |
| 6,635 | 3,054 |
| 10,619 | 2,994 |
| 13,937 | 2,947 |
| 17,921 | 2,892 |
| 21,240 | 2,850 |
| 25,223 | 2,795 |
| 28,543 | 2,752 |
| 32,884 | 2,697 |
| 35,932 | Pump on |
| 39,332 | Pump on |
| 39,435 | 3,550 |
| 43,318 | 3,445 |
| 46,636 | 3,350 |
| 49,953 | 3,260 |
| 53,936 | 3,167 |
| 57,254 | 3,087 |
| 60,574 | 3,012 |
| 64,554 | 2,927 |
| 68,535 | 2,842 |
| 71,854 | 2,757 |
| 75,021 | 2,697 |
| 79,254 | Pump on |
| 82649 | Pump on |
| 85,968 | 3,475 |
| 89,953 | 3,397 |
| 93,270 | 3,340 |

To contrast explicative and empirical models, consider the data shown below in Figure 6.1. If the modeler's assumptions lead to the expectation of a quadratic explicative model, then a parabola would be fit to the data points, as illustrated in Figure 6.2. However, if the modeler has no reason to expect a model of a particular type, a smooth curve may be passed through the data points to serve as an empirical model, as illustrated in Figure 6.3.

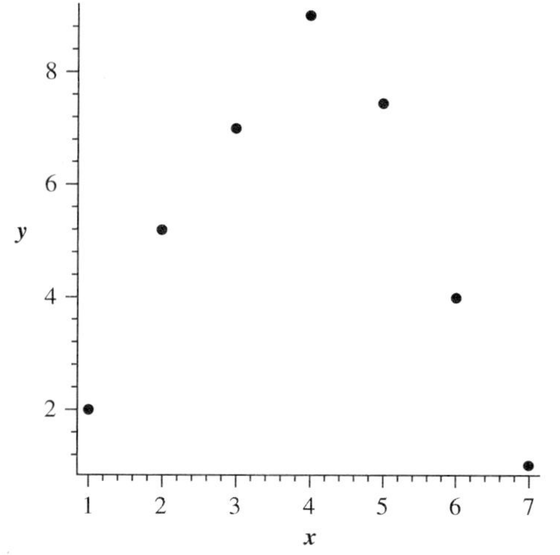

**FIGURE 6.1**
Random data

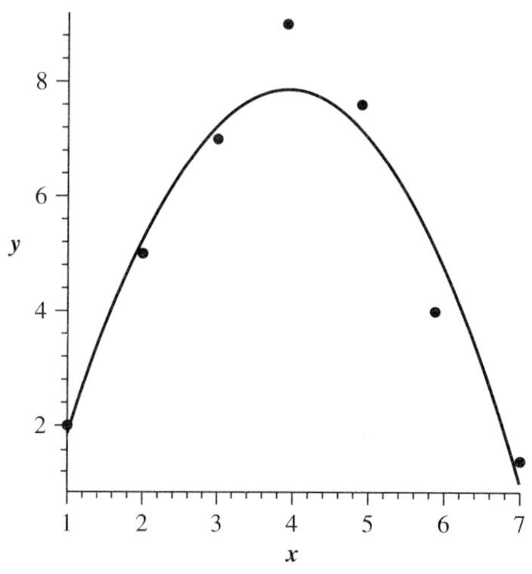

**FIGURE 6.2**
Quadratic fit to
the random data

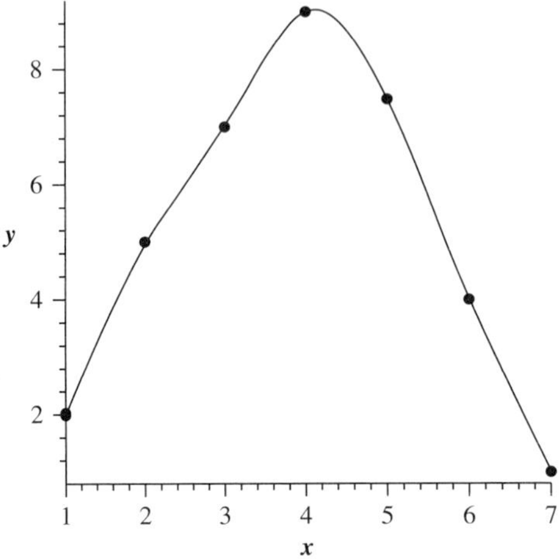

**FIGURE 6.3**
Smooth curve fitting all of the data

This chapter describes the construction of empirical models. Although there are many techniques to assist in model construction, we present a logical procedure for determining an appropriate empirical model and present only a few techniques such as using the ladder of powers and other transformations for one-term models (such as $y = kx$), polynomial fitting, and cubic splines.

We provide the following steps in the empirical modeling process.

**Step 1**    Examine a scatterplot of the data for trends.

**Step 2**    Examine outliers and replace or discard them.

**Step 3**    Find a one-term model and, if possible, fit the chosen one-term model; if the fit is adequate, then end the process.

**Step 4**    Otherwise, attempt another one-term model or consider a polynomial model.

**Step 5**    From a divided difference table, fit a low-order polynomial; if adequate, then end.

**Step 6**    Otherwise, if the data is extremely accurate, then construct a cubic spline.

First, examine if a trend in the data is discernible. If so, begin with the simplest technique available (a one-term model) and then proceed through the above steps until an acceptable empirical model is developed that satisfies the requirements of the particular application. As a general rule, we select the simplest adequate model that captures the trend of the data.

## 6.1    SIMPLE ONE-TERM MODELS

We commence our study on empirical model building with the presentation of simple one-term models. These models have several advantages. One obvious and powerful advantage is their simplicity. Another is that a one-term model may capture the trend of the data better than any other empirical model (such as a polynomial). This feature is particularly important in situations when the modeler intends to make a predication by extrapolating for values outside the intervals of the observations—for instance, when the modeler intends to predict a future value based on historical information.

When constructing an empirical model, always begin with a careful analysis of the collected data. Investigate whether the data suggests the existence of a trend. Are there data points that fail to follow the trend? If such **outliers** may exist, you may wish to discard them. Or if the data were obtained experimentally, then you may choose to repeat the experiment to check the accuracy of the data to preclude any data-collection error.

If a trend is observed to exist, the use of one-term models is investigated by the matching of the trend's concavity using the suggested ladder of transformations in Figure 6.4. We deal strictly with data whose concavity trends are either concave down [Figure 6.4(a)] or concave up [Figure 6.4 (b)].

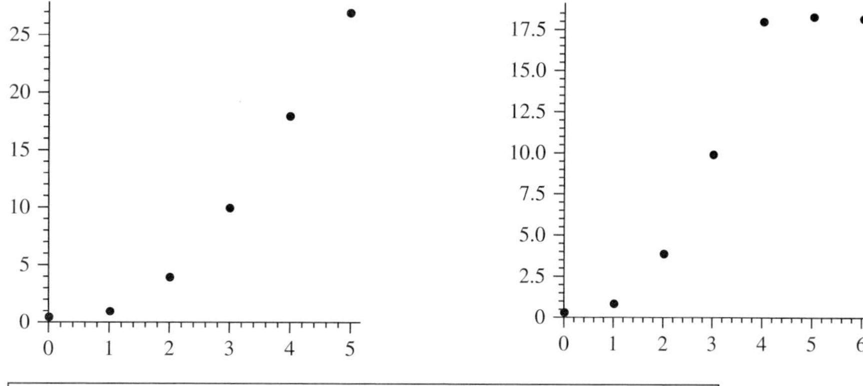

Ladder by:
squeezing the right-hand tail downward (change $y$ to $\sqrt{y}$, log $y$, $\frac{1}{\sqrt{y}}$, $etc$) or

stretching the right-hand tail to the right (change $x$ to $x^2$, $x^3$, etc.)

Table 6.2 shows the ladder of powers for the transformation to either the $x$ variable or the $y$ variable. Table 6.3 displays the details to the use of the ladder and the concavity trends. Let's illustrate with an example.

| Example 1 | Bass Fishing Derby |

Consider the bass fishing derby, where we have data recorded for the length of given bass fish and its corresponding weight in ounces. We want a simple model that allows fisherman to estimate the weight so that they can decide whether or not to keep the fish that they caught. In this example, the scatterplot of the original data (see Figure 6.5) suggests a slight trend that is concave up.

$len:=[14.5,12.5,17.25,14.5,12.625,17.75,14.125,12.625];$
$wt:=[27,17,41,26,17,49,23,16];$

Thus, we select the $z^2$ transformation and plot weight ($wt$) versus length$^2$ ($len$)$^2$. Because the plot is reasonably linear, we fit the model analytically (using least squares) and plot the residuals. Because there is an evident trend in the residuals, we investigate $wt$ versus $len^3$ in a similar manner. The two resulting models are: $wt = 0.13108 \, len^2$ and $wt = 0.00844 \, len^3$. The residuals

**Table 6.2** Ladder of Transformations

.
.
.

$Z^3$

$Z^2$

$Z$ (no change)

$\sqrt{Z}$

Log $Z$

$\dfrac{-1}{\sqrt{z}}$

$\dfrac{-1}{z}$

$\dfrac{-1}{z^2}$

.
.

**Table 6.3** Variable Transformations Based on Concavity

| Data Trend | Variable to Be Transformed | Ladder of Transformations |
|---|---|---|
| Concave up | $x$ | $x \rightarrow Z^2, \ldots$ |
| Concave up | $y$ | $y \rightarrow \sqrt{Z}, \ldots$ |
| Concave down | $x$ | $x \rightarrow \sqrt{Z}, \ldots$ |
| Concave down | $y$ | $y \rightarrow Z^2, \ldots$ |

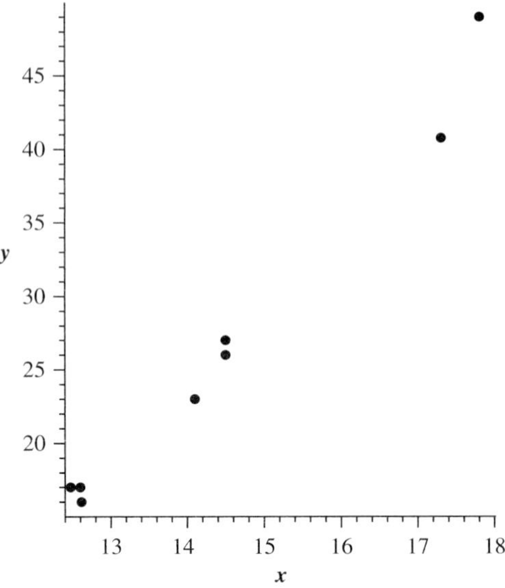

**FIGURE 6.5**

Scatterplot of weight (*y*-axis) versus length (*x*-axis), suggesting a slight concave upward trend

for the model, $W$ versus $len^2$, appear to show a trend that suggests that an improvement in the model can be made. Note that the residuals of $wt$ versus $len^3$ are more random about the origin, indicating that the second model appears to be the better model. Below is the Maple output for our analysis using a program available on the companion Web site called SLR (Simple Linear Regression) that produces a least-squares fit with an ANOVA table.

wt versus $ln^2$
Estimated Regression Equation: $y = .1310819784*x^2$
ANOVA Table
($F^* = 42.0085$)

| | degrees of freedom df | sum of squares SS | mean square MS |
|---|---|---|---|
| Regression | 1 | 890.7725 | 890.7725 |
| Error | 6 | 127.2275 | 21.2046 |
| Total | 7 | 1018.0000 | |

Although the fit appears reasonable (see Figure 6.6), we clearly see a trend in the residual plot in Figure 6.7, regardless of the $R^2$ value,

$$(1 - SSE/SST) = 0.8750$$

**FIGURE 6.6**

Plot of the data and the regression equation $y=0.1310819784\,x^2$

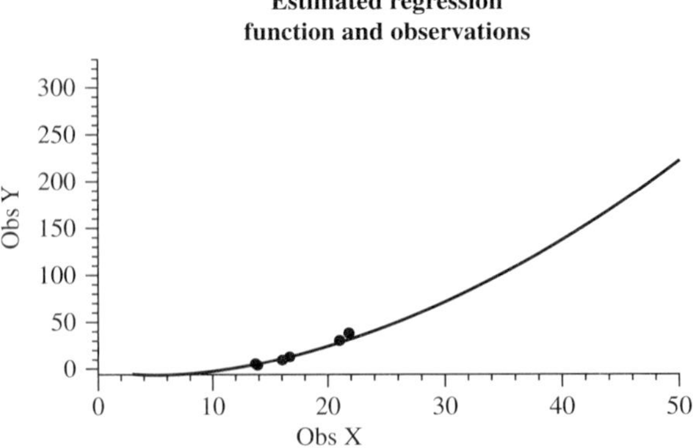

**FIGURE 6.7**
Residual plot

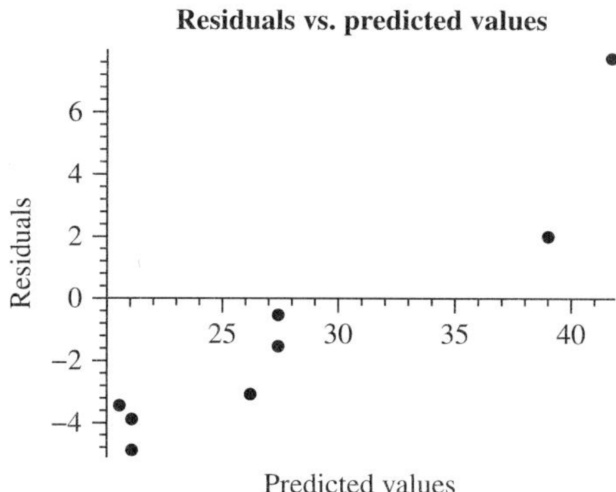

or how "smooth" the fit appears. Graphically, the trends indicate that this model is not as adequate as other models might be. Therefore, we continue up the ladder to repeat the analysis and see if we can obtain a more random pattern in the residuals.

Estimated Regression Equation: $y = 0.0008436671276$.
ANOVA Table
($F^* = 495.6325$)

|  | df | SS | MS |
| --- | --- | --- | --- |
| Regression | 1 | 1005.8238 | 1005.8238 |
| Error | 6 | 12.1762 | 2.0294 |
| Total | 7 | 1018.0000 | |

Figure 6.8 shows a reasonable fit, and we see that there appear to be no trends in the residual plot, Figure 6.9, thus indicating an adequate model. The $R^2$ happens to equal 0.998. However, because we do not see a trend in the residual plot, we conclude that our model, $wt = .008436671276 * len^3$, is adequate.

**FIGURE 6.8**
Plot of data and regression equation
$y = 0.008436671276\,x^3$

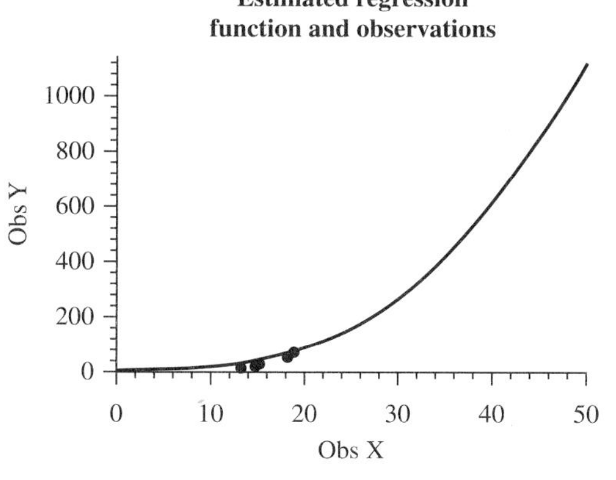

**FIGURE 6.9**
Residual plot

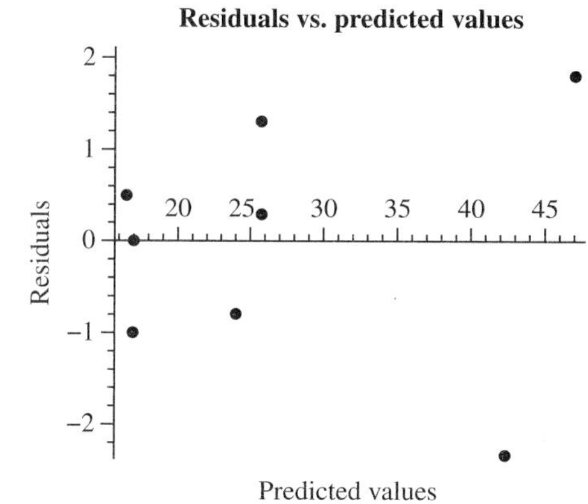

## Example 2    Terror Bird Revisited

Recall the terror bird example from Chapters 2 and 5. We will now use simple one-term models to obtain an empirical model using the bird data displayed in Table 6.4.

We obtain a scatterplot of the raw data and interpret it as increasing and concave up (see Figure 6.10).

**Table 6.4** Bird Data for Modeling the Terror Bird

| Femur Circumference (cm) ($x$) | Weight (kg) ($y$) |
| --- | --- |
| .7943 | .0832 |
| .7079 | .0912 |
| 1.000 | .1413 |
| 1.122 | .1479 |
| 1.6982 | .2455 |
| 1.2023 | .2818 |
| 1.9953 | .7943 |
| 2.2387 | 2.5119 |
| 2.5119 | 1.4125 |
| 2.5119 | .8913 |
| 3.1623 | 1.9953 |
| 3.5481 | 4.2658 |
| 4.4668 | 6.3096 |
| 5.8884 | 11.2202 |
| 6.7608 | 19.9500 |
| 15.136 | 141.25 |
| 15.850 | 158.4893 |

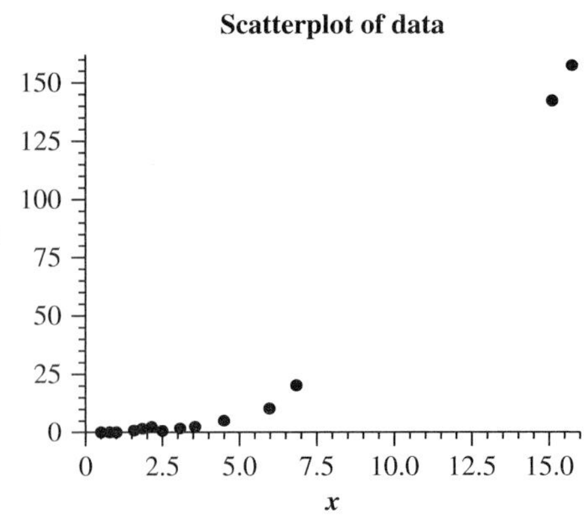

**FIGURE 6.10**
Scatterplot of raw data indicating increasing and concave up



We decide to transform $x$ going up the ladder in an attempt to linearize the data for a simple one-term model. Our plot of $y$ versus $x^3$ is the most linear (see Figure 6.11).

We use least squares to obtain the parameters and the model.

> *xobs:=[.7079,.7943,1,1.122,1.2023,1.6982,1.9953,2.2387,2.5119,*
> *2.5120,3.1623,3.5481,4.4668,5.8884,6.7608,15.136,15.850]:*
> *yobs:=[.0912,.082,.1413,.1479,.2818,.2455,.7943,2.5119,.8913,1.4125,*
> *1.9953,4.2658,6.3096,11.2202,19.95,141.25,158.4893]:*
> *dim:=vectdim(xobs):*

User-provided regression model equation and domain over which estimated regression function is plotted:

> *fcn:='y=a1*x³':plotLB:=0:plotUB:=20:#fcn:='y=a*x³':plotLB:=0:plotUB:=20:*

**FIGURE 6.11**
Plot of *y* versus *x³* appears linear

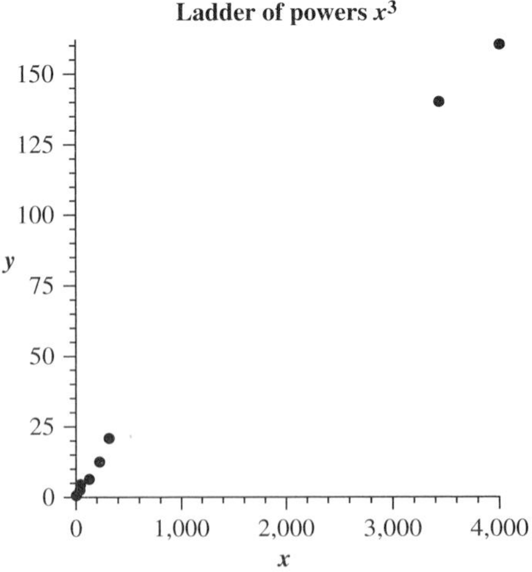

Invoke procedure SLR and display:

1. least-squares estimated regression function and observed values, and

2. plot of residuals against predicted values

```
> SLR(fcn,xobs,yobs,dim,plotLB,plotUB):
> display(ObsPoints,LR_PredCurve);
> printf("\n\n\n");display(CheckErrors);
```

Estimated Regression Equation: $y = .4032417679e-1*x\text{\textasciicircum}3$
ANOVA Table
($F^* = 6358.5947$)

|  | df | SS | MS |
|---|---|---|---|
| Regression | 1 | 38,366.5522 | 38,366.5522 |
| Error | 15 | 90.5072 | 6.0338 |
| Total | 16 | 38,457.0594 |  |

Our model to predict the weight of the terror bird is $y = 0.0403\ x^3$. Thus, at a femur circumference of 21 cm, we evaluate and find $y(21) = .0403(21^3) = 371.22$ kg. Our mode predicts the weigh of the terror bird as 371.22 kg. The $R^2$ for our model is 0.9976. Figure 6.12 shows a reasonable fit. However, the residual plot in Figure 6.13 does not appear to be random about $y = 0$. Thus, we might conclude the model is not as adequate as we might like.

We decide to try another transformation to this data and see how our model does. The scatterplot, being increasing and concave up, suggests trying a ln–ln transformation.

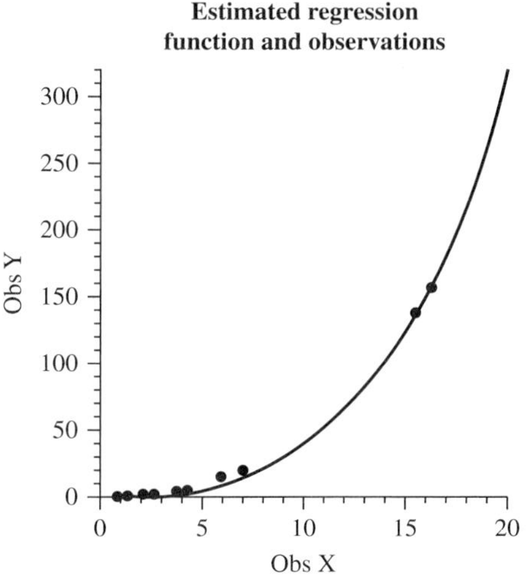


**FIGURE 6.12**
Plot of data and regression equation, $y = 0.04032417679\ x^3$

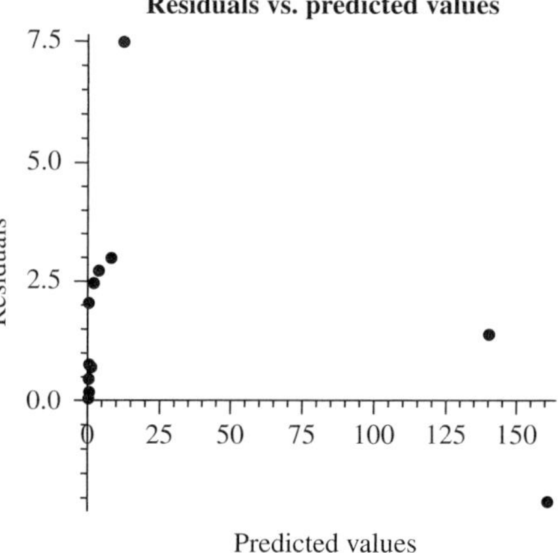


**FIGURE 6.13**
Residual plot

---

| **Example 3** | ## Cost of a Postage Stamp |

We examine the cost of a postage stamp from 1963 through 2009. We desire to build a model to determine the future costs of postage. We set $t = 0$ for 1963, and we use cents rather than decimals for the rate (5 cents versus 0.05).

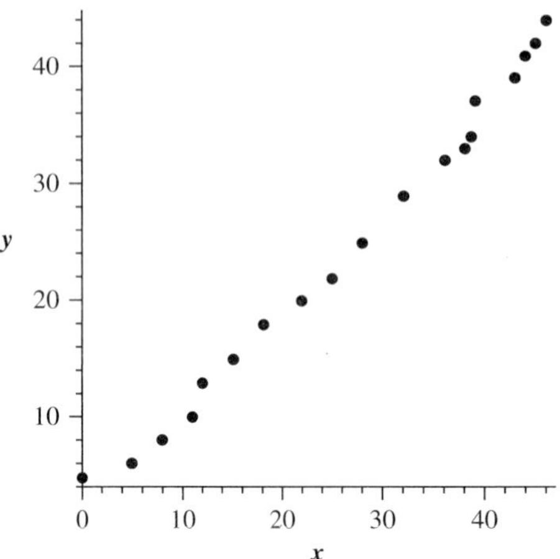

**FIGURE 6.14**
Scatterplot of postage
stamp data

*rate:=[5,6,8,10,13,15,18,20,22,25,29,32,33,34,37,
39,41,42,44] :year:=[0,5,8,11,12,15,18,22,25,28,
32,36,38,38.6,39,43,44,45,46];*

$$year := [0, 5, 8, 11, 12, 15, 18, 22, 25, 28,$$
$$32, 36, 38, 38.6, 39, 43, 44, 45, 46]$$

*> pointplot({seq([year[i], rate[i]],i=1..19)});*

Our plot in Figure 6.14 looks reasonably linear so we use the model, *rate= a*year + b*. We fit the model to find the parameters {*a,b*}

*> eq_fit:=fit[leastsquare[[x,y], y=a*x+b]]([year, rate]);*

$$eq\_fit := y = 0.8649269478\ x + 1.878575535$$

*> eq_fit:=Fit(a*x+b,year,rate,x);*

$$eq\_fit := 0.8649269478\ x + 1.878575535$$

We transform the results using the fit command, but we do not have to transform if we use the Fit command. We will also find the percent relative error and plot the residuals.

*> eq_function:=unapply(rhs(eq_fit),x);*

$$eq\_function := x \to 0.8649269478\ x + 1.878575535$$

and then give the predicted values:

*> Yvaluespredicted:=map(x–>year,eq_fit);transform[apply[eq_function]](year);*

$$Yvaluespredicted$$
$$:= [1.878575535, 6.203210274, 8.797991117, 11.39277196,$$
$$12.25769890, 14.85247976, 17.44726060, 20.90696838,$$
$$23.50174924, 26.09653008, 29.55623786, 33.01594566,$$
$$34.74579956, 35.26475572, 35.61072650, 39.07043430,$$
$$39.93536124, 40.80028818, 41.66521514]$$

Find the residuals:

*> Residuals:=seq(rate[i]–Yvaluespredicted[i], i=1..19);*
or
*> Residuals:=zip((x,y)–>x–y,rate, Yvaluespredicted);*

$$Residuals := [3.121424465, -.203210274, -.797991117, -1.39277196,$$
$$0.74230110, 0.14752024, 0.55273940, -.90696838,$$
$$-1.50174924, -1.09653008, -.55623786, -1.01594566,$$
$$-1.74579956, -1.26475572, 1.38927350, -0.07043430,$$
$$1.06463876, 1.19971182, 2.33478486]$$

> *#% relative error*
> *%relerr:=zip((x,y)->100*(x-y)/y,rate. Yvaluespredicted);*

$$\%relerr := [166.1591140, -3.275888855, -9.070151429,$$
$$-12.22504905, 6.055794860, 0.9932364318, 3.168058371,$$
$$-4.338115233, -6.389946657, -4.201823295, -1.881964351,$$
$$-3.077136334, -5.024490966, -3.586458191, 3.901278172,$$
$$-.1802751908, 2.665904920, 2.940449378, 5.603678877]$$

We could also use the following command

> *Yvaluespredicted:=[seq(0.8649269478*year[i]+1.878575535, i=*
> *#Plot the residuals*
> *pointplot{seq([Yvaluespredicted[i], Residuals[i]], i= 1 ..19)},*
> *title = 'Residual Plot');*
> *# Comment on residual pattern*

The residual plot in Figure 6.15 reveals no trend, so we accept the model as adequate. We notice that the first estimate is a terrible percent error, but the percent errors toward the latter years are all less than 10 percent, indicating a good fit. We use the model to make predictions about the future.

To predict the rate in year 2023, we substitute $x = 60$ into our model to obtain prediction of 53.77 cents, or most likely 54 cents. To determine when the cost might be \$1, we solve the equation to find the year to be about 2076.

### Modeling from Ln–Ln Transformations

Another approach to simple one-term models is to try a ln–ln transformation. We transform the original data into ln $x$ and ln $y$ and see if that plot is linear. If it appears linear, then we can

**FIGURE 6.15**

Residual plot from postage
stamp model

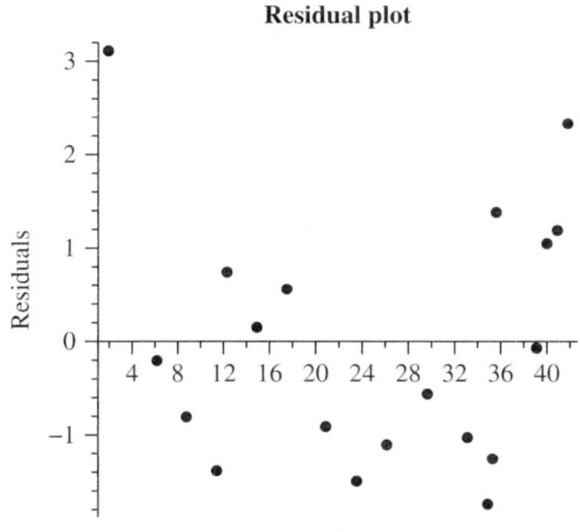

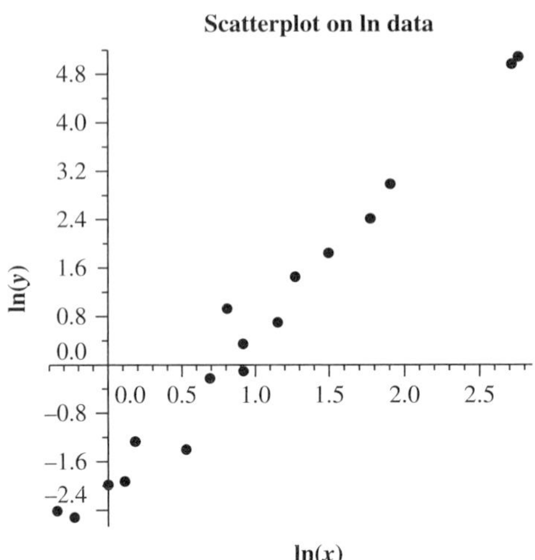



**FIGURE 6.16**

The ln–ln plot appears very linear

find the slope and the intercept for the ln–ln space (see Figure 6.16). We then return the ln–ln model into its original *xy* space as a power model to determine the adequacy.

Let's apply this ln–ln transformation to the terror bird problem.

> *lnx:=map(x−>ln(x),xobs):lny:=map(y−>ln(y),yobs):*

We fit with basic least squares using the Fit command:

> *eq:=Fit(a*x+b,lnx,lny,x);*

$$eq := 2.516167640\,x - 1.944089036$$

Remember, this is really ln $(y)$= 2.516167640 * ln$(x)$ − 1.944089036.

We now transform the model back into its original space and examine both the model fit (visually in Figure 6.17) and the residual plot (see Figure 6.18). We return the model by making both sides of the equation the exponent of the exponential, *e*.

So,

$$e^{\ln(y)} = e^{2.516167640\,*\,\ln(x)\,-\,1.944089036}$$

We have the power model:

$$y = .1431175389\,x^{2.516167640}$$

There appears to be a linear trend in the residual plot, indicating the model is not adequate.

---

## 6.1  EXERCISES

Requirements for Exercises 1-4:

**Step 1**  Scatterplot the original data. List the description and trends of the data. If the data appears either concave up or concave down (single concavity), then go to step 2. Example: Data is increasing and concave up.

**Step 2**  Either go "up" or "down" the ladder of powers in an attempt to linearize the data. This is done graphically. Stop when you are satisfied that you obtained a linear plot.

**Step 3**  Use least squares and fit the model from the ladder of powers. For example, if we went to $z^2$ using the x variable to get a linear graph, then our model will be $y = kx^2$. We fit the parameter, $k$, using least squares.

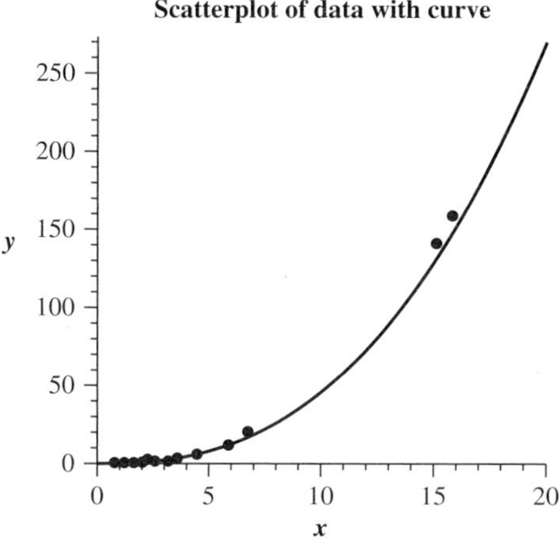





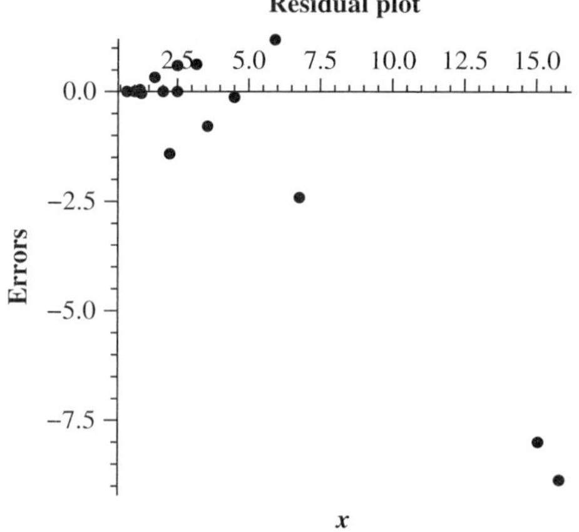



**Step 4**   Perform a residual plot analysis for the adequacy of your simple one-term model. If you are dissatisfied with this model, then move to step 5.

**Step 5**   Try a ln–ln plot of the data. If it looks linear, then use least squares to fit the linear model:

$$\ln y = b + a \ln x$$

Transform this back into the original space to find and plot the residual analysis.

**1.** Population

**Population data**

| Year 1910 = 10 | Population |
| --- | --- |
| 10 | 1,766 |
| 20 | 15,673 |
| 30 | 56,214 |
| 40 | 139,123 |
| 50 | 280,974 |
| 60 | 498,984 |
| 70 | 810,903 |
| 80 | 1,234,934 |
| 90 | 1,789,675 |
| 100 | 2,494,079 |

**2.** Bass Fishing

**Bass fishing data**

| Length | 12.5 | 12.625 | 14.125 | 14.5 | 17.25 | 17.75 |
|---|---|---|---|---|---|---|
| Weight | 17 | 16.5 | 23 | 26.5 | 41 | 49 |

**3.** Telecommunications: $t = 1$ corresponds to 1997, telecom in millions

**Telecom data**

| Year, $t$ | 1 | 2 | 3 | 4 | 5 | 6 | 7 | 8 | 9 |
|---|---|---|---|---|---|---|---|---|---|
| Telecom | 9.2 | 9.6 | 10 | 10.4 | 10.6 | 11 | 11.1 | 11.2 | 11.3 |

**4.** We want to build a mathematical model to determine the length of a bird's wingspan (in meters) as a function of the bird's weight (in grams). We collected the following data from six birds.

**Wingspan of birds**

| Weight | 0.25 | 0.5 | 1.5 | 2.0 | 2.5 | 3.0 |
|---|---|---|---|---|---|---|
| Wingspan | 0.4 | 0.77 | 1.1 | 1.22 | 1.33 | 1.40 |

**5.** Use the following U.S. population data to build a model and predict the results of the 2010 U.S. Census.

**U.S. population**

| Year | Population (Millions) |
|---|---|
| 1998 | 275.854 |
| 1999 | 279.040 |
| 2000 | 282.171 |
| 2001 | 285.039 |
| 2002 | 287.726 |
| 2003 | 290.210 |
| 2004 | 292.892 |
| 2005 | 295.560 |
| 2006 | 298.362 |
| 2007 | 301.290 |
| 2008 | 304.059 |
| 2009 | 307.006 |

## 6.1   FURTHER READING

Giordano, F. R., W. P. Fox, S. Horton, and Maurice D. Weir, *A First Course in Mathematical Modeling.* 4th ed. Belmont, CA: Cengage, 2009.

## 6.2   FITTING AN *N*–1 ORDER POLYNOMIAL TO *N* DATA POINTS

Because of the inherent simplicity, one-term models facilitate model analysis, including sensitivity analysis, optimization, estimation of rates of change and area under a curve applications, and so forth. However, precisely because of this simplicity, one-term models are not likely to

**Table 6.5** Data Points for a Polynomial

| $x$ | 1 | 2 | 3 |
|-----|---|---|----|
| $y$ | 5 | 8 | 25 |

**Table 6.6** The System of Equations

| $a$ | $+$ | $1b$ | $+$ | $1c$ | $=$ | 5 |
|-----|-----|------|-----|------|-----|----|
| $a$ | $+$ | $2b$ | $+$ | $2^2 c$ | $=$ | 8 |
| $a$ | $+$ | $3b$ | $+$ | $3^2 c$ | $=$ | 25 |

capture the trend of every arbitrary data set. In many cases, models with more than one term must be considered. The remainder of this chapter considers one type of multiterm model—namely, the polynomial. Because polynomials are easy to integrate and differentiate, they are especially popular to use. Consider passing a quadratic $P_2(x) = a + bx + cx^2$ through the following data points in Table 6.5.

Requiring that $P_2(x_i) = y_i$ yields the following system of equations in Table 6.6:

To solve the above system conveniently using Maple, we consider the linear system in the form of the matrix equation $AX = B$, which has the solution $X = A^{-1}B$, provided that $A$ is invertible. Thus, the above system can be written in matrix form as

$$\begin{bmatrix} 1 & 1 & 1 \\ 1 & 2 & 4 \\ 1 & 3 & 9 \end{bmatrix} \cdot \begin{bmatrix} a \\ b \\ c \end{bmatrix} = \begin{bmatrix} 5 \\ 8 \\ 25 \end{bmatrix} \text{ where } X = \begin{bmatrix} a \\ b \\ c \end{bmatrix} \tag{6.1}$$

and, because the coefficient matrix is invertible, the solution is

$$\begin{bmatrix} a \\ b \\ c \end{bmatrix} = \begin{bmatrix} 3 & -3 & 1 \\ -2.5 & 4 & -1.5 \\ 0.5 & -1 & 0.5 \end{bmatrix} \cdot \begin{bmatrix} 5 \\ 8 \\ 25 \end{bmatrix} = \begin{bmatrix} 16 \\ -18 \\ 7 \end{bmatrix} \tag{6.2}$$

or, $P_2(x) = 16 - 18x + 7x^2$.

In general, the requirement that an $n - 1$ degree polynomial pass through $n$ distinct data points yields a system of $n$ linear algebraic equations in $n$ unknowns. It is important to realize that large systems of equations can be difficult to solve with great accuracy, and small round-off errors in computer arithmetic can cause large oscillations to occur because the higher-order terms are present.

The process of solving for $a$, $b$, and $c$ by inverting $A$ and multiplying it by $B$ is computed by Maple using the LinearAlgebra package and the ReducedRowEchelonForm command. Often in using high-order polynomials, oscillations in the curve between data pairs and near end points occur. Therefore, although we can build $(n–1)$st-order polynomials, we might want to consider capturing the trend with lower-order polynomials. We illustrate with the following examples.

---

**Example 1**

## An (*N*–1)-Degree Polynomial

The Maple linear algebra package is capable of in-depth matrix manipulations. Specifically, Maple quickly solves the type of problem discussed above with one command. Reviewing that earlier example:

$$\begin{bmatrix} 1 & 1 & 1 \\ 1 & 2 & 4 \\ 1 & 3 & 9 \end{bmatrix} \cdot \begin{bmatrix} a \\ b \\ c \end{bmatrix} = \begin{bmatrix} 5 \\ 8 \\ 25 \end{bmatrix} \tag{6.3}$$

*> with( LinearAlgebra ):*
*> B := ⟨⟨1, 1, 1, 5⟩|⟨1, 2, 4, 8,⟩|⟨1, 3, 9, 25⟩⟩;*

$$B := \begin{bmatrix} 1 & 1 & 1 \\ 1 & 2 & 3 \\ 1 & 4 & 9 \\ 5 & 8 & 25 \end{bmatrix}$$

**FIGURE 6.19**

The quadratic curve passes through the points

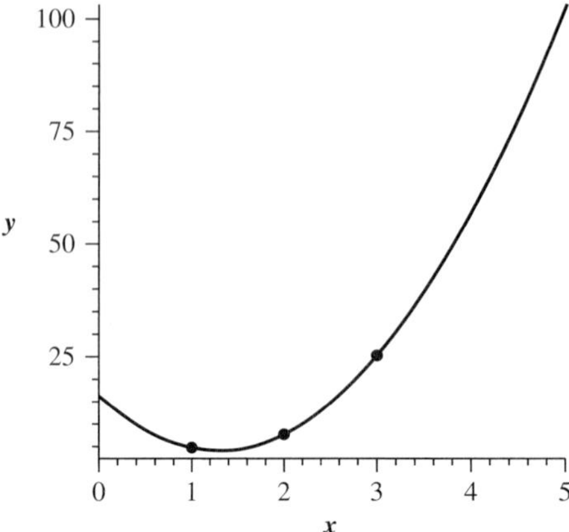

> $B1:=Transpose\ (B)$;

$$B1 := \begin{bmatrix} 1 & 1 & 1 & 5 \\ 1 & 2 & 4 & 8 \\ 1 & 3 & 9 & 25 \end{bmatrix}$$

> $ReducedRowEchelonFrom(B1)$;

$$\begin{bmatrix} 1 & 0 & 0 & 16 \\ 0 & 1 & 0 & -18 \\ 0 & 0 & 1 & 7 \end{bmatrix}$$

This provides the coefficients for the solution of $P_2(x) = 16 - 18x + 7x^2$. The graphical presentation demonstrates the fit of the solution, presented in Figure 6.19.

> $x:=[1,2,3]$:
$y:=[5,8,25]$ :
$xy:=\{seq([x[i],y[i]],i=1.3)\}$:
$with\ (plots)$ :
$plot1:=plot(xy,\ style = point,\ symbol=circle)$ :
$plot2:=plot(16{-}18*z + 7*z^2,z=0..5)$ :
$display(\{plot1,plot2\})$;

### Problems with Higher-Order Polynomials

There are problems that might exist in higher-order polynomial fitting. As we said earlier, sometimes in fitting the data, the curve has oscillations, and snaking occurs near the end points. This oscillating and snaking behavior makes the polynomial very poor for interpolation near the end points and further makes predictions outside the end points pointless. We provide an illustrative example where we fit a complete polynomial. Note that we have lost the trend of the data (although we fit each datum exactly), and at the end points we have oscillations. Therefore, we suggest a close examination of all higher-order polynomials before accepting any.

Here are the data and the scatterplot (see Figure 6.20) suggesting a smooth curve with concave-up trends.

> $Xvalues:=[.55,1.2,2,4,6.5,12,16]; Yvalues:=[.13,.64,5.8,102,210,2030,3900]$;

$$Xvalues := [\,0.55,\ 1.2,\ 2,\ 4,\ 6.5,\ 12,\ 16\,]$$
$$Yvalues := [\,0.13,\ 0.64,\ 5.8,\ 102,\ 210,\ 2030,\ 3900\,]$$

> $pointplot(zip((x,y){-}{>}[x,y],Xvalues,Yvalues))$;

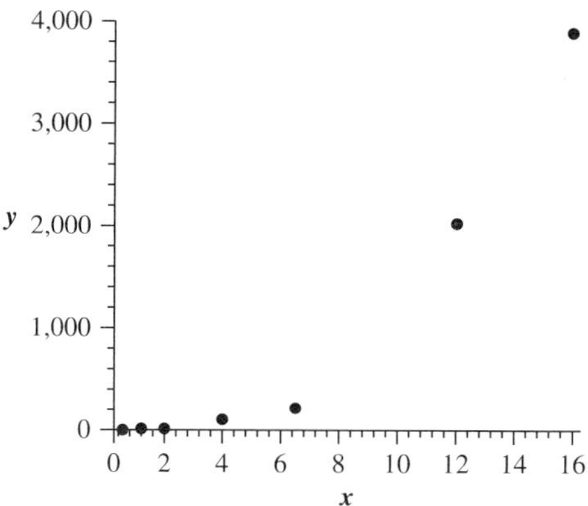

**FIGURE 6.20**
Scatterplot of data

Next, we fit a sixth-order polynomial ($n = 7$), and then plot the polynomial. We will use the *interp* command to obtain a *n*-1st order polynomial.

> *with( CurveFitting ):Fi:=PolynomialInterpolation( [.55,1.2,2,4,6.5,12,16 ], [.13,.64,5.8,102,210,2030,3900 ], z );*

$$FI := -0.01383726235\, z^6 + 0.5084246673\, z^5 - 18.09506969 - 6.437923862\, z^4 + 64.31279044\, z + 34.85731000\, z^3 - 73.99155349\, z^2$$

> *plot( −.1383726235e−1*z^6+.5084246673*z^5−18.09506969−6.437923862*z^4 +64.31279044*z+34.85731000*z^3−73.99155349*z^2,z=0..16 );*

In Figure 6.21, we note the oscillation at the end points. Thus, although the higher-order ($n$–1)st-order polynomial gives a perfect fit through the data points, its oscillations make interpolation and prediction less accurate.

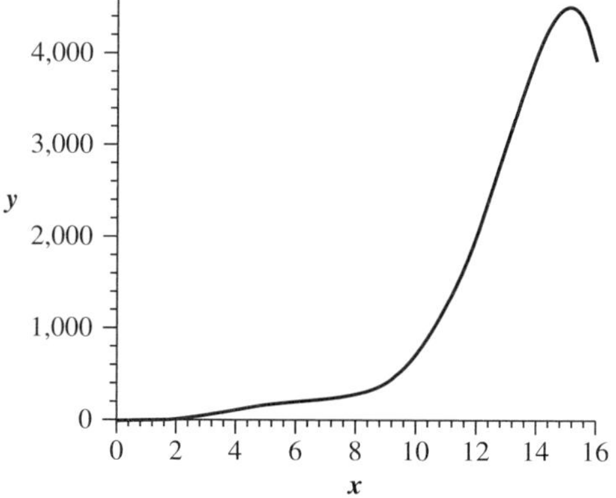

**FIGURE 6.21**
High-order polynomial fit (sixth order) showing oscillations at the end points.

---

## Example 2    Fitting a Fifth-Order Polynomial Using Least Squares

Given a set of data points, see Table 6.7 an analyst decides to attempt to fit a curve to the data using a high-order polynomial. Table 6.8 displays the commands required, and Figure 6.22 displays the polynomial curve superimposed on the data.

Do we have to use a fifth-order polynomial? The answer lies in the need for and use of the model as well as how well the trend is captured regardless of the perfect fit. There is a trade off between a perfect fit and a model that fits well but yields some error. The modeler and user must decide which is best.

**Table 6.7** The Data

| $x$ | 1 | 2 | 3 | 4 | 5 | 6 |
|---|---|---|---|---|---|---|
| $y$ | 305 | 266 | 135 | −16 | 125 | 1,230 |

**Table 6.8**

```
> with(stats)with(plots):with(Statistics):
> xdata:=[1,2,3,4,5,6];
```

$$xdata:=[1,2,3,4,5,6]$$

```
> ydata:=[305,266,135,−16,125,1230];
```

$$ydata:=[305, 266, 135, -16, 125, 1230]$$

```
> xyfit:=Fit(a*x^5+b*x^4+c*x^3+d*x^2+e*x+f,xdata,ydata,x);
```

$$xyfit:=x^5 - 5x^4 - 3x^3 + 7x^2 + 5x + 300$$

```
> f:=map(x->xdata,xyfit);
```

$$f:= [305, 266, 135, -16, 125, 1230];$$

```
> xy:={seq([xdata[i],ydata[i]], i=1 ..6)};
```

$$xy := \{[1,305], [2,266], [3, 135], [4,-16], [5, 125], [6, 1230\}$$

```
> c1:=pointplot[{seq][xdata[i],ydata[i]], i=1..6)},thickness=3) :

c2:=plot(xyfit,x=0..6):
display({c1,c2});
```

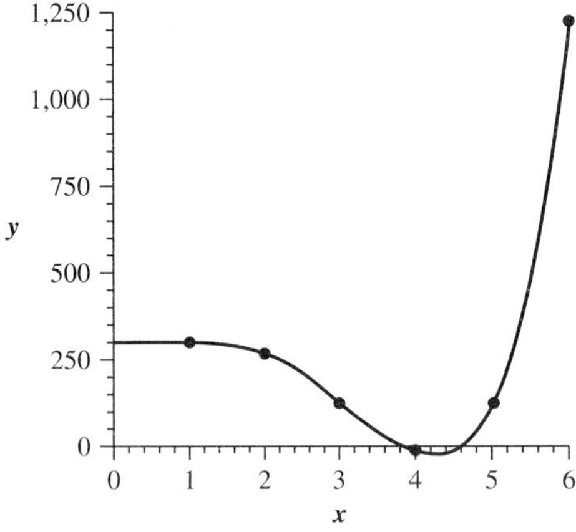

**FIGURE 6.22**

The plot for fitting a fifth-order polynomial

Determine the higher-order polynomial to exactly fit the following data. If possible, plot the functions.

## 6.2 | EXERCISES

**1.**

| $x$ | 1 | 2 | 3 | 4 |
|---|---|---|---|---|
| $y$ | 7 | 16 | 23 | 36 |

**2.** Use the following population data.

| Year 1998 = 10 | Population |
|---|---|
| 10 | 1,766 |
| 20 | 15,673 |
| 30 | 56,214 |
| 40 | 139,123 |
| 50 | 280,974 |
| 60 | 498,984 |
| 70 | 810,903 |
| 80 | 1,234,934 |
| 90 | 1,789,675 |
| 100 | 2,494,079 |

**3.** Use the following bass fishing data.

| Length | 12.5 | 12.625 | 14.125 | 14.5 | 17.25 | 17.75 |
|---|---|---|---|---|---|---|
| Weight | 17 | 16.5 | 23 | 26.5 | 41 | 49 |

**4.** Use the following telecommunications data ($t = 1$ corresponds to 1997; telecom in millions)

| Year, $t$ | 1 | 2 | 3 | 4 | 5 | 6 | 7 | 8 | 9 |
|---|---|---|---|---|---|---|---|---|---|
| Telecom | 9.2 | 9.6 | 10 | 10.4 | 10.6 | 11 | 11.1 | 11.2 | 11.3 |

**5.** We want to build a mathematical model to determine the length of a bird's wingspan (in meters) as a function of the bird's weight (in grams). We collected the following data from six birds.

| Weight | 0.25 | 0.5 | 1.5 | 2.0 | 2.5 | 3.0 |
|---|---|---|---|---|---|---|
| Wingspan | 0.4 | 0.77 | 1.1 | 1.22 | 1.33 | 1.40 |

**6.** Use the following U.S. population data to build a model and predict the results of the 2010 census.

| Year | Population (Millions) |
|---|---|
| 1998 | 275.854 |
| 1999 | 279.040 |
| 2000 | 282.171 |
| 2001 | 285.039 |
| 2002 | 287.726 |
| 2003 | 290.210 |
| 2004 | 292.892 |
| 2005 | 295.560 |
| 2006 | 298.362 |
| 2007 | 301.290 |
| 2008 | 304.059 |
| 2009 | 307.006 |

## 6.3    POLYNOMIAL SMOOTHING

Smoothing with low-order polynomials is an attempt to retain the advantages of polynomials as empirical models while simultaneously reducing the tendencies of higher-order polynomials to snake and oscillate. The idea of smoothing is illustrated graphically in Figure 6.23.

Rather than forcing a polynomial to pass through all the data points, a low-order polynomial is chosen and fit to the data. This choice normally results in a situation in which the number of data points exceeds the number of constants necessary to determine the polynomial. Smoothing with polynomials requires two decisions:

1. Is a low-order polynomial appropriate? If so, what should the order be?

2. What are the parameters of the model according to some criterion of best fit, such as the least-squares criterion?

We suggest two methods to qualitatively determine whether or not to try a low-order polynomial. First, a visual determination from the scatterplot might be helpful. We can access whether the data appear to follow part of a parabola (so we can try a quadratic polynomial) or whether a change in concavity is seen, suggesting a cubic polynomial. We can also refer back to our one-term models to see the order of the one-term fit and use that to expand the polynomial. For example, if we found that $y = kx^2$ was the best one-term model, we could decide to try the following polynomials:

$$y = ax^2 + bx + c$$
$$y = ax^2 + bx$$
$$y = ax^2 + c$$

Another qualitative method that we could use is by using divided difference tables.

### Divided Difference Tables

We offer a detailed discussion of how to answer these questions using divided difference tables as a qualitative method to access low-order polynomial trends in the data.

Let's assume we have a polynomial of the form:

$$y = ax^2 + bx + c$$

Now,

$$\frac{dy}{dx} = 2ax + b,$$

$$\frac{d^2y}{dx^2} = 2a,$$

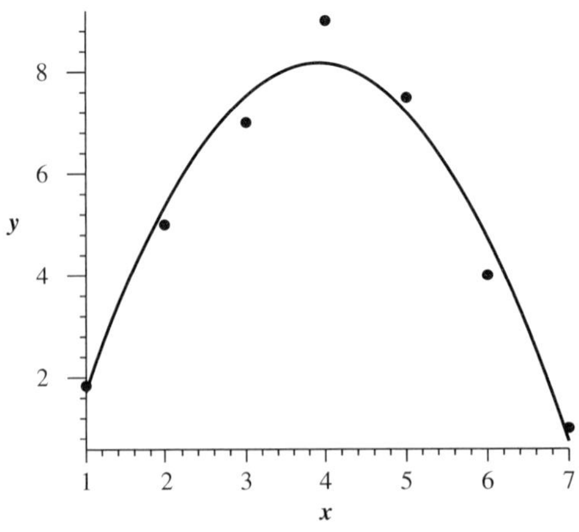

**FIGURE 6.23**
Polynomial smoothing: A quadratic function of the form $y = ax^2 + bx + c$ smoothes data because it is not required to pass through all seven data points

**Table 6.9** Data for Polynomial Fit Example

| $x$ | 1 | 2 | 3 | 4 | 5 |
|---|---|---|---|---|---|
| $y$ | 6 | 11 | 18 | 27 | 38 |

**Table 6.10** Divided Difference Table for Our Polynomial Fit Data

| $x$ | $y$ | $\Delta^1$ | $\Delta^2$ | $\Delta^3$ | $\Delta^4$ |
|---|---|---|---|---|---|
| 1 | 6 | | | | |
| | | 5 | | | |
| 2 | 11 | | 1 | | |
| | | 7 | | 0 | 0 |
| 3 | 18 | | 1 | | |

**Table 6.11** Generic Divided Difference Formulas

| $X$ | $Y$ | $\Delta^1$ | $\Delta^2$ |
|---|---|---|---|
| $X1$ | $Y1$ | | |
| | | $\dfrac{(Y2-Y1)}{(X2-X1)}$ | |
| $X2$ | $Y2$ | | $\dfrac{\frac{(Y3-Y2)}{(X3-X2)}-\frac{(Y2-Y1)}{(X2-X1)}}{(X3-X1)}$ |
| | | $\dfrac{(Y3-Y2)}{(X3-X2)}$ | |
| $X3$ | $Y3$ | | |

$$\frac{d^3y}{dx^3} = 0$$

We have driven the third derivative to zero.

Now, consider a data set that comes directly from this polynomial, such as the data in Table 6.9. Recall that the definition of the derivative is

$$\frac{dy}{dx} = \lim_{\Delta x \to 0} \frac{\Delta y}{\Delta x}$$

Now, a slope may not be a very good approximation to $dy/dx$, but if we take the difference between successive functional values, we can gain insight into what the derivative is doing. Our goal is to find the successive functional differences that might approximate zero. Let's illustrate.

In Table 6.10, $\Delta^1$ are all positive, indicating an increasing function. All $\Delta^2$ are positive, indicating concave up. All $\Delta^3$ are 0, indicating the third derivative is approximately 0. This corresponds to what we saw earlier when we took successive derivatives of the function $y = ax^2 + bx + c$.

The divided difference table (see Table 6.11) is composed of two columns of our original data and successive columns of divided differences that qualitatively (not numerically) give information about the derivatives.

We have created a Maple program that takes the input $x$ and $y$ data and produces a divided difference table. The modeler must interpret the divided difference table. This divided difference program with example is provided on the Web site.

---

| **Example 1** | ## Simple Squared Data |
|---|---|

```
> with(stats):with(LinearAlgebra):with(Statistics):
> xdata:=[0,2,4,6,8]:
```

140    Chapter 6   Empirical Model Construction

> *ydata:=[0,4,16,36,64]:*
> *ddproc(xdata,ydata);*

$$\begin{bmatrix} 0 & 0 & 0 & 0 & 0 \\ 2 & 4 & 2 & 0 & 0 \\ 4 & 16 & 6 & 1 & 0 \\ 6 & 36 & 10 & 1 & 0 \\ 8 & 64 & 14 & 1 & 0 \end{bmatrix}$$

The first two columns are the data. We find that the $\Delta^3$ is zero, so we can try a quadratic polynomial and use least squares to fit this data to the quadratic.

---

**Example 2**          # Vehicular Stopping Distance

The divided difference table assists in determining what order polynomial should be used to approximate a set of data. This can be demonstrated using the vehicular stopping-distance example.

> *xdata:=[20,25,30,35,40,45,50,55,60,65,70,75,80]:*
> *ydata:=[42,56,73.5,91.5,116,142.5,173,209.5,248,292.5,343,401,464]:*
> *ddproc(xdata,ydata);*

| | | | | | | | | | | | |
|---|---|---|---|---|---|---|---|---|---|---|---|
| 20. | 42. | 0. | 0. | 0. | 0. | 0. | 0. | 0. | 0. | 0. | 0. |
| 25. | 56. | 2.800 | 0. | 0. | 0. | 0. | 0. | 0. | 0. | 0. | 0. |
| 30. | 73.5 | 3.500 | 0.07000 | 0. | 0. | 0. | 0. | 0. | 0. | 0. | 0. |
| 35. | 91.5 | 3.600 | 0.01000 | $-0.004000$ | 0. | 0. | 0. | 0. | 0. | 0. | 0. |
| 40. | 116. | 4.900 | 0.1300 | 0.008000 | 0.0006000 | 0. | 0. | 0. | 0. | 0. | 0. |
| 45. | 142.5 | 5.300 | 0.04000 | $-0.006000$ | $-0.0007000$ | $-0.00005200$ | 0. | 0. | 0. | 0. | 0. |
| 50. | 173. | 6.100 | 0.08000 | 0.002667 | 0.0004333 | 0.00004533 | 0. | 0. | 0. | 0. | 0. |
| 55. | 209.5 | 7.300 | 0.1200 | 0.002667 | 0. | $-0.00001733$ | 0. | 0. | 0. | 0. | 0. |
| 60. | 248. | 7.700 | 0.04000 | $-0.005333$ | $-0.0004000$ | $-0.00001600$ | 0. | 0. | 0. | 0. | 0. |
| 65. | 292.5 | 8.900 | 0.1200 | 0.005333 | 0.0005333 | 0.00003733 | 0. | 0. | 0. | 0. | 0. |
| 70. | 343. | 10.10 | 0.1200 | 0. | $-0.0002667$ | $-0.00003200$ | 0. | 0. | 0. | 0. | 0. |
| 75. | 401. | 11.60 | 0.1500 | 0.002000 | 0.0001000 | 0.00001467 | 0. | 0. | 0. | 0. | 0. |
| 80. | 464. | 12.60 | 0.1000 | $-0.003333$ | $-0.0002667$ | $-0.00001467$ | 0. | 0. | 0. | 0. | 0. |

We notice this table differs from previous tables. The third column is small but has both positive and negative values. Negative values within a divided difference table indicate that we might not be able to use the qualitative assessment of the table. However, in this case the values are so small that we can assume they are all very close to zero (ignoring the signs). We might conclude a quadratic polynomial based on this assessment. However, if we accepted the changing signs, then all future columns cannot be used. We might conclude that the tables did not provide us qualitative information for a low-order polynomial. We again demonstrate the Maple solution of fitting the quadratic polynomial using the least-squares criterion (see Figures 6.24 and 6.25).

> *xdata:=[20,25,30,35,40,45,50,55,60,65,70,75,80]:*
> *ydata:=[42,56,73.5,91.5,116,142.5,173,209.5,248,292.5,343,401,464]:*
> *xyfit:=Fit(ax²+bx+c,xdata,ydata,x);*

$$xyfit := 0.0885914085914085576\, x^2 - 1.97012987012986840\, x + 50.0594405594405387$$

**Figure 6.24**
Plot of function and data

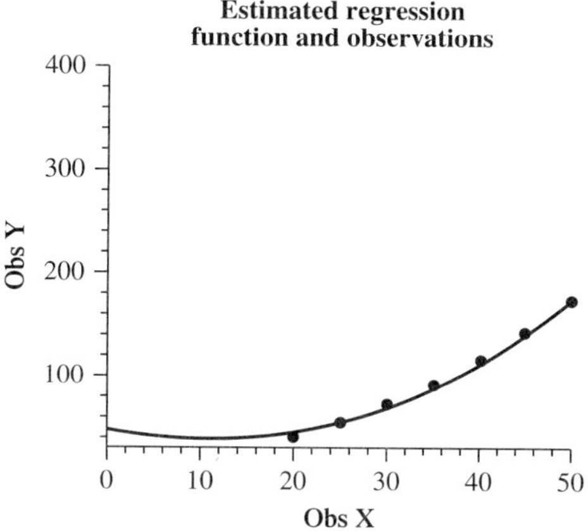


**Figure 6.25**
Residual plot with
curved trend

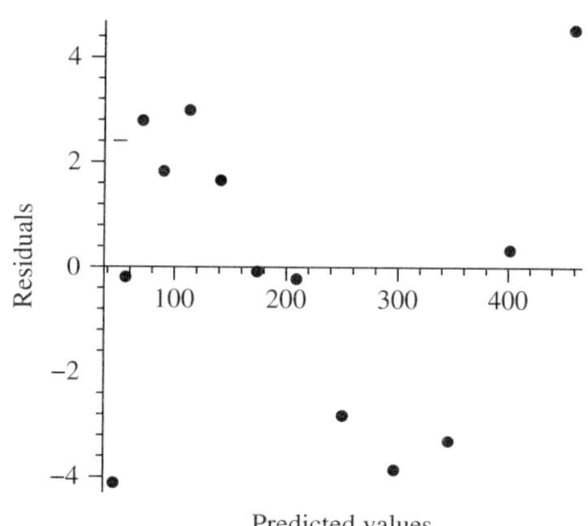


Estimated Regression Equation: $y = 0.08859140859*x\text{\textasciicircum}2 - 1.970129870*x + 50.05944056$
ANOVA Table
($F^* = 26467.9511$)

|  | df | SS | MS |
|---|---|---|---|
| Regression | 1 | 225756.4071 | 225756.4071 |
| Error | 11 | 93.8237 | 8.5294 |
| Total | 12 | 225850.2308 | |

We conclude that our quadratic model is not adequate.

## 6.3 | EXERCISES

Find a low-order polynomial that best fits this data in Exercises 1–5.

**1.** Use the following data.

| $X$ | 1 | 2 | 3 | 4 |
|---|---|---|---|---|
| $Y$ | 7 | 16 | 23 | 36 |

**2.** Use the following population data.

| Year (10 = 1910) | Population |
|---|---|
| 10 | 1,766 |
| 20 | 15,673 |
| 30 | 56,214 |
| 40 | 139,123 |
| 50 | 280,974 |
| 60 | 498,984 |
| 70 | 810,903 |
| 80 | 1,234,934 |
| 90 | 1,789,675 |
| 100 | 2,494,079 |

**3.** Use the following bass fishing data.

| Length | 12.5 | 12.625 | 14.125 | 14.5 | 17.25 | 17.75 |
|---|---|---|---|---|---|---|
| Weight | 17 | 16.5 | 23 | 26.5 | 41 | 49 |

**4.** Telecommunications: $t = 1$ corresponds to 1997, telecom in millions.

| Year, $t$ | 1 | 2 | 3 | 4 | 5 | 6 | 7 | 8 | 9 |
|---|---|---|---|---|---|---|---|---|---|
| Telecom | 9.2 | 9.6 | 10 | 10.4 | 10.6 | 11 | 11.1 | 11.2 | 11.3 |

**5.** We want to build a mathematical model to determine the length of a bird's wingspan (in meters) as a function of the bird's weight (in grams). We collected the following data from six birds.

| Weight | 0.25 | 0.5 | 1.5 | 2.0 | 2.5 | 3.0 |
|---|---|---|---|---|---|---|
| Wingspan | 0.4 | 0.77 | 1.1 | 1.22 | 1.33 | 1.40 |

**6.** Using the following U.S. population data, build a model and predict the results of the 2010 U.S. Census.

| Year | Population (Millions) |
|---|---|
| 1998 | 275.854 |
| 1999 | 279.040 |
| 2000 | 282.171 |
| 2001 | 285.039 |
| 2002 | 287.726 |
| 2003 | 290.210 |
| 2004 | 292.892 |
| 2005 | 295.560 |
| 2006 | 298.362 |
| 2007 | 301.290 |
| 2008 | 304.059 |
| 2009 | 307.006 |

## 6.4   THE CUBIC-SPLINE MODEL

In this section, we introduce cubic-spline interpolation as an alternative method for constructing empirical models. By using difference cubic polynomials between successive pairs of data points and by connecting the cubic polynomials together in a smooth fashion, we can capture the trend of the data, regardless of the underlying relationships. Simultaneously, we will reduce the tendency toward oscillation and the sensitivity of the coefficients to changes in the data.

For a full development of the cubic-spline model, with a discussion of obtaining the equations to solve for all the cubic coefficients, see Chapter 11 (Section 11.2).

Earlier versions of Maple had an internal spline command, and the later versions of Maple use *with(CurveFitting)*. We will show both initially. Typing the command *readlib(spline):* activates the ability to use the *spline* command. The *spline* command produces the group of polynomials that define the curve. The command *s:=unapply(",x);* assigns to *s* the value of the curve at a particular point, which is displayed with two arbitrary points: $s(1,4859)$ and $s(2.3)$.

The following is the command description for spline in Maple.

*spline – compute a natural spline*
*Calling Sequence*
*spline(X, Y, z, d)*
*Parameters*
*X, Y – two vectors or two lists*
*z – name*
*d – (optional) positive integer or name*

---

**Example 1**   ## Obtaining Linear and Cubic Splines

> *with(CurveFitting):*
> *Spline([0,1,2,3],[0,1,4,3],x,linear);*

$$\begin{bmatrix} x & x < 1 \\ -2 + 3x & x < 2 \\ 6 - x & otherwise \end{bmatrix}$$

or
> *with(CurveFitting):*
> *Spline([0,1,2,3],[1,2,4,3],x,degree=3);*

$$\begin{bmatrix} \dfrac{1}{5}x + \dfrac{4}{5}x^3 & x < 1 \\ \dfrac{14}{5} - \dfrac{41}{5}x + \dfrac{42}{5}x^2 - 2x^3 & x < 2 \\ -\dfrac{114}{5} + \dfrac{151}{5}x - \dfrac{54}{5}x^2 + \dfrac{6}{5}x^3 & otherwise \end{bmatrix}$$

---

**Example 2**   ## Fruit Fly Population

The following data in Table 6.12 represents the population of fruit flies in a lab measured weekly for six weeks.

**Table 6.12** Fruit Fly Data

| Time | 7 | 14 | 21 | 28 | 35 | 42 |
|---|---|---|---|---|---|---|
| Number of fruit flies | 8 | 41 | 133 | 250 | 280 | 297 |

Now we will fit the data using natural cubic splines.

```
> with(CurveFitting):
> dx:=[7,14,21,28,35,42]:,dy:=[8,41,133,250,280,297):
```
or
```
> Spline(dx,dy,x,degree=3);
> readlib(spline):
> spline([7,14,21,28,35,42],[8,41,133,250,280,297],x,cubic);
```

$$
\begin{cases}
-25 + \dfrac{12085}{1463}x - \dfrac{7782}{10241}x^2 + \dfrac{2594}{71687}x^3 & x < 14 \\[2ex]
\dfrac{20639}{209} - \dfrac{26711}{1463}x + \dfrac{1056}{931}x^2 - \dfrac{639}{71687}x^3 & x < 21 \\[2ex]
\dfrac{196274}{209} - \dfrac{202346}{1463}x + \dfrac{10023}{1463}x^2 - \dfrac{376}{3773}x^3 & x < 28 \\[2ex]
-\dfrac{632590}{209} + \dfrac{419302}{1463}x - \dfrac{85251}{10241}x^2 + \dfrac{5807}{71687}x^3 & x < 35 \\[2ex]
\dfrac{170535}{209} - \dfrac{8939}{209}x + \dfrac{11124}{10241}x^2 - \dfrac{618}{71687}x^3 & \text{otherwise}
\end{cases}
$$

```
> s:=unapply(%,x);
```

$$
\begin{aligned}
s := x \rightarrow \text{piecewise}\Bigg( & x < 14,\ -25 + \frac{12085}{1463}x - \frac{7782}{10241}x^2 + \frac{2594}{71687}x^3,\ x < 21, \\
& \frac{20639}{209} - \frac{26711}{1463}x + \frac{1056}{931}x^2 - \frac{639}{71687}x^3,\ x < 28, \\
& \frac{196274}{209} - \frac{202346}{1463}x + \frac{10023}{1463}x^2 - \frac{376}{3773}x^3,\ x < 35, \\
& -\frac{632590}{209} + \frac{419302}{1463}x - \frac{85251}{10241}x^2 + \frac{5807}{71687}x^3,\ \frac{170535}{209} - \frac{8939}{209}x + \frac{11124}{10241}x^2 - \frac{618}{71687}x^3\Bigg)
\end{aligned}
$$

```
> with(plots):
> dxy :=pointplot({seq([dx[i],dy[i]],i=1..6)}):
> curve := plot (s (x), x=0..42):
> display ({dxy, curve});
```

Figure 6.26 shows that the five cubic equations smoothly passing through the six data pairs. We can use the cubic equations to evaluate points that fall between their respective intervals.

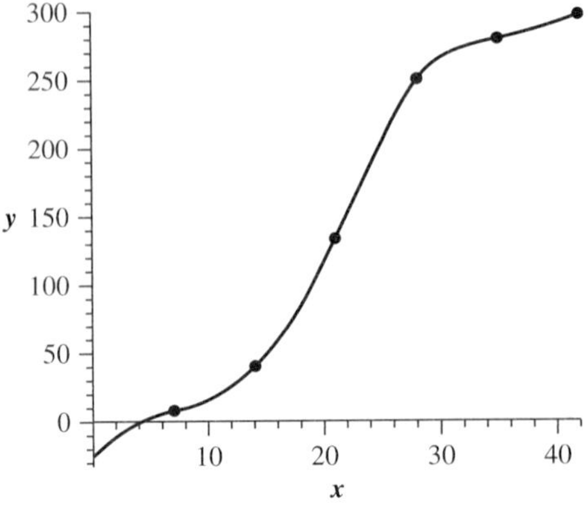

**Figure 6.26**
Cubic-spline fit for fruit fly data

Evaluating points $x = 14.56$ and $x = 33.69$ yields the following results:

> *s(14.56);*

$$45.86257246$$

> *s(33.69);*

$$278.053883$$

---

| **Example 3** | ## Vehicle Stopping Distance |
|---|---|

Because our quadratic model was not adequate, we might try the cubic spline.

> *with( CurveFitting ):*

> *Spline( [20, 25, 30, 35, 40, 45, 50, 55, 60, 65, 70, 75, 80],*
>        *[42, 56, 73.5, 91.5, 116, 142.5, 173, 209.5, 248, 292.5, 343,*
>        *401, 464], x, degree=3 );*

$$
\begin{aligned}
&-9.91986184 + 2.59599309200000006\,x + 8.88178419700125282\ 10^{-17}\,(x-20)^2 \\
&+\ 0.0081602763399999968\,(x-20)^3 \quad x < 25 \\
&-24.20034542 + 3.20801381699999988\,x + 0.122404145077720197\,(x-25)^2 \\
&-\ 0.0128013816900000004\,(x-25)^3 \quad x < 30 \\
&-30.65854920 + 3.47195164099999998\,x - 0.0696165803108808184\,(x-30)^2 \\
&+\ 0.0190452504300000006\,(x-30)^3 \quad x < 35 \\
&-55.64628670 + 4.20417961999999967\,x + 0.216062176165803088\,(x-35)^2 \\
&-\ 0.0153796200400000000\,(x-35)^3 \quad x < 40 \\
&-92.4531952 + 5.21132987900000000\,x - 0.0146321243523316030\,(x-40)^2 \\
&+\ 0.0064732297069999984\,(x-40)^3 \quad x < 45 \\
&-107.2725388 + 5.55050086299999990\,x + 0.0824663212435232940\,(x-45)^2 \\
&+\ 0.0054867012100000012\,(x-45)^3 \quad x < 50 \\
&-166.3333334 + 6.78666666700000044\,x + 0.164766839378238384\,(x-50)^2 \\
&-\ 0.0124200345500000004\,(x-50)^3 \quad x < 55 \\
&-203.1557858 + 7.50283246999999953\,x - 0.0215336787564766874\,(x-55)^2 \\
&+\ 0.0121934369599999998\,(x-55)^3 \quad x < 60 \\
&-244.1202072 + 8.20200345399999975\,x + 0.161367875647668385\,(x-60)^2 \\
&-\ 0.0043537132969999962\,(x-60)^3 \quad x < 65 \\
&-324.2949913 + 9.48915371300000032\,x + 0.0960621761658031342\,(x-65)^2 \\
&+\ 0.0052214162369999992\,(x-65)^3 \quad x < 70 \\
&-415.8967183 + 10.8413816900000004\,x + 0.174383419689119190\,(x-70)^2 \\
&-\ 0.0045319516400000026\,(x-70)^3 \quad x < 75 \\
&-517.3989640 + 12.2453195200000007\,x + 0.106404145077720196\,(x-75)^2 \\
&-\ 0.0070936096730000020\,(x-75)^3 \quad otherwise
\end{aligned}
$$

> *s := unapply(%, x);*

$s := x \rightarrow piecewise$
$(x < 25, -9.91986184 + 2.59599309200000006x + 8.88178419700125282\ 10^{-17}\,(x-20)^2$
$+\ 0.0081602763399999968\,(x-20)^3, x < 30, -24.20034542 + 3.20801381699999988$
$x + 0.122404145077720197\,(x-25)^2 - 0.0128013816900000004\,(x-25)^3, x < 35,$
$-30.65854920 + 3.47195164099999998\,x - 0.0696165803108808184\,(x-30)^2$
$+\ 0.0190452504300000006\,(x-30)^3, x < 40, -55.64628670 + 4.20417961999999967\,x$
$+\ 0.216062176165803088\,(x-35)^2 - 0.0153796200400000000\,(x-35)^3, x < 45,$
$-92.4531952 + 5.21132987900000000\,x - 0.0146321243523316030\,(x-40)^2$
$+\ 0.0064732297069999984\,(x-40)^3, x < 50, -107.2725388 + 5.55050086299999990\,x$
$+\ 0.0824663212435232940\,(x-45)^2 + 0.0054867012100000012\,(x-45)^3, x < 55,$

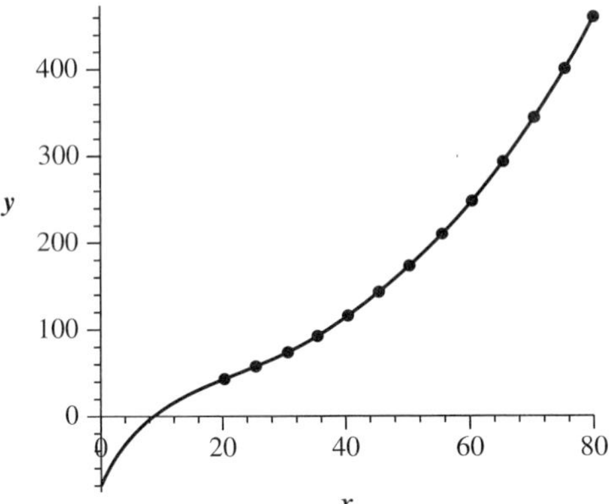

**Figure 6.27**
Cubic spline fits the vehicle-stopping data

$$-166.3333334 + 6.78666666700000044\,x + 0.164766839378238384\,(x-50)^2$$
$$-0.0124200345500000004\,(x-50)^3,\ x < 60,\ -203.1557858 + 7.50283246999999953\,x$$
$$-0.0215336787564766874\,(x-55)^2 + 0.0121934369599999998\,(x-55)^3,\ x < 65,$$
$$-244.1202072 + 8.20200345399999975\,x + 0.161367875647668385\,(x-60)^2$$
$$-0.00435371329699999962\,(x-60)^3,\ x < 70,\ -324.2949913 + 9.48915371300000032\,x$$
$$+ 0.0960621761658031342\,(x-65)^2 + 0.00522141623699999992\,(x-65)^3,\ x < 75,$$
$$-415.8967183 + 10.8413816900000004\,x + 0.174383419689119190\,(x-70)^2$$
$$-0.00453195164000000026\,(x-70)^3,\ -517.3989640 + 12.2453195200000007\,x$$
$$+ 0.106404145077720196\,(x-75)^2 - 0.00709360967300000020\,(x-75)^3)$$

```
> with(plots):
> dxy:=pointplot( {seq( [xdata[i], ydata[i]],i=1..13)}):
> curve:=plot(s(x),x=0..80) :
> display( {dxy, curve} );
```

Figure 6.27 clearly shows that the cubic equations capture the trends between the data and why we do not want to extrapolate with cubic splines, because as $x$ gets small we might not expect $y$ to be negative or as $x$ gets large we might expect $x$ to flatten out and not grow without bound.

---

| Example 4 | Determining the Cost of Postage Stamps Using Cubic Splines |

## Determining the Cost of Postage Stamps Using Cubic Splines

Between 1963 and 2009, the cost of postage stamps has increased on more than 20 occasions. An analyst might like to be able to model the cost of the stamp. Because the dates and the cost are extremely accurate, we can use splines to model this. We also modify the years and use $x = 0$ to represent year 1963 so that $x = 46$ represents 2009. We also convert the rate to cents so 0.44 is 44 cents and show this data in Table 6.13.

```
> with(CurveFitting):

> rate :=[5,6,8,10,13,15,18,20,22,25,29,32,33,34,37,
        39,41,42,44]:year:=[0,5,8,11,12,15,18,22,25,28,
        32,36,38,38.6,39,43,44,45,46];
```

$$year := [0, 5, 8, 11, 12, 15, 18, 22, 25, 28, 32, 36, 38, 38.6, 39,$$
$$43, 44, 45, 46]$$

**Table 6.13** Cost of a Postage Stamp

| Date Introduced | Rate for first ounce (USD) |
| --- | --- |
| January 7, 1963 | .05 |
| January 7, 1968 | .06 |
| May 16, 1971 | .08 |
| March 2, 1974 | .10 |
| December 31, 1975 | .13 |
| May 29, 1978 | .15 |
| March 22, 1981 | .18 |
| November 1, 1981 | .20 |
| February 17, 1985 | .22 |
| April 3, 1988 | .25 |
| February 3, 1991 | .29 |
| January 1, 1995 | .32 |
| January 10, 1999 | .33 |
| January 7, 2001 | .34 |
| June 30, 2002 | .37 |
| January 8, 2006 | .39 |
| May 14, 2007 | .41 |
| May 12, 2008 | .42 |
| May 11, 2009 | .44 |

Postage rate information from http://en.wikipedia.org/wiki/History_of_United_States_postage_rates.uu

> *Spline ( [0, 5, 8, 11, 12, 15, 18, 22, 25, 28, 32, 36, 38, 38.6, 39, 43, 44, 45, 46],*
> *[5, 6, 8, 10, 13, 15, 18, 20, 22, 25, 29, 32, 33, 34, 37, 39, 41*
> *42, 44], x, degree=3);*

$$5. \; -0.0467543872999999986\,x \; - \; 8.88178419700125282\,10^{-17}\,x^2$$
$$+\;0.0098701754930000034\,x^3 \qquad x < 5$$
$$2.532456127 \; + \; 0.693508774599999933\,x \; + \; 0.148052632388911282\,(x-5)^2$$
$$-0.0523333339000000009\,(x-5)^3 \qquad x < 8$$
$$6.649403576 \; + \; 0.168824552999999988\,x \; - \; 0.322947372740860106\,(x-8)^2$$
$$+\;0.162964914699999991\,(x-8)^3 \qquad x < 11$$
$$-\;18.94312312 \; + \; 2.63119301100000014\,x \; + \; 1.14373685857452911\,(x-11)^2$$
$$-\;0.774929869699999996\,(x-11)^3 \qquad x < 12$$
$$-\;18.12652543 \; + \; 2.59387711900000006\,x \; - \; 1.18105275037365320\,(x-12)^2$$
$$+\;0.179549755300000002\,(x-12)^3 \qquad x < 15$$
$$9.668939810 \; + \; 0.355404012699999994\,x \; + \; 0.434895048138232210\,(x-15)^2$$
$$-\;0.0733432396300000012\,(x-15)^3 \qquad x < 18$$
$$0.27887705 \; + \; 0.984506830299999970\,x \; - \; 0.225194108845942154\,(x-18)^2$$
$$+\;0.0260168503099999990\,(x-18)^3 \qquad x < 22$$
$$10.50121896 \; + \; 0.431762774500000002\,x \; + \; 0.0870080948571233387\,(x-22)^2$$
$$-\;0.00290226582199999998\,(x-22)^3 \qquad x < 25$$
$$0.11374584 \; + \; 0.875450166499999960\,x \; + \; 0.0608877024613472818\,(x-25)^2$$
$$-\;0.00645703042300000038\,(x-25)^3 \qquad x < 28$$
$$-\;4.86022368 \; + \; 1.06643656000000008\,x \; + \; 0.00277442863082087104\,(x-28)^2$$
$$-\;0.00484589214000000040\,(x-28)^3 \qquad x < 32$$
$$1.60706669 \; + \; 0.856029166000000008\,x \; - \; 0.0553762770538835034\,(x-32)^2$$
$$+\;0.00721724638699999984\,(x-32)^3 \qquad x < 36$$
$$4.65991606 \; + \; 0.759446776099999998\,x \; + \; 0.031230679584713 1329\,(x-36)^2$$
$$-\;0.0804770338299999888\,(x-36)^3 \qquad x < 38$$
$$36.09148662 \; - \; 0.0813549110000000020\,x \; - \; 0.451631523400511770\,(x-38)^2$$
$$+\;5.60833470000000034\,(x-38)^3 \qquad x < 38.6$$

$$\begin{aligned}
&-\,175.7403852 + 5.43368873699999977\,x + 9.64337093918872590\,(x-38.6)^2 \\
&-\,11.1939819600000004\,(x-38.6)^3 \qquad x < 39 \\
&-\,266.2356918 + 7.77527414999999956\,x - 3.78940741084286126\,(x-39)^2 \\
&+\,0.492647218300000023\,(x-39)^3 \qquad x < 43 \\
&-\,8.60449784 + 1.10708134500000011\,x + 2.12235920993542227\,(x-43)^2 \\
&-\,1.22944055500000005\,(x-43)^3 \qquad x < 44 \\
&-\,32.19303636 + 1.66347809899999998\,x - 1.56596245598277894\,(x-44)^2 \\
&+\,0.902484356700000045\,(x-44)^3 \qquad x < 45 \\
&-\,13.75528156 + 1.23900625700000000\,x + 1.14149061399569463\,(x-45)^2 \\
&-\,0.380496871299999984\,(x-45)^3 \qquad otherwise
\end{aligned}$$

$> s := unapply(\%, x);$

$s := x \rightarrow piecewise$
$(x < 5,\ 5. - 0.0467543872999999986\,x - 8.88178419700125282\ 10^{-17}x^2$
$+\ 0.0098701754930000034\,x^3,\ x < 8,\ 2.532456127 + 0.693508774599999933x$
$+\ 0.148052632388911282\,(x-5)^2 - 0.0523333339000000009\,(x-5)^3,\ x < 11,$
$6.649403576 + 0.168824552999999988x - 0.322947372740860106\,(x-8)^2$
$+\ 0.162964914699999991\,(x-8)^3,\ x < 12,\ -18.94312312 + 2.63119301100000014x$
$+\ 1.14373685857452911\,(x-11)^2 - 0.774929869699999996\,(x-11)^3,\ x < 15,$
$-18.12652543 + 2.59387711900000006x - 1.18105275037365320\,(x-12)^2$
$+\ 0.179549755300000002\,(x-12)^3,\ x < 18,\ 9.668939810 + 0.355404012699999994x$
$+\ 0.434895048138232210\,(x-15)^2 - 0.0733432396300000012\,(x-15)^3,\ x < 22,$
$0.27887705 + 0.984506830299999970x - 0.225194108845942154\,(x-18)^2$
$+\ 0.0260168503099999990\,(x-18)^3,\ x < 25,\ 10.50121896 + 0.431762774500000002\,x$
$+\ 0.0870080948571233387\,(x-22)^2 - 0.00290226582199999998\,(x-22)^3,\ x < 28,$
$0.11374584 + 0.875450166499999960\,x + 0.0608877024613472818\,(x-25)^2$
$-0.00645703042300000038\,(x-25)^3,\ x < 32,\ -4.86022368 + 1.06643656000000008x$
$+\ 0.00277442863082087104\,(x-28)^2 - 0.00484589214000000040\,(x-28)^3,\ x < 36,$
$1.60706669 + 0.856029166000000008x - 0.0553762770538835034\,(x-32)^2$
$+\ 0.00721724638699999984\,(x-32)^3,\ x < 38,\ 4.65991606 + 0.759446776099999998x$
$+\ 0.0312306795847131329\,(x-36)^2 - 0.0804770338299999888\,(x-36)^3,\ x < 38.6,$
$36.09148662 - 0.0813549110000000020\,x - 0.451631523400511770\,(x-38)^2$
$+\ 5.60833470000000034\,(x-38)^3,\ x < 39,\ -175.7403852 + 5.43368873699999977\,x$
$+\ 9.64337093918872590\,(x-38.6)^2 - 11.1939819600000004\,(x-38.6)^3,\ x < 43,$
$-266.2356918 + 7.77527414999999956\,x - 3.78940741084286126\,(x-39)^2$
$+\ 0.492647218300000023\,(x-39)^3,\ x < 44,\ -8.60449784 + 1.10708134500000011x$
$+\ 2.12235920993542227\,(x-43)^2 - 1.22944055500000005\,(x-43)^3,\ x < 45,$
$-32.19303636 + 1.66347809899999998\,x - 1.56596245598277894\,(x-44)^2$
$+\ 0.902484356700000045\,(x-44)^3,\ -13.75528156 + 1.23900625700000000\,x$
$+\ 1.14149061399569463\,(x-45)^2 - 0.380496871299999984\,(x-45)^3)$

$> with\ (plots):$
$> dxy := pointplot(\{seq([year[i],\ rate[i]], i=1..19)\}):$
$> curve := plot(s(x), x=0..50, thickness=2):$
$> display(\{dxy,\ curve\});$

Figure 6.28 displays the spline curve superimposed over the original data points. Estimations of the value of the function can be taken from the graph or by using the function $s(x)$. Notice that, by the year 2001, the spline approximates the cost of the postage stamp to be 36 cents, a reasonable estimation. Note that the cubic spline is dependent on the data points; approximations outside of the end points can act erratically. An approximation of the cost of the postage stamp in 2023 is –966 cents: The government would pay each of us to use the mail system, an unlikely event. Table 6.14 summarizes these results.

**FIGURE 6.28**
Cubic-spline plot of
postage stamp data

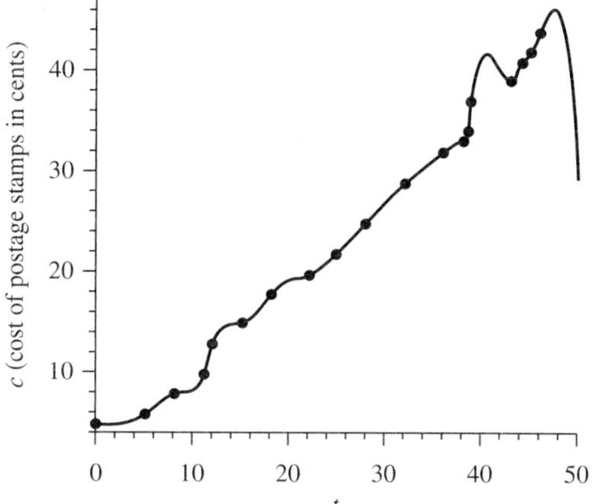


**Table 6.14** Summary Cubic Spline Output for the
Cost of a Postage Stamp

| Command | Output |
| --- | --- |
| $s(21)$; | 19.62 in year 1984 |
| $s(48)$; | 36 cents in year 2008 |
| $s(49)$; | 37 cents in year 2009 |
| $s(60)$; | −966.75 cents in year 2023 |

## 6.4 | EXERCISES

Use cubic splines on Exercises 1–6 from Section 6.3.

## PROJECTS

In this special problem, you will construct a predictive model based on one of the following scenarios:

> Population of the United States I
> Population of the United States II
> Cost of Medicare
> Postage Stamps
> Terror Bird, Revisited
> Telemetry
> Water Commission

In addition to questions asked in each specific scenario, the following will be required as part of the written report:

1. Executive summary of modeling results (1 page or less)

2. A concise problem statement

3. Discussion of the modeling assumptions used

4. A scatterplot of the original data. Discuss any trends. Discuss any potential outliers.

5. Discuss the data briefly. How good do you think the data are? In particular, how confident are you in the accuracy of the data points? How confident are you that the data represent what you are trying to model?

6. Obtain a model by *each* of the techniques below. Show all work, and turn in computer outputs for each technique used to build the model.

> Simple one-term model
> Log–log (or ln–ln) transformation models
> Interpolating polynomial
> Low-order polynomial suggested by divided difference table
> Cubic splines

7. Which technique or model do you consider best for answering the questions in your scenario? Why?

## Population of the United States I

Consider the data in the following table. These data give the population of the United States on July 1 of each given year. Use the procedures described in empirical model building to capture the trends of this data.

| Year | Population* |
|---|---|
| 1790 | 3,929,000 |
| 1800 | 5,308,000 |
| 1860 | 31,443,000 |
| 1900 | 76,212,000 |
| 1920 | 106,461,000 |
| 1930 | 123,203,000 |
| 1940 | 132,122,446 |
| 1950 | 152,271,417 |
| 1960 | 180,671,158 |
| 1970 | 205,052,174 |
| 1980 | 227,224,681 |
| 1990 | 249,464,396 |
| 1999 | 279,040,168 |
| 2000 | 282,171,936 |
| 2005 | 295,560,599 |
| 2009 | 307,006,550 |

*Data from U.S. population census Web site and the *World Almanac*.

### Additional Questions

1. Would you eliminate any data points? Why or why not?

2. If forced to predict, what do you expect the population of the United States to be on July 1, 2010? How about July 1, 2020?

3. What would you expect the population to have been on July 1, 1995? Find an almanac and compare for fun. Why did you choose the model you did to answer the question? How comfortable are you with your prediction?

4. What is the population for July 1, 1956, expressed from each model type developed? What do you think the population was on July 1, 1956? Why?

## Population of the United States II

Population of the United States II
   Source: Population Estimates Program, Population Division, U.S. Census Bureau
   Internet Release Date: March 5, 2009

The data represent the average annual population for their respective years.

| Date | National Population | Population Change |
|---|---|---|
| July 1, 2009 | 307,006,550 | 2,944,826 |
| July 1, 2008 | 304,059,724 | 2,769,392 |
| July 1, 2007 | 301,290,332 | 2,927,359 |
| July 1, 2006 | 298,362,973 | 2,802,374 |
| July 1, 2005 | 295,560,599 | 2,668,432 |
| July 1, 2004 | 292,892,127 | 2,681,213 |
| July 1, 2003 | 290,210,914 | 2,484,267 |
| July 1, 2002 | 287,726,647 | 2,686,844 |
| July 1, 2001 | 285,039,803 | 2,867,866 |
| July 1, 2000 | 282,171.937 | 3,131,769 |
| July 1, 1999 | 279,040,168 | 2,442,810 |

Use your models and determine the population for July 1, 1979. Use your models and predict the population in this current year. When will the population break 300 million?

| Year | Gross Debt in Billions (Undeflated) | As % of GDP | Debt Held by Public ($Billions) | As % of GDP |
|---|---|---|---|---|
| 1910 | 2.6 | Unknown | 2.6 | Unknown |
| 1920 | 25.9 | Unknown | 25.9 | Unknown |
| 1928 | 18.5 | Unknown | 18.5 | Unknown |
| 1930 | 16.2 | Unknown | 16.2 | Unknown |
| 1940 | 50.6 | 52.4 | 42.8 | 44.2 |
| 1950 | 256.8 | 94.0 | 219.0 | 80.2 |
| 1960 | 290.5 | 56.0 | 236.8 | 45.6 |
| 1970 | 380.9 | 37.6 | 283.2 | 28.0 |
| 1980 | 909.0 | 33.4 | 711.9 | 26.1 |
| 1990 | 3,206.3 | 55.9 | 2,411.6 | 42.0 |
| 2000 | 5,628.7 | 58.0 | 3,409.8 | 35.1 |
| 2001 | 5,769.9 | 57.4 | 3,319.6 | 33.0 |
| 2002 | 6,198.4 | 59.7 | 3,540.4 | 34.1 |
| 2003 | 6,760.0 | 62.6 | 3,913.4 | 35.1 |
| 2004 | 7,354.7 | 63.9 | 4,295.5 | 37.3 |
| 2005 | 7,905.3 | 64.6 | 4,592.2 | 37.5 |
| 2006 | 8,451.4 | 65.0 | 4,829.0 | 37.1 |
| 2007 | 8,950.7 | 65.6 | 5,035.1 | 36.9 |
| 2008 | 9,985.8 | 70.2 | 5,802.7 | 40.8 |
| 2009 | 12,311.4 | | | |

Data from http://en.wikipedia.org/wiki/United_States_public_debt

1. Would you eliminate any data points? Why or why not?

2. If forced to make a prediction, what would you predict the debt to be in 2010? 2015? 2020? Are you comfortable with your predictions? Which model did you use and why?

3. What debt does each of your models predict occurred in 2010? Which prediction would you use and why?

## Postage Stamps

Sometime in the future, the U.S. public will see another increase in postage. The following data are provided.

| Year | Cost |
|------|------|
| 1885 | .02 |
| 1917 | .03 (wartime increase) |
| 1919 | .02 (restored by Congress) |
| 1932 | .03 |
| 1958 | .04 |
| 1963 | .05 |
| 1968 | .06 |
| 1971 | .08 |
| 1974 | .10 |
| 1975 | .13 (temporary) |
| 1976 | .13 |
| 1978 | .15 |
| 1981 | .18 |
| 1981 | .20 |
| 1985 | .22 |
| 1987 | .25 |
| 1991 | .29 |
| 1995 | .32 |
| 1999 | .33 |
| 2001 | .34 |
| 2002 | .37 |
| 2006 | .34 |
| 2007 | .41 |
| 2008 | .42 |
| 2009 | .44 |

### Additional Questions

1. Would you eliminate any data points? Why?

2. What does each model predict the cost of postage to be in 2011 and 2015? Which prediction would you use? Why?

3. Realistically, after this year, when do you expect the next postage hike? What will be the cost? Provide your rationale.

4. Predict the year in which the cost of a postage stamp will reach $1.00. Which model did you use and why?

# The Terror Bird, Revisited

As you recall from Chapters 2 and 5, *Titanis walleri,* the terror bird, lived about 2 million to 3 million years ago on the oak and grass savannas of what is now Florida. A giant, flightless predatory bird, it probably ambushed and then devoured its prey in tall grasslands.

Scientists believe that bones give a good indication of size (height) but not reliable body weight (weight is not easily fossilized). We want to build a mathematical model to predict its weight.

The following data sets are available.

### Bird Data

| Femur circumference (cm) | Weight (kg) |
|------|------|
| .7943 | .0832 |
| .7079 | .0912 |
| 1.000 | .1413 |

### Bird Data (continued)

| Femur circumference (cm) | Weight (kg) |
| --- | --- |
| 1.122 | .1479 |
| 1.6982 | .2455 |
| 1.2023 | .2818 |
| 1.9953 | .7943 |
| 2.2387 | 2.5119 |
| 2.5119 | 1.4125 |
| 2.5119 | .8913 |
| 3.1623 | 1.9953 |
| 3.5481 | 4.2658 |
| 4.4668 | 6.3096 |
| 5.8884 | 11.2202 |
| 6.7608 | 19.9500 |
| 15.136 | 141.25 |
| 15.850 | 158.4893 |

### Dinosaur Data

| Name | Femur Circumference(mm) | Weight (kg) |
| --- | --- | --- |
| Hypsilophodontidae | 103 | 55 |
| Ornithomimdae | 136 | 115 |
| Thescelosauridae | 201 | 311 |
| Ceratosauridae | 267 | 640 |
| Allosauridae | 348 | 1,230 |
| Hadrosauridae-1 | 400 | 1,818 |
| Hadrosauridae-2 | 504 | 3,300 |
| Hadrosauridae-3 | 512 | 3,500 |
| Tyrannosauridae | 534 | 4,000 |

Build a mathematical model and test with each data set. Predict the size (weight) of the terror bird by each empirical modeling method. Which of the models (from the data sets) would you use to present your answer to a scientist?

# Telemetry

1. **Situation.** Your mathematical modeling team is responsible for collecting, analyzing, and coordinating the use of certain telemetry data from an experimental ground-to-air rocket. The data represent the scaled velocity of the rocket. From rocket launching through the first 6.5 seconds, these data are transmitted by the rocket back to the ground at varying time intervals. Occasionally, radio interference causes some of the transmitted data to be garbled and consequently lost. The data from one experiment are listed in enclosure 1.

2. **Task.** Based on the table in enclosure 1, perform the following tasks:

   a. Estimate what the telemetry data should be for the two garbled transmissions.
   b. Estimate the telemetry data at seven seconds. Is this a valid estimate?
   c. Estimate the flight time for telemetry data values at 50 fps and 800 fps.

**3. Specific Requirements**

**a.** First, construct a simple one-term model, using the ladder of power for the data up to but not including the first garbled data transmission. Specifically state this model.

**b.** Build a ln–ln model for this same data (up to but not including the first garbled transmission). Specifically state this model.

**c.** Construct a Lagrange interpolating polynomial (Section 4.2) for the given data using only four data points *centered* on each garbled transmission. Reduce each polynomial to an equation of the form

$$P(x) = a_0 + a_1 x + a_2 x^2 + a_3 x^3$$

where $a_i$, i = 0, 1, 2, 3, are constants containing at least four significant digits. Use the polynomial to estimate the missing data value.

**d.** Construct a divided difference table for the two garbled transmissions. Use as many or few data points as necessary in your table. *Do not center* the data about the first garbled transmission. Form two tables, one using the point just prior to the garbled transmission and subsequent values after the garbled transmission, and the other using the point just after the garbled transmission and the remaining points from prior to the garbled transmission. Center the data about the second garbled transmission. From the tabulated values, decide on a low-order polynomial, if any. Use least squares to fit the polynomial. Analyze the residuals and comment on the model's adequacy.

**e.** Use cubic splines to determine the values for the garbled transmission.

**f.** Obtain a graphical representation of your simple one-term model, your interpolating polynomials, and your low-order polynomial fit that resulted from analyzing a divided difference table.

**g.** One of the important performance parameters of this ground-to-air rocket is the flight time when the rocket reaches specific speeds. Use an appropriate interpolating polynomial to determine the flight time (seconds) to an accuracy of $10^{-2}$ for the rocket to reach a scaled velocity of:

(1) 50 fps and
(2) 800 fps

**h.** Estimate what the speed will be at time $t = 7$ seconds. Is this kind of extrapolation valid/accurate?

**4. Submission.** Submit to your instructor, in the following in order:

**a.** Executive summary describing the problem, methodology, and summarizing your results. Reference supporting materials in your tabbed appendices.

**b.** Table of contents. The rest are tabbed appendixes.

**c.** The interpolating polynomials, including all work used in their development.

**d.** The interpolating polynomials chosen for determination of flight times for scaled velocities of 50 fps and 800 fps and all calculations for determination of the flight time.

**e.** Final answers and conclusions.

(1) Discussion of the most appropriate interpolating polynomial for this problem and your reasons for selection; include discussion of what order polynomial yields the best approximation and why.

(2) Your graphical representation of the data and interpolating polynomial.

(3) Discussion of computations leading to flight times for scaled velocities and the extrapolated velocity.

**f.** Thought question: Is there a better way to fit a curve to these data that would provide a better approximation?

**Flight Data**

| Flight Time (Seconds) | Telemetry Data (Feet per Second) |
| --- | --- |
| 0.0 | 0.0000 |
| 0.25 | 1.295 |
| 0.50 | 1.955 |
| 0.75 | 3.288 |
| 1.00 | 5.695 |
| 1.25 | 9.579 |
| 1.50 | 15.34 |
| 1.75 | 23.37 |
| 2.00 | 34.05 |
| 2.25 | 47.83 |
| 2.50 | Garbled |
| 2.75 | 86.20 |
| 3.00 | 111.7 |
| 3.25 | 141.6 |
| 3.57 | 187.4 |
| 3.70 | 208.6 |
| 3.86 | 240.8 |
| 4.16 | 296.7 |
| 4.40 | 351.4 |
| 4.73 | 436.7 |
| 4.93 | 495.0 |
| 5.13 | 557.1 |
| 5.43 | 662.1 |
| 5.65 | 746.4 |
| 6.11 | Garbled |
| 6.20 | 987.8 |
| 6.43 | 1,102.1 |

# Water Commission

Consider the following problem: California's State Water Commission is requiring data from La Mesa housing on the rate of water use, in gallons per hour, and the total amount of water used each day. The Department of Engineering and Housing (DEH) in Monterey does not have the sophisticated equipment that measures the flow of water in or out of the main water tank. Instead, DEH can measure only the level of water in the tank, within 0.5 percent accuracy, every hour. More importantly, whenever the level in the tank drops below some minimum level $L$, a pump fills the tank up to the maximum level, $H$, but there is no measurement of the pump flow at these times, either. Thus, one cannot readily relate the level in the tank to the amount of water used while the pump is working, which occurs once or twice per day, for a couple of hours each time. The following table contains the time, in seconds, since the first measurement, and the level of water in the tank in hundredths of a foot. For example, after 3,316 seconds, the depth of the water in the tank reached 31.10 feet. The tank is a vertical circular cylinder with a height of 40 feet and a diameter of 57 feet. Usually, the pump starts filling the tank when the level drops to about 27.00 feet, and the pump stops when the level rises back to about 35.50 feet.

# Additional Requirements

1. Build a mathematical model for the DEH.
2. Use your model and estimate the flow out of the tank, $f(t)$, at all times, even when the pump is working.
3. Use your model and estimate the total amount of water used in one day (24 hours).

| Time (Sec) | Level (0.01 ft) |
| --- | --- |
| 0 | 3,175 |
| 3,316 | 3,110 |
| 6,635 | 3,054 |
| 10,619 | 2,994 |
| 13,937 | 2,947 |
| 17,921 | 2,892 |
| 21,240 | 2,850 |
| 25,223 | 2,795 |
| 28,543 | 2,752 |
| 32,884 | 2,697 |
| 35,932 | Pump on |
| 39,332 | Pump on |
| 39,435 | 3,550 |
| 43,318 | 3,445 |
| 46,636 | 3,350 |
| 49,953 | 3,260 |
| 53,936 | 3,167 |
| 57,254 | 3,087 |
| 60,574 | 3,012 |
| 64,554 | 2,927 |
| 68,535 | 2,842 |
| 71,854 | 2,757 |
| 75,021 | 2,697 |
| 79,254 | Pump on |
| 82,649 | Pump on |
| 85,968 | 3,475 |
| 89,953 | 3,397 |
| 93,270 | 3,340 |

7

# Modeling with Linear Programming

## Introduction

Consider planning the shipment of needed items, from the warehouses where they are manufactured and stored to the distribution centers where they are needed.

There are three warehouses at different cities: Detroit, Pittsburgh, and Buffalo. They have 250, 130, and 235 tons of paper, respectively. There are four publishers in different cities: Boston, New York, Chicago, and Indianapolis. They ordered 75, 230, 240, and 70 tons of paper, respectively, to publish new books.

The following table shows the costs in dollars of transporting one ton of paper from one city to another.

| From\To | Boston (BS) | New York (NY) | Chicago (CH) | Indianapolis (IN) |
|---|---|---|---|---|
| Detroit (DT) | 15 | 20 | 16 | 21 |
| Pittsburgh (PT) | 25 | 13 | 5 | 11 |
| Buffalo (BF) | 15 | 15 | 7 | 17 |

Your boss wants you to minimize the shipping costs while meeting demand. This problem deals with the allocation of resources and can be modeled as a linear programming problem, as we will discuss.

Consider starting a new diet, which needs to be healthy. A nutritionist gives you lots of information on foods and recommends that you stick to six different foods—bread, milk, cheese, fish, potatoes, and yogurt—and provides you a table of information that includes the average cost of the items.

| | Bread | Milk | Cheese | Potato | Fish | Yogurt |
|---|---|---|---|---|---|---|
| Cost ($) | 2.0 | 3.5 | 8.0 | 1.5 | 11.0 | 1.0 |
| Protein (g) | 4.0 | 8.0 | 7.0 | 1.3 | 8.0 | 9.2 |
| Fat (g) | 1.0 | 5.0 | 9.0 | 0.1 | 7.0 | 1.0 |
| Carbohydrates (g) | 15.0 | 11.7 | 0.4 | 22.6 | 0.0 | 17.0 |
| Calories (Cal) | 90 | 120 | 106 | 97 | 130 | 180 |

The nutritionist also recommends that our diet contain not less than 150 calories, not more than 10 g of protein, not less than 10 g of carbohydrates, and not less than 8 g of fat. We also conclude that our diet should include at least 0.5 g of fish and not more than 1 cup of milk. Also, we decide that our diet should have **minimal cost**. Again, this is an allocation of resources problem in which we want the optimal diet at minimal cost. We have six unknown variables that define the weight of the food. There is a lower bound for fish (0.5 g). There is an upper bound for milk (1 cup). To model and solve this problem, we can use linear programming.

Modern linear programming was the result of a research project undertaken by the U.S. Air Force Office of Statistical Control. As the number of fronts in the Second World War increased, it became more and more difficult to coordinate troop supplies effectively. Mathematicians looked for ways to use the new computers being developed to perform calculations quickly. One SCOOP (Scientific Computation of Optimum Programs) team member, George Dantzig, developed the simplex algorithm for solving simultaneous linear programming problems. The simplex method has several advantageous properties: It is highly efficient, allowing the solution of problems with many variables, and it uses methods from linear algebra that are readily solvable.

In January 1952, the first successful solution to a linear programming (LP) problem was found using a high-speed electronic computer on the National Bureau of Standards SEAC machine. Today, most LPs are solved via high-speed computers. Computer-specific software programs such as LINDO, Excel Solver, GAMS have been developed to help analyze and solve LP problems. We will use the power of Maple to solve our linear programming problems.

To provide a framework for our discussions, we offer the following basic model:

$$\text{Maximize (or minimize) } f(X)$$

$$\text{Subject to } g_i(X) \begin{Bmatrix} \geq \\ = \\ \leq \end{Bmatrix} b_i \text{ for all } i.$$

Now let's explain this notation. The various component of the vector $X$ are called the *decision variables* of the model. These are the variables that can be controlled or manipulated. The function, $f(X)$, is called the *objective function*. By "Subject to" (ST) we connote that certain side conditions, resource requirements, or resource limitations must be met. These conditions are called *constraints*. The constant $b_i$ represents the level that is associated with constraint $g(X_i)$ and is called the *right-hand side* in the model.

Linear programming is a method for solving linear problems, which occur frequently in almost every modern industry. In fact, diverse areas use linear programming, including defense, health, transportation, manufacturing, advertising, and telecommunications. The reason for this is the classic economic problem: A company wants to maximize output, but it is competing for limited resources. The *linear* in linear programming means that, in the case of production, the quantity produced is proportional to the resources used and also the revenue generated. The coefficients are constants, and no products of variables are allowed.

To use this technique, the company must identify a number of constraints that will limit the production or transportation of their goods; these may include factors such as labor hours, energy, or raw materials. Each constraint must be quantified in terms of one unit of output, because the problem-solving method relies on the constraints being used.

An optimization problem that satisfies the following five properties is said to be a linear programming problem.

1. There is a unique objective function, $f(X)$.

2. Whenever a decision variable, $X$, appears in either the objective function or a constraint function, it must appear with an exponent of 1, possibly multiplied by a constant.

3. No terms contain products of decision variables.

**4.** All coefficients of decision variables are constants.

**5.** Decision variables are permitted to assume fractional as well as integer values.

Linear problems, by the nature of the many unknowns, are very hard to solve by human inspection, but methods have been developed to use the power of computers to do the hard work quickly. We will illustrate with two variables, graphically.

## 7.1 FORMULATING LINEAR PROGRAMMING PROBLEMS

A linear programming problem is a problem that requires an objective function to be maximized or minimized subject to resource constraints. The key to formulating a linear programming problem is recognizing the decision variables. The objective function and all constraints are written in terms of these decision variables.

The conditions for a mathematical model to be a linear program are:

- All variables are continuous (i.e., they can take fractional values).

- The model has a single objective (minimize or maximize).

- The objective and constraints are linear (i.e., any term is either a constant or a constant multiplied by an unknown).

- The decision variables must be nonnegative.

Many practical problems can be formulated as LPs. There also exists an algorithm (called the *simplex* algorithm) that enables us to solve LPs numerically relatively easily.

We will return later to the simplex algorithm for solving LPs, but for the moment we will concentrate on formulating them.

Some of the major application areas to which LP can be applied are:

- blending,

- production planning,

- oil-refinery management,

- distribution,

- financial and economic planning,

- human resources planning,

- blast-furnace burdening, and

- farm planning.

In the following, we consider specific examples of the types of problem that can be formulated as LPs. Note here that the key to formulating them is *practice*. However, a useful hint is that common objectives for LPs are *minimize cost and maximize profit*.

---

## Example 1    Production Mix of New Drinks

Consider the following problem statement:

A company wants to can two new different drinks for the holiday season. It takes 2 hours to can one gross of drink A, and it takes 1 hour to label the cans. It takes 3 hours to can one gross of drink B, and it takes 4 hours to label the cans. The company makes $10 profit on one gross of drink A and a $20 profit of one gross of drink B. Given that we have 20 hours to devote to canning the drinks and 15 hours to devote to labeling cans per week, how many cans of each type of drink should the company package to maximize profits?

# Required Submission for Formulation Solution

## Problem Identification

Maximize the profit of selling these new drinks.

## Define Variables

$X_1$ = the number of gross cans produced for drink A per week
$X_2$ = the number of gross cans produced for drink B per week

## Objective Function

$$Z = 10X_1 + 20X_2$$

## Constraints

**1.** Canning with only 20 hours available per week

$$2X_1 + 3X_2 \leq 20$$

**2.** Labeling with only 15 hours available per week

$$X_1 + 4X_2 \leq 15$$

**3.** Nonnegativity restrictions

$$X_1 \geq 0 \text{ (nonnegativity of the production items)}$$
$$X_2 \geq 0 \text{ (nonnegativity of the production items)}$$

## The Complete Formulation

$$\text{Maximize } Z = 10X_1 + 20X_2$$
Subject to
$$2X_1 + 3X_2 \leq 20$$
$$X_1 + 4X_2 \leq 15$$
$$X_1 \geq 0$$
$$X_2 \geq 0$$

We will see in the next section how to solve these two-variable problems graphically.

---

## Example 2    Financial Planning

A bank makes four kinds of loans to its personal customers, and these loans yield the following annual interest rates to the bank:

- First mortgage, 14 percent
- Second mortgage, 20 percent
- Home improvement, 20 percent
- Personal overdraft, 10 percent

The bank has a maximum foreseeable lending capability of $250 million and is further constrained by the following policies:

**1.** First mortgages must be at least 55 percent of all mortgages issued and at least 25 percent of all loans issued (in $ terms).

**2.** Second mortgages cannot exceed 25 percent of all loans issued (in $ terms).

**3.** To avoid public displeasure and the introduction of a new windfall tax, the average interest rate on all loans must not exceed 15 percent.

Formulate the bank's loan problem as an LP so as to maximize interest income while satisfying the policy limitations.

Note here that these policy conditions, while potentially limiting the profit that the bank can make, also limit its exposure to risk in a particular area. It is a fundamental principle of risk reduction that risk is reduced by spreading money (appropriately) across different areas.

## Financial Planning Formulation

Note here that, as in *all* formulation exercises, we are translating a verbal description of the problem into an *equivalent* mathematical description.

A useful tip when formulating LPs is to express the variables, constraints, and objective in words before attempting to express them in mathematics.

## Variables

Essentially, we are interested in the amount (in dollars) the bank has loaned to customers in each of the four different areas (not in the actual number of such loans). Hence, let $x_i$ = amount loaned in area $i$ in million of dollars (where $i = 1$ corresponds to first mortgages, $i = 2$ corresponds to second mortgages, etc.) and note that each $x_i \geq 0$ ($i = 1,2,3,4$).

Note here that it is conventional in LPs to have all variables $\geq 0$. Any variable ($X$, say) that can be positive *or* negative can be written as $X_1 - X_2$ (the difference of two new variables) where $X_1 \geq 0$ and $X_2 \geq 0$.

## Constraints

**1.** Limit on amount lent:

$$x_1 + x_2 + x_3 + x_4 \leq 250$$

**2.** Policy condition 1:

$$x_1 \geq 0.55(x_1 + x_2)$$

that is, first mortgages $> = 0.55$(total mortgage lending) and also

$$x_1 \geq 0.25(x_1 + x_2 + x_3 + x_4)$$

that is, first mortgages $\geq 0.25$(total loans)

**3.** Policy condition 2:

$$x_2 < 0.25(x_1 + x_2 + x_3 + x_4)$$

**4.** Policy condition 3: We know that the total annual interest is $0.14x_1 + 0.20x_2 + 0.20x_3 + 0.10x_4$ on total loans of $(x_1 + x_2 + x_3 + x_4)$. Hence, the constraint relating to policy condition (3) is

$$0.14x_1 + 0.20x_2 + 0.20x_3 + 0.10x_4 \leq 0.15(x_1 + x_2 + x_3 + x_4)$$

## Objective Function

To maximize interest income (which is given above)—that is,

$$\text{Maximize } Z = 0.14x_1 + 0.20x_2 + 0.20x_3 + 0.10x_4$$

## Example 3

## Blending Problem and Formulation

Consider the example of a manufacturer of animal feed who is producing feed mix for dairy cattle. In our simple example, the feed mix contains two active ingredients. One kg of feed mix must contain a minimum quantity of each of the following four nutrients:

| Nutrient | A | B | C | D |
|---|---|---|---|---|
| Grams | 90 | 50 | 20 | 2 |

The ingredients have the following nutrient values and cost:

| | A | B | C | D | Cost/kg |
|---|---|---|---|---|---|
| Ingredient 1 (gram/kg) | 100 | 80 | 40 | 10 | 40 |
| Ingredient 2 (gram/kg) | 200 | 150 | 20 | 0 | 60 |

What should be the amounts of active ingredients in one kg of feed mix that minimizes cost?

## Blending Problem Solution

### Variables

To solve this problem, it is best to think in terms of 1 kilogram of feed mix. That kilogram is made up of two parts, ingredient 1 and ingredient 2:

$$x_1 = \text{amount (kg) of ingredient 1 in 1 kg of feed mix}$$
$$x_2 = \text{amount (kg) of ingredient 2 in 1 kg of feed mix}$$

where $x_1 \geq 0$, $x_2 \geq 0$.

Essentially, these variables ($x_1$ and $x_2$) can be thought of as the recipe telling us how to make up 1 kilogram of feed mix.

### Constraints

- Nutrient constraints

$$100x_1 + 200x_2 \geq 90 \text{ (nutrient A)}$$
$$80x_1 + 150x_2 \geq 50 \text{ (nutrient B)}$$
$$40x_1 + 20x_2 \geq 20 \text{ (nutrient C)}$$
$$10x_1 \geq 2 \text{ (nutrient D)}$$

- Balancing constraint (an *implicit* constraint due to the definition of the variables)

$$x_1 + x_2 = 1$$

### Objective Function

Presumably to minimize cost—that is,

$$\text{Minimize } Z = 40x_1 + 60x_2$$

This gives us our complete LP model for the blending problem.

## Example 4

## Production Planning Problem

A company manufactures four variants of the same table, and in the final part of the manufacturing process there are assembly, polishing, and packing operations. The following table shows the time (in minutes) required for these operations for each variant, as is the profit per unit sold.

|         |   | Assembly | Polish | Pack | Profit (\$) |
|---------|---|----------|--------|------|------------|
| Variant | 1 | 2 | 3 | 2 | 1.50 |
|         | 2 | 4 | 2 | 3 | 2.50 |
|         | 3 | 3 | 3 | 2 | 3.00 |
|         | 4 | 7 | 4 | 5 | 4.50 |

- Given the current state of the labor force, the company estimates that each year it has 100,000 minutes of assembly time, 50,000 minutes of polishing time, and 60,000 minutes of packing time available. How many of each variant should the company make per year, and what is the associated profit?

## Variables

Let:

$$x_i \text{ be the number of units of variant } i \; (i = 1, 2, 3, 4) \text{ made per year}$$

where $x_i \geq 0$, $i = 1, 2, 3, 4$.

## Constraints

Resources for the operations of assembly, polishing, and packing

$$2x_1 + 4x_2 + 3x_3 + 7x_4 \leq 100{,}000 \; (assembly)$$
$$3x_1 + 2x_2 + 3x_3 + 4x_4 \leq 50{,}000 \; (polishing)$$
$$2x_1 + 3x_2 + 2x_3 + 5x_4 \leq 60{,}000 \; (packing)$$

## Objective Function

$$\text{Maximize } Z = 1.5x_1 + 2.5x_2 + 3.0x_3 + 4.5x_4$$

## 7.1 | EXERCISES

Formulate the following scenarios 1–7 as linear programming problems.:

1. A company wants to can two different drinks for the holiday season. It takes 3 hours to can one gross of drink A, and it takes 2 hour to label the cans. It takes 2.5 hours to can one gross of drink B, and it takes 2.5 hours to label the cans. The company makes \$15 profit on one gross of drink A and an \$18 profit of one gross of drink B. Given that we have 40 hours to devote to canning the drinks and 35 hours to devote to labeling cans per week, how many cans of each type drink should the company package to maximize profits?

2. The Mariners Toy Company wishes to make three models of ships to maximize its profits. They found that a model steamship takes the cutter 1 hour, the painter 2 hours, and the assembler 4 hours of work; it produces a profit of \$6. The sailboat takes the cutter 3 hours, the painter 3 hours, and the assembler 2 hours. It produces a \$3 profit.

The submarine takes the cutter 1 hour, the painter 3 hours, and the assembler 1 hour. It produces a profit of $2. The cutter is only available for 45 hours per week, the painter for 50 hours, and the assembler for 60 hours. Assume that they sell all the ships that they make and formulate this LP to determine how many ships of each type that Mariners should produce.

3. To produce 1,000 tons of nonoxidizing steel for BMW engine valves, at least the following units will be needed weekly: 10 units of manganese, 12 units of chromium, and 14 units of molybdenum (1 unit is 10 lbs). These materials are obtained from a dealer who markets these metals in three sizes: small (S), medium (M), and large (L). One S case costs $9 and contains two units of manganese, two units of chromium, and one unit of molybdenum. One M case costs $12 and contains two units of manganese, three units of chromium, and one unit of molybdenum. One L case costs $15 and contains one unit of manganese, one unit of chromium, and five units of molybdenum. How many cases of each kind (S, M, L) should be purchased weekly so that we have enough manganese, chromium, and molybdenum at the smallest cost?

4. The Superbowl Advertising Agency wishes to plan an advertising campaign in three different media—television, radio, and magazines. The purpose or goal is to reach as many potential customers as possible. Results of a marketing study are shown in the following table:

| | Daytime TV | Prime-Time TV | Radio | Magazines |
|---|---|---|---|---|
| Cost of advertising unit | $40,000 | $75,000 | $30,000 | $15,000 |
| Number of potential customers reached per unit | 400,000 | 900,000 | 500,000 | 200,000 |
| Number of woman customers reached per unit | 300,000 | 400,000 | 200,000 | 100,000 |

The company does not want to spend more than $800,000 on advertising. It further requires (1) at least 2 million exposures take place among woman, (2) TV advertising be limited to $500,000, (3) at least three advertising units be bought on daytime TV and 2 units on prime-time TV, and (4) the number of radio and magazine advertisement units should each be between five and 10 units.

5. A tomato cannery has 5,000 pounds of grade A tomatoes and 10,000 pounds of grade B tomatoes, from which the cannery will make whole canned tomatoes and tomato paste. Whole tomatoes must be composed of at least 80 percent grade A tomatoes, where as tomato paste must be made with at least 10 percent grade A tomatoes. Whole tomatoes sell for $0.08 per pound, and grade B tomatoes sell for $0.05 per pound. Maximize revenue of the tomatoes.

   *Hint*: Let $X_{wa}$ = pounds of grade A tomatoes used to whole tomatoes and $X_{wb}$ = pounds of grade B tomatoes used to whole tomatoes. The total number of whole tomato cans produced is the sum of $X_{wa} + X_{wb}$ after each is found. Also, remember a percent is a fraction of the whole times 100 percent.

6. The McCow Butchers company is a large-scale distributor of dressed meats for Myrtle Beach restaurants and hotels. Ryan's orders meat for meat loaf (mixed ground beef, pork, and veal) for 1,000 pounds according to the following specifications:
   a. Ground beef is to be no less than 400 lbs and no more than 600 lbs.
   b. The ground pork is to be between 200 and 300 pounds.
   c. The ground veal must weigh between 100 and 400 lbs.
   d. The weigh of the ground pork must be no more than one and one-half (3/2) times the weight of the veal.

   The contract calls for Ryan's to pay $1,200 for the meat. The cost per pound for the meat is $0.70 for hamburger, $0.60 for pork, and $0.80 for veal. How can this be modeled?

**Table 7.1**

| Bond Name | Bond Type | Moody's Quality Scale | Bank's Quality Scale | Years to Maturity | Yield at Maturity (%) | After-Tax Yield (%) |
|-----------|-----------|------------------------|------------------------|--------------------|------------------------|----------------------|
| A | Municipal | Aa | 2 | 9 | 4.3 | 4.3 |
| B | Agency | Aa | 2 | 15 | 5.4 | 2.7 |
| C | Govt 1 | Aaa | 1 | 4 | 5 | 2.5 |
| D | Govt 2 | Aaa | 1 | 3 | 4.4 | 2.2 |
| E | Local | Ba | 5 | 2 | 4.5 | 4.5 |

**7.** Portfolio Investments. A portfolio manager in charge of a bank wants to invest
$10 million. The securities available for purchase, as well as their respective quality
ratings, maturate, and yields, are shown below in Table 7.1.

  The bank places certain policy limitations on the portfolios manager's actions:

  **a.** Government and agency bonds must total at least $4 million.

  **b.** The average quality of the portfolios cannot exceed 1.4 on the bank's quality scale.
  Note that a low number means high quality.

  **c.** The average years to maturity must not exceed five years.

  Assume the objective is to maximize after-tax earning on the investment.

## 7.2   GRAPHICAL SIMPLEX FOR LINEAR PROGRAMMING PROBLEMS WITH TWO VARIABLES

Many applications in business and economics involve a process called **optimization**. In optimization problems, you are asked to find the minimum or the maximum result. This section illustrates the strategy in graphical simplex of linear programming. We will restrict ourselves in this graphical context to two dimensions. Variables in the simplex method are restricted to positive variables (for example $x \geq 0$).

A two-dimensional linear programming problem consists of a linear *objective function* and a system of linear inequalities called **constraints**. The objective function gives the linear quantity that is to be maximized (or minimized). The constraints determine the set of **feasible solutions**.

### Example 1        Memory Chips for CPUs

Let's start with a manufacturing example. Suppose a small business wants to know how many of two types of high-speed computer chips to manufacturer weekly to maximize its profits. First, we need to define our decision variables. Let

$$x_1 = \text{number of high speed chip type A to produce weekly}$$
$$x_2 = \text{number of high speed chip type B to produce week}$$

The company reports a profit of $140 for each type A chip and $120 for each type B chip sold. The production line reports the following information:

|  | **Chip A** | **Chip B** | **Quantity Available** |
|---|---|---|---|
| Assembly time (hours) | 2 | 4 | 1,400 |
| Installation time (hours) | 4 | 3 | 1,500 |
| Profit ( per unit) | 140 | 120 |  |

The constraint information from the table becomes inequalities that are written mathematically as:

$$2x_1 + 4x_2 \leq 1{,}400 \text{ (assembly time)}$$
$$4x_1 + 3x_2 \leq 1{,}500 \text{ (installation time)}$$
$$x_1 \geq 0, \ x_2 \geq 0$$

The profit equation is:

$$\text{Maximize Profit} = 140x_1 + 120x_2$$

### The Feasible Region

The constraints of a linear program, which include any bounds on the decision variables, essentially shape the region in the $x$–$y$ plane that will be the domain for the objective function before any optimization is performed. Every inequality constraint that is part of the formulation divides the entire space defined by the decision variables into two parts: the portion of the space containing points that violate the constraint and the portion of the space containing points that satisfy the constraint.

It is very easy to determine which portion will contribute to shaping the domain. We can simply substitute the value of some point in either *half-space* into the constraint. Any point will do, but the origin is particularly appealing. Because there is only one origin, if it satisfies the constraint, then the half-space containing the origin will contribute to the domain of the objective function.

When we do this for each of the constraints in the problem, the result is an area representing the intersection of all the half-spaces that satisfied the constraints *individually*. This intersection is the domain for the objective function for the optimization. Because it contains points that satisfy all the constraints *simultaneously*, these points are considered feasible to the problem. Naturally, the common name for this domain is the *feasible region*.

Consider our Example 1 constraints:

$$2x_1 + 4x_2 \leq 1{,}400 \text{ (assembly time)}$$
$$4x_1 + 3x_2 \leq 1{,}500 \text{ (installation time)}$$
$$x_1 \geq 0, \ x_2 \geq 0$$

For our graphical work, we use the constraints $x_1 \geq 0$, $x_2 \geq 0$ to set the region. Here we are strictly in the $x_1 x_2$ plane (the first quadrant).

Let's first take constraint 1 (assembly time) in the first quadrant: $2x_1 + 4x_2 \leq 1{,}400$.

In Figure 7.1, we see the shaded region for constraint 1 that makes the inequality true.

**FIGURE 7.1**
Shaded inequality

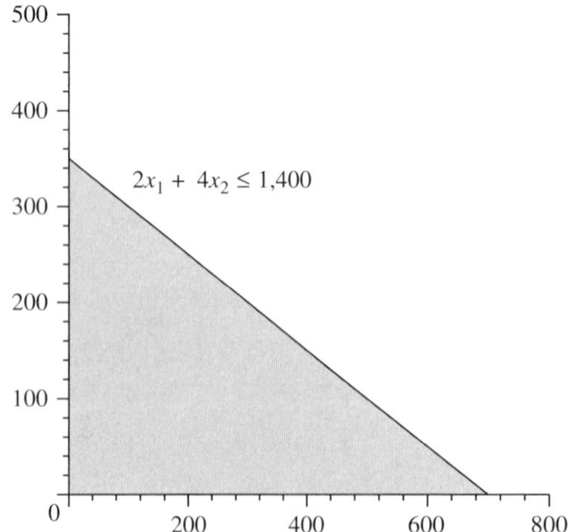

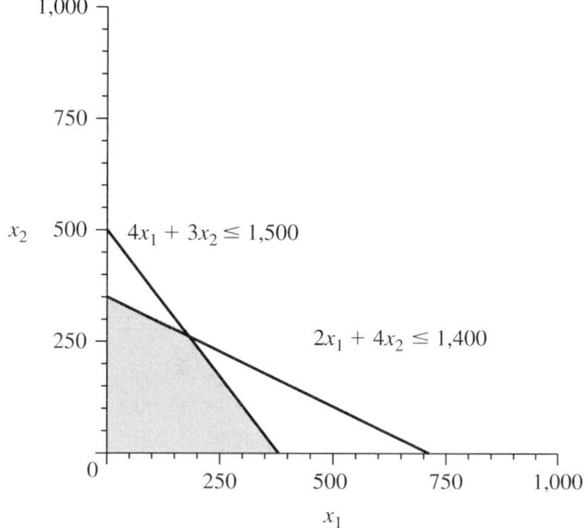


We repeat this process for all the other constraint to obtain the results shown in Figure 7.2.

Figure 7.2 shows a plot of (1) the assembly hours constraint and (2) the installation hours constraint in the first quadrant. Along with the nonnegativity restrictions on the decision variables, the intersection of the half-spaces defined by these constraints is the feasible region shown in yellow. This area represents the domain for the objective function optimization.

<table>
<tr><td>**Example 2**</td><td></td></tr>
</table>

## Finding the Feasible Region

Shade in the feasible region defined by the following set of constraints:

$$x_1 + 2x_2 \leq 20$$
$$2x_1 + x_2 \leq 20$$
$$x_1 \geq 0, x_2 \geq 0$$

## Solution

The feasible region is the set of ordered pairs $(x_1, x_2)$ that satisfy all four constraints simultaneously. They are points that lie below $x_1 + 2x_2 \leq 20$, below $2x_1 + x_2 \leq 20$, and above $x_2 = 0$ and to the right of $x_1 = 0$. We note that the nonnegativity constraints, $x_1 \geq 0$, $x_2 \geq 0$, restrict the feasible region to the first quadrant.

If the problem is well behaved, this should be a closed and bounded polyhedral shape, called a *polyhedron*, such as the one shown in yellow in Figure 7.3. It does not have to be so. Sometimes the orientation and location of the constraints fail to hold back the objective function in the direction of the optimization. When this happens, the problem is *unbounded*; the objective function value goes off to positive or negative infinity. Can you draw a sketch of a situation in which this will happen?

Other times, the intersection of the half-spaces is an empty set. In this case, the problem is *infeasible*; there are no possible solutions that will satisfy the requirements of all the constraints simultaneously. Can you draw a sketch of a situation in which this will happen?

**FIGURE 7.3**
Shaded feasible region

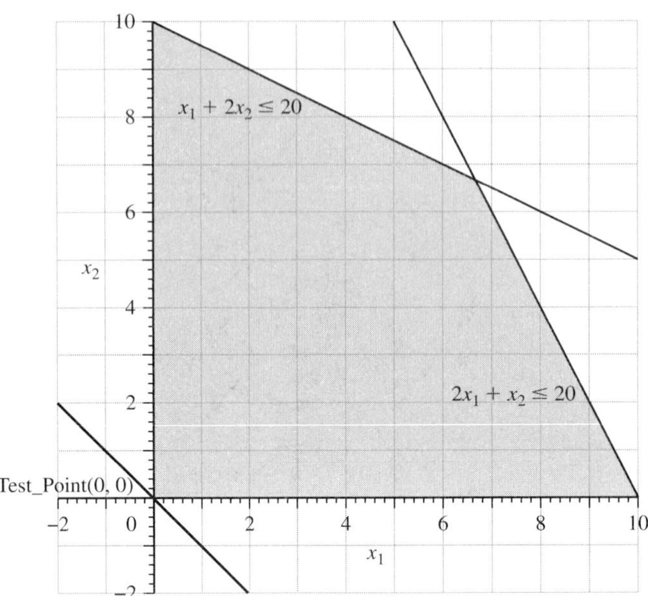


### Solving a Linear Programming Problem Graphically

Recall that we have decision variables defined and an objection function that is to be maximized or minimized. Although all points inside the feasible region provide feasible solutions, the solution, if one exists, occurs according to the fundamental theorem of linear programming:

> If the optimal solution exists, then it occurs at a corner point of the feasible region

Notice the various corners formed by the intersections of our constraints and the constraints with each axis and the axis themselves in our example above. These points are of great importance to us. There is a cool theorem (didn't know there were any of these, huh?) in linear optimization that states, "If an optimal solution exists, then an optimal corner point exists." The result of this is that any algorithm searching for the optimal solution to a linear program should have some mechanism of heading toward the corner point where the solution will occur. If the search procedure stays on the outside border of the feasible region while pursuing the optimal solution, it is called an *exterior point* method. If the search procedure cuts through the inside of the feasible region, it is called an *interior point* method.

Thus, in a linear programming problem, if there exists a solution, it must occur at a corner point of the set of feasible solutions (these are the vertices of the region). Note that in Figure 7.3 the corner points of the feasible region are the coordinates: (0,0), (0,10) (10, 0), and (20/3, 20/3).

How did we get the point (20/3, 20/3)?

This point is the intersection of the lines: $x_1 + 2x_2 = 20$ and $2x_1 + x_2 = 20$. You have solved these problems before. In the second constraint, let $x_2 = 20 - 2x_1$ and substitute $(20 - 2x_1)$ for $x_2$ in the first equation, so $x_1 + 2(20 - 2x_1) = 20$ and solve for $x_1$. We find $3x_1 = 20$ or $x_1 = 20/3$. Because $x_2 = 20 - 2x_1$, we substitute $x_1 = 20/3$ for $x_1$ and solve for $x_2$. Now, $x_2 = 20/3$.

Now that we have all the possible solution coordinates for $(x_1, x_2)$, we need to know which is the **optimal solution**. Here is how we determine that:

We evaluate the objective function at each point and choose the best solution.

Assume our objective function is to maximize $Z = 2x_1 + 2x_2$. We can set up a table of coordinates and corresponding Z-values as follows.

**FIGURE 7.4**
ISO-profit lines added

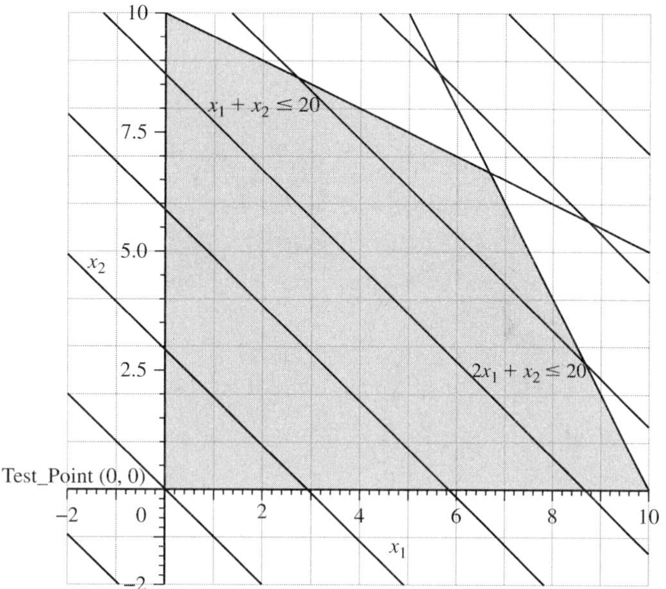


| Coordinate of Corner Point | $Z = 2x_1 + 2x_2$ |
| --- | --- |
| (0,0) | $Z = 0$ |
| (0,10) | $Z = (2)(0) + (2)(10) = 20$ |
| (20/3, 20/3) | $Z = (2)(20/3) + (2)(20/3) = 80/3*$ |
| (10,0) | $Z = (2)(10) + (2)(0) = 20$ |
| **Best solution is (20/3, 20/3)** | $Z = 80/3 = 26.666$ |

Graphically, we see the result by plotting the objective function line, $Z = 2x_1 + 2x_2$, with the feasible region. Determine the parallel direction for the line to maximize (in this case) $Z$.

Move the line parallel until it crosses the last point in the feasible set. That point is the solution. The line that goes through the origin at a slope of $-2/2$ is called the *ISO-profit line*. We have provided this in Figure 7.4.

Here are the steps for solving a linear programming problem involving only two variables.

1. Sketch the region corresponding to the system of constraints. The points satisfying all constraints make up the feasible solution.

2. Find all the corner points (or intersection points in the feasible region).

3. Test the objective function at each corner point and select the values of the variables that optimize the objective function. For bounded regions, both a maximum and a minimum will exist. For an unbounded region, if a solution exists, it will exist at a corner.

## Example 3　Minimization Problem

$$\text{Minimize } Z = 5x + 7y$$
$$\text{Subject to } 2x + 3y \geq 6$$
$$3x - y \leq 15$$
$$-x + y \leq 4$$
$$2x + 5y \leq 27$$
$$x \geq 0$$
$$y \geq 0$$

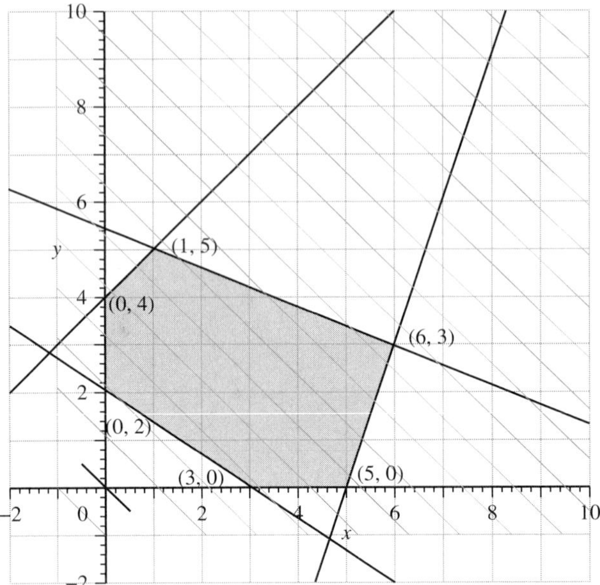



The corner points shown in Figure 7.5 are (0,2), (0,4,) (1,5), (6,3), (5,0), and (3,0). See if you can solve and find all of these corner points.

If we evaluate $Z = 5x + 7y$ at each of these points, we find the following:

| Corner Point | $Z = 5x + 7y$ **(MINIMIZE)** |
|---|---|
| (0,2) | $Z = 14$ |
| (1,5) | $Z = 40$ |
| (6,3) | $Z = 51$ |
| (5,0) | $Z = 25$ |
| (3,0) | $Z = 15$ |
| (0,4) | $Z = 28$ |

The minimum value occurs at (0,2) with a $Z$ value of 14. Notice in our graph that the blue ISO-profit line will last cross the point (0,2) as it moves out of the feasible region in the direction that minimizes $Z$.

## Example 4    Unbounded Feasible Region

Let's examine the concept of an unbounded feasible region as in Figure 7.6. Look at the constraints:

$$x + 2y \geq 4$$
$$3x + y \geq 7$$
$$x \geq 0 \text{ and } y \geq 0$$

Note that the corner points are (0,7), (2,1) and (4,0) and that the region is unbounded. If our solution is to minimize $Z = x + y$, then our solution is (2,1) with $Z = 3$.

Determine why there is no solution to the LP to Maximize $Z = x + y$.

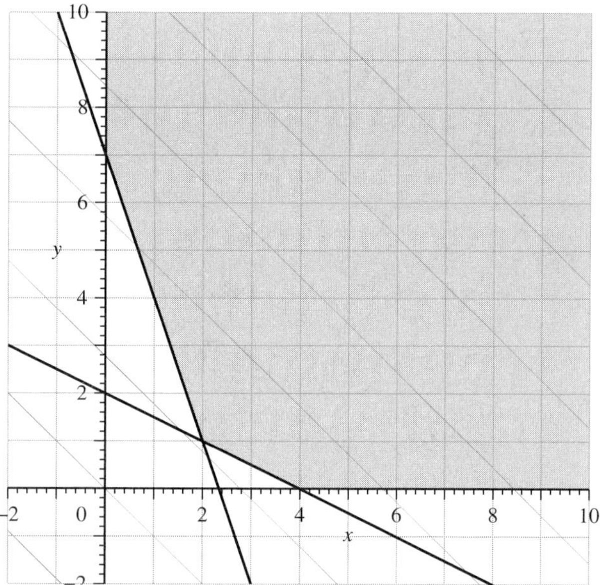

**FIGURE 7.6**
Unbounded feasible region

---

## 7.2 | EXERCISES

Find the maximum and minimum solution. Assume we have $x \geq 0$ and $y \geq 0$ for each problem.

1. $Z = 2x + 3y$

   subject to:
   $2x + 3y \geq 6$
   $3x - y \leq 15$
   $-x + y \leq 4$
   $2x + 5y \leq 27$

2. $Z = 6x + 4y$

   subject to:
   $-x + y \leq 12$
   $x + y \leq 24$
   $2x + 5y \leq 80$

3. $Z = 6x + 5y$

   subject to:
   $x + y \geq 6$
   $2x + y \geq 9$

4. $Z = x - y$

   subject to:
   $x + y \geq 6$
   $2x + y \geq 9$

5. $Z = 5x + 3y$

   subject to:
   $1.2x + 0.6y \leq 24$
   $2x + 1.5y \leq 80$

# 7.2 | PROJECTS

1. With the rising cost of gasoline and increasing prices to consumers, the use of additives to enhance gasoline performance is being considered. Consider two additives: additive 1 and additive 2. The following conditions must hold for their use:
   - Harmful carburetor deposits must not exceed 1/2 lb for each car's gasoline tank.
   - The quantity of additive 2 plus twice the quantity of additive 1 must be at least 1/2 lb for each car's gasoline tank.
   - One pound of additive 1 will add 10 octane units per tank, and 1 lb of additive 2 will add 20 octane units per tank. The total number of octane units added must not be less than six (6).
   - Additives are expensive: \$1.53/lb for additive 1 and \$4/lb for additive 2.

   We want to determine the quantity of each additive that will meet the above restrictions and will minimize their cost while meeting the following requirements.
   a. List the decision variables and define them.
   b. List the objective function.
   c. List the resources that constrain this problem.
   d. Graph the feasible region.
   e. Label all intersection points of the feasible region.
   f. Plot the objective function in a different color (highlight the objective function line, if necessary) and label it the ISO-profit line.
   g. Clearly indicate the point that is the optimal solution on the graph.
   h. List the coordinates of the optimal solution and the value of the objective function.
   i. Assume now that a manufacturer of additives has the opportunity to sell you a nice special TV deal to deliver at least 0.5 lb of additive 1 and at least 0.3 lb of additive 2. Use graphical LP methods to help determine whether you should buy this TV offer. Support your recommendation.
   j. Write a one-page cover letter to the boss of your company that summarizes the results that you found.

2. A farmer has 30 acres on which to grow tomatoes and corn. Each 100 bushels of tomatoes require 1,000 gallons of water and 5 acres of land. Each 100 bushels of corn require 6,000 gallons of water and 2½ acres of land. Labor costs are \$1 per bushel for both corn and tomatoes. The farmer has available 30,000 gallons of water and \$750 in capital. He knows that he cannot sell more than 500 bushels of tomatoes or 475 bushels of corn. He estimates a profit of \$2 on each bushel of tomatoes and \$3 for each bushel of corn. How many bushels of each should he raise to maximize profits? Meet the following requirements:
   a. List the decision variables and define them.
   b. List the objective function.
   c. List the resources that constrain this problem.
   d. Graph the feasible region.
   e. Label all intersection points of the feasible region.
   f. Plot the objective function in a different color (highlight the objective function line, if necessary) and label it the ISO-profit line.
   g. Clearly indicate on the graph the point that is the optimal solution.
   h. List the coordinates of the optimal solution and the value of the objective function.
   i. Assume now that farmer has the opportunity to sign a nice contract with a grocery store to grow and deliver at least 300 bushels of tomatoes and at least 500 bushels of corn. Use graphical LP methods to help recommend a decision to the farmer. Support your recommendation.
   j. If the farmer can obtain an additional 10,000 gallons of water for a total cost of \$50, is it worth it to obtain the additional water? Determine the new optimal solution caused by adding this level of resource.
   k. Write a one-page cover letter to your boss that summarizes the result that you found.

3. Firestone Tires, headquartered in Akron, Ohio, has a plant in Florence, South Carolina, that manufactures two types of tires: SUV 225 radials and SUV 205 radials. Demand is high because of the recent recall of tires. Each 100 lot quantity of SUV 225 radials requires 100 gallons of synthetic plastic and 5 lb of rubber. Each 100-lot quantity of SUV 205 radials require 60 gallons synthetic plastic and 2½ lb of rubber. Labor costs are $1 per tire for each type tire. The manufacturer has weekly quantities available of 660 gallons of synthetic plastic, $750 in capital, and 300 lbs. of rubber. The company estimates a profit of $3 on each SUV 225 radial and $2 of each SUV 205 radial. How many of each type of tire should the company manufacture to maximize its profits?

**Required:**

a. List the decision variables and define them.
b. List the objective function.
c. List the resources that constrain this problem.
d. Graph the feasible region.
e. Label all intersection points of the feasible region.
f. Plot the objective function in a different color (highlight the objective function line, if necessary) and label it the ISO-profit line.
g. Clearly indicate on the graph the point that is the optimal solution.
h. List the coordinates of the optimal solution and the value of the objective function.
i. Assume now that manufacturer has the opportunity to sign a nice contract with a tire outlet store to deliver at least 500 SUV 225 radial tires and at least 300 SUV 205 radial tires. Use graphical LP methods to help recommend a decision to the manufacturer. Support your recommendation.
j. If the manufacturer can obtain an additional 1,000 gallons of synthetic plastic for a total cost of $50, is it worth it to obtain this amount? Determine the new optimal solution caused by adding this level of resource.
k. If the manufacturer can obtain an additional 20 lb of rubber for $50, should it obtain the rubber? Determine the new solution caused by adding this amount.
l. Write a one-page cover letter to your boss of the company that summarizes the results that you found.

4. Consider a toy maker who carves wooden soldiers. The company specializes in two types: Confederate soldiers and Union soldiers. The estimated profit for each is $28 and $30, respectively. A Confederate soldier requires 2 units of lumber, 4 hours of carpentry, and 2 hours of finishing to complete. A Union soldier requires 3 units of lumber, 3.5 hours of carpentry, and 3 hours of finishing to complete. Each week, the company has 100 units of lumber delivered. The workers can provide at most 120 hours of carpentry and 90 hours of finishing. Determine the number of each type of wooden soldiers to produce to maximize weekly profits. Formulate and then solve this linear programming graphically.

<br>



# GRAPHICAL SENSITIVITY ANALYSIS

One of the most important topics in linear programming is sensitivity analysis. In this section, we illustrate the concept of sensitivity analysis through a graphical example. Sensitivity analysis is concerned with how changes in the parameters (coefficient and right-hand-side values) affect the LP's optimal solution. Very often, we can ascertain whether the optimal solution variables remain the same (perhaps with different solution values) or whether the variables will change.

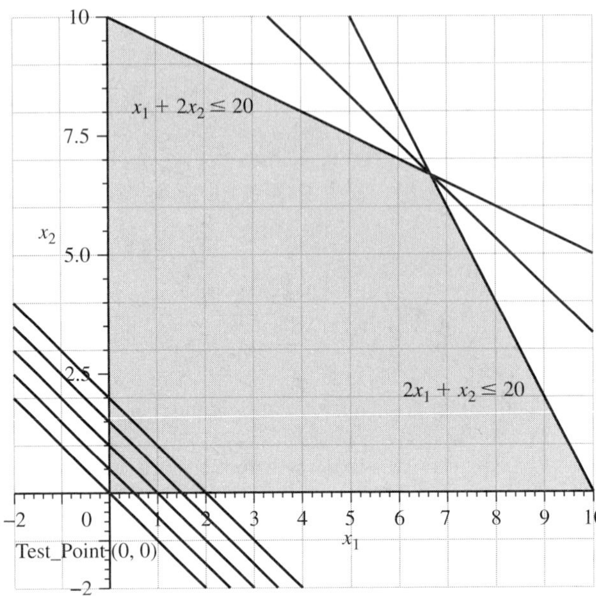



Reconsider the example from Section 7.2.

$$\text{Maximize } 2x_1 + 2x_2$$
$$\text{Subject to:}$$
$$x_1 + 2x_2 \leq 20$$
$$2x_1 + x_2 \leq 20$$
$$x_1 \geq 0, \, x_2 \geq 0$$

The feasible region and objective function contours are shown in Figure 7.7. Note that we have shown the last contour in the direction of increase and that it passes through the corner point (20/3, 20/3), making it in the optimal solution. The corners of the feasible region with the respective objective function values are shown in the following table.

| Point | Coordinates | Z-Value |
|---|---|---|
| A | (0,0) | $Z = 0$ |
| B | (10,0) | $Z = 20$ |
| C | (0,10) | $Z = 20$ |
| D | (20/3,20/3) | $Z = 80/3 = 26.66666$ Optimal |

The optimal solution is currently $x_1 = 20/3$, $x_2 = 20/3$, and $Z = 80/3$.

### Graphical Analysis of the Effect of an Objective Function Coefficient

Recall our objective function:

$$Z = 2x_1 + 2x_2$$

The slope of this objective function line is $-1$. The solution is currently at point D, the coordinates (20/3,20/3). Recall that the optimal solution, if it exists, must lie at the corner point of the feasible region. The next closest points to D are B and C. B lies on constraint (2), and C lies on constraints (1). The slope of constraint (1) is –1/2, and the slope of constraint (2) is –2.

You can see that if we change the coefficients—let's say, $Ax_1 + 2x_2$—its slope is –A/2. Currently, A is 2 with a slope of –1. The bounds for the slope to retain point D as a solution is between $-2 \leq$ slope $< -1/2$ (note that at equality we will have alternate optimal solutions). We can easily solve for the values of A that keep the slope within its bounds: $1 \leq A \leq 4$. Because A is currently 2, it can decrease by one unit or increase by two units and still keep point D optimal.

Let's try $2x_1 + bx_2$ with its slope $-2/b$. Currently, B is 2 with a slope of –1. The bounds for the slope to retain point D as a solution is between $-2 \leq$ slope $\leq -1/2$ (note that at equality we will have alternate optimal solutions). We can easily solve for the values of B that keep the slope within its bounds: $1 \leq B \leq 4$. Because B is currently 2, it can decrease by one unit or increase by two units and still keep point D optimal.

### Graphical Analysis of the Effect of a Change in a Right-Hand-Side Coefficient

The right-hand-side value for each constraint controls the $y$-intercept of the problem. The slopes will remain the same. Changing the right-hand side yields a parallel line to the original constraint. Point D is the intersection of constraints (1) and (2). Points B and C are the intersections of constraints (1) and (2), each with nonnegativity.

Let's consider constraint (1): $x_1 + 2x_2 \leq$ B.

The value of B is currently 20. If it is reduced, then the line moves down until it intersects point (10,0). Using (10,0) in the equation yields a B value of 10. If we increase B, then we move up to an original infeasible point (0,20), which will now become feasible. Using (0,20) in the equation yields a B value of 40.

Thus, $10 \leq B \leq 40$. This is a decrease of 10 units and an increase of 20 units from its original value. We see this displayed in Figures 7.8 for the decrease and 7.9 for the increase.

Let's consider constraint (2): $2x_1 + x_2 \leq$ C.

The value of C is currently 20. If it is reduced, then the line moves down until it intersects point (0,10). Using (0,10) in the equation yields a C value of 10. If we increase C, then we move up to an original infeasible point (20,0), which will now become feasible. Using (20,0) in the equation yields a C value of 40.

Thus, $10 \leq C \leq 40$. This is a decrease of 10 units and an increase of 20 units.

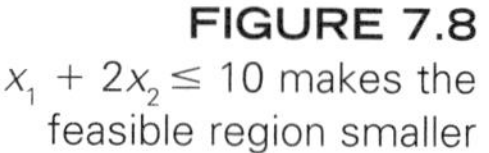

**FIGURE 7.8**
$x_1 + 2x_2 \leq 10$ makes the feasible region smaller

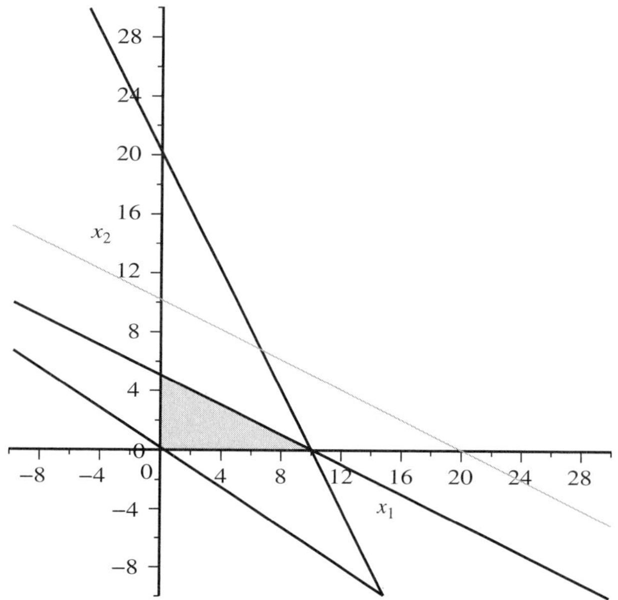

**FIGURE 7.9**
$x_1 + 2 \cdot x_2 \le 30$ makes the feasible region larger

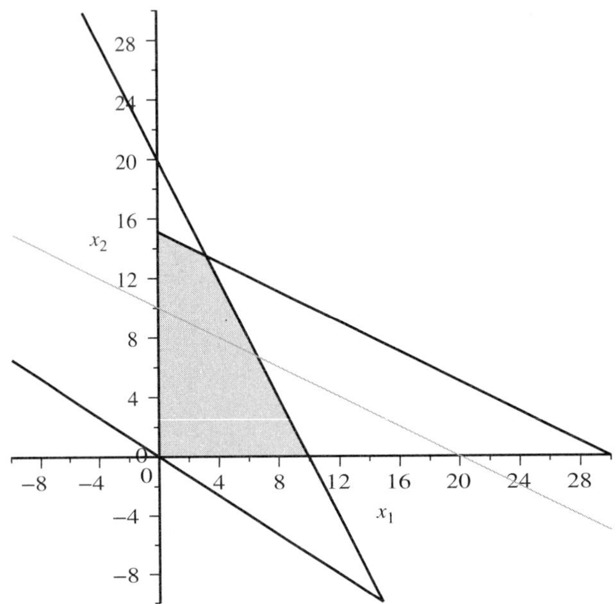

---

| **Example 1** | ## Manufacturing Problem Sensitivity Analysis |

$$\text{Maximize } Z = 20x_1 + 30x_2$$
Subject to
$$2x_1 + 4x_2 \le 1{,}400 \text{ (assembly time)}$$
$$4x_1 + 3x_2 \le 1{,}500 \text{ (installation time)}$$

The solution is (180,260), $Z = 11{,}400$ as seen in Figure 7.10

Let's see the effects of changing the coefficients of the objective function because management has some leeway with the revenues and costs:

$$Z = 20x_1 + 30x_2$$

**FIGURE 7.10**
Manufacturing example

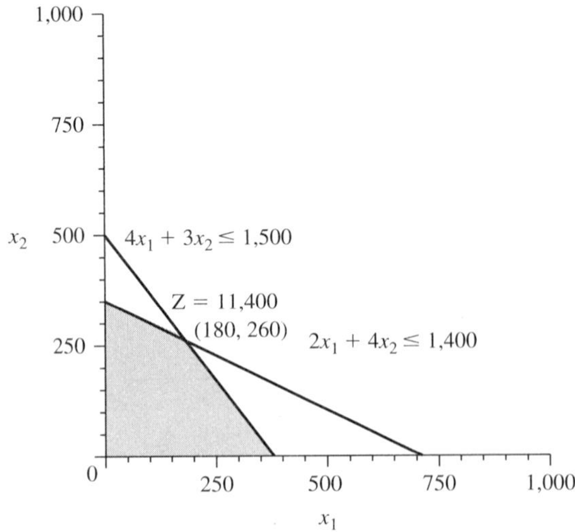

The slope of this objective function line is $-2/3$. The solution is currently at point D, the coordinates (180,260). Recall that the optimal solution, if it exists, must lie at the corner point of the feasible region. The next closest points to D are B and C. B lies on constraint (2), and C lies on constraints (1). The slope of constraint (1) is $-1/2$, and the slope of constraint (2) is $-4/3$.

You can see that if we change the coefficients—let's say, $Ax_1 + 30x_2$—its slope is $-A/30$. Currently, A is 20 with a slope of $-2/3$. The bounds for the slope to retain point D has a solution between $-4/3 \le$ slope $\le -1/2$. We can easily solve for the values of A that keep the slope within its bounds: $15 \le A \le 40$. Because A is currently 20, it can decrease by five units or increase by 20 units and still keep point D optimal.

Now, let's try changing the coefficient of the other variable: $20x + By$ with its slope $-20/B$. Currently, B is 30 with a slope of $-2/3$. The bounds for the slope to retain point D as a solution is between $-4/3 \le$ slope $\le -1/2$ We can easily solve for the values of B that keep the slope within its bounds.: $15 \le B \le 40$. Because B is currently 30, it can decrease by 15 units or increase by 10 units and still keep point D optimal.

---

## 7.4    THE SIMPLEX METHOD

In the previous sections we discussed formulating linear programming problems, solving two-dimensional linear programming problems by graphical methods, and graphical sensitivity analysis. The graphical method illustrates some key concepts but is only practical for problems with two variables. As you see, linear programming problems often have more than two variables. With problems with more than two variables, an algebraic method may be used. This method is called the **simplex method**. The method was developed by George Dantzig in 1947 and incorporates both *optimality* and *feasibility* tests to find the optimal solution(s) to a linear program (if one exists).

An **optimality test** shows whether or not an intersection point corresponds to a value of the objective function better than the best value found so far.

A **feasibility test** determines whether the proposed intersection point is feasible. It does not violate any of the constraints.

The simplex method starts with the selection of a corner point (usually the origin if it is a feasible point) and then, in a systematic method, moves to adjacent corner points of the feasible region until the optimal solution is found or it can be shown that no solution exists.

### Steps of the Simplex Method

**Step 1: Tableau Format**  Place the linear program in *tableau format*, as explained below.

$$\text{Maximize } Z = 25x_1 + 30x_2$$
$$\text{Subject to}$$
$$20x_1 + 30x_2 \le 690$$
$$5x_1 + 4x_2 \le 120$$
$$x_1, x_2, \ge 0$$

To begin the simplex method, we start by converting the inequality constraints (of the form $\le$) to equality constraints. This is accomplished by adding a unique, nonnegative variable, called a *slack variable*, to each constraint. For example, the inequality constraint $20x_1 + 30x_2 \le 690$ is converted to an equality constraint by adding the slack variable $S_1$ to obtain $20x_1 + 30x_2 + S_1 = 690$, where $S_1 \ge 0$.

The inequality $20x_1 + 30x_2 \le 690$ states that the sum $20x_1 + 30x_2$ is less than or equal to 690. The slack variable "takes up the slack" between the values used for $x_1$ and $x_2$ and the value 690. For example, if $x_1 = x_2 = 0$, the $S_1 = 690$. If $x_1 = 24$, $x_2 = 0$, then $20(24) + 30(0) + S_1 = 690$, so $S_1 = 210$.

A unique slack variable must be added to each inequality constraint.

$$\text{Maximize } Z = 25x_1 + 30x_2$$
$$\text{Subject to}$$
$$20x_1 + 30x_2 + S_1 = 690$$
$$5x_1 + 4x_2 + S_2 = 120$$
$$x_1 \geq 0,\ x_2 \geq 0,\ S_1 \geq 0,\ S_2 \geq 0$$

Adding slack variables makes the constraint set a system of linear equations. We write these with all variables on the left side of the equation and all constants on the right-hand side.

We will even rewrite the objective function by moving all variables to the left-hand side. Maximize $Z = 25x_1 + 30x_2$ is written as

$$Z - 25x_1 - 30x_2 = 0$$

Now these can be written in the following form:

$$Z - 25x_1 - 30x_2 = 0$$
$$20x_1 + 30x_2 + S_1 = 690$$
$$5x_1 + 4x_2 + S_2 = 120$$

or more simply in a matrix. This matrix is called the *simplex tableau.*

| $Z$ | $x_1$ | $x_2$ | $S_1$ | $S_2$ | | RHS |
|---|---|---|---|---|---|---|
| 1 | $-25$ | $-30$ | 0 | 0 | $=$ | 0 |
| 0 | 20 | 30 | 1 | 0 | $=$ | 690 |
| 0 | 5 | 4 | 0 | 1 | $=$ | 120 |

**Step 2: Initial Extreme Point**  The simplex method begins with a known extreme point, usually the origin $(0,0)$ for many of our examples. The requirement for a basic feasible solution gives rises to special simplex methods such as Big M and two-phase simplex, which can be studied in a linear programming course.

The tableau previously shown contains the corner point $(0, 0)$ as our initial solution.

| $Z$ | $x_1$ | $x_2$ | $S_1$ | $S_2$ | | RHS |
|---|---|---|---|---|---|---|
| 1 | $-25$ | $-30$ | 0 | 0 | $=$ | 0 |
| 0 | 20 | 30 | 1 | 0 | $=$ | 690 |
| 0 | 5 | 4 | 0 | 1 | $=$ | 120 |

The basic variables have value and are indicated by columns with an identity matrix column below the variable. We also know that all non-basic variables have value zero. Thus, we can read our solution as follows:

$$x_1 = 0$$
$$x_2 = 0$$
$$S_1 = 690$$
$$S_2 = 120$$
$$Z = 0$$

Let's continue to define a few of these variables further. We have five variables $\{Z, x_1, x_2, S_1, S_2\}$ and three equations. We can have at most three solutions. $Z$ will always be a solution by convention of our tableau. We have two nonzero variables among $\{x_1, x_2, S_1, S_2\}$. These non-zero variables are called the **basic variables**. The remaining variables are called the **nonbasic variables**. The corresponding solutions are called the **basic feasible solutions (BFSs)** and correspond to corner points. The complete step of the simplex method produces a solution that corresponds to a corner point of the feasible region. These solutions are read directly from the tableau matrix.

We also note the basic variables are variables that have a column consisting of a single 1; the rest are zeros in their column. We will add a column to label these as shown in the following table.

| Basic Variables | $Z$ | $x_1$ | $x_2$ | $S_1$ | $S_2$ | | RHS |
|---|---|---|---|---|---|---|---|
| $Z$ | 1 | $-25$ | $-30$ | 0 | 0 | $=$ | 0 |
| $S_1$ | 0 | 20 | 30 | 1 | 0 | $=$ | 690 |
| $S_2$ | 0 | 5 | 4 | 0 | 1 | $=$ | 120 |

**Step 3: Optimality Test**  We need to determine if an adjacent intersection point improves the value of the objective function. If not, the current extreme point is optimal. If an improvement is possible, the optimality test determines which variable currently in the independent set (having value zero) should *enter* the dependent set as a basic variable and become nonzero. For our maximization problem, we look at the $Z$-row (the row marked by the basic variable $Z$). If any coefficients in that row are negative, then we select the variable with the most negative coefficient as the entering variable.

| Basic Variables | $Z$ | $x_1$ | $x_2$ | $S_1$ | $S_2$ | | RHS |
|---|---|---|---|---|---|---|---|
| $Z$ | 1 | $-25$ | $-30$ | 0 | 0 | $=$ | 0 |
| $S_1$ | 0 | 20 | 30 | 1 | 0 | $=$ | 690 |
| $S_2$ | 0 | 5 | 4 | 0 | 1 | $=$ | 120 |

In the $Z$-row, the coefficients are:

| Basic Variables | $Z$ | $x_1$ | $x_2$ | $S_1$ | $S_2$ |
|---|---|---|---|---|---|
| $Z$ | 1 | $-25$ | $-30$ | 0 | 0 |

The variable with the most negative coefficient is $x_2$ with value $-30$. Thus, $x_2$ wants to become a basic variable. We can only have three basic variables in this example (because we have three equations), so one of the current basic variables $\{S_1, S_2\}$ must be replaced by $x_2$.

Let's proceed to see how we determine which variable exits being a basic variable.

**Step 4: Feasibility Test**  To find a new intersection point, one of the variables in the basic variable set must *exit* to allow the entering variable from step 3 to become basic. The feasibility test determines which current dependent variable to choose for exiting, ensuring that we stay inside the feasible region. We will use the minimum positive ratio test as our feasibility test. The minimum positive ratio test is the $\text{MIN}(\text{RHS}_j / a_j \geq 0)$. Make a quotient of the $\frac{rhs_j}{a_j}$.

| | | | Most Negative Coefficient $(-30)$ | | | | | |
|---|---|---|---|---|---|---|---|---|
| Basic Variables | $Z$ | $x_1$ | $x_2$ | $S_1$ | $S_2$ | | RHS | Ratio Test Quotient |
|---|---|---|---|---|---|---|---|---|
| $Z$ | 1 | $-25$ | $-30$ | 0 | 0 | $=$ | 0 | |
| $S_1$ | 0 | 20 | 30 | 1 | 0 | $=$ | 690 | $690/30 = 23$ |
| $S_2$ | 0 | 5 | 4 | 0 | 1 | $=$ | 120 | $120/4 = 30$ |

Note that we will always disregard all quotients with either 0 or negative values in the denominator. In our example, we compare {23,30} and select the smallest nonnegative value. This gives the location of the matrix pivot that we will perform.

**Most Negative Coefficient ($-30$)**

| Basic Variables | $Z$ | $x_1$ | $x_2$ | $S_1$ | $S_2$ | | RHS | Ratio Test<br>Quotient |
|---|---|---|---|---|---|---|---|---|
| $Z$ | 1 | $-25$ | $-30$ | 0 | 0 | = | 0 | |
| $S_1$ | 0 | 20 | 30 Pivot | 1 | 0 | = | 690 | $690/30 = 23$ |
| $S_2$ | 0 | 5 | 4 | 0 | 1 | = | 120 | $120/4 = 30$ |

**Step 5: Pivot**  We can form a new equivalent system by using row operations to change the pivot element to a 1 and all other numbers in the pivot column to zero. We do the row operations by adding a suitable multiple of the pivot row to a multiple of each row in the tableau, thus eliminating the new basic variable. We then set the new nonbasic variables to zero in the new system to find the values of the new basic variables, thereby determining an intersection point.

**Pivot Column**

| Basic Variables | $Z$ | $x_1$ | $x_2$ | $S_1$ | $S_2$ | | RHS | |
|---|---|---|---|---|---|---|---|---|
| $Z$ | 1 | $-25$ | $-30$ | 0 | 0 | = | 0 | |
| $S_1$ | 0/30 | 20/30 | 30/30 | 1/30 | 0/30 | = | 690/30 | Pivot Row |
| $S_2$ | 0 | 5 | 4 | 0 | 1 | = | 120 | |

Make the entry in the intersection of the pivot column and pivot row equal to 1.

**Pivot Column**

| Basic Variables | $Z$ | $x_1$ | $x_2$ | $S_1$ | $S_2$ | | RHS | |
|---|---|---|---|---|---|---|---|---|
| $Z$ | 1 | $-25$ | $-30$ | 0 | 0 | = | 0 | |
| $S_1$ | 0 | 2/3 | 1 | 1/30 | 0 | = | 23 | Pivot Row |
| $S_2$ | 0 | 5 | 4 | 0 | 1 | = | 120 | |

Using row operations make all other entries in the pivot column equal to 0.

**Pivot Column**

| Basic Variables | $Z$ | $x_1$ | $x_2$ | $S_1$ | $S_2$ | | RHS | |
|---|---|---|---|---|---|---|---|---|
| $Z$ | 1 | $-5$ | 0 | 1 | 0 | = | 690 | $30R2 + R1 \rightarrow R1$ |
| $x_2$ | 0 | 2/3 | 1 | 1/30 | 0 | = | 23 | |
| $S_2$ | 0 | 7/3 | 0 | $-4/30$ | 1 | = | 28 | $-4R2 + R3$<br>$\rightarrow R3$ |

Let's interpret our current basic feasible solution.

$$\text{Basic variables}$$
$$x_2 = 23$$
$$S_2 = 28$$
$$Z = 690$$
$$\text{Nonbasic variables: } x_1 = 0, \ S_1 = 0$$

**Step 6: Repeat Steps 3–5**  Repeat the steps until an **optimal extreme point** is found.

We note that $x_1$ has a coefficient of $-5$ in the Z-Row. Therefore, we are not optimal.

**Step 3**

|  |  |  |  |  | Negative Coefficient |  |  |
|---|---|---|---|---|---|---|---|
| **Basic Variables** | **Z** | $x_1$ | $x_2$ | $S_1$ | $S_2$ |  | **RHS** |
| Z | 1 | $-5$ | 0 | 1 | 0 | = | 690 |
| $x_2$ | 0 | 2/3 | 1 | 1/30 | 0 | = | 23 |
| $S_2$ | 0 | 7/3 | 0 | $-4/30$ | 1 | = | 28 |

**Step 4**

|  |  |  | Pivot Column |  |  |  |  |  |
|---|---|---|---|---|---|---|---|---|
| **Basic Variables** | **Z** | $x_1$ | $x_2$ | $S_1$ | $S_2$ |  | **RHS** | **Ratio Test Quotient** |
| Z | 1 | $-5$ | 0 | 1 | 0 | = | 690 |  |
| $x_2$ | 0 | 2/3 | 1 | 1/30 | 0 | = | 23 | $69/2 = 34.5$ |
| $S_2$ | 0 | 7/3 | 0 | $-4/30$ | 1 | = | 28 | $84/7 = 12^*$ Min |

The minimum nonnegative quotient is 12. This indicates that to remain in the feasible region, $x_1$ enters as a basic variable, and $S_2$ leaves as a basic variable.

|  |  |  | Pivot Column |  |  |  |  |  |
|---|---|---|---|---|---|---|---|---|
| **Basic Variables** | **Z** | $x_1$ | $x_2$ | $S_1$ | $S_2$ |  | **RHS** | **Ratio Test Quotient** |
| Z | 1 | $-5$ | 0 | 1 | 0 | = | 690 |  |
| $x_2$ | 0 | 2/3 | 1 | 1/30 | 0 | = | 23 |  |
| $S_2$ | 0 | 7/3 | 0 | $-4/30$ | 1 | = | 28 | Pivot Row |

**Step 5**

We make the highlighted position a 1 and all other column entries 0 for the column of $x_1$. We divide the entire $S_2$ row by 7/3.

| **Basic Variables** | **Z** | $x_1$ | $x_2$ | $S_1$ | $S_2$ |  | **RHS** |
|---|---|---|---|---|---|---|---|
| Z | 1 | $-5$ | 0 | 1 | 0 | = | 690 |
| $x_2$ | 0 | 2/3 | 1 | 1/30 | 0 | = | 23 |
| $S_2$ | 0 | 1 | 0 | $-12/210$ | 3/7 | = | 12 |

| **Basic Variables** | **Z** | $x_1$ | $x_2$ | $S_1$ | $S_2$ |  | **RHS** |  |
|---|---|---|---|---|---|---|---|---|
| Z | 1 | 0 | 0 | 5/7 | 15/7 | = | 750 | $5R3 + R1 \rightarrow R1$ |
| $x_2$ | 0 | 0 | 1 | 1/14 | $-2/7$ | = | 15 | $-2/3R3 + R2 \rightarrow R2$ |
| $x_1$ | 0 | 1 | 0 | $-2/35$ | 3/7 | = | 12 |  |

The current solution is read as follows:

$$\text{Basic variables}$$
$$x_2 = 15$$
$$x_1 = 12$$
$$Z = 750$$
$$\text{Nonbasic variables}$$
$$S_1 = S_2 = 0$$

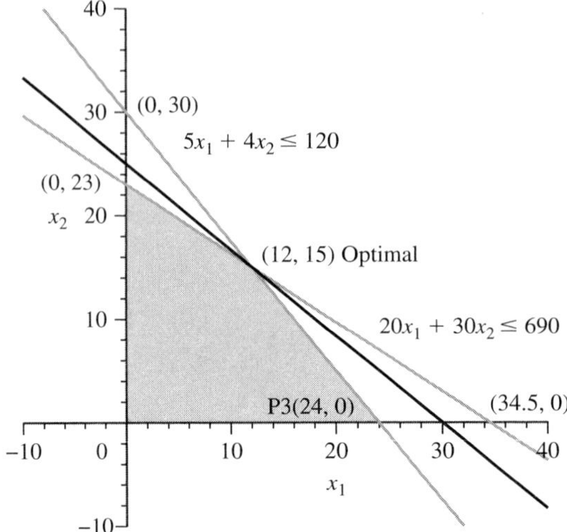



**FIGURE 7.11**
The set of points satisfying the constraint of this linear programming problem (the convex set as a shaded region)

There are no negative coefficients in the $Z$-row, so we are optimal, as we see in our graphical simplex (see Figure 7.11).

|   | $Z$ | $x_1$ | $x_2$ | $S_1$ | $S_2$ |
|---|-----|-------|-------|-------|-------|
| $Z$ | 1 | 0 | 0 | 5/7 | 15/7 |

Each solution found in the tableau corresponds to a corner point. We went from corner (0,0) to corner (0,23) to corner (12,15).

Although Maple has several commands that directly solve linear programming problems and which we illustrate later in this chapter, we have written an interactive program to do the simplex procedure. You are prompted to provide information regarding columns number and the row number that are involved in simplex operations.

We provide the procedure, called tableau, and a repeat of this example.

TABLEAU GENERATOR PROGRAM
LPS from an educational perspective.

```
> restart;
> Tableau:=proc( A::matrix,r1::posint,c1::posint)
> local R,C,S,R2,Q,B,B1,A1;
> global label;
> with(linalg):with(LinearAlgebra):
> A1:=Matrix(r1,c1,A);
> print(A1);
> #print(A);
> label[1]:
> print(A1);
> printf('Your choices.\n');
> printf('Enter 1 for Pivot and Row operations, 2 to stop.\n');
> S:=readstat("Operation #?");
> if S=1then
> printf('row number and column number.\n');
> R:=readstat("Row?");
> C:=readstat("Column?");
> printf('divide row number by number');
> R2:=readstat("divide row");
```

```
> Q:=readstat("by the value");
> print(R2,Q);
> print(R,C);
> B:=Matrix(3,6,A1):print(B);
> B:=evalm(RowOperation((Matrix(3,6,A1)),R2,(1/Q))):Matrix(3,6,B);
> print(B);
> #B1:=with(LinearAlgebra):Pivot(Matrix(3,6,B),R,C);
> B1:=evalm(Pivot((Matrix(3,6,B)),R,C)):Matrix(3,6,B1);
> print(B1);
> A1:=Matrix(3,6,B1);
> goto(label[1]) ;
> else
> if S=2 then
> printf('Optimal Tableau previously found');
> return([A1]);
> else
> end if;
> end if;
> end;
```

```
Tableau := proc(A:matrix, r1:posint, c1:posint)
    local R, C, S,R2, Q, B, B1, A1;
    global label;
    with(linalg);
    with(LinearAlgebra);
    A1 := Matrix(r1,c1,A);
    print (A1);
    label[1]:
    print (A1);
    printf('Your choices.'); printf('Enter 1 for Pivot and Row operations,2 to stop.');
    S := readstat ("Operation #?");
    if S = 1 then printf('row number and column number.');
    R := readstat ("Row?");
    C := readstat ("Column?");
    printf ('divide row number by number');
    R2 := readstat("divide row");
    Q := readstat ("by the value");
    print(R2, Q);
    print (R,C);
    B := Matrix(r1,c1,A1):
    print(B);
    B :=evalm(LinearAlgebra:-RowOperation (Matrix(r1,c1,A1),R2, 1/Q));
    Matrix(r1,c1,B);
    print(B);
    B1 :=evalm(LinearAlgebra:-Pivot(Matrix(r1,c1,B),R,C));
    Matrix(r1,c1,B1);
    print (B1);
    A1 := Matrix(r1,c1,B1);
    goto(label[1])
else
    if S = 2 then
        printf ('Optimal Tableau previously found'); return [A1]
    else
    end if
end if
end proc
```

> $A := matrix(3, 6, [1, -25, -30, 0, 0, 0, 0, 20, 30, 1, 0, 690, 0, 5, 4, 0, 1, 120])$;

| | **Z** | **$x_1$** | **$x_2$** | **$s_1$** | **$s_2$** | **RHS** |
|---|---|---|---|---|---|---|
| $A :=$ | 1 | $-25$ | $-30$ | 0 | 0 | 0 |
| | 0 | 20 | 30 | 1 | 0 | 690 |
| | 0 | 5 | 4 | 0 | 1 | 120 |

> $Tableau(A, 3, 6)$;

| **Z** | **x1** | **x2** | **s1** | **s2** | **RHS** |
|---|---|---|---|---|---|
| 1 | $-25$ | $-30$ | 0 | 0 | 0 |
| 0 | 20 | 30 | 1 | 0 | 690 |
| 0 | 5 | 4 | 0 | 1 | 120 |

Your choices:

Enter 1 for Pivot and Row operations, 2 to stop.

Enter row number and column number separated by a comma.

Enter the row number to be divided and the value to divided by separated by a comma.

$$2, 30$$
$$2, 3$$
$$2, 3$$

$$\begin{bmatrix} 1 & -25 & -30 & 0 & 0 & 0 \\ 0 & 20 & 30 & 1 & 0 & 690 \\ 0 & 5 & 4 & 0 & 1 & 120 \end{bmatrix}$$

$$\begin{bmatrix} 1 & -25 & -30 & 0 & 0 & 0 \\ 0 & \dfrac{2}{3} & 1 & \dfrac{1}{30} & 0 & 23 \\ 0 & 5 & 4 & 0 & 1 & 120 \end{bmatrix}$$

$$\begin{bmatrix} 1 & -5 & 0 & 1 & 0 & 690 \\ 0 & \dfrac{2}{3} & 1 & \dfrac{1}{30} & 0 & 23 \\ 0 & \dfrac{7}{3} & 0 & -\dfrac{2}{15} & 1 & 28 \end{bmatrix}$$

Your choices:

Enter 1 for Pivot and Row operations, 2 to stop.

Enter row number and column number separated by a comma.

Enter the row number to be divided and the value to divided by separated by a comma.

$$3, \dfrac{3}{7}$$
$$3, 2$$

$$\begin{bmatrix} 1 & -5 & 0 & 1 & 0 & 690 \\ 0 & \dfrac{2}{3} & 1 & \dfrac{1}{30} & 0 & 23 \\ 0 & \dfrac{7}{3} & 0 & -\dfrac{2}{15} & 1 & 28 \end{bmatrix}$$

$$\begin{bmatrix} 1 & -5 & 0 & 1 & 0 & 690 \\ 0 & \dfrac{2}{3} & 1 & \dfrac{1}{30} & 0 & 23 \\ 0 & \dfrac{49}{9} & 0 & -\dfrac{14}{45} & \dfrac{7}{3} & \dfrac{196}{3} \end{bmatrix}$$

$$\begin{bmatrix} 1 & 0 & 0 & \dfrac{5}{7} & \dfrac{15}{7} & 750 \\ 0 & 0 & 1 & \dfrac{1}{14} & -\dfrac{2}{7} & 15 \\ 0 & \dfrac{49}{9} & 0 & -\dfrac{14}{45} & \dfrac{7}{3} & \dfrac{196}{3} \end{bmatrix}$$

Your choices:

Enter 1 for Pivot and Row operations, 2 to stop.

Enter row number and column number separated by a comma.

Enter the row number to be divided and the value to divided by separated by a comma.

3, 5.444444444444

3, 2

$$\begin{bmatrix} 1 & 0 & 0 & \dfrac{5}{7} & \dfrac{15}{7} & 750 \\ 0 & 0 & 1 & \dfrac{1}{14} & -\dfrac{2}{7} & 15 \\ 0. & 1.000000000 & 0. & -0.05714285715 & 0.4285714286 & 12.00000000 \end{bmatrix}$$

$$\begin{bmatrix} 1. & 0. & 0. & 0.7142857143 & 2.142857143 & 750. \\ 0. & 0. & 1. & 0.07142857143 & -0.2857142857 & 15. \\ 0. & 1.000000000 & 0. & -0.05714285715 & 0.4285714286 & 12.00000000 \end{bmatrix}$$

Your choices:

Enter 1 for Pivot and Row operations, 2 to stop.

Optimal Tableau previously found.

| > Z | $x_1$ | $x_2$ | $s_1$ | $s_2$ | RHS |
|---|---|---|---|---|---|
| 1. | 0. | 0. | 0.7142857143 | 2.142857143 | 750. |
| 0. | 0. | 1. | 0.07142857143 | −0.2857142857 | 15. |
| 0. | 1.000000000 | 0. | −0.05714285715 | 0.4285714286 | 12.00000000 |

We have found the optimal tableau. We easily read our solution as $Z = 750$, $x_1 = 12$, and $x_2 = 15$.

---

**Example 1**   ## Maximization Problem with Simplex

$$\text{Maximize } Z = 3x_1 + x_2$$
Subject to
$$2x_1 + x_2 \le 6$$
$$x_1 + 3x_2 \le 9$$
$$x_1, x_2 \ge 0$$

The tableau format with slack variables $y_1$, $y_2$:

| Basic Variables | Z | $x_1$ | $x_2$ | $y_1$ | $y_2$ | RHS |
|---|---|---|---|---|---|---|
| Z | 1 | $-3$ | $-1$ | 0 | 0 | 0 |
| $Y_1$ | 0 | 2 | 1 | 1 | 0 | 6 |
| $Y_2$ | 0 | 1 | 3 | 0 | 1 | 9 |

Basic variable: $\{Z, y_1, y_2\}$
Nonbasic variable: $\{x_1, x_2\}$
Extreme point (0,0): corresponds to the values of $(x_1, x_2)$
Value of objective function: $Z = 0+$

## Optimality Test

The entering variable is $x_1$ (corresponding to $-3$ in the Z-row).

## Feasibility Test

Compute the ratios of the RHS divided by the column labeled $x_1$ to determine the minimum positive ratio.

| Basic Variables | Z | $x_1$ | $x_2$ | $y_1$ | $y_2$ | RHS | Quotient |
|---|---|---|---|---|---|---|---|
| Z | 1 | $-3$ | $-1$ | 0 | 0 | 0 | |
| $y_1$ | 0 | 2 | 1 | 1 | 0 | 6 | $6/2 = 3$ |
| $y_2$ | 0 | 1 | 3 | 0 | 1 | 9 | $9/1 = 9$ |

Choose $y_1$ to leave because it corresponds to the minimum positive ratio test value of 3.

## Pivot

Divide the row containing the exiting variable (the first row in this case) by the coefficient of the entering variable in that row (the coefficient of $x_1$ in this case), giving a coefficient of 1 for the entering variable in this row. Then eliminate the entering variable $x_1$ from the remaining rows (which do not contain the exiting variable $y_1$ and have a zero coefficient for it). The results are summarized in the next tableau.

| Basic Variables | Z | $x_1$ | $x_2$ | $y_1$ | $y_2$ | RHS |
|---|---|---|---|---|---|---|
| Z | 1 | 0 | 0.5000 | 1.5 | 0 | 9 |
| $x_1$ | 0 | 1 | 0.5000 | 0.5 | 0 | 3 |
| $y_2$ | 0 | 0 | 2.5000 | $-0.5$ | 1 | 6 |

Basic variable: $\{Z, x_1, y_2\}$
Nonbasic variable: $\{y_1, x_2\}$
Extreme point (3,0): corresponds to the values of $(x_1, x_2)$
Value of objective function: $Z = 9$

The pivot determines that the dependent variables have the values $x_1 = 3$, $y_2 = 6$, and $Z = 9$.

## Optimality Test

There are no negative coefficients in the $Z$-row. Thus, $x_1 = 3$ (a basic variable) and $x_2 = 0$ (a nonbasic variable) is an extreme point giving the optimal objective function value $Z = 9$.

*> Tableau( A,3,6 );*

$$
\begin{array}{c c c c c c}
Z & x_1 & x_2 & y_1 & y_2 & \mathbf{RHS}
\end{array}
$$

$$
\begin{bmatrix}
1 & -3 & -1 & 0 & 0 & 0 \\
0 & 2 & 3 & 1 & 0 & 6 \\
0 & 1 & 3 & 0 & 1 & 9
\end{bmatrix}
$$

Your choices:
Enter 1 for Pivot and Row operations, 2 to stop.
Enter row number and column number separated by a comma.
Enter the row number to be divided and the value to divided by separated by a comma.

2, 2
2, 2

$$
\begin{bmatrix}
1 & -3 & -1 & 0 & 0 & 0 \\
0 & 2 & 3 & 1 & 0 & 6 \\
0 & 1 & 3 & 0 & 1 & 9
\end{bmatrix}
$$

$$
\begin{bmatrix}
1 & -3 & -1 & 0 & 0 & 0 \\
0 & 1 & \dfrac{3}{2} & \dfrac{1}{2} & 0 & 3 \\
0 & 1 & 3 & 0 & 1 & 9
\end{bmatrix}
$$

$$
\begin{bmatrix}
1 & 0 & \dfrac{7}{2} & \dfrac{3}{2} & 0 & 9 \\
0 & 1 & \dfrac{3}{2} & \dfrac{1}{2} & 0 & 3 \\
0 & 0 & \dfrac{3}{2} & -\dfrac{1}{2} & 1 & 6
\end{bmatrix}
$$

Your choices:
Enter 1 for Pivot and Row operations, 2 to stop.
Optimal Tableau previously found.

$$
\begin{array}{c c c c c c}
Z & x_1 & x_2 & s_1 & s_2 & \mathbf{RHS}
\end{array}
$$

$$
\begin{bmatrix}
1 & 0 & \dfrac{7}{2} & \dfrac{3}{2} & 0 & 9 \\
0 & 1 & \dfrac{3}{2} & \dfrac{1}{2} & 0 & 3 \\
0 & 0 & \dfrac{3}{2} & -\dfrac{1}{2} & 1 & 6
\end{bmatrix}
$$

We read the solution as $Z = 9$, $x_1 = 3$, and $s_2 = 6$.

**Remarks:** We have assumed that the origin is a feasible extreme point. If it is not, then an extreme point must be found before the simplex method presented here can be used. We have also assumed that the linear program is not "degenerate" in the sense that no more than two constraints intersect at the same point. These and other topics are studied in more advanced optimization courses.

**Example 2**

## Wooden Soldier Revisited

Let's revisit our wooden soldier problem.

## Problem Identification

Maximize profits from making and selling wooden soldiers.

## Assumptions

We sell all the soldiers that we make each week. We also assume that we can make a fractional number of soldiers. Unfinished soldiers are completed in the following week.

## Define Variables

$$X_1 = \text{number of Confederate soldiers made per week}$$
$$X_2 = \text{number of Union soldiers made per week}$$

## Objective Function

$$\text{Maximize } Z = 28X_1 + 30X_2$$

## Resource Constraints

$$\text{Units of lumber: } 2X_1 + 3X_2 \le 100$$
$$\text{Hours of carpentry: } 4X_1 + 3.5X_2 \le 120$$
$$\text{Hours of finishing: } 2X_1 + 3X_2 \le 90$$
$$\text{Nonnegativity: } X_1 \ge 0,\ X_2 \ge 0$$

The initial tableau with slack variables $Y_1$, $Y_2$, and $Y_3$ added:

| Basic Variables | $Z$ | $X_1$ | $X_2$ | $Y_1$ | $Y_2$ | $Y_3$ | RHS |
|---|---|---|---|---|---|---|---|
| $Z$ | 1 | $-28$ | $-30$ | 0 | 0 | 0 | 0 |
| $Y_1$ | 0 | 2 | 3 | 1 | 0 | 0 | 100 |
| $Y_2$ | 0 | 4 | 3.5 | 0 | 1 | 0 | 120 |
| $Y_3$ | 0 | 2 | 3 | 0 | 0 | 1 | 90 |

Basic variable: $\{Z, Y_1, Y_2, Y_3\}$
Nonbasic variable: $\{X_1, X_2\}$
Extreme point (0,0): corresponds to the values of $(X_1, X_2)$
Value of objective function: $Z = 0$

## Optimality Test

The entering variable is $X_2$ (corresponding to $-30$ in the $Z$-row).

## Feasibility Test

Compute the ratios for the RHS divided by the coefficients in the column labeled $X_2$ to determine the minimum positive ratio.

| Basic Variables | $Z$ | $X_1$ | $X_2$ | $Y_1$ | $Y_2$ | $Y_3$ | RHS | Quotient |
|---|---|---|---|---|---|---|---|---|
| $Z$ | 1 | $-28$ | $-30$ | 0 | 0 | 0 | 0 | |
| $Y_1$ | 0 | 2 | 3 | 1 | 0 | 0 | 100 | 33.333 |
| $Y_2$ | 0 | 4 | 3.5 | 0 | 1 | 0 | 120 | 34.286 |
| $Y_3$ | 0 | 2 | 3 | 0 | 0 | 1 | 90 | 30 |

Choose $Y_3$ to it corresponding to the minimum positive ratio (90/3 = 30).

## Pivot

| Basic Variables | $Z$ | $X_1$ | $X_2$ | $Y_1$ | $Y_2$ | $Y_3$ | RHS |
|---|---|---|---|---|---|---|---|
| $Z$ | 1 | $-8$ | 0 | 0 | 0 | 10 | 900 |
| $Y_1$ | 0 | 0 | 0 | 1 | 0 | $-1$ | 10 |
| $Y_2$ | 0 | 1.6667 | 0 | 0 | 1 | $-1.16667$ | 15 |
| $X_2$ | 0 | 0.66667 | 1 | 0. | 0 | 0.3333 | 30 |

Basic variable: $\{Z,\ Y_1,\ Y_2,\ X_2\}$
Nonbasic variable: $\{X_1,\ Y_3\}$
Extreme point (0,30): corresponds to the values of $(X_1, X_2)$
Value of objective function: $Z = 900$

## Optimality Test

The entering variable is $X_1$ (corresponding to –8 in the $Z$-row).

## Feasibility Test

Compute the ratios for the RHS divided by the coefficients in the column labeled $X_1$ to determine the minimum positive ratio.

| Basic Variables | $Z$ | $X_1$ | $X_2$ | $Y_1$ | $Y_2$ | $Y_3$ | RHS | Quotient |
|---|---|---|---|---|---|---|---|---|
| $Z$ | 1 | $-8$ | 0 | 0 | 0 | 10 | 900 | |
| $Y_1$ | 0 | 0 | 0 | 1 | 0 | $-1$ | 10 | Undefined |
| $Y_2$ | 0 | 1.6667 | 0 | 0 | 1 | $-1.16667$ | 15 | 9 |
| $X_2$ | 0 | 0.66667 | 1 | 0. | 0 | 0.3333 | 30 | 45 |

Choose $Y_2$ corresponding to the minimum positive ratio 9 as the exiting variable.

## Pivot

| Basic Variable | $Z$ | $X_1$ | $X_2$ | $Y_1$ | $Y_2$ | $Y_3$ | RHS |
|---|---|---|---|---|---|---|---|
| $Z$ | 1 | 0 | 0 | 0 | 4.79999 | 4.39999 | 972 |
| $Y_1$ | 0 | 0 | 0 | 1 | 0 | $-1$ | 10 |
| $X_1$ | 0 | 1 | 0 | 0 | 0.59999 | $-0.7$ | 9 |
| $X_2$ | 0 | 0 | 1 | 0 | $-0.40000$ | 0.8 | 24 |

Basic variable: $\{Z, Y_1, X_1, X_2\}$
Nonbasic variable: $\{Y_2, Y_3\}$
Extreme point (9,24): corresponds to the values of $(X_1, X_2)$
Value of objective function: $Z = 972$

## Optimality Test

Because there are no negative coefficients in the $Z$-row, we are optimal: $X_1 = 9$ and $X_2 = 24$ give the optimal solution $Z = \$972$ for the objective function.

We state that we make 9 Confederate soldiers and 24 Union soldiers, yielding a profit of $972.

We also have information about the resources. The fact that $Y_1 = 10$ (because $Y_2 = Y_3 = 0$) indicates that we have 10 units of lumber left over; that is, we are only truly constrained by hours of carpentry and hours of finishing.

> $A:=matrix(4,7,[1,-28,-30,0,0,0,0,0,2,3,1,0,0,100,0,4,3.5,0,1,0,120,0,2,3,0,0,1,90]);$

|   | Z | $x_1$ | $x_2$ | $y_1$ | $y_2$ | $y_3$ | RHS |
|---|---|---|---|---|---|---|---|
| $A :=$ | 1 | −28 | −30 | 0 | 0 | 0 | 0 |
|   | 0 | 2 | 3 | 1 | 0 | 0 | 100 |
|   | 0 | 4 | 3.5 | 0 | 1 | 0 | 120 |
|   | 0 | 2 | 3 | 0 | 0 | 1 | 90 |

> $Tableau(A,4,7);$

$$\begin{bmatrix} 1 & -28 & -30 & 0 & 0 & 0 & 0 \\ 0 & 2 & 3 & 1 & 0 & 0 & 100 \\ 0 & 4 & 3.5 & 0 & 1 & 0 & 120 \\ 0 & 2 & 3 & 0 & 0 & 1 & 90 \end{bmatrix}$$

Your choices:
Enter 1 for Pivot and Row operations, 2 to stop.
Enter row number and column number separated by a comma.
Enter the row number to be divided and the value to divided by separated by a comma.

4, 3
4, 3

$$\begin{bmatrix} 1 & -28 & -30 & 0 & 0 & 0 & 0 \\ 0 & 2 & 3 & 1 & 0 & 0 & 100 \\ 0 & 4 & 3.5 & 0 & 1 & 0 & 120 \\ 0 & 2 & 3 & 0 & 0 & 1 & 90 \end{bmatrix}$$

$$\begin{bmatrix} 1 & -28 & -30 & 0 & 0 & 0 & 0 \\ 0 & 2 & 3 & 1 & 0 & 0 & 100 \\ 0 & 4 & 3.5 & 0 & 1 & 0 & 120 \\ 0 & \frac{2}{3} & 1 & 0 & 0 & \frac{1}{3} & 30 \end{bmatrix}$$

$$\begin{bmatrix} 1 & -8 & & 0 & 0 & 0 & 10 & 900 \\ 0 & 0 & & 0 & 1 & 0 & -1 & 10 \\ 0. & 1.666666667 & 0. & 0. & 1. & -1.166666667 & 15.0 \\ 0 & \frac{2}{3} & & 1 & 0 & 0 & \frac{1}{3} & 30 \end{bmatrix}$$

Your choices.
Enter 1 for Pivot and Row operations, 2 to stop.
Enter row number and column number separated by a comma.
Enter the row number to be divided and the value to divided by separated by a comma.

3, 1.66666666'
3, 2

$$\begin{bmatrix} 1 & -8 & 0 & 0 & 0 & 10 & 900 \\ 0 & 0 & 0 & 1 & 0 & -1 & 10 \\ 0. & 1.666666667 & 0. & 0. & 1. & -1.166666667 & 15.0 \\ 0 & \dfrac{2}{3} & & 1 & 0 & 0 & \dfrac{1}{3} & 30 \end{bmatrix}$$

$$\begin{bmatrix} 1 & -8 & 0 & 0 & 0 & 10 & 900 \\ 0 & 0 & 0 & 1 & 0 & -1 & 10 \\ 0. & 1.000000000 & 0. & 0. & 0.5999999999 & -0.7000000001 & 8.999999998 \\ 0 & \dfrac{2}{3} & & 1 & 0 & 0 & \dfrac{1}{3} & 30 \end{bmatrix}$$

$$\begin{bmatrix} 1 & 0 & 0 & 0 & 4.799999999 & 4.399999999 & 972 \\ 0 & 0 & 0 & 1 & 0 & -1 & 10 \\ 0 & 1 & 0 & 0 & 0.599999999 & -0.700000001 & 8.99999998 \\ 0 & 0 & 1 & 0 & -0.400000000 & 0.800000001 & 24.00000000 \end{bmatrix}$$

$$\begin{bmatrix} 1 & 0 & 0 & 0 & 4.799999999 & 4.399999999 & 972 \\ 0 & 0 & 0 & 1 & 0 & -1 & 10 \\ 0 & 1 & 0 & 0 & 0.599999999 & -0.700000001 & 8.99999998 \\ 0 & 0 & 1 & 0 & -0.400000000 & 0.800000001 & 24.00000000 \end{bmatrix}$$

Your choices.
Enter 1 for Pivot and Row operations, 2 to stop.
Optimal Tableau previously found.

$$\begin{bmatrix} 1 & 0 & 0 & 0 & 4.799999999 & 4.399999999 & 972 \\ 0 & 0 & 0 & 1 & 0 & -1 & 10 \\ 0 & 1 & 0 & 0 & 0.599999999 & -0.700000001 & 8.99999998 \\ 0 & 0 & 1 & 0 & -0.400000000 & 0.800000001 & 24.00000000 \end{bmatrix}$$

## 7.4 | EXERCISES

Use the simplex procedure. Find the maximum and minimum solution. Assume we have $x \geq 0$ and $y \geq 0$ for each problem.

**1.** $Z = 2x + 3y$

Subject to
$2x + 3y \geq 6$
$3x - y \leq 15$
$-x + y \leq 4$
$2x + 5y \leq 27$

**2.** $Z = 6x + 4y$

Subject to
$-x + y \leq 12$
$x + y \leq 24$
$2x + 5y \leq 80$

**3.** $Z = 6x + 5y$

Subject to
$x + y \leq 6$
$2x + y \leq 9$

**4.** $Z = x - y$

Subject to
$x + y \geq 6$
$2x + y \geq 9$

**5.** $Z = 5x + 3y$

Subject to
$1.2x + 0.6y \leq 24$
$2x + 1.5y \leq 80$

## 7.4 | PROJECTS

Set up and solve projects 1–4 from Section 7.2 using the simplex method, and set up and solve the exercises 1–7 from Section 7.1 using the simplex method.

**1.** With the rising cost of gasoline and increasing prices to consumers, the use of additives to enhance gasoline performance is being considered. Consider two additives: additive 1 and additive 2. The following conditions must hold for their use:
   - Harmful carburetor deposits must not exceed 1/2 lb for each car's gasoline tank.
   - The quantity of additive 2 plus twice the quantity of additive 1 must be at least 1/2 lb for each car's gasoline tank.
   - One pound of additive 1 will add 10 octane units per tank, and 1 lb of additive 2 will add 20 octane units per tank. The total number of octane units added must not be less than six (6).
   - Additives are expensive: \$1.53/lb for additive 1 and \$4/lb for additive 2.

   We want to determine the quantity of each additive that will meet the above restrictions and will minimize their cost while meeting the following requirements.
   **a.** List the decision variables and define them.
   **b.** List the objective function.
   **c.** List the resources that constrain this problem.
   **d.** Graph the feasible region.
   **e.** Label all intersection points of the feasible region.
   **f.** Plot the objective function in a different color (highlight the objective function line, if necessary) and label it the ISO-profit line.
   **g.** Clearly indicate the point that is the optimal solution on the graph.
   **h.** List the coordinates of the optimal solution and the value of the objective function.
   **i.** Assume now that a manufacturer of additives has the opportunity to sell you a nice special TV deal to deliver at least 0.5 lb of additive 1 and at least 0.3 lb of additive 2. Use graphical LP methods to help determine whether you should buy this TV offer. Support your recommendation.
   **j.** Write a one-page cover letter to the boss of your company that summarizes the results that you found.

**2.** A farmer has 30 acres on which to grow tomatoes and corn. Each 100 bushels of tomatoes require 1,000 gallons of water and 5 acres of land. Each 100 bushels of corn require 6,000 gallons of water and 2½ acres of land. Labor costs are \$1 per bushel for both corn and tomatoes. The farmer has available 30,000 gallons of water and \$750 in capital. He knows that he cannot sell more than 500 bushels of tomatoes or 475 bushels of corn. He estimates a profit of \$2 on each bushel of tomatoes and \$3 for each bushel of corn.

How many bushels of each should he raise to maximize profits? Meet the following requirements:

**a.** List the decision variables and define them.
**b.** List the objective function.
**c.** List the resources that constrain this problem.
**d.** Graph the feasible region.
**e.** Label all intersection points of the feasible region.
**f.** Plot the objective function in a different color (highlight the objective function line, if necessary) and label it the ISO-profit line.
**g.** Clearly indicate on the graph the point that is the optimal solution.
**h.** List the coordinates of the optimal solution and the value of the objective function.
**i.** Assume now that farmer has the opportunity to sign a nice contract with a grocery store to grow and deliver at least 300 bushels of tomatoes and at least 500 bushels of corn. Use graphical LP methods to help recommend a decision to the farmer. Support your recommendation.
**j.** If the farmer can obtain an additional 10,000 gallons of water for a total cost of $50, is it worth it to obtain the additional water? Determine the new optimal solution caused by adding this level of resource.
**k.** Write a one-page cover letter to your boss that summarizes the result that you found.

3. Firestone Tires, headquartered in Akron, Ohio, has a plant in Florence, South Carolina, that manufactures two types of tires: SUV 225 radials and SUV 205 radials. Demand is high because of the recent recall of tires. Each 100 lot quantity of SUV 225 radials requires 100 gallons of synthetic plastic and 5 lb of rubber. Each 100-lot quantity of SUV 205 radials require 60 gallons synthetic plastic and 2½ lb of rubber. Labor costs are $1 per tire for each type tire. The manufacturer has weekly quantities available of 660 gallons of synthetic plastic, $750 in capital, and 300 lbs. of rubber. The company estimates a profit of $3 on each SUV 225 radial and $2 of each SUV 205 radial. How many of each type of tire should the company manufacture to maximize its profits?

   **Required:**

   a. List the decision variables and define them.
   b. List the objective function.
   c. List the resources that constrain this problem.
   d. Graph the feasible region.
   e. Label all intersection points of the feasible region.
   f. Plot the objective function in a different color (highlight the objective function line, if necessary) and label it the ISO-profit line.
   g. Clearly indicate on the graph the point that is the optimal solution.
   h. List the coordinates of the optimal solution and the value of the objective function.
   i. Assume now that manufacturer has the opportunity to sign a nice contract with a tire outlet store to deliver at least 500 SUV 225 radial tires and at least 300 SUV 205 radial tires. Use graphical LP methods to help recommend a decision to the manufacturer. Support your recommendation.
   j. If the manufacturer can obtain an additional 1,000 gallons of synthetic plastic for a total cost of $50, is it worth it to obtain this amount? Determine the new optimal solution caused by adding this level of resource.
   k. If the manufacturer can obtain an additional 20 lb of rubber for $50, should it obtain the rubber? Determine the new solution caused by adding this amount.
   l. Write a one-page cover letter to your boss of the company that summarizes the results that you found.

4. Consider a toy maker that carves wooden soldiers. The company specializes in two types: Confederate soldiers and Union soldiers. The estimated profit for each is $28 and $30, respectively. A Confederate soldier requires 2 units of lumber, 4 hours of carpentry, and 2 hours of finishing to complete. A Union soldier requires 3 units of lumber, 3.5 hours of carpentry, and 3 hours of finishing to complete. Each week the company has 100 units of

lumber delivered. The workers can provide at most 120 hours of carpentry and 90 hours of finishing. Determine the number of each type of wooden soldiers to produce to maximize weekly profits. Formulate and then solve this linear programming graphically.

5. A company wants to can two different drinks for the holiday season. It takes 3 hours to can one gross of drink A, and it takes 2 hour to label the cans. It takes 2.5 hours to can one gross of drink B, and it takes 2.5 hours to label the cans. The company makes $15 profit on one gross of drink A and an $18 profit of one gross of drink B. Given that we have 40 hours to devote to canning the drinks and 35 hours to devote to labeling cans per week, how many cans of each type drink should the company package to maximize profits?

6. The Mariners Toy Company wishes to make three models of ships to maximize its profits. They found that a model steamship takes the cutter 1 hour, the painter 2 hours, and the assembler 4 hours of work; it produces a profit of $6. The sailboat takes the cutter 3 hours, the painter 3 hours, and the assembler 2 hours. It produces a $3 profit. The submarine takes the cutter 1 hour, the painter 3 hours, and the assembler 1 hour. It produces a profit of $2. The cutter is only available for 45 hours per week, the painter for 50 hours, and the assembler for 60 hours. Assume that they sell all the ships that they make and formulate this LP to determine how many ships of each type that Mariners should produce.

7. To produce 1,000 tons of nonoxidizing steel for BMW engine valves, at least the following units will be needed weekly: 10 units of manganese, 12 units of chromium, and 14 units of molybdenum (1 unit is 10 lbs). These materials are obtained from a dealer who markets these metals in three sizes small (S), medium (M), and large (L). One S case costs $9 and contains two units of manganese, two units of chromium, and one unit of molybdenum. One M case costs $12 and contains two units of manganese, three units of chromium, and one unit of molybdenum. One L case costs $15 and contains one unit of manganese, one unit of chromium, and five units of molybdenum. How many cases of each kind (S, M, L) should be purchased weekly so that we have enough manganese, chromium, and molybdenum at the smallest cost?

8. The Superbowl Advertising Agency wishes to plan an advertising campaign in three different media—television, radio, and magazines. The purpose or goal is to reach as many potential customers as possible. Results of a marketing study are shown in the following table:

|  | Daytime TV | Prime-Time TV | Radio | Magazines |
| --- | --- | --- | --- | --- |
| Cost of advertising unit | $40,000 | $75,000 | $30,000 | $15,000 |
| Number of potential customers reached per unit | 400,000 | 900,000 | 500,000 | 200,000 |
| Number of woman customers reached per unit | 300,000 | 400,000 | 200,000 | 100,000 |

The company does not want to spend more than $800,000 on advertising. It further requires (1) at least 2 million exposures take place among woman, (2) TV advertising be limited to $500,000, (3) at least three advertising units be bought on daytime TV and 2 units on prime-time TV, and (4) the number of radio and magazine advertisement units should each be between five and 10 units.

9. A tomato cannery has 5,000 pounds of grade A tomatoes and 10,000 pounds of grade B tomatoes, from which the cannery will make whole canned tomatoes and tomato paste. Whole tomatoes must be composed of at least 80 percent grade A tomatoes, whereas tomato paste must be made with at least 10 percent grade A tomatoes. Whole tomatoes sell for $0.08 per pound, and grade B tomatoes sell for $0.05 per pound. Maximize revenue of the tomatoes.

*Hint:* Let $X_{wa}$ = pounds of grade A tomatoes used to whole tomatoes and $X_{wb}$ = pounds of grade B tomatoes used to whole tomatoes. The total number of whole tomato cans produced is the sum of $X_{wa} + X_{wb}$ after each is found. Also remember a percent is a fraction of the whole times 100 percent.

10. The McCow Butchers company is a large-scale distributor of dressed meats for Myrtle Beach restaurants and hotels. Ryan's orders meat for meat loaf (mixed ground beef, pork, and veal) for 1,000 pounds according to the following specifications:
    a. Ground beef is to be no less than 400 lbs and no more than 600 lbs.
    b. The ground pork is to be between 200 and 300 pounds.
    c. The ground veal must weigh between 100 and 400 lbs.
    d. The weigh of the ground pork must be no more than one and one-half (3/2) times the weight of the veal.
       The contract calls for Ryan's to pay \$1,200 for the meat. The cost per pound for the meat is \$0.70 for hamburger, \$0.60 for pork, and \$0.80 for veal. How can this be modeled?

11. Portfolio Investments. A portfolio manager in charge of a bank wants to invest \$10 million. The securities available for purchase, as well as their respective quality ratings, maturate, and yields, are shown in the following table.

| Bond Name | Bond Type | Moody's Quality Scale | Bank's Quality Scale | Years to Maturity | Yield at Maturity (%) | After-Tax Yield (%) |
|---|---|---|---|---|---|---|
| A | Municipal | Aa | 2 | 9 | 4.3 | 4.3 |
| B | Agency | Aa | 2 | 15 | 5.4 | 2.7 |
| C | Govt 1 | Aaa | 1 | 4 | 5 | 2.5 |
| D | Govt 2 | Aaa | 1 | 3 | 4.4 | 2.2 |
| E | Local | Ba | 5 | 2 | 4.5 | 4.5 |

The bank places certain policy limitations on the portfolios manager's actions:
a. Government and agency bonds must total at least \$4 million.
b. The average quality of the portfolios cannot exceed 1.4 on the bank's quality scale. Note that a low number means high quality.
c. The average years to maturity must not exceed five years.
   Assume the objective is to maximize after-tax earning on the investment.

## 7.5  LINEAR PROGRAMMING IN MAPLE

Maple provides a suite of powerful, robust routines for solving optimization problems, including LPs. Using Maple's flexible mathematical programming language, it is simple to conduct thorough sensitivity studies on solutions to optimization problems. Maple has intrinsic built-in features that solve linear programming problems. We will present several of these features.

Method 1 uses the following commands:

**Command**

```
with(simplex):
Maximize(objective function, constraints,
NONNEGATIVE)
Subs( result, objective function)
```

| **Example 1** | ## Maximizing a Linear Programming Problem in Maple |
|---|---|

$$\text{Maximize } Z = x + y$$
$$\text{Subject to}$$
$$x + 3y \leq 9$$
$$2x + y \leq 8$$
$$x \geq 0$$
$$y \geq 0$$

## Method 1

We evoke the Maple library package, with (simplex). Then we enter the set of constraints and call the set **cnsts**. Note we did not have to enter the nonnegativity constraints. Then we enter the objective function and call it **obj**. We maximize the objective function with the constraints set and now tell Maple that the variables are nonnegative. The results of the decision variables that satisfy the requirements are returned by Maple. We then substitute those results back into the objective function to get the optimal value for the problem.

```
> with(simplex):
> cnsts:={x+3*y<=9,2*x+y<=8}:
obj:=x+y:
> maximize(obj,cnsts,NONNEGATIVE);
```

$$\{y = 2, x = 3\}$$

```
> subs({y=2,x=3},obj);
```

$$5$$

The solution is $x = 3$, $y = 2$, and $Z = 5$.

## Method II

This method evokes the use of the new optimization package in Maple and requires us only to enter the same inputs as in method I, but input them in the LP solver. We present the LPSolve command.

**Optimizational [LP Solve]** – solve a linear program

**Calling Sequence**
LP Solve (**obj, constr, bd, opts**)

**Parameters**

obj – **algebraic**; linear objective function
constr – (optional) **set(relation)** or **list(relation)**; linear constraints
bd – (optional) sequence of **name = range**; bounds for one or more variables
opts – (optional) equation(s) of the form **option = value** where **option** is one of **assume, binaryvariables, depthlimit, feasibilitytolerance, infinitebound, initialpoint, integertolerance, integervariables, interationlimit, maximize, nodelimit** or **output**; specify options for the **LPSolve** command

| **Example 2** | ## Method 2 in Maple with Maximization Problem from Example 1 |
|---|---|

We enter the constraints and objective function as we did in method 1.

```
> cnsts:={x+3*y≤9,2*x+y≤8};obj:=x+y;
```

$$cnsts := \{x + 3y \leq 9, 2x + y \leq 8\}$$
$$obj := x + y$$

> *with(Optimization) :*
> *LPSolve(obj, cnsts, assume=nonnegative, maximize=true);*

$$[5., [x = 3., y = 1.99999999999999978]]$$

The output-provided solution is $Z = 5$, is $x = 3$, $y = 1.99999999999999978$. Note that we would round this value to 2.

## Method III

This method evokes the use of the optimization package in Maple and requires the use and knowledge of linear algebra. The coefficient matrix, A, the coefficient of the constraints, is entered as a matrix A. The right-hand-side values b are entered as a column matrix B. The cost coefficients (or profit coefficients) of the objective function are entered as a column matrix C. We then use the LPSolve command.

> *restart;*
> *with (Optimization):with(LinearAlgebra):*
> *A:=<<1,2>|<3,1>>;*

$$A := \begin{bmatrix} 1 & 3 \\ 2 & 1 \end{bmatrix}$$

> *B:=<9,8>;*

$$B := \begin{bmatrix} 9 \\ 8 \end{bmatrix}$$

> *C:=<1,1>:*

$$C := \begin{bmatrix} 1 \\ 1 \end{bmatrix}$$

> *mysol:=LPSolve(C,[A,B],assume=nonnegative,maximize);*

$$mysol := \left[ 5., \begin{bmatrix} 3. \\ 1.999999999999978 \end{bmatrix} \right]$$

Note the optimal answer is read as "The value of the objective function is 5 when $x = 3$ and $y = 1.999999999999978$."

In addition, the CD has two additional features: a tableau generator and a Maplet generator for a maximization LP. These can be used to see and discuss the tableau feature.

## 7.5 | EXERCISES

Solve the exercises from Section 7.4 in Maple.

## 7.5 | PROJECTS

Solve the projects in Section 7.4 using Maple.

## 7.6  SENSITIVITY ANALYSIS IN MAPLE (OPTIONAL)

We have looked at Maple to solve linear programming problems and to generate tableaus for further analysis on problems. In this section, we explore using Maple to generate sensitivity analysis. This sensitivity analysis is essential for good decision making. We will present methods to obtain sensitivity analysis on parameters of (1) coefficient of nonbasic variables (2) coefficient of basic variables , and (3) changes in the resources (the RHS values). We will examine these one at a time.

We must reduce the analysis to formula form using the following matrix notation for our analysis:

$CBV$: matrix of the original cost coefficients of the basic variables
$B^{-1}$: the inverse of the matrix of the basic variables in the order in which they enter the basis.
$a_j$: the original column of the resources coefficient of the $j$th variable.
$cj$: the original column of the cost coefficient of the $j$th variable.
$b_j$: = the resource limit of the $j$th constraint.

Let's start with the linear programming problem given by

$$\text{Maximize } Z = 60d + 30t + 20c$$
$$\text{Subject to}$$
$$8d + 6t + c \le 48$$
$$4d + 2t + 1.5c \le 20$$
$$2d + 1.5t + 0.5c \le 8$$
$$d,\ t,\ c \ge 0$$

We solve this LP to obtain the following the solution and also to create the tableaus.

*> with (Optimization):with( LinearAlgebra):with(simplex):with (linalg):*
*> obj :=60·d+30·t+20·c;*

$$obj := 60d + 30\ t + 20c$$

*> constr :={8·d+6·t+c≤48,4·d+2·t+1.5·c≤20,2·d+1.5·t+0.5·c≤8};*

$$constr := \{8d + 6t + c \le 48,\ 4d + 2t + 1.5\ c \le 20,\ 2d + 1.5t + 0.5c < 8\}$$

*> LPSolve (obj, {8·d+ 6·t+c≤48,4·d+2·t+1.5·c≤20,2·d+1.5·t+0.5·c≤8},maximize,
assume=nonnegative)*

$$[280.,\ [t = 0.,\ d = 2.,\ c = 7.99999999999999912]]$$

*> m :=LPSolve(obj, {8·d+6·t+c≤48,4·d+2·t+1.5·c≤20,2·d+1.5·t+0.5·c≤8},maximize,
assume=nonnegative, output=solutionmodule);*

$$m := module\ (\ )\ export\ Results,\ Settings;\ end\ module$$

*> m:-Results();*

$$[\text{"objectivevalue"}= 280.,\ \text{"solutionpoint"}= ([t = 0.,d = 2.,$$
$$c = 7.99999999999999912]),\ \text{"iterations"}= 2]$$

Using the Tableau method from earlier to type in the initial and final tableaus.

*> initTab :=matrix (4,8,[1,−60,−30,−20, 0, 0, 0, 0, 0, 8, 6, 1, 1, 0, 0, 48, 0, 4, 2, 1.50,1,0,20,0,2,
1.5, .5, 0,0, 1,8]);*

$$initTab := \begin{bmatrix} 1 & -60 & -30 & -20 & 0 & 0 & 0 & 0 \\ 0 & 8 & 6 & 1 & 1 & 0 & 0 & 48 \\ 0 & 4 & 2 & 1.5 & 0 & 1 & 0 & 20 \\ 0 & 2 & 1.5 & 0.5 & 0 & 0 & 1 & 8 \end{bmatrix}$$

> *finalTabl :=matrix(4, 8, [1, 0, 5, 0, 0, 10, 10, 280, 0, 0,–2, 0, 1, 2,–8, 24, 0, 0,–2, 1, 0, 2,–4, 8, 0, 1,1.25, 0,0,–.5, 1.5, 2]);*

$$finalTabl := \begin{bmatrix} 1 & 0 & 5 & 0 & 0 & 10 & 10 & 280 \\ 0 & 0 & -2 & 0 & 1 & 2 & -8 & 24 \\ 0 & 0 & -2 & 1 & 0 & 2 & -4 & 8 \\ 0 & 1 & 1.25 & 0 & 0 & -.5 & 1.5 & 2 \end{bmatrix}$$

The solution is $Z = 280$ when $s_1 = 24$, $d = 2$, and $c = 8$. The importance of the initial and final tableau are that they provide information about the variables needed in the sensitivity analysis.

**Coefficient of Nonbasic Variables**

We will reduce the analysis to formula form using the following matrix notation. We will use

$$\bar{c}_j = c_{BV}\, B^{-1} a_j - c_j$$

For the current solution basis to remain the optimal basis, all nonbasic variable coefficients must remain "greater than or equal to" zero in the final tableau (for a maximization problem). Thus, $\bar{c}_j \geq 0$.

Change of Nonbasic Variables Coefficients

> *cbv:=⟨⟨0⟩|⟨–20⟩|⟨–60⟩⟩;*

$$cbv := [0\ -20\ -60]$$

> *cbvl :=convert(cbv, Matrix);*

$$cbvl := [0\ -20\ -60]$$

> *NewB:=⟨⟨1,0,0⟩|⟨1,1.5,.5⟩|⟨8,4,2⟩⟩;*

$$NewB := \begin{bmatrix} 1 & 1 & 8 \\ 0 & 1.5 & 4 \\ 0 & 0.5 & 2 \end{bmatrix}$$

> *NewBinv:=MatrixInverse(NewB);*

$$NewBinv := \begin{bmatrix} 1. & 2. & -8. \\ 0. & 1.99999999999999956 & -3.99999999999999912 \\ 0. & -.49999999999999888 & 1.49999999999999978 \end{bmatrix}$$

> *CbvBinv:=MatrixMatrixMultiply(–1·cbvl, NewBinv);*

$$CbvBinv := [0.\quad 10.\quad 10.]$$

> *a2:=⟨⟨6, 2, 1.5⟩⟩;*

$$a2 := \begin{bmatrix} 6 \\ 2 \\ 1.5 \end{bmatrix}$$

> *MatrixMatrixMultiply (CbvBinv, a2);*

$$[35.]$$

> *c2:=–⟨⟨30+delta⟩⟩;*

$$c2 := [-30-\delta]$$

> *solve(35–30–delta≥0,delta);*

$$RealRange(-\infty, 5)$$

This means that if the coefficient of table, $t$, that is currently at a value of 30 is less than 35, indicating that the variable, tables, will never be made because the coefficient is not large enough to ever enter the basis.

### Change in a Basic Variable Cost Coefficient

First, change coefficient for $d$ and then the coefficient for $c$.

$$\text{Change in } d$$

```
> cbv:=⟨⟨0>⟩|⟨–20⟩|⟨–60–delta⟩⟩;
```

$$cbv: = \begin{bmatrix} 0 & -20 & -60 & -\delta \end{bmatrix}$$

```
> CbvBinvl:=MatrixMatrixMultiply(–1·cbv, NewBinv);
```

$$CbvBinv1 := [0., 10.00000000 - 0.499999999999999888\ \delta,$$
$$10.00000000 + 1.49999999999999978\ \delta]$$

```
> a1:= ⟨⟨8, 4, 2⟩⟩;
```

$$a1 := \begin{bmatrix} 8 \\ 4 \\ 2 \end{bmatrix}$$

```
> a3:= ⟨⟨1, 1.5, .5⟩⟩;
```

$$a3 := \begin{bmatrix} 1 \\ 1.5 \\ 0.5 \end{bmatrix}$$

```
> t1:=MatrixMatrixMultiply(CbvBinv1,a1);
```

$$t1 := \begin{bmatrix} 60.00000000 + 1.000000000\ \delta \end{bmatrix}$$

```
> t2:=MatrixMatrixMultiply(CbvBinv1,a2);
```

$$t2 := [\ 35.00000000 + 1.250000000\ \delta\ ]$$

```
> t3:=MatrixMatrixMultiply(CbvBinv1, a3);
```

$$t3 := [\ 20.00000000\ ]$$

```
> solve( {10–.5·delta≥0, 10+1.5·delta≥0,5+1.25·delta≥}, {delta});
```

$$\{-4. \leq \delta, \delta \leq 20.\}$$

$$\text{Change in Coefficient for } t$$

```
> cbv:=⟨⟨0⟩|⟨–20–delta⟩|⟨–60⟩⟩;
```

$$cbv := [\ 0 \ -20 \ -\delta \ -60]$$

```
> CbvBinv1:=MatrixMatrixMultiply(–1.cbv, NewBinv);
```

$$CbvBinv1 := [0., 10.00000000 + 1.99999999999999956\ \delta,$$
$$10.00000000 - 3.99999999999999912\ \delta]$$

```
> t1:=MatrixMatrixMultiply(CbvBinv1, a1);
```

$$t1 := [\ 60.00000000]$$

```
> t2:=MatrixMatrixMultiply(CbvBinv1, a2);
```

$$t2 := [\ 35.00000000 - 2.000000000\ \delta]$$

> $t3:=Matrix\,Matrix\,Multiply\,(Cbv\,Binv1,a3);$

$$t3 := [20.00000000 + 1.000000000\,\delta]$$

> $solve(\,\{5-2\cdot delta \geq 0, 10+2\cdot delta \geq 0, 10-4\cdot delta \geq 0\}, \{delta\}\,);$

$$\left\{-5 \leq \delta, \delta \leq \frac{5}{2}\right\}$$

Let's interpret these results. First, for the coefficient of $d$, we find that if the value is between 56 and 80, then $d$ will remain a basic variable. If the coefficient of $c$ is between 15 and 22.5, then $c$ remains a basic variable.

### Change in RHS Values

Test with a change in b1, originally 48.

> $b:=\langle\langle 48+delta,20,8\rangle\rangle;$

$$b := \begin{bmatrix} 48 + \delta \\ 20 \\ 8 \end{bmatrix}$$

> $Matrix\,Matrix\,Multiply\,(New\,Binv,\ b);$

$$\begin{bmatrix} 24. + 1.\delta \\ 8.00000000 \\ 2.00000000 \end{bmatrix}$$

> $solve(24+delta \geq 0, delta);$

$$Real\,Range(-24, \infty)$$

This is interpreted as "If the resource of constraint 1 currently at 48 units is at least 24, then $d$ and $c$ remain as basic variables."

## 7.6 | EXERCISES

Perform sensitivity analysis using Maple for problems 1–6 from Section 7.4.

## 7.7 | MODELING OF RANKING UNITS USING DATA ENVELOPMENT ANALYSIS

Data envelopment analysis (DEA), which is occasionally called *frontier analysis*, was first put forward by Charnes, Cooper, and Rhodes in 1978. It is a performance-measurement technique that, as we shall see, can be used to evaluate the *relative efficiency* of *decision-making units* (DMUs) in organizations. Here a DMU is a distinct unit within an organization that has flexibility with respect to some of the decisions it makes but not necessarily complete freedom with respect to these decisions.

Examples of such units to which DEA has been applied are banks, police stations, hospitals, tax offices, prisons, defense bases (Army, Navy, Air Force), schools, and university departments. Note here that one advantage of DEA is that it can be applied to not-for-profit organizations.

Since the technique was first proposed, much theoretical and empirical work has been done. Many studies have been published that deal with applying DEA in real-world situations. Obviously, there are many more unpublished studies—for example, those done internally by companies and those done by external consultants.

We will initially illustrate DEA by means of a small example. More about DEA can be found online using Google and the search term "data envelopment analysis." Note here that much of what follows is a graphical (pictorial) approach to DEA. This is very useful when DEA is explained to those who are less technically qualified (such as people in the military and the world of management). There is, however, a mathematical approach to DEA that can be adopted. We will present the single measure first to demonstrate the idea and then move to multiple measures and use linear programming methodology from our course.

## Example 1    Ranking Banks

Consider a number of bank branches. For each branch, we have a single output measure (number of personal transactions completed) and a single input measure (number of staff).

The data we have are as follows:

| Branch | Personal Transactions (1,000s) | Number of Staff |
|---|---|---|
| 1 | 125 | 18 |
| 2 | 44 | 16 |
| 3 | 80 | 17 |
| 4 | 23 | 11 |

For example, for branch 2 in one year, there were 44,000 transactions relating to personal accounts and 16 staff members were employed.

How then can we compare these branches and measure their performance using this data?

## Ratios

A commonly used method is *ratios*. Typically, we take some output measure and divide it by some input measure. Note the terminology here: We view branches as taking *inputs* and converting them (with varying degrees of efficiency, as we shall see below) into *outputs*.

For our bank branch example, we have a single input measure, the number of staff, and a single output measure, the number of personal transactions. Hence we have:

| Branch | Personal Transactions per Staff Member (1,000s) |
|---|---|
| 1 | 6.94 |
| 2 | 2.75 |
| 3 | 4.71 |
| 4 | 2.09 |

Here we can see that branch 1 has the highest ratio of personal transactions per staff member, whereas branch 4 has the lowest ratio of personal transactions per staff member.

Because branch 1 has the highest ratio (6.94), we can compare all other branches to it and calculate their *relative efficiency* with respect to branch 1. To do this, we divide the ratio for any branch by 6.94 (multiply by 100 to convert to a percentage). This gives:

| Branch | Relative Efficiency |
|--------|--------------------|
| 1 | $100(6.94/6.94) = 100\%$ |
| 2 | $100(2.75/6.94) = 40\%$ |
| 3 | $100(4.71/6.94) = 68\%$ |
| 4 | $100(2.09/6.94) = 30\%$ |

The other branches do not compare well with branch 1, so they presumably are performing less well—that is, they are relatively less efficient at using their given input resource (staff members) to produce output (number of personal transactions).

We could, if we wish, use this comparison with branch 1 to set *targets* for the other branches. For example, we could set a target for branch 4 of continuing to process the same level of output but with one fewer staff member. This is an example of an **input target** as it deals with an input measure.

An example of an **output target** would be for branch 4 to increase the number of personal transactions by 10 percent (e.g., by obtaining new accounts).

Plainly, in practice, we might well set a branch a mix of input and output targets that we want it to achieve. We can use linear programming.

# Linear Programming Example of DEA

| **Example 2** | ## Ranking Banks with LP |
|---|---|

Typically, we have more than one input and one output. For the bank branch example, suppose now that we have two output measures (number of personal transactions completed and number of business transactions completed) and the same single input measure (number of staff) as before.

The data we have are as follows:

| Branch | Personal Transactions (1,000) | Business Transactions (1,000) | Number of Staff |
|--------|-------------------------------|-------------------------------|-----------------|
| 1 | 125 | 50 | 18 |
| 2 | 44 | 20 | 16 |
| 3 | 80 | 55 | 17 |
| 4 | 23 | 12 | 11 |

We start by scaling (via ratios) the inputs and outputs to reflect the ratio of 1 unit.

| Branch | Personal Transactions (1,000) | Business Transactions (1,000) | Per Employee or Staff |
|--------|-------------------------------|-------------------------------|-----------------------|
| 1 | $125/18 = 6.94$ | $50/18 = 2.78$ | $18/18 = 1$ |
| 2 | $44/16 = 2.75$ | $20/16 = 1.25$ | $16/16 = 1$ |
| 3 | $80/17 = 4.71$ | $55/17 = 3.24$ | $17/17 = 1$ |
| 4 | $23/11 = 2.09$ | $12/11 = 1.09$ | $11/11 = 1$ |

Pick a DMU to maximize: $E_1$, $E_2$, $E_3$, or $E_4$

Let $W_1$ and $W_2$ be the personal and business transactions at the branch.

In this example, we choose to maximize branch 2, $E_2$.

Here is the LP formulation:

$$\text{Maximize } E_2$$
$$\text{Subject to}$$
$$E_1 = 6.94 \ W_1 + 2.78 \ W_2$$
$$E_2 = 2.75 \ W_1 + 1.25 \ W_2$$
$$E_3 = 4.71 \ W_1 + 3.24 \ W_2$$
$$E_4 = 2.09 \ W_1 + 1.09 \ W_2$$
$$E_1 \leq 1$$
$$E_2 \leq 1$$
$$E_3 \leq 1$$
$$E_4 \leq 1$$

```
> with(Optimization) :
> obj:=E2;
```

$$obj := E2$$

```
> const:={E1-6.94·W1-2.78·W2=0,E2-2.75·W1-1.25·W2=0,E3-4.17·W1-3.24·W2=0,
    E4-2.09·W1-1.09·W2=0,E1≤1,E2≤1,E3≤1,E4≤1};
```

$$const := \{E1 - 6.94 \ W1 - 2.78 \ W2 = 0, \ E2 - 2.75 \ W1 - 1.25 \ W2 = 0, \ E3 - 4.71$$
$$W1 - 3.24 \ W2 = 0, \ E4 - 2.09 \ W1 - 1.09 \ W2 = 0, \ E1 \leq 1, \ E2 \leq 1,$$
$$E3 \leq 1, \ E4 \leq 1\}$$

```
> LPSolve(obj, const, maximize);
```

$$[0.43149343043932, \ [E2 = 0.431493430439319481, \ E1 = 1.,$$
$$W1 = 0.0489788964841672560, \ W2 = 0.237441172086287788,$$
$$E3 = 0.99999999999999988, \ E4 = 0.3611767711225963034]]$$

Now, what did we learn from this? If we rank order the branches on efficiency performance of our inputs and outputs, we find the following:

| | |
|---|---|
| Branch 1 | 100% |
| Branch 3 | 100% |
| Branch 2 | 43.2% |
| Branch 4 | 36.2% |

We know we need to improve on branch 2 and branch 4 performances while not losing our efficiency in branches 1 and 3. A better interpretation could be that if the practices and procedures used by the other branches were to be adopted by branch 4, then it could improve its performance. Would things have changed if we had optimized different variables, like E1? No, the results would be the same.

This invokes issues of highlighting and disseminating examples of best practices. Equally, there are issues relating to identification of poor practices.

In DEA, the concept of the reference set can be used to identify the best-performing branches for comparison with poorly performing branches. If you use this procedure, use it wisely.

## 7.7 | EXERCISES

**1.** Consider rating departments at a college. The following table is provided:

| DMU Departments | Inputs No. Faculty | Outputs Student Credit Hours | Number of Students | Total Degrees (MS and PhD) |
|---|---|---|---|---|
| Unit 1 | 25 | 18,341 | 9,086 | 63 |
| Unit 2 | 15 | 8,190 | 4,049 | 23 |
| Unit 3 | 10 | 2,857 | 1,255 | 31 |
| Unit 4 | 33 | 22,277 | 6,102 | 31 |
| Unit 5 | 12 | 6,830 | 2,910 | 19 |

Formulate and solve the DEA model and rank order the five departments.

**2.** Consider ranking companies within a task force. For simplification reasons, we will consider only six companies.

| Companies | Inputs (No. Size of Unit) | Output 1 | 2 | 3 |
|---|---|---|---|---|
| Unit 1 | 120 | 18,341 | 9,086 | 63 |
| Unit 2 | 110 | 8,190 | 4,049 | 23 |
| Unit 3 | 100 | 2,857 | 1,255 | 31 |
| Unit 4 | 135 | 22,277 | 6,102 | 31 |
| Unit 5 | 120 | 6,830 | 2,910 | 19 |
| Unit 6 | 95 | 5,050 | 1835 | 12 |

## FURTHER READING

Bazarra, M. S., J. J. Jarvis, and H. D. Sheralli, *Linear Programming and Network Flows.* New York. John Wiley & Sons, 1990.

Ecker, J., and M. Kupperschmid, *Introduction to Operations Research.* New York. John Wiley and Sons, 1988.

Hiller, F. S., and G. J. Liberman, *Introduction to Mathematical Programming.* New York: McGraw Hill Publishing Company, 1990.

Winston, W. L., *Introduction to Mathematical Programming Applications and Algorithms*, 4th edition. Belmont, CA. Duxbury Press. 2002.

8

# Modeling with Single-Variable Unconstrained Optimization

## Introduction

Consider an oil-drilling rig that is 8.5 miles offshore. The rig is required to be connected by underwater pipe to a pumping station. The pumping station is connected by land-based pipe to a refinery, which is 14.7 miles down the shoreline from the drilling rig (see Figure 8.1). The underwater pipe costs \$31,575 per mile, and the land-based pipe costs \$13,342 per mile. We are to determine where to place the pumping station to minimize the cost of the pipe.

In this chapter, we will discuss models that require single-variable calculus to solve. We will review the calculus concepts for optimization and then apply them to this application. We will use Maple to assist us.

## 8.1  SINGLE-VARIABLE BASIC THEORY

We want to solve problems of the following form:

$$\max \text{ (or min) } f(x)$$
$$x \in (a, b)$$

If $a = -\infty$ and $b = \infty$, then we are looking at $R^2$, the $xy$ plane. If either $a$ or $b$ or both $a$ and $b$ are restricted, then we must consider possible end points in our solution. We will examine the following three cases.

**Case 1.**  Points where $a < x < b$ and $f'(x) = 0$.

**Case 2.**  Points where $f'(x)$ does not exist.

**Case 3.**  End points $a$ and $b$ of the interval $[a, b]$.

**FIGURE 8.1**
Oil rig and pipeline schematic
(not to scale)

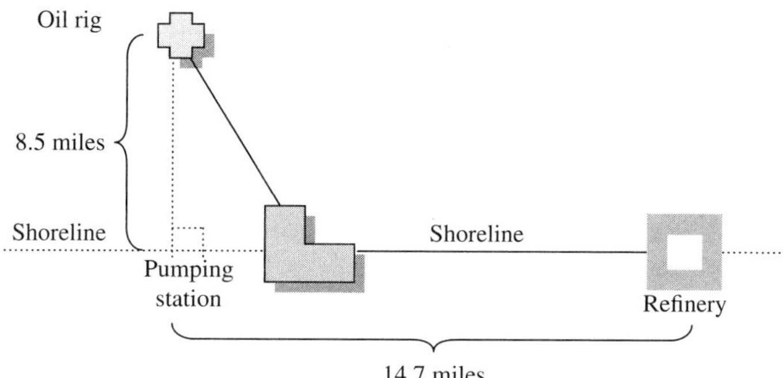


In addition, we will define points where $f'(x) = 0$ as critical points or stationary points. The following additional definitions and theorems from calculus might be useful.

**Definition** | A function $f$ has a maximum (global) at a point $c$ if $f(c) \geq f(x)$ for all $x$ in the domain of $f$, and a function $f$ has a minimum (global) at a point $c$ if $f(c) \leq f(x)$ for all $x$ in the domain of $f$.

**Extreme Value Theorem** | If $f$ is continuous on a closed interval $[a, b]$, then $f$ has both a global maximum and a global minimum over the interval.

Recall from your study of calculus the analysis of the first derivative test:

If $f'(x) > 0$ to the left of $x^*$ and $f'(x) < 0$ to the right of $x^*$, then $x^*$ is a local maximum and we find $f(x)$ increasing to the left and decreasing to the right of point $x^*$.

Also recall the second derivative test. If $f'(x_0) = 0$, then we compute $f''(x_0)$.

If $f''(x_0) < 0$, then $x_0$ is a local maximum.

If $f''(x_0) > 0$, then $x_0$ is a local minimum.

If $f''(x_0) = 0$, then $x_0$ might be an inflection point.  ■

**Theorem 8.1** | If $f'(x_0) = 0$ and $f''(x_0) < 0$, then $x_0$ is a local maximum.
If $f'(x_0) = 0$ and $f''(x_0) > 0$, then $x_0$ is a local minimum.  ■

**Theorem 8.2** | If $f''(x_0) = 0$, and

1. if the first nonzero derivative at $x_0$ occurs at an odd-order derivative, then $x_0$ is neither a local maximum nor a local minimum;

2. if the first nonzero derivative is positive and occurs at an even-order derivative, then $x_0$ is a local minimum;

3. if the first nonzero derivative is negative and occurs at an even-order derivative, then $x_0$ is a local maximum.  ■

# Case 2. Points Where $f'(x)$ Does Not Exist

If $f(x)$ does not have a derivative at $x_0$, then $x_0$ might be a local maximum, a local minimum, or neither. In this case, we test points near $x_0$ and evaluate the function at those neighboring points where $x_1 < x_0 < x_2$.

| Relationship Between $f(x_0)$ and Close Neighbors | $x_0$ Classification |
| --- | --- |
| $f(x_0) > f(x_1); f(x_0) < f(x_2)$ | Not a local extremum |
| $f(x_0) < f(x_1); f(x_0) > f(x_2)$ | Not a local extremum |
| $f(x_0) \geq f(x_1); f(x_0) \geq f(x_2)$ | Local maximum |
| $f(x_0) \leq f(x_1); f(x_0) \leq f(x_2)$ | Local minimum |

# Case 3. End points $a$ and $b$

From Figure 8.2(a)–(d), we see the following:

1. If $f'(a) > 0$, then $a$ is a local minimum, see Figure 8.2(a).
2. If $f'(a) < 0$, then $a$ is a local maximum, see Figure 8.2(b).
3. If $f'(b) < 0$, then $b$ is a local minimum, see Figure 8.2(c).
4. If $f'(b) > 0$, then $b$ is a local maximum, see Figure 8.2(d).

If both $f'(a) = 0$ and $f'(b) = 0$, then draw a sketch and test neighboring points to determine if $a$ or $b$ (or both) is an extremum.

## Example: Minimize $1.5x^2$

Solution:

$$f(x) = 1.5x^2$$

$$\frac{df(x)}{dx} = 3x$$

We solve $\dfrac{dx}{dt} = 0$, so

$$3x = 0$$
$$x = 0$$

We take the 2nd derivative and evaluate at the critical point, $x = 0$.

$$\frac{d^2 f(x)}{dx^2}\Big|_{x=0} = 3$$

The first derivative is equal to 0 at $x = 0$. The second derivative is positive at $x = 0$, so the critical point yields a local minimum (which is also a global minimum) of the function.

## Example: Minimize $f(x) = x^3$

where we restrict the value of $x$ to be in the following interval: $-1 \leq x \leq 3$.

### Solution

We find $f'(x) = 0$ at $x = 0$. The second-derivative test yields $f''(0) = 0$, so we have an inflection point at $x = 0$. We also note that the $f(0) = 0$. Next, we test the end points:

$$f(-1) = -1$$
$$f(3) = 27$$

We find a minimum at $x = -1$.

**FIGURE 8.2**

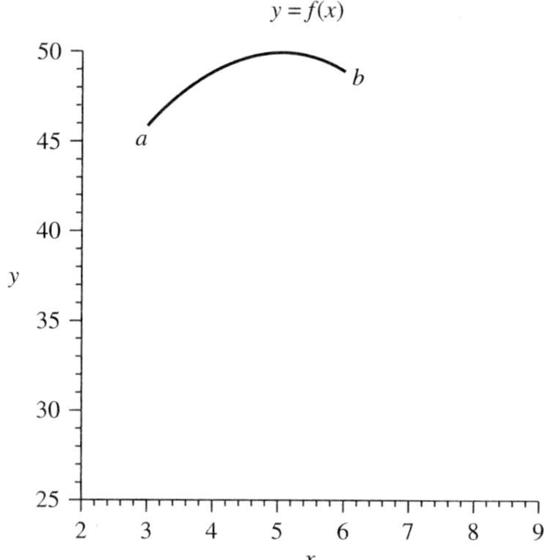
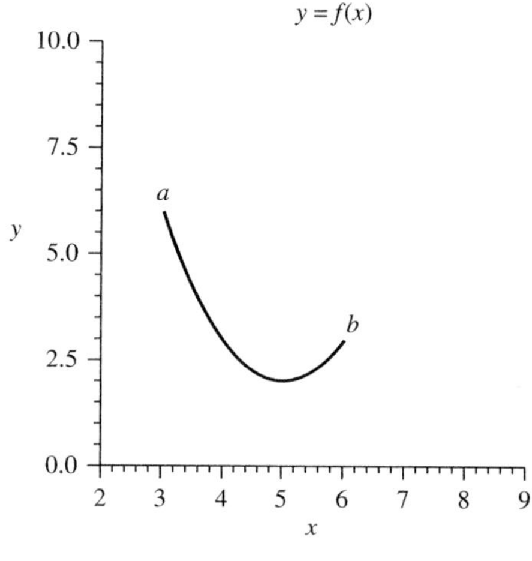

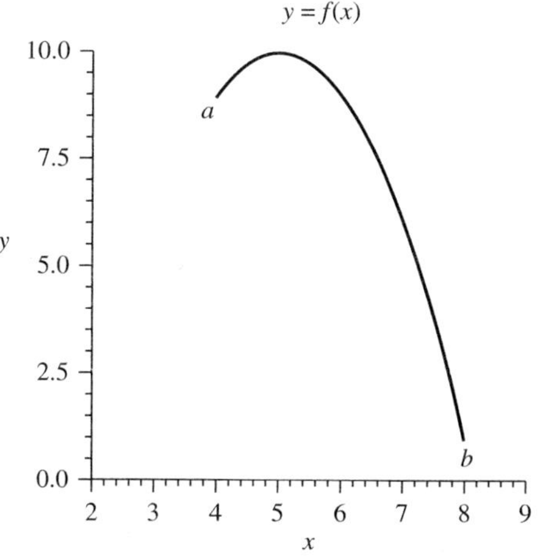
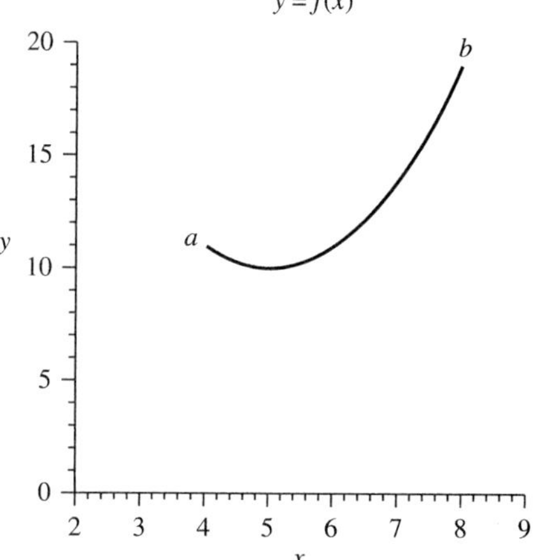

## 8.1 | EXERCISES

Solve the following optimization problems.

1. Maximize $f(x) = -2x^2 + 10$ for $x$ in the interval $[0,2]$.
2. Maximize $f(x) = -2x^2 + 10$ for $x$ in the interval $[0,4]$.
3. Minimize $f(x) = 2.1x^2 - 16.8x + 5.1$.
4. Find all local maximum and minimum of $f(x) = 3x^4 - 4x^3$.
5. Find the maximum and minimum for $f(x) = 2x^2 - 3x$.
6. What is the location of the maximum and minimum of $f(x) = \sqrt{2 + x} - \sqrt{2 - x}$?
7. Find the maximum of $f(x) = \dfrac{4}{x^2 - 4}$ in the interval $(-2, 2)$.

<table>
<tr><td>**8.2**</td><td># MODELS WITH BASIC APPLICATIONS OF MAX–MIN THEORY</td></tr>
</table>

## Example 1 — Chemical Sales

A chemical-manufacturing company sells sulfuric acid at a price of $100 per unit. If the daily total production cost in dollars for $x$ units is

$$C(x) = 100{,}000 + 50x + 0.0025x^2$$

and the daily production is at most 7,000 units, then how many units of sulfuric acid should the manufacturer produce to maximize daily profits?

Solution: Profit = revenue − cost

$$P = 100x - (100{,}000 + 50x + 0.0025x^2)$$
$$P = -100{,}000 + 50x - 0.0025x^2, \text{ for } 0 \le x \le 7{,}000$$
$$\frac{dP}{dx} = 50 - 0.005x = 0$$
$$x = 10{,}000 \text{ units with } P(10{,}000) = \$150{,}000$$
$$\frac{d^2P}{dx^2} = -0.005.$$

Because $P'' < 0$, we have a local maximum.

We must also check the end points.

$$x = 0, \, P = -\$100{,}000$$
$$x = 7{,}000, \, P = \$127{,}500$$

Our solution is $x = 7{,}000$ and $P = \$127{,}500$ at the end point because $x^*$ is outside of the domain of production, even though its mathematical results are better: $x^* = 10{,}000$ units with $P(10{,}000) = \$150{,}000$.

In Maple, we will use commands that we have previously seen in Chapter 1:

---

diff or Diff  - **Differentiation or Partial Differentiation**

**Calling Sequence**

   diff(**f, x1, ..., xj**)

   diff(**f, [x1\$n]**)

   diff(**f, x1\$n, [x2\$n, x3], ... xi, [xj\$m]**)

solve - **solve one or more equations**

Calling Sequence

   solve(**equations, variables**)

---

```
> cost:=100*x−(100000+50*x+0.0025*x^2);
```

$$cost := 50x - 100000 - 0.0025x^2$$

```
> dc:=diff(cost,x);
```

$$dc := 50 - 0.0050x$$

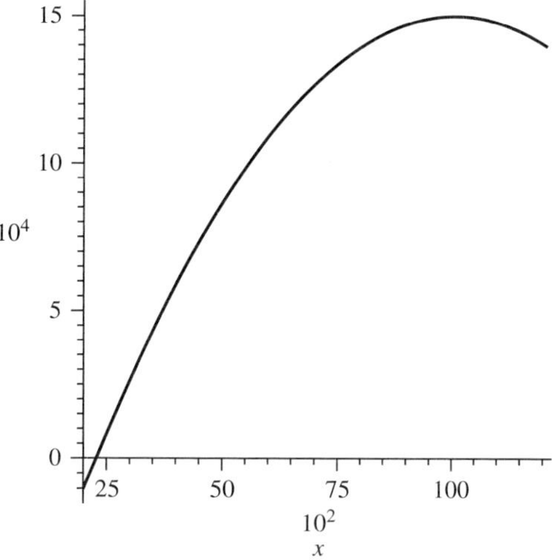

**FIGURE 8.3**
Profit function from chemical sales

> *ans:=solve(dc=0,x);*

$$ans := 10000.$$

> *ddc:=diff(dc,x);*

$$ddc := -0.0050$$

> *subs(x=ans,cost);*

$$150,000$$

> *plot(cost,x=0..12000);*

From Figure 8.3, we see that there is no local maximums within our interval. Thus, we check the end points:

> *subs(x=0,cost);*

$$-10000.$$

> *subs(x=7000,cost);*

$$12750$$

The end-point solution is selected because the critical point is outside the domain.

---

## Example 2    Inventory Model: Manufacturing and Storage

A company wants to minimize the costs involved with manufacturing and storing items for future sales.

Let $x$ = the number of batches of the item produced per year.
Let $s$ = the storage costs of one unit for one year.
Let $F$ = the fixed setup costs to produce the item (includes insurance, machines, labor, etc.).
Let $v$ = the variable cost in producing one unit of the item.
Let $T$ = the total number of units produced annually.

$$\text{Total production cost, } M(x) = \left(F + \frac{vT}{x}\right)x$$

$$\text{Average storage cost, } s(x) = \frac{kT}{2x}$$

$$\text{Total cost function, } C(x) = \left(F + \frac{vT}{x}\right)x + \frac{kT}{2x}$$

## Solution

$$C(x) = \left(F + \frac{vT}{x}\right)x + \frac{kT}{2x}$$

$$\frac{dC(x)}{x} = F - \frac{kT}{2x^2} = 0$$

$$x^* = \sqrt{\frac{kT}{2F}}$$

$$\frac{d^2 C(x)}{dx^2} = \frac{kT}{x^3} > 0 \text{ for real values of } k, \, T, \text{ and } x.$$

Because the second derivative is greater than zero, $\dfrac{kT}{x^3} > 0$, the critical point $x^* = \sqrt{\dfrac{kT}{2x}}$ represents a minimum.

Maple does symbolic mathematics as well.

```
> cost:=(F+v*T/x)*x+(k*T)/(2*x);
```

$$cost := \left(F + \frac{vT}{x}\right)x + \frac{kT}{2x}$$

```
> dcost:=diff(cost,x);
```

$$dcost := F - \frac{kT}{2x^2}$$

```
> solve(dcost=0,x);
```

$$\frac{\sqrt{2}\,\sqrt{F\,k\,T}}{2F}, \, -\frac{\sqrt{2}\,\sqrt{F\,k\,T}}{2F}$$

We only need the positive of the two solutions.

```
> subs(x=1/2/F*2^(1/2)*(F*k*T)^(1/2),diff(dcost,x));
```

$$\frac{2\,k\,T\,F^3\,\sqrt{2}}{(F\,k\,T)^{(3/2)}}$$

The same arguments hold for the analysis of the critical points. Since $k$, $T$, and $F$ are always positive, then our second derivative is positive and we have the minimum of the function.

---

<table><tr><td>**Example 3**</td><td>## Producing the SP6 Computer</td></tr></table>

A company spends $200 in variable costs to produce a SP6 computer, plus a fixed cost of $5,000 if any SP6 computers are produced. If the company spends $x$ dollars on advertising its new SP6 computer, it can sell $x^{1/2}$ units at $500 per computer. How many SP6 computers should the company produce to maximize profits?

## Solution

$$\text{Maximize profit} = \text{revenue} - \text{cost}$$
$$\text{Cost} = \text{fixed} + \text{variable} + \text{advertising costs} = 5{,}000 + 200 \cdot x^{1/2} + x$$

$$\text{Revenue} = 500 \cdot x^{1/2}$$
$$\text{Maximize } P = 500 \cdot x^{1/2} - (5{,}000 + 200 \cdot x^{1/2} + x)$$

$$\frac{dP}{dx} = \frac{d(500 \cdot x^{1/2} - (5000 + 200 \cdot x^{1/2} + x))}{dx} = -\frac{-150 + \sqrt{x}}{\sqrt{x}} = \frac{150}{\sqrt{x}} - 1$$

$$\text{Set } \frac{dP}{dx} = 0, \text{ yields } \frac{150}{\sqrt{x}} - 1 = 0, \; x = 22{,}500$$

$$500 \cdot x^{1/2} - (5000 + 200 \cdot x^{1/2} + x)$$

$$\frac{d^2 P}{dx^2} = \frac{-75}{x^{3/2}} < 0 \text{ for all } x > 0.$$

Therefore, we have found a maximum.

We can use the following Maple commands to obtain our solution.

> *restart;*
> *profit:=5000*sqrt(x)–(5000+200*sqrt(x)+x);*

$$\textit{profit} := 300 \sqrt{x} - 5000 - x$$

> *dp:=diff(profit,x);*

$$dp := \frac{150}{\sqrt{x}} - 1$$

> *ans:=solve(dp=0,x);*

$$\textit{ans} := 22500$$

> *evalf(subs(x=ans,profit));*

$$22000.00000$$

> *evalf(subs(x=ans,diff(dp,x)));*

$$-0.00002222222222$$

Because $p'' < 0$, we found the maximum (see Figure 8.4).

> *plot(profit, x=0..30000, title='Profit',thickness=3);*

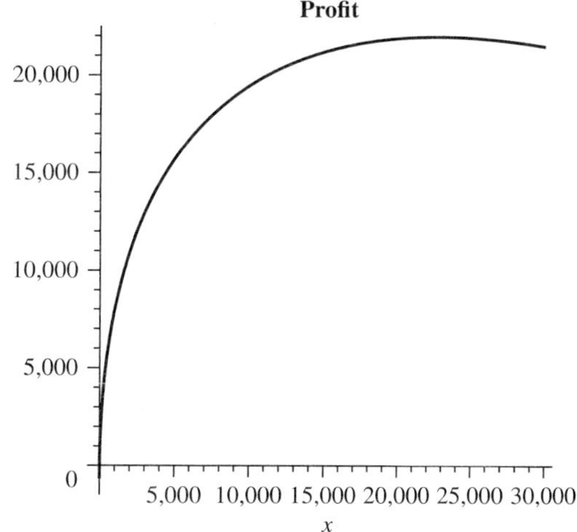

**FIGURE 8.4**
Profit function of
SP6 computers

## 8.2 | EXERCISES

1. Each morning during rush hour, 10,000 people travel from New Jersey to New York City. If a person takes the subway, the trip lasts 40 minutes. If $x$ thousand people drive to New York City, it takes $20 + 5x$ minutes to make the trip.
   a. Formulate the problem with the objective of minimizing the average travel time per person. Let $x$ = number of people (in 1,000s) who drive to New York City from New Jersey.
   b. Find the optimal number of people who drive so that the average time per person is minimized.

2. Use calculus to find the optimal solution to

$$\text{Maximize } f(x) = 2x^3 - 1$$
$$-1 \le x \le 1$$

3. Given the following plot of the function: $f(x) = 0.5x^3 - 8x^2$

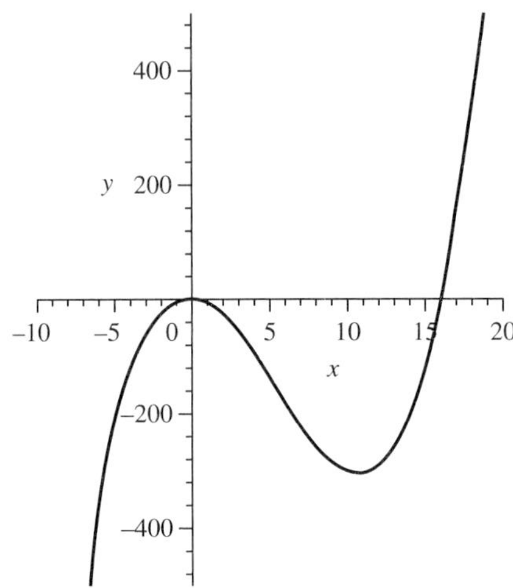

   a. Using analytical techniques (i.e., calculus), find and classify all **extrema** for $f(x)$.
   b. Apply the definition of convexity to show that the function $f(x)$ is *concave* over the interval between $x_1 = -6$ and $x_2 = 5$, using $c = 0.5$. (We're fixing the value of $c$ here to keep the algebra from getting ugly.)
   c. Confirm your response to part (b) using the second-derivative test to show that $f(x)$ is concave over this entire interval.
   d. Over what interval is the above function convex? Concave?
   e. Why is knowing the concavity important in optimization?

4. Find the optimal solution to the bounded single-variable optimization problem. (You may use calculus here.)

$$\text{Maximize } f(x) = 2x^3 - 2$$
$$\text{in the interval } -1 \le x \le 1$$

5. Dr. E. N. Throat has been taking X-rays of the trachea contracting during coughing. He has found that the trachea appears to contract by 33 percent of its normal size. He has asked the department of mathematics to confirm or deny his claim. You perform initial research and find that under reasonable assumptions about the elasticity of the tracheal wall and about how air near the wall is slowed by friction, the average flow of velocity $v$ can be modeled by the equation

$$v = c(r_o - r)r^2 \text{ cm/sec, between } r_o/2 \le r \le r_o$$

where $c$ is a positive constant (let $c = 1$) and $r_o$ is the resting radius of the trachea in centimeters.

Find the value of $r$ that maximizes $v$ and then *support* or *deny* the claim.

# 8.3  APPLIED SINGLE-VARIABLE OPTIMIZATION MODELS

### Example 1    An Inventory Problem: Revisit Minimizing the Cost of Delivery and Storage

You have been hired as a consultant by a chain of gasoline stations to determine how often and how much gasoline should be delivered to various stations. After some questioning, you determine that each time gasoline is delivered stations incur a charge of $d$ dollars, which is in addition to the cost of the gasoline and is independent of the amount delivered.

Costs are also incurred when the gasoline is ordered. One such cost is capital tied up in inventory—that is, money that is invested in the stored gasoline and cannot be used elsewhere. The cost is normally computed by multiplying the cost of the gasoline to the company by the current interest rate for the period the gasoline was stored. Other costs include amortization of the tanks and the equipment necessary to store the gasoline, insurance, taxes, security measures, and so forth.

The gasoline stations are located near interstate highways, where demand is fairly constant throughout the week. Records are available that indicate the gallons sold daily for each station.

## Problem Identification

Assume the company wishes to maximize its profits and that demand and price are constant in the short run. Thus, because total revenue is constant, total profit can be maximized by minimizing the total costs. There are many components of total costs such as overhead, employee costs, and the like. If these costs are affected by the amount and the time of the deliveries, they should be considered. Let's assume the costs are not so affected and focus our attention on the following problem: *Minimize the average daily cost of delivering and storing sufficient gasoline at each station to meet consumer demand.* Intuitively, we expect such a minimum to exist. If the delivery charge is very high and the storage cost is very low, then we would expect very large orders of gasoline to be delivered infrequently. On the other hand, if the delivery cost is very low and the storage costs are very high, then we would expect small orders of gasoline to be delivered very frequently.

## Assumptions

In the following presentation, we consider some factors that are important in deciding how large an inventory to maintain. Obvious factors to consider are delivery costs, storage costs, and demand rate for the product. Perishability of the product being stored also may be of paramount concern. In the case of gasoline, condensation also may become increasingly important as the gasoline level gets lower in the tank.

The market stability of the selling price of the product and the cost of raw materials need to be considered. For example, if the market price of the product is volatile, the seller would be reluctant to store large quantities of the product. On the other hand, an expected large increase in the price of raw materials in the near future argues for large inventories. The stability of consumer demand is another factor to be taken into account. There are many seasonal fluctuations in the demand for the product. In addition, a technological breakthrough could cause the product to become obsolete. The time horizon being considered also can be extremely important. In the short run, contracts may be signed for warehouse space, some of which will not be needed in the long run. Another consideration is the importance to the owner of an occasional unsatisfied demand (stock outs). Some owners would opt for a more costly inventory strategy to ensure that they never run out of gas. From this discussion, we can see that any inventory decision is not an easy one. We restrict our initial model here to the following variables:

$$\text{average daily cost} = f(\text{storage costs, delivery costs, demand rate})$$

# The Submodels

## Storage Costs

We need to consider how the storage cost per unit varies with the number of units being stored. Are we renting space and receiving a discount when storage exceeds certain levels or do we rent the cheapest storage first (adding more space as needed)? Do we need to rent an entire warehouse or floor? If so, the per unit price is likely to decrease as the quantity stored increases until another warehouse or floor needs to be rented. Does the company own its own storage facilities? If so, what alternative use can be made of them? These are all valid options. To keep it simple, we will assume that per unit storage is a constant in our model.

## Delivery Costs

In many cases, the delivery charge depends on the amount delivered. For example, if a larger truck or an extra flatcar is needed, an additional charge is made. In our model, we consider a constant delivery charge to be independent of the amount delivered.

## Demand

Even though we realize that the demands occur in discrete time periods, for purposes of simplification we take a continuous submodel for demand. This continuous submodel is a line through the origin with a positive slope, where the slope of the line represents the constant demand rate. Notice the importance of our assumptions in producing the linear submodel.

## Model Formulation

We use the following notation for constructing our model:

$$s = \text{storage costs, per gallon, per day}$$
$$d = \text{delivery cost, in dollars, per delivery}$$
$$r = \text{demand rate, in gallons, per day}$$
$$Q = \text{quantity of gasoline, in gallons}$$
$$T = \text{time, in days}$$

Now suppose an amount of gasoline—say, $Q = q$—is delivered at time $T = 0$ and that the gasoline is used up after $T = t$ days. The same cycle is then repeated, as illustrated in the sawtooth graph in Figure 8.5. The slope of each line segment is $-r$ (the negative of the demand rate). The problem is to determine an order quantity $Q^*$ and a time between order $T^*$ that minimizes the delivery and storage costs.

We seek an expression for the average daily cost, so consider the delivery and storage costs for a cycle of length $t$ days. The delivery costs are the constant amount, $d$, because only one

**FIGURE 8.5**
An inventory cycle

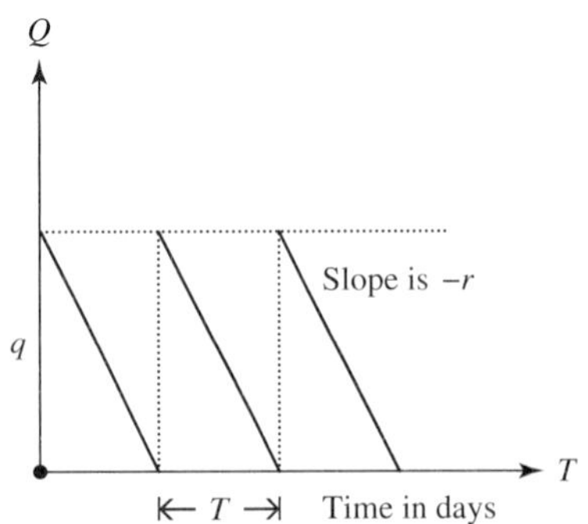

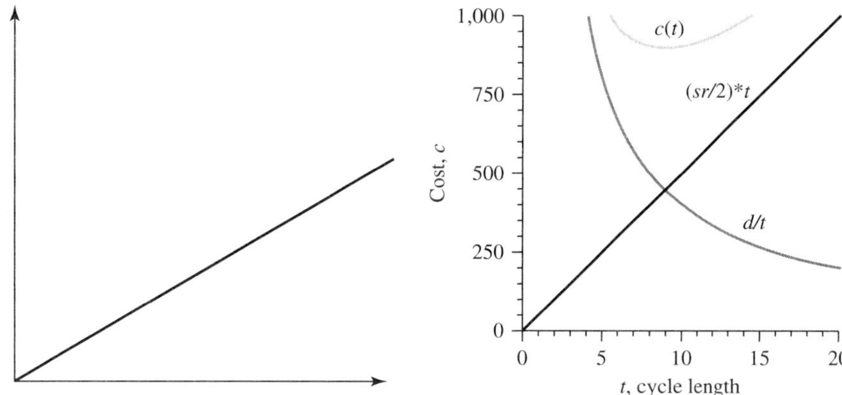



**FIGURE 8.6**
The average daily cost curve as a sum of a hyperbola and a line

delivery is made over the single time period. To compute the storage cost, take the average daily inventory, $q/2$, multiply by the number of days in storage, $t$, and multiply that by the storage cost per item per day, $s$. In our notation, this gives:

$$\text{Cost per cycle} = d + s\frac{q}{2}t$$

On division by $t$, this yields the average daily cost: $c = \dfrac{d}{t} + \dfrac{sq}{2}$

## Model Solution

Apparently, the cost function to be minimized has two independent variables, $q$ and $t$. For a single cyclic period, the amount delivered equals the amount demanded. This translates to $q = rt$. Substitution into the average daily cost equation yields $c = \dfrac{d}{t} + \dfrac{srt}{2}$.

Equation (8.1) is the sum of a hyperbola and a linear function. The situation is depicted in Figure 8.6.

Let's find the time between orders $T^*$ that minimizes the average daily cost. Differentiating $c$ with respect to $t$ and setting $c' = 0$ yields $c' = -\dfrac{d}{t^2} + \dfrac{sr}{2} = 0.$

This last equation gives the (positive) critical point: $T^* = \left(\dfrac{2d}{sr}\right)^{1/2}.$

This critical point provides a relative minimum for the cost function because the second derivative $c'' = \dfrac{2d}{t^3}$ is always positive for positive values of $t$. It is clear from Figure 8.6 that $T^*$ gives a global minimum as well. And note that $\dfrac{d}{T^*} = \dfrac{sr}{2}T^*$ so that $T^*$ is the point at which the linear function and the hyperbola intersect in Figure 8.6.

---

<table><tr><td>

**Example 2**

</td><td>

## Oil-Rig Location Problem

</td></tr></table>

Consider an oil-drilling rig that is 8.5 miles offshore. The rig is to be connected by underwater pipe to a pumping station. The pumping station is connected by land-based pipe to a refinery, which is 14.7 miles down the shoreline from the drilling rig (refer to Figure 8.1 at the beginning of the chapter). The underwater pipe costs \$31,575 per mile, and the land-based pipe costs \$13,342 per mile. You are to determine where to place the pumping station to minimize the cost of the pipe.

## Problem Identification

Find a relationship between the location of the pumping station and cost of the pipe installation.

## Assumptions and Variables

First, we assume no cost saving for the pipe if we purchase in larger lot sizes. We further assume no additional costs are incurred in preparing the terrain to lay the pipe.

## Variables

$x$ = the location of the pumping station along the horizontal distance from $x = 0$ to $x = 14.7$ miles

$TC$ = total cost of the pipe for both underwater and onshore piping

## Model Construction

We use Pythagoras's theorem for the underwater distance of the pipe, which is the hypotenuse of the right triangle with height 8.5 miles and base $= x$. The hypotenuse is $\sqrt{8.5^2 + x^2}$. The length of the pipe onshore is $14.7 - x$.

$$\text{Total cost} = 31{,}575 \sqrt{8.5^2 + x^2} + 13{,}342\,(14.7 - x).$$

```
> TC:=31575*sqrt(8.5^2+x^2)+13342*(14.7–x);
```

$$TC := 31575 \sqrt{72.25 + x^2} + 196127.4 - 13342x$$

```
> cp:=diff(TC,x);
```

$$cp := \frac{31575\,x}{\sqrt{72.25 + x^2}} - 13342$$

```
> xstar:=solve(cp=0,x);
```

$$xstar := 3.962829844$$

```
> TC_at_xstar:=subs(x=xstar,TC);
```

$$TC\_at\_xstar := 439377.6878$$

```
> plot(TC,x=0..7, thickness=3,title='Total_Cost');
```

Thus, if the pumping station is located at $14.7 - 3.963 = 10.737$ miles from the refinery, we will minimize the total cost at a cost of \$439,377.69 (see also Figure 8.7).

**FIGURE 8.7**
Plot of total cost

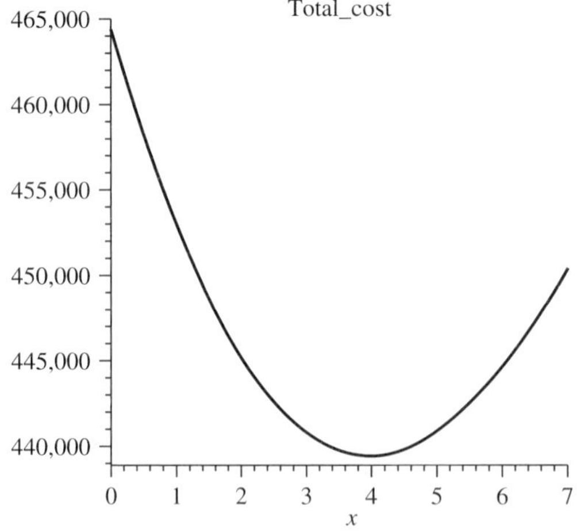

# 8.3 | EXERCISES

1. Consider an industrial situation in which an assembly line must be set up. Suppose that each time the line is set up a cost $c$ is incurred. Assume $c$ is in addition to the cost of producing any item and is independent of the amount produced. Suggest submodels for the production rate. Now assume a constant production rate $k$ and a constant demand rate $r$. What assumptions are implied by the model in the following figure? Assume a storage cost of $s$ (in dollars per unit per day) and compute the optimal length of the production run $P^*$ in order to minimize the costs. List all of your assumptions. How sensitive is the average daily cost to the optimal length of the production run?

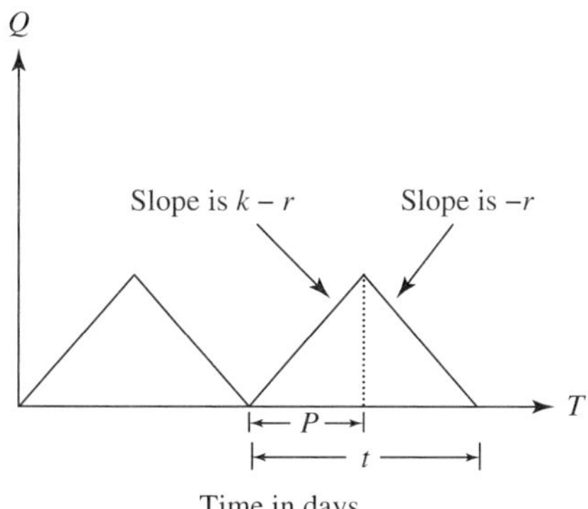


2. Consider a company that allows back-ordering. In other words, the company notifies customers that a temporary stock out exists and that their orders will be filled shortly. What conditions might argue for such a policy? What effect does such a policy have on storage costs? Should costs be assigned to stock outs? Why? How would we make such an assignment? What assumptions are implied by the model in the following figure? Suppose a "loss of goodwill cost" of $w$ dollars per unit per day is assigned to each stock out. Compute the optimal order quantity $Q^*$ and interpret your model.

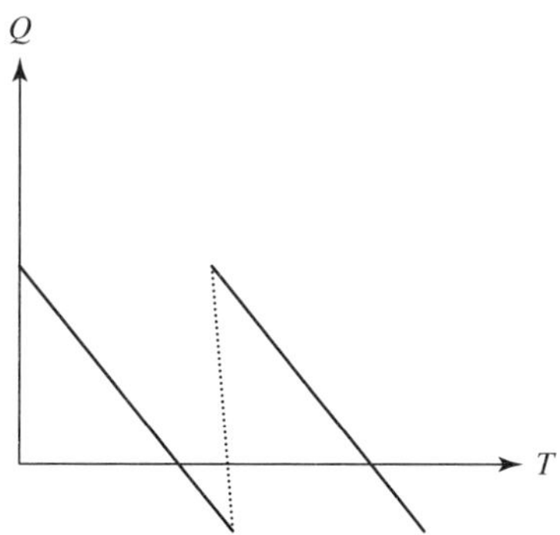


3. In the inventory model discussed in the text, we assumed a constant delivery cost that is independent of the amount delivered. Actually, in many cases, the cost varies in discrete amounts, depending on the size of the truck needed, the number of platform cars required, and so forth. How would you modify the model in the text to take into account these changes? We also assumed a constant cost for raw materials. However, bulk-order discounts are often given. How would you incorporate these discounts into the model?

4. What is the optimal speed and safe following distance that allows the maximum flow rate (cars per unit time)? The solution to the problem would be of use in controlling traffic in

tunnels, for roads under repair, or for other congested areas. In the following schematic, $l$ is the length of a typical car and $d$ is the distance between cars.

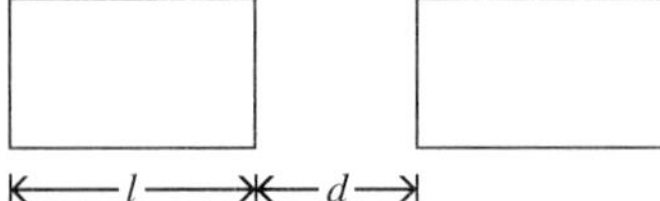

$$\text{Justify that the flow rate is given by } f = \frac{velocity}{distance}.$$

Let's assume a car length of 15 ft. The safe stopping distance $d = 1.1v + 0.054v^2$ was determined, where $d$ is measured in feet and $v$ in miles per hour. Find the velocity in miles per hour and the corresponding following distance $d$ that maximizes traffic flow. How sensitive is the solution to changes in $v$? Can you suggest a practical rule? How would you enforce it?

5. Consider an athlete competing in the shot put. What factors influence the length of his or her throw? Construct a model that predicts the distance thrown as a function of the initial velocity and angle of release. What is the optimal angle of release? If the athlete cannot maximize the initial velocity at the angle of release you propose, should he or she be more concerned with satisfying the angle of release or generating a high initial velocity? What are the trade-offs?

6. John Martin is responsible for periodically buying new trucks to replace older trucks in his company's fleet of vehicles. He is expected to determine the time a truck should be retained so as to minimize the average cost of owning the truck. Assume the purchase price of a new truck is \$9,000 with a trade-in. Also assume the maintenance cost (in dollars) per truck for $t$ years can be expressed analytically by the following empirical model: $C(t) = 640 + 180(t + 1)t$ where $t$ is the time in years the company owns the truck.

## 8.3 │ PROJECT

1. You are a volunteer working with the Peace Corps in a country in Central America. Your group has just finished building a flight strip that will be used to bring supplies to a village. There is another village 100 km away that is inaccessible by vehicle from the flight strip (see the following diagram).

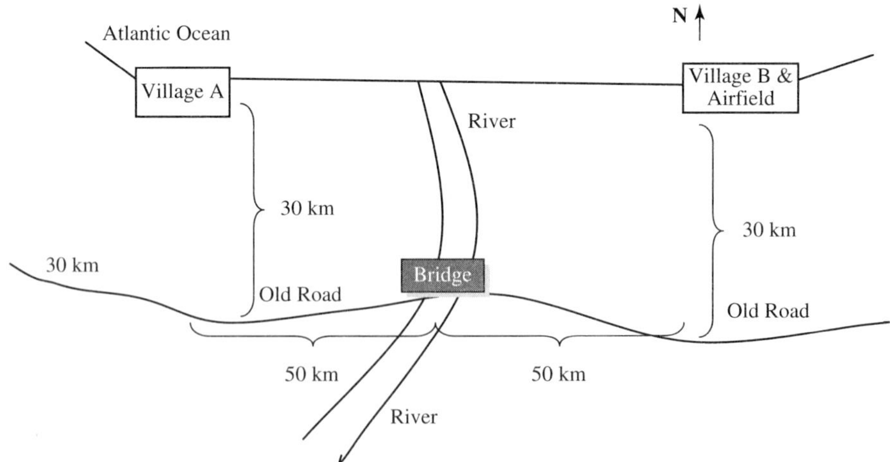


Your group has been asked to build a road connecting the two villages. *You must cross the existing bridge.* Currently, an unimproved dirt road 30 km south of both villages connects with the bridge. Assume the Old Road is straight and in an east–west direction. You want to build the most cost-effective road connecting the two villages.

After surveying the Old Road and the terrain between the two villages, you estimate that the cost of materials and equipment to improve the Old Road is $120,000 per kilometer. Constructing a new road, on the other hand, will cost $220,000 per kilometer.

**a.** Draw a diagram that includes definitions of any variables you choose to use.

**b.** Analysis of the problem should include at least the answers to the following questions (all of this can be neatly handwritten):

- What is the cost function for the roads?
- What are the lengths of the newly constructed road and the refurbished Old Road that will minimize the cost of the road complex that connects the two villages?
- What is the minimum cost of the road complex?
- If the intersection (new road to Old Road) closest to village B cannot be positioned at the site that provides the minimum cost, would it be better to move the intersection toward village B or toward village A along the Old Road? And why?

## 8.4 SINGLE-VARIABLE SEARCH TECHNIQUES WITH MAPLE

The basic approach of most numerical methods in optimization is to produce a sequence of improved approximations to the optimal solution according to a specific scheme. We will examine both elimination (dichotomous, golden section, and Fibonacci) and interpolation methods (Newton's and bisection).

In numerical methods of optimization, a procedure is used to obtain values of the objective function at various combinations of the decision variables; conclusions are then drawn regarding the optimal solution. The elimination methods can be used to find an optimal solution for even discontinuous functions. An important relationship (assumption) must be made to use these elimination methods. The function must be unimodal. A unimodal function is one that has only one peak (maximum) or one valley (minimum). This can be stated mathematically as follows:

A function $f(x)$ is unimodal if (1) $x_2 < x^*$ implies that $f(x_2) < f(x_1)$, and (2) $x_1 > x^*$ implies that $f(x_1) < f(x_2)$ where $x^*$ is a minimum and $x_1 < x_2$.

A function $f(x)$ is unimodal if (1) $x_2 > x^*$ implies that $f(x_2) > f(x_1)$, and (2) $x_1 < x^*$ implies that $f(x_1) > f(x_2)$ where $x^*$ is a maximum and $x_1 < x_2$.

Examples of unimodal functions are shown in Figures 8.8 (a)–(c). Thus, as seen, unimodal functions may or may not be differentiable.

A function can be a nondifferentiable (corners) or even a discontinuous function. If a function is known to be unimodal in a given interval, then the optimum (maximum or minimum) can be found as a smaller interval.

In this section, we will learn many techniques for numerical searches. For the elimination methods, we accept an interval answer. If a single value is required, then we usually evaluate the function at each end point of the final interval and the midpoint of the final interval and take the optimum of those three values to approximate our single value.

### Unrestricted Search

The method called *unrestricted search* is used when the optimum needs to be found but we have no known interval of uncertainty. This will involve a search with a fixed step size. This method is not very computationally effective.

1. Start with a guess—say, $x_1$.
2. Find $f_1 = f(x_1)$.

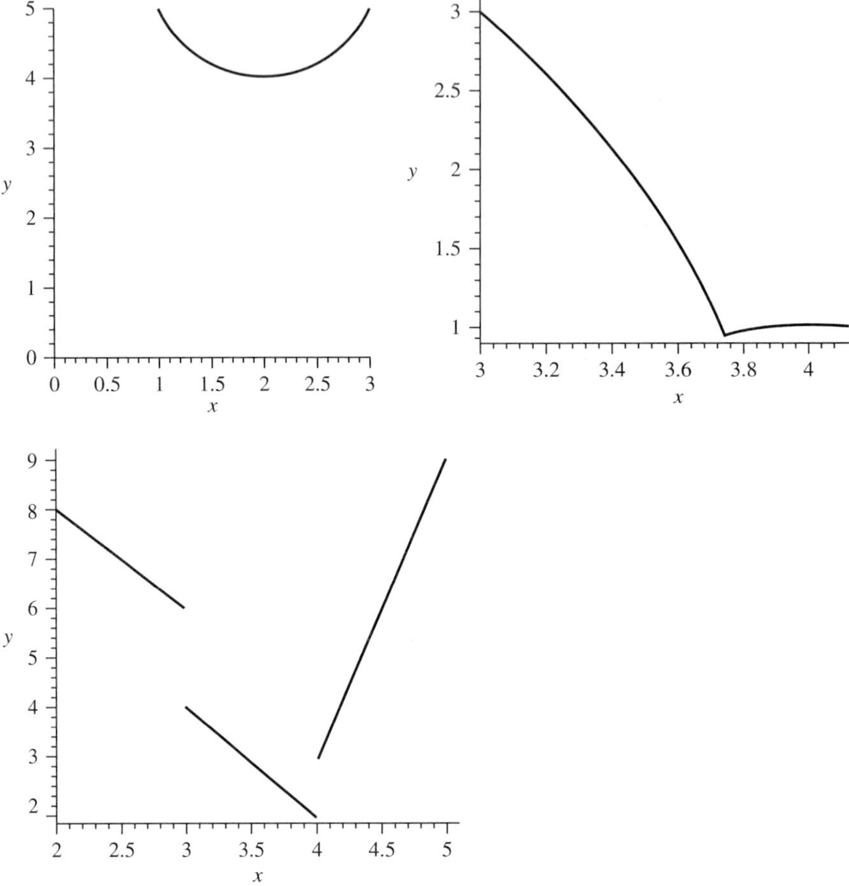

**FIGURE 8.8**
Unimodal functions

3. Assume a step size, $S$, and find $x_2 = x_1 + S$.
4. Find $f_2 = f(x_2)$.
5. For a minimization problem: If $f_2 < f_1$, then the solution interval cannot possibly lie in $x < x_1$, so we find points $x_3, x_4, \ldots x_n$ until an increase of the function is found.
6. The search is terminated with the interval $x_i$ and $x_{i-1}$.
7. If $f_2 > f_1$, then the search goes in the reverse direction.
8. If $f_1 = f_2$, then either $x_1$ or $x_2$ is optimum.
9. If the search proceeds too slowly an acceleration step size, a constant $c$ times $S$, $cS$, can be used.

## Example 1   Minimizing a Step Function

Find the minimum of the following function:

$$f(x) = \begin{cases} -x/2 & x \leq 2 \\ x - 3 & x > 2 \end{cases}$$

Use an unrestricted search method with an initial guess as 1.0 and a step size = 0.4.

$$x_1 = 1 \text{ and } f(x_1) = f_1 = -0.5$$
$$x_2 = x_1 + S = 1 + 0.4 = 1.4, \ f(1.4) = f_2 = -0.7$$
$$f_1 > f_2$$

so

$$x_3 = 1.8, f_3 = -0.9, f_2 > f_3$$
$$x_4 = 2.2, f_4 = -0.8, f_4 > f_3$$
$$\text{Stop.}$$

Thus, the optimum (minimum) must lie between 1.8 and 2.2. If a single value is required, we evaluate:

$$f(1.8) = -0.9$$
$$f(2.2) = -0.8$$
$$f((1.8 + 2.2)/2) = f(2) = -1$$

Because $f(2)$ is the smallest value of $f(x)$ among those tested, we will use $x = 2$ as our approximation to the minimum yielding a value of $f(2) = -1$.

### Dichotomous Search

This search method is a simultaneous search method in which all the experiments are conducted before any judgment is made concerning the location of the optimum point.

The following is the dichotomous algorithm:

1.  Initialize a distinguishable constant $2\varepsilon > 0$ ($e$ is a very small number such as 0.01).
2.  Select a length of uncertainty for the final interval, $t > 0$ ($t$ is also small).
3.  Calculate the number of iterations required ($n$) using:

$$0.5^n = t/(b - a)$$

4.  Let $k = 1$.

### Main Steps

1.  If $(b - a) < t$, then stop because $(b - a)$ is the final interval.

    If $(b - a) > t$, then let

$$x_1 = \frac{(a + b)}{2} - e, \ x_2 = \frac{(a + b)}{2} + e$$

2.  Perform comparisons of function values at these points.

    Minimization Problem

| If $f(x_1) < f(x_2)$ | If $f(x_1) > f(x_2)$ |
|---|---|
| $a = a$ | $a = x_1$ |
| $b = x_2$ | $B = b$ |
| $k = k + 1$ | $k = k + 1$ |
| Return to main step 1. | Return to main step 1. |

Example 2

## Dichotomous Search to Find a Minimum

Minimize $f(x) = x^2 + 2x$ over the interval $[-3, 6]$ using dichotomous search. Since we want a minimum, we have several options. We can reprogram our code or merely maximize the negative of the function. We choose to maximize the negative of the function. We wrote a program named Dichotomous to perform the iterative calculations.

$$\text{Let } t = 0.2 \text{ and } e = 0.01.$$
$$\text{Find } n \text{ using } 0.5^n = t/(b - a).$$

$$0.5^n = 0.2/[6 - (-3)] = 0.2/9$$
$$0.5^n = (0.2/9)$$
$$n \ln(0.5) = \ln(0.2/9)$$
$$n = 5.49 \text{ or } 6 \text{ (rounding up)}$$

*f:=x–>–x^2–2*x;*

$$f := x \rightarrow -x^2 - 2\,x$$

*> DICHOTOMOUS(f,–3,6,.2,.01);*

The interval [a,b] is [−3.00, 6.00] and user specified tolerance level is .20000.
The first 2 experimental end points are x1 = 1.490 and x2 = 1.510.

| Iteration | $x(1)$ | $x(2)$ | $f(x1)$ | $f(x2)$ | Interval |
|---|---|---|---|---|---|
| 1 | 1.4900 | 1.5100 | −5.2001 | −5.3001 | [−3.0000, 6.00] |
| 2 | −.7550 | −.7350 | .9400 | .9298 | [−3.0000, 1.51] |
| 3 | −1.8775 | −1. | .2300 | .2647 | [−3.0000, −.735] |
| 4 | −1.3163 | −1.2963 | .9000 | .9122 | [−1.8775, −.735] |
| 5 | −1.0356 | −1.0156 | .9987 | .9998 | [−1.3163, −.735] |
| 6 | −.8953 | −.8753 | .9890 | .9845 | [−1.0356, −.7350] |
| 7 | −.9655 | −.9455 | .9988 | .9970 | [−1.0356, −.8753] |

The midpoint of the final interval is −.990547 and f(midpoint) = 1.000.
The minimum function is .999 and the $x$ value = −1.035625

The final interval is [−1.035, −0.8753]. The length of this interval is 0.1603, which is less than our tolerance of $t = 0.2$. The value of $n$ refers to the number of $x$ pairs observed.

### Golden Section Search

Golden section search is a search procedure that utilizes the *golden ratio*. To better understand the golden ratio, consider a line segment over the interval that is divided into two separate regions, as shown in Figure 8.9. These segments are divided into the golden ratio if the length of the whole line is to the length of the larger part as the length of the larger part is to the length of the smaller part of the line. Symbolically, this can be written as $\dfrac{1}{r} = \dfrac{r}{(1-r)}$.

Algebraic manipulation of the golden ratio relationship yields $r^2 + r - 1 = 0$. Solving this function for its roots (using the quadratic formula) gives us two real solutions:

$$r_1 = \frac{\sqrt{5} - 1}{2}, r_2 = \frac{-\sqrt{5} - 1}{2}$$

Only the positive root, $r_1$, satisfies the requirement of residing on the given line segment. The numerical value of $r_1$ is 0.618, which is known as the *golden ratio*. Among its properties, this ratio is the limiting value for the ratio of the consecutive Fibonacci sequences, which we will see in the next method. It is noted here because there is also a Fibonacci search method that could be used in lieu of the golden section method.

To use the golden section search procedure, we must ensure that the following assumptions hold:

1. The function must be unimodal over a specified interval.

2. The function must have an optimal solution over a known interval of uncertainty.

3. We must accept an interval solution because the exact optimum cannot be found by this method.

**FIGURE 8.9**
Golden section schematic

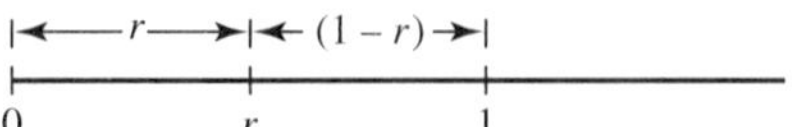

Only an interval solution known as the *final interval of uncertainty* can be found using this technique. The length of this final interval is controllable by the user and can be made arbitrarily small by the selection of a *tolerance value*. The final interval is guaranteed to be less than this tolerance level.

Line search procedures use an initial interval of uncertainty to iterate to the final interval of uncertainty. As shown earlier, the procedure is based on solving for the unique positive root of the quadratic equation, $r^2 + r = 1$. The positive root form using the quadratic formula with $a = 1, b = 1$, and $c = -1$ is:

$$r = \frac{-b \pm \sqrt{b^2 - 4ac}}{2a} \approx 0.618$$

**Finding the Maximum of a Function Over an Interval with the Golden Section**  This search procedure to find a maximum is iterative, requiring evaluations of $f(x)$ at experimental points $x_1$ and $x_2$, where $x_1 = b - r(b - a)$ and $x_2 = a + r(b - a)$. These experimental points will lie between the original interval $[a, b]$. These experimental points are used to help determine the new interval of search. If $f(x_1) < f(x_2)$, then the new interval is $[x_1, b]$; and if $f(x_1) > f(x_2)$, then the new interval is $[a, x_2]$. The iterations continue in this manner until the final interval length is less than our imposed tolerance. Our final interval contains the optimum solution. It is the size of this final interval that determines our accuracy in finding the approximate optimum solution. The number of iterations required to achieve this accepted interval length can be found as the smallest integer greater than $k$, where $k$ equals:

$$k = \frac{\ln\left(\frac{tolerance}{(b - a)}\right)}{\ln(0.618)}$$

Often, we are required to provide a point solution instead of the interval solution. When this occurs, the method of selecting a point is to evaluate the function $f(x)$ at the end points of the final interval and at the midpoint of this final interval. For maximization problems, we select the value of $x$ that yields the largest $f(x)$ solution. For minimization problems, we select the value of $x$ that yields the smallest $f(x)$ solution.

The algorithm used is shown in Figure 8.10.

**FIGURE 8.10**
Golden section algorithm

To find a maximum solution to a given function $f(x)$ on the interval $[a, b]$ where the function $f(x)$ is unimodal.

**INPUT:**   endpoints $a, b$; tolerance, $t$
**OUTPUT:**   final interval $[a_i, b_i]$, $f$(midpoint)
**Step 1.**   Initialize the tolerance, $t > 0$.
**Step 2.**   Set $r = 0.618$ and define the test points

$$x_1 = a + (1 - r)(b - a)$$
$$x_2 = a + r(b - a)$$

**Step 3.**   Calculate $f(x_1)$ and $f(x_2)$
**Step 4.**   Compare $f(x_1)$ and $f(x_2)$
   **a.** If $f(x_1) \leq f(x_2)$, then the new interval is $[x_1, b]$:

     $a$ becomes the previous $x_1$
     $b$ does not change
     $x_1$ becomes the previous $x_2$

     Find the new $x_2$ using the formula in step 2.
   **b.** If $f(x_1) > f(x_2)$, then the new interval is $[a, x_2]$:

     $a$ does not change
     $b$ becomes the previous $x_2$
     $x_2$ becomes the previous $x_1$

     Find the new $x_1$ using the formula in step 2.
**Step 5.**   If the length of the new interval from step 4 is less than the tolerance specified, then stop. Otherwise go back to step 3.
**Step 6.**   Estimate $x^*$ as the midpoint of the final interval and compute $f(x^*)$, the estimated maximum of the function.

**STOP**

Although the golden section method can be used with any unimodal function to find the maximum (or minimum) over a specified interval, its main advantage comes when normal calculus procedures fail. Consider the following example:

$$\text{Maximize } f(x) = -|2 - x| - |5 - 4x| - |8 - 9x| \text{ over the interval } 0 \le x \le 3.$$

In calculus, absolute values are not differentiable because they have corner points. Thus, taking the first derivative and setting it equal to zero is not an option. Another method needs to be used to find the solution. We use the golden section to solve this problem and other examples. We wrote a Maple proc, GOLD, to perform the iterations of golden section for us. We will use it in Example 3.

## Example 3    Using Golden Section Search to Maximize a Function That Does Not Have a Derivative

$$\text{Maximize } f(x) = -|2 - x| - |5 - 4x| - |8 - 9x| \text{ over the interval } 0 \le x \le 3.$$

> f:= x−>−(abs(2−x)+abs(5−4*x)+abs(8−9*x));

$$f := x \rightarrow - |2 - x| - |5 - 4x| - |8 - 9x|$$

> GOLD(f,0,3,.1);

The interval [a,b] is [0.00, 3.00] and user specified tolerance level is 0.10000.
The first 2 experimental end points are x1 = 1.146 and x2 = 1.854.

| Iteration | $x(1)$ | $x(2)$ | f(x1) | f(x2) | Interval |
|---|---|---|---|---|---|
| 1 | 1.1460 | 1.8540 | −3.5840 | −11.2480 | [0.0000, 1.8540] |
| 2 | 0.7082 | 1.1460 | −5.0848 | −3.5840 | [0.7082, 1.8540] |
| 3 | 1.1460 | 1.4163 | −3.5840 | −5.9958 | [0.7082, 1.4163] |
| 4 | 0.9787 | 1.1460 | −2.9149 | −3.5840 | [0.7082, 1.1460] |
| 5 | 0.8755 | 0.9787 | −2.7436 | −2.9149 | [0.7082, 0.9787] |
| 6 | 0.8116 | 0.8755 | −3.6382 | −2.7436 | [0.8116, 0.9787] |
| 7 | 0.8755 | 0.9149 | −2.7436 | −2.6594 | [0.8755, 0.9787] |
| 8 | 0.9149 | 0.9393 | −2.6594 | −2.7571 | [0.8755, 0.9393] |

The midpoint of the final interval is 0.907364 and f(midpoint) = −2.629.
The maximum of the function is −2.629 and the $x$ value = 0.907364

In this example, we want a specific point as our solution. The midpoint yields the maximum value of $f(x)$. Thus, we will use 0.907364 as the value of $x$ that maximizes this function.

## Example 4    Maximizing a Transcendental Function with Golden Section Search

Maximize the function $f(x) = 1 - \exp(-x) + 1/(1 + x)$ over the interval [0,20].

> f:=x−>1−exp(−x)+(1/(1+x));

$$f := x \rightarrow 1 - e^{-x} + \frac{1}{1 + x}$$

> GOLD(f,0,20,.001);

The interval [a,b] is [0.00, 20.00] and user specified tolerance level is 0.00100.
The first 2 experimental end points are x1 = 7.640 and x2 = 12.360.

| Iteration | $x(1)$ | $x(2)$ | $f(x1)$ | $f(x2)$ | Interval |
|---|---|---|---|---|---|
| 1 | 7.6400 | 12.360 | 1.1153 | 1.0748 | [0.0000, 12.3600] |
| 2 | 4.7215 | 7.6400 | 1.1659 | 1.1153 | [0.0000, 7.6400] |
| 3 | 2.9185 | 4.7215 | 1.2012 | 1.1659 | [0.0000, 4.7215] |
| 4 | 1.8036 | 2.9185 | 1.1920 | 1.2012 | [1.8036, 4.7215] |
| 5 | 2.9185 | 3.6069 | 1.2012 | 1.1899 | [1.8036, 3.6069] |
| 6 | 2.4925 | 2.9185 | 1.2036 | 1.2012 | [1.8036, 2.9185] |
| 7 | 2.2295 | 2.4925 | 1.2021 | 1.2036 | [2.2295, 2.9185] |
| 8 | 2.4925 | 2.6553 | 1.2036 | 1.2033 | [2.2295, 2.6553] |
| 9 | 2.3921 | 2.4925 | 1.2034 | 1.2036 | [2.3921, 2.6553] |
| 10 | 2.4925 | 2.5548 | 1.2036 | 1.2036 | [2.3921, 2.5548] |
| 11 | 2.4543 | 2.4925 | 1.2036 | 1.2036 | [2.4543, 2.5548] |
| 12 | 2.4925 | 2.5164 | 1.2036 | 1.2036 | [2.4925, 2.5548] |
| 13 | 2.5164 | 2.5310 | 1.2036 | 1.2036 | [2.4925, 2.5310] |
| 14 | 2.5072 | 2.5164 | 1.2036 | 1.2036 | [2.5072, 2.5310] |
| 15 | 2.5164 | 2.5219 | 1.2036 | 1.2036 | [2.5072, 2.5219] |
| 16 | 2.5128 | 2.5164 | 1.2036 | 1.2036 | [2.5072, 2.5164] |
| 17 | 2.5107 | 2.5128 | 1.2036 | 1.2036 | [2.5107, 2.5164] |
| 18 | 2.5128 | 2.5142 | 1.2036 | 1.2036 | [2.5107, 2.5142] |
| 19 | 2.5120 | 2.5128 | 1.2036 | 1.2036 | [2.5120, 2.5142] |
| 20 | 2.5128 | 2.5134 | 1.2036 | 1.2036 | [2.5120, 2.5134] |
| 21 | 2.5125 | 2.5128 | 1.2036 | 1.2036 | [2.5125, 2.5134] |

The midpoint of the final interval is 2.512961 and f(midpoint) = 1.204.
The maximum of the function is 1.204 and the $x$ value = 2.512961

Again, assuming that we desire a specific numerical value as the solution, our solution is $x$ = 2.512705 with $f(2.512705)$ = 1.204.

### Fibonacci Search

The Fibonacci search is a procedure that utilizes the ratio of Fibonacci numbers to set up experimental points in a sequence. The Fibonacci numbers are a sequence that follows the rules:

$$F_0 = 1$$
$$F_1 = 1$$
$$F_i = F_{i-1} + F_{i-2}$$

This generates the sequence {1, 1, 2, 3, 5, 8, 13, 21, 34, 55, 89, ...}.

The limiting value for the ratio of the consecutive Fibonacci sequences is the golden ratio 0.618. It is noted here because the golden section search method could be used in lieu of the Fibonacci method. However, the Fibonacci search converges faster than the golden section method.

To use the Fibonacci search procedure, we must ensure that the following key assumptions hold:

1. The function must be unimodal over a specified interval.

2. The function must have an optimal solution over a known interval of uncertainty.

3. We must accept an interval solution because the exact optimum cannot be found by this method.

Only an interval solution, known as the *final interval of uncertainty*, can be found using this technique. The length of this final interval is controllable by the user and can be made arbitrarily small by the selection of a *tolerance value*. The final interval is guaranteed to be less than this tolerance level.

Line search procedures use an initial interval of uncertainty to iterate to the final interval of uncertainty.

**Finding the Maximum of a Function Over an Interval with the Fibonacci Method**  This search procedure to find a maximum is iterative, requiring evaluations of $f(x)$ at experimental points $x_1$ and $x_2$, where $x_1 = a + (F_{n-2}/F_n)(b - a)$ and $x_2 = a + (F_{n-1}/F_n)(b - a)$. These experimental points will lie between the original interval $[a, b]$. These experimental points are used to help determine the new interval of search. If $f(x_1) < f(x_2)$, then the new interval is $[x_1, b]$; and if $f(x_1) > f(x_2)$, then the new interval is $[a, x_2]$. The iterations continue in this manner until the final interval length is less than our imposed tolerance. Our final interval contains the optimum solution. It is the size of this final interval that determines our accuracy in finding the approximate optimum solution. The number of iterations required to achieve this accepted interval length can be found as the smallest Fibonacci number from the sequence that satisfies the inequality,

$$F_k > \frac{(b - a)}{tolerance}.$$

Often we are required to provide a point solution instead of the interval solution. When this occurs, the method of selecting a point is to evaluate the function $f(x)$ at the end points of the final interval and at the midpoint of this final interval. For maximization problems, we select the value of $x$ that yields the largest $f(x)$ solution. For minimization problems, we select the value of $x$ that yields the smallest $f(x)$ solution.

The algorithm used is shown in Figure 8.11.

**FIGURE 8.11**

Fibonacci algorithm

To find a maximum solution to a given function $f(x)$ on the interval $[a, b]$ where the function $f(x)$ is unimodal.

| | |
|---|---|
| **INPUT:** | endpoints $a$, $b$; tolerance, $t$, Fibonacci sequence |
| **OUTPUT:** | final interval $[a_i, b_i]$, $f$(midpoint) |
| **Step 1.** | Initialize the tolerance, $t > 0$. |
| **Step 2.** | Set $F_n > (b - a)/t$ as the smallest $F_n$ and define the test points |

$$x_1 = a + (F_{n-2}/F_n)(b - a)$$
$$x_2 = a + (F_{n-1}/F_n)(b - a)$$

**Step 3.** Calculate $f(x_1)$ and $f(x_2)$

**Step 4.** Compare $f(x_1)$ and $f(x_2)$

 **a.** If $f(x_1) \leq f(x_2)$, then the new interval is $[x_1, b]$:

   $a$ becomes the previous $x_1$
   $b$ does not change
   $x_1$ becomes the previous $x_2$

   $$n = n - 1$$

   Find the new $x_2$ using the formula in step 2.

 **b.** If $f(x_1) > f(x_2)$, then the new interval is $[a, x_2]$:

   $a$ does not change
   $b$ becomes the previous $x_2$
   $x_2$ becomes the previous $x_1$

   $$n = n - 1$$

   Find the new $x_1$ using the formula in step 2.

**Step 5.** If the length of the new interval from step 4 is less than the tolerance specified, then stop. Otherwise go back to step 3.

**Step 6.** Estimate $x^*$ as the midpoint of the final interval and compute $f(x^*)$, the estimated maximum of the function.

**STOP**

Although Fibonacci can be used with any unimodal function to find the maximum (or minimum) over a specified interval, its main advantage comes when normal calculus procedures fail. Consider the following example:

$$\text{Maximize } f(x) = -|2 - x| - |5 - 4x| - |8 - 9x| \text{ over the interval } 0 \le x \le 3.$$

In calculus, absolute values are not differentiable because they have corner points. Thus, taking the first derivative and setting it equal to zero is not an option. Another method needs to be used to find the solution. We use the Fibonacci to solve this problem and other examples. As before, we wrote a proc, FIBSearch, to perform the iterations based on the algorithm. We illustrate this in Example 5.

---

## Example 5 — Maximizing a Function That Does Not Have a Derivative with Fibonacci Search

$$\text{Maximize } f(x) = -|2 - x| - |5 - 4x| - |8 - 9x| \text{ over the interval } 0 \le x \le 3.$$

```
> f:=x->-(abs(2-x)+abs(5-4*x)+abs(8-9*x));
```

$$f := (x) \rightarrow -|2 - x| - |5 - 4x| - |8 - 9x|$$

```
> FIBSearch(f,0,3,.1);
```

The interval [a,b] is [0.00, 3.00] and user specified tolerance level is 0.10000.
The first 2 experimental end points are x1 = 1.147 and x2 = 1.853.

| Iteration | $x(1)$ | $x(2)$ | f(x1) | f(x2) | Interval |
|---|---|---|---|---|---|
| 1 | 1.1471 | 1.8529 | −3.5882 | −11.2353 | [0.0000, 1.8529] |
| 2 | 0.7059 | 1.1471 | −5.1176 | −3.5882 | [0.7059, 1.8529] |
| 3 | 1.1471 | 1.4118 | −3.5882 | −5.9412 | [0.7059, 1.4118] |
| 4 | 0.9706 | 1.1471 | −2.8824 | −3.5882 | [0.7059, 1.1471] |
| 5 | 0.8824 | 0.9706 | −2.6471 | −2.8824 | [0.7059, 0.9706] |
| 6 | 0.7941 | 0.8824 | −3.8824 | −2.6471 | [0.7941, 0.9706] |

The midpoint of the final interval is 0.882353 and f(midpoint) = −2.647.
The actual maximum of the function is −2.882 and the $x$ value = 0.970588

In this example, we want a specific point as our solution. The end point yields the maximum value of $f(x)$. Thus, we will use $x = 0.970588$ with $f(0.970588) = -2.882$ as our solution.

---

## Example 6 — Maximizing a Transcendental Function with Fibonacci Search

$$\text{Maximize the function } f(x) = 1 - \exp(-x) + 1/(1 + x) \text{ over the interval } [0, 20].$$

```
> f:=(x)->1-exp(-x)+1/(1+x);
```

$$> f := x \rightarrow 1 - e^{-x} + \frac{1}{1 + x}$$

```
> FIBSearch(f,0,20,.1);
```

The interval [a,b] is [0.00, 20.00] and user specified tolerance level is 0.10000.
The first 2 experimental end points are $x_1 = 7.639$ and $x_2 = 12.361$.

| Iteration | $x(1)$ | $x(2)$ | $f(x1)$ | $f(x2)$ | Interval |
|---|---|---|---|---|---|
| 1 | 7.6395 | 12.3605 | 1.1153 | 1.0748 | [0.0000, 12.3605] |
| 2 | 4.7210 | 7.6395 | 1.1659 | 1.1153 | [0.0000,  7.6395] |
| 3 | 2.9185 | 4.7210 | 1.2012 | 1.1659 | [0.0000,  4.7210] |
| 4 | 1.8026 | 2.9185 | 1.1919 | 1.2012 | [1.8026,  4.7210] |
| 5 | 2.9185 | 3.6052 | 1.2012 | 1.1900 | [1.8026,  3.6052] |
| 6 | 2.4893 | 2.9185 | 1.2036 | 1.2012 | [1.8026,  2.9185] |
| 7 | 2.2318 | 2.4893 | 1.2021 | 1.2036 | [2.2318,  2.9185] |
| 8 | 2.4893 | 2.6609 | 1.2036 | 1.2033 | [2.2318,  2.6609] |
| 9 | 2.4034 | 2.4893 | 1.2034 | 1.2036 | [2.4034,  2.6609] |
| 10 | 2.4893 | 2.5751 | 1.2036 | 1.2036 | [2.4034,  2.5751] |

The midpoint of the final interval is 2.489270 and f(midpoint) = 1.204.
The maximum of the function is 1.204 and the $x$ value = 2.403433

Again, assuming that we desire a specific numerical value as the solution, our solution is $x = 2.489270$ with $f(2.489270) = 1.204$.

### Interpolation with Derivatives: Newton's Method for Nonlinear Optimization

**Finding the Critical Points (Roots) of a Function**  Newton's method has been adapted to solve nonlinear optimization problems. For a function of a single variable, the adaptation is straightforward. Newton's method is applied to the *derivative* of the function we wish to optimize because the function's critical points occur where the derivative's roots are found. When finding the critical points of the function, Newton's method is based on the derivative of the quadratic approximation of the function $f(x)$ at the point $x_k$ :

$$q(x) = f(x_k) + f'(x_k)(x - x_k) + \tfrac{1}{2}f''(x_k)(x - x_k)^2$$

The result, $q'(x)$, is a linear approximation of $f'(x)$ at the point $x_k$. Setting $q'(x) = 0$ and solving for $x$ yields the formula

$$x_{k+1} = x_k - \frac{f'(x_k)}{f''(x_k)}$$

Newton's method can be terminated when $|x_{k+1} - x_k| < \varepsilon$, where $\varepsilon$ is a prespecified scalar tolerance or when $|f'(x)| < \tilde{\varepsilon}$.

To use Newton's method to find the critical points of a function, the function's first and second derivatives must exist in the neighborhood of interest. Also note that when the second derivative at $x_k$ is zero, the point $x_{k+1}$ cannot be computed.

It is important to first master the computations required in the algorithm. Note also that Newton's method finds only the approximate critical value; it does not know whether it is finding a maximum or a minimum. The sign of the second derivative may be used to determine if we have a maximum or a minimum.

**The Basic Application**  Consider any simple polynomial, such as $f(x) = 5x - x^2$, whose critical point can easily be found by calculus, taking the first derivative and setting it equal to zero. We find that the critical point $x = 2.5$ yields a maximum of the function. Applying Newton's method to find critical points requires finding $f'(x)$ and $f''(x)$ and then using a computation device to perform the iterations:

$$f'(x) = 5 - 2x \text{ and } f''(x) = -2$$

Newton's method uses $x_{k+1} = x_k - \dfrac{f'(x_k)}{f''(x_k)}$ or $x_{k+1} = x_k - \dfrac{(5 - 2x_k)}{(-2)}$.

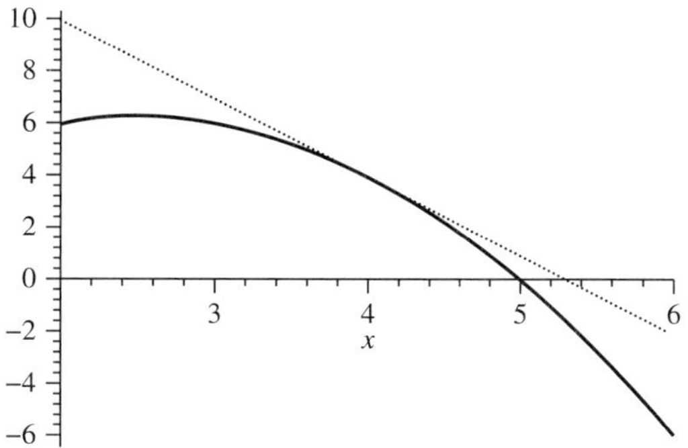



**FIGURE 8.12**
A graph of the function $f(x) = 5x - x^2$ and its linear approximation at the point $x_k = 4$

Starting at $x_0 = 1$ yields the following:

| $k$ | $x_k$ | $f'(x)$ | $f''(x)$ | $x_{k+1}$ | $|x_k - x_{k+1}|$ |
|---|---|---|---|---|---|
| 0 | 1 | 3 | $-2$ | 2.5 | 1.5 |
| 1 | 2.5 | 0 | $-2$ | 2.5 | 0 |

Starting at other values also yields $x = 2.5$. Because this simple quadratic function has a derivative that is a linear function, the linear approximation of the derivative will be exact regardless of the starting point, and the answer will be confirmed at the second iteration. Newton's method produces the critical values of $f'(x)$ without regard to the point $x_k$ being a maximum or a minimum. We know we have found a maximum by looking at the entries in the table for $f''(x)$. We have found a maximum at $x = 2.5$ because $f''(x)$ at $x = 2.5$ is $-2$, which is less than or equal to 0.

Note that the slope of the linear approximation of the function at the point $x_k$ is precisely the slope of the function at that point, so the linear approximation is tangent to the function at the point $x_k$ (see Figure 8.12).

---

## Example 7    Newton's Method to Minimize $f(x) = x^2 + 2x$

We will start our guess with $x = 4$ and use a stopping criteria of $\varepsilon = 0.01$.

$$f(x) = x^2 + 2x$$
$$f'(x) = 2x + 2$$
$$f''(x) = 2$$
$$x_2 = x_1 - f'(x_1)/f''(x_1)$$
$$x_2 = 4 - (10/2) = 4 - 5 = -1$$
$$x_3 = x_2 - f'(x_2)/f''(x_2)$$
$$x_2 = -1 - (0/2) = -1$$
Stop.

$$|f'(x)| = 0 < 0.01 \text{ or } |x_2 - x_1| = 0 < 0.01$$

---

## Example 8    Maximize $f(x) = -2x^3 + 10x - 10$

We will begin at $x = 1$ and use a stopping criteria of $\varepsilon = 0.01$.

$$f'(x) = -6x^2 + 10$$
$$f''(x) = -12x$$
$$\text{At } x = 1, f'(1) = 4, f''(1) = -12$$
$$x_2 = 1 - (-4/12) = 1.33333$$

**Step 1.**  Find two values $a$ and $b$ where $f'(a)$ and $f'(b)$ have opposite signs.
**Step 2.**  Fix a tolerance for the final interval of the solution $[a_f, b_f]$ so that
$|b_f - a_f| <$ tolerance.
**Step 3.**  Find the midpoint, $m_i = (b_i + a_i)/2$
**Step 4.**  Compute $f'(a_i), f'(b_i),$ and $f'(m_i).$
**Step 5.**  Determine if $f'(a_i)*f'(m_i) < 0.$
If true, then

$a_i = a_i$
$b_i = m_i$

Otherwise,

$a_i = m_i$
$b_i = b_i$

**Step 6.**  If $f'(m_i) \neq 0$ and the new $|b_i - a_i| >$ tolerance, then using the new interval
$[a_i, b_i]$ go back to step 3 and repeat the process. Otherwise,

**STOP.**

Neither $|f'(x)|$ nor $|x_{k+1} - x_k|$ are less than $\varepsilon$, so we continue. We summarize in the following table.

| $k$ | $x$ | $f'(x)$ | $f''(x)$ | $x(k + 1)$ |
|---|---|---|---|---|
| 1 | 1 | 4 | $-12$ | 1.333333 |
| 2 | 1.333333 | $-0.66667$ | $-16$ | 1.291667 |
| 3 | 1.291667 | $-0.01042$ | $-15.5$ | 1.290995 |
| 4 | 1.290995 | $-2.7E-06$ | $-15.4919$ | 1.290994 |

Because $f'(x) = |-2.7E - 06| < 0.01$, we stop. Our critical point is $x = 1.290994$. Because $f''(x) < 0$, then we have found a maximum.

### The Bisection Method with Derivatives

To utilize bisection method with derivatives correctly, certain properties must hold for the given function:

Let $f'$ be a continuous function that has opposite sign values at each end of some specified interval. Then, by the *intermediate-value property* (IVP) of continuous functions (mentioned in many basic college algebra texts), we are guaranteed to find a root between the end points of the given interval. Specifically, the IVP states that given two points $(x_1, y_1)$ and $(x_2, y_2)$ with $y_1 \neq y_2$ on a graph of the continuous function $f$, the function $f$ takes on every value between $y_1$ and $y_2$. Thus, with values having opposite signs, then there must be a value for which $f'(x) = 0$.

The algorithm is listed in Figure 8.13.

## Example 9    Bisection Method to Minimize the Function $f(x)$

Minimize $f(x) = x^2 + 2x, -3 \leq x \leq 6$. Let $\varepsilon = 0.2$.

The number of required observations is found by solving the formula $(0.5)^n = t/(b - a)$. We find that $n = 6$ observations.

$$f'(x) = 2x + 2$$
$$k = 1$$

1. $x_{mp} = 0.5(6 - 3) = 1.5, f'(1.5) = 5, f'(1.5) > 0$

   so
   $a = -3$
   $b = 1.5$
   $x_{mp} = -0.75$

2. $f'(-0.75) = 0.5 > 0$

   so
   $a = -3$
   $b = -0.75$
   $x_{mp} = -1.875$

3. $f'(-1.875) = -1.75 < 0$

   so
   $a = -1.875$
   $b = -75$
   $x_{mp} = -1.3125$

4. $f'(-1.3125) = -0.625 < 0$

   so
   $a = -1.3125$
   $b = -0.75$
   $x_{mp} = -1.03125$

5. $f'(-1.03125) = -0.0625 < 0$

   so
   $a = -1.30125$
   $b = -0.75$
   $x_{mp} = -0.89025$

6. $f'(-0.89025) = 0.2195 > 0$

   $a = 1.030125$
   $b = -0.89025$

The final interval is less than 0.2, so we stop. Our solution is between $[-1.030125, -0.89025]$. As previously discussed, if we need a single value we can evaluate the derivative of the function $f'$ at $a$, at $b$, and at the midpoint, selecting the value that is closest to $f'(x) = 0$ from those points.

$$f'(-1.030125) = -0.0625$$
$$f'(-0.89025) = 0.2195$$
$$f'(-0.96075) = 0.0705$$
$$x = -1.030125 \text{ with } f'(x) = -0.0625 \text{ is best}$$

The exact value, via single-variable calculus, for the maximization of $f$ is the solution $x = -1$.

## 8.4 | EXERCISES

Use the golden section, Fibonacci's method, Newton's method, and the bisection method to solve the following:

1. Maximize $f(x) = -x^2 - 2x$ on the closed interval $[-2,1]$, using a tolerance for the final interval of **0.6**. (*Hint*: Start Newton's method at $x = -0.5$.)

2. Maximize $f(x) = -x^2 - 3x$ on the closed interval $[-3,1]$, using a tolerance for the final interval of **0.6**. (*Hint*: Start Newton's method at $x = 1$.)

3. Minimize $f(x) = x^2 + 2x$ on the closed interval $[-3,1]$, using a tolerance for the final interval of **0.5**. (Start Newton's at $x = -3$.)

4. Minimize $f(x) = -x + e^x$ over the interval $[-1,3]$ using a tolerance of 0.1. (Start Newton's at $x = -1$.)

5. List at least two assumptions required by both the golden section and Fibonacci's search methods.

6. Consider minimizing $f(x) = -x + e^x$ over the interval $[-1,3]$. Assume your final interval yielded a solution within the tolerance of $[-0.80,0.25]$. Report a single best value of $x$ to minimize $f(x)$ over the interval.

## 8.4 | PROJECTS

1. We are considering buying the new e-phone system, which is computer compatible with the new e-computer system. The company is concerned with how often it will need to replace the machines and the cost involved. The company's R&D team estimates that when the e-phone is $t$ years old, it will allow the user to earn revenue at a rate of $e^{-t}$ per year. After $t$ years of use, the e-phone can be sold to a third-world company for $1/(1 + t)$ dollars. Maintenance costs after $t$ years are estimated as $0.01t$ dollars. Build a model for the company and determine how long it should keep the e-phone system before replacing it. Make a recommendation to the CEO as to what you would do based on your modeling.

2. Previously we discussed the least-squares method to fit the parameters to a proposed model. Another choice to fit the model is minimize the sum of the absolute deviations between the proposed model and the data. For example, consider the model using $W = KL^3$ and the following available data.

| Length | 12.5 | 12.625 | 12.625 | 14.125 | 14.5 | 14.5 | 17.27 | 17.75 |
|--------|------|--------|--------|--------|------|------|-------|-------|
| Weight | 17 | 16 | 17 | 23 | 26 | 27 | 43 | 49 |

A model that we could use to find the slope is a search method. Find the value of $k$ that minimizes the function $S$.

$$\text{Minimize } S = |17 - k*12.5^3| + |16 - k*12.625^3| + \dots + |49 - k*17.75^3|$$

Use any of our numerical techniques and find the value of $k$.

3. An economist estimates the profit function in millions as $y = x^2 \cdot \cos(x) \cdot e^x$. Plot the profit function as a function of $x$. Pick an interval for the maximum and use the golden section to solve. State your golden section experiments on the interval, giving their location, the corresponding value of the profit function, and the length and location of the final interval of uncertainty. Determine the optimal solution to a tolerance of 0.25, 0.1, 0.01?

4. An economic analysis of a proposed facility is being conducted in order to select an operating life such that the maximum uniform annual income is achieved. A short life results in high annual amortization costs, but the maintenance costs become excessive for a long life. The annual income after deducting all operating costs, except maintenance costs, is $180,000. The installed cost of the facility, $C$, is $500,000 borrowed at 10-percent interest compounded annually. The maintenance charges on an annual basis are evaluated using the product of the gradient present-worth factor and the capital-recovery factor. In the gradient present-worth method, there are no maintenance charges the first year, a cost $M$ for the second year, $2M$ for the third year, $3M$ for the fourth year, and so on. The second-year cost, $M$, for the problem is $10,000. The annual profit is given by the following equation:

$$P = 180,000 - \{[(i + 1)^N - 1 - iN] / i[(1 + i)^N - 1]\} M - \{i(1 + i)^N / [(1 + i)^N - 1]\} C$$

where $i$ is the interest rate and $N$ is the number of years. Determine the number of years, $N$, that give the maximum uniform annual income, $P$. For your convenience, the following table gives the values of the coefficients of $M$ and $C$ for $i = 0.1$ as a function of $N$.

| | Coefficient of | | | | Coefficient of | |
| Year | $M$ | $C$ | | Year | $M$ | $C$ |
|---|---|---|---|---|---|---|
| 1 | 0 | 1.10 | | 11 | 4.05 | 0.154 |
| 2 | 0.476 | 0.576 | | 12 | 4.39 | 0.147 |
| 3 | 0.909 | 0.403 | | 13 | 4.69 | 0.141 |
| 4 | 1.30 | 0.317 | | 14 | 5.00 | 0.136 |
| 5 | 1.80 | 0.264 | | 15 | 5.28 | 0.131 |
| 6 | 2.21 | 0.230 | | 16 | 5.54 | 0.128 |
| 7 | 2.63 | 0.205 | | 17 | 5.80 | 0.125 |
| 8 | 2.98 | 0.188 | | 18 | 6.05 | 0.122 |
| 9 | 3.38 | 0.174 | | 19 | 6.29 | 0.120 |
| 10 | 3.71 | 0.163 | | 20 | 6.51 | 0.117 |

5. It is proposed to recover the waste heat from exhaust gases leaving a furnace (flow rate, $m = 60{,}000$ lb/hr; heat capacity, $c_p = 0.25$ BTU/lb°F) at a temperature of $T_{in} = 500°$F by installing a heat exchanger (overall heat-transfer coefficient, $U = 4.0$ BTU/hr,ft$^2$,°F) to produce steam at $T_s = 220°$F from saturated liquid water at 220°F. The value of heat in the form of steam is $p = \$0.75$ per million BTUs, and the installed cost of the heat exchanger is $c = \$5$ per ft$^2$ of gas side area. The life of the installation is $n = 5$ years, and the interest rate is $i = 8.0\%$. The following equation gives the net profit $P$ for the 5-year period from the sale of the steam and the cost of the heat exchanger. The exhaust gas temperature $T_{out}$ can be between the upper and lower limits of 500°F and 220°F.

$$P = pqn - cA(1 + i)n$$

where

$$q = mc_p(T_{in} - T_{out}) = UADT_{LM}$$

$$\Delta T_{LM} = \frac{(T_{in} - T_s) - (T_{out} - T_s)}{\ln\left(\dfrac{T_{in} - T_s}{T_{out} - T_s}\right)}$$

a. Derive the following equation for this design.

$$P = 91{,}137 - 492.75 T_{out} + 27{,}550 \ln(T_{out} - 220)$$

b. Use a Fibonacci search with seven experiments to locate the optimal outlet temperature $T_{out}$ to maximize the profit $P$ on the interval of $T_{out}$ from 220°F to 500°F. Find the largest value of the profit and the size and location of the final interval of uncertainty for the fractional resolution based on the initial interval of $e = 0.01$.

c. Use another numerical method and compare your results.

## 8.4  FURTHER READING

Bazarra, M., C. Shetty, and H. D. Scherali, *Nonlinear Programming: Theory and Applications*. New York: Wiley. 1993.

Fox, W. P. "Teaching Nonlinear Programming with Minitab." *COED Journal*. Vol. II No. 1, January–March 1992, pages 80–84.

Fox, W. P. "Using Microcomputers in Undergraduate Nonlinear Optimization." *Collegiate Microcomputer*. VOL XI(3). Pages 214–218. 1993.

Fox, W. P., and Margie Witherspoon, "Single Variable Optimization When Calculus Fails: Golden Section Search Methods in Nonlinear Optimization Using Maple," *COED*. VOL XI (2), 2001, pages 50–56.

Fox, W. P., F. Giordano, and Maurice Weir, *A First Course in Mathematical Modeling*, 3rd edition. Monterey, CA: Brooks/Cole. 2003.

Phillips, D. T., A. Ravindran, and J. Solberg, 1976. *Operations Research*. New York: John Wiley and Sons.

Rao, S. S., *Optimization: Theory and Applications*. New Delhi, India: Wiley Eastern Limited. 1979.

Winston, W., *Introduction to Mathematical Programming: Applications and Algorithms*. 4th edition. Belmont. CA: Duxbury Press, ITP. 2002.

# 9

# Models Using Unconstrained Optimization: Maximization and Minimization with Several Variables

## Introduction

Consider a small company that is planning to install a central computer with cable links to five new departments. According to the floor plan, the peripheral computers for the five departments will be situated as shown by the dark circles in Figure 9.1. The company wishes to locate the central computer so that the minimal amount of cable will be used to link to the five peripheral computers. Assuming that cable can be strung over the ceiling panels in a straight line from a point above any peripheral to a point above the central computer, the distance formula may be used to determine the length of cable needed to connect any peripheral to the central computer. Ignore all lengths of cable from the computer itself to a point above the ceiling panel immediately over that computer. In other words, work only with lengths of cable strung over the ceiling panels.

The coordinates of the locations of the five peripheral computers are listed in Table 9.1.

Assume the central computer will be positioned at coordinates $(m, n)$ where $m$ and $n$ are *integers* in the grid representing the office space. Determine the coordinates $(m, n)$ for placement of the central computer that minimize the total amount of cable needed. Report the total number of feet of cable needed for this placement along with the coordinates $(m, n)$.

To model and solve problems like this, we need to learn about multivariable unconstrained optimization.

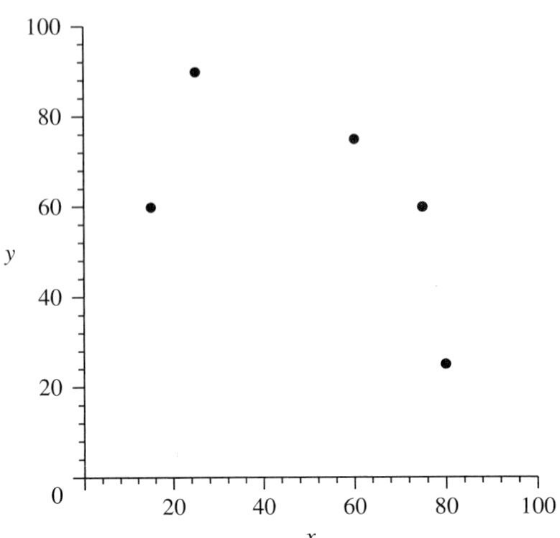



**FIGURE 9.1**
The Grid for the Five Departments

**Table 9.1**  Grid Coordinates of Five Departments

| X | Y |
|---|---|
| 15 | 60 |
| 25 | 90 |
| 60 | 75 |
| 75 | 60 |
| 80 | 25 |

## 9.1  BASIC THEORY

We will discuss how to find an optimal solution (if one exists) for the following unconstrained nonlinear optimization problem:

$$\text{Maximize (or minimize)} \, f(x_1, x_2, \ldots, x_n) \text{ over } R^n \tag{9.1}$$

We assume that the first and second partial derivatives of $f(x_1, x_2, \ldots, x_n)$ exist and are continuous at all points in the domain of $f$. Let

$$\frac{\partial f(x_1, x_2, \ldots x_n)}{\partial x_i}$$

be the partial derivative of $f(x_1, x_2, \ldots, x_n)$ with respect to $x_i$. Candidate critical points (stationary points) are found where

$$\frac{\partial f(x_1, x_2, \ldots x_n)}{\partial x_i} = 0, \text{ for } i = 1, 2, \ldots, n$$

This sets up a system of equations that, when solved, yields the critical point (if one or more is found) that satisfies all partial derivatives.

**Theorem 1:**  If $x$ is a local extremum, then $x$ satisfies $\dfrac{\partial f(x_1, x_2, \ldots x_n)}{\partial x_i} = 0$, for $i = 1, 2, \ldots, n$.  ∎

We have previously defined all points that satisfy $\dfrac{\partial f(x_1, x_2, \ldots x_n)}{\partial x_i} = 0$, for $i = 1, 2, \ldots, n$ as critical points (stationary points). Not all critical points (stationary points) are local extrema. If a stationary point is not a local extremum (a maximum or a minimum), then it is called a **saddle point**.

## 9.2  THE HESSIAN MATRIX

How do we determine the convexity of functions of more than one variable?

**Definition:** A function $f(x)$ is convex if

$$f(x^{(1)} + \lambda (x^{(2)} - x^{(1)})) \leq f(x^{(1)}) + \lambda (f(x^{(2)}) - f(x^{(1)}))$$

for every $x^{(1)}$ and $x^{(2)}$ in its domain and every $\lambda \in [0,1]$. Similarly, $f(x)$ is concave if

$$f(x^{(1)} + \lambda (x^{(2)} - x^{(1)})) \geq f(x^{(1)}) + \lambda (f(x^{(2)}) - f(x^{(1)}))$$

for every $x^{(1)}$ and $x^{(2)}$ in its demain and every $\lambda \in [0,1]$.

We introduce the Hessian matrix that allows us to determine the convexity of multivariable functions. As we will see, the Hessian matrix provides us with additional information about the critical points as well.

**Definition:** The Hessian matrix is the $n \times n$ matrix of the second partial derivatives of a multivariable function $f(x_1, x_2, \ldots , x_n)$ where the $ij$th entry is $\dfrac{\partial^2 f}{\partial x_i x_j}$.

$$H = \begin{bmatrix} \dfrac{\partial^2 f}{\partial x_1^2} & \dfrac{\partial^2 f}{\partial x_1 x_2} & \dfrac{\partial^2 f}{\partial x_1 x_3} & \cdots & \dfrac{\partial^2 f}{\partial x_1 x_n} \\[2mm] \dfrac{\partial^2 f}{\partial x_2 x_1} & \dfrac{\partial^2 f}{\partial x_2^2} & \cdots & & \dfrac{\partial^2 f}{\partial x_2 x_n} \\[2mm] \dfrac{\partial^2 f}{\partial x_3 x_1} & \cdots & \cdots & & \dfrac{\partial^2 f}{\partial x_3 x_n} \\[2mm] \cdots & & \cdot & \cdot & \cdot \\[2mm] \dfrac{\partial^2 f}{\partial x_n x_1} & \dfrac{\partial^2 f}{\partial x_n x_2} & \cdots & & \dfrac{\partial^2 f}{\partial x_n^2} \end{bmatrix}$$

In the $2 \times 2$ case,

$$H = \begin{bmatrix} \dfrac{\partial^2 f}{\partial x_1^2} & \dfrac{\partial^2 f}{\partial x_1 x_2} \\[3mm] \dfrac{\partial^2 f}{\partial x_2 x_1} & \dfrac{\partial^2 f}{\partial x_2^2} \end{bmatrix}$$

We note that the mixed partials are always equal: $\dfrac{\partial^2 f}{\partial x_1 \partial x_2} = \dfrac{\partial^2 f}{\partial x_2 \partial x_1}$ as you will see in our examples.

---

**Example 1**     Finding the Hessian for $f(x_1, x_2) = x_1^2 + 3x_2^2$

If $f(x_1, x_2) = x_1^2 + 3x_2^2$, find the Hessian.

$$\frac{\partial f}{\partial x_1} = 2x_1, \ \frac{\partial f}{\partial x_2} = 6x_2^2$$

$$\frac{\partial^2 f}{\partial x_1^2} = 2$$

$$\frac{\partial^2 f}{\partial x_2^2} = 6$$

$$\frac{\partial^2 f}{\partial x_1 x_2} = \frac{\partial^2 f}{\partial x_2 x_1} = 0$$

$$H = \begin{bmatrix} 2 & 0 \\ 0 & 6 \end{bmatrix}$$

---

**Example 2**     Finding the Hessian for $f(x_1, x_2) = -x_1^2 - 3x_2^2 + 3x_1 \cdot x_2$

If $f(x_1, x_2) = -x_1^2 - 3x_2^2 + 3x_1 \cdot x_2$, find the Hessian.

$$\frac{\partial f}{\partial x_1} = -2x_1 + 3x_2, \ \frac{\partial f}{\partial x_2} = -6x_2 + 3x_1$$

$$\frac{\partial^2 f}{\partial x_1^2} = -2$$

$$\frac{\partial^2 f}{\partial x_2^2} = -6$$

$$\frac{\partial^2 f}{\partial x_1 x_2} = \frac{\partial^2 f}{\partial x_2 x_1} = 3$$

$$H = \begin{bmatrix} -2 & 3 \\ 3 & -6 \end{bmatrix}$$

---

**Definition:** | The $i$th leading principal minor of an $n \times n$ matrix is the determinant of any $i \times i$ matrix obtained by deleting $n - i$ rows and the corresponding $n - i$ columns of the matrix.

---

**Example 3**     Finding the Principal Minors of a 3 × 3 Hessian Matrix

Given the 3×3 Hessian matrix,

$$H = \begin{bmatrix} 2 & 0 & 4 \\ 0 & 1 & 5 \\ 4 & 5 & 3 \end{bmatrix}$$

1. There are three first leading principal minors ($i = 1$, so eliminate $3 - 1 = 2$ rows and 2 columns).

   a. Eliminate rows 2,3 and columns 2,3 to yield the matrix [2]:

$$\text{Det}[2] = 2$$

**b.**   Eliminate rows 1,3 and columns 1,3 to yield the matrix [1]:

$$\text{Det }[1] = 1$$

**c.**   Eliminate rows 1,2 and columns 1,2 to yield the matrix [3]:

$$\text{Det }[3] = 3$$

Note: These first leading principal minors are the entries of the main diagonal.

**2.**   There are three second leading principal minors ($i = 2$, so we eliminate $3 - 2 = 1$ row and 1 column).

    **a.**   Eliminate row 3 and column 3 to yield the matrix

$$\begin{bmatrix} 2 & 0 \\ 0 & 1 \end{bmatrix}, \text{Det} \begin{bmatrix} 2 & 0 \\ 0 & 1 \end{bmatrix} = 2$$

    **b.**   Eliminate row 2 and column 2 to yield the matrix

$$\begin{bmatrix} 2 & 4 \\ 4 & 3 \end{bmatrix}, \text{Det} \begin{bmatrix} 2 & 4 \\ 4 & 3 \end{bmatrix} = 6 - 16 = -10$$

    **c.**   Eliminate row 1 and column 1 to yield the matrix

$$\begin{bmatrix} 1 & 5 \\ 5 & 3 \end{bmatrix}, \text{Det} \begin{bmatrix} 1 & 5 \\ 5 & 3 \end{bmatrix} = 3 - 25 = -22$$

**3.**   There is only one third leading principal minor ($i = 0$, so eliminate no rows or columns):

$$\text{Det} \begin{bmatrix} 2 & 0 & 4 \\ 0 & 1 & 5 \\ 4 & 5 & 3 \end{bmatrix} = -60$$

---

**Definition:**   The $k$th-leading principal minor of an $n \times n$ matrix is the determinant of the $k \times k$ matrix obtained by deleting the last $n - k$ rows and $n - k$ columns of the matrix.

---

## Example 4 — Determinants of Principal Minors

Given the $3 \times 3$ Hessian matrix,

$$H = \begin{bmatrix} 2 & 0 & 4 \\ 0 & 1 & 5 \\ 4 & 5 & 3 \end{bmatrix}$$

**1.**   The first leading principal minor is the determinant of the $1 \times 1$ matrix, obtained by deleting the last $3 - 1 = 2$ rows and columns, which yields the matrix [2], with det [2] = 2.

**2.**   The second leading principal minor is the determinant of the $2 \times 2$ matrix, obtained by deleting the last $3 - 2 = 1$ rows and columns, which yields the matrix:

$$\det \begin{bmatrix} 2 & 0 \\ 0 & 1 \end{bmatrix} = 2$$

**3.**   The third leading principal minor is the determinant of the $3 \times 3$ matrix, obtained by deleting the last $3 - 3 = 0$ rows and columns, which yields the matrix:

$$\det \begin{bmatrix} 2 & 0 & 4 \\ 0 & 1 & 5 \\ 4 & 5 & 3 \end{bmatrix} = -60$$

Notice that if you examine the matrix, $H$, the leading principal minors are just the determinant of the square matrices along the main diagonal.

So, how do we use all these determinants that give the principal minor and leading principal minors of the Hessian matrix to determine the convexity of the multivariate function?

**Theorem 2a:** | Let $f(x_1, x_2, \ldots x_n)$ be a function with continuous second-order partial derivatives for every point in the domain of $f$. Then $f(x_1, x_2, \ldots x_n)$ is a convex function if all the leading principal minors of the Hessian matrix are nonnegative.  ∎

**Theorem 2b:** | Let $f(x_1, x_2, \ldots x_n)$ be a function with continuous second-order partial derivatives for every point in the domain of $f$. $f(x_1, x_2, \ldots x_n)$ is a concave function if all the nonzero leading principal minors of the Hessian matrix follow the sign of $(-1)^k$, where $k$ represents the order of the principal minors ($k = 1,2,3$, etc).  ∎

**Theorem 2c:** | If the leading principal minors do not follow either Theorem 2a or Theorem 2b, then $f(x_1, x_2, \ldots x_n)$ is neither a convex function nor a concave function.  ∎

## Example 5    A Convex Function

Determine the convexity of the function using the Hessian matrix, $f(x_1, x_2) = x_1^2 + 3x_2^2$.

$$H = \begin{bmatrix} 2 & 0 \\ 0 & 6 \end{bmatrix}$$

The first leading principal minors are det $[2] = 2 > 0$, det$[6] = 6 > 0$.

The second leading principal minor is det $\begin{bmatrix} 2 & 0 \\ 0 & 6 \end{bmatrix} = 12 > 0$.

Because all principal minors are nonnegative, we can classify $f(x_1, x_2) = x_1^2 + 3x_2^2$ as a convex function by Theorem 2a. The graph is shown in Figure 9.2.

**FIGURE 9.2**
Graph of
$f(x_1, x_2) = x_1^2 + 3x_2^2$

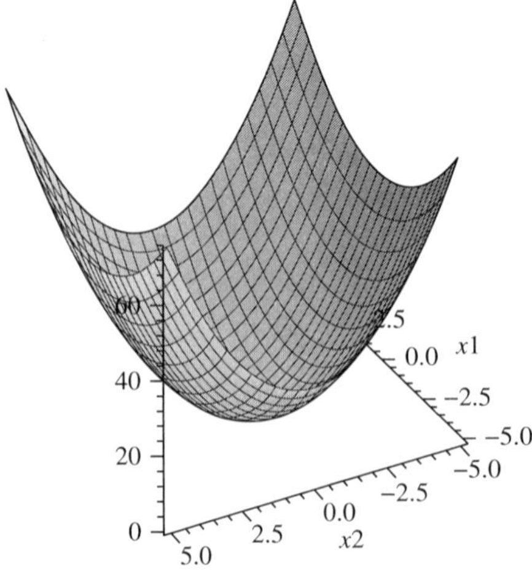

## Example 6

## A Concave Function

Determine the convexity of the function using the Hessian matrix $f(x_1, x_2) = -x_1^2 - 3x_2^2 + x_1 \cdot x_2$.

$$H = \begin{bmatrix} -2 & 1 \\ 1 & -6 \end{bmatrix}$$

The first leading principal minors are $\det[-2] = -2 < 0$ and $\det[-6] = -6 < 0$, both of which follow the signs of $(-1)^k$ when $k = 1 = (-1)^1 = -1$ (which is negative).

The second leading principal minor is $\det \begin{bmatrix} -2 & 1 \\ 1 & -6 \end{bmatrix} = 12 - 1 = 11 > 0$, which follows the sign of $(-1)^k$ when $k = 2 = (-1)^2 = 1$ (which is positive).

So we can classify the function $f(x_1, x_2) = -x_1^2 - 3x_2^2 + x_1 \cdot x_2$ as a concave function according to Theorem 2b. The graph is shown in Figure 9.3.

**FIGURE 9.3**
Graph of
$f(x_1, x_2) = -x_1^2 - 3x_2^2 + x_1 \cdot x_2$

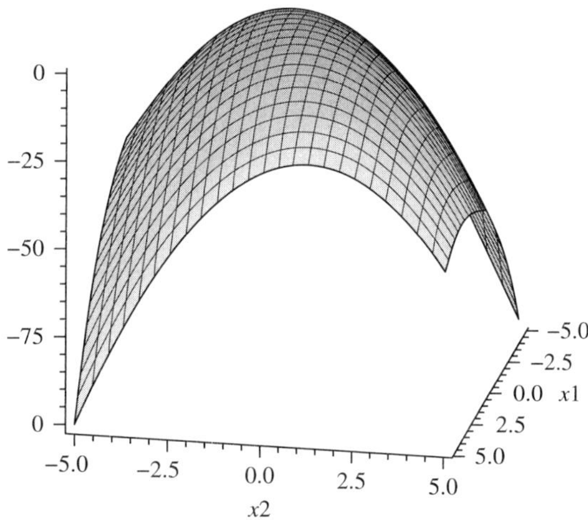


## Example 7

## Neither a Convex nor a Concave Function

Determine the convexity of the function $f(x_1, x_2) = x_1^2 - 3x_2^2 + x_1 \cdot x_2$.

$$H = \begin{bmatrix} 2 & 1 \\ 1 & -6 \end{bmatrix}$$

The first leading principal minors are $\det[2] = 2 > 0$ and $\det[-6] = -6 < 0$. Neither follows Theorem 2a or Theorem 2b; because the signs along the main diagonal change, we found one negative and one positive. We can stop finding the principal minors (because the first principal minors rules were violated) and conclude that the function is neither convex nor concave according to Theorem 2c. This Hessian matrix is called *indefinite*—the Hessian did not follow either theorem. The graph for $f(x_1, x_2) = x_1^2 - 3x_2^2 + x_1 \cdot x_2$ is shown in Figure 9.4.

```
> z2:=x1^2−3·x2^2+x1·x2;
```

$$z2 := x1^2 - 3\,x2^2 + x1\,x2$$

```
h :=Hessian(z2,[x1,x2]);
```

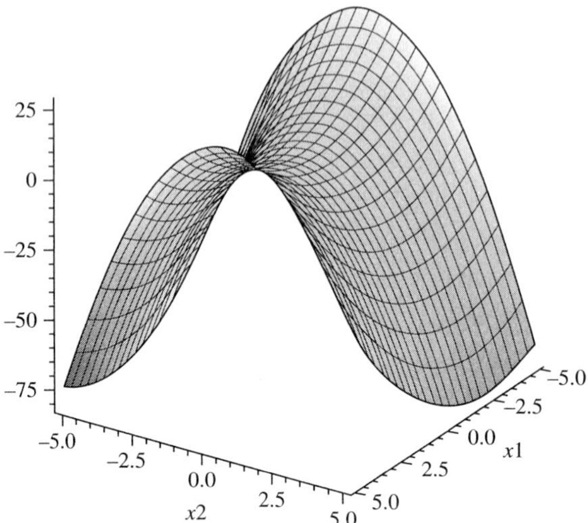

**FIGURE 9.4**
Graph of
$f(x_1, x_2) = x_1^2 - 3x_2^2 + x_1 \cdot x_2$



$$h := \begin{bmatrix} 2 & 1 \\ 1 & -6 \end{bmatrix}$$

> *with( LinearAlgebra ):*
> *IsDefinite( h, 'query'='indefinite' );*

$$true$$

Therefore, the function is neither convex nor concave because the Hessian is indefinite.

Let's return to Examples 1 and 2.

In Example 1, the Hessian matrix that follows Theorem 2a is called *positive definite*. In Example 2, the Hessian matrix that follows Theorem 2b is called *negative definite*. Maple can be used to test and verify definiteness.

> *With( LinearAlgebra ):with( linalg ):with( VectorCalculus ):*
> *z1:=x1²+3·x2²;*

$$z1 := x1^2 + 3\,x2^2$$

> *h1:=Hessian( z1, [x1,x2] );*

$$h1 := \begin{bmatrix} 2 & 0 \\ 0 & 6 \end{bmatrix}$$

> *IsDefinite( h1, 'query'='positive_definite' );*

$$true$$

> *z2:=-x1²-3·x2²+3·x1·x2;*

$$z2 := -x1^2 - 3\,x2^2 + 3x1\,x2$$

> *h2:=Hessian( z2, [x1,x2] );*

$$h2 := \begin{bmatrix} -2 & 3 \\ 3 & -6 \end{bmatrix}$$

> *IsDefinite( h2, 'query'='negative_definite' );*

$$true$$

Three library packages can be used to obtain the Hessian matrix from Maple: Linear-Algebra, and VectorCalculus. We suggest consulting the Help menu in Maple to decide which one you might need to use.

As discussed earlier, the Hessian matrix provides information concerning the nature or classification of the stationary points so that we may classify these points as maximum, minimum, or saddle points.

$$H(X) = \begin{bmatrix} \dfrac{\partial^2 f}{\partial x_1^2} & \dfrac{\partial^2 f}{\partial x_1 x_2} & \dfrac{\partial^2 f}{\partial x_1 x_3} & \cdots & \dfrac{\partial^2 f}{\partial x_1 x_n} \\[2ex] \dfrac{\partial^2 f}{\partial x_2 x_1} & \dfrac{\partial^2 f}{\partial x_2^2} & \cdots & & \dfrac{\partial^2 f}{\partial x_2 x_n} \\[2ex] \dfrac{\partial^2 f}{\partial x_3 x_1} & \cdots & \cdots & & \dfrac{\partial^2 f}{\partial x_3 x_n} \\[2ex] \cdots & & \cdot & \cdot & \cdot \\[2ex] \dfrac{\partial^2 f}{\partial x_n x_1} & \dfrac{\partial^2 f}{\partial x_n x_2} & \cdots & & \dfrac{\partial^2 f}{\partial x_n^2} \end{bmatrix}$$

$$H(x) = \begin{bmatrix} F_{xx} & F_{xy} & \cdot & \cdot & F_{xz} \\ F_{yx} & F_{yy} & \cdot & \cdot & F_{yz} \\ \cdot & \cdot & \cdot & \cdot & \cdot \\ \cdot & \cdot & \cdot & \cdot & \cdot \\ F_{zx} & F_{zy} & \cdot & \cdot & F_{zz} \end{bmatrix}$$

The Hessian matrix is a symmetric $n \times n$ matrix, where $n$ is the number of independent variables. For example, $z = f(x,y)$ would yield a $2 \times 2$ symmetric matrix. The matrix is symmetric because $F_{xy} = F_{yx}$. The function is $x^T H x$. This notation is used to describe the matrix as one of the following:

| | |
|---|---|
| Positive definite | $> 0$ |
| Positive semidefinite | $\geq 0$ |
| Negative definite | $< 0$ |
| Negative semidefinite | $\leq 0$ |
| Indefinite | Otherwise |

These also relate to the signs of the leading principal minors (PMs), which were previously described.

Positive definite: All leading principal minors are $> 0$.

Positive semidefinite: All leading principal minors are nonnegative (some are zero).

Negative definite: The leading principal minors follow the signs of $(-1)^k$, where $k$ represents the order of the leading principal minor. For example, with $k = 1, 2, 3$, the signs of the leading principal minors are minus, positive, minus.

Negative semidefinite: All nonzero valued leading principal minors follow the signs of $(-1)^k$, where k represents the order of the leading principal minor. For example, with $k = 1, 2, 3$, the signs of the leading principal minors are minus, positive, minus. Some leading principal minors have value zero.

Indefinite: Some leading principal minors do not follow any of the rules for positive definite, positive semidefinite, negative definite, or negative semidefinite defined here.

The Hessian matrix is not always a matrix of all constants. If the Hessian is a function of the independent variables, then its definiteness might vary from one value of $x$ to another. To test the definiteness of the Hessian at a point $x^*$, it is necessary to evaluate the Hessian at the point $x^*$. For example, consider the following Hessian:

$$H(X) = \begin{bmatrix} 2x_1 & x_2 \\ x_2 & 4 \end{bmatrix}$$

The values of $x_1$ and $x_2$ determine whether the matrix is positive definite, positive semidefinite, negative definite, negative semidefinite, or indefinite.

There exists a relationship between the Hessian matrix definiteness and the classification of stationary points (extrema) as maximum, minimum, or saddle points, or as inconclusive.

The following table summarizes these results, with $k$ indicating the order of the leading principal minors of the Hessian. The $i$th PM is found by eliminating the $n - i$ rows and corresponding columns of the matrix. The first leading PMs are always the main diagonal of the original Hessian matrix.

| Determinants: $k$, Leading Principal Minors (PMs) | Results | Conclusions about Stationary Points |
|---|---|---|
| $H_k > 0$ | Positive definite, $f$ convex | Minima |
| $H_k \geq 0$ | Positive semidefinite, $f$ convex | Local minima |
| $H_k$ follows the signs of $(-1)^k$ | Negative definite, $f$ concave | Maxima |
| $H_k$ either 0 (not all 0) or follows $(-1)^k$ | Negative definite, $f$ concave | Local maxima |
| $H_k$ not all 0 and none of the above | Indefinite, $f$ neither | Saddle point |
| $H_k$ all 0″s | Indefinite | Inconclusive |

---

## Example 8    A Negative Definite Hessian Matrix

Suppose the Hessian of the function is given by:

$$H(x) = \begin{bmatrix} -2 & -1 \\ -1 & -4 \end{bmatrix}$$

The first leading PMs are $-2$, $-4$, and the second leading PM is the determinant of the 2×2 matrix, 7. Because the first leading PMs are negative and the second leading PM is positive, we find that they follow the rule that the leading PM follows the sign of $(-1)^k$, where $k$ represents the order of the leading principal minor. We would conclude that the function is concave and that any corresponding stationary point found was a maximum. The Hessian matrix is negative definite (ND).

> $h:=\langle\langle -2,-1\rangle|\langle -1,-4\rangle\rangle;$

$$h := \begin{bmatrix} -2 & -1 \\ -1 & -4 \end{bmatrix}$$

> $IsDefinite(\,h,'query'='negative\_definite'\,);$

$$true$$

---

## Example 9    A Positive Semidefinite Hessian Matrix

$$H(x) = \begin{bmatrix} 3 & 2 & 1 \\ 6 & 5 & 4 \\ 9 & 8 & 7 \end{bmatrix}$$

The first leading PMs are 3, 5, and 7. Note they are the entries of the main diagonal. The second leading PMs are the determinants found by eliminating the row and column containing the first PM one at a time. This yields three 2×2 submatrices.

Eliminating the row and column with the 3 yields the 2×2 matrix

$$A = \begin{bmatrix} 5 & 4 \\ 8 & 7 \end{bmatrix}$$

with a determinate value of $35 - 32$, or 3. Eliminating the row and column with the 5 yields the 2×2 matrix

$$B = \begin{bmatrix} 3 & 1 \\ 9 & 7 \end{bmatrix}$$

with a determinate value of $21 - 9$, or 12.

Eliminating the row and column with the 7 yields the 2×2 matrix

$$C = \begin{bmatrix} 3 & 2 \\ 6 & 5 \end{bmatrix}$$

with a determinate value of $15 - 12$, or 3.

The third leading PM is the determinant of the 3×3 original matrix. This value is zero (0).

The Hessian follows the positive semidefinite form where all leading principal minors are greater than or equal to zero. The function would be convex, and any corresponding stationary points would be local minima.

This is easier in Maple.

> $h := \langle\langle 3,6,9 \rangle | \langle 2,5,8 \rangle | \langle 1,4,7 \rangle\rangle;$

$$h := \begin{bmatrix} 3 & 2 & 1 \\ 6 & 5 & 4 \\ 9 & 8 & 7 \end{bmatrix}$$

> $IsDefinite(h, \text{'query'} = \text{'negative\_definite'});$

$$false$$

> $IsDefinite(h, \text{'query'} = \text{'positive\_definite'})$

$$false$$

> $IsDefinite(h, \text{'query'} = \text{'positive\_semidefinite'})$

$$True$$

<table><tr><td>**Example 10**</td><td>**Maple Checking "Definiteness"**</td></tr></table>

$$H(x) = \begin{bmatrix} -2 & -4 \\ -4 & -3 \end{bmatrix}$$

has the first leading PM of $-2$ and $-3$, whereas the second leading PM is $-10$. This is an indefinite Hessian, and any stationary point corresponding to this Hessian would be a saddle point.

> $nh := \langle\langle -2,-4 \rangle | \langle -4,-3 \rangle\rangle;$

$$nh := \begin{bmatrix} -2 & -4 \\ -4 & -3 \end{bmatrix}$$

> *IsDefinite( nh,'query'='indefinite');*

$$true$$

---

## 9.3    UNCONSTRAINED OPTIMIZATION

We put the theorems and definitions from the previous sections to work in this section.

---

### Example 1    Finding and Classifying Stationary Points

Find and classify all the stationary points of $f(x,y) = 55 \cdot x - 4 \cdot x^2 + 135 \cdot y - 15 \cdot y^2 - 100$.

$$\frac{\partial f}{\partial x} = 55 - 8x = 0$$

$$\frac{\partial f}{\partial x} = 135 - 30y = 0$$

These solve as $x = \dfrac{55}{8}$ and $y = \dfrac{135}{30}$. There is only stationary point.

$$H\left(\frac{55}{8}, \frac{135}{30}\right) = \begin{bmatrix} -8 & 0 \\ 0 & -30 \end{bmatrix}$$

The first PM are $-8, -30$, which follows $(-1)^1$.
The second PM is $240 > 0$, which follows $(-1)^2$.

The function, $f$, is concave at $\left(\dfrac{55}{8}, \dfrac{135}{30}\right)$, therefore $\left(\dfrac{55}{8}, \dfrac{135}{30}\right)$ represents the maximum of $f$.

$$f\left(\frac{55}{8}, \frac{135}{30}\right) = 392.81$$

The commands in Maple are:

> *z:=55*x−4*x^2+135*y−15*y^2−100;*

$$z := 55x - 4x^2 + 135\,y - 15y^2 - 100$$

> *zx:=diff (z,x);*

$$zx := 55 - 8x$$

> *zy:=diff (z,y);*

$$zy := 135 - 30y$$

> *solve( {zx=0,zy=0}, {x,y});*

$$\{y = \frac{9}{2}, x = \frac{55}{8}\}$$

> *evalf (%);*

$$\{y = 4.500000000, x = 6.875000000\}$$

```
> with( VectorCalculus):
> h:=Hessian(z,[x,y]);
```

$$h := \begin{bmatrix} -8 & 0 \\ 0 & -30 \end{bmatrix}$$

```
> evalf ( subs( {y=9/2,x=55/8},z ) );
```

$$392.8125000$$

```
> IsDefinite(h,'query'='negative_definite');
```

$$true$$

---

<table><tr><td>

**Example 2**

</td><td>

## Finding and Classifying All Extreme Points

Find all the local maxima, local minima, and saddle points for $f(x_1, x_2) = x_1 x_2^2 + x_1^3 x_2 - x_1 x_2$.

We find the partial derivatives and set them equal to zero to find the stationary points.

$$\frac{\partial f}{\partial x_1} = x_2^2 + 3x_1^2 x_2 - x_2 = 0$$

$$\frac{\partial f}{\partial x_2} = 2x_1 x_2 + x_1^3 x_2 - x_1 = 0$$

$$x_2^2 + 3x_1^2 - x_2 = 0 \text{ or } x_2(x_2 + 3x_1^2 - 1) = 0$$

$$2x_1 x_2 + x_1^3 - x_1 = 0 \text{ or } x_1(2x_2 + x_1^2 - 1) = 0$$

The candidate points are:

(1)   $x_2 = 0$ or $x_1 = \sqrt{\dfrac{1 - x_2}{3}}$

(2)   $x_1 = 0$ or $x_2 = \dfrac{1 - x_1^2}{2}$

Thus, the following points will work for equations (1) and (2), alternating between $x_2 = 0$ and $x_1 = 0$:

(0, 0)

(1, 0)

(0, 1)

(−1, 0)

Also, we need to find solutions when $x_1 = \sqrt{\dfrac{1 - x_2}{3}}$ and $x_2 = \dfrac{1 - x_1^2}{2}$.

By substitution, $x_2 = \dfrac{1 - \left(\sqrt{\dfrac{1-x_2}{3}}\right)^2}{2} = \dfrac{1 - \left(\dfrac{1-x_2}{3}\right)}{2}$.

$$2x_2 = 1 - \frac{1}{3} + \frac{x_2}{3}$$

$$\frac{5}{3}x_2 = \frac{2}{3}$$

$$x_2 = \frac{2}{5}$$

</td></tr></table>

$$\text{If } x_2 = \frac{2}{5}, \text{ then } x_1 = \sqrt{\frac{1}{5}} = \pm\frac{\sqrt{5}}{5}.$$

$$\left(\frac{\sqrt{5}}{5}, \frac{2}{5}\right)$$

$$\left(-\frac{\sqrt{5}}{5}, \frac{2}{5}\right)$$

$$(0, 0)$$
$$(1, 0)$$
$$(0, 1)$$
$$(-1, 0)$$

$$\left(\frac{\sqrt{5}}{5}, \frac{2}{5}\right)$$

$$-\left(\frac{\sqrt{5}}{5}, \frac{2}{5}\right)$$

We have six stationary points that need to be tested with the Hessian to see if they are local maxima, local minima, or saddle points. We need the Hessian matrix.

$$\frac{\partial f}{\partial x_1} = x_2^2 + 3x_1^2\, x_2 - x_2 = 0$$

$$\frac{\partial f}{\partial x_1} = 2x_1 x_2 + x_1^3 - x_1 = 0$$

$$\frac{\partial f}{\partial x_1} = 6x_1 x2$$

$$\frac{\partial f}{\partial x_1} = 2x_1$$

$$H = \begin{bmatrix} 6x_1 x_2 & 2x_2 + 3x_1^2 - 1 \\ 2x_2 + 3x_1^2 - 1 & 2x_1 \end{bmatrix}$$

$$H(0,0) = \begin{bmatrix} 0 & -1 \\ -1 & 0 \end{bmatrix}$$

$$H(1,0) = \begin{bmatrix} 0 & 2 \\ 2 & 1 \end{bmatrix}$$

$$H(0,1) = \begin{bmatrix} 0 & 1 \\ 1 & 0 \end{bmatrix}$$

$$H(-1,0) = \begin{bmatrix} 0 & 2 \\ 2 & -2 \end{bmatrix}$$

$$H\left(\frac{\sqrt{5}}{5}, \frac{2}{5}\right) = \begin{bmatrix} 12\frac{\sqrt{5}}{25} & \frac{2}{5} \\ \frac{2}{5} & 2\frac{\sqrt{5}}{5} \end{bmatrix}, \text{ determinant: } \frac{4}{5}$$

$$H\left(-\frac{\sqrt{5}}{5}, \frac{2}{5}\right) = \begin{bmatrix} -12\frac{\sqrt{5}}{25} & \frac{2}{5} \\ \frac{2}{5} & -2\frac{\sqrt{5}}{5} \end{bmatrix}, \text{ determinant: } \frac{4}{5}$$

| Point | First PM | Second PM | Classification and Result |
|---|---|---|---|
| (0,0) | 0,0, both nonnegative | $-1$ | Neither, saddle point |
| (1,0) | 0,0, both nonnegative | $-4$ | Neither, saddle point |
| (0,1) | 0,0, both nonnegative | $-1$ | Neither, saddle point |
| (−1,0) | $0, -2$, follows $(-1)^1$ | $-4$ | Neither, saddle point |
| $\left(\dfrac{\sqrt{5}}{5}, \dfrac{2}{5}\right)$ | $12\dfrac{\sqrt{5}}{25}, 2\dfrac{\sqrt{5}}{5}$, both nonnegative | $\dfrac{4}{5}$ | $f$ is convex, local minimum |
| $-\left(\dfrac{\sqrt{5}}{5}, \dfrac{2}{5}\right)$ | $-12\dfrac{\sqrt{5}}{25}, -2\dfrac{\sqrt{5}}{5}$, follows $(-1)^1$ | $\dfrac{4}{5}$ (follows $(-1)^2$) | $f$ is concave, local maximum |

> $z:=x2\^2*x1+x1\^3*x2-x1*x2;$

$$z := x2^2 x1 + x1^3 x2 - x1\,x2$$

> $zx1:=diff(z,x1);$

$$zx1 := x2^2 + 3x1^2\,x2 - x2$$

> $zx2:=diff(z,x2);$

$$zx2 := 2x1\,x2 + x1^3 - x1$$

> $solve(\{zx1=0,zx2=0\},\{x1,x2\});$

$$\{x1 = 0, x2 = 0\}, \{x1 = 0, x2 = 1\}, \{x1 = 1, x2 = 0\}, \{x1 = -1, x2 = 0\},$$
$$\left\{x2 = \frac{2}{5}, x1 = RootOf(5\_Z^2 - 1,\ label = \_L6)\right\}$$

> $solve(5*x\^2-1,x);$

$$\frac{\sqrt{5}}{5}, -\frac{\sqrt{5}}{5}$$

> $h:=Hessian(z,[x1,x2]);$

$$h := \begin{bmatrix} 6x1x2 & 2x2 + 3x1^2 - 1 \\ 2x2 + 3x1^2 - 1 & 2x1 \end{bmatrix}$$

> $subs(\{x1=0,x2=0\},h);$

$$\begin{bmatrix} 0 & -1 \\ -1 & 0 \end{bmatrix}$$

> $subs(\{x1=0,x2=1\},h);$

$$\begin{bmatrix} 0 & 1 \\ 1 & 0 \end{bmatrix}$$

> $subs(\{x2=1,x1=0\},h);$

$$\begin{bmatrix} 0 & 1 \\ 1 & 0 \end{bmatrix}$$

> $subs(\{x1=-1,x2=0\},h);$

$$\begin{bmatrix} 0 & 2 \\ 2 & -2 \end{bmatrix}$$

> $subs(\{x1=sqrt(5)/5,x2=2/5\},h);$

$$\begin{bmatrix} \dfrac{12\sqrt{5}}{25} & \dfrac{2}{5} \\ \dfrac{2}{5} & \dfrac{2\sqrt{5}}{5} \end{bmatrix}$$

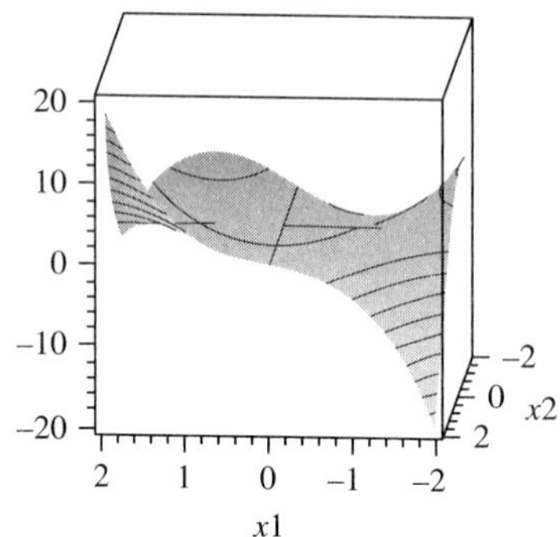

**FIGURE 9.5**
Graph of
$x_1 x_2^2 + x_1^3 x_2 - x_1 x_2$

> $subs(\{x1=-sqrt(5)/5,x2=2/5\},h);$

$$
\begin{bmatrix}
-\dfrac{12\sqrt{5}}{25} & \dfrac{2}{5} \\[2ex]
\dfrac{2}{5} & -\dfrac{2\sqrt{5}}{5}
\end{bmatrix}
$$

> $evalf(subs(\{x1=sqrt(5)/5,x2=2/5\},z));$

$$-0.07155417526$$

> $evalf(subs(\{x1=-sqrt(5)/5,x2=2/5\},z));$

$$0.07155417526$$

The graph of $x_1 x_2^2 + x_1^3 x_2 - x_1 x_2$ is shown in Figure 9.5.

---

**Example 3**

## Finding Stationary Points for $f(x,y) = 2xy + 4x + 6y - 2x^2 - 2y^2$

Find and classify all the stationary points of the function, $f(x,y) = 2xy + 4x + 6y - 2x^2 - 2y^2$.

$$f(x,y) = 2xy + 4x + 6y - 2x^2 - 2y^2$$

$$\frac{\partial f}{\partial x} = 2y + 4 - 4x$$

$$\frac{\partial f}{\partial x} = 2x + 6 - 4y$$

To find where $\dfrac{\partial f}{\partial x} = \dfrac{\partial f}{\partial y} = 0$, we solve

$$-4x + 2y = -4$$
$$2x - 4y = -6$$

$$
\begin{bmatrix}
-4 & 2 & -4 \\
2 & -4 & -6
\end{bmatrix},\ \text{row echelon}
\begin{bmatrix}
1 & 0 & \dfrac{7}{3} \\[1ex]
0 & 1 & \dfrac{8}{3}
\end{bmatrix}
$$

$$x = \frac{7}{3},\ y = \frac{8}{3}$$

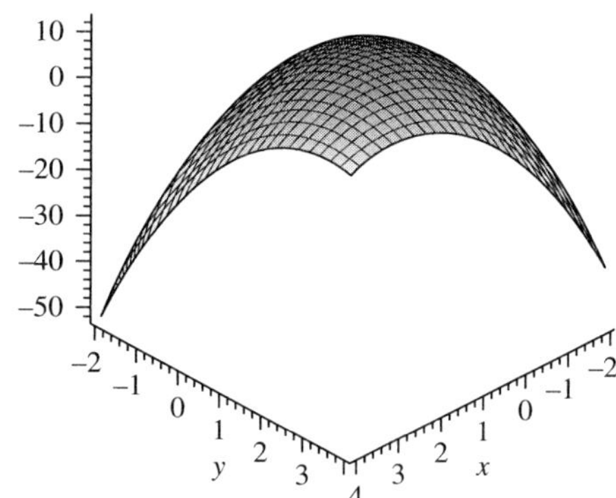


**FIGURE 9.6**
Graph of
$z := 2xy + 4x + 6y - 2x^2 - 2y^2$

$$H = \begin{bmatrix} -4 & 2 \\ 2 & -4 \end{bmatrix}$$

First leading PM is $-4$, $-4$, following $(-1)^k$, $k = 1$
Second leading PM is $16 - 4 = 12 > 0$, following $(-1)^k$, $k = 2$

Thus, $H$ is negative definite, $f$ is concave, and $(x^*, y^*) = (7/3, 8/3)$ is a global maximum. This is clearly seen in Figure 9.6.

```
> z:=2*x*y+4*x+6*y−2*x^2−2*y^2;
```

$$z := 2xy + 4x + 6y - 2x^2 - 2y^2$$

```
> zx:=diff(z,x);
```

$$zx := 2y + 4 - 4x$$

```
> zy:=diff(z,y);
```

$$zy := 2x + 6 - 4y$$

```
> fsolve({zx=0,zy=0},{x,y});
```

$$\{x = 2.333333333, \ y = 2.666666667\}$$

```
> h:=Hessian(z,[x,y]);
> IsDefinite(h,'query'='negative_definite');
```

$$true$$

```
> subs({x = 2.333333333, y = 2.666666667},z);
```

$$12.66666665$$

```
> with(plots):
> plot3d(z,x=−2..4,y=−2..4);
```

---

## Example 4    Least-Squares Model

In a least-squares model fitting a line, $y = a + bx$, we want to minimize the sum of squared error,

$$f(a,b) = \sum_{i=1}^{i=n} (y_i - a - bx_i)^2$$

$$\frac{\partial f}{\partial a} = 2\left(\sum_{i=1}^{i=n}(y_i - a - bx_i)(-1)\right) = 0$$

$$\frac{\partial f}{\partial b} = 2\left(\sum_{i=1}^{i=n}(y_i - a - bx_i)(-x_i)\right) = 0$$

$$\frac{\partial f}{\partial a} = \frac{\partial f}{\partial b} = 0$$

hold when we find $(a,b)$ such that the following hold:

$$na + b\sum_{i=1}^{i=n} x_i = \sum_{i=1}^{i=n} y_i$$

$$a\sum_{i=1}^{i=n} x_i + b\sum_{i=1}^{i=n} x_i^2 = \sum_{i=1}^{i=n} x_i y_i$$

These are called the *normal equations for least squares*:

$$H = \begin{bmatrix} 2n & 2\sum_{i=1}^{i=n} x_i \\ 2\sum_{i=1}^{i=n} x_i & 2\sum_{i=1}^{i=n} x_i^2 \end{bmatrix}$$

First leading PMs are $2n$, $2\sum_{i=1}^{i=n} x_i^2$ both are $> 0$.

Second leading PM is $4n - 4 > 0$.

$H$ is positive definite, and $f$ is strictly convex, so $(a,b)^*$ will be a global minimum.
Now let's look at this with the following data:

| $x$ | 1 | 2 | 3 |
|---|---|---|---|
| $y$ | 2 | 4.8 | 7 |

$$\text{Min } S = (2 - a - b)^2 + (4{\cdot}8 - 2a - b)^2 + (7 - 3a - b)^2$$

`> S:=(2–a–b)²+(4.8–2·a–b)²+(7–3·a–b)²;`

$$S := (2-a-b)^2 + (4.8-2a-b)^2 + (7-3a-b)^2$$

`> pda:=diff(S,a);`

$$pda := -65.2 + 28a + 12b$$

`> pdb:=diff(S,b);`

$$pdb := -27.6 + 12a + 6b$$

`> cp:=solve({pda=0,pdb=0},{a,b});`

$$cp := \{b = -.4000000000,\ a = 2.500000000\}$$

`> with(VectorCalculus):`
`> h:=Hessian(S,[a,b]);`

$$h := \begin{bmatrix} 28 & 12 \\ 12 & 6 \end{bmatrix}$$

The Hessian matrix is positive definite, so we have found the minimum.

## Example 5

## Finding the Island

Consider the following problem. We want to see if an island exists in a harbor. The following function represents the topographic underwater region of the harbor that we see in Figure 9.7 and the contour plot in Figure 9.8.

> $f:=-300*y^3-695*y^2+7*y-300*x^3-679*x^2-235*x+570;$

$$f:=-300y^3 -695y^2 + 7y-300x^3 -679x^2 -235x + 570$$

> $with\ (plots):$

> $contourplot(f, x =-2\ ..2,\ y =-2\ ..2,\ contours = 50);$

> $pdx := diff(f, x);$

$$pdx := -900.0x^2 -1358.0x -235.0$$

> $pdy := diff(f,y);$

$$pdy := -900.0y^2 -1390.0y + 7.0$$

> $with(VectorCalculus):$

> $solve(\{pdx = 0, pdy = 0\},\ \{x,y\});$

$$\{y = 0.005019656649, x = -.1993991232\}, \{y = 0.00501965664,$$
$$x = -1.309489766\}, \{x = -.1993991232, y = -1.549464101\}$$
$$\{x = -1.309489766, y = -1.549464101\}$$

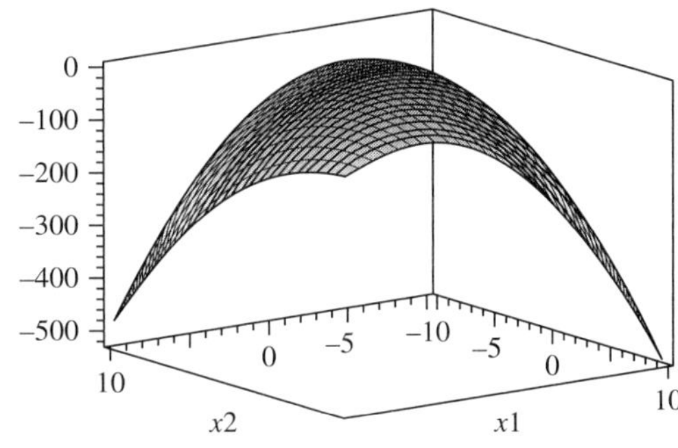

**FIGURE 9.7**

The function $Z = f(x_1, x_2) = 2x_1x_2 + 2x_2 - x_1^2 - 2x_2^2$



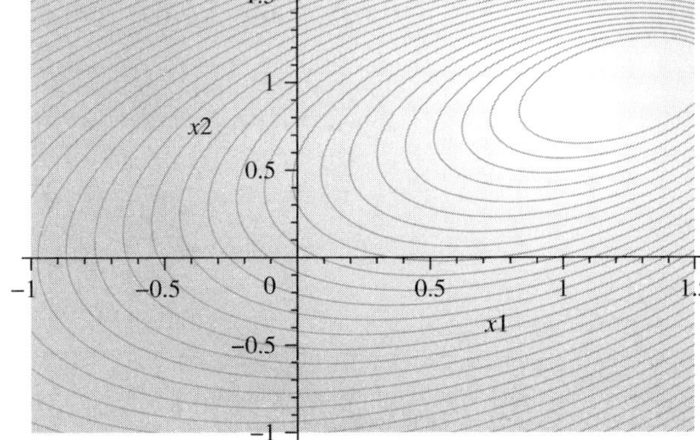

**FIGURE 9.8**

Contour plot of $2x_1x_2 + 2x_2 - x_1^2 - 2x_2^2$

```
> subs({x=-.1993991232,y=-1.549464101},f);
```

$$28.8149796$$

```
> h :=Hessian(f,[x,y]);
```

$$h := \begin{bmatrix} -1800x-1358 & 0 \\ 0 & -1800y-1390 \end{bmatrix}$$

```
> subs({x=-.1993991232, y=-1.549464101},Hessian(f,[x,y]);
```

$$\begin{bmatrix} -999.0815782 & 0 \\ 0 & 1399.035382 \end{bmatrix}$$

```
> subs({y=0.005019656649,x=-.1993991232},f);
```

$$592.2577681$$

```
> subs({y=0.005019656649,x=-.1993991232},
    Hessian (f, [x,y]));
```

$$\begin{bmatrix} -999.0815782 & 0 \\ 0 & -1399.035382 \end{bmatrix}$$

```
> subs({y=0.005019656649,x=-1.309489766},f);
```

$$387.0628573$$

```
> subs({y=0.005019656649,x=-1.309489766},
    Hessian(f, [x,y]));
```

$$\begin{bmatrix} 999.081579 & 0 \\ 0 & -1399.035382 \end{bmatrix}$$

```
> subs({y=-1.549464101,x=-1.309489766},f);
```

$$-176.3799310$$

```
> subs({y=1.549464101,x=-1.309489766},Hessian(f, [x,y])
```

$$\begin{bmatrix} 999.081579 & 0 \\ 0 & 1399.035382 \end{bmatrix}$$

We find that at $x = -0.1993991232$ and $y = 0.0050019656649$ with a maximum height of $f(x,y) = 592.2577681$.

The Hessian is negative definite at that point, indicating that we found the maximum.

If we assume that $Z = 0$ is sea level, then an island exists in the harbor that is 592.2577681 feet above sea level.

## 9.3 | EXERCISES

1. Indicate both the "definiteness" {positive definite, positive semidefinite, negative definite, negative semidefinite, and indefinite} of the following Hessian matrices and indicate the concavity of the function from which these Hessians were derived.

Assume the Hessian, $H$, is:

**a.** $\begin{bmatrix} 2 & 3 \\ 3 & 5 \end{bmatrix}$

**b.** $\begin{bmatrix} 4 & 3 \\ 3 & 2 \end{bmatrix}$

**c.** $\begin{bmatrix} -2 & 1 \\ 1 & -2 \end{bmatrix}$

**d.** $\begin{bmatrix} -3 & 4 \\ 4 & -5 \end{bmatrix}$

**e.** $\begin{bmatrix} 6x & 0 \\ 0 & 2x \end{bmatrix}$

**f.** $\begin{bmatrix} x^2 & 0 \\ 0 & 2y \end{bmatrix}$

**g.** $\begin{bmatrix} 2 & 2 & 3 \\ 2 & 6 & 4 \\ 3 & 4 & 4 \end{bmatrix}$

**h.** $\begin{bmatrix} 1 & 2 & 2 \\ 2 & 1 & 2 \\ 2 & 2 & 1 \end{bmatrix}$

2. Using the Hessian matrix, $H$, determine the convexity and then find the critical points and classify them for the following:

   **a.** $f(x,y) = x^2 + 3xy - y^2$

   **b.** $f(x,y) = x^2 + y^2$

   **c.** $f(x,y) = -x^2 - xy - 2y^2$

   **d.** $f(x,y) = 3x + 5y - 4x^2 + y^2 - 5xy$

   **e.** $f(x,y,z) = 2x + 3y + 3z - xy + xz - yz - x^2 - 3y^2 - z^2$

   **f.** Determine the values of $a$, $b$, and $c$ such that $ax^2 + bxy + cy^2$ is convex and concave.

3. Find and classify all the extreme points for the following:

   **a.** $f(x,y) = x^2 + 3xy - y^2$

   **b.** $f(x,y) = x^2 + y^2$

   **c.** $f(x,y) = -x^2 - xy - 2y^2$

   **d.** $f(x,y) = 3x + 5y - 4x^2 + y^2 - 5xy$

   **e.** $f(x,y,z) = 2x + 3y + 3z - xy + xz - yz - x^2 - 3y^2 - z^2$

4. Find and classify all critical points of $f(x,y) = e^{(x-y)} + x^2 + y^2$.

5. Find and classify all critical points of the function $f(x,y) = (x^2 + y^2)^{1.5} - 4(x^2 + y^2)$.

6. Consider a small company that is planning to install a central computer with cable links to five new departments. According to their floor plan, the peripheral computers for the five departments will be situated as shown by the dark circles in Figure 9.6. The company wishes to locate the central computer so that the minimal amount of cable will be used to link to the five peripheral computers. Assuming that cable can be strung over the ceiling panels in a straight line from a point above any peripheral to a point above the central computer, the distance formula may be used to determine the length of cable needed to connect any peripheral to the central computer. Ignore all lengths of cable from the

computer itself to a point above the ceiling panel immediately over that computer. In other words, work only with lengths of cable strung over the ceiling panels. This was displayed earlier in Figure 9.1.

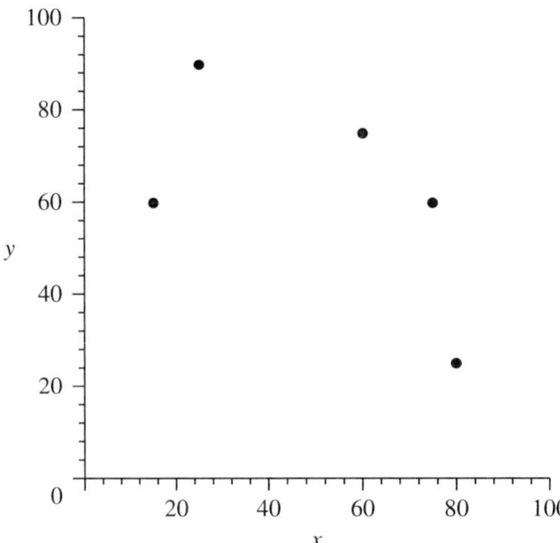

**FIGURE 9.1**
The Grid for the Five
Departments

The coordinates of the locations of the five peripheral computers are listed in Table 9.1.

**Table 9.1** Grid Coordinates of Five Departments

| X | Y |
|---|---|
| 15 | 60 |
| 25 | 90 |
| 60 | 75 |
| 75 | 60 |
| 80 | 25 |

Assume the central computer will be positioned at coordinates $(m, n)$, where $m$ and $n$ are *integers* in the grid representing the office space. Determine the coordinates $(m,n)$ for a placement of the central computer that minimizes the total amount of cable needed. Report the total number of feet of cable needed for this placement along with the coordinates $(m,n)$.

7. Find all the extrema and then classify the extrema for the following functions:

   **a.** $f(x,y) = x^3 - 3xy^2 + 4y^4$

   **b.** $w(x,y,z) = x^2 + 2xy - 4z + yz^2$

8. Three oil fields are located according to a rectangular coordinate system. Each field produces an equal amount of oil. A pipeline is to be laid from each oil field to a centrally located refinery. If the oil wells are located at coordinates $(0,0)$, $(12,6)$, and $(10, 20)$, where should the refinery be located to minimize the following total squared Euclidean distance?

$$\sum_{i=1}^{3} (x-a_i)^2 + (y-b_i)^2$$

9. The first partials of a function $f(x,y)$ evaluated at the point $(0,0)$ are $(-5,1)$ and the Hessian matrix, $H(x)$, is as follows:

$$H(x) = \begin{bmatrix} 6 & -1 \\ -1 & 2 \end{bmatrix}$$

Using your knowledge of partial derivatives and Hessians, determine the values of $(x,y)$ that minimize the function. Provide the value of $f(x,y)$. Show all work.

**10.** Find and classify all stationary points of the function:

$$k(x,y) = -5 + 3x^3 + 7x^2 + 2x + 7y^2 - y + 3y^3.$$

## 9.4  MULTIVARIABLE NUMERICAL SEARCH METHODS

In the previous sections, we discussed analytical techniques to solve the unconstrained nonlinear programming problem (NLP):

$$\text{Max } z = f(x_1, x_2, x_3, \ldots, x_n) \text{ over } R^n \tag{9.2}$$

In many problems, it is quite difficult to find the stationary points (critical points) and then use them to determine the nature of the stationary point. In this chapter, we will discuss several numerical techniques to either maximize or minimize a multivariable function as expressed in Equation 9.2.

### Gradient Search Methods

Suppose we want to solve the following unconstrained NLP using Equation 9.2:

$$\text{Maximize } z = f(x_1, x_2, x_3, \ldots, x_n)$$

In calculus, if Equation 9.2 is a concave function, then the optimal solution (if there is one) will occur at a stationary point $x^*$ having the following property:

$$\frac{\partial f(x^*)}{\partial x_1} = \frac{\partial f(x^*)}{\partial x_2} = \cdots = \frac{\partial f(x^*)}{\partial x_n} = 0$$

In many problems, it is not an easy task to find the stationary point. Thus, the method of steepest ascent (maximization problems) and the method of steepest descent (minimization problems) offer alternatives to finding an approximate stationary point. We will continue to discuss the gradient method for the steepest ascent.

Given a function like the one in Figure 9.9, assume that we want to find the maximum point of the function. If we started at the bottom of the hill, then we might proceed by finding the

**FIGURE 9.9**
Plot of
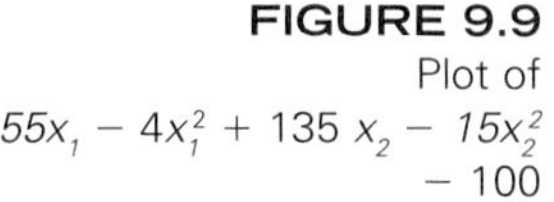
$-100$

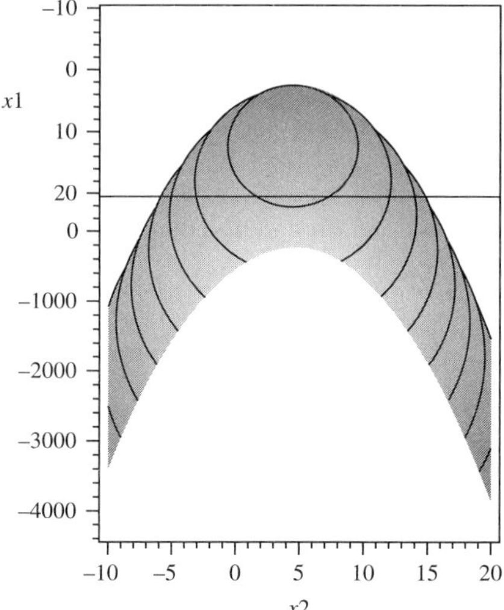

**FIGURE 9.10**

Steepest Ascent Algorithm

To find a maximum solution to a given multivariable unconstrained function, $f(x)$

**INPUT:**     Starting point $\mathbf{x}_0$; tolerance, $t$

**OUTPUT:**    Approximate $x^*$ and $f(x^*)$

**Step 1.**    Initialize the tolerance, $t > 0$.

**Step 2.**    Set $\mathbf{x} = \mathbf{x}_0$ and define the gradient at that point.

$$\nabla f(x_0)$$

**Step 3.**    Calculate the maximum of the new function $f[x_i + t_i\, \nabla f(x_i)]$, where $t_i \geq 0$, by finding the value of $t_i$.

**Step 4.**    Find the new $x_i$ point by substituting $t_i$ into

$$x_{i+1} = x_i + t_i\, \nabla f(x_i)$$

**Step 5.**    If the length (magnitude) of $\mathbf{x}$, defined by

$$\| \mathbf{x} \| = (x_1^2 + x_2^2 + \ldots + x_n^2)^{\frac{1}{2}},$$

(difference between 2 successive points) is less than the tolerance specified or if the **absolute magnitude of gradient is less than our tolerance** (derivative approximately zero), then **continue**. Otherwise, go back to step 3.

**Step 6.**    Use $x^*$ as the approximate stationary point and compute $f(x^*)$, the estimated maximum of the function.

**STOP**

gradient. The gradient is the vector of the partial derivatives that points "up the hill." We define the gradient vector as follows:

$$\nabla f(x) = \left[ \frac{\partial f(x^*)}{\partial x_1}, \frac{\partial f(x^*)}{\partial x_2}, \ldots, \frac{\partial f(x^*)}{\partial x_n} \right]$$

If we were lucky, the gradient would point all the way to the top of the function, but the contours of functions rarely cooperate. Thus, the gradient points up hill—but for how far? We need to find the distance along the gradient to travel that maximizes the height of the function in that direction. From that new point, we recompute a new gradient vector to find a new direction that points up hill. We continue this method until we get to the top of the hill.

From a starting point, we move in the direction of the gradient as long as we continue to increase the value of $f$. At that point, we move in the direction of a newly calculated gradient as far as we can so long as it continues to improve $f$. This continues until we achieve our maximum value within some specific tolerance (or margin of acceptable error). Figure 9.10 displays an algorithm for the method of steepest ascent using the gradient. We illustrate some examples of the gradient search.

## Example 1        Maximization with Gradient Search

$$\text{Maximize } f(x_1, x_2) = 2x_1 x_2 + 2x_2 - x_1^2 - 2x_2^2$$

The gradient of $f(x_1, x_2)$, $\nabla f$, is found using the partial derivatives as shown in the last chapter. The gradient is the vector $[2x_2 - 2x_1, 2x_1 + 2 - 4x_2]$. If we evaluate the gradient at $(0,0)$, we have

$Vf(0,0) = [0,2]$. From (0,0) we move along (up) the $x_2$-axis in the direction of [0,2]. How far do we go? We need to maximize the function starting at the point (0,0) using the function $f[x_i + t_i Vf(x_i)] = f(0 + 0t, x0 + 2t) = 2(2t) - 2(2t)^2 = 4t - 8t^2$.

This function can be maximized by using any of the one-dimensional search techniques that we discussed in Chapter 8 (single-variable optimization). This function can also be maximized by simple single-variable calculus:

$$\frac{df}{dt} = 0 = 4 - 16t = 0, t = 0.25$$

Iteration 1: The new point is found by substitution into $x_{i+1} = x_i + t_i Vf(x_i)$.
So, $x_1 = [0 + 0(.25), 0 + 2(.25)] = [0, 0.5]$.
The magnitude of $x_1$ is 0.5, which is not less than our tolerance of 0.01 (chosen arbitrarily). Because we are not optimal, we continue. We now repeat the calculations from the new point [0, 0.5].

## Iteration 2

The gradient vector is $[2x_2 - 2x_1, 2x_1 + 2 - 4x_2]$.

$Vf(0,0.5) = [1,0]$. From (0,0.5) we move in the direction of [1,0]. How far do we go? We need to maximize the function starting at the new point (0,0.5) using the function $f[x_i + t_i Vf(x_i)] = f(0 + 1t, 0.5 + 0t) = 2(t)(0.5) + 2(0.5) - t^2 - 2(0.5)^2 = -t^2 + t + 0.5$.

This function can also be maximized be using any of the one-dimensional search techniques that we discussed in Chapter 8, or it can be maximized by simple single-variable calculus:

$$\frac{df}{dt} = 0 = -2t + 1 = 0, t = 0.50$$

The new point is found by substitution into $x_{i+1} = x_i + t_i Vf(x_i)$.
So, $x_1 = [0 + 1(0.5), 0.5 + 0(0.5)], [0.5, 0.5]$

The magnitude of $x_1$ is $\sqrt{.5} = 0.707$, which is not less than our tolerance of 0.01 (chosen arbitrarily) . The magnitude of $Vf = 1$, which is also not less than 0.01. Because we are not optimal, we continue. We repeat the calculations from the new point [0.5,0.5].

We use Maple to complete the process. We wrote a program to perform the iterations for us.

$> f := 2*x1*x2 + 2*x2 - x1\wedge2 - 2*x2\wedge2;$

$$f := 2x1x2 + 2x2 - x1^2 - 2\,x2^2$$

$> (kt, MP, z1, z2, z3) := STEEPEST(50, .05, 1., 1., f);$

---

**Initial Condition: ( 0.0000, 0.0000)**

| Iteration | Gradient Vector | G magnitude | G x[k] | Step Length |
|---|---|---|---|---|
| 1 | ( 0.0000, 2.0000) | 2.0000 | ( 0.0000, 0.0000) | .25 |
| 2 | ( 1.0000, 0.0000) | 1.0000 | ( 0.0000,  .5000) | .50 |
| 3 | ( 0.0000, 1.0000) | 1.0000 | ( .5000,  .5000) | .25 |
| 4 | ( .5000, 0.0000) | .5000 | ( .5000,  .7500) | .50 |
| 5 | ( 0.0000,  .5000) | .5000 | ( .7500,  .7500) | .25 |
| 6 | ( .2500, 0.0000) | .2500 | ( .7500,  .8750) | .50 |
| 7 | ( 0.0000,  .2500) | .2500 | ( .8750,  .8750) | .25 |
| 8 | ( .1250, 0.0000) | .1250 | ( .8750,  .9375) | .50 |
| 9 | ( 0.0000,  .1250) | .1250 | ( .9375,  .9375) | .25 |
| 10 | ( .0625, 0.0000) | .0625 | ( .9375,  .9688) | .50 |
| 11 | ( 0.0000,  .0625) | .0625 | ( .9688,  .9688) | .25 |
| 12 | ( .0313, 0.0000) | .0313 | ( .9688,  .9844) | .50 |

*( Continued )*

*(Continued)*

| Iteration | Gradient Vector | G magnitude | G x[k] | Step Length |
|---|---|---|---|---|
| 13 | ( 0.0000,  .0313) | .0313 | ( .9844, .9844) | .25 |
| 14 | ( .0156, 0.0000) | .0156 | ( .9844, .9922) | .50 |
| 15 | ( 0.0000,  .0156) | .0156 | ( .9922, .9922) | .25 |
| 16 | ( .0078,  .0000) | .0078 | | |

Approximate Solution: ( .9922, .9961)

Maximum Functional Value: 1.0000

Number gradient evaluations: 17

Number function evaluations: 16

The solution, via calculus, is as follows:

$$\text{Maximize } f(x_1,x_2) = 2x_1x_2 + 2x_2 - x_1^2 - 2x_2^2$$

$$\frac{\partial f}{\partial x_1} = 0 = 2x_2 - 2x_1$$

$$\frac{\partial f}{\partial x_2} = 0 = 2x_1 + 2 - 4x_2$$

Solving these two equations simultaneously yields:

$$x_1 = 1$$
$$x_2 = 1$$
$$f(x_1,x_2) = 1$$

The Hessian matrix, $\begin{bmatrix} -2 & 2 \\ 2 & -4 \end{bmatrix}$, is negative definite, so the point $x^*$ is a maximum. Note that our approximate solutions, $x$ (0.9922, 0.9961) and $f(x) = 1.000$, are close to the exact value of $x^*$, (1,1), and $f(x^*) = 1$. To get a closer approximation, we should make our tolerance smaller. A look at the contour plot confirms a hill at approximately (1,1) (see Figure 9.11).

**FIGURE 9.11**
Hill at approximately (1,1) in the plot of $2x_1x_2 + 2x_2 - x_1^2 - 2x_2^2$

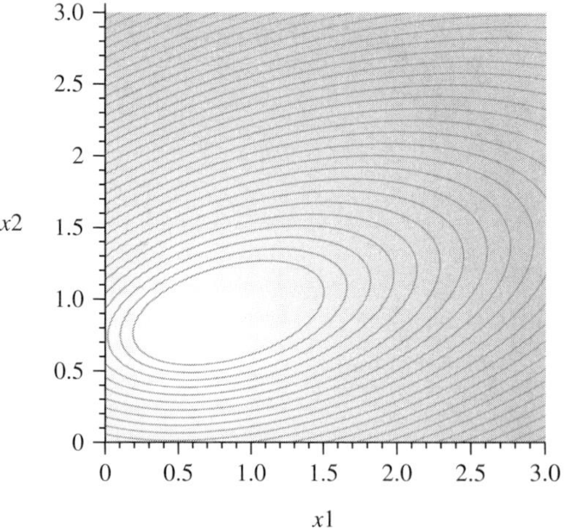


---

**Example 2**

## Maximizing the Function $f(x_1, x_2) = 55x_1 - 4x_1^2 + 135x_2 - 15x_2^2 - 100$

Maximize $f(x_1, x_2) = 55x_1 - 4x_1^2 + 135x_2 - 15x_2^2 - 100$ from the point (1,1).

We will maximize $f(x_1, x_2)$ starting from the point (1,1) and using a tolerance of 0.01. Figure 9.12 is provided as a visual reference.

**FIGURE 9.12**

3-D plot of $f(x_1, x_2) = 55\,x_1 - 4\,x_1^2 + 135\,x_2 - 15\,x_2^2 - 100$

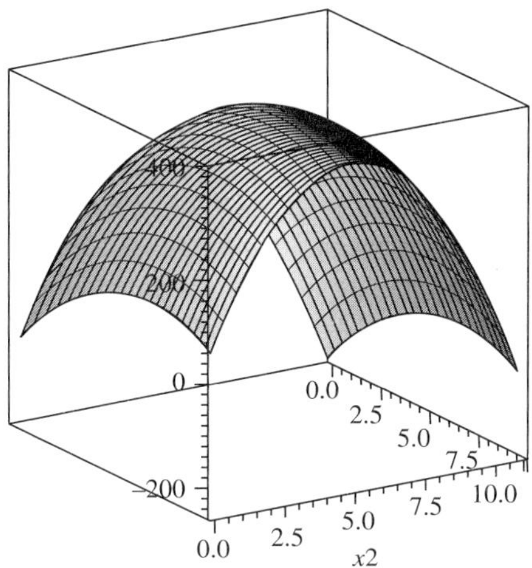

## Iteration 1

The gradient vector is $[55 - 8x_1,\ 135 - 30x_2]$.

$\nabla f(1,1) = [47,105]$. From $(1,1)$ we move in the direction of $[47,105]$. How far do we go? We need to maximize the function starting at the point $(1,1)$ using the function

$$f[x_i + t_i\,\nabla f(x_i)] = f(1 + 47t,\ 1 + 105t)$$
$$= 55(1 + 47t) - 4(1 + 47t)^2 + 135(1 + 105t) - 15(1 + 105t)^2 - 100$$
$$= 71 + 13234t - 174211t^2$$

This function can also be maximized be using any of the one-dimensional search techniques that we discussed in Chapter 8 or by simple single-variable calculus:

$$\frac{df}{dt} = 0 = 13234 - 348422t,\ t = 0.0379$$

The new point is found by substitution into $x_{i+1} = x_i + t_i\,\nabla f(x_i)$.

$$\text{So, } x_1 = [1 + 47(0.0379),\ 1 + 105(.0379)],\ [2.785,\ 4.98].$$

The magnitude of the gradient is 115.73, which is not less than our tolerance of 0.01. Because we are not optimal, we continue. We repeat the calculations from the new point [2.785, 4.98].

## Iteration 2

The gradient vector is $[55 - 8x_1,\ 135 - 30x_2]$.

$\nabla f(2.785, 4.98) = [32.72, -14.4]$. From $(2.785, 4.98)$, we move in the direction of $[32.72, -14.4]$. How far do we go? We need to maximize the function starting at the new point $(2.785, 4.98)$ using the function

$$f[x_i + t_i\,\nabla f(x_i)] = f(2.785 + 32.72t,\ 4.98 - 14.4t)$$
$$= 322.444100 + 1277.95840\,t - 7392.7936\,t^2$$

This function can also be maximized by simple single-variable calculus:

$$\frac{df}{dt} = 0 \text{ when } t = .0864$$

The new point is found by substitution into $x_{i+1} = x_i + t_i\,\nabla f(x_i)$.

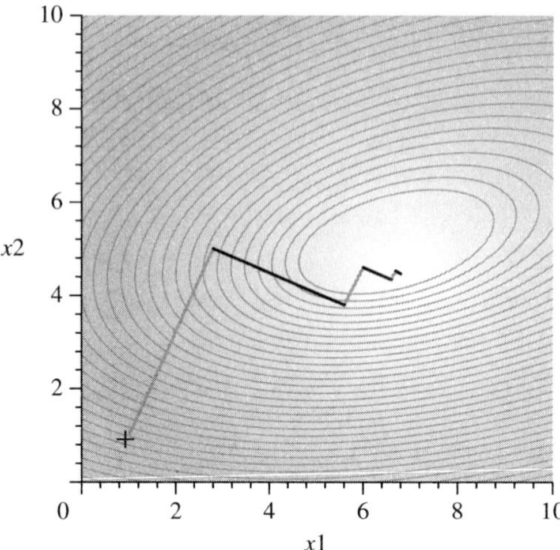



**FIGURE 9.13**
Zigzagging to a solution

So, $x_1 = [5.612, 3.736]$.

Because we are not optimal, we continue. We now repeat the calculations from the new point [5.612,3.736].

We use technology to complete the solution process.

```
> f:=55*x1-4x1^2+135*x2-15*x2^2-100;
> (ky,MP,z1,z2,z3):=STEEPEST(50,0.01,1.,1.,f);
```

**Initial Condition: ( 1.0000, 1.0000)**

| Iteration | Gradient Vector | G magnitude | G x[k] | Step Length |
|---|---|---|---|---|
| 1 | ( 47.0000,105.0000) | 115.0391 | ( 1.0000, 1.0000) | .0380 |
| 2 | ( 32.7185, −14.6454) | 35.8468 | ( 2.7852, 4.9882) | .0857 |
| 3 | ( 10.2936, 22.9964) | 25.1951 | ( 5.5883, 3.7335) | .0380 |
| 4 | ( 7.1658, −3.2075) | 7.8509 | ( 5.9793, 4.6069) | .0857 |
| 5 | ( 2.2544, 5.0365) | 5.5180 | ( 6.5932, 4.3321) | .0380 |
| 6 | ( 1.5694, −.7025) | 1.7195 | ( 6.6788, 4.5234) | .0857 |
| 7 | ( .4938, 1.1031) | 1.2085 | ( 6.8133, 4.4632) | .0380 |
| 8 | ( .3437, −.1539) | .3766 | ( 6.8320, 4.5051) | .0857 |
| 9 | ( .1081, .2416) | .2647 | ( 6.8615, 4.4919) | .0380 |
| 10 | ( .0753, −.0337) | .0825 | ( 6.8656, 4.5011) | .0857 |
| 11 | ( .0237, .0529) | .0580 | ( 6.8720, 4.4982) | .0380 |
| 12 | ( .0165, −.0074) | .0181 | ( 6.8729, 4.5002) | .0857 |
| 13 | ( .0052, .0116) | .0127 | ( 6.8744, 4.4996) | .0380 |
| 14 | ( .0036, −.0016) | .0040 | ( 6.8745, 4.5001) | |

Approximate Solution: ( 6.8745, 4.5001)
Maximum Functional Value: 392.8125
Number gradient evaluations: 15
Number function evaluations: 14

We point out that the solution process might zigzag and converge more slowly to an optimal solution (see Figure 9.13).

## Example 3    Gradient Search When Calculus Fails

We desire to maximize a transcendental multivariable function and find that using only calculus fails. Instead, we will use gradient search to find our maximum.

$$f(x_1, x_2) = 2x_1x_2 + 2x_2 - e^{x_1} - e^{x_2} + 10$$

These partial derivative equations,

$$2x_2 - e^{x_1} = 0$$

$$2x_1 + 2 - e^{x_2} = 0$$

are not solvable in closed form for $(x_1, x_2)$ without numerical methods. So, we will use the gradient method to approximate the solution. Here we have no analytical solution for comparison.

We will start at the point (0.5, 0.5).

The gradient is $\nabla f = [2x_2 - e^{x_1}, 2x_1 + 2 - e^{x_2}]$.

$$\nabla f(0.5, 0.5) = [-0.6487, 1.3513] \text{ with magnitude} = 1.4989$$

A new point is found by maximizing

$$f[0.5 - 0.6487t, 0.5 + 1.3513t] = 2(0.5 - 0.6487)(0.5 + 1.3513t)$$
$$+ 2(0.5 + 1.3513t) - e^{0.5 - 0.6487t} - e^{0.5 + 1.3513t}$$

$$df/dt = 0 = 3.40520 - 3.50635324\, t + 0.6487\, e^{(0.5 - 0.6487t)} - 1.3513\, e^{(0.5 + 1.3513t)}$$

$$t = 0.2873853295 = 0.2874 \text{ (rounded to 4 decimals)}$$

This moves us to the next point (0.3136, 0.8883). We continue the search using technology.

```
> f:=2*x1*x2+2*x2-exp(x10-exp(x2)+10;
```

$$f := 2x1x2 + 2x2 - e^{x1} - e^{x2} + 10$$

```
> (kt,MP,z1,z2,z3):=STEEPEST(50,0.05,.5.,5.,f);
```

**Initial Condition: ( .5000, .5000)**

| Iteration | Gradient Vector | G magnitude | G x[k] | Step Length |
|---|---|---|---|---|
| 1 | ( −.6487, 1.3513) | 1.4989 | ( .5000,  .5000) | .2874 |
| 2 | ( .4084,  .1961) | .4530 | ( .3136,  .8883) | 1.7041 |
| 3 | ( −.2993,  .6235) | .6917 | ( 1.0095, 1.2224) | .2001 |
| 4 | ( .1097,  .0526) | .1216 | ( .9496, 1.3472) | .7345 |
| 5 | ( −.0298,  .0621) | .0689 | ( 1.0302, 1.3859) | .1868 |
| 6 | ( .0090,  .0043) | .0099 | ( 1.0246, 1.3975) | |

Approximate Solution: ( 1.0246, 1.3975)
Maximum Functional Value: 8.8277
Number gradient evaluations: 7
Number function evaluations: 6

## Modified Newton's Method

An alternative search method is the Newton−Raphson numerical method, illustrated in two variables. This numerical method appears to do a more efficient and faster job in converging to the near optimal solution. It is an iterative root-finding technique using the partial derivatives of the function as the new system of equations. The algorithm uses Cramer's rule to find the solution of the system of equations.

Newton's method for multivariable optimization searches is based on Newton's single-variable algorithm for finding the roots and the Newton−Raphson method for finding roots of the first derivative, given a $x_0$, iterate $x_{n+1} = x_n - f'(x_n)/f''(x_n)$ until $|x_{n+1} - x_n|$ is less than some small tolerance. In several variables, we may use a vector $\mathbf{x}_0$ or two variables $(x_0, y_0)$. The algorithm is expanded to include partial derivatives with respect to each variable's dimension. In two variables $(x, y)$, this would yield a system of equations where $F$ is the derivative of $f(x, y)$ with respect to $x$ and $G$ is the derivative of $f(x, y)$ with respect to $y$. Thus, we need to find both $F = 0$ and $G = 0$ simultaneously.

**INPUT:** $x(0)$, $y(0)$, $N$, toleranace

**OUTPUT:** $x(n)$, $y(n)$

**Step 1.** For $n = 1$ to $N$ do

**Step 2.** Calculate the new estimate for $x(n)$ and $y(n)$ as follows:

$$\frac{\partial F}{\partial x}(x(n-1), y(n-1)) \rightarrow q$$

$$\frac{\partial F}{\partial y}(x(n-1), y(n-1)) \rightarrow r$$

$$\frac{\partial G}{\partial x}(x(n-1), y(n-1)) \rightarrow s$$

$$\frac{\partial G}{\partial x}(x(n-1), y(n-1)) \rightarrow t$$

$$-F(x(n-1), y(n-1)) \rightarrow u$$

$$-G(x(n-1), y(n-1)) \rightarrow v$$

$$qt - rs \rightarrow D$$

$$x(n-1) + (ut-vr)/D \rightarrow x(n)$$

$$y(n-1) + (qv-su)/D \rightarrow y(n)$$

**Step 3.** If $((x(n)-x(n-1))^2 + (y(n)-y(n-1))^2)^{1/2} <$ tolerance,

Then Stop

Else, Go back to Step 2.

**STOP**

This yields a matrix equation $\sum_{j=1}^{N}\alpha_{ij}\delta x_j = \beta_i$, where

$$\alpha_{ij} = \frac{\partial f_i}{\partial x_i}, \beta_i = -f_i.$$

The matrix equation can be solved by LU decomposition or in the case of a 2×2 by Cramer's rule. The corrections are then added to the solution vector

$$x_i^{new} = x_i^{old} + \delta x_i, i = 1,...N$$

and iterated until it converges within a tolerance.

### The Modified Newton with Technology

The algorithm is seen in Figure 9.14.

Let's repeat our examples with technology using a modified Newton's method with a Maple program we wrote to perform the iterations.

---

## Example 1

## Newton's Method to Find a Maximum

Maximize $f(x_1,x_2) = 2x_1x_2 + 2x_2 - x_1^2 - 2x_2^2$ starting at the point $(0,0)$ with a tolerance of 0.01.

```
> f:=2*x1*x2+2*x2-x1^2-2*x2^2;
```

$$f := 2\ x1\ x2 + 2\ x2 - x1^2 - 2\ x2^2$$

```
> f1:=(x1,x2)->2*x2-2*x1;
```

$$f1 := (x1, x2) \rightarrow 2\ x2 - 2\ x1$$

```
> f2:=(x1,x2)->2*x1+2-4*x2;
```

$$f2 := (x1, x2) \rightarrow 2x1 + 2 - 4x2$$

```
> (z1,z2,z3):=Newtons(f,f1,f2,100,.05,0,0);
```

| Hessian: | $[-2.000$ | $2.000\ ]$ |
| | $[2.000$ | $-4.000\ ]$ |
| eigenvalues: | $-5.236$ | $-.764$ |
| pos def: false | | |
| new | $x = 1.000$ | new $y = 1.000$ |
| Hessian: | $[-2.000$ | $2.000\ ]$ |
| | $[2.000$ | $-4.000\ ]$ |
| eigenvalues: | $-5.236$ | $-.764$ |
| pos def: false | | |
| new | $x = 1.000$ | new $y = 1.000$ |
| final new | $x = 1.000$ | final new $y = 1.000$ |

final fvalue is 1.000

It converges to the point (1,1) after 2 iterations.

---

## Example 2 — Maple with Newton's Method

Maximize $f(x_1, x_2) = 55 \cdot x_1 - 4 \cdot x_1^2 + 135 \cdot x_2 - 15 \cdot x_2^2 - 100$

```
> f:=55*x1–4*x1^2+135*x2–15*x2^2–100;
```

$$f := 55\,x1 - 4\,x1^2 + 135\,x2 - 15\,x2^2 - 100$$

Using technology:

```
> (z1,z2,z3):=Newtons(f,100,.01,1,1);
```

| Hessian: | $[-8.000$ | $0.000\ ]$ |
| | $[0.000$ | $-30.000\ ]$ |
| eigenvalues: | $-30.000$ | $-8.000$ |
| pos def: false | | |
| new | $x = 6.875$ | new $y = 4.500$ |
| Hessian: | $[-8.000$ | $0.000\ ]$ |
| | $[0.000$ | $-30.000\ ]$ |
| eigenvalues: | $-30.000$ | $-8.000$ |
| pos def: false | | |
| new | $x = 6.875$ | new $y = 4.500$ |
| final new | $x = 6.875$ | final new $y = 4.500$ |
| final fvalue is 392.813 | | |

$$z1,\ z2,z3 := \frac{55}{8},\ \frac{9}{2},\ 392.8125000$$

---

## Example 3 — Maximize $2x_1 x_2 + 2x_2 - e^{x1} - e^{x2} + 10$

```
> f:=2*x1*x2+2*x2–exp(x1)–exp(x2)+10;
```

$$f := 2\,x1\,x2 + 2\,x2 - e^{x1} - e^{x2} + 10$$

```
> df1:=diff(f,x1);
```

$$df1 := 2\,x2 - e^{x1}$$

```
> df2:=diff(f,x2);
```

$$df2 := 2\,x1 + 2 - e^{x2}$$

> *f1:=unapply(df1,x1,x2);*

$$f1 := (x1, x2) \rightarrow 2\,x2 - e^{x1}$$

> *f2:=unapply(df2,x1,x2);*

$$f2 := (x1,x2) \rightarrow 2x1 + 2 - e^{x2}$$

> *(z1,z2,z3):=Newtons(f,f1,f2,100,.1,0,0);*

---

```
pos def:
Hessian:                        [−1.000            2.000 ]
                                [2.000            −1.000 ]
eigenvalues:                    −3.000             1.000
pos def: false
new                             x = −.333      new y = .333
Hessian:                        [−.717             2.000 ]
                                [2.000            −1.396 ]
eigenvalues:                    −3.085             .973
pos def: false
new                             x = −.269      new y = .381
final new                       x = −.269  final new y = .381
final fvalue is 8.329
```

---

The maximum is found at $(-0.269, 0.381)$ with a function value of 8.329.

### Comparisons of Methods

We compared these two routines and found that Newton's method converges faster than the gradient method. This is displayed in the following table.

---

### Function 1

|  | Initial Condition | Iterations | Feval | gevals | Solution | Max $F$ |
|---|---|---|---|---|---|---|
| Steepest ascent | (0,0) | 16 | 16 | 17 | $x = 0.9922$ $y = 0.9961$ | 1.0 |
| Newton's method | (0,0) | 2 | 2 |  | $x = 1$ $y = 1$ | 1.00000 |

### Function 2

|  | Initial Condition | Iterations | Feval | gevals | Solution | Max $F$ |
|---|---|---|---|---|---|---|
| Steepest ascent | (0,0) | 4 | 5 | 4 | $x = -0.26638,$ $y = 0.3853$ | 8.3291 |
| Newton's method | (0,0) | 2 |  |  | $x = -0.269,$ $y = 0.381$ | 8.329 |

---

## 9.4 | EXERCISES

1. Given: Maximize $f(x,y) = 2xy - 2x^2 - y^2$. Assume our tolerance for the magnitude of the gradient is 0.10.

   a. Start at the point $(x,y) = (1,1)$. Perform two complete iterations of a gradient search. For each iteration, clearly show $X_n$, $X_{n+1}$, $\nabla f(X_n)$, and $t^*$. Justify how we will eventually find the approximate maximum.

   **b.** Use Newton's method to find the maximum starting at $(x,y) = (1,1)$. Clearly show $X_n$, $X_{n+1}$, $\nabla f(X_n)$, and $H^{-1}$ for each iteration. Clearly indicate when the stopping criterion is achieved.

2. Given: Maximize $f(x,y) = 3xy - 4x^2 - 2y^2$. Assume our tolerance for the magnitude of the gradient is 0.10.

   **a.** Start at the point $(x,y) = (1,1)$. Perform two complete iterations of gradient search. For each iteration, clearly show $X_n$, $X_{n+1}$, $\nabla f(X_n)$, and $t^*$. Justify how we will eventually find an approximate maximum.

   **b.** Use Newton's method to find the maximum starting at $(x,y) = (1,1)$. Clearly show $X_n$, $X_{n+1}$, $\nabla f(X_n)$, and $H^{-1}$ for each iteration. Clearly indicate when a stopping criterion is achieved.

3. Apply the modified Newton's method (multivariable) to find the following:

   **a.** Maximize $f(x,y) = -x^3 + 3x + 84y - 6y^2$; start at $(1,1)$. Why can't we start at $(0,0)$?

   **b.** Minimize $f(x,y) = -4x + 4x^2 - 3y - y^2$; start at $(0,0)$.

   **c.** Perform three iterations to minimize $f(x,y) = (x - 2)^4 + (x - 2y)^2$; start at $(0,0)$. Why is this problem not converging as quickly as problem 3(b)?

4. Use the gradient search to find the approximate minimum to $f(x,y) = (x - 2)^2 + x + y^2$; start at $(2.5, 1.5)$.

## 9.4 | PROJECTS

1. Write a computer program in Maple that uses a one-dimensional search algorithm—say, the golden section search—instead of calculus to perform iterations of gradient search. Use your code to find the maximum of

$$f(x,y) = xy - x^2 - y^2 - 2x - 2y + 4$$

2. Write a computer program in Maple that uses a one-dimensional search algorithm—say, Fibonacci search—instead of calculus to perform iterations of gradient search. Use your code to find the maximum of

$$f(x,y) = xy - x^2 - y^2 - 2x - 2y + 4$$

## 9.4 | FURTHER READING


Bazarra, M., C. Shetty, and H. D. Scherali. *Nonlinear Programming: Theory and Applications*. New York: Wiley. 1993.

Fox, W. "Teaching Nonlinear Programming with Minitab." *COED Journal*. Vol. II No. 1, January–March 1992, pages 80–84.

Fox, W. P. "Using Microcomputers in Undergraduate Nonlinear Optimization." *Collegiate Microcomputer*. VOL XI(3), 1993, pages 214–218.

Fox, W., and J. Appleget, "Some Fun with Newton's Method." *COED Journal*, Vol X(4), October–December 2000, pages 38–43.

Fox, W. P., F. Giordano, S. Maddox, and M. Weir. *Mathematical Modeling with Minitab*. Monterey, CA: Brooks/Cole. 1987.

Fox, W. P., F. Giordano, and M. Weir. *A First Course in Mathematical Modeling*, 2nd Edition. Monterey, CA: Brooks/Cole. 1997.

Fox, W. P., and W. Richardson. *Mathematical Modeling with Least Squares Using Maple.* Maple Application Center, Nonlinear Mathematics, October 2000.

Meerschaert, M. *Mathematical Modeling.* San Diego: Academic Press. 1993.

Phillips, D. T., A. Ravindran, and J. Solberg, 1976. *Operations Research.* New York: John Wiley & Sons.

Press, W. H., B. Flannery, S. Teukolsky, and W. Vetterling. 1987). *Numerical Recipes.* New York: Cambridge University Press, pages 269−271.

Rao, S. S. *Optimization: Theory and Applications.* New Delhi: Wiley Eastern Limited. 1979.

Winston, W. *Introduction to Mathematical Programming: Applications and Algorithm.* 4th Edition. Belmont, CA: Duxbury Press. 2002.

## Project:  PLANNING AND PRODUCTION CONTROL

## Introduction

Many optimization problems require the simultaneous consideration of a number of independent variables. In planning and producing items, one must consider many factors that affect the process. The company might desire to maximize profit, minimize cost, maximize production levels, improve efficiency, and minimize shipping time, among a host of other options. Many of these can be solved by the following techniques:

Differential calculus

Lagrange multipliers

Linear programming

Dynamic programming

We will cover most of these methods during this course except linear programming; we devote an entire course to the subject.

During this block, we studied only two numerical search techniques for multivariable functions:

**1.** Gradient search (ascent and descent)

**2.** Newton−Raphson method

## Key Definitions and Variables

- The term $z$ is the measure of performance.
- Inputs $x_1, x_2, \ldots, x_n$ affect z.
- Optimum point: The values of $x_1, x_2, \ldots, x_n$ maximize or minimize $z$.
- Optimal value: The value of $z$ for the optimum point.
- Unimodal: Most search strategies rely on the assumption that the surface is unimodal, that is, it only has one peak over the region of concern.

## Part I

You are hired as a consultant and asked to optimize all facets of the company's planning and production.

RABA manufactures 15-inch color TV sets. The company plans to improve its production of color TVs by adding a new chip to the circuitry that will improve each TV's reception and survivability. The new chip is extremely sensitive and must be continuously monitored. The monitoring process is assumed to be modeled by the expression:

$$Y = Ax + B/x$$

where the value of $A$ is assumed to remain constant throughout the process at a value of $A = 68$. However, the value of $B$ fluctuates slowly because of gradual environmental changes.

1. The process was recently measured in terms of $(x,y)$ at (0.5, 79). Find the value of $B$ at that instant.

2. Using the value of B from question 1, determine the value of $x$ that will minimize $y$. Assume $y$ measures the production cost of monitoring. What is this value of $y$?

3. The preceding result seems too unrealistic for you, so you collect the following data over a 12-week period. Find the values of $A$ and $B$ that minimize the model $S = \sum (Y_i - (Ax_i + (B/x_i)))^2$.

| Week | 1 | 2 | 3 | 4 | 5 | 6 | 7 | 8 | 9 | 10 | 11 | 12 |
|---|---|---|---|---|---|---|---|---|---|---|---|---|
| $y$ | 76.6 | 78 | 97.5 | 120.5 | 145 | 170.5 | 196 | 222 | 248 | 274 | 301 | 328 |
| $x$ | 1.1 | 2.1 | 3 | 4.5 | 5 | 6 | 7.1 | 8 | 9 | 10 | 11.1 | 12 |

4. If A and B are settings on a machine, then what settings would you use and why?

# Part II

The TV manufacturer is planning the introduction of two new products: a 19-inch stereo color set with a manufacturer's suggested retail price (MSRP) of $339 and a 21-inch stereo color set with a MSRP of $399. The cost to the company is $195 per 19-inch set and $225 per 21-inch set and an additional $400,000 in fixed costs of initial parts, initial labor, and machinery.

In the competitive market in which the company desires to sell the sets, the number of sales per year will affect the average selling price. It is estimated that for each type of set, the average selling price drops by one cent for each additional unit sold. Furthermore, sales of 19-inch sets will affect the sales of 21-inch sets and vice versa. It is estimated that the average selling price for the 19-inch set will be reduced by an additional 0.3 cents for each 21-inch set sold, and the price for the 21-inch set will decrease by 0.4 cents for each 19-inch set sold. We desire to find the optimal number of units of each type of sets to produce and to determine the expected profits. Recall that profit is revenue minus cost: $P = R - C$.

**Required Information**

1. Formulate the model to maximize profits. Be sure that you have accounted for all revenues and costs. Define all your variables.

2. Solve for the optimal levels of 21-inch and 19-inch sets to be manufactured using
   a. classical optimization—that is, calculus;
   b. a combination of 3-D surface plot and contour plot (estimated answers are all right here); and
   c. the Newton–Raphson method (briefly explain why you can use this technique).

3. Using Maple, obtain a contour plot of the function. Color or identify the optimal point in some other manner. Illustrate the gradient-search technique, starting from the initial point (0,0). Only perform two or three iterations, showing the gradient, the distance traveled, and the new point. There is no requirement to obtain the optimal solution using this method.

4. Comment about the accuracy and rate of convergence (number of iterations or difficulty) in obtaining a result in questions 2 and 3 above.

5. Comment on the solution in terms of the scenario. Did we use an appropriate technique? Should there be any constraints that were not considered? Make your recommendation to the CEO concerning these TV sets.

## Appendix A   MAPLE CODE FOR STEEPEST ASCENT METHOD

## Gradient Method-Steepest Ascent

Let's use a gradient search method to illustrate.

The algorithm inputs are:

$n$ = maximum number of iterations

$tol$ = tolerance to stop when magnitude of gradient is less than tolerance $ptx1, ptx2$ are the initial $(x_0, y_0)$ point $f$ is the multivariable function to be maximized

Outputs are the approximate optimal point, $(x,y)$, and the functional value at that point, $f(x,y)$.

If you need to minimize a function, multiply $f$ by $-1$ and use the following algorithm:

```
> restart:
> with(linalg):
>UP:= proc(n,tol,ptx1,ptx2,f)
> local
numIter,numfeval,numgeval,f1,p1,p2,rv,x1pt,x2pt,temp,max,mg,v1,v2,newt,1am,nv1,nv2,Fvalue;
> f1:=f;temp:=grad(f1,vector([x1,x2]));
> p1 := unapply(temp[1],x1,x2);p2 := unapply(temp[2],x1,x2);
> x1pt:=ptx1;
> x2pt:=ptx2;
> rv:=vector([p1(x1pt,x2pt),p2(x1pt,x2pt)]):
> numIter:=1;numgeval:=1;numfeval:=1;
> printf("\n\n----------------------------------------");
> printf("----------------------------------------");
> printf("\n\n InitialCondition:");
> printf("(%8.4f,%8.4f)\n\n",x1pt,x2pt);
> printf(" Iter Gradient Vector G magnitude G");
> printf(" x[k] Step Length\n\n");
> label_7;
> rv:=vector([p1(x1pt,x2pt),p2(x1pt,x2pt)]):
> numgeval:=numgeval+1;
> printf("%5d (%8.4f,%8.4f)",numIter,rv[1],rv[2]);
> max:=n;
> mg:=convert(sqrt(dotprod(rv,rv)),float);
> printf("%12.4f",mg);
> if(mg<tol or numIter>=max) then
> goto(label_6);
> else
> numIter:=numIter+1;
> fi;
> v1:=x1pt+t*rv[1];
> v2:=x2pt+t*rv[2];
```

```
> newt:=evalf(subs({x1=v1,x2=v2},f1));
> numfeval:=numfeval+1;
> lam:=fsolve(diff(newt,t)=0,t,maxsols=1);
> nv1:=evalf(subs({t=lam},v1));
> nv2:=evalf(subs({t=lam},v2));
> printf("(%8.4f,%8.4f)%13.4f\n",x1pt,x2pt,lam);
> x1pt:=nv1;
> x2pt:=nv2;
> goto(label_7);
> label_6;
> printf("\n\n----------------------------------------");
> printf("----------------------------------------");
> printf("\n\n Approximate Solution:");
> printf(" (%8.4f,%8.4f)\n",x1pt,x2pt);
> Fvalue:=evalf(subs(x1=x1pt,x2=x2pt,f));
> printf(" Maximum Functional Value:");
> printf("%21.4f",Fvalue);
> printf("\n Number gradient evaluations:");
> printf("%22d",numgeval);
> printf("\n Number function evaluations:");
> printf("%22d",numfeval);
> printf("\n\n----------------------------------------");
> printf("----------------------------------------");
> end:
```

## Appendix B     NEWTON'S METHOD FOR OPTIMIZATION

## Newton's Method

```
Steepest:=proc(f,f1::procedure,f2::procedure,n::posint,tol::numeric,x11::numeric,x22::numeric):
> t1:=tol;
> print(tol);
> x:=x11;
> y:=x22;
> label_6:
> dq:=D[1](f1):q:=evalf(dq(x,y)):#print(q);
> dr:=D[2](f1):r:=evalf(dr(x,y)):#print(r);
> ds:=D[1](f2):s:=evalf(ds(x,y)):#print(s);
> dt:=D[2](f2):t:=evalf(dt(x,y)):#print(t);
> printf("\nHessian:[ %8.3f %8.3f]\n",q,r);
> printf(" [ %8.3f %8.3f ]\n",s,t);
> A := array([[q,r],[s,t]]);
> printf("eigenvalues:%8.3f %8.3f\n", evalf(Eigenvals(A))[1],evalf(Eigenvals(A))[2]);
> A := matrix(2,2, [q,r,s,t]);
> printf("pos def: %s\n",definite(A, 'positive_def'));
> u:=-1*f1(x,y):#print(u);
> v:=-1*f2(x,y):#print(v);
> dv:=q*t-r*s:#print(dv);
> newx:=x+((u*t-v*r)/dv);
> newy:=y+((q*v-s*u)/dv);
> xx:=x;
> yy:=y;
> x:=newx;
```

```
> y:=newy;
> printf("new x=%8.3f new y=%8.3f\n",newx,newy);
> if (evalf(sqrt((newx–xx)^2+(newy–yy)^2)) < t1) then
> printf("\n\nfinal new x=%8.3f final new y=%8.3f\n",newx,newy);
> printf("final fvalue is %8.3f",evalf(subs({x1=newx,x2=newy},f)));
> else
> goto(label_6);
> fi;
> #print(newx,newy);
> end:
```

# 10

# Modeling Optimization with Constraints

A company manufactures new e-phones that are supposed to capture the market by storm. The two main inputs components of the new e-phone are the circuit board and the relay switches that make the phone faster and smarter and give it more memory. The number of e-phones to be produced is estimated to equal $E = 200\, a^{\frac{1}{2}}\, b^{\frac{1}{4}}$, where $E$ is the number of phones produced, while $a$ and $b$ are the number of circuit board hours and the number of relay hours worked, respectively. Such a function is known to economists as a *Cobb–Douglas function*. Laborers are paid by the type of work they do: the circuit boards and the relays for \$5 and \$10 an hour, respectively. We want to maximize the number of e-phones to be made if we have \$150,000 to spend on these components in the short run.

Problems such as this can be modeled using constrained optimization. We begin our discussion with equality constrained optimization, and then we discuss inequality constrained optimization

 ## EQUALITY CONSTRAINTS METHOD OF LAGRANGE MULTIPLIERS

Lagrange multipliers can be used to solve *nonlinear optimization problems* (NLPs) in which all the constraints are equality constrained. We consider the following type of NLPs as shown by Equation 10.1:

$$\text{Maximize (Minimize) } z = f(x_1, x_2, x_3, \ldots, x_n) \tag{10.1}$$

Subject to
$$g_1(x_1, x_2, \ldots, x_n) = b_1$$
$$g_2(x_1, x_2, \ldots, x_n) = b_2$$
$$\vdots$$
$$g_m(x_1, x_2, \ldots, x_n) = b_m$$

In our e-phones example, we find we can build an equality constrained model. We want to maximize

$$E = 200a^{\frac{1}{2}}b^{\frac{1}{4}}$$

subject to the constraint

$$5a + 10b = 150{,}000$$

### Introduction and Basic Theory

To solve NLPs in the form of Equation 10.1, we associate a Lagrangian multiplier, $\lambda_i$, with the $i$th constraint and form the Lagrangian equation to get Equation 10.2:

$$L(X,\lambda) = f(X) + \sum_{i=1}^{m} \lambda_i (b_i - g_i(X)) \tag{10.2}$$

The computational procedure for Lagrange multipliers requires that all the partials of this Lagrangian function, Equation 10.2, must equal zero. These partials are the *necessary conditions* of the NLP problem. These are the conditions required for $\mathbf{x} = \{x_1, x_2, \ldots, x_n\}$ to be a solution to Equation 10.1.

$$\text{The Necessary Conditions} \tag{10.3}$$
$$\partial L/-\partial Xj = 0 \,(\, j = 1, 2, \ldots, n \text{ variables}) \tag{10.3a}$$
$$\partial L/-\partial \lambda_i = 0 \,(i = 1, 2, \ldots, m \text{ constraints}) \tag{10.3b}$$

**Definition** | Definition: $\mathbf{x}$ is a regular point if and only if $\nabla g_i(x), i = 1, 2, \ldots, m$ are linearly independent.

**Theorem 10a** | **(a)**  Let (1) be a maximization problem. If $f$ is a concave function and each $g_i(x)$ is a linear function, then any point satisfying Equation 10.1 will yield an optimal solution.

**(b)**  Let (1) be a minimization problem. If $f$ is a convex function and each $g_i(x)$ is a linear function, then any point satisfying Equation 10.1 will yield an optimal solution.  ∎

Recall from Chapter 9 that we used the Hessian matrix to determine if a function was convex, concave, or neither. We also note that the above theorem limits our constraints to linear functions. What if we have nonlinear constraints?

We can use the bordered Hessian in the sufficient conditions. Given the bivariate Lagrangian function, as in

$$L(x_1, x_2, \lambda) = f(x_1, x_2) + \sum_{i=1}^{m} \lambda_i (b_i - g(x_1, x_2))$$

The bordered Hessian is defined from the 2nd partial derivatives of the Lagrangian function as follows:

$$BdH = \begin{bmatrix} 0 & g_1 & g_2 \\ g_1 & f_{11} - \lambda g_{11} & f_{12} - \lambda g_{12} \\ g_2 & f_{21} - \lambda g_{21} & f_{22} - \lambda g_{22} \end{bmatrix}$$

We find the determinant of this bordered Hessian as Equation 10.4:

$$|BdH| = \det \begin{bmatrix} 0 & g_1 & g_2 \\ g_1 & f_{11} - \lambda g_{11} & f_{12} - \lambda g_{12} \\ g_2 & f_{21} - \lambda g_{21} & f_{22} - \lambda g_{22} \end{bmatrix} = g_1 g_2 (f_{21} - \lambda g_{21}) + g_2 g_1 (f_{12} - \lambda g_{12})$$

$$- g_2^2 (f_{11} - \lambda g_{11}) - g_1^2 (f_{22} - \lambda g_{22})$$

(10.4)

The sufficient condition for a *maximum,* in the bivariate case with one constraint, is that the determinant of its bordered Hessian is positive when evaluated at the critical point.

The sufficient condition for a *minimum,* in the bivariate case with one constraint, is that the determinant of its bordered Hessian is negative when evaluated at the critical point.

If $\mathbf{x}$ is a regular point and $g_i(\mathbf{x}) = 0$ (constraints are satisfied) then $M = \{y|\ \nabla g_i(\mathbf{x}) \cdot y = 0\}$ defines a plane tangent to the feasible region at $\mathbf{x}$.

**Lemma** | If $x$ is regular, $g_i(\mathbf{x}) = 0$, and $\nabla g_i(\mathbf{x}) \cdot y = 0$, then $\nabla f(\mathbf{x}) \cdot y = 0$.

Note that the Lagrange multiplier conditions are exactly the same for a minimization problem as a maximization problem. This is why these conditions alone are not sufficient conditions. Thus, a given solution can either be a maximum or a minimum. To determine whether the point found is a maximum, minimum, or saddle point, we will use the Hessian.

The Lagrange multiplier, $\lambda$, has an important modeling interpretation. It is the "shadow price" for scarce resources. Thus, $\lambda_i$ is the shadow price of the $i$th constraint. Thus, if the right-hand side (RHS) is increased by a small amount $\Delta$ in a max or a min problem, then the optimal solution will change by $\lambda_i \Delta$ We will illustrate the shadow price both graphically and computationally.

### Graphical Interpretation of Lagrange Multipliers

The method of a Lagrange multiplier is based on its geometric interpretation. This geometric interpretation involves the gradients of both the function and the constraints. Initially, let's consider only one constraint,

$$g(x_1, x_2, \ldots, x_n) = b$$

so that the Lagrangian equation simplifies to

$$\nabla f = \lambda \nabla g$$

The solution is the point in $\mathbf{x}$ where the gradient vector, $\nabla g(\mathbf{x})$, is perpendicular to the surface. The gradient vector, $\nabla f$, always points in the direction in which $f$ increases fastest. At both maximums and minimums, this direction must also be perpendicular to $S$. Thus, because both $\nabla f$ and $\nabla g$ both point along the same perpendicular line, then $\nabla f = \lambda \nabla g$.

In the case of multiple constraints, the geometrical arguments are similar (see Figure 10.1). Let's preview a graphical solution to our first computational example.

**FIGURE 10.1**
One equality constraint

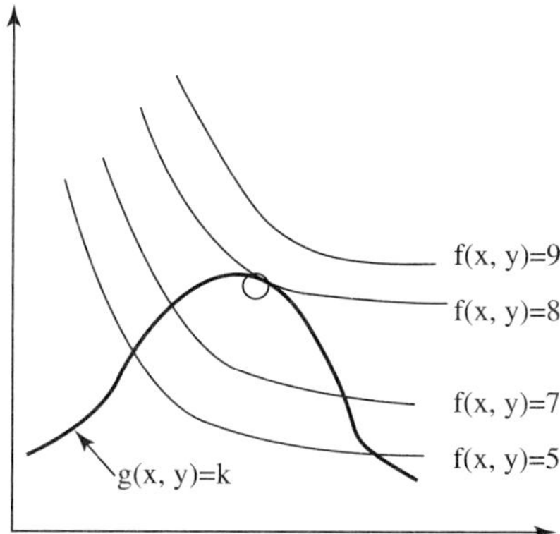

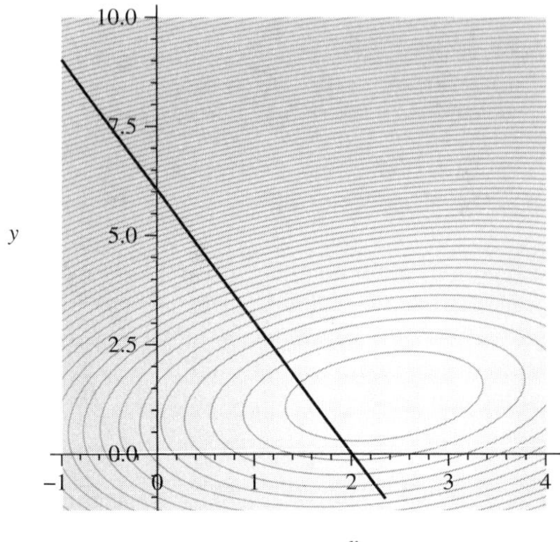

**FIGURE 10.2**

The contour plot of function, *f*, and the equality constraint example $g(x,y) = 3x + y = 6$

$$\text{Maximize} \quad z = -2x^2 - 2y^2 + xy + 8x + 3y$$
$$\text{S.T.} \quad 3x + y = 6$$

We obtained a contour plot of $z$ from Maple and overlaid the single constraint onto the contour plot (see Figure 10.2). What information can we obtain from this graphical representation? First, we note that the unconstrained optimal does not lie on the constraint. We can estimate the unconstrained optimal $(x^*, y^*) = (2.3, 1.3)$. The optimal constrained solution lies at the point where the constraint is tangent to a contour of the function, $f$. This point is estimated as $(1.8, 1.0)$. We see that the resource constraint does not pass through the unconstrained maximum, and thus if feasible, the amount of the resource might be increased until it is tangent to the unconstrained optimum. At that point, we would no longer add (or subtract) any more resources (see Figure 10.3). We can gain valuable insights about the problem if we are able to plot the information.

### Computational Method of Lagrange Multipliers with Maple

Consider the set of equations in Equation 10.3. This gives $m + n$ equations in the $m + n$ unknowns $(x_j, \lambda_i)$. Generally speaking this is a difficult problem to solve without a computer except for simple problems. Also, because the Lagrange multipliers are necessary conditions only (not sufficient), we may find solutions $(x_j, \lambda_i)$ that are not optimal for our NLP. We

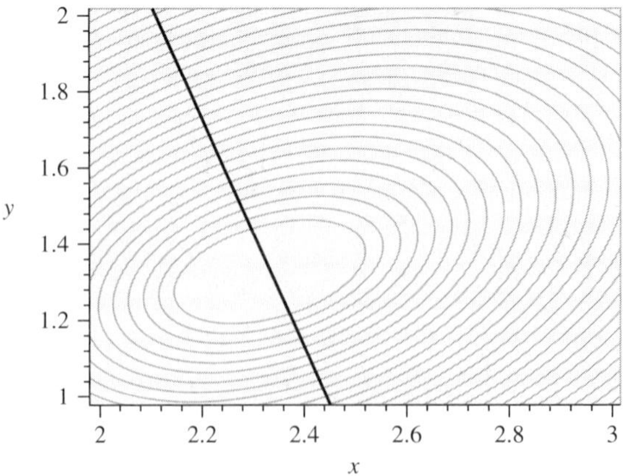

**FIGURE 10.3**

Added value to a resource, $g(x) = 8.45 = 3x + y$

need to be able to determine the classification of points found in the solution to the necessary conditions. Commonly used methods as justification include:

1. the Hessian matrix and

2. the bordered Hessian, Det [BH] = 0.

We will illustrate these, where feasible, in the following examples with Maple.

---

| **Example 1** | ## Minimization with Lagrange Multipliers |

$$\text{Minimize} \quad z = -2x^2 - 2y^2 + xy + 8x + 3y$$
$$\text{S.T.} \qquad 3x + y = 6$$

We enter these into Maple.

$> f := -2 \cdot x^2 - 2 \cdot y^2 + x \cdot y + 8 \cdot x + 3 \cdot y;$

$$f := -2x^2 - 2y^2 + xy + 8x + 3y$$

$> c1 := 6 - 3 \cdot x - y;$

$$c1 := 6 - 3\,x - y$$

We obtain the following 3-D plot (if it's available) from Maple, shown in Figure 10.4.

```
> p:=plot3d(f,x=−10..10,y=−10..10):
> q :=contourplot(({f, c1}),x=−10..10,y=−10..10):
> t :=transform((x,y)→[x,y,−150]):
> display({p, t(q)});
```

Within Maple, you can rotate the 3-D figure to inspect it from many different perspectives. Try this in Maple to see how useful this feature is.

The Lagrangian function is:

$$L(x, y, \lambda) = -2x^2 - y^2 + xy + 8x + 3y + \lambda[6 - 3x - y]$$

The necessary conditions are:

$$L_x = -4x - 4y - 8 - 3\lambda = 0$$
$$L_y = -4y + x + 3 - \lambda = 0$$
$$L_\lambda = 3x + y + 6 = 0$$

We have three equations and three unknowns to solve. Several commands are available in Maple to obtain solutions, depending on the output we are looking to view. We will

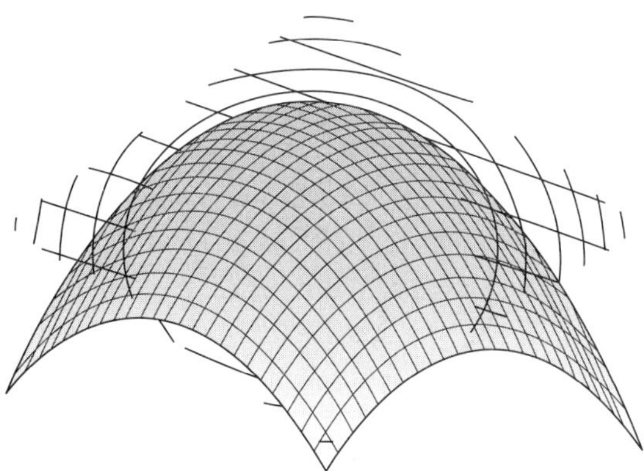

**FIGURE 10.4**

Display of 3D view and contours together

illustrate various methods, including with(Student[MultivariateCalculus]), with(Optimization), with(Maximization), and with(Minimization), as well as a set of commands to use the computer algebra system (CAS) procedures in Maple.

---

Student[MultivariateCalculus][LagrangeMultipliers] - solve types of optimization

problems using the method of Lagrange multipliers

Calling Sequence

LagrangeMultipliers(**f(x,y,..)**, **[g(x,y,..), h(x,y,..),..]**, **[x,y,..]**, **opts**)

---

Parameters

f(x,y,..) − algebraic expression; objective function

[g(x,y,..), h(x,y,..),..] − algebraic expression; constraint functions, assumed equal

to 0

[x,y,..] − list of names; independent variables

opts - (optional) equation(s) of the form **option=value** where **option** is one of

**constraintoptions,   levelcurveoptions,   pointoptions,   output,   showconstraints,**

**showlevelcurves, showpoints, title,** or **view**; output options

---

The Maple command *with( Optimization )* is used. Under the optimization options, we find the following:

---

Optimization[NLPSolve] − solve a nonlinear program

Calling Sequence

NLPSolve(**obj, constr, bd, opts**)

NLPSolve(**opfobj, ineqcon, eqcon, opfbd, opts**)

---

Optimization[Minimize] − minimize an objective function, possibly subject to

constraints

Optimization[Maximize] − maximize an objective function, possibly subject to

constraints

Calling Sequence

Minimize(**obj, constr, bd, opts**)

Maximize(**obj, constr, bd, opts**)

Minimize(**opfobj, ineqcon, eqcon, opfbd, opts**)

Maximize(**opfobj, ineqcon, eqcon, opfbd, opts**)

---

We might also just use Maple as a computational tool. We will illustrate all three methods. We also point out that a critical element of the solutions with the Lagrange multiplier is the value and interpretation of the multiplier, $\lambda$.

We can set up in Maple a system of equations and obtain a solution.

**Method 1**  Set up the equations and use Maple as a computational tool.

> *L:=obj1+l·(6−3·x−y);*

$$L := -2x^2 - 2y^2 + xy + 8x + 3y + l(6 - 3x - y)$$

> *lx:=diff(L, x);*

$$lx := -4x + y + 8 - 3l$$

> *ly:=diff(L, y);*

$$ly := -4y + x + 3 - l$$

> *ll:=diff(L, l);*

$$ll := 6 - 3x - y$$

> *fsolve({lx=0,ly=0,ll=0}, {x,y,l});*

$$\{l = 0.7608695652,\ x = 1.673913043,\ y = 0.9782608696\}$$

> *subs({l=0.7608695652,x=1.673913043,y=0.9782608696},(obj1);*

$$10.44565217$$

We find the solution to all variables and the Lagrange multiplier.

**Method 2**  Enter the equations and use NLPSolve.

*NLPSolve(obj1, {3·x+y=6},maximize,assume=nonnegative);*

$$[10.4456521739130430, [x = 1.67391304347826096, y = 0.978260869565217406]]$$

We obtain the solution, but we do not capture the value of the Lagrange multiplier.
*Caution*: If we try to fix the problem by placing the Lagrangian function into the solver, we do not get the same value of $\lambda$.

> *NLPSolve(L, {3·x+y=6},maximize,assume=nonnegative);*

$$[10.4456521739130430, [x = 1.67391304347826074,$$
$$y = 0.978260869565217406, l = 0.99999999999999944]]$$

**Method 3**  Enter the equations and use the maximization or minimization function.

> *Maximize(obj1, {3·x+y=6});*

$$[10.445652173913, [x = 1.67391304347826074, y = 0.978260869565217184]]$$

We obtained the correction solution without the value of the multiplier, $\lambda$.
*Caution*: Again, if we try to remedy this, we do not get the correct value of $\lambda$.

> *Maximize(L, {3·x+y=6},assume=nonnegative);*

Warning: problem appears to be unbounded.

$$[10.445652173913, [x = 1.67391304347826096, y = 0.978260869565217184, l = 0.]]$$

We recommend setting up the Lagrangian and solving using Maple as a computational tool for these problems or using the with(Student[MultivariateCalculus]) routine.

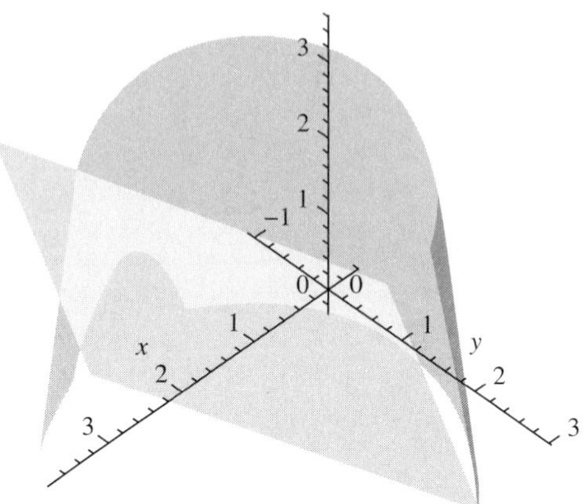

**Method 4** Enter the equations and use the with(Student)[MultivariateCalculus]).

The command is *Lagrange Multiplier(function, constraints, variables, output=detailed);*

> *with( Student[ MultivariateCalculus ]) :*
> *Lagrange Multipliers( obj1, [3x+y−6 ], [x,y ], output=detailed )*

$$\left[ x = \frac{77}{46}, y = \frac{45}{46}, \lambda_1 = \frac{35}{46}, -2x^2 - 2y^2 + xy + 8x + 3y = \frac{961}{92} \right]$$

> *evalf (%)*

$$[x = 1.673913043, y = 0.9782608696, \lambda_1 = 0.7608695652, -2 \cdot x^2 - 2 \cdot y^2 + xy + 8 \cdot x + 3 \cdot y = 10.44565217]$$

We may use this to obtain a plot as shown in Figure 10.5.

> *with( Student[ MultivariateCalculus ]):*
> *Lagrange Multipliers( obj1, [3x+y−6 ], [x,y ], output=plot, showlevelcurves=true )*

Thus, our solution is

$$x^* = 1.673913043$$
$$y^* = 0.978260896$$
$$\lambda^* = 0.7608695652$$

We evaluate the function to obtain its value of 10.44565217:

$$f(x^*, y^*) = 10.4456$$

We have a solution, but we need to know whether this solution represents the maximum or the minimum of the Lagrangian.

We use the Hessian matrix in our justification. We could use either the Hessian or the bordered Hessian described in the following to justify that we have found the correct solution to our problem to maximize L.

***Hessian*** If you have *with(linalg)*, use the following commands:

> *h:=hessian( L,[x,y ]);*

$$h := \begin{bmatrix} -4 & 1 \\ 1 & -4 \end{bmatrix}$$

> *det(h);*

The Hessian is negative definite for all values of (x,y) so the regular point also called the stationary point (x*,y*) is a maximum.

Using with(VectorCalculus):

*> with( VectorCalculus ):*
*> h1:=Hessian( L,[x,y] );*

$$h1 := \begin{bmatrix} -4 & 1 \\ 1 & -4 \end{bmatrix}$$

The determinant is 15, so the Hessian is negative definite.

*> det(h1);*

$$15$$

### Bordered Hessian

*> bdh:=matrix([[−4, 1,−3],[1,−4,−1],[−3,−1,0]]);*

$$bdh := \begin{bmatrix} -4 & 1 & -3 \\ 1 & -4 & -1 \\ -3 & -1 & 0 \end{bmatrix}$$

*> det(bdh);*

$$46$$

Because the determinant is positive, we have found the maximum at the critical point.

Either method works in this example to determine that we have found the maximum of L. Now, let's interpret the shadow price, $\lambda = 0.76$. If the right-hand side of the constraint is increased by a small amount $\Delta$, then the function will increase by 0.76 $\Delta$. Because this is a maximization problem, we would add to the resource if possible: It improves the value of the objective function. From the graph, we can see that the incremental change must be small or the objective function will begin to increase. If we increase the RHS by one unit so that $g(x) = 3x + y = 7$, the solution at the new point $(x^{**},y^{**})$ should yield a functional value, $f(x^{**},y^{**}) \approx$ old $f + \lambda = 5.810625 + 5.075 = 10.885625$. In actuality, changing the constraints yields an actual solution of 11.04347826.

*> L:=f+l1*(c1+1);*

$$L := -2x^2 - 2y^2 + xy + 8x + 3y + l1\,(7 - 3x - y)$$

*> nc :=grad( L,[x,y,l1 ]);*

$$nc := [-4x + y + 8 - 3l1, -4y + x + 3 - l1, 7 - 3x - y]$$

*> lgsol: =solve({−4*x+y+8−3*l1,−4*y+x+3−l1,7−3*x−y}, {x,y,l1 });*

$$lgsol := \{x = \frac{45}{23}, y = \frac{26}{23}, l1 = \frac{10}{23}\}$$

*> evalf(%);*

$$\{x = 1.956521739, l1 = .4347826087, y = 1.130434783\}$$

*> subs({x=1.956521739,l1=.4347826087,y=1.130434783},L);*

$$11.04347826$$

The increase was about 0.60.

| **Example 2** | Lagrange Multipliers with Multiple Constraints |
|---|---|

$$\text{Minimize } w = x^2 + y^2 + 3z$$
$$\text{S.T. } x + y = 3$$
$$x + 3y + 2z = 7$$

We will illustrate only two methods: CAS and with(Student[MultivariateCalculus]).

## Method 1 Entering the Equations and Solving a System of Equations with CAS

$$L(x,y,z,\lambda_1,\lambda_2) = x^2 + y^2 + 3z + \lambda_1[x + y - 3] + \lambda_2[x + 3y + 2z - 7]$$

$> f1:=x^2+y^2+3z;$

$$f1 := x^2 + y^2 + 3z$$

$> c1:=x+y-3$

$$c1 := x + y - 3$$

$> c2:=7-(x+3\cdot y+2\cdot z);$

$$c2 := 7 - x - 3y - 2z$$

$> L:=f1+l1\cdot c1+l2\cdot c2;$

$$L := x^2 + y^2 + 3z + l1\,(x + y - 3) + l2\,(7 - x - 3y - 2z)$$

$> nc:=grad(L,[x,y,z,l1,l2]);$

$$nc := [[2x + l1 - l2, 2y + l1 - 3\,l2, 3 - 2\,l2, x + y - 3, 7 - x - 3y - 2z]]$$

$> lgsol:=solve(\{2\cdot x+l1+l2,2\cdot y+l1+3\cdot l2,3+2\cdot l2,x+y-3,x+3\cdot y+2\cdot z-7\}, \{x,y,z,l1,l2\});$

$$lgsol := \left\{ x = \frac{3}{4},\ y = \frac{9}{4},\ l1 = 0,\ z = -\frac{1}{4},\ l2 = -\frac{3}{2} \right\}$$

$> evalf(\%);$

$$\{x = 0.7500000000,\ y = 2.250000000,\ l1 = 0.,\ z = -.2500000000,\ l2 = -1.500000000\}$$

$> subs(\{l2=1.500000000,z=-.2500000000,l1=0.,y=2.250000000,x=0.7500000000\},L);$

$$4.875000000$$

Justification with the Hessian:

$> h2:=Hessian(L,[x,y,z]);$

$$h2 := \begin{bmatrix} 2 & 0 & 0 \\ 0 & 2 & 0 \\ 0 & 0 & 0 \end{bmatrix}$$

The Hessian is always positive semidefinite. The function is convex, and our critical point is a minimum.

## Revisit the previous problem using the: with(Student[MultivariateCalculus])

> *with( Student[ MultivariateCalculus ]) :*
> *LagrangeMultipliers(f1,[cons1,cons2],[x,y,z],output=detailed);*

$$\left[ x = \frac{3}{4},\ y = \frac{9}{4},\ z = -\frac{1}{4},\ \lambda_1 = 0,\ \lambda_2 = -\frac{3}{2},\ x^2 + y^2 + 3z = \frac{39}{8} \right]$$

> *evalf(%);*

$$[x = 0.7500000000,\ y = 2.250000000,\ z = -.2500000000,\ \lambda_1 = 0.,$$
$$\lambda_2 = -1.500000000,\ x^2 + y^2 + 3.\ z = 4.875000000]$$

Justification with the Hessian:

> *h2:=Hessian( L,[x,y,z] );*

$$h2 := \begin{bmatrix} 2 & 0 & 0 \\ 0 & 2 & 0 \\ 0 & 0 & 0 \end{bmatrix}$$

The Hessian is always positive semidefinite. The function is convex, and our critical point is a minimum.

Let us interpret the shadow prices, $\lambda_1$ and $\lambda_2$. If we could only spend an extra dollar on one of the two resources, which one would we spend it on? The values of the shadow prices are 0 and 1.5, respectively. Because the shadow price is $\partial w/\partial b$, we would not spend an extra dollar on resource 2: It will cause the objective function to increase by approximately \$1.50.

---

**Modeling and Applications with Lagrange Multipliers**

---

## Example 3          Cobb-Douglas Function

Recall the problem suggested in the introduction of the chapter. A company manufactures new e-phones that are supposed to capture the market by storm. The two main inputs components to the e-phone are the circuit board and the relay switches. The number of e-phones produced is estimated to equal $E = 200a^{\frac{1}{2}} b^{\frac{1}{4}}$, where $E$ is the number of phones produced and $a$ and $b$ are the number of circuit broad hours and the number of relay hours worked, respectively. Such a function is known to economists as a *Cobb–Douglas function*. Our laborers are paid by the type work they do: the circuit boards and the relays for \$5 and \$10 an hour. We want to maximize the number of e-phones to make if we have \$150,000 to spend on these components in the short run.

> *LG2:=200· A·⁵· B ·²⁵+lam·( 150000−5·A−10·B );*

$$LG2 := 200\,A^{0.5}\,B^{0.25} + lam\,(150000 - 5A - 10B)$$

> *l1:=diff( LG2,A );*

$$l1 := \frac{100.0B^{0.25}}{A^{0.5}} - 5\,lam$$

> *l2:=diff(LG2, B);*

$$l2 := \frac{50.00 A^{0.5}}{B^{0.75}} - 10 \, lam$$

> *l3:=diff(LG2, lam);*

$$l3 := 150000 - 5A - 10B$$

> *fsolve({l1=0,l2=0,l3=0}, {A,B,lam});*

$$\{A = 20000.00000, \ B = 5000.000000, \ lam = 1.189207115\}$$

> *subs({A=20000.00000,B=5000.000000,lam=1.189207115}, LG2);*

$$2.378414230 \ 10^5$$

We find we can make 237,841.42 e-phones using 20,000 relays and 5,000 circuit boards hours of labor.

## Revisit the Cobb-Douglas Function Using the with(Student[MultivariateCalculus]) Method

> *CDF:=200·A^{0.5}·B^{0.25};*

$$CDF := 200 \ A^{0.5} B^{0.25}$$

> *consCDF:=150000−5·A−10·B;*

$$consCDF := 150000 - 5 A - 10 B$$

> *with(Student[MultivariateCalculus]) :*

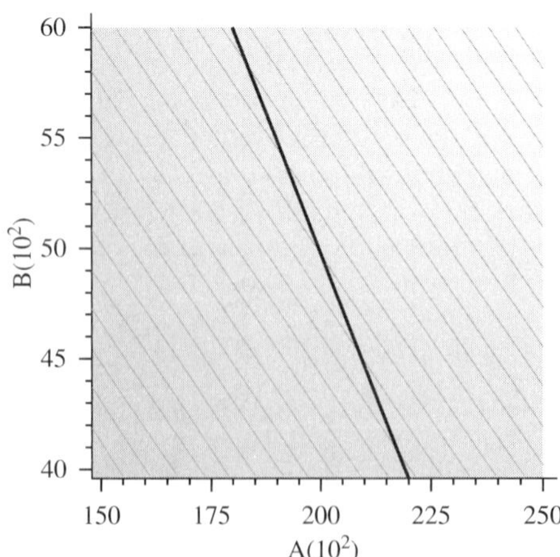



**FIGURE 10.6**
Cobb–Douglas function with constraint

> *LagrangeMultipliers( CDF,[ −1·consCDF],[ A,B],output=detailed);*

$$[A = 20000., B = 5000., \lambda_1 = 1.189207115, 200\, A^{0.5} B^{0.25} = 2.378414230\ 10^5]$$

> *evalf(%);*

$$[A = 20000., B = 5000., \lambda_1 = 1.189207115, 200.\, A^{0.5} B^{0.25} = 2.378414230\ 10^5]$$

> *p1:=contourplot( CDF,A=15000..25,000,B=4000..6000,contours=50):*
> *p2:=contourplot( consCDF,A=15000..25000,B=4000..6000,contours=[0],thickness=3):*
> *display( p1,p2);*

We obtained the solution, as well as the plot shown in Figure 10.6.

---

## Example 4     Oil Transfer

You are employed as a consultant for a small oil-transfer company. The management desires a minimum cost policy because of restricted tank-storage space. Historical records have been studied, and a formula has been derived that describes system costs:

$$f(X) = \sum_{n=1}^{N} \frac{(A_n B_n)}{X_n} + \frac{(H_n X_n)}{2}$$

where:

$A_n$ is the fixed costs for the $n$th item,
$B_n$ is the withdrawal rate per unit time for the $n$th item, and
$H_n$ is the holding costs per unit time for the $n$th item.

The tank-space constraint is given by:

$$g(x) = \sum_{n=1}^{N} t_n X_n = T$$

where

$t_n$ is the space required for the $n$th item (in correct units), and
$T$ is the available tank space (in correct units).

You determine the following information:

| Item ($n$) | $A_n$ ($) | $B_n$ | $H_n$ ($) | $t_n$ (cubic feet) |
|---|---|---|---|---|
| 1 | 9.6 | 3 | 0.47 | 1.4 |
| 2 | 4.27 | 5 | 0.26 | 2.62 |
| 3 | 6.42 | 4 | 0.61 | 1.71 |

You measure the storage tanks and find only 22 cubic feet of space available. We want to find the optimal solution as a minimum cost policy.

First, we will solve the unconstrained problem.

If we assume that $\lambda = 0$, we find an unconstrained optimal solution.

$$> Lf := (9.6)*(3)/x + .47*x/2 + (4.27)*(5)/y + .26*y/2 + (6.42)*(4)/z + .61*z/2;$$

$$Lf := 28.8\,\frac{1}{x} + .2350000000x + \frac{21.35}{y} + .1300000000y + \frac{25.68}{z} + .3050000000z$$

$$> pd1 := diff(Lf, x);$$

$$pd1 := -28.8\,\frac{1}{x^2} + .2350000000$$

$$> pd2 := diff(Lf, y);$$

$$pd2 := -21.35\,\frac{1}{y^2} + .1300000000$$

$$> pd3 := diff(Lf, z);$$

$$pd3 := -25.68\,\frac{1}{z^2} + .3050000000$$

$$> solve(\{pd1=0, pd2=0, pd3=0\}, \{x, y, z\});$$

$$\{z = -9.175877141,\ y = -12.81525533,\ x = -11.07037450\},$$
$$\{x = 11.07037450,\ z = -9.175877141,\ y = -12.81525533\},$$
$$\{z = 9.175877141,\ y = -12.81525533,\ x = -11.07037450\},$$
$$\{z = 9.175877141,\ x = 11.07037450,\ y = -12.81525533\},$$
$$\{y = 12.81525533,\ z = -9.175877141,\ x = -11.07037450\},$$

$$\{x = 11.07037450,\ y = 12.81525533,\ z = -9.175877141\},$$
$$\{z = 9.175877141,\ y = 12.81525533,\ x = -11.07037450\},$$
$$\{z = 9.175877141,\ x = 11.07037450,\ y = 12.81525533\}$$

The only useful solution is where $x$, $y$, $z \geq 0$

$$x = 11.07037450,\ y = 12.81525533,\ z = 9.175877141$$

This solution is $(x^*, y^*, z^*) = (11.07, 12.82, 9.176)$. This solution provides an unconstrained upper bound because those values do not satisfy the constraint

$$1.4x + 2.62y + 1.71z = 22$$

We set up the constrained model as:

$$\text{Let } x = \text{item 1}$$
$$y = \text{item 2}$$
$$z = \text{item 3}$$

$$L(x,y,z,\lambda) = (9.6)(3)/x + 0.47x/2 + (4.27)(5)/y + 0.26y/2 + (6.42)(4)/z + 0.61z/2$$
$$+ \lambda\,[1.4x + 2.62y + 1.71\,z - 22]$$
$$L_x = -28.8x^{-2} + 0.235 + 1.4\,\lambda = 0$$
$$L_y = -21.35\,y^{-2} + 0.13 + 2.62\lambda = 0$$
$$L_z = -25.68z^{-2} + 0.305 + 1.71\lambda = 0$$
$$L_\lambda = 1.4x + 2.62y + 1.71z - 22 = 0$$

$$> L := (9.6)*(3)/x + .47*x/2 + (4.27)*(5)/y + .26*y/2 + (6.42)*(4)/z + .61*z/2 + l1*$$
$$(1.4*x + 2.62*y + 1.71*z - 22);$$

$$L := 28.8\,\frac{1}{x} + .2350000000x + \frac{21.35}{y} + .1300000000y + \frac{25.68}{z} + .3050000000z$$
$$+ l1\,(1.4x + 2.62y + 1.71z - 22)$$

> $nc:=grad(L,[x,y,z,l1]);$

$$nc := \left[ -28.8\,\frac{1}{x^2} + .2350000000 + 1.4\,l1,\ -21.35\,\frac{1}{y^2} + .1300000000 + 2.62\,l1, \right.$$
$$\left. -25.68\,\frac{1}{z^2} + .3050000000 + 1.71\,l1,\ 1.4x + 2.62y + 1.71z - 22 \right]$$

> $solve(\{-28.8*1/(x^2)+.2350000000+1.4*l1,-21.35*1/(y^2)+.1300000000+2.62*l1,$
$-25.68*1/(z^2)+.3050000000+1.71*l1,\ 1.4*x+2.62*y+1.71*z-22\},\{x,y,z,l1\});$

The answer selected from the many solutions in Maple is:

$$\{y = 3.213131453,\ z = 4.044536153,\ x = 4.761027695,\ l1 = .7396771332\}$$

> $subs(\{y=3.213131453,z=4.044536153,x=4.761027695,l1=.7396771332\},L);$

$$21.81316118$$

Do we have the minimum?
The Hessian matrix, $H$, is:

> $h:=hessian(L,[x,y,z]);$

$$h := \begin{bmatrix} 57.6\,\dfrac{1}{x^3} & 0 & 0 \\[2ex] 0 & 42.70\,\dfrac{1}{y^3} & 0 \\[2ex] 0 & 0 & 51.36\,\dfrac{1}{z^3} \end{bmatrix}$$

> $h1:=det(h);$

$$h1 := 126320.9472\,\frac{1}{x^3\,y^3\,z^3}$$

> $subs(\{y=3.213131453,z=4.044536153,x=4.761027695,l1=.7396771332\},h1);$

$$.5333123053$$

The Hessian is positive definite at our critical point $x = 4.761027695$, $y = 3.213131453$, $z = 4.044536153$, $x = 4.761027695$. Therefore, the solution found is the minimum for this convex function.

Should we add storage space? We know from the unconstrained solution that, if possible, we would add storage space to decrease the costs. In addition, we have found the value of $\lambda = 0.7396771332$, which suggests that any small increase ($\Delta$) in the RHS of the constraint causes the objective function to decrease by approximately $0.74\Delta$. The cost of the extra storage tank would have to be less than the savings incurred by adding the tank.

## 10.1 | EXERCISES

1. Solve the following constrained problems.
   a. Minimize $x^2 + y^2$
      subject to $x + 2y = 4$
   b. Maximize $(x - 3)^2 + (y - 2)^2$
      subject to $x + 2y = 4$

   **c.** Maximize $x^2 + 4xy + y^2$
      subject to $x^2 + y^2 = 1$
   **d.** Maximize $x^2 + 4xy + y^2$
      subject to $x^2 + y^2 = 4$
      $x + 2y = 4$

**2.** Maximize $3X^2 + Y^2 + 2XY + 6X + 2Y$

   S.T.       $2X - Y = 4$

Did you find the maximum? Explain.

**3.** Find and classify the extrema for

   $f(x,y,z) = x^2 + y^2 + z^2$

   S.T.       $x^2 + 2y^2 - z^2 = 1$

**4.** Given that two manufacturing processes both use resource $b$, we want to find the maximum of $f_1(x_1) + f_2(x_2)$ subject to $x_1 + x_2 = b$.

   If $f_1(x_1) = 50 - (x_1 - 2)^2$ and $f_2(x_2) = 50 - (x_2 - 2)^2$, analyze this process to
   **a.** determine the amount of $x_1$ and $x_2$ to use to maximize the process, and
   **b.** determine the amount of resource $b$ to use.

**5.** Maximize $Z = -2x^2 - y^2 + xy + 8x + 3y$

   S.T.       $3x + y = 10$

            $x^2 + y^2 = 16$

**6.** Use the method of LaGrange multipliers to find the maxima for the following:

   Maximize $f(x,y,w) = xyw$

   S.T.       $2x + 3y + 4w = 36$

Determine how much $f(x,y,w)$ would change if one more unit was added to the constraint.

## 10.1 | PROJECTS

**1.** Suppose a newspaper publisher must purchase three kinds of paper stock but must minimize its costs in the process. It decides to use an economic lot size model to assist in its decisions. Given an economic order quantity (EOQ) model with constraints where the total cost is the sum of the individual quantity costs:

$$C(Q_1,Q_2,Q_3) = C(Q_1) + C(Q_2) + C(Q_3)$$
$$C(Q_i) = a_i d_i / Q_i + h_i Q_i / 2$$

where

      $d$ is the order rate.
      $h$ is the cost per unit time (storage).
      $Q/2$ is the average amount on hand, and
      $a$ is the order cost.

The constraint is the amount of storage area available to the publisher so that it can have the three kinds of paper on hand for use. The items cannot be stacked, but they can be laid side by side. They are constrained by the available storage area, $S$.
The following data are collected:

|   | Type I | Type II | Type III |
|---|---|---|---|
| $d$ | 32 rolls/week | 24 | 20 |
| $a$ | \$25 | \$18 | \$20 |
| $h$ | \$1/roll/week | \$1.5 | \$2.0 |
| $S$ | 4 sqft/roll | 3 | 2 |

You have 200 sq ft of storage space available.

Required:

  **a.** Find the levels of quantity that are the unconstrained minimum total cost, and show that these values would not satisfy the constraint. What purpose do these values serve?

  **b.** Find the constrained optimal solution by using the Lagrange multipliers, assuming we will use all 200 sq feet.

  **c.** Find and interpret the shadow prices.

**2.** Resolve the tank-storage problem to determine whether it is better to have a cylindrical storage space or a rectangular storage space of 50 cubic units.

**3.** Suppose you want to use the Cobb–Douglass function—$P(L,K) = A\,L^a\,K^b$—to predict output in thousands, based on the amount of capital and labor used. Suppose you know the price of capital and labor per year is \$10,000 and \$7,000 respectively. Your company estimates the following values: $A = 1.2$, $a = 0.3$, and $b = 0.6$. Your total cost is assumed to be $T = PL*L + Pk*k$, where $PL$ and $Pk$ are the price of capital and labor. There are three possible funding levels: \$63,940, \$55,060, and \$71,510. Determine which budget yields the best solution for your company. Interpret the Lagrange multiplier.

## 10.2  INEQUALITY CONSTRAINTS: KUHN–TUCKER NECESSARY AND SUFFICIENT CONDITIONS

### Introduction to Kuhn-Tucker Condition (KTC)

In the previous sections, we investigated procedures to solve problems with equality constraints. The method of Lagrange multipliers provided a methodology to solve NLP problems as shown by Equations 10.1–10.3.

However, in the most realistic problems, many of the constraints are inequalities. These constraints form the boundaries for the solution. The generic form of the NLP we will study in this section is shown by Equation 10.5:

$$\text{Maximize (minimize) } z = f(x_1, x_2, x_3, \ldots, x_n)$$
$$\text{Subject to}$$
$$g_1(x_1, x_2, \ldots, x_n) \geq b_1$$
$$g_2(x_1, x_2, \ldots, x_n) \geq b_2 \tag{10.5}$$
$$\cdot$$
$$\cdot$$
$$\cdot$$
$$g_m(x_1, x_2, \ldots, x_n) \geq b_m$$

One method for solving NLPs of this type, Equation 10.5, is the Kuhn–Tucker condition (KTC). In this chapter, we describe the KTC first graphically and then analytically. We discuss the necessary and sufficient conditions for $\mathbf{X} = \{x_1, x_2, \ldots, x_n\}$ to be an optimal solution to the NLP of Equation 10.5. We illustrate how to use Maple to solve these KTC problems. We then present some applications using the KTC solution methodology.

### Basic Theory of Constrained Optimization

During these sections on KTC, we are concerned with problems of the form:

$$\text{Maximize (minimize) } f(X)$$
$$\text{Subject to}$$
$$g_i(X) \begin{Bmatrix} \geq \\ \leq \\ = \end{Bmatrix} b_i \quad i = 1, 2, \ldots, m \tag{10.5a}$$

We allow the constraints to be either "less than or equal to" or "greater than or equal to." We have previously completed a block on Lagrange multipliers to solve problems with equality constraints. Recall that during the Lagrange block the optimal solution actually fell on one constraint or at an intersection of several constraints. With the inequality constraints, the solution no longer must lie on a constraint or at an intersection point of constraints. This concept poses new problems. We need a method to account for the position of the optimal solution relative to each constraint. This KTC procedure involves setting up a Lagrangian function of the decision variables $\mathbf{X}$, the Lagrange multipliers $\lambda$ and the slack or surplus variables $\mathbf{U}_i^2$. The $\mathbf{X}_j$ are the decision variables $(x_1, x_2, \ldots, x_n)$, the $-\lambda_i$ are the shadow prices for the $i$th constraint, and the $\mathbf{U}_i^2$ are either added (slack variables from $\leq$ constraints) or subtracted (surplus variables from $\geq$ constraints). Thus, with the sign of $\mathbf{U}_i^2$ we are able to accommodate both $\leq$ and $\geq$ constraints.

We set up this generic Lagrangian function:

$$L(X,\lambda,U^2) = f(X) + \sum_{i=1}^{m} \lambda_i(g(X_i) + (\pm U_i^2) - b_i) \qquad (10.6)$$

Note: $U_i^2$ depends on the type of inequality constraint, as we will explain more in detail. The computational procedure for the KTC requires that all the partials of this Lagrangian function equal zero. These partials are the *necessary conditions* of the NLP problem. These are the conditions required for $\mathbf{x} = \{x_1, x_2, \ldots, x_n\}$ to be a solution to Equation 10.6.

| | | |
|---|---|---|
| The Necessary Conditions | | (10.7) |
| $\partial L/\partial X_j = 0$ | $(j = 1, 2, \ldots, n)$ | (10.7a) |
| $\partial L/\partial \lambda_i = 0$ | $(i = 1, 2, \ldots, m)$ | (10.7b) |
| $\partial L/\partial U_i = 0$ or $2U_i \lambda_i = 0$ | $(i = 1, 2, \ldots, m)$ | (10.7c) |

The following two theorems give the sufficient conditions for $x^* = \{x_1, x_2, \ldots, x_n\}$ to be an optimal solution to the NLP given in Equation 10.7.

**The Sufficient Conditions [10.8]**

(10.8a) **Minimum**: If $f(x)$ is a convex function and each of the $g_i(x)$ are convex functions, then any point that satisfies the necessary conditions is an optimal solution. An optimal point is a point that minimizes the function subject to the constraints. $\lambda_i$ is greater than or equal to zero for all $i$.

(10.8b) **Maximum**: If $f(x)$ is a concave function and each of the $g_i(x)$ are convex functions, then any point that satisfies the necessary conditions is an optimal solution. An optimal point is a point that maximizes the function subject to the constraints. $\lambda_i$ is less than or equal to zero for all $i$.

If these above conditions are not completely satisfied, then we may use another method to check the nature of a potential stationary or regular point, such as the bordered Hessian.

**Bordered Hessian**  The bordered Hessian is a symmetric matrix of second partials of the Lagrangian:

$$H_B = \partial^2 L/\partial(X_j^2, \lambda_k^2) \ \{j = 1, 2, \ldots, n, k = 1, 2, \ldots, m\}$$

We can determine, if possible, the nature of the stationary point by classifying the bordered Hessian. This method is only valid leading to max or min points. If the bordered Hessian is indefinite, then another method should be used.

**Complementary Slackness**  The KTC computational solution process uses these necessary conditions for solving $2^m$ possible cases, where $m$ equals the number of constraints. The value 2 comes from the possible conditions placed on $\lambda_i$: it either equals zero or does not equal zero. There is actually more to this process because it really involves the complementary slackness condition embedded in the necessary condition, $2\,U_i\lambda_i = 0$. Thus, either $U_i$ equals zero and $\lambda_i$ does not equal zero or $U_i$ is greater than or equal to zero and $\lambda_i$ is equal to zero. This ensures that the complementary slackness conditions are satisfied while the other necessary conditions from Equations 10.7a and 10.7b are solved.

It is these complementary slackness necessary conditions, Equation 17.6c, that leads to the solution process on which we focus our computational and geometric interpretation. We have defined $U_i^2$ as a slack or surplus variable. Therefore, if $U_i^2$ equals zero, then our point is on the $i$th constraint; and if $U_i^2$ is greater than zero, then the point does not lie on the $i$th constraint. Furthermore, if the value of $U_i^2$ is undefined because it equals a negative number, then the point of concern is infeasible. Figures 10.7 through 10.9 illustrate these conditions.

### Geometric Interpretation of KTC

We begin with a geometric illustration. It is a basic two variable linear problem. The purpose of this geometric interpretation is to lay a foundation for both further geometric interpretation and computational results.

**Spanning Cones (Optional)**  Let's consider the following example:

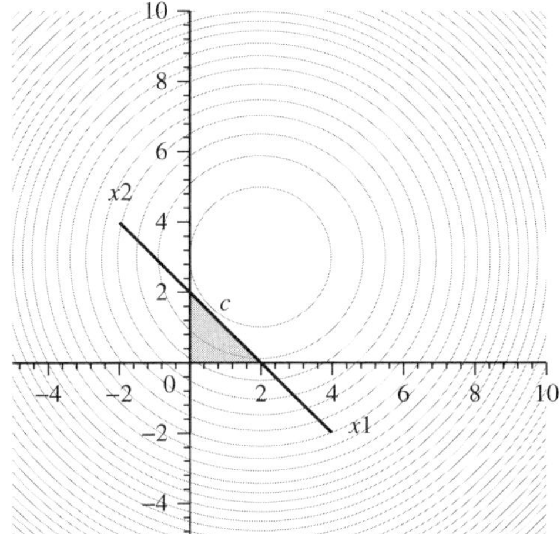

**FIGURE 10.7**

$U^2 = 0$: PT C is on the constraint

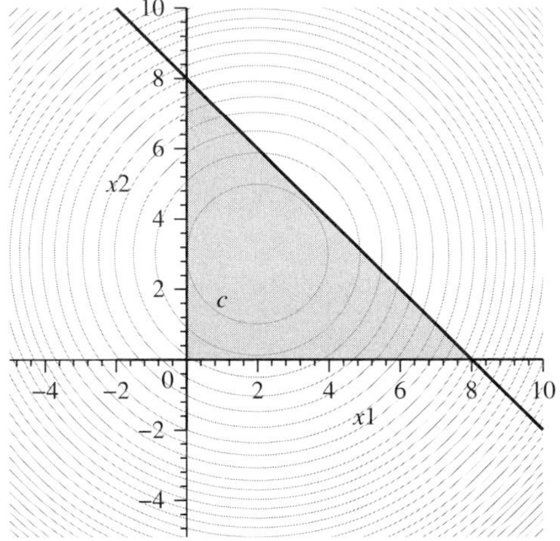

**FIGURE 10.8**

The point C is inside the feasible region, $U^2 > 0$

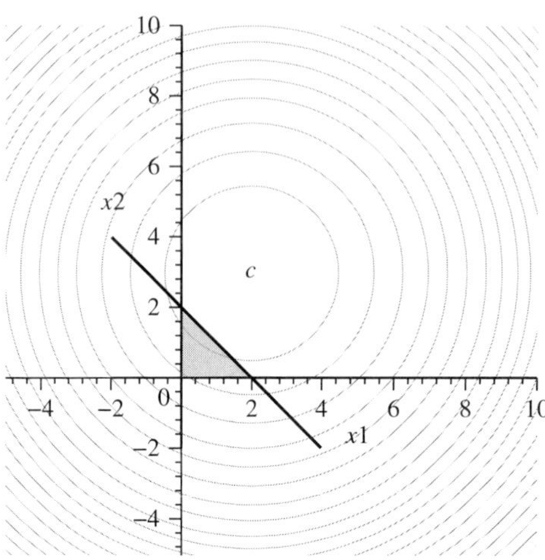



**FIGURE 10.9**
The point C is not in the feasible region, $U^2 < 0$

---

<table><tr><td>**Example 1**</td><td>

## Maximizing with Inequality Constraints

</td></tr></table>

$$\text{Maximize } Z = x_1 + x_2$$
$$\text{Subject to}$$
$$x_1 + 3x_2 \le 9$$
$$2x_1 + x_2 \le 8$$
$$x_1, x_2 \ge 0$$

The gradient of the objective function (we will call this *gradient vector* **C**) is the vector of partial derivatives of $Z$ with respect to $x_1$ and $x_2$. Because these are linear functions, they are the coefficients of $x_1$ and $x_2$. In our example, **C** = [1,1]. The vector **C** points in the direction of greatest local increase of the objective function.

Each intersection of the feasible region is found as the intersection of the *binding constraints*, or constraints that are satisfied at equality without slack or surplus.

### Spanning Cones (Optional)

The cone spanned by a set of vectors is the set of all nonnegative linear combinations of all the vectors. The coefficients of the linear combination are called the *multipliers* of the cone. In a plane, a cone has the appearance that its name suggests (see Figure 10.10).

Thus, the binding constraints are used to find the gradient vectors of the feasible region's boundary or corner points. There are four constraints (counting the nonnegativity), and we will refer to their gradients as $A_1$, $A_2$, $A_3$ and $A_4$. *Note:* All constraints need to be in the form $\le$ RHS, including the nonnegativity constraints:

$$A_1 = [1,3]$$
$$A_2 = [2,1]$$
$$A_3 = [-1,0]$$
$$A_4 = [0,-1]$$

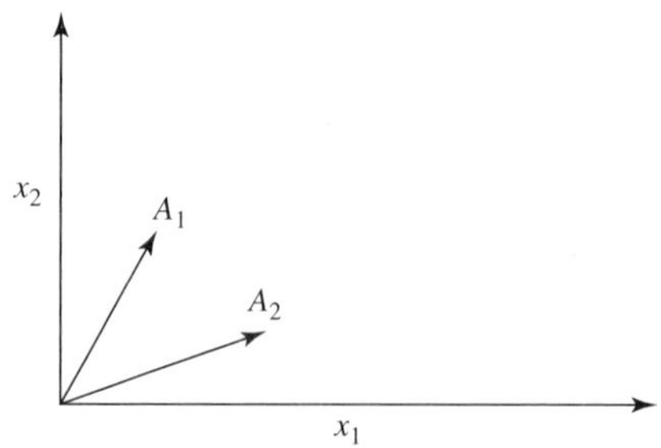



**FIGURE 10.10**
Cone spanned by $A_1$ and $A_2$

*Note*: We have corner points P1, P2, P3, and P4. The optimal solution is located at P2, and this is where the gradient of **C** is contained within the cone spanned by their binding constraints (see Figures 10.11 through 10.14). Each is shown as a spanning cone at points P1–P4. They were drawn separately for clarity.

**FIGURE 10.11**
Spanning cone illustration at P1 (opt vector not in span)

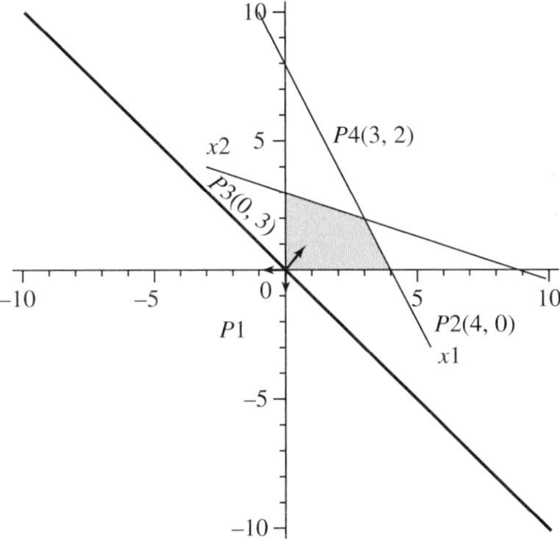

**FIGURE 10.12**
Spanning cones at P3 (opt vector not in span)

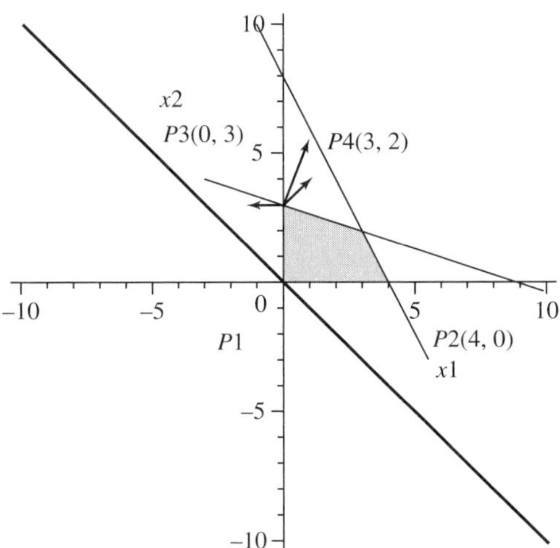

**FIGURE 10.13**
Spanning cones at P2 (opt vector not in span)

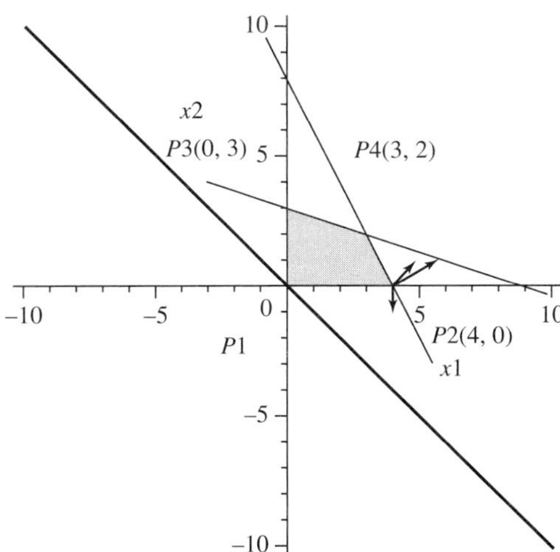

**FIGURE 10.14**

Spanning cone at P4 (optimal vector is in span of binding constraints)

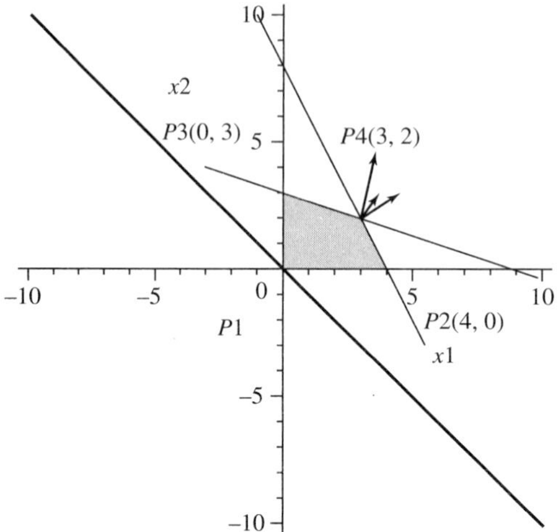


```
> L:=obj2+l1·(x1+3·x2+u1²−9)+l2·(2·x1+x2+u2²−8);
```

$$L := x1 + x2 + l1\,(x1 + 3\,x2 + u1^2 - 9) + l2\,(2\,x1 + x2 + u2^2 - 8)$$

```
> lx1:=diff(L,x1);
```

$$lx1 := 1 + l1 + 2\,l2$$

```
> lx2:=diff(L,x2);
```

$$lx2 := 1 + 3\,l1 + l2$$

```
> ll1 :=diff(L,l1);
```

$$ll1 := x1 + 3\,x2 + u1^2 - 9$$

```
> ll2:=diff(L,l2);
```

$$ll2 := 2x1 + x2 + u2^2 - 8$$

```
> lu1 :=diff(L, u1);
```

$$lu1 := 2\,l1\,u1$$

```
> lu2:=diff(L,u2);
```

$$lu2 := 2\,l2\,u2$$

```
> solve({lx1=0,lx2=0,ll1=0,ll2=0,lu1=0,lu2=0},{x1,x2,l1,l2, u1, u2});
```

$$\left\{ u1 = 0,\ u2 = 0,\ x1 = 3,\ x2 = 2,\ l1 = -\frac{1}{5},\ l2 = -\frac{2}{5} \right\}$$

### Computational KTC with Maple

Again we will illustrate two methods that can help us obtain all the parameters and variable values. Method 1 will be with the CAS solving the systems of equations and, and method 2 with the with(Student[MultivariateValculus]) command.

## Example 2

## Two Variable-Two Constraint Linear Problem

Consider the problem:

$$\text{Maximize } 3x + 2y$$
$$\text{Subject to}$$
$$2x + y \leq 100$$
$$x + y \leq 80$$

Putting this problem into our generalized Lagrangian form, Equation 10.6 we obtain the following:

$$L(x, y, \lambda, U) = 3x + 2y + \lambda_1 (2x + y + U_1^2 - 100) + \lambda_2 (x + y + U_2^2 - 80)$$

The six necessary conditions are:

1. $3 + 2\lambda_1 + \lambda_2 = 0$
2. $2 + \lambda_1 + \lambda_2 = 0$
3. $2x + y + U_1^2 - 100 = 0$
4. $x + y + U_2^2 - 80 = 0$
5. $2U_1\lambda_1 = 0$
6. $2U_2\lambda_2 = 0$

Because there are two constraints, recognize that there are four ($2^2$) cases required to solve for the optimal solution. These four cases stem from necessary conditions listed in equations $2U_1\lambda_1 = 0$ and $2U_2\lambda_2 = 0$ (see the following table).

| Cases | Condition Imposed | Condition Inferred |
|-------|-------------------|--------------------|
| I | $\lambda_1 = \lambda_2 = 0$ | $U_2^2 \neq 0,\ U_1^2 \neq 0$ |
| II | $\lambda_1 = 0,\ \lambda_2 \neq 0$ | $U_2^2 = 0,\ U_1^2 \neq 0$ |
| III | $\lambda_2 = 0,\ \lambda_1 \neq 0$ | $U_1^2 = 0,\ U_2^2 \neq 0$ |
| IV | $\lambda_1 \neq 0,\ \lambda_2 \neq 0$ | $U_2^2 = 0,\ U_1^2 = 0$ |

For simplicity, we have arbitrarily made both $x$ and $y \geq 0$ for this maximization problem. We provide a graphical representation in Figure 10.15.

**FIGURE 10.15**
Region for KTC example Max $3x + 2y$, subject to $2x + y \leq 100$, $x + y \leq 80$, $x \geq 0$, $y \geq 0$

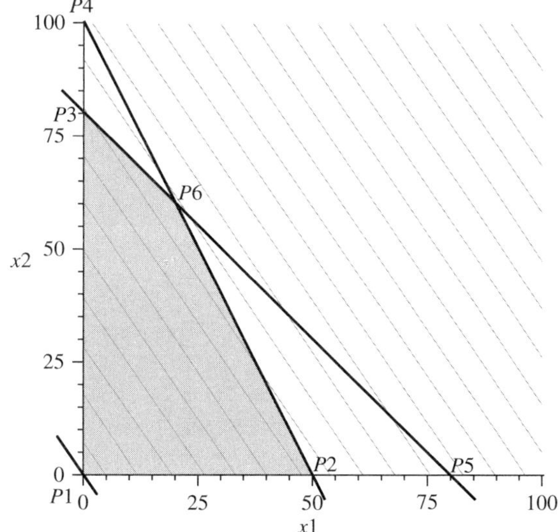

Returning to cases I–IV, we observe the following:

1. Case I considers slack in both the first and second constraints, so we do not fall exactly on either of the constraints. This corresponds to the intersection point labeled P1 at (0,0) because only intersection points can lead to linear optimization solutions. This point is feasible but clearly not optimal because we need to move out from (0,0) to be optimal. This case will not yield an optimal solution.

2. Case II places the possible solution point on the second constraint but not on the first. There exist two possible solutions: points 3 and 5 from Figure 10.15. Point 5 is infeasible, and point 3 is feasible but not an optimal solution. This case will not yield an optimal solution.

3. Case III places the possible solution on the first constraint but not on the second. There exist two possible solutions: points 2 and 4 from Figure 10.15. Point 4 is infeasible, and point 2 is feasible but not the optimal solution. Again, this case does not yield an optimal solution.

4. Case IV places the possible solution on both constraints 1 and 2 simultaneously. This corresponds to point 6 on Figure 10.15. Point 6 is the optimal solution, the point in the feasible region tangent to the contours of the objective function in the direction of increased value for the objective function. This case will computationally yield the optimal solution to the problem.

Sensitivity analysis is also enhanced by geometric interpretation. Thus, it is clear from Figure 10.16 that if the right-hand side of either or both constraints increases, then the feasible region will be extended and the value of the objective function will increase. We should find this through the computational process and the solution of $\lambda$. This computational sensitivity analysis will be shown through the value of the shadow price $(-\lambda_i)$.

Because both our objective function and constraints are linear, convex, and concave functions, the sufficient conditions are also satisfied.

Solving these cases shows that case IV yields the optimal solution, confirming the graphical solution.

## Computational Results

**Case I:**   $\lambda_1 = \lambda_2 = 0$. This case violates the first two of the necessary conditions that we listed earlier because $2 \neq 0$ and $3 \neq 0$. Case I also implies that $U_2^2 \neq 0$ and $U_1^2 \neq 0$.

**Case II:**   $\lambda_1 = 0, \lambda_2 \neq 0$, implying that $U_2^2 = 0$ and $U_1^2 \neq 0$. This case violates both the first two necessary conditions listed earlier because $\lambda_2$ cannot equal both $-2$ and $-3$ at the same time.

FIGURE 10.16<br>Geometric sensitivity analysis

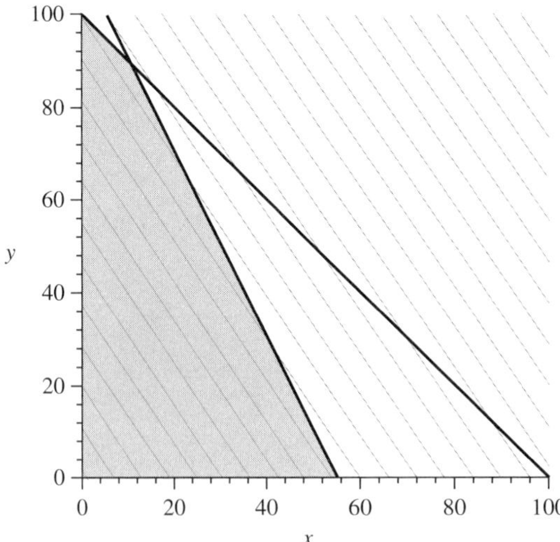

**Case III:**    $\lambda_2 = 0$, $\lambda_1 \neq 0$, implying that $U_1^2 = 0$ and $U_2^2 \neq 0$. This case also violates the first two necessary conditions listed earlier because $\lambda_1$ cannot equal $-2$ and $-3/2$ at the same time.

**Case IV:**    $\lambda_1 \neq 0$, $\lambda_2 \neq 0$, implying that $U_2^2 = 0$ and $U_1^2 = 0$.

Imposing case IV yields the following sets of equations that we refer to as Equations 10.9 through 10.12:

$$3 + 2\lambda_1 + \lambda_2 = 0 \tag{10.9}$$
$$2 + \lambda_1 + \lambda_2 = 0 \tag{10.10}$$

and

$$2x + y - 100 = 0 \tag{10.11}$$
$$x + y - 80 = 0 \tag{10.12}$$

---

Solving these pairs simultaneously yields the following optimal solution.

The solutions are $x^* = 20$, $y^* = 60$, $f(x^*,y^*) = 180$, $\lambda_1 = -1$, $\lambda_2 = -1$, $U_1^2 = U_2^2 = 0$. The shadow prices indicate for a small change, $\Delta$ in the right-hand side value of either constraint 1 or constraint 2, the objective value will increase by approximately *one x*$\Delta$. The geometric interpretation reinforces the computational results and gives them meaning. It fully shows the effect of binding constraints (where $U_i^2 = 0$) on the solution.

**Method 1** Solve by a system of equations using CAS.

> *obj3:=3·x+2·y;*

$$obj3 := 3x + 2y$$

> *L:=obj3+l1·(2·x+y+u1²−100)+l2·(x+y+u2²−80);*

$$L := 3x + 2y + l1\,(2x + y + u1^2 - 100)$$
$$+ l2(x + y + u2^2 - 80)$$

> *lx1:=diff(L,x);*

$$lx1 := 3 + 2l1 + l2$$

> *lx2:=diff(L,y);*

$$lx2 := 2 + l1 + l2$$

> *ll1:=diff(L,l1);*

$$ll1 := 2x + y + u1^2 - 100$$

> *ll2:=diff(L,l2);*

$$ll2 := x + y + u2^2 - 80$$

> *lu1:=diff(L,u1);*

$$lu1 := 2\,l1\,u1$$

> *lu2:=diff(L, u2);*

$$lu2 := 2\,l2\,u2$$

> *solve({lx1=0,lx2=0,ll1=0,ll2=0,lu1=0,lu2=0}, {x,y,l1,l2,u1,u2});*

$$\{l2 = -1,\ u2 = 0,\ l1 = -1,\ y = 60,\ x = 20,\ u1 = 0\}$$

The optimal solution is $f(20,60) = 180$.

How do we know we found a maximum? Recall the rules for finding the maximum or minimum.

**Minimum**: If $f(x)$ is a convex function and each of the $g_i(x)$ are convex functions, then any point that satisfies the necessary conditions is an optimal solution. An optimal point is a point that minimizes the function subject to the constraints. $\lambda_i$ is greater than or equal to zero for all $i$.

**Maximum**: If $f(x)$ is a concave function and each of the $g_i(x)$ are convex functions, then any point that satisfies the necessary conditions is an optimal solution. An optimal point is a point that maximizes the function subject to the constraints. $\lambda_i$ is less than or equal to zero for all $i$.

The objective function in linear and is both convex and concave as are the constraints. Because the values of $\lambda_i$ are negative, we have found the maximum.

---

## Example 3

## Two-Variable, Three-Constraints Linear Problem

In this example, we merely added one constraint to the previous problem, $x \le 40$. The addition of one constraint has caused the number of solution cases to grow from $2^2$ to $2^3$ or 8 cases. The problem with all three constraints is shown in Figure 10.17. Again, for simplicity, we arbitrarily force both $x$ and $y \ge 0$.

A summary of the graphical interpretation is displayed in Table 10.1. The optimal solution is found using case IV. Again, the computational solution merely looks for the point where all the necessary conditions are either met or violated. The geometric interpretation reinforces why the other cases do not yield an optimal solution.

The optimal solution will be found only in case IV, which geometrically shows that the solution is binding on constraints 1 and 2 and not binding on constraint 3 (slack still exists). The optimal solution is found computationally using case IV (similar to the previous example): $f(x^*,y^*) = f(20,60) = 180, \lambda_1 = -1, \lambda_2 = -1, \lambda_3 = 0, U_1^2 = 0, U_2^2 = 0,$ and $U_3^2 = 20$.

The geometric interpretation takes the mystery out of the case-wise solutions. You can visually see why we can or cannot achieve optimality conditions in each specific case. If possible, make and obtain a quick graph and analyze the graph to eliminate as many cases as possible before doing the computational solution procedures. Let's apply this procedure to another example.

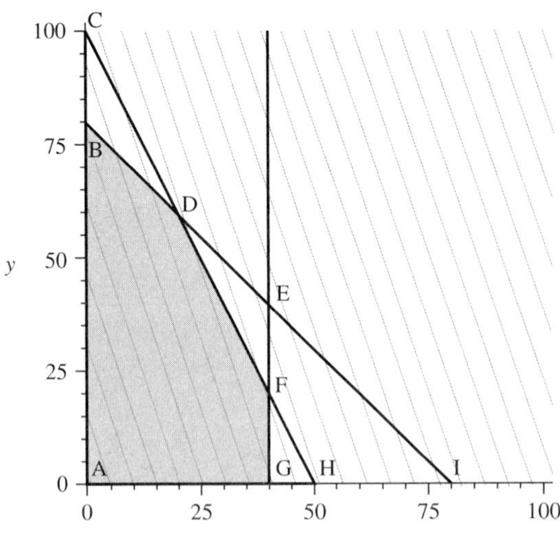


**FIGURE 10.17**

Contours of $3x + 2y$ with three constraints

**Table 10.1** Example 3 Summary

| Case | Condition Imposed | Point Number | Feasible | Optimal |
|---|---|---|---|---|
| I | $\lambda_1 = \lambda_2 = \lambda_3 = 0$ (all constraints have slack) | A | Yes | No |
| II | $\lambda_1 = 0, \lambda_2 \neq 0, \lambda_3 \neq 0$ (on constrain 2 and 3 , not on constraint 1) | E | No | No |
| III | $\lambda_2 = 0, \lambda_1 \neq 0, \lambda_3 \neq 0$ (on constraint 1 and 3, not on 2) | F | Yes | No |
| IV | $\lambda_3 = 0, \lambda_2 \neq 0, \lambda_1 \neq 0$ (not on 3, on 1 and 2) | D | Yes | Yes |
| V | $\lambda_1 = \lambda_2 = 0, \lambda_3 \neq 0$ (on constraint 3 , not on 1 or 2) | G | Yes | No |
| VI | $\lambda_1 = \lambda_3 = 0, \lambda_2 \neq 0$ (on 2, not on 1 or 3) | B<br>I | Yes<br>No | No<br>No |
| VII | $\lambda_2 = \lambda_3 = 0, \lambda_1 \neq 0$ (on 1, not on 2 or 3) | C or H | No | No |
| VIII | $\lambda_1 \neq 0, \lambda_2 \neq 0, \lambda_3 \neq 0$ (on all three constraints) | Does not exist | No | No |

## Example 4

## Geometric Three-Variable Nonlinear Constrained Problem

$$\text{Minimize } z = (x - 14)^2 + (y - 11)^2$$
$$\text{Subject to}$$
$$(x - 11)^2 + (y - 13)^2 \leq 49$$
$$x + y \leq 19 \tag{10.13}$$

We use Maple to generate the plots, contour plots, and constraints to obtain a geometrical interpretation shown in Figure 10.18. As shown, the optimal solution is the point where *the level curve of the objective function is tangent to the constraint*, $x + y = 19$ in the direction of increase for the contours. The solution merely satisfies the other constraint $(x - 11)^2 + (y - 13)^2 \leq 49$. This corresponds to the case where constraint 2, ($x + y \leq 19$), is binding, and constraint 1, $(x - 11)^2 + (y - 13)^2 \leq 49$, is not binding. This is written as $\lambda_2 \neq 0$ $(U_2^2 = 0)$ and $\lambda_1 = 0$ $(U_1^2 \neq 0)$. We can eyeball an estimate for the solution from the plot.

We use the fact that constraint 2 is binding and constraint 1 is merely satisfied to directly solve this case to find the optimal solution. Graphically, we can obtain a good approximation but cannot obtain the shadow prices that are invaluable in sensitivity analysis.

**FIGURE 10.18**
Contour plot of $(x - 14)^2 + (y - 11)^2$ with constraints

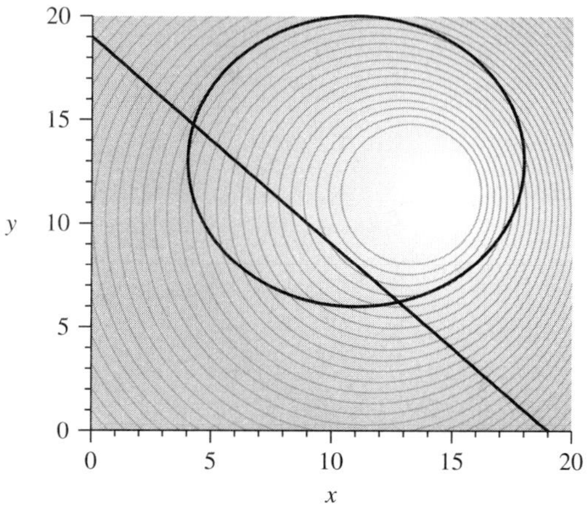

To obtain the Figure 10.18 we use the following commands:

```
> with(plots):
> f1:=contourplot(f,x=0..20,y=-10..20,contours=35):
> f2:=contourplot(g1,x=0..20,y=-10..20,contours=[0], thickness=3):
> f3:=contourplot(g2,x=0..20,y=-10..20,contours=[0], thickness=3):
> display({f1,f2,f3});
```

## Case: $\lambda_2 \neq 0$ $(U_2{}^2 = 0)$ and $\lambda_1 = 0$ $(U_1{}^2 \neq 0)$

Necessary conditions with this case applied are:

$$2x - 28 + \lambda_2 = 0 \tag{10.14}$$
$$2y - 22 + \lambda_2 = 0 \tag{10.15}$$
$$(x - 11)^2 + (y - 13)^2 + U_1{}^2 - 49 = 0 \tag{10.16}$$
$$x + y = 19 \tag{10.17}$$

```
> f:=(x-14)^2+(y-11)^2;
```

$$f := (x - 14)^2 + (y - 11)^2$$

```
> g1:=(x-11)^2+(y-13)^2-49;
```

$$g1 := (x - 11)^2 + (y - 13)^2 - 49$$

```
> g2:=(x+y-19);
```

$$g2 := x + y - 19$$

```
L:=f+l1*(-g1-u1^2)+l2*(-g2-u2^2);
```

$$L := (x - 14)^2 + (y - 11)^2 + l1\,(-(x - 11)^2 - (y - 13)^2 + 49 - u1^2)$$
$$+\, l2(-x - y + 19 - u2^2)$$

```
> Lx:=diff(L,x);
```

$$Lx := 2\,x - 28 + l1\,(-2\,x + 22) - l2$$

```
> Ly:=diff(L,y);
```

$$Ly := 2\,y - 22 + l1\,(-2\,y + 26) - l2$$

```
> Ll1:=diff(L,l1);
```

$$Ll1 := -(x - 11)^2 - (y - 13)^2 + 49 - u1^2$$

```
> Ll2:=diff(L,l2);
```

$$Ll2 := -x - y + 19 - u2^2$$

```
> Lu1:=u1*l1;
```

$$Lu1 := u1\,l1$$

```
> Lu2:=u2*l2;
```

$$Lu2 := u2\,l2$$

```
> solve({Lx=0,Ly=0,Ll1=0,Ll2=0,Lu1=0,Lu2=0}, {x,y,l1,l2,u1,u2});
```

$$\{y = \mathrm{RootOf}(\_Z^2 - 21\,\_Z + 92),\ x = -\mathrm{RootOf}(\_Z^2 - 21\,\_Z + 92) + 19,$$

$$l1 = -\frac{32}{73} + \frac{10}{73}\,\mathrm{RootOf}(\_Z^2 - 21\_Z + 92),\ l2 = \frac{50}{73}$$

$$\mathrm{RootOf}(\_Z^2 - 21\,\_Z + 92) - \frac{598}{73},$$

$$u1 = 0,\ u2 = 0\},\ \{l2 = 0,\ u2 = \mathrm{RootOf}(276 + 130\,\_Z^2 + 13\,\_Z^4),$$

$$y = 23 + 2\,\mathrm{RootOf}(276 + 130\,\_Z^2 + 13\,\_Z^4)^2,$$

$$l1 = \frac{114}{49} + \frac{13}{49}\,\mathrm{RootOf}(276 + 130\,\_Z^2 + 13\,\_Z^4)^2,$$

$$x = -4 - 3\,\mathrm{RootOf}(276 + 130\,\_Z^2 + 13\,\_Z^4)^2,\ u1 = 0\},$$

$$\{l2 = -6,\ u1 = 2\,\mathrm{RootOf}(-6 + \_Z^2,\ label = \_L5),\ l1 = 0,\ y = 8,\ u2 = 0,\ x = 11\},$$

$$\{l2 = 0,\ u2 = \mathrm{RootOf}(6 + \_Z^2,\ label = \_L4),\ l1 = 0,\ u1 = 6,\ y = 11,\ x = 14\},$$

$$\{l2 = 0,\ u2 = \mathrm{RootOf}(6 + \_Z^2,\ label = \_L4),\ l1 = 0,\ u1 = -6,\ y = 11,\ x = 14\}$$

```
> fsolve({Lx=0,Ly=0,Ll1=0,Ll2=0,Lu1=0,Lu2=0}, {x,y,l1,l2,u1, u2});
```

$$\{l1 = 0.,\ x = 11.00000000,\ u2 = 0.,\ u1 = -4.898979486,\ y = 8.000000000,\ l2 = -6.000000000\}$$

This is not the solution because the value of $u1$ is negative. We find these following values positive.

```
> solve(276+130*z1^2+13*z1^4,z1);
```

$$-\frac{1}{13}\sqrt{-845 + 91\sqrt{13}},\ \frac{1}{13}\sqrt{-845 + 91\sqrt{13}},\ -\frac{1}{13}\sqrt{-845 - 91\sqrt{13}},\ \frac{1}{13}\sqrt{-845 - 91\sqrt{13}}$$

```
> fsolve(Z^2-21*Z+92,Z);
```

$$6.227998127,\ 14.77200187$$

We extract from the Maple output the solution that satisfies all of our necessary and sufficient conditions is as follows:

$$x^* = 11,\ y^* = 8,\ \lambda_1 = 0,\ \lambda_2 = 6,\ U_2^2 = 0,\ \text{and } U_1^2 = 24$$

The value of the objective function is $f(x^*, y^*) = 18$. This is the optimal solution to Equation (13).

The interpretation of the shadow prices shows that if we have more resource for constraint 2, then our objective function will decrease. If we add $\Delta$ to the right-hand side of constraint 2, then the objective function value will decrease by approximately $6\Delta$. If we changed the right-hand side of the constraint from 19 to 20, the optimal solution becomes $x^* = 11.5,\ y^* = 8.5,\ f(x^*, y^*) = 12.5$ or a decrease of 5.5 units (the work is left to the student in the exercise set).

## We show only Method 2 for this example

```
> obj4:=(x-14)²+(y-11)²;
```

$$obj4 := (x-14)^2 + (y-11)^2$$

```
> cons14:=(x-11)²+(y-13)²+U1²-49;
```

$$cons\ 14 := (x-11)^2 + (y-13)^2 + U1^2 - 49;$$

```
> cons24:=x+y+U2²-19;
```

$$cons\ 24 := x + y + U2^2 - 19$$

```
> with(Student[MultivariateCalculus]) :
> LagrangeMultipliers(obj4,[cons14,cons24],[x,y,U1,U2],output=detailed);
```

$$[x = 11, y = 8, U1 = 2\,\mathrm{RootOf}(-6 + \_Z^2, \mathit{label} = \_L32),\ U2 = 0,$$
$$\lambda_1 = 0,\ \lambda_2 = -6,\ (x - 14)^2 + (y - 11)^2 = 18],\ [x = 14, y = 11,$$
$$U1 = 6, U2 = \mathrm{RootOf}(\_Z^2 + 6, \mathit{label} = \_L31),\ \lambda_1 = 0,\ \lambda_2 = 0,$$
$$(x-14)^2 + (y - 11)^2 = 0],\ [x = 14,\ y = 11,\ U1 = -6,$$
$$U2 = \mathrm{RootOf}(\_Z^2 + 6, \mathit{label} = \_L31),\ \lambda_1 = 0,\ \lambda_2 = 0,$$
$$(x-14)^2 + (y - 11)^2 = 0],\ x = -\mathrm{RootOf}(\_Z^2 + 92 - 21\,\_Z)$$
$$+ 19, y = \mathrm{RootOf}(\_Z^2 + 92 - 21\,\_Z),\ U1 = 0, U2 = 0,$$
$$\lambda_1 = -\frac{32}{73} + \frac{10}{73}\,\mathrm{RootOf}(\_Z^2 + 92 - 21\,\_Z),$$
$$\lambda_2 = \frac{50}{73}\,\mathrm{RootOf}(\_Z^2 + 92 - 21\,\_Z) - \frac{598}{73},$$
$$(x - 14)^2 + (y - 11)^2 = (-\mathrm{RootOf}(\_Z^2 + 92 - 21\,\_Z) + 5)^2$$
$$+ (\mathrm{RootOf}(\_Z^2 + 92 - 21\,\_Z) - 11)^2],\ [x = -3$$
$$\mathrm{RootOf}(13\,\_Z^4 + 130\_Z^2 + 276)^2 - 4,$$
$$y = 2\,\mathrm{RootOf}(13\_Z^4 + 130\_Z^2 + 276)^2 + 23,\ U1 = 0,$$
$$U2 = \mathrm{RootOf}(13\_Z^4 + 130\_Z^2 + 276),$$
$$\lambda_1 = \frac{13}{49}\,\mathrm{RootOf}(13\_Z^4 + 130\_Z^2 + 276)^2 + \frac{114}{49},\ \lambda_2 = 0$$
$$(x-14)^2 + (y-11)^2 = (-3\,\mathrm{RootOf}(13\_Z^4 + 130\_Z^2 + 276)^2 - 18)^2$$
$$+ (2\,\mathrm{RootOf}(13\_Z^4 + 130\_Z^2 + 276)^2 + 12)^2]$$

Now we must go through each possible solution and solve the RootOF equations for any solution that has a possibility of working. We can exclude all possible solutions whose solutions for any $U_i$ is negative.

We determine the solution that satisfies our conditions is:

$$x^* = 11,\ y^* = 8,\ \lambda_1 = 0,\ \lambda_2 = 6,\ U_2^2 = 0,\ \text{and } U_1^2 = 24$$

The value of the objective function is $f(x^*, y^*) = 18$, which is a minimum because $f(x)$ is convex, the $g(x)$ are convex, and the $\lambda_i$ are nonnegative.

---

### Necessary and Sufficient Conditions for Computational KTC

We have shown how the use of visual interpretation can reduce the amount of work required to solve the problem. By seeing the plot, you can interpret the conditions involved at the optimal point and then solve directly for that point. However, sometimes we cannot obtain the graphical interpretation, so we must rely on the computational method alone. When this occurs, we must solve all the cases and interpret the results. Let's illustrate with a few examples.

---

## Example 5   Revisiting Example 4

Let's revisit the previous NLP, Equation (10.13). The Lagrangian form of the NLP is:

$$L(x,y,\lambda,U) = (x - 14)^2 + (y - 11)^2 + \lambda_1[(x - 11)^2 + (y - 13)^2 + U_1^2 - 49]$$
$$+ \lambda_2[x + y + U_2^2 - 19] \tag{10.18}$$

The necessary conditions are:

$$2(x - 14) + 2\lambda_1(x - 11) + \lambda_2 = 0 \tag{10.19}$$
$$2(y - 11) + 2\lambda_1(y - 13) + \lambda_2 = 0 \tag{10.20}$$
$$(x - 11)^2 + (y - 13)^2 + U_1^2 - 49 = 0 \tag{10.21}$$
$$x + y + U_2^2 - 19 = 0 \tag{10.22}$$
$$2\lambda_1 U_1 = 0 \tag{10.23}$$
$$2\lambda_2 U_2 = 0 \tag{10.24}$$

## Case I: $\lambda_1 = \lambda_2 = 0$

This case finds $x = 14$ and $y = 11$. These values violate Equation (10.22) because $U_2^2 \neq -5$, indicating this solution is infeasible.

## Case II: $\lambda_1 = 0, \lambda_2 \neq 0$, implying that $U_2^2 = 0$ and $U_1^2 \neq 0$

This case yields the optimal solution as follows.

$$2(x - 14) + \lambda_2 = 0 \tag{10.19}$$
$$2(y - 11) + \lambda_2 = 0 \tag{10.20}$$
$$(x - 11)^2 + (y - 13)^2 + U_1^2 - 49 = 0 \tag{10.21}$$
$$x + y + -19 = 0 \tag{10.22}$$

Because $y = 19 - x$ from (10.22), we can substitute for $y$ into (10.20) and then solve equations (10.19) and (10.20) simultaneously for $x$ and $\lambda_2$. All the necessary conditions are satisfied at the point $x^* = 11$, $y^* = 8$, $\lambda_1 = 0$, $\lambda_2 = 6$, $U_2^2 = 0$ and $U_1^2 = 24$. The value of the function at this point is 18. It is left to the student to show that this point is an optimal point.

## Case III: $\lambda_2 = 0$, $\lambda_1 \neq 0$, Implying That $U_1^2 = 0$ and $U_2^2 \neq 0$

This case yields values for $y$ as either 9.117 or 16.8829, for $x$ as either 16.824 or 5.15 and for 11 as either $-0.484$ or $-1.514$ both violate the vale of 11 for our solution.

## Case IV: $\lambda_1 \neq 0$, $\lambda_2 \neq 0$, Implying That $U_2^2 = 0$ and $U_1^2 = 0$

Imposing case IV yields the following sets of equations:

$$2(x - 14) + 2\lambda_1(x - 11) + \lambda_2 = 0 \tag{10.19}$$
$$2(y - 11) + 2\lambda_1(y - 13) + \lambda_2 = 0 \tag{10.20}$$
$$(x - 11)^2 + (y - 13)^2 - 49 = 0 \tag{10.21}$$
$$x + y - 19 = 0 \tag{10.22}$$

Solving (10.21) and (10.22) for $x$ and $y$ yields two results. We use these results to solve equations (10.19) and (10.20) for the following results:

$$x = 12.772, \ y = 6.228, \ \lambda_1 = -0.5465, \ \lambda_2 = 0.3439$$

and

$$x = 4.23, \ y = 14.77, \ \lambda_1 = -1.49, \ \lambda_2 = -0.15314$$

The functional values are:

$$f(12.722, 6.228) = 24.279 \text{ and } f(4.23, 14.77) = 109.665$$

These are not optimal values because they do not satisfy the sufficient conditions for $\lambda_i$ in the case for a relative minimum. You should show that $f(4.23, 14.77)$ is a relative maximum. Why?

---

**Example 6**     ## Minimization with Two Inequality Constraints

Minimize $2x^2 - 8x - 6y + y^2$
Subject to
$x + y \leq 4$
$y \leq 2$
$L(x, y, \lambda_1, \lambda_2) = 2x^2 - 8x - 6y + y^2 + \lambda_1[x + y + U_1^2 - 4] + \lambda_2[y + U_2^2 - 2]$

# Necessary Conditions

$$L_x = 4x - 8 + \lambda_1 = 0 \tag{10.23}$$
$$L_y = -6 + 2y + \lambda_1 + \lambda_2 = 0 \tag{10.24}$$
$$L_{\lambda I} = x + y + U_I^2 - 4 = 0 \tag{10.25}$$
$$L_{\lambda 2} = y + U_2^2 - 2 = 0 \tag{10.26}$$

## Case I: $\lambda_1 = \lambda_2 = 0$

$x = 2$ from (10.23)
$y = 3$ from (10.24)
$U_1^2 = -1$ violates $U_1^2 \geq 0$ for (10.25).

## Case II: $\lambda_1 = 0, \lambda_2 \neq 0$

$x = 2$ from (10.23)
$y = 2$ from (10.26)
$U_1^2 = 0$
$\lambda_2 = 2$

All necessary conditions have been satisfied. Because the objective function is convex, the constraints are linear, and $\lambda_i$ for all binding constraints are positive ($\lambda_2 = 2$), the sufficient conditions are met.

Thus, the optimal solution must be among those solutions that satisfy all necessary conditions because all first partials are continuous. This solution is the only solution that satisfies all of the necessary conditions.

## Case III: $\lambda_1 \neq 0, \lambda_2 = 0$

$\lambda_1 = 4/3$ from (10.23), (10.24), and (10.25).
$y = 2.667$ from back substitution into (10.24).
$x = 1.667$ from back substitution into (10.23).
$U_2^2 = -0.667$, which violates the fact that $U_2^2 \geq 0$ to be a real solution in (10.26).

## Case IV: $\lambda_1 \neq 0, \lambda_2 \neq 0$

So $U_1^2 = 0$ and $U_2^2 = 0$. Case IV yields the same solution as case II.

---

**Modeling and Applications of KTC**

---

| **Example 7** | ## Maximizing Profit from Perfume Manufacturing |

A company manufactures perfumes and can purchase as much as 1,925 oz of the main chemical ingredient for \$10 per oz. At a cost of \$3 per oz, the chemical can be manufactured into an ounce of perfume 1; at a cost of \$5 per oz, the chemical can be manufactured into an ounce of the higher-priced perfume 2. An advertising firm estimates that if $x$ ounces of perfume 1 are manufactured, it will sell for $30 - 0.01x$ per ounce. If $y$ ounces of perfume 2 are produced, it can sell for $50 - 0.02y$ per ounce. The company wants to maximize its profits.

## Formulation

$x =$ ounces of perfume 1 produced
$y =$ ounces of perfume 2 produced
$z =$ ounces of main chemical purchased

$$\text{Max } f(x,y,z) = x(30 - 0.01x) + y(50 - 0.02y) - 3x - 5y - 10z$$

Subject to

$$x + y \leq z$$
$$z \leq 1925$$

## Solution

Set up the Lagrangian function, $L$.

$$L(x,y,z,\lambda_1,\lambda_2) = x(30 - 0.01x) + y(50 - 0.02y) - 3x - 5y - 10z + \lambda_1$$
$$[x + y - z + U_1^2] + \lambda_2[z + U_2^2 - 1925]$$

The necessary conditions are:

$$L_x = 30 - 0.02x - 3 + \lambda_1 = 0 \tag{10.27}$$
$$L_y = 50 - 0.04y - 5 + \lambda_1 = 0 \tag{10.28}$$
$$L_z = -10 - \lambda_1 + \lambda_2 = 0 \tag{10.29}$$
$$L_{\lambda 1} = x + y - z + U_1^2 = 0 \tag{10.30}$$
$$L_{\lambda 2} = z + U_2^2 - 1925 = 0 \tag{10.31}$$

We set this up and solve using Maple.

```
> L:=x·(30−.01·x)+y·(50−.02·y)−3·x−5·y−10·z+l1·(x+y−z+U1²)+l2·(z+U2²−1925);
```

$$L := x\,(30 - 0.01\,x) + y\,(50 - 0.02y) - 3x - 5y - 10z + l1\,(x + y - z + U1^2) + l2$$
$$(z + U2^2 - 1925)$$

```
> Lx:=diff(L,x);
```

$$Lx := 27. - 0.02x + l1$$

```
> Ly:=diff(L,y);
```

$$Ly := 45. - 0.04y + l1$$

```
> Lz:=diff(L,z);
```

$$Lz := -10 - l1 + l2$$

```
> Ll1:=diff(L,l1);
```

$$Ll1 := x + y - z + U1^2$$

```
> Ll2:=diff(L,l2);
```

$$Ll2 := z + U2^2 - 1925$$

```
> LU1:=diff(L,U1);
```

$$LU1 := 2\,l1\,U1$$

```
> LU2:=diff(L,U2);
```

$$LU2 := 2l2\,U2$$

```
> solve({Lx=0,Ly=0,Lz=0,Ll1=0,Ll2=0,LU1=0,LU2=0},{x,y,z,l1,l2,U1,U2});
```

$$\{l2 = 2.666666667, x = 983.3333333, l1 = -7.333333333,$$
$$y = 941.6666667, U1 = 0., z = 1925., U2 = 0.\}, \{l1 = 0.,$$
$$l2 = 10., x = 1350., y = 1125., U1 = 23.45207880I, z\ 1925.,$$
$$U2 = 0.\}, \{l1 = 0., l2 = 10., x = 1350., y = 1125.,$$
$$U1 = -23.45207880I, z = 1925., U2 = 0.\}, \{U2 = 14.14213562,$$
$$U1 = 0., l2 = 0., z = 1725., l1 = -10., x = 850., y = 875.\},$$

$$\{U1 = 0., U2 = -14.14213562, l2 = 0., z = 1725.,$$
$$l1 = -10., x = 850., y = 875.\}$$

We disregard all solutions where either $U_1 < 0$ or $U_2 < 0$ because this violates $U_1$ and $U_2$ having to be real values.

We use the four cases as before and put the results in a table such as the following.

| Cases | $x$ | $y$ | $z$ | $\lambda_1$ | $\lambda_2$ | $U_1$ | $U_2$ | Remarks |
|---|---|---|---|---|---|---|---|---|
| I | 1,350 | 1,125 | 1,925 | 0 | 10 | 23.45I | 0 | $U_1$ not real |
| II | 850 | 875 | 1,725 | –10 | 0 | 0 | 14.1421 | Candidate solution |
| III | 1,350 | 1,125 | 1,925 | 0 | 10 | –23.45I | 0 | $U_1$ not real |
| IV | 983.33 | 941.67 | 1,925 | –7.33 | 2.667 | 0 | 0 | (10.27) and (10.28) led to two different signs of $\lambda_i \rightarrow$ violation. |

We have a concave function and linear constraints, and $\lambda_i\,(\lambda_1 = -10)$ is negative for all binding constraints. Thus, we have met the sufficient conditions for the point (850, 875) to be the optimal solution. The optimal manufacturing strategy is to purchase 1,725 ounces of the chemical and produce 850 ounces of perfume 1 and 875 ounces of perfume 2.

> *subs({U1=0.,U2=−14.14213562,l2=0.,z=1725.,l1=−10.,x=850.,y=875.},L);*

$$22537.50$$

This yields a profit of \$22,537.50.

Consider the significance of the shadow price for $\lambda_1$. How do we interpret the shadow price in terms of this scenario? If we could obtain an extra ounce ($\Delta = 1$) of chemical at no cost, it would improve the profit to about \$22,537.50 + \$10.

---

<table>
<tr><td>**Example 8**</td><td>

## Minimum Variance of Expected Investment Returns

</td></tr>
</table>

A new company has \$5,000 to invest, but it needs to earn about 12 percent interest. A stock expert has suggested three mutual funds (A, B, and C) in which the company could invest. Based on the previous year's returns, these funds appear relatively stable. The expected return, variance on the return, and covariance between funds are shown as follows.

| Expected Value | A | B | C |
|---|---|---|---|
| | 0.14 | 0.11 | 0.10 |
| Variance | A | B | C |
| | 0.2 | 0.08 | 0.18 |
| Covariance | AB | AC | BC |
| | 0.05 | 0.02 | 0.03 |

### Formulation

We use laws of expected value, variance, and covariance in our model. Let $x_j$ be the number of dollars invested in funds $j$ ( $j = 1,2,3$).

$$\begin{aligned}
\text{Minimize } V_I &= \text{var}(Ax_1 + Bx_2 + Cx_3) = x_1^2\,\text{var}(A) + x_2^2\,\text{var}(B) + x_3^2\,\text{var}(C)\\
&\quad + 2x_1x_2\,\text{cov}(AB) + 2x_1x_3\,\text{cov}(AC) + 2x_2x_3\,\text{cov}(BC)\\
&= 0.2x_1^2 + 0.08x_2^2 + 0.18x_3^2 + 0.10x_1x_2 + 0.04x_1x_3 + 0.06\,x_2x_3
\end{aligned}$$

Our constraints include the following:

1. We expect to achieve at least the expected return of 12 percent from the sum of all the expected returns:

$$0.14x_1 + 0.11x_2 + 0.10x_3 \geq (0.12 \times 5{,}000) \text{ or}$$
$$0.14x_1 + 0.11x_2 + 0.10x_3 \geq 600$$

2. The sum of all investments must not exceed the \$5,000 capital:

$$x_1 + x_2 + x_3 \leq \$5{,}000$$

## Solution:

We set up the Lagrangian function, $L$:

$$\text{Min } L = 0.2x_1^2 + 0.08x_2^2 + 0.18x_3^2 + 0.10x_1x_2 + 0.04x_1x_3 + 0.06x_2x_3 + \lambda_1$$
$$[0.14x_1 + 0.11x_2 + 0.10x_3 - U_1^2 - 600] + \lambda_2[x_1 + x_2 + x_3 + U_2^2 - \$5{,}000]$$

## Necessary Conditions

$$L_{x_1} = 0.4x_1 + 0.10x_2 + 0.04x_3 + 0.14\lambda_1 + \lambda_2 = 0 \tag{10.32}$$
$$L_{x_2} = 0.16x_2 + 0.10x_1 + 0.06x_3 + 0.11\lambda_1 + \lambda_2 = 0 \tag{10.33}$$
$$L_{x_3} = 0.36x_3 + 0.04x_1 + 0.06x_2 + 0.10\lambda_1 + \lambda_2 = 0 \tag{10.34}$$
$$L_{\lambda 1} = 0.14x_1 + 0.11x_2 + 0.10x_3 - U_1^2 - 600 = 0 \tag{10.35}$$
$$L_{\lambda 2} = x_1 + x_2 + x_3 + U_2^2 - \$5{,}000 = 0 \tag{10.36}$$

There are only two constraints, so we need to consider four cases.

The solution is found in the case where both $\lambda_1$ and $\lambda_2$ do not equal zero. This is the case where the solution lies at the intersection of the constraints. Both constraints are binding.

```
> obj7:=0.2*x1^2+.08*x2^2+.18*x3^2+.1*x1·x2+.04*x1*x3+.06*x2*x3;
```

$$obj7 := 0.2x1^2 + 0.08\ x2^2 + 0.18\ x3^2 + 0.1\ x1\ x2 + 0.04x1\ x3 + 0.06\ x2\ x3$$

```
> cons17:=.14·x1+.11·x2+.1·x3−U1²−600;
```

$$cons17 := 0.14x1 + 0.11\ x2 + 0.1\ x3 - U1^2 - 600$$

```
> cons27:=x1+x2+x3+U2²−5000;
```

$$cons27 := x1 + x2 + x3 + U2^2 - 5000$$

```
> with(Student[MultivariateCalculus]):
> LagrangeMultipliers(obj7,[cons17,cons27],[x1,x2,x3,U1,U],output=detailed);
```

$$[x1 = 1904.761905, x2 = 2380.952381, x3 = 714.2857143, U1 = 0.,$$
$$U2 = 0., \lambda_1 = 13809.52381, \lambda_2 = -904.7619048, 0.2\ x1^2 + 0.08\ x2^2$$
$$+ 0.18\ x3^2 + 0.1\ x1\ x2 + 0.04x1\ x3 + 0.06x2x3 = 1.880952381\ 10^6],$$
$$[x1 = 702.2471910, x2 = 3117.977528, x3 = 1179.775281,$$
$$U1 = 6.382032363\ I, U2 = 0., \lambda_1 = 0, \lambda_2 = 639.8876404, 0.2\ x1^2$$
$$+ 0.08\ x2^2 + 0.18\ x3^2 + 0.1\ x1\ x2 + 0.04x1\ x3 + 0.06x2x3 = 1.599719101\ 10^6],$$
$$[x1 = 702.2471910, x2 = 3117.977528, x3 = 1179.775281,$$
$$U1 = -6.382032363\ I, U2 = 0., \lambda_1 = 0., \lambda_2 = 639.8876404,$$
$$0.2\ x1^2 + 0.08\ x2^2 + 0.18\ x3^2 + 0.1\ x1\ x2 + 0.04x1\ x3$$
$$+ 0.06x2\ x3 = 1.599719101\ 10^6], [x1 = 1250.108970,$$
$$x2 = 2929.125621, x3 = 1027.809258, U1 = 0., U2 = 14.38901837\ I,$$
$$\lambda_1 = 5957.632290, \lambda_2 = 0., 0.2\ x1^2 + 0.08\ x2^2 + 0.18\ x3^2$$
$$+ 0.1\ x1\ x2 + 0.04x1\ x3 + 0.06x2\ x3 = 1.787289686\ 10^6],$$
$$[x1 = 1250.108970, x2 = 2929.125621, x3 = 1027.809258,$$
$$U1 = 0., U2 = -14.38901837\ I, \lambda_1 = 5957.632290, \lambda_2 = 0.,$$
$$0.2\ x1^2 + 0.08\ x2^2 + 0.18\ x3^2 + 0.1\ x1\ x2 + 0.04x1\ x3$$

$$+\ 0.06\ x2\ x3 = 1.787289686\ 10^6],\ [x1 = 0.,\ x2 = 0.,\ x3 = 0.,$$
$$U1 = 24.49489743\ I,\ U2 = 70.71067812,\ \lambda_1 = 0.,\ \lambda_2 = 0.,$$
$$0.2\ x1^2 + 0.08x2^2 + 0.18\ x3^2 + 0.1\ x1\ x2 + 0.04x1\ x3$$
$$+\ 0.06\ x2\ x3 = 0.],\ [x1 = 0.,\ x2 = 0.,\ x3 = 0.,$$
$$U1 = -24.49489743\ I,\ U2 = 70.71067812,\ \lambda_1 = 0.,\ \lambda_2 = 0.,$$
$$0.2\ x1^2 + 0.08\ x2^2 + 0.18\ x3^2 + 0.1\ xl\ x2 + 0.04x1\ x3$$
$$+\ 0.06\ x2\ x3 = 0.],\ [x1 = 0.,\ x2 = 0.,\ x3 = 0.,$$
$$U1 = 24.49489743\ I,\ U2 = -70.71067812,\ \lambda_1 = 0.,\ \lambda_2 = 0.,$$
$$0.2\ x1^2 + 0.08\ x2^2 + 0.18\ x3^2 + 0.1\ x1\ x2 + 0.04x1\ x3$$
$$+\ 0.06\ x2\ x3 = 0.],\ [x1 = 0.,\ x2 = 0.,\ x3 = 0.,$$
$$U1 = -24.49489743\ I,\ U2 = -70.71067812,\ \lambda_1 = 0.,\ \lambda_2 = 0.,$$
$$0.2\ x1^2 + 0.08\ x2^2 + 0.18\ x3^2 + 0.1\ x1\ x2 + 0.04x1\ x3$$
$$+\ 0.06\ x2\ x3 = 0.]$$

$$> h := Hessian(obj7, [x1, x2, x3])$$

$$h := \begin{bmatrix} 0.4 & 0.1 & 0.04 \\ 0.1 & 0.16 & 0.06 \\ 0.04 & 0.06 & 0.36 \end{bmatrix}$$

We solve the following system of equations equal to the right-hand side matrix [0 0 0 600 5000].

| $x_1$ | $x_2$ | $x_3$ | $\lambda_1$ | $\lambda_2$ |
|---|---|---|---|---|
| 0.4 | 0.10 | 0.04 | 0.14 | 1 |
| 0.1 | 0.16 | 0.06 | 0.11 | 1 |
| 0.04 | 0.06 | 0.36 | 0.10 | 1 |
| 0.14 | 0.11 | 0.10 | 0 | 0 |
| 1 | 1 | 1 | 0 | 0 |

The solution is

$$x_1 = 1{,}904.80$$
$$x_2 = 2{,}381.00$$
$$x_3 = 714.20$$
$$\lambda_1 = -1{,}3809.50$$
$$\lambda_2 = 904.80$$
$$z = \$1{,}880{,}942.29 \text{ or a standard deviation of } \$1{,}371.50.$$

The expected return is 12 percent as found by

$$0.14(1{,}904.8) + 0.11(2381) + 0.1(714.2)/5{,}000$$

This solution is optimal. The Hessian matrix, $H$, has all positive leading principal minors. Therefore, because $H$ is always positive definite, our solution is the optimal minimum.

$$H = \begin{bmatrix} .4 & .1 & .04 \\ .1 & .16 & .06 \\ .04 & .06 & .36 \end{bmatrix}$$

## 10.2   EXERCISES

1. Use the KTC conditions to find the optimal solution to the following nonlinear problem:

$$\text{Maximize } f(x,y) = -x^2 - y^2 + xy + 7x + 4y$$
$$\text{subject to: } 2x + 3y \le 24$$
$$-5x + 12y \ge 20$$

2. Use the KTC conditions to find the optimal solution to the following nonlinear problem:

$$\text{Maximize } f(x,y) = -x^2 - y^2 + xy + 7x + 4y$$
$$\text{subject to: } 2x + 3y \geq 16$$
$$-5x + 12y \leq 20$$

3. Use the KTC conditions to find the optimal solution to the following nonlinear problem:

$$\text{Minimize } f(x,y) = 2x + xy + 3y$$
$$\text{subject to: } x^2 + y \leq 3$$
$$2.5 - 0.5x - y \leq 0$$

4. Use the KTC conditions to solve

$$\text{Minimize } f(x,y) = 2x + xy + 3y$$
$$\text{subject to: } x^2 + y \geq 3$$
$$x + 0.5 \geq 0$$
$$y \geq 0$$

5. Solve the following:

$$\text{Max } -(x - 0.5)^2 - (y - 5)^2$$
$$\text{Subject to } -x + 2y \leq 4$$
$$x^2 + y^2 \leq 14$$
$$0 \leq x, 0 \leq y$$

6. Minimize $x^2 + y^2$

$$\text{Subject to}$$
$$2x + y \leq 100$$
$$x + y \leq 80$$

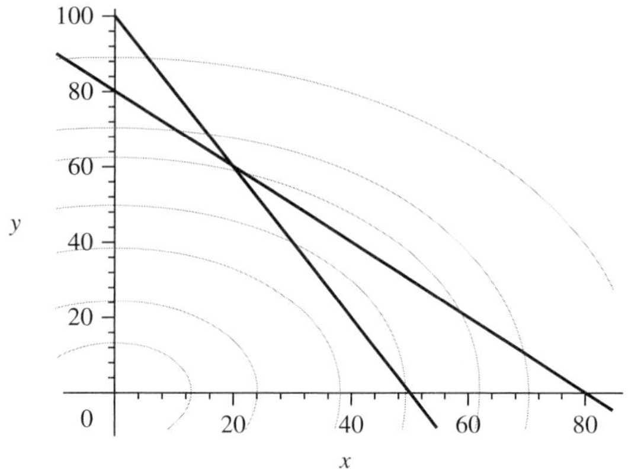

7. Maximize: $-(x - 4)^2 + xy - (y - 4)^2$

$$\text{Subject to}$$
$$2x + 3y \leq 18$$
$$2x + y \leq 8$$

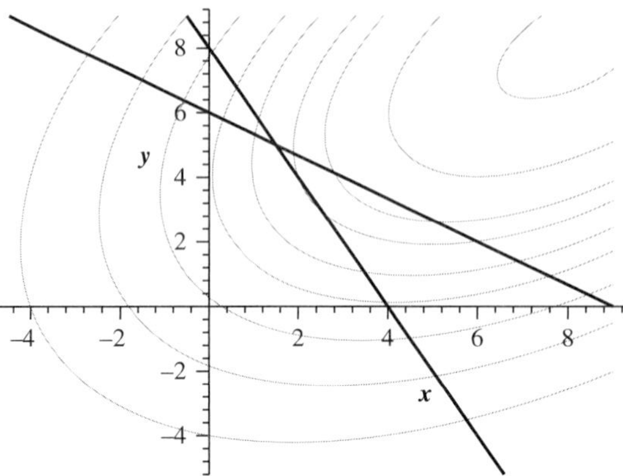

## 10.2 | PROJECT

## Manufacturing

### Part I

The manufacturer of a new plant is planning the introduction of two new products, a 19-inch stereo color set with a manufacturer's suggested retail price (MSRP) of $339 and a 21-inch stereo color set with a MSRP of $399. The cost to the company is $195 per 19-inch set and $225 per 21-inch set, plus an additional $400,000 in fixed costs of initial parts, initial labor, and machinery. In a competitive market in which the company desires to sell the sets, the number of sales per year will affect the average selling price. For each type of set, it is estimated that the average selling price drops by one cent for each additional unit sold. Furthermore, sales of 19-inch sets will affect the sales of 21-inch sets and vice versa. It is estimated that the average selling price for the 19-inch set will be reduced by an additional 0.3 cents for each 21-inch set sold, and the price for the 21-inch set will decrease by 0.4 cents for each 19-inch set sold. We want to find the optimal number of units of each type set to produce and to determine the expected profits. Recall that profit is revenue minus cost:

$$P = R - C$$

### Part II

1. Formulate the model to maximize profits. Be sure you have accounted for all revenues and costs. Define all of your variables.

2. Solve for the optimal levels of 21-inch and 19-inch sets to be manufactured using
   **a.** classical optimization—calculus,
   **b.** a combination of 3-D surface plot and contour plot (an estimated answer is all right here), and
   **c.** the Newton–Raphson method (briefly explain why you can use this technique).

3. Using Maple, obtain a contour plot of the function. Identify the optimal point with color or in some other manner. Illustrate the gradient search technique, starting from the initial point (0,0). Perform only two or three iterations, showing the gradient, the distance traveled, and the new point. There is no requirement to obtain the optimal solution using this method.

### Part III

Previously, we assumed that the company has the potential to produce any number of TV sets per year. Now we realize that there is a limit on production capacity. Consideration of these two products came about because the company planned to discontinue manufacturing its black-and-white sets, thus providing excess capacity at its assembly plants. This excess capacity could be used to increase production of other existing product lines, but the company felt that these new products would be more profitable. It is estimated that the available production capacity is sufficient to produce 10,000 sets per year (about 200 per week). The company has ample supply of 19-inch and 21-inch color tubes, chassis, and other standard components; however, circuit assemblies are in short supply. Also, the 19-inch TV requires different circuit assemblies than the 21-inch TV. The supplier can deliver 8,000 boards per year for the 21-inch model and 5,000 boards per year for the 19-inch model. Taking this new information into account, what should the company now do?

**Required**

1. Solve the above as a LaGrange multiplier problem with equality (=) constraints. Consider only the 10,000 sets per year as the constraint. Interpret the shadow prices.

2. Assuming the above constraints are each inequalities ($\leq$), solve the constraints using KTC. Consider all constraints above. Interpret the shadow prices.

3. Compare this solution to your solution to the constraints as equality in the previous chapter.

## 10.2 | FURTHER READING

Bazarra, M. S., and C. M. Shetty, *Nonlinear Programming Theory and Algorithms*. New York. John Wiley & Sons, 1993.

Ecker, J., and M. Kupperschmid, *Introduction to Operations Research*. John Wiley and Sons, 1988.

Hiller, F. S, and G. J. Liberman, *Introduction to Mathematical Programming*. McGraw Hill Publishing Co., 1990.

Winston, W L., *Introduction to Mathematical Programming Applications and Algorithms*, 4th Edition. Belmont, CA: Duxbury Press, 2002.

# 11

# Modeling with Linear Systems of Equations Using Linear Algebra Techniques

## Introduction

Trusses are lightweight structures capable of carrying heavy loads. In civil engineering bridge design, individual members of the truss are connected with rotatable pin joints that permit forces to be transferred from one member of the truss to another. Figure 11.1 shows a truss that is held stationary at the lower left end point 1 is permitted to move horizontally at the lower right end point 4 and has pin joints at 1, 2, 3, and 4 as shown. A load of 10 kilonewtons (kN) is placed at joint 3, and the forces on the members of the truss have magnitudes given by $f_1, f_2, f_3, f_4,$ and $f_5$ as shown. The stationary support member has both a horizontal force $F_1$ and a vertical force $F_2$, but the movable support member has only the vertical force $F_3$.

If the truss is in static equilibrium, then the forces at each joint must add to the zero vector, so the sum of the horizontal and vertical components of the forces at each joint must be zero. We can model this producing a system of linear equations. We will build and solve this model later in this chapter.

## 11.1 INTRODUCTION TO SYSTEMS OF EQUATIONS

In this chapter, we illustrate the use of systems of equations to solve real-world applications from the sciences, engineering, and the economic sciences. Previously, we have seen examples of this in systems of discrete dynamical systems and in model fitting with least squares, to name a few examples. Maple has a linear algebra package—With(LinearAlgebra)—and we illustrate several features from this package that are useful in solving these problems.

**FIGURE 11.1**
Bridge truss

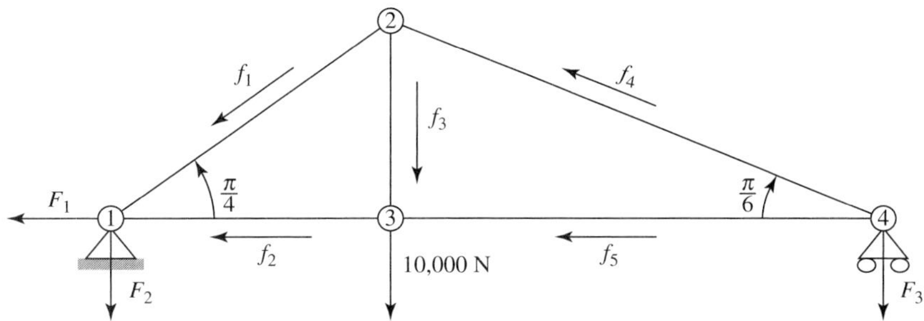


Linear systems of equations fall into three cases: unique solutions, infinite solutions, and no solutions. Consider the following system of equations:

$$a_1 x + b_1 y = c_1$$
$$a_2 x + b_2 y = c_2$$

We will use the augmented matrix form:

$$\begin{bmatrix} a_1 & b_1 & c_1 \\ a_2 & b_2 & c_2 \end{bmatrix}$$

If we row reduce this augmented matrix, we will find one of following three results:

**1.** Unique Solution Result

$$\begin{bmatrix} 1 & 0 & d \\ 0 & 1 & e \end{bmatrix}$$

where $d$ and $e$ are real numbers.

**2.** Infinite Solution

$$\begin{bmatrix} 1 & 0 & f \\ 0 & 0 & 0 \end{bmatrix} \quad or \quad \begin{bmatrix} 0 & 0 & 0 \\ 0 & 1 & g \end{bmatrix} \quad or \quad \begin{bmatrix} 0 & 0 & 0 \\ 0 & 0 & 0 \end{bmatrix}$$

where $f$ and $g$ are real numbers for one of the two variables while the other variable takes on infinitely many different values. It is possible for all variables to take on infinitely many values.

**3.** No Solution

$$\begin{bmatrix} 1 & 0 & h \\ 0 & 0 & 1 \end{bmatrix}$$

Because it is impossible for $0x + 0y = 1$, there are no solutions.

In two-dimensional space these are represented visually by Figures 11.2, 11.3, and 11.4.

We will rely on the matrix forms to recognize our solutions in this chapter.

The LinearAlgebra package is an efficient and robust suite of routines for doing computational linear algebra. Full rectangular and sparse matrices are fully supported at the data structure level, as well as upper and lower triangular matrices, unit triangular matrices, banded matrices, and a variety of others. Further, *symmetric, skew-symmetric, hermitian,* and *skew-hermitian* are known qualifiers that are used appropriately to reduce storage and select among algorithms.

The LinearAlgebra package has been designed to accommodate different sets of usage scenarios: casual use and programming use. Correspondingly, there are functions and notations designed for easy casual use (sometimes at the cost of some efficiency) and some functions designed for maximal efficiency (sometimes at the cost of ease of use). In this way, the LinearAlgebra facilities scale easily from first-year classroom use to heavy industrial usage, emphasizing the different qualities that each type of use needs.

**FIGURE 11.2**
Intersecting lines leads to a
unique solution

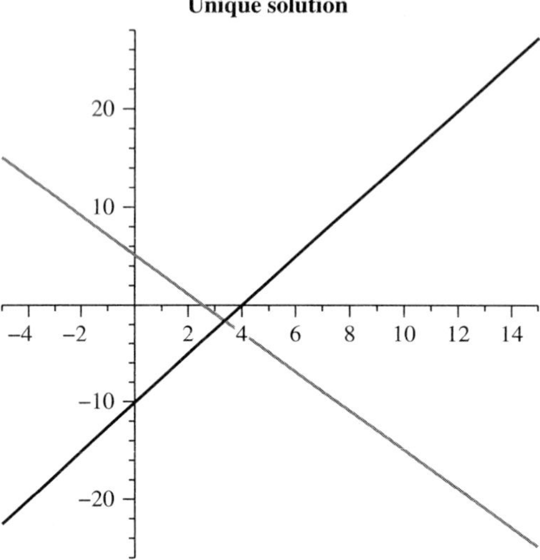


**FIGURE 11.3**
Lines $y = 5 - 2x$ and
$2y + 4x = 10$ are the same
line and provide infinite
solutions.

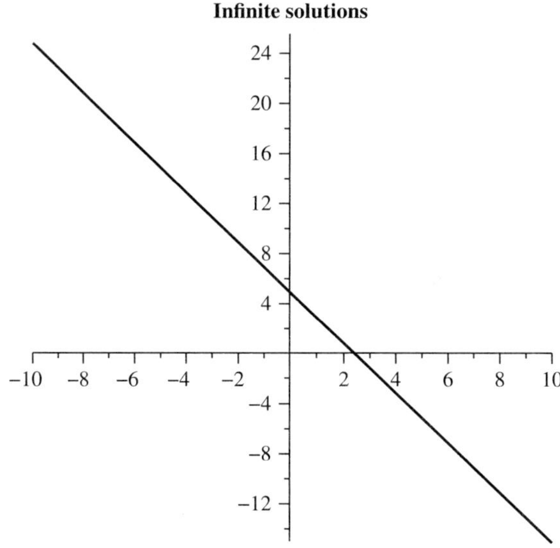


**FIGURE 11.4**
The lines $y = 5 - 2x$ and
$y = 10 - 2x$ are parallel and
do not intersect, so there
is no solution common to
both lines.

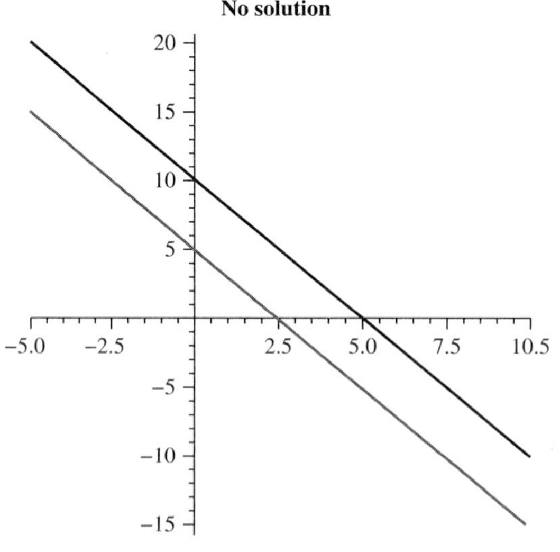

We will illustrate the following commands in this chapter:

- LinearAlgebra[GaussianElimination] performs Gaussian elimination on a matrix.
- LinearAlgebra[ReducedRowEchelonForm] performs Gauss−Jordan elimination on a matrix.
- LinearAlgebra[MatrixInverse] computes the inverse of a square matrix or the Moore−Penrose pseudo-inverse of a nonsquare matrix.

We begin with a simple $3 \times 3$ system of linear equations.

## Example 1

## Solving a System of Equations

Given the following systems of equations, determine the solution.

$$2x + 4y + 4z = 4$$
$$x + 3y + z = 4$$
$$-x + 3y + 2z = -1$$

We will open the linear algebra package, enter the matrix in augmented form, solve using RREF (Gaussian elimination), and interpret the results.

> *with(LinearAlgebra):*

We enter the system of equations in augmented matrix form.

> *A:=<<2,1,−1>|<4,3,3>|<4,1,2>|<4,4,−1>>;*

$$A := \begin{bmatrix} 2 & 4 & 4 & 4 \\ 1 & 3 & 1 & 4 \\ -1 & 3 & 2 & -1 \end{bmatrix}$$

We row reduce using Maple elimination method.

> *ReducedRowEchelonForm(A);*

$$\begin{bmatrix} 1 & 0 & 0 & 2 \\ 0 & 1 & 0 & 1 \\ 0 & 0 & 1 & -1 \end{bmatrix}$$

We interpret the results as a unique solution: $x = 2$, $y = 1$, and $z = -1$.

## 11.1 | EXERCISES

Use Maple to solve the following system of equations and interpret the solution.

1. $x - 5y = -154$
   $x - 3y = -84$

2. $11x - 6y = 494$
   $x + 7y = -23$

3. $9x + y = 56$
   $6x - 5y = 128$

4. $6x + y = 50$
   $18x + 3y = 150$

**5.** $x + 2y + 3z = 5$
$x - y + 6z = 1$
$3x - 2y = 4$

**6.** $x + y + 3z - 4t = 12$
$3x + y - 2z - t = 0$

**7.** $2x + 3y + 4z = 5$
$x - y + 2z = 6$
$3x - 5y - z = 0$

**8.** $2x - 3y + 2z = 21$
$x + 4y - z = 1$
$-x + 2y + z = 17$

**9.** $3x - 4y + 5z - 4t = 12$
$3x - 4y + 3z - 6t = 0$
$2x + y + 2z + 3t = 52$
$2x - 3y + 2z - t = 4$

## 11.1   PROJECT

Model a solution methodology for $Ax = b$ using linear algebra and illustrate the methodology using Maple commands.

## 11.2   MODELS WITH UNIQUE SOLUTIONS USING SYSTEMS OF EQUATIONS

### Example 1   A Bridge Too Far

Trusses are lightweight structures capable of carrying heavy loads. In civil engineering bridge design, the individual members of the truss are connected with rotatable pin joints that permit forces to be transferred from one member of the truss to another. Figure 11.5 shows a truss that is held stationary at the lower-left end point 1; it is permitted to move horizontally at the lower-right end point 4 and has pin joints at 1, 2, 3, and 4 as shown. A load of 10 kilonewtons (kN) is placed at joint 3, and the forces on the members of the truss have magnitudes given by $f_1, f_2, f_3, f_4$, and $f_5$ as shown in Figure 11.5. The stationary support member has both a horizontal force, $F_1$, and a vertical force, $F_2$, but the movable support member has only the vertical force $F_3$.

If the truss is in static equilibrium, then the forces at each joint must add to the zero vector, so the sum of the horizontal and vertical components of the forces at each joint must be zero. This produces the system of linear equations shown in Table 11.1.

According to Newton's law for equilibrium of forces, $\Sigma F = 0$. We will examine each joint and calculate both the horizontal and vertical components of the applied forces. Let's begin with joint 1. We find that $f_1$ and $f_2$ are forces pushing into joint 1, and $F_1$ and $F_2$ are forces pulling away from joint 1. Force $f_1$ is acting from an angle of $\pi/4$ radians. It has two components: a horizontal ($x$) component and a vertical ($y$) component. Thus, the forces acting on joint 1 are in equilibrium when both the horizontal and vertical components are zero, or

**FIGURE 11.5**
Bridge truss

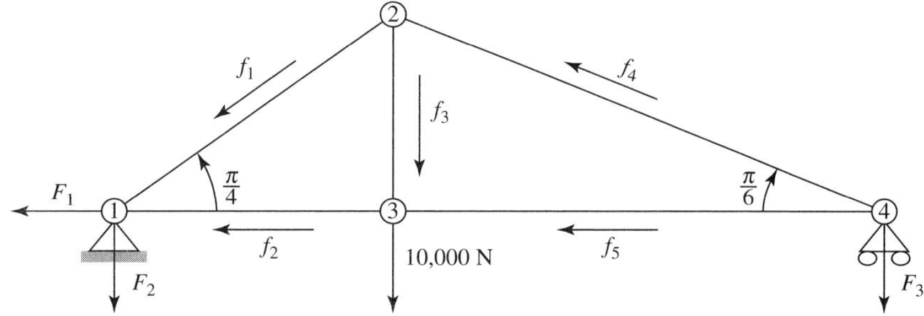


$$-F_1 + \frac{\sqrt{2}}{2}f_1 + f_2 = 0$$

and

$$\frac{\sqrt{2}}{2}f_1 - F_2 = 0$$

We completely model the forces at each joint and place the force vector equations from Table 11.1. You should verify that these are correct.

We will write this model as a system of equations in matrix notation, $Ax = b$, with eight equations and eight unknowns:

$$\begin{bmatrix} -1 & 0 & 0 & \frac{\sqrt{2}}{2} & 1 & 0 & 0 & 0 \\ 0 & -1 & 0 & \frac{\sqrt{2}}{2} & 0 & 0 & 0 & 0 \\ 0 & 0 & 0 & \frac{-\sqrt{2}}{2} & 0 & 0 & \frac{\sqrt{3}}{2} & 0 \\ 0 & 0 & 0 & \frac{-\sqrt{2}}{2} & 0 & -1 & .5 & 0 \\ 0 & 0 & 0 & 0 & -1 & 0 & 0 & 1 \\ 0 & 0 & 0 & 0 & 0 & 1 & 0 & 0 \\ 0 & 0 & 0 & 0 & 0 & 0 & \frac{-\sqrt{3}}{2} & -1 \\ 0 & 0 & -1 & 0 & 0 & 0 & 0.5 & 0 \end{bmatrix} \begin{bmatrix} F_1 \\ F_2 \\ F_3 \\ f_1 \\ f_2 \\ f_3 \\ f_4 \\ f_5 \end{bmatrix} = \begin{bmatrix} 0 \\ 0 \\ 0 \\ 0 \\ 0 \\ 10000 \\ 0 \\ 0 \end{bmatrix}$$

We use Maple to enter the augmented matrix, solve using RREF, and interpret the results.

**Table 11.1**

| Joint | Horizontal Component | Vertical Component |
|---|---|---|
| 1 | $-F_1 + \frac{\sqrt{2}}{2}f_1 + f_2 = 0$ | $\frac{\sqrt{2}}{2}f_1 - F_2 = 0$ |
| 2 | $-\frac{\sqrt{2}}{2}f_1 + \frac{\sqrt{3}}{2}f_4 = 0$ | $-\frac{\sqrt{2}}{2}f_1 - f_3 + \frac{1}{2}f_4 = 0$ |
| 3 | $-f_2 + f_5 = 0$ | $f_3 - 10{,}000 = 0$ |
| 4 | $-\frac{\sqrt{3}}{2}f_4 - f_5 = 0$ | $-\frac{1}{2}f_4 - F_3 = 0$ |

> *with( LinearAlgebra ):*
> *B:=<<−1,0,0,0,0,0,0,0>|<0,−1,0,0,0,0,0,0>|<0,0,−1,0,0,0,0,0>|*
> *< sqrt(2)/2,sqrt(2)/2,0,−sqrt(2)/2,0,0,−sqrt(2)/2,0>|<1,0,0,0,−1,0,0,0>|*
> *< 0,0,0,−1,0,1,0,0>|<0,0,.5,.5,0,0,sqrt(3)/2,−sqrt(3)/2>|*
> *< 0,0,0,0,1,0,0,−1>|<0,0,0,0,0,10000,0,0>>;*

$$
B := \begin{bmatrix}
-1 & 0 & 0 & \frac{\sqrt{2}}{2} & 1 & 0 & 0 & 0 & 0 \\
0 & -1 & 0 & \frac{\sqrt{2}}{2} & 0 & 0 & 0 & 0 & 0 \\
0 & 0 & -1 & 0 & 0 & 0 & 0.5 & 0 & 0 \\
0 & 0 & 0 & -\frac{\sqrt{2}}{2} & 0 & -1 & 0.5 & 0 & 0 \\
0 & 0 & 0 & 0 & -1 & 0 & 0 & 1 & 0 \\
0 & 0 & 0 & 0 & 0 & 1 & 0 & 0 & 10000 \\
0 & 0 & 0 & -\frac{\sqrt{2}}{2} & 0 & 0 & \frac{\sqrt{3}}{2} & 0 & 0 \\
0 & 0 & 0 & 0 & 0 & 0 & -\frac{\sqrt{3}}{2} & -1 & 0
\end{bmatrix}
$$

> *C:=ReducedRowEchelonForm( B );*

$$
C := \begin{bmatrix}
1 & 0 & 0 & 0 & 0 & 0 & -0. & -0. & -0. \\
0 & 1 & 0 & 0 & 0 & 0 & -0. & -0. & -\dfrac{5000.000000\,\sqrt{3}}{\frac{\sqrt{3}}{2}-0.5} \\
0 & 0 & 1 & 0 & 0 & 0 & -0. & -0. & -\dfrac{5000.0}{\frac{\sqrt{3}}{2}-0.5} \\
0 & 0 & 0 & 1 & 0 & 0 & -0. & -0. & -\dfrac{5000.000000\,\sqrt{3}\,\sqrt{2}}{\frac{\sqrt{3}}{2}-0.5} \\
0 & 0 & 0 & 0 & 1 & 0 & -0. & -0. & -\dfrac{5000\,\sqrt{3}}{\frac{\sqrt{3}}{2}-0.5} \\
0 & 0 & 0 & 0 & 0 & 1 & 0. & 0. & 10000. \\
0 & 0 & 0 & 0 & 0 & 0 & 1 & 0 & -\dfrac{10000}{\frac{\sqrt{3}}{2}-0.5} \\
0 & 0 & 0 & 0 & 0 & 0 & 0 & 1 & \dfrac{5000\,\sqrt{3}}{\frac{\sqrt{3}}{2}-0.5}
\end{bmatrix}
$$

> *convert( C,float );*

$$
\begin{bmatrix}
1. & 0. & 0. & 0. & 0. & 0. & -0. & -0. & -0. \\
0. & 1. & 0. & 0. & 0. & 0. & -0. & -0. & -23660.25403 \\
0. & 0. & 1. & 0. & 0. & 0. & -0. & -0. & -13660.25403 \\
0. & 0. & 0. & 1. & 0. & 0. & -0. & -0. & -33460.65214 \\
0. & 0. & 0. & 0. & 1. & 0. & -0. & -0. & 23660.25403 \\
0. & 0. & 0. & 0. & 0. & 1. & 0. & 0. & 10000. \\
0. & 0. & 0. & 0. & 0. & 0. & 1. & 0. & -27320.50806 \\
0. & 0. & 0. & 0. & 0. & 0. & 0. & 1. & 23660.25403
\end{bmatrix}
$$

Note that we used the *convert* command to obtain decimal equivalents.
We interpret the unique solution of the forces as

$F_1 = 0$, $F_2 = -23{,}660$, $F_3 = -13{,}660$, $f_1 = -33{,}461$, $f_2 = 23{,}660$, $f_3 = 10{,}000$, $f_4 = -27{,}321$,
and $f_5 = 23{,}660$

This is continued as a student project at the end of the section.

---

**Example 2**

## The Leontief Input–Output Economic Model as a Linear System of Equations

Wassily Leontief (1906–1999), a Russian-born American economist, explained his input–output model in the April 1965 issue of *Scientific American*. Leontief organized the 1958 American economy into an 81 × 81 matrix. Each of the 81 sectors of the economy—such as steel, agriculture, manufacturing, transportation, and utilities—represented resources that rely on input from the output of other resources. For example, the production of clothing requires an input from manufacturing, transportation, agriculture, and other manufacturing. The following is a brief example of the Leontief model and its solution.

Let's consider a production model. In the model, let's produce 1 unit considering the following specific requirements:

- Petroleum requires 0.2 units of transportation, 0.4 units of chemicals, and 0.1 unit of itself.
- Textiles require 0.4 units of petroleum, 0.1 units of textiles, 0.15 units of transportation, 0.3 unit of chemicals, and 0.35 units of manufacturing.
- Transportation requires 0.6 units of petroleum, 0.1 units of itself, and 0.25 units of chemicals.
- Chemical requires 0.2 units of petroleum, 0.1 units of textiles, 0.3 units of transportation, 0.1 unit of manufacturing, and 0.2 units of itself.
- Manufacturing requires 0.1 units of petroleum, 0.3 units of transportation, and 0.2 units of itself.

Units are usually measured in dollars.
The following table shows the technology matrix that represents this model.

| | Petroleum | Textiles | Transportation | Chemicals | Manufacturing |
|---|---|---|---|---|---|
| Petroleum | 0.1 | 0.4 | 0.6 | 0.2 | 0.1 |
| Textiles | 0.0 | 0.1 | 0.0 | 0.1 | 0.0 |
| Transportation | 0.2 | 0.15 | 0.1 | 0.3 | 0.3 |
| Chemicals | 0.4 | 0.3 | 0.25 | 0.2 | 0.0 |
| Manufacturing | 0.0 | 0.35 | 0.0 | 0.1 | 0.2 |

If the economy produces \$900 million of petroleum, \$300 million of textiles, \$850 million of transportation, \$800 million of chemicals, and \$750 million of manufacturing, how much of this production is internally consumed by the economy?

We begin by setting this up as five equations and five unknowns in a system of equations in matrix form: $Ax = b$.

$$\begin{bmatrix} 0.1 & 0.4 & 0.6 & 0.2 & 0.1 \\ 0 & 0.1 & 0 & 0.1 & 0 \\ 0.2 & 0.15 & 0.1 & 0.3 & 0.3 \\ 0.4 & 0.3 & 0.25 & 0.2 & 0 \\ 0 & 0.35 & 0 & 0.1 & 0.2 \end{bmatrix} \begin{bmatrix} x_1 \\ x_2 \\ x_3 \\ x_4 \\ x_5 \end{bmatrix} = \begin{bmatrix} 900 \\ 300 \\ 850 \\ 800 \\ 750 \end{bmatrix}$$

The Leontief exchange input−output model or production equation is

$$x = Cx = d$$

where  $x =$ amount produced,
$Cx =$ intermediate demand, and
$d =$ final demand.

We put $x$ into a matrix by multiplying it by the identity matrix, $I$. Thus, $Ix - Cx = d$.

To achieve equilibrium, we need to set up $(I - C)x = d$, where $I$ is the identity matrix. So we subtract our matrix from the $5 \times 5$ identity matrix and obtain the new system of equations that we will need to solve:

$$\begin{bmatrix} 0.9 & -0.4 & -0.6 & -0.2 & -0.1 \\ 0 & 0.9 & 0 & -0.1 & 0 \\ -0.2 & -0.15 & 0.9 & -0.3 & -0.3 \\ -0.4 & -0.3 & -0.25 & 0.8 & 0 \\ 0 & -0.35 & 0 & -0.1 & 0.8 \end{bmatrix} \begin{bmatrix} x_1 \\ x_2 \\ x_3 \\ x_4 \\ x_5 \end{bmatrix} = \begin{bmatrix} 900 \\ 300 \\ 850 \\ 800 \\ 750 \end{bmatrix}$$

We use Maple to input the augmented matrix and solve:

```
> C:=<<.9,0,-.2,-.4,0>|<-.4,.9,-.15,-.3,-.35>|<-.6,0,.9,-.25,0>|<-.2,-.1,-.3,.8,-.1>|<-.1,
0,-.3,0,.8>|<900,300,850,800,750>>;
```

$$C := \begin{bmatrix} 0.9 & -0.4 & -0.6 & -0.2 & -0.1 & 900 \\ 0 & 0.9 & 0 & -0.1 & 0 & 300 \\ -0.2 & -0.15 & 0.9 & -0.3 & -0.3 & 850 \\ -0.4 & -0.3 & -0.25 & 0.8 & 0 & 800 \\ 0 & -0.35 & 0 & -0.1 & 0.8 & 750 \end{bmatrix}$$

```
> CC:=ReducedRowEchelonForm(C);
```

$$CC := \begin{bmatrix} 1 & 0 & 0 & 0 & 0 & 6944.2 \\ 0 & 1 & 0 & 0 & 0 & 1070.0 \\ 0 & 0 & 1 & 0 & 0 & 5620.6 \\ 0 & 0 & 0 & 1 & 0 & 6629.7 \\ 0 & 0 & 0 & 0 & 1 & 2234.4 \end{bmatrix}$$

We interpret the unique solution as needed to produce the following amounts of petroleum, textiles, transportation, chemicals, and manufacturing:

$$x_1 = 6{,}944.2, \ x_2 = 1{,}070, \ x_3 = 5{,}620.6, \ x_4 = 6{,}629.7, \ \text{and} \ x_5 = 2{,}234.4$$

---

**Example 3**

# Least-Squares Models as Systems of Equations

Consider fitting the model $y = ax^2 + bx + c$ to the following data

| $x$ | 0 | 1 | 2 | 3 | 4 |
|---|---|---|---|---|---|
| $y$ | 62 | 50 | 39 | 18 | 2 |

The method of least squares in Chapter 4 and the calculus of least squares in Chapter 9 sought to minimize the sum of squares error:

$$\text{Minimize } S = \sum (y_i - ax_i^2 - bx_i - c)^2$$

As discussed in Chapter 9, we take the partial derivatives and set them equal to zero and obtain the following normal equations:

$$\frac{\partial S}{\partial a} = 0 = \sum y_i x_i^2 - a \cdot \sum x_i^4 - b \sum x_i^3 - c \sum x_i^2$$

$$\frac{\partial S}{\partial b} = 0 = \sum y_i x_i - a \cdot \sum x_i^3 - b \sum x_i^2 - c \sum x_i$$

$$\frac{\partial S}{\partial c} = 0 = \sum y_i - a \cdot \sum x_i^2 - b \sum x_i - c \sum 1$$

We rewrite these as a system of linear equations:

$$\sum y_i x_i^2 = a \cdot \sum x_i^4 + b \sum x_i^3 + c \sum x_i^2$$

$$\sum y_i x_i = a \cdot \sum x_i^3 + b \sum x_i^2 + c \sum x_i$$

$$\sum y_i = a \cdot \sum x_i^2 + b \sum x_i + c \sum 1$$

We substitute the following:

$$\sum x = 10, \ \sum x^2 = 30, \ \sum x^3 = 100, \ \sum x^4 = 354, \ \sum_{i=1}^{5} 1 = 5$$

$$\sum y = 171, \ \sum yx = 190, \ \sum yx^2 = 400$$

and obtain the following equations:

$$400 = 354a + 100b + 30c$$
$$190 = 100a + 30b + 10c$$
$$171 = 30a + 10b + 5c$$

We can input this system of equations into Maple and solve using reduced row echelon form.

```
> with(LinearAlgebra):with(linalg):
> A:=<<354,100,30,400>|<100,30,10,190>|<30,10,5,171>>;
```

$$A := \begin{bmatrix} 354 & 100 & 30 \\ 100 & 30 & 10 \\ 30 & 10 & 5 \\ 400 & 190 & 171 \end{bmatrix}$$

> *A1:=Transpose(A);*

$$A1 := \begin{bmatrix} 354 & 100 & 30 & 400 \\ 100 & 30 & 10 & 190 \\ 30 & 10 & 5 & 171 \end{bmatrix}$$

> *A2:=GaussianElimination(A1);*

$$A2 := \begin{bmatrix} 354 & 100 & 30 & 400 \\ 0 & \dfrac{310}{177} & \dfrac{90}{59} & \dfrac{13630}{177} \\ 0 & 0 & \dfrac{35}{31} & \dfrac{2171}{31} \end{bmatrix}$$

> *A3:= ReducedRowEchelonForm(A1);*

$$A3 := \begin{bmatrix} 1 & 0 & 0 & -\dfrac{9}{7} \\ 0 & 1 & 0 & -\dfrac{352}{35} \\ 0 & 0 & 1 & \dfrac{2171}{35} \end{bmatrix}$$

The output yields $a = -\dfrac{9}{7}$, $b = -\dfrac{352}{35}$, and $c = \dfrac{2171}{35}$, so

$$y = \left(-\frac{9}{7}\right)x^2 - \left(\frac{352}{35}\right)x + \left(\frac{2171}{35}\right)$$

> *xdata:=[0,1,2,3,4];*

$$xdata := [0, 1, 2, 3, 4]$$

> *ydata:=[62,50,39,18,2];*

$$ydata := [62, 50, 39, 18, 2]$$

> *p1:=pointplot({seq([xdata[i],ydata[i]],i=1..5)}):*

> *p2:=plot$\left(\left(-\dfrac{9}{7}\right)x^2-\left(\dfrac{352}{35}\right)x+\left(\dfrac{2171}{35}\right), x=0..6, thickness=2\right)$:*

> *display(p1,p2);*

In Figure 11.6, we note the fit visually and that it captures the trends of the data.

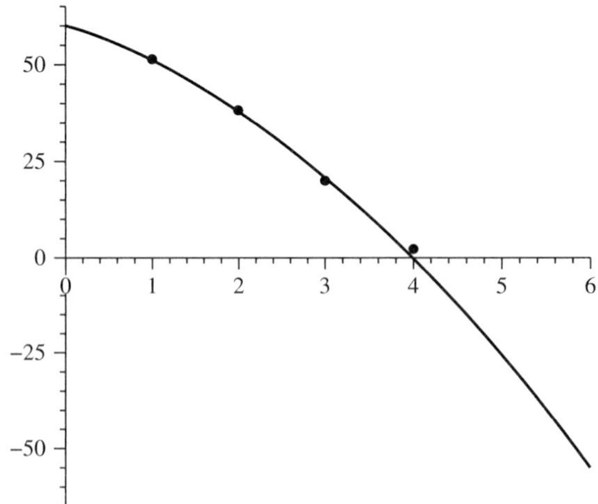

## Example 4

## Natural Cubic Splines as a System of Equations

In modeling natural cubic splines, we want to fit a third-order polynomial between every pair of data points. Let's assume that we have the following four data pairs:

| $x$ | 7 | 14 | 21 | 28 | 35 | 42 |
|---|---|---|---|---|---|---|
| $y$ | 125 | 275 | 800 | 1,200 | 1,700 | 1,650 |

We will need five third-order equations:

$$[7-14],\ S_1 = a_3x^3 + a_2x^2 + a_1x + a_0$$
$$[14-21],\ S_2 = b_3x^3 + b_2x^2 + b_1x + b_0$$
$$[21-28],\ S_3 = c_3x^3 + c_2x^2 + c_1x + c_0$$
$$[28-35],\ S_4 = d_3x^3 + d_2x^2 + d_1x + d_0$$
$$[35-42],\ S_5 = e_3x^3 + e_2x^2 + e_1x + e_0$$

We note that there are 20 unknowns $(a_3, a_2, \ldots, e_1, e_0)$, and we need 20 equations to uniquely solve for the these unknowns.

By substituting in the $(x,y)$ data pairs, we obtain 10 equations. We still lack 10 equations.

We force both the first derivative (slope) and second derivatives (concavity) at the interior to match. For $i \geq 0$ and $i < n - 1$,

$$\frac{dS_{i+1}}{dx} = \frac{dS_{i+2}}{dx}$$

$$\frac{d^2S_{i+1}}{dx^2} = \frac{d^2S_{i+2}}{dx^2}$$

Because there are four interior points, this gives eight more equations. Matching the derivatives ensures smoothness of the curves. The last two equations concern the end points. Under the natural cubic spline, we want the slope at the end points to be constants that force the second derivatives to equal zero. This yields two more equations, and we now have 20 equations. *Note*: If we had clamped cubic splines, then the first derivatives at the two end points would equal specific constants and again we would have 20 equations.

We now present the 20 equations:

1. $343a_3 + 49a_2 + 7a_1 + a_0 = 125$
2. $2{,}744a_3 + 196a_2 + 14a_1 + a_0 = 275$
3. $2{,}744b_3 + 196b_2 + 14b_1 + b_0 = 275$
4. $9{,}621b_3 + 441b_2 + 21b_1 + b_0 = 800$
5. $9{,}621c_3 + 441c_2 + 21c_1 + c_0 = 800$
6. $2{,}1952c_3 + 784c_2 + 28c_1 + c_0 = 1{,}200$
7. $21{,}952d_3 + 784d_2 + 28d_1 + d_0 = 1{,}200$
8. $42{,}875d_3 + 1{,}225d_2 + 35d_1 + d_0 = 1{,}700$
9. $42{,}875e_3 + 1{,}225e_2 + 35e_1 + e_0 = 1{,}700$
10. $74{,}088e_3 + 1{,}764e_2 + 42e_1 + e_0 = 1{,}650$
11. $588a_3 + 28a_2 + a_1 = 588b_3 + 28b_2 + b_1$
12. $1{,}323b_3 + 42b_2 + b_1 = 1{,}323c_3 + 42c_2 + c_1$
13. $2{,}352c_3 + 56c_2 + c_1 = 2{,}352d_3 + 56d_2 + d_1$

14. $3{,}675d_3 + 70d_2 + d_1 = 3{,}675e_3 + 70e_2 + e_1$

15. $84a_3 + 2a_2 = 84b_3 + 2b_2$

16. $126b_3 + 2b_2 = 126c_3 + 2c_2$

17. $168c_3 + 2c_2 = 168d_3 + 2d_2$

18. $210d_3 + 2d_2 = 210e_3 + 2e_2$

19. $42a_3 + 2a_2 = 0$

20. $252e_3 + 2e_2 = 0$

We will use Maple to solve this system of equations. We increase the accuracy in the number of digits as well as increase the user interface for the matrix size. (See also Figure 11.7.)

> *digits :=50;*

$$digits := 50$$

> *interface (rtablesize = infinity);*

$$\infty$$

$$
B := \begin{bmatrix}
343 & 49 & 7 & 1 & 0 & 0 & 0 & 0 & 0 & 0 & 0 & 0 & 0 & 0 & 0 & 0 & 0 & 0 & 0 & 0 & 125 \\
2744 & 196 & 14 & 1 & 0 & 0 & 0 & 0 & 0 & 0 & 0 & 0 & 0 & 0 & 0 & 0 & 0 & 0 & 0 & 0 & 275 \\
0 & 0 & 0 & 0 & 2744 & 196 & 14 & 1 & 0 & 0 & 0 & 0 & 0 & 0 & 0 & 0 & 0 & 0 & 0 & 0 & 275 \\
0 & 0 & 0 & 0 & 9261 & 441 & 21 & 1 & 0 & 0 & 0 & 0 & 0 & 0 & 0 & 0 & 0 & 0 & 0 & 0 & 800 \\
0 & 0 & 0 & 0 & 0 & 0 & 0 & 0 & 9261 & 441 & 21 & 1 & 0 & 0 & 0 & 0 & 0 & 0 & 0 & 0 & 800 \\
0 & 0 & 0 & 0 & 0 & 0 & 0 & 0 & 21952 & 784 & 28 & 1 & 0 & 0 & 0 & 0 & 0 & 0 & 0 & 0 & 1200 \\
0 & 0 & 0 & 0 & 0 & 0 & 0 & 0 & 0 & 0 & 0 & 0 & 21952 & 784 & 28 & 1 & 0 & 0 & 0 & 0 & 1200 \\
0 & 0 & 0 & 0 & 0 & 0 & 0 & 0 & 0 & 0 & 0 & 0 & 42875 & 1225 & 35 & 1 & 0 & 0 & 0 & 0 & 1700 \\
0 & 0 & 0 & 0 & 0 & 0 & 0 & 0 & 0 & 0 & 0 & 0 & 0 & 0 & 0 & 0 & 42875 & 1225 & 35 & 1 & 1700 \\
0 & 0 & 0 & 0 & 0 & 0 & 0 & 0 & 0 & 0 & 0 & 0 & 0 & 0 & 0 & 0 & 74088 & 1764 & 42 & 1 & 1650 \\
588 & 28 & 1 & 0 & -588 & -28 & -1 & 0 & 0 & 0 & 0 & 0 & 0 & 0 & 0 & 0 & 0 & 0 & 0 & 0 & 0 \\
0 & 0 & 0 & 0 & 1323 & 42 & 1 & -1323 & -42 & -1 & 0 & 0 & 0 & 0 & 0 & 0 & 0 & 0 & 0 & 0 & 0 \\
0 & 0 & 0 & 0 & 0 & 0 & 0 & 0 & 2352 & 56 & 1 & 0 & -2352 & -56 & -1 & 0 & 0 & 0 & 0 & 0 & 0 \\
0 & 0 & 0 & 0 & 0 & 0 & 0 & 0 & 0 & 0 & 0 & 0 & 3765 & 70 & 1 & 0 & -3765 & -70 & -1 & 0 & 0 \\
84 & 2 & 0 & 0 & -84 & -2 & 0 & 0 & 0 & 0 & 0 & 0 & 0 & 0 & 0 & 0 & 0 & 0 & 0 & 0 & 0 \\
0 & 0 & 0 & 0 & 126 & 2 & 0 & 0 & -126 & -2 & 0 & 0 & 0 & 0 & 0 & 0 & 0 & 0 & 0 & 0 & 0 \\
0 & 0 & 0 & 0 & 0 & 0 & 0 & 0 & 168 & 2 & 0 & 0 & -168 & -2 & 0 & 0 & 0 & 0 & 0 & 0 & 0 \\
0 & 0 & 0 & 0 & 0 & 0 & 0 & 0 & 0 & 0 & 0 & 0 & 210 & 2 & 0 & 0 & -210 & -2 & 0 & 0 & 0 \\
42 & 2 & 0 & 0 & 0 & 0 & 0 & 0 & 0 & 0 & 0 & 0 & 0 & 0 & 0 & 0 & 0 & 0 & 0 & 0 & 0 \\
0 & 0 & 0 & 0 & 0 & 0 & 0 & 0 & 0 & 0 & 0 & 0 & 0 & 0 & 0 & 0 & 252 & 2 & 0 & 0 & 0
\end{bmatrix}
$$

C := convert(ReducedRowEchelonForm(B),float);

$$B := \begin{bmatrix}
1 & 0 & 0 & 0 & 0 & 0 & 0 & 0 & 0 & 0 & 0 & 0 & 0 & 0 & 0 & 0 & 0 & 0 & 0 & 0 & 0.3285032886 \\
0 & 1 & 0 & 0 & 0 & 0 & 0 & 0 & 0 & 0 & 0 & 0 & 0 & 0 & 0 & 0 & 0 & 0 & 0 & 0 & -6.898569060 \\
0 & 0 & 1 & 0 & 0 & 0 & 0 & 0 & 0 & 0 & 0 & 0 & 0 & 0 & 0 & 0 & 0 & 0 & 0 & 0 & 53.62189371 \\
0 & 0 & 0 & 1 & 0 & 0 & 0 & 0 & 0 & 0 & 0 & 0 & 0 & 0 & 0 & 0 & 0 & 0 & 0 & 0 & -25. \\
0 & 0 & 0 & 0 & 1 & 0 & 0 & 0 & 0 & 0 & 0 & 0 & 0 & 0 & 0 & 0 & 0 & 0 & 0 & 0 & -0.549221982 \\
0 & 0 & 0 & 0 & 0 & 1 & 0 & 0 & 0 & 0 & 0 & 0 & 0 & 0 & 0 & 0 & 0 & 0 & 0 & 0 & 29.96589232 \\
0 & 0 & 0 & 0 & 0 & 0 & 1 & 0 & 0 & 0 & 0 & 0 & 0 & 0 & 0 & 0 & 0 & 0 & 0 & 0 & -462.4805656 \\
0 & 0 & 0 & 0 & 0 & 0 & 0 & 1 & 0 & 0 & 0 & 0 & 0 & 0 & 0 & 0 & 0 & 0 & 0 & 0 & 2383.478143 \\
0 & 0 & 0 & 0 & 0 & 0 & 0 & 0 & 1 & 0 & 0 & 0 & 0 & 0 & 0 & 0 & 0 & 0 & 0 & 0 & 0.410658693 \\
0 & 0 & 0 & 0 & 0 & 0 & 0 & 0 & 0 & 1 & 0 & 0 & 0 & 0 & 0 & 0 & 0 & 0 & 0 & 0 & -30.50659023 \\
0 & 0 & 0 & 0 & 0 & 0 & 0 & 0 & 0 & 0 & 1 & 0 & 0 & 0 & 0 & 0 & 0 & 0 & 0 & 0 & 807.4415678 \\
0 & 0 & 0 & 0 & 0 & 0 & 0 & 0 & 0 & 0 & 0 & 1 & 0 & 0 & 0 & 0 & 0 & 0 & 0 & 0 & -6505.97679 \\
0 & 0 & 0 & 0 & 0 & 0 & 0 & 0 & 0 & 0 & 0 & 0 & 1 & 0 & 0 & 0 & 0 & 0 & 0 & 0 & -0.4373436113 \\
0 & 0 & 0 & 0 & 0 & 0 & 0 & 0 & 0 & 0 & 0 & 0 & 0 & 1 & 0 & 0 & 0 & 0 & 0 & 0 & 40.7333735 \\
0 & 0 & 0 & 0 & 0 & 0 & 0 & 0 & 0 & 0 & 0 & 0 & 0 & 0 & 1 & 0 & 0 & 0 & 0 & 0 & -1187.277417 \\
0 & 0 & 0 & 0 & 0 & 0 & 0 & 0 & 0 & 0 & 0 & 0 & 0 & 0 & 0 & 1 & 0 & 0 & 0 & 0 & 12111.4004 \\
0 & 0 & 0 & 0 & 0 & 0 & 0 & 0 & 0 & 0 & 0 & 0 & 0 & 0 & 0 & 0 & 1 & 0 & 0 & 0 & 0.247496114 \\
0 & 0 & 0 & 0 & 0 & 0 & 0 & 0 & 0 & 0 & 0 & 0 & 0 & 0 & 0 & 0 & 0 & 1 & 0 & 0 & -31.18451036 \\
0 & 0 & 0 & 0 & 0 & 0 & 0 & 0 & 0 & 0 & 0 & 0 & 0 & 0 & 0 & 0 & 0 & 0 & 1 & 0 & 1290.479268 \\
0 & 0 & 0 & 0 & 0 & 0 & 0 & 0 & 0 & 0 & 0 & 0 & 0 & 0 & 0 & 0 & 0 & 0 & 0 & 1 & -15877.14509
\end{bmatrix}$$

> *xdata:=[7,14,21,28,35,42];*

$$xdata := [7, 14, 21, 28, 35, 42]$$

> *ydata:=[125,275,800,1200,1700,1650];*

$$ydata:=[125,275,800,1200,1700,1650]$$

```
> P1:=pointplot({seq([xdata[i],ydata[i]],i=1..6)}):
> p2:=plot(C[1,21]·x³+C[2,21]·x²+C[3,21]·x+C[4,21],x=7..14):
> p3:=plot(C[5,21]·x³+C[6,21]·x²+C[7,21]·x+C[8,21],x=14..21):
> p4:=plot(C[9,21]·x³+C[10,21]·x²+C[11,21]·x+C[12,21],x=21..28):
> p5:=plot(C[13,21]·x³+C[14,21]·x²+C[15,21]·x+C[16,21],x=28..35):
> p6:=plot(C[17,21]·x³+C[18,21]·x²+C[19,21]·x+C[20,21],x=35..42):
> display({p1, p2, p3, p4, p5, p6});
```

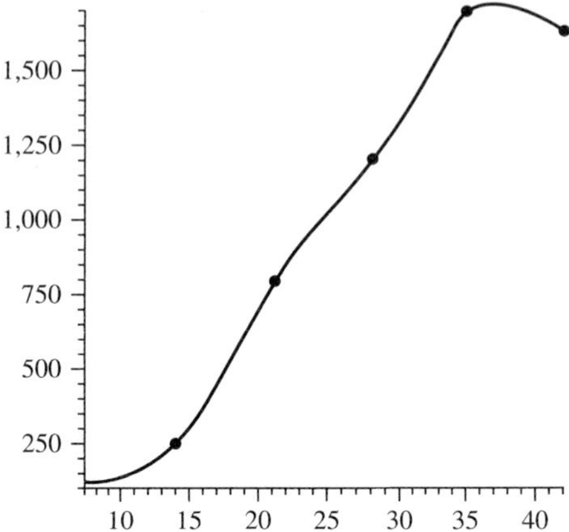

**FIGURE 11.7**
Cubic-spline model

We note that, in general, with $N$ pairs of data we will need $N - 1$ cubic equations. To set up the system of equations:

1. We will need $2N - 2$ cubic equations (two for each successive pairs of points).

2. For each interior point ($N - 2$) points, we obtain two equations: One equation is the first derivative being set equal at each interior point, and the second is the second derivative being set equal at each interior point. This yields another $2(N - 2)$ equations.

3. Finally, if we need clamped cubic splines, then we use the fact that the first derivatives at the two end points equal specific constants to obtain the last two equations. If we need natural cubic splines, we assume the slopes are unknown but constant and set the second derivatives equal to zero at the two end points to obtain the last two equations.

## 11.2 | EXERCISES

1. A Bridge Too Far (Revisited). Set up a model and solve if the angle at joint 4 is $\dfrac{\pi}{3}$.

2. A Bridge Too Far (Revisited). Set up a model and solve if the angle at joint 4 is $\dfrac{\pi}{8}$.

3. A Bridge Too Far (Revisited). Set up a model and solve if the angles at joints 1 and 4 are both $\dfrac{\pi}{3}$. If this does not solve, explain.

4. Consider the Leontief model in Example 2.
   a. Determine the solution with the following technology matrix.

|  | Petroleum | Textiles | Transportation | Chemicals | Manufacturing |
|---|---|---|---|---|---|
| Petroleum | 0.2 | 0.3 | 0.6 | 0.2 | 0.1 |
| Textiles | 0.0 | 0.2 | 0.0 | 0.1 | 0.0 |
| Transportation | 0.2 | 0.15 | 0.2 | 0.3 | 0.3 |
| Chemicals | 0.4 | 0.35 | 0.25 | 0.2 | 0.0 |
| Manufacturing | 0.0 | 0.35 | 0.0 | 0.1 | 0.2 |

   b. If the economy now produces \$1 billion in petroleum, \$400 million in textiles, \$950 million in transportation, \$750 million in chemicals, and \$950 million in manufacturing, how much of this production is internally consumed by the economy?

**5.** Use least squares and fit the model $y = kx^2 + bx + c$ using the following data:

| $x$ | 0.5 | 1.0 | 1.5 | 2.0 | 2.5 |
|---|---|---|---|---|---|
| $y$ | 0.7 | 3.4 | 7.2 | 12.4 | 20.1 |

**6.** Use least squares and fit the model $W = aL^3 + bL^2 + cL + d$ using the following data.

| Length, $L$ | 12.5 | 12.625 | 12.625 | 14.125 | 14.5 | 14.5 | 17.27 | 17.75 |
|---|---|---|---|---|---|---|---|---|
| Weight, $W$ | 17 | 16 | 17 | 23 | 26 | 27 | 43 | 49 |

**7.** Use cubic-spline interpolation to obtain the cubic equations for the following data:

**a.**

| $X$ | 1 | 2 | 3 |
|---|---|---|---|
| $Y$ | 9 | 27 | 48 |

**b.**

| $X$ | 0.5 | 1.0 | 1.5 | 2.0 | 2.5 |
|---|---|---|---|---|---|
| $Y$ | 0.7 | 3.4 | 7.2 | 12.4 | 20.1 |

# 11.2 PROJECTS

**1.** One decision that manufacturers must make is how much of something to produce. Considering that an economy consists of manufacturers of many things, you can see how this problem can get really complex. Leontief's work on this problem earned him the Nobel Prize. What we will discuss now comes from his work, hence the name *Leontief models.* Here is an example of such a model.

First we divide an economy into certain sectors. In reality, there are hundreds of sectors, but to keep things simple we will say for this example that there are three: manufacturing (M), electronics (E), and agriculture. We must decide how many units of each sector to produce. We can put this into a **production matrix**:

$$X = \begin{matrix} M \\ E \\ A \end{matrix} \begin{bmatrix} ?? \\ ?? \\ ?? \end{bmatrix}$$

Now let's say that the public wants 100 units of manufacturing, 200 units of electronics, and 300 units of agriculture. We can put this into an **(external) demand matrix:**

$$D = \begin{matrix} M \\ E \\ A \end{matrix} \begin{bmatrix} 100 \\ 200 \\ 300 \end{bmatrix}$$

Now, one might say that if this is what the people want, then this is what should be produced (i.e., $X = D$). The problem, however, is that the production of certain resources actually *uses up* resources as well. In other words, it takes stuff to make stuff. How much stuff it takes to make stuff can be expressed by an *input−output matrix.*

$$\begin{array}{c} \text{output} \\ \text{M} \quad \text{E} \quad \text{A} \end{array}$$

$$T = .\overset{\text{input}}{\underset{}{\begin{array}{c} \text{M} \\ \text{E} \\ \text{A} \end{array}}} \begin{bmatrix} .1 & .2 & .3 \\ .2 & .1 & .3 \\ .1 & .1 & .2 \end{bmatrix}$$

This matrix says that the production of 1 unit of manufacturing uses up 0.1 units of manufacturing, 0.2 units of electronics, and 0.1 units of agriculture. The production of 1 unit of electronics uses up 0.2 units of manufacturing, 0.1 units of electronics, and 0.1 units of agriculture. Finally, the production of 1 unit of agriculture uses up 0.3 units of manufacturing, 0.3 units of electronics, and 0.1 units of agriculture. So, not only must we account for what the people want, but we must also make up for what is used up in the process of making what the people want. This is called *internal demand*, and it is given by the matrix product $TX$. Hence, what we produce needs to satisfy both internal demand and external demand—that is,

$$X = TX + D$$

2. Aside from developing highly sophisticated economic theories, Wassily Leontief also enjoyed trout fishing, ballet, and fine wines. He won the 1973 Nobel Prize in Economics for his work in creating mathematical models to describe various economic phenomena. In the remainder of this lab, we will look at a very simple special case of his work called a *closed exchange model*. Here is the premise:

Suppose in a far away land of Eigenbazistan, in a small country town called Matrixville, there lived a Farmer, a Tailor, a Carpenter, a Coal Miner, and Slacker Bob. The Farmer produced food; the Tailor, clothes; the Carpenter, housing; the Coal Miner supplied energy; and Slacker Bob made High-Quality 100-Proof Moonshine, half of which he drank himself. Let us make the following assumptions:

- Everyone buys from and sells to the central pool (i.e., there is no outside supply and demand).
- Everything produced is consumed.

For these reasons, this is called a *closed exchange model*. Next we must specify what fraction of each of the goods is consumed by each person in our town: See the following table.

| | Food | Clothes | Housing | Energy | High-Quality 100-Proof Moonshine |
|---|---|---|---|---|---|
| Farmer | 0.25 | 0.15 | 0.25 | 0.18 | 0.20 |
| Tailor | 0.15 | 0.28 | 0.18 | 0.17 | 0.05 |
| Carpenter | 0.22 | 0.19 | 0.22 | 0.22 | 0.10 |
| Coal Miner | 0.20 | 0.15 | 0.20 | 0.28 | 0.15 |
| Slacker Bob | 0.18 | 0.23 | 0.15 | 0.15 | 0.50 |

So, for example, the Carpenter consumes 22 percent of all food, 19 percent of all clothes, 22 percent of all housing, 22 percent of all energy and 10 percent of all High-Quality 100-Proof Moonshine.

If $I - T$ is invertible, then this equation can be solved for $X$:

$$X - TX = D$$
$$(I - T)X = D$$
$$(I - T)^{-1}[(I - T)X] = (I - T)^{-1}D$$
$$[(I - T)^{-1}(I - T)]X = (I - T)^{-1}D$$
$$IX = (I - T)^{-1}D$$
$$X = (I - T)^{-1}D$$

In this example,

$$I - T = \begin{bmatrix} 1 & 0 & 0 \\ 0 & 1 & 0 \\ 0 & 0 & 1 \end{bmatrix} - \begin{bmatrix} .1 & .2 & .3 \\ .2 & .1 & .3 \\ .1 & .1 & .2 \end{bmatrix} = \begin{bmatrix} .9 & -.2 & -.3 \\ -.2 & .9 & -.3 \\ -.1 & -.1 & .8 \end{bmatrix}.$$

The inverse of this matrix is (approximately):

$$(I - T)^{-1} \approx \begin{bmatrix} 1.255 & .345 & .6 \\ .345 & 1.255 & .6 \\ .2 & .2 & 1.4 \end{bmatrix}.$$

And so

$$X = (I - T)^{-1} D = \begin{bmatrix} 1.255 & .345 & .6 \\ .345 & 1.255 & .6 \\ .2 & .2 & 1.4 \end{bmatrix} \begin{bmatrix} 100 \\ 200 \\ 300 \end{bmatrix} \approx \begin{bmatrix} 375 \\ 465 \\ 480 \end{bmatrix}$$

In other words, we want to produce 375 units of manufacturing, 465 units of electronics, and 480 units of agriculture.

The internal demand is now easy to calculate. Although we could do it by finding the matrix product $TX$, we can just say it is total production minus external demand, that is, $X - D$.

$$\text{internal demand} = X - D = \begin{bmatrix} 375 \\ 465 \\ 480 \end{bmatrix} - \begin{bmatrix} 100 \\ 200 \\ 300 \end{bmatrix} = \begin{bmatrix} 275 \\ 265 \\ 180 \end{bmatrix}$$

In reality, the numbers in $T$ will be much smaller, and so the internal demand is usually very small compared to the external demand.

## 11.3 MODELS WITH INFINITE SOLUTIONS USING SYSTEMS OF EQUATIONS

### Basic Chemical Balancing

During your life, you have witnessed numerous chemical reactions. How would you describe them to someone else? How could you obtain quantitative information about what went on in the reaction? Chemists use *stoichiometric chemical equations* to answer these questions.

By definition, a chemical equation is a written representation of a chemical reaction that shows the reactants and products, their physical states, and the direction in which the reaction proceeds. In addition, many chemical equations designate the conditions necessary (such as high temperature) for the reactions to occur. A chemical equation provides stoichiometric information about a chemical reaction only if it is balanced.

For a chemical equation to be balanced, the same number of each kind of atom must be present on both sides of the chemical equation. French chemist Antoine Lavoisier introduced the law of conservation of matter during the latter half of the 18th century. This law states that matter can be neither created nor destroyed. Lavoisier's principles for naming chemical substances are still used today.

John Dalton (1766−1844) developed the first useful atomic theory of matter around 1803. In the course of his studies on meteorology, Dalton concluded that evaporated water exists in air as an independent gas. Solid bodies cannot occupy the same space at the same time, but obviously water and air can. If water and air were composed of discrete particles, Dalton reasoned, then evaporation might be viewed as a mixing of water particles with air particles.

He performed a series of experiments on mixtures of gases to determine what effect properties of the individual gases had on the properties of the mixture as a whole. While trying to explain the results of those experiments, Dalton developed the hypothesis that the sizes of the particles making up different gases must be different. He later wrote that "it became an object to determine the relative sizes and weights, together with the relative numbers of atoms entering into such combinations. . . . Thus a train of investigation was laid for determining the number and weight of all chemical elementary particles which enter into any sort of combination one with another." Dalton was the first to associate the ancient idea of atoms with stoichiometry.

From Dalton's atomic theory, we know that all substances are composed of atoms. During a chemical reaction atoms may be combined, separated, or rearranged but not created or destroyed. Dalton proposed his atomic theory of matter in 1803. While these statement largely remain true, they might not be completely true in the light of later discoveries. The postulates included the following.

1. All matter is composed of atoms. These indivisible and indestructible objects are the ultimate chemical particles.

2. All atoms of a given element are identical, both in mass and in properties. Atoms of different elements have different masses and different properties.

3. Compounds are formed by combination of two or more different kinds of atoms. Atoms combine in the ratio of small whole numbers.

4. Atoms are the units of chemical change. A chemical reaction involves only the combination, separation, or rearrangement of atoms.

Let's examine the meaning of chemical equations and compare them to mathematical equations to gain some insights. A mathematical equation might be of the following form:

$$x + 2x = 3x$$

A chemical equation identifies the starting and finishing chemical as reactants and products:

$$\text{reactants} \rightarrow \text{products}$$

For example, the combustion of propane is written as:

$$C_3H_8 + 5O_2 \rightarrow 3CO_2 + 4H_2O$$

A chemical equation is balanced when it reflects the conservation of matter.

### Chemical Equations Versus Mathematical Equations

We usually think of an equation such as $x + 2x = 3x$ as purely mathematical, even if $x$ represents a physical quantity such as distance or mass. A chemical equation may look like a mathematical equation, but it describes experimental observations: the quantities and kinds of reactants and products for a particular chemical reaction. Reactants appear on the left-hand side of a chemical equation; products on the right. The products, which are the result of combining the reactants, are known from experimental observations—they cannot be derived mathematically. In fact, combining the same reactants at different concentrations or temperatures often produces different products from the same reactants.

First-year chemistry students cannot predict these effects, and they are not generally asked to predict these effects. On the other hand, a chemical equation is similar to a mathematical equation in that there are certain restrictions on what may appear on the left- and right-hand sides of a chemical reaction. These mathematical rules represent the effects of the conservation of matter on the reaction. The principle of the conservation of matter says that no atoms are destroyed or created during a chemical reaction.

### How a Chemist Approaches Balancing an Equation

Many chemical equations, in the view of the chemists, can be balanced by inspection, that is, the process of "trial and error." The objectives of a chemist are as follows:

- recognize a balanced equation,
- recognize an unbalanced equation,
- balance by inspecting chemical equations with given reactants and products, and
- write the unbalanced equation when given compound names for reactants and products.

According to chemistry textbooks, here is a stepwise procedure to balance equations:

**Step 1**   Determine what reaction is occurring: Know the reactants, the products, and the physical states.

**Step 2**   Write the unbalanced equation that summarizes the reaction described in step 1.

**Step 3**   Balance the equation by inspection; start with the most complicated molecules. Do not change the identities of any reactant or product.

Thus, balancing the equation is done by inspection, a trial-and-error process that some students catch on to and some students do not.

### Balancing Equations with Systems of Equations

Many students are frustrated with the trial-and-error method and their inability to balance chemical equations. Balancing chemical equations can be an application of solving a linear system of equations. Placing variables as the multipliers for each compound and making equations for each type of atom results in a system of linear equations. This system is usually underdetermined, meaning that there are more variables than equations. This leads to infinitely many solutions. Our goal is to provide a procedure to find one of these solutions in terms of integers. This method is best grasped through an example. It also lays the foundation for balancing the more complicated oxidation−reduction equations.

---

**Example 1**   Simple Chemical Balancing: Balancing $S_6 + O_2 \rightarrow SO_2$

**Step 1**   Introduce multipliers for each compound. We have three compounds, so we identify three multipliers $\{x_1, x_2, x_3\}$.

$$x_1 S_6 \,(s) + x_2 O_2 \,(g) \rightarrow x_3 SO_2 \,(g)$$

**Step 2**   Identify the chemicals: sulfur (S) and oxygen (O) . Set up equations for each element involving the multipliers and the subscripts.

$$S: 6x_1 = x_3$$
$$O: 2x_2 = 2x_3$$

**Step 3**   Put all variables on the same side of the equation and set equal to zero.

$$S: 6x_1 - x_3 = 0$$
$$O: 2x_2 - 2x_3 = 0$$

**Step 4**   Put into matrix notation, $A\mathbf{x} = \mathbf{b}$, where $A$ is the matrix of coefficients for variables $x_1$, $x_2$, and $x_3$, $\mathbf{x}$ is the column vector $\begin{bmatrix} x_1 \\ x_2 \\ x_3 \end{bmatrix}$, and $\mathbf{b}$ is the column vector $\begin{bmatrix} 0 \\ 0 \end{bmatrix}$.

**Step 5**   Place the linear system into an augmented matrix, $[A \,|\, b]$. The augmented matrix is a shorthand summary of the system of equations in steps 3 and 4:

$$[A|b] = \begin{bmatrix} 6 & 0 & -1 & 0 \\ 0 & 2 & -2 & 0 \end{bmatrix}$$

**Step 6**    Use Gaussian elimination (reduction) to simplify the matrix $A\mathbf{x} = \mathbf{b}$ to $H\mathbf{x} = \mathbf{c}$, where $H$ is in row-reduced echelon form, using a sequence of elementary row operations. After several sequences, we achieve:

$$[H|c] = \begin{bmatrix} 1 & 0 & -1/6 & 0 \\ 0 & 1 & -1 & 0 \end{bmatrix}$$

Steps 5 and 6 can be done from Maple. We enter the matrix and then obtain the reduced row echelon form.

> *with(Linear Algebra):*
> $B := \langle\langle 6,0\rangle|\langle 0,2\rangle|\langle -1,-2\rangle|\langle 0,0\rangle\rangle;$

$$B := \begin{bmatrix} 6 & 0 & -1 & 0 \\ 0 & 2 & -2 & 0 \end{bmatrix}$$

> *Reduced Row Echelon Form ($\langle B\rangle$);*

$$\begin{bmatrix} 1 & 0 & -\dfrac{1}{6} & 0 \\ 0 & 1 & -1 & 0 \end{bmatrix}$$

This represents the following equations:

$$x_1 - 1/6x_3 = 0$$
$$x_2 - x_3 = 0$$

or

$$x_1 = 1/6x_3$$
$$x_2 = x_3$$

Balancing the chemical equation means finding the *smallest whole numbers* $x_1$, $x_2$, and $x_3$. Because there are more equations than unknowns, we have infinite solutions. We note that $c$ is part of every equation, and we can let $c$ equal anything. Yet we want whole numbers and have $1/6$ as a coefficient in one of our equations, so we let $x_3 = 6$.

**Step 7**    Solve for the multipliers $\{x_1, x_2, x_3\}$.

$$x_3 = 6$$
$$x_1 = 1$$
$$x_2 = 6$$

**Step 8**    Write the balanced equation:

$$1S_6 + 6O_2 \rightarrow 6SO_2$$

Checking our work:

$$\text{Sulfur: } 6 = 6$$
$$\text{Oxygen: } 12 = 12$$

---

## Example 2        Balancing a More Complicated Equation

We are presented with the following more-complicated unbalanced equation: where s and g stand for solid and gas.

$$C_2H_8N_2(s) + N_2O_4(g) \rightarrow N_2(g) + CO_2(g) + H_2O(g)$$

This could be tough by inspection, so let's use a system of equations.

**Step 1**    Introduce five multipliers: $\{a,b,c,d,e\}$.

$$a\mathrm{C_2H_8N_2}(s) + b\mathrm{N_2O_4}(g) \rightarrow c\mathrm{N_2}(g) + d\mathrm{CO_2}(g) + e\mathrm{H_2O}(g)$$

**Step 2**    Set up the following equations:

$$
\begin{aligned}
\text{C: } & 2a = d \\
\text{H: } & 8a = 2e \\
\text{N: } & 2a + 2b = 2c \\
\text{O: } & 4b = 2d + e
\end{aligned}
$$

**Step 3**

$$
\begin{aligned}
\text{C: } & 2a - d = 0 \\
\text{H: } & 8a - 2e = 0 \\
\text{N: } & 2a + 2b - 2c = 0 \\
\text{O: } & 4b - 2d - e = 0
\end{aligned}
$$

**Step 4**

$$
\begin{bmatrix}
2 & 0 & 0 & -1 & 0 \\
8 & 0 & 0 & 0 & -2 \\
2 & 2 & -2 & 0 & 0 \\
0 & 4 & 0 & -2 & -1
\end{bmatrix}
\begin{bmatrix} a \\ b \\ c \\ d \\ e \end{bmatrix}
=
\begin{bmatrix} 0 \\ 0 \\ 0 \\ 0 \\ 0 \end{bmatrix}
$$

**Step 5**

$$
[A|b] =
\begin{bmatrix}
2 & 0 & 0 & -1 & 0 & 0 \\
8 & 0 & 0 & 0 & -2 & 0 \\
2 & 2 & -2 & 0 & 0 & 0 \\
0 & 4 & 0 & -2 & -1 & 0
\end{bmatrix}
$$

**Step 6**    Gaussian elimination yields:

$$
[A] =
\begin{bmatrix}
1 & 0 & 0 & 0 & -1/4 & 0 \\
0 & 1 & 0 & 0 & -1/2 & 0 \\
0 & 0 & 1 & 0 & -3/4 & 0 \\
0 & 0 & 0 & 1 & -1/2 & 0
\end{bmatrix}
$$

$$
[A|b] =
\begin{bmatrix}
2 & 0 & 0 & -1 & 0 & 0 \\
8 & 0 & 0 & 0 & -2 & 0 \\
2 & 2 & -2 & 0 & 0 & 0 \\
0 & 4 & 0 & -2 & -1 & 0
\end{bmatrix}
$$

> $B := \langle\langle 2,8,2,0\rangle | \langle 0,0,2,4\rangle | <0,0,-2,0> | \langle -1,0,0,-2\rangle | <0,-2,0,-1> | \langle 0,0,0,0\rangle\rangle;$

$$
B :=
\begin{bmatrix}
2 & 0 & 0 & -1 & 0 & 0 \\
8 & 0 & 0 & 0 & -2 & 0 \\
2 & 2 & -2 & 0 & 0 & 0 \\
0 & 4 & 0 & -2 & -1 & 0
\end{bmatrix}
$$

> *Reduced Row Echelon Form* $(\langle B \rangle)$;

$$\begin{bmatrix} 1 & 0 & 0 & 0 & -\frac{1}{4} & 0 \\ 0 & 1 & 0 & 0 & -\frac{1}{2} & 0 \\ 0 & 0 & 1 & 0 & -\frac{3}{4} & 0 \\ 0 & 0 & 0 & 1 & -\frac{1}{2} & 0 \end{bmatrix}$$

**Step 7**    Choose $e = 4$

$$a = 1/4e, \text{ so } a = 1$$
$$b = 1/2e, \text{ so } b = 2$$
$$c = 3/4e, \text{ so } c = 3$$
$$d = 1/2e, \text{ so } d = 2$$

**Step 8**    $1C_2H_8N_2(s) + 2N_2O_4(g) \rightarrow 3N_2(g) + 2CO_2(g) + 4H_2O(g)$

Check:

$$C: 2 = 2$$
$$H: 8 = 8$$
$$N: 2 + 4 = 6, 6 = 6$$
$$O: 8 = 4 + 4, 8 = 8$$

### Balancing More Complicated Oxidation–Reduction (Redox Reaction) Equations

These equations are the more difficult to balance. Chemistry textbooks suggest a methodology that also involves inspection. There are two types of oxidation–reduction equations: acidic and basic solutions.

The steps in oxidation–reduction reactions in acidic solution via standard chemistry textbooks are as follows:

**Step 1**    Write separate equations for oxidation and reduction half-reactions.

**Step 2**    For each equation:

    **a.**    Balance all elements except hydrogen and oxygen.

    **b.**    Balance oxygen using $H_2O$.

    **c.**    Balance hydrogen using $H^+$.

    **d.**    Balance the charge.

Thus, in acidic solution, we balance H and O atoms with $H_2O$ and charge with $H^+$.

**Step 3**    If necessary, multiply balanced equations by integers to equalize electrons.

**Step 4**    Add the half-reaction and cancel identical species.

**Step 5**    Check that the elements and charges are balanced.

The steps in oxidation–reduction reactions in basic solution from a standard chemistry textbook are as follows:

**Step 1**    Use the half-reaction method as previously specified for acidic solutions to obtain a balanced equation *as if $H^+$ ions were present.*

**Step 2**    To both sides of the equation obtained above, add a number of $OH^-$ ions that is equal to the number of $H^+$ ions. In basic solution, $H_2O$ balances H and O atoms and $OH^-$ balances the charge.

**Step 3**   Form $H_2O$ on the side containing both $H^+$ and $OH^-$ ions, and eliminate the number of $H_2O$ molecules that appear on both sides.

**Step 4**   Check that the elements and charges are balanced

**Balancing Equations with the Conservation of Mass and Charge: Oxidation–Reduction Equations, with Systems of Equations**

**Step 1**   Apply appropriate variables for either acidic or basic solutions
For acid:

$$(\text{Reactants}) + H^+ \rightarrow (\text{Products}) + H_2O$$

For base:

$$(\text{Reactants}) + H_2O \rightarrow (\text{Products}) + OH^-$$

**Step 2**   Create equations that balance atoms.

**Step 3**   Create charge conservation equation.

**Step 4**   Place into augmented coefficient matrix.

**Step 5**   Use Gaussian elimination (use a reduced-row echelon form).

**Step 6**   Select the value of the free variables to create integer solutions.

**Step 7**   Write the balanced equation. Step 3 requires the student to model a new equation for the oxidation–reduction reaction using oxidation numbers. An atom's oxidation number signifies the number of charges the atom would have in a molecule if all electrons were completely transferred. We will use the following rules to assign oxidation numbers:

   **a.**   In free elements, each atom has an oxidation number of zero. Thus, each atom in a stand-alone reaction (such as Na, $H_2$, Br, K, and $S_4$) has the same oxidation number: zero.

   **b.**   The oxidation number of a monotonic ion is the same as the charge. For example, the oxidation number of a $Na^+$ ion is $+1$ and of a $Cl^-$ ion is $-1$.

   **c.**   In compounds, fluorine is always assigned an oxidation number of $-1$.

   **d.**   Oxygen is always assigned an oxidation number of $-2$ in its compounds.

   **e.**   In compounds with nonmetals, hydrogen is assigned an oxidation number of $+1$.

   **f.**   The sum of the oxidation states *must* equal zero for an electronically neutral compound.

   We illustrate these rules to obtain the oxidation equations in the following examples.

---

## Example 3        Balancing a Redox Equation

Consider the following redox equation with elements copper (Cu), hydrogen (H), nitrogen (N), and oxygen (O):

$$Cu + HNO_{3(aq)} \rightarrow Cu_{(aq)}^{2+} + NO_{2(g)} + NO_{3(aq)}^{1-} + H_2O_{(l)}$$

If we add our multipliers, we have six unknowns and only four chemical elements:

$$x_1Cu + x_2HNO_3 \rightarrow x_3Cu^{2+} + x_4NO_2 + x_5NO_3^{1-} + x_6H_2O$$

$$
\begin{aligned}
\text{Copper:} \quad & x_1 = x_3 \\
\text{Hydrogen:} \quad & x_2 = 2x_6
\end{aligned}
$$

Nitrogen: $\quad x_2 = x_4 + x_5$

Oxygen: $\quad 3x_2 = 2x_4 + 3x_5 + x_6$

It might appear as though we cannot use the system of equations method because of the six equations and only four elements. However, charge must be conserved as well, giving rise to another equation. We list the charges with each element. These are the oxidation numbers. They are found in the tables of chemistry textbooks.

$$\overset{+0}{Cu} + \overset{+1\,+5\,-2}{HNO_3} \rightarrow \overset{+2}{Cu^{2-}} + \overset{+2\,-2}{NO} + \overset{+5\,-2}{NO_3^{1-}} + \overset{+1\,-2}{H_2O}$$

or with our set of multipliers

$$x_1 \cdot 0 + x_2 \cdot 5 = x_3 \cdot 2 + x_4 \cdot 2 + x_5 \cdot 5 + x_6 \cdot 0$$

This generates the following equation:

$$x_1 \overset{+0}{Cu} + x_2 \overset{+1\,+5\,-2}{HNO_3}$$

$$\rightarrow$$

$$x_3 \overset{+2}{Cu^{2-}} + x_4 \overset{+2\,-2}{NO} + x_5 \overset{+5\,-2}{NO_3^{1-}} + x_6 \overset{+1\,-2}{H_2O}$$

This generates the following equation:

$$x_1 \cdot 0 + x_2 \cdot 5 = x_3 \cdot 2 + x_4 \cdot 2 + x_5 \cdot 5 + x_6 \cdot 0$$

Notice that this equation uses only the oxidation numbers of atoms whose oxidation numbers are changing.

The equation may be rewritten as:

$$0x_1 + 5x_2 - 2x_3 - 2x_4 - 5x_5 - 0x_6 = 0$$

We put all the equations in matrix form.

> $A:=\langle\langle 1,0,0,0,0\rangle|\langle 0,1,1,3,5\rangle|\langle -1,0,0,0,-2\rangle|<0,0,-1,-2,-2>$
  $|\langle 0,0,-1,-3,-5\rangle|<0,-2,0,-1,0>|<0,0,0,0,0>>;$

$$A := \begin{bmatrix} 1 & 0 & -1 & 0 & 0 & 0 & 0 \\ 0 & 1 & 0 & 0 & 0 & -2 & 0 \\ 0 & 1 & 0 & -1 & -1 & 0 & 0 \\ 0 & 3 & 0 & -2 & -3 & -1 & 0 \\ 0 & 5 & -2 & -2 & -5 & 0 & 0 \end{bmatrix}$$

> *Reduced Row Echelon Form* $(\langle A \rangle)$;

$$\begin{bmatrix} 1 & 0 & 0 & 0 & 0 & -\frac{3}{2} & 0 \\ 0 & 1 & 0 & 0 & 0 & -2 & 0 \\ 0 & 0 & 1 & 0 & 0 & -\frac{3}{2} & 0 \\ 0 & 0 & 0 & 1 & 0 & -1 & 0 \\ 0 & 0 & 0 & 0 & 1 & -1 & 0 \end{bmatrix}$$

We obtain the solution with a free variable.

$$
\begin{bmatrix} x_1 \\ x_2 \\ x_3 \\ x_4 \\ x_5 \\ x_6 \end{bmatrix} = x_6 \begin{bmatrix} \frac{3}{2} \\ 2 \\ \frac{3}{2} \\ 1 \\ 1 \\ 1 \end{bmatrix}
$$

We choose the solution to the free variable to be the smallest integer value that clears all of the following fractions:

$$
\begin{aligned}
x_1 &= 3 \\
x_2 &= 4 \\
x_3 &= 3 \\
x_4 &= 2 \\
x_5 &= 2 \\
x_6 &= 2
\end{aligned}
$$

Thus, we choose to let $x_6 = 2$, thus obtaining the following values for our variables: We write the balanced form as:

$$
\overset{+0}{3Cu} + \overset{+1+5-2}{4HNO_3} \rightarrow \overset{+2}{3Cu^{2-}} + \overset{+2-2}{2NO} + \overset{+5-2}{2NO_3^{1-}} + \overset{+1\ -2}{2H_2O}
$$

We check our balanced equation.

|     | Left Side | Right Side |
| --- | --- | --- |
| Cu | 3 | 3 |
| H | 4 | 4 |
| N | 4 | 4 |
| O | 12 | 12 |

We also check to see that the electrons balance:

$$
\text{Electrons: } 20 = 20
$$
$$
\text{Because } 20 = 20, \text{ our electrons balance.}
$$

Thus, we have a balanced equation.

---

## Example 4     Balancing an Oxidation–Reduction Equation

Consider the following oxidation–reduction equation:

$$
OH^- + SnO_2^{2-} + Bi(OH)_3 \rightarrow Bi + SnO_3^{2-} + H_2O
$$

We multiply each by a multiplier ($x_1$, $x_2$, $x_3$, $x_4$, $x_5$, and $x_6$) to obtain:

$$
x_1OH^- + x_2SnO_2^{2-} + x_3Bi(OH)_3 \rightarrow x_4Bi + x_5SnO_3^{2-} + x_6H_2O
$$

Next the oxidation numbers are added:

$$
x_1OH^- + x_2\overset{+2-2}{SnO_2^{2-}} + x_3\overset{+3\ -3}{Bi(OH)_3} \rightarrow x_4Bi + x_5\overset{+4\ -4}{SnO_3^{2-}} + x_6H_2O
$$

The oxidation of Bi goes from $+3$ to $0$, and the oxidation of Sn goes from $+2$ to $+4$. We add in the oxidation numbers above to obtain an equation:

$$0x_1 + 2x_2 + 3x_3 - 0x_4 - 4x_5 + 0x_6 = 0$$

We notice that only Bismuth (Bi) and Tin (Sn) oxidation numbers change, so the preceding equation only deals with those atoms.

The multipliers are $x_1$, $x_2$, $x_3$, $x_4$, $x_5$, and $x_6$.

The equations:

$$\text{Sn: } x_2 = x_5$$
$$\text{O: } 4x_1 + 2x_2 + x_3 = 0x_4 + 4x_5 + x_6 = 0$$
$$\text{Bi :} x_3 = x_4$$
$$\text{H} x_1 + 3x_3 = 2x_6$$
$$\text{Oxidation } 0x_1 + 2x_2 + 3x_3 = 0x_4 + 4x_5 + 0x^6$$

We put these equations into matrix form:

$$\begin{bmatrix} 0 & 1 & 0 & 0 & -1 & 0 & 0 \\ 1 & 2 & 3 & 0 & -3 & -1 & 0 \\ 0 & 0 & 1 & -1 & 0 & 0 & 0 \\ 1 & 0 & 3 & 0 & 0 & -2 & 0 \\ 0 & 2 & 3 & 0 & -4 & 0 & 0 \end{bmatrix}$$

We reduce this system using Gaussian elimination:

$$\begin{bmatrix} 1 & 0 & 0 & 0 & 0 & 0 & 0 \\ 0 & 1 & 0 & 0 & 0 & -1 & 0 \\ 0 & 0 & 1 & 0 & 0 & -\dfrac{2}{3} & 0 \\ 0 & 0 & 0 & 1 & 0 & -\dfrac{2}{3} & 0 \\ 0 & 0 & 0 & 0 & 1 & -1 & 0 \end{bmatrix}$$

We obtain the solution with a free variable.

$$\begin{bmatrix} x_1 \\ x_2 \\ x_3 \\ x_4 \\ x_5 \\ x_6 \end{bmatrix} = x_6 \begin{bmatrix} 0 \\ 1 \\ 2/3 \\ 2/3 \\ 1 \\ 1 \end{bmatrix}$$

We choose the solution for the free variable to be the smallest integer value that clears the fractions. Thus, we choose to let $x_6 = 3$, thus obtaining the following:

$$x_1 = 0$$
$$x_2 = 3$$
$$x_3 = 2$$
$$x_4 = 2$$
$$x_5 = 3$$
$$x_6 = 3$$

We write the balanced form as:

$$(0) \cdot OH^- + 3 \cdot SnO_2^{2-} + 2 \cdot Bi(OH)_3 \rightarrow 2 \cdot Bi + 3 \cdot SnO_3^{2-} + 3 \cdot H_2O$$

We also note that $x_1 = 0$, so OH is eliminated.
We check our balanced equation.

|  | Left Side | Right Side |
|---|---|---|
| O | 12 | 12 |
| H | 6 | 6 |
| Sn | 3 | 3 |
| Bi | 2 | 2 |
| Oxidation (electrons) | 12 | 12 |

## Example 5

## Oxidation–Reduction: Balancing $CH_3CH_2OH + Cr_2O_7^{2-} + H^+ \rightarrow CH_3CO_2H + Cr^{3+} + H_2O$

Consider the following unbalanced oxidation–reduction reaction:

$$CH_3CH_2OH + Cr_2O_7^{2-} + H^+ \rightarrow CH_3CO_2H + Cr^{3+} + H_2O$$

We obtain the oxidation numbers and model only those elements whose oxidation numbers actually change:

$$\overset{-3}{C}H_3 \, \overset{+1\,-1\,+1}{C}H_2 \, \overset{-2\,+1}{O}H + \overset{+6}{C}r_2 \, \overset{-2}{O_7^{2-}} + \overset{-1}{H^+} \rightarrow \overset{-3\,+1}{C}H_3 \, \overset{+3\,-2\,+1}{C}O_2 H + \overset{+6^{3+}}{C}r + \overset{+\,-2}{H_2O}$$

We look at what oxidation levels change and build an appropriate equation:

$$\overset{-1}{C} \rightarrow \overset{+3}{C}$$

$$\overset{+6}{Cr} \rightarrow \overset{+3}{Cr}$$

We obtain the oxidation equation as

$$-x_1 + 12x_2 + 0x_3 - 3x_4 - 3x_5 + 0x_6 = 0$$

| | |
|---|---|
| Carbon: | $2x_1 - 2x_4 = 0$ |
| Hydrogen: | $6x_1 + x_3 - 4x_4 - 2x_6 = 0$ |
| Chromium: | $2x_2 - x_5 = 0$ |
| Oxygen: | $1x_1 + 7x_2 - 2x_3 - x_6 = 0$ |
| Oxidation: | $-x_1 + 12x_2 + 0x_3 - 3x_4 - 3x_5 + 0x_6 = 0$ |

We put into matrix form and then solve

$$A := \begin{bmatrix} 2 & 0 & 0 & -2 & 0 & 0 & 0 \\ 6 & 0 & 1 & -4 & 0 & -2 & 0 \\ 1 & 7 & 0 & -2 & 0 & -1 & 0 \\ 0 & 2 & 0 & 0 & -1 & 0 & 0 \\ -1 & 12 & 0 & -3 & -3 & 0 & 0 \end{bmatrix}$$

We use Gaussian elimination in Maple to obtain:

$$B := \begin{bmatrix} 1 & 0 & 0 & 0 & 0 & \frac{-3}{11} & 0 \\ 0 & 1 & 0 & 0 & 0 & \frac{-2}{11} & 0 \\ 0 & 0 & 1 & 0 & 0 & \frac{-16}{11} & 0 \\ 0 & 0 & 0 & 1 & 0 & \frac{-3}{11} & 0 \\ 0 & 0 & 0 & 0 & 1 & \frac{-4}{11} & 0 \end{bmatrix}$$

We allow the free variable, $x_6 = 11$, to clear the fractions at the lowest value. This makes $x_1 = 3$, $x_2 = 2$, $x_3 = 16$, $x_4 = 3$, $x_5 = 4$, and $x_6 = 11$.

$$3(CH_3CH_2OH) + 2(Cr_2O_7^{2-},+) 16H^+ \rightarrow 3(CH_3CO_2H) + 4Cr^{3+} + 11(H_2O)$$

Checking the balancing:

| | |
|---|---|
| Carbon: | $6 = 6$ |
| Hydrogen: | $34 = 34$ |
| Oxygen: | $17 = 17$ |
| Chromium: | $4 = 4$ |
| Electrons: | $21 = 21$ |

We have balanced the oxidation–reduction equation.

## SUMMARY

Oxidation–reduction, or *redox*, equations involve acidic or basic solutions in their reactions. Use your knowledge of chemistry and the following information before starting the balancing procedure.

For an acid:

$$(\text{Reactants}) + H^+ \rightarrow (\text{Products}) + H_2O$$

For a base:

$$(\text{Reactants}) + H_2O \rightarrow (\text{Products}) + OH^-$$

In summary, for the more complicated oxidation–reduction equations involving conservation of mass or charge, we apply the following steps to accurately balance the equation:

- Apply variables.
- Create equations that balance atoms.
- Create charge-conservation equation.
- Place into augmented coefficient matrix.
- Use Gaussian elimination (use a reduced-row echelon form).
- Select the value of the free variables to create integer solutions.
- Write and check the balanced equation.

## 11. 3 | EXERCISES

In problems 1−4, balance basic chemical reactions.

1. $Cu + AgNO_3 \rightarrow Ag + Cu(NO_3)_2$
   Copper (Cu), silver (Ag), nitrogen (N), and oxygen (O) are the elements.

2. $Zn + HCl \rightarrow ZnCl_2 + H_2$
   Zinc (Zn), hydrogen (H), and chlorine (Cl) are the elements.

3. $Ca(OH)_2 + H_3PO_4 \rightarrow H_2O + Ca_3(PO_4)_2$
   Calcium (Ca), hydrogen (H), oxygen (O), and phosphorus (P) are the elements.

4. $FeO + O_2 \rightarrow Fe_2O_3$
   Iron (Fe) and oxygen (O) are the elements.

5. Balance the following oxidation−reduction equation:

$$H + Mn^{2+} + NaBiO_3 \rightarrow H_2O + MnO_4^- + Bi^{3+} + Na$$

## 11.3 | PROJECTS

1. Explain why we need to use the lowest integer solution in the chemical reactions. Provide an example.

2. Sometimes oxidation−reduction equations do not require the oxidation (electron changes) equations in order to balance. The equation is redundant to the solution process. Explain this and provide an example.

## 11.3 | FURTHER READING

Fox, W. P., K. Varanzo, and J. Croteau. "Mathematical Modeling of Chemical Stoichiometry," *Primus*, XVII(4), 301−315. 2007.

Fox, W. P., K. Varanzo, and J. Croteau. "Oxidation−Reduction Chemical Equation Balancing with Maple," *Computers in Education Journal* (COED), Vol XII (2), April−June 2007, pages 50−57.

# 12

# Modeling First-Order Ordinary Differential Equations (ODEs)

## Introduction

Consider the sports parachutist shown here. We want to build a model on how far the parachutist free-falls before he opens his parachute. We might initially assume the parachutist is falling from rest and choose to neglect all resistive forces. Or we might choose to neglect a variable because we want to investigate the problem without the influence of that variable. For instance, we might want to study the falling body by neglecting buoyancy but considering the effects of drag within Earth's atmosphere. We can simplify the submodel for the resistive force by making a reasonable assumption that the drag force equals some constant times the speed of the falling body, or we might decide to try a constant times the speed squared of the falling body. Next, we want to construct submodels for propulsion, drag, and buoyancy forces. The propulsion force acting on the falling body from rest is the result of gravity. This gravitational attraction in turn depends on the mass of the falling body and its distance above Earth's surface. Thus,

$$
\begin{aligned}
\text{Propulsion force} &= F_p \\
&= \text{gravitational attraction} \\
&= f(\text{mass, distance})
\end{aligned}
$$

Next, consider the resistive force, which is the sum of the drag and buoyancy forces:

$$\text{Drag} = F_d$$

$= f(\text{speed, air density, cross-sectional area of the body, shape of the body})$
Buoyancy $= F_b$
$= f(\text{air density, density of the body})$
A free-body diagram, such as that shown in Figure 12.1, displays all the forces acting on the body with a coordinate system indicating directions.

From the free-body diagram, we can see that the total force $F$ acting on the body is given by

$$F = F_p + F_d + F_b$$

We assume that Newton's second law applies: $\Sigma F = ma$ (itself a mathematical model). If we assume to neglect all resisting forces, we find

$$F = ma = -mg + 0 + 0$$

or $my'' = mg$, where $y''$ is a second derivative and subject to initial conditions:

$$y(0) = h, y'(0) = 0$$

Here $m$ is the mass of the body, $g$ is the acceleration due to gravity, and $y$ is the distance of the body measured from Earth's surface. The acceleration $a$ is the second derivative of $y$ with respect to time, and it is this idea of change that led to a differential equation. Because the propulsion force is acting down, it is negative.

If we were to include the force from air resistance as a constant times speed, then $F_r = kv$. We then would have the model

$$F = F_p + F_d + F_b$$
$$F = ma = my'' = -mg + kv$$

with the same initial conditions of $y(0) = h$, $y'(0) = 0$.

In this chapter, we will consider a variety of models, concepts, and techniques that give the reader some of the basic tools needed in solving and analyzing first-order differential

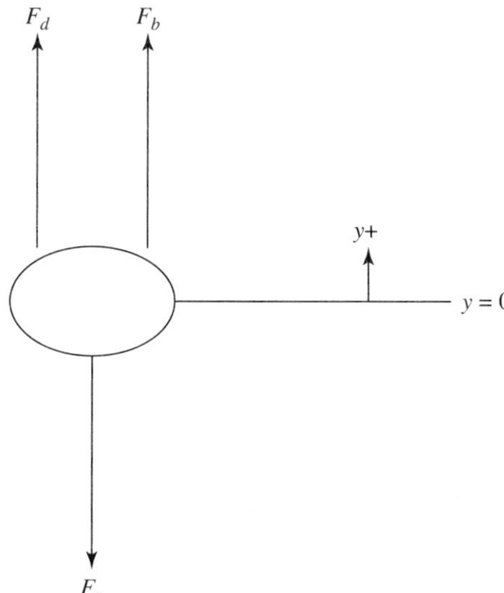


**FIGURE 12.1**
Free-body diagram

equations. Because differential equations are used often in mathematical modeling, we devote a more thorough approach in this chapter.

Section 12.1 provides some models that we will solve and analyze. These models come from a variety of disciplines in science and engineering, including chemistry, physics, biology, fluids mechanics, Newtonian mechanics, environmental engineering, and financial mathematics. Modeling techniques are discussed that help determine the necessary coefficients in the models.

Section 12.2 introduces slope fields. Slope fields provide qualitative viewing of solutions of first-order differential equations. Recall that we said this book discusses the modeling of change. Slope fields provide information about rates of change because they are derived from derivative information.

Section 12.3 covers the analytical method of solution from Maple. Existence and uniqueness are addressed from conceptual and theoretical basis. The focus is on the initial-value problem and the question as to whether or not the initial value problem has a unique solution and whether the new functions can be defined by the initial-value problem.

Section 12.4 is devoted to various numerical techniques. Euler, improved Euler, and Runge–Kutta techniques are introduced as techniques to approximate solutions. Because problems may arise in numerical methods, some errors are discussed as well as methods to compare long-term behavior of the solutions to determine qualitative accuracy of solutions.

## 12.1 APPLIED FIRST-ORDER MODELS

In this section, we introduce many mathematical models from a variety of disciplines. Our emphasis here is in building the mathematical model and its associated differential equation, which will be solved later in the chapter. Recall our previous discussions of the modeling process. In this section, we will confine ourselves to the first three steps: (1) identifying the problem, (2) assumptions and variables, and (3) building the model.

---

**Example 1**      Radioactive Decay

Lord Ernest Rutherford (1871–1937), a British physicist, established that an atom consists of a nucleus surrounded by electrons. The nucleus itself consists of protons and neutrons and two nuclei with identical numbers of protons and identical numbers of neutrons belong to the same nuclear species, or **nuclide**.

The accidental discovery of radioactivity by Henri Becquerel in 1896 was followed quite closely by that of X-rays. Becquerel had left a substance containing uranium on a photographic plate wrapped in black paper. After the plate was developed, it showed images of crystals of uranium compounds. The uranium had emitted penetrating rays, and Becquerel coined the word *radioactive* to describe this property. Later research by Becquerel, Marie Curie, and Rutherford led to the discovery that there are three different kinds of radioactive emissions: alpha, beta, and gamma radiation. Ordinarily, it is of interest or concern to know how long it will take for a certain radioactive substance to lose some fraction of its total mass through the emission process. Let's construct a model to provide these answers.

Assumptions: Assume we have an object that will emit radiation if exposed to a certain substance. Assume that the emissions are all the same kind and that they all decay by the same process (alpha, beta, gamma emissions). The decaying process is such that once a particular nucleus decays, it cannot repeat the process again. Moreover, the probability of its decaying in any time interval is the same. Suppose that at time $t$ there are $N$ undecayed nuclei present. Then the number of nuclei that decay during the time interval from $t$ to $t + \Delta t$ must be proportional to the product of $N$ and $\Delta t$; that is,

$$\Delta N = -kN\Delta t$$

where $k > 0$ is the proportionality coefficient, called the *decay constant*. Let us assume that the number of undecayed nuclei is a continuous function of time. Then dividing both sides of the last equation by $\Delta t$ and passing to the limit as $\Delta t \to 0$ results in the following differential equation:

$$\frac{dN}{dt} = -kN$$

For a given radioactive element, the number $k$ is determined from experimental data. We desire information concerning the radioactive decay of an element.

## Problem Identification

Determine a relationship between time and amount of radiation emitting from a specific substance after exposure.

## Assumptions

We assume the rate in which an item decays is proportional to the amount of that item at any time, $t$. The model is accurate over a short period of time. Half-life is the measure of stability of a radioactive substance. The half-life is simply the time, $t_i$, that it takes one-half of the atoms to disintegrate. The longer the half-life, the more stable the item is.

## Model Construction

Ordinarily, it is of interest to know how long it takes an object emitting radiation to lose some fraction of its total mass or how much of its mass is remaining after a specified amount of time. Many radioactive materials disintegrate at a rate proportional to the amount present. For example, if $x$ is the radioactive material and $Q(t)$ is the amount of radioactive material present at time, $t$, then the rate of change of $Q(t)$ with respect to $t$ is given by $\frac{dQ}{dt} = -kQ$, where $k$ represents the proportionality constant and $k > 0$. Let us refer to $Q(0) = Q_0$ as the initial quantity of radioactive material at any arbitrary time, $t = 0$.

Find the proportionality constant $k$ from the following data.

| Time | Grams of Radioactive Material |
|---|---|
| 0 | 100 |
| 1 | 58.5 |
| 2 | 34.2 |
| 3 | 20 |
| 4 | 11.7 |
| 5 | 7 |
| 6 | 4 |
| 7 | 2.3 |
| 8 | 1.4 |

**Estimation Using the Law of Radioactive Decay**

The law of radioactive decay leads to the following solution:

$$Q(t) = Q_0 e^{-kt}$$

From the data, we know initially that $Q(0) = 100$ grams. After 8 hours, 1.4 grams remain. So,

$$Q(8) = 1.4 = 100e^{-k(8)}$$
$$0.014 = e^{(-8k)}$$
$$\ln(0.014)/-8 = k$$
$$k = 0.5336$$

If we use this method, then we might use the model $Q(t) = Q_o e^{0.5336kt}$.

If we use all the data, then we must create the change in the number of nuclei and then plot versus $N$. The slope (using linear regression without an intercept) will be our value for $k$. Use the following data.

| Time | Grams of Radioactive Material | Difference |
|---|---|---|
| 0 | 100 | −41.5 |
| 1 | 58.5 | −24.3 |
| 2 | 34.2 | −14.2 |
| 3 | 20 | −8.3 |
| 4 | 11.7 | −4.7 |
| 5 | 7 | −3 |
| 6 | 4 | −1.7 |
| 7 | 2.3 | −0.9 |
| 8 | 1.4 | |

We plot the difference versus $N$ and find the slope as $-0.4152$. If we use this approach, our model is

$$Q(t) = Q_0 e^{(-0.4152t)}$$

From a modeling perspective, if you have all the data, or a substantial amount of data, then build the proportionality model to find the slope. If you only have a few data points, then use the model form to estimate the value of $k$.

---

**Example 2**      # Newton's Law of Cooling

You are working as an assistant for crime scene investigators at a possible homicide. A body was found in a trash Dumpster, and you were called to the site immediately. The ambient temperature is approximately 68°F. The body temperature is 77 degrees. After 1 hour, the body temperature is 74°F.

## Problem Identification

Build a mathematical model relating a dead body's cooling temperature to time in order to find the time of death.

## Assumptions

A simplified list of relevant factors could include normal body temperature, average seasonal outdoor temperatures, constant indoor temperature, and no other outside sources of heat or cooling.

## Model Construction

Newton's law of cooling states that the surface temperature of an object changes at a rate proportional to its relative temperature—that is, the difference between its temperature and the surrounding environment temperature. Let $T(t) =$ the temperature of the object at time $t$.

According to Newton's law, $\dfrac{dT}{dt} = k(T - T_s)$, where $k$ is the proportionality constant.

## Example 3

## Mixtures

Consider an initial amount of a substance—say, barley measured in pounds—is dissolved in a large vat in order to make beer. The solution is pumped in at one rate, mixed well, and pumped out at the same rate. We want to know the concentration of the substance in the vat after time.

### Problem Identification

Build a mathematical modeling relating the concentration of barley in the vat to time $t$.

### Assumptions

We know the size of the vat: 300 gallons. The initial amount dissolved is 50 pounds. The pumping rate is constant both into and out of the vat. No other substances are interacting with the vat and the pumps.

### Model Construction

The rate of change of concentration is measured as the difference between the rate pumped in and the mixture being pumped out of the vat. Let $C(t)$ be the concentration or amount of barley in the vat after time $t$, $\dfrac{C(t)}{dt} = R_{in} - R_{out}$.

## Example 4

## Population Models

Consider an experiment measuring the growth of a yeast culture. An analysis yields the growth rate as approximately 0.6. We might try $\dfrac{dP}{dt} = 0.6P$ as our ordinary differential equation (ODE).

Note that this model will predict a population that increases forever. We might think that is not possible.

**Model Refinement: Modeling Births, Deaths, and Resources**

Both births and deaths during a period are proportional to the population, so the change in population itself should be proportional to the population, as illustrated earlier. However, certain resources (food, for instance) can support only a maximum population level rather than one that increases indefinitely. As these maximum levels are approached, growth should slow and move toward *carrying capacity*.

Suppose we estimate the carrying capacity to be 665. Then, as $P$ approaches 665, the change does slow considerably. Because $665 - P$ does get smaller as $P_n$ approaches 665, consider the following model: $\dfrac{dP}{dt} = k(665 - P)P$.

If we accept the proportionality argument, then we can estimate the slope of the line approximating the data from a proportionality graph that we will see later to be $k \approx 0.0008271$, which gives the model

$$P_{n+1} - P_n = 0.0008271(665 - P_n)P_n$$

Converting to an ODE yields

$$\frac{dp}{dt} = 0.0008271 \cdot P \cdot (665 - P)$$

## Example 5    The Spread of a Contagious Disease

Suppose there are 400 students in a college dormitory and that one or more of the students has a severe case of the flu. Let $i_n$ represent the number of infected students after $n$ time periods. Assume that some interaction between those who are infected and those who are not is required to spread the disease. If all are susceptible to the disease, then $(400 - i_n)$ represents those susceptible but not yet infected. If those infected remain contagious, then we may model the change in the number of infected as a proportionality to the product of those infected by those susceptible but not yet infected, or

$$\Delta i_n = i_{n+1} - i_n = k i_n (400 - i_n)$$

In this model, the product $i_n(400 - i_n)$ represents possible interactions between those infected and those not infected. A fraction $k$ would now become infected. There are many refinements to the above model. For example, we can consider that a segment of the populations is not susceptible to the disease, that the infection period is limited, that infected students are removed from the dorm to prevent interaction with uninfected students, and so forth.

$$\frac{di}{dt} = ki(400 - i), \ k > 0$$

**Some Classical First-Order Models in Differential Equations**

Decay: $\dfrac{dQ}{dt} = -kQ, \ k > 0$

Growth: $\dfrac{dQ}{dt} = kQ, \ k > 0$

Newton's law of cooling: $\dfrac{dT}{dt} = k(T - T_0), \ k > 0$

Chemical mixtures, drug dissemination, dialysis: $\dfrac{dA}{dt} = R_{in} - R_{out}$

L-R series circuit: $L\dfrac{dI}{dt} + RI = E(t)$

Falling bodies, we use Newton's law of $\Sigma$ Forces $= 0$

$$m\frac{dv}{dt} = F_{weight} + F_{resistance} = mg - kv, \ k > 0$$

Chemical saturation: $\dfrac{dX}{dt} = k(\alpha - X)(\beta - X)$ where $k$ is a saturation rate, $X$ is the substance being saturated, $\alpha$ is current level and $\beta$ is the saturation level.

Spread of a disease or a rumor: $\dfrac{dN}{dt} = kN(L - N), \ k > 0, \ L > 0$

Tank with cross-sectional area $A$ filled from the top at rate $k$ while draining from the bottom through an orifice of area $\alpha$ with height of the tank $h$:

$$\frac{dh}{dt} = k - \frac{\left(k - \alpha a \sqrt{2gh}\right)}{A}$$

## 12.2    SLOPE FIELDS AND QUALITATIVE ASSESSMENTS OF AUTONOMOUS FIRST-ORDER ODEs

We have used the modeling process to create a number of models that are important to differential equations. However, sometimes we can predict outcomes of models graphically without actually solving them. This is true when we are looking for behavior rather than specific values. For example, "Will a specific population survive?" is a question that can be answered graphically.

In calculus, we used derivative information to analyze a function or just to gain more information. Knowing where a function is increasing or decreasing, locations of extrema, concavity, and so forth can be important information.

Interpreting the derivative as the slope of the line tangent to the graph of the function is useful in gaining information about the solution to the differential equation. From this information, it is possible to sketch qualitatively the solution curves to the equations from any real initial condition. These graphs provide considerable information regarding the solution as well as provide piratical benefits. Because most real differential equations cannot be solved analytically, other techniques such as graphical analysis are needed to analyze the behavior of solutions.

**Slope Fields**

Suppose the first-order differential equation $\frac{dy}{dt} = f(t,y)$ is given. Each time an initial value $y(t_0) = y_0$ is specified, the solution curve is required to pass through the coordinate point $(t_0, y_0)$, and we also know the value of slope at that point: $\frac{dy}{dt} = f(t_0, y_0)$. Therefore, it is possible to draw a short line segment with the correct slope through each point $(t,y)$ in the $t - y$ plane. The result is known as the *slope field of the first-order differential equation.* We can say that the slope field of a differential equation is the set of all line segments with the correct slope or minitangent lines.

Let's sketch the possible solution curves of the differential equation, $\frac{dy}{dt} = \frac{(t-y)}{2}$, without solving. Consider the information in the following table, which provides specific points in the plane and the slope at each point.

| Point | Slope |
|---|---|
| (0,1) | $-\frac{1}{2}$ |
| (0,2) | $-1$ |
| (1,0) | $\frac{1}{2}$ |
| (2,0) | $1$ |
| (−1,0) | $-\frac{1}{2}$ |
| (−2,0) | $-1$ |
| (1,1) | $0$ |
| (−1,1) | $-1$ |
| (2,2) | $0$ |
| (1,2) | $-\frac{1}{2}$ |

In Maple, we must first call the library with(DEtools). Then we enter the differential equation and enter the commands to get a specific plot.

**Command Sequence**

- *with(DEtools):*
- *deq:=diff(y(t),t)=(t−y(t))/2;*
- *DEplot(deq,y(t), t = a..b, y=c..d);*

A complete slope field over the region $-2 \le t \le 2$ is seen in Figure 12.2.

```
> example_1:=diff(y(t),t)=(t−y(t))/2;
> DEplot(example_1,y(t),t=−2..2,y=0..2);
```

$$example\_1 := \frac{d}{dt}\,y(t) = \frac{t}{2} - \frac{1}{2}\,y(t)$$

We use Maple to produce the slope-field plot. Although knowledge of how to graph slope fields is important, you will not want to graph complete slope fields without technology. A

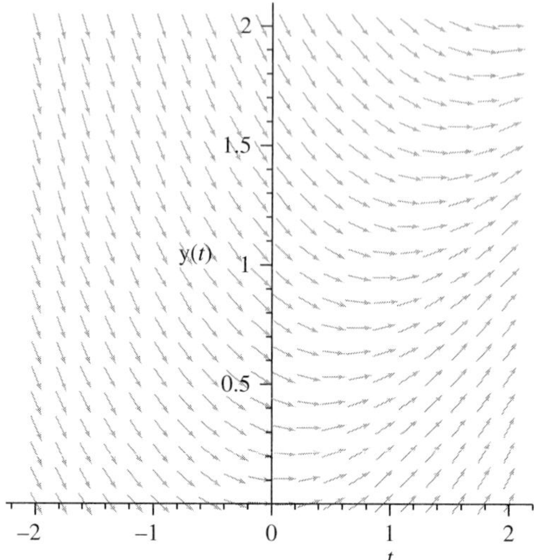

**FIGURE 12.2**
Slope field for
$$\frac{dy}{dt} = \frac{(t - y)}{2}$$

solution curve is then tangent to the slope line at each point through which the curve passes. Thus, the slope field gives a visual representation of what a family of possible solution curves of the differential equation looks like. Several solution curves are now depicted in Figure 12.3 for $\frac{dP}{dt} = 0.0008271 \cdot P \cdot (665 - P)$ with starting points (0,2) and (0,900). The possible solution curves are seen in Figure 12.3.

We examine the slope field and notice that the solution curves do not cross tangent lines.

Below are the Maple commands to obtain the slope fields and possible solution curves for the following first-order differential equations:

$$y' = y - t^2$$

> *example1:=diff(y(t),t)=y(t)–t^2;*

$$example1 := \frac{d}{dt}\, y(t) = y(t) - t^2$$

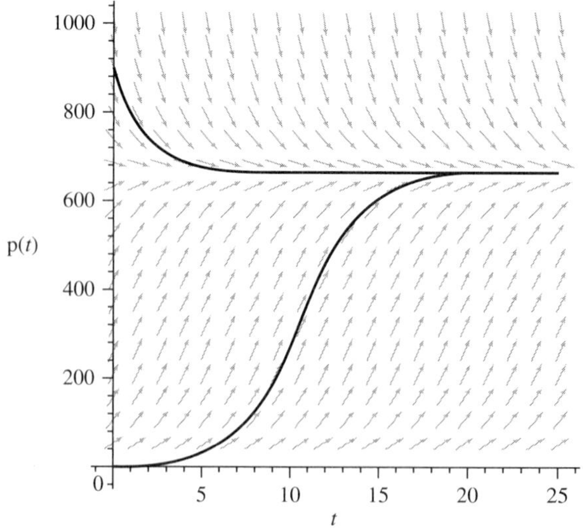

**FIGURE 12.3**
Slope Field for
$$\frac{dP}{dt} = 0.0008271 \cdot P \cdot (665 - P)$$
with starting points (0,2) and (0.900).

> *DEplot(example1,y(t),t=0..5,y=0..10,{[0,2],[3,3]});*

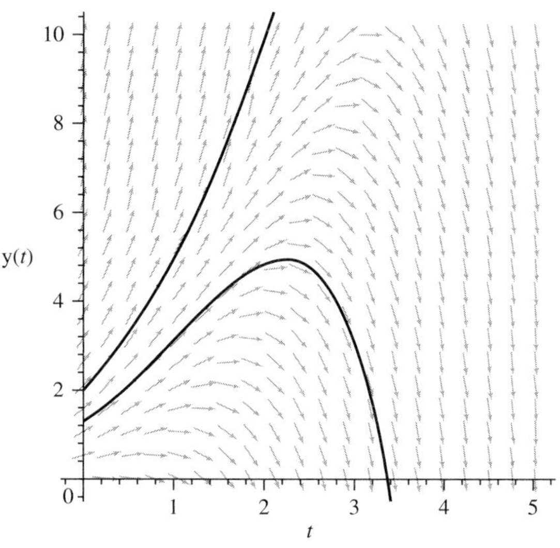

$$y' = y \cdot (2 - y)$$

> *example2:=diff(y(t),t)=y(t)*(2−y(t));*
> *DEplot(example2,y(t),t=0..5,y=0..5,{[0,.5],[0,2],[0,4]});*

$$example2 := \frac{d}{dt}\,y(t) = y(t)\,(2 - y(t))$$

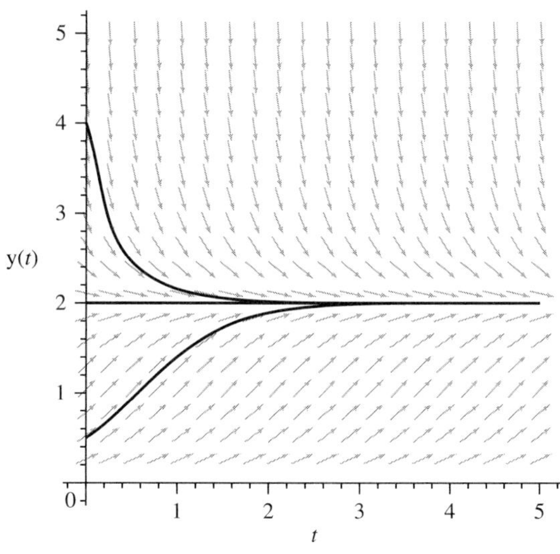

$$y' = sin\,(t + y)$$

> *example3:=diff(y(t),t)=sin(t+y(t));*
> *DEplot(example3,y(t),t=0..5,y=0..7,{[0,.5],[0,4],[0,6]});*

$$example3 := \frac{d}{dt}\,y(t) = \sin\,(t + y(t))$$

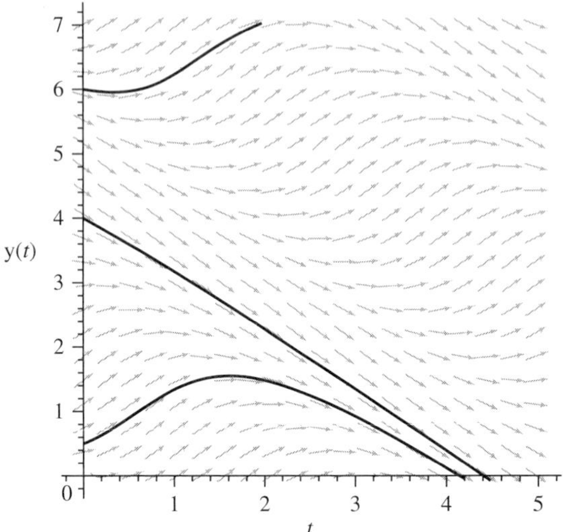

Slope fields provide us with possible solution curves for a differential equation.

### Autonomous Qualitative Assessment

In many modeling applications, the instantaneous rate of change $\frac{dy}{dt}$ is assumed to be proportional to some function of the dependent variable $y$ alone. These are called *autonomous differential equations*, $\frac{dy}{dt} = f(y)$. In these cases, it is possible to analyze the nature of the solution curve by analyzing $\frac{dy}{dt}$ versus $y$. Let's explore this qualitative assessment.

### Rest Points

Qualitative information concerning the solutions can often be obtained from the differential equation. For example, consider the differential equation describing the Malthusian model of population growth discussed earlier: $\frac{dP}{dt} = kP,\ k > 0$. Thus, the Malthusian model assumes a simple proportionality between growth rate and population (see Figures 12.4 and 12.5). Because $P > 0$ and $k > 0$, you can see that $\frac{dP}{dt} > 0$, which means that the population curve $P(t)$ is everywhere increasing. Moreover, the smaller the value of $k$, the less rapid is the growth in the population over time. For a fixed value of $k$, as $P$ increases, so does its rate of change $\frac{dP}{dt}$. Thus, the solution curves must appear qualitatively as depicted in Figure 12.4. Assuming $k$ is constant, at each population level $P$ the derivative $\frac{dP}{dt}$ is constant. Therefore, all the solution

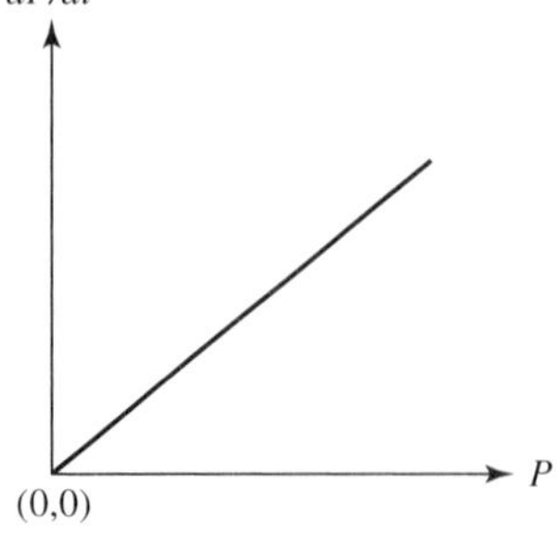

**FIGURE 12.4**
Graph of *dP/dt* versus *P*

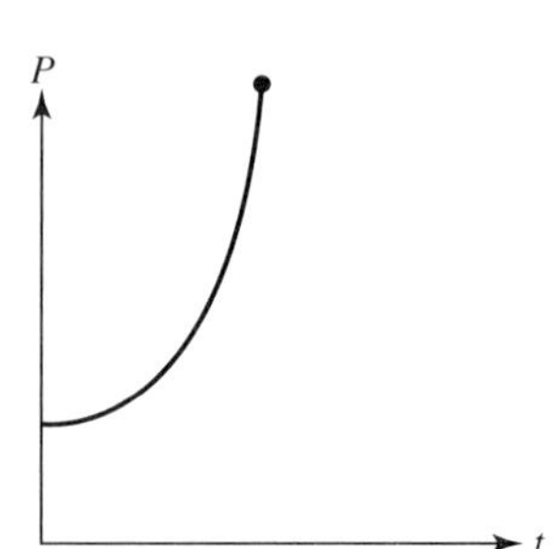

**FIGURE 12.5**
Plot of *P* versus *t*
from Figure 12.4

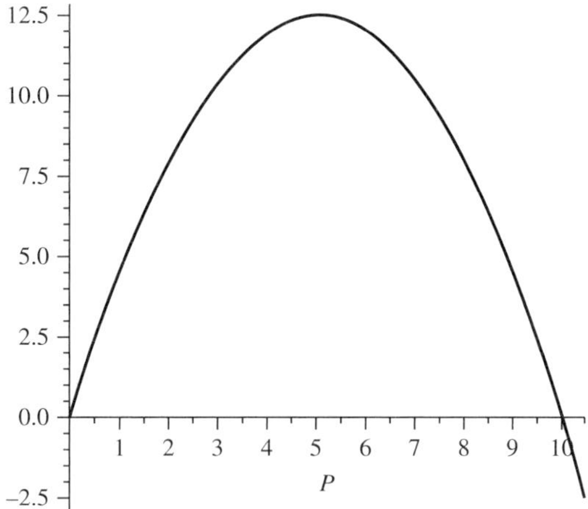

curves are horizontal translations of one another. The initial population $P(0) = P_0$ distinguishes these various curves.

Next, recall the refined population model from Section 12.1: $\dfrac{dP}{dt} = r\,(M - P)P$. The graph of $\dfrac{dP}{dt}$ versus P is the parabola shown in Figure 12.6. Because the second derivative $P'' = rP'$ $(M - 2P)$ is zero when $P = \dfrac{M}{2}$, positive for $P < \dfrac{M}{2}$, and negative for $P > \dfrac{M}{2}$, we conclude that $\dfrac{dP}{dt}$ is at a maximum at the population level. Note that for $0 < P < M$, the derivative $\dfrac{dP}{dt}$ is positive and the population $P(t)$ is increasing; for $M < P$, $\dfrac{dP}{dt}$ is negative and $P(t)$ is decreasing. Moreover, if $P < \dfrac{M}{2}$, then the second derivative $P''$ is positive and the population curve is concave upward for $P > \dfrac{M}{2}$, $P''$ is negative, and the population curve is concave downward.

The parabola in Figure 12.6 has zeros at the population levels $P = 0$ and $P = M$. At those levels, $\dfrac{dP}{dt} = 0$, so no change in the population $P$ can occur. In other words, if the population level is at 0, then it will remain there for all time; if it is at the level $M$, then it will remain there for all time. Points for which the derivative $\dfrac{dP}{dt}$ at 0 are called **rest points**, **equilibrium points**, **stationary points**, or **critical points** of the differential equation. The behavior of solutions near rest points are of significant interest to us. Let's examine what happens when the population $P$ is near the rest points $P = M$ and $P = 0$.

Suppose the population $P$ is slightly less than $M$. Then $\dfrac{dP}{dt}$ is positive, so the population increases and gets closer to $M$. On the other hand, if $P > M$, $\dfrac{dP}{dt} < 0$, the population will decrease toward $M$. Thus, no matter what positive value is assigned to the starting population, it will tend to the limiting value $M$ as time tends to infinity. We say that $M$ is an **asymptotically stable rest point**, because whenever the population level is perturbed away from that level it tends to return there again. However, if the starting population is not at the level $M$, the population $P(t)$ cannot reach M in a finite amount of time. This fact follows from the property that $P = M$ is a solution to the limited growth equation and the two solutions cannot cross. Look at the slope field plots that we have previously shown to see that trajectories never cross.

Next, consider the rest point $P = 0$ in Figure 12.6. If $P$ is perturbed slightly away from 0 so that $M > P > 0$, then $\dfrac{dP}{dt}$ is positive and the population increases toward $M$. In this situation, we say that $P = 0$ is an **unstable rest point**, because any population not starting at that level tends to move away from it.

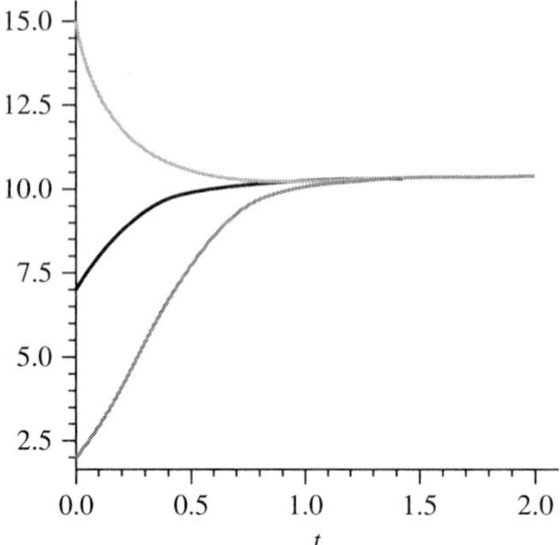

**FIGURE 12.7**

Possible plots of $P$ versus $t$ with $M = 10$

In general, equilibrium solutions to differential equations are classified as stable or unstable according to whether, graphically, nearby solutions stay close to or converge to the equilibrium or whether they diverge away from the equilibrium, respectively, as the independent variable $t$ tends to infinity.

At $P = \dfrac{M}{2}$, $P''$ is zero and the derivative $\dfrac{dP}{dt}$ is at a maximum (see Figure 12.6). Thus, the population $P$ is increasing most rapidly when $P = \dfrac{M}{2}$, and a point of inflection occurs in the graph there. These features give each solution curve its characteristic "S" or sigmoid shape. From the information we have obtained, the family of solutions to the limited growth model must appear (approximately) as shown in Figure 12.5. Notice again that at each population level $P$, the derivative $\dfrac{dP}{dt}$ is constant, so that the solution curves are horizontal translations of one another. The curves are distinguished by the initial population level $P(0) = P_0$. In particular, note the solution curve when $\dfrac{M}{2} < P_0 < M$ in Figure 12.7.

## 12.2 | EXERCISES

**1.** Construct a direction field for the following differential equations and sketch in a solution curve.

**a.** $\dfrac{dy}{dx} = y$

**b.** $\dfrac{dy}{dx} = x$

**c.** $\dfrac{dy}{dx} = x + y$

**d.** $\dfrac{dy}{dx} = x - y$

**e.** $\dfrac{dy}{dx} = xy$

**f.** $\dfrac{dy}{dx} = \dfrac{1}{y}$

**2.** Sketch a number of solutions to the following equations showing the correct slope, concavity, and any points of inflection.

**a.** $\dfrac{dy}{dx} = (y + 2)(y - 3)$

**b.** $\dfrac{dy}{dx} = y^2 - 4$

**c.** $\dfrac{dy}{dx} = y^3 - 4$

**d.** $\dfrac{dy}{dx} = x - 2y$

3. Analyze graphically the equation $\dfrac{dy}{dt} = ry$, when $r < 0$. What happens to any solution curve as $t$ becomes large?

4. Develop graphically the following models. First graph $\dfrac{dP}{dt}$ versus $P$, then obtain various graphs of $P$ versus $t$ by selecting different initial values $P(0)$ (as in our population example in the text). Identify and discuss the nature of the equilibrium points in each model.

   **a.** $\dfrac{dP}{dt} = a - bP,\ a,b > 0$

   **b.** $\dfrac{dP}{dt} = P(a - bP),\ a,b > 0$

   **c.** $\dfrac{dP}{dt} = k(M - P)(P - m),\ k, M, m > 0$

   **d.** $\dfrac{dP}{dt} = kP(M - P)(P - m),\ k, M, m > 0$

5. The department of fish and game in a certain state is planning to issue deer-hunting permits. It is known that if the deer population falls below a certain level, $m$, then the deer will become extinct. It is also known that if the deer population goes above the maximum carrying capacity, $M$, the population will decrease to $M$.

   **a.** Discuss the reasonableness of the following model for the growth rate of the deer population as a function of time: $\dfrac{dP}{dt} = kP(M - P)(P - m)$, where $P$ is the population of the deer and $k$ is a constant of proportionality. Include a graph of $\dfrac{dP}{dt}$ versus $P$ as part of your discussion.

   **b.** Explain how this growth-rate model differs from the logistic model $\dfrac{dP}{dt} = kP(M - P)$. Is it better or worse than the logistic model?

   **c.** Show that if $P > M$ for all $t$, then the limit $P(t)$ as $t \to \infty$ is $M$.

   **d.** Discuss what happens if $P < m$ for all $t$.

   **e.** Assuming that $m < P < M$ for all $t$, explain briefly the steps you would use to solve the differential equation. Do not attempt to solve the differential equation.

   **f.** Graphically discuss the solutions to the differential equation. What are the equilibrium points of the model? Explain the dependence of the equilibrium level of $P$ on the initial conditions. How many deer-hunting permits should be issued?

## 12.3  ANALYTICAL SOLUTION TO FIRST-ORDER ODEs

### Separable ODEs

The differential equation of the form $\dfrac{dy}{dt} = f(t,y)$ is called *separable* if $f(t,y) = h(t) \cdot g(y)$; that is, $\dfrac{dy}{dt} = h(t) \cdot g(y)$. You need to able to write the function $f(t,y)$ as $h(t) \cdot g(y)$.

**Example 1**      ### Recognizing a Separable ODE

$$\frac{dy}{dt} = t \cdot y$$

This is separable because the function $f(t,y) = h(t) \cdot g(y) = t \cdot y$.

### Example 2    Recognizing a Nonseparable ODE

$$\frac{dy}{dt} = t + 9y$$

This is not separable because $f(t,y) = t + 9y$ cannot be written as the product of two functions, $h(t){\cdot}g(y)$.

Sometimes this is more complicated to see, such as in the following example.

### Example 3    A Separable ODE after Manipulation

$$\frac{dy}{dt} = 8ty - 12y - 2t + 3$$

This can be written as $\dfrac{dy}{dt} = (2t - 3) \cdot (4y - 1)$, and it is separable in this form.

To solve separable ODEs, we perform the following steps:

1. Separate the differential equation into a clear form of

$$\frac{dy}{dt} = h(t){\cdot}g(y)$$

2. Rewrite the differential equation as

$$\frac{dy}{g(y)} = h(t)dt$$

3. Integrate both sides of the expression and include the constant of integration, $C$,

$$\int \frac{dy}{g(y)} = \int h(t) \cdot dt$$

to obtain the general solution

$$G(y) = H(t) + C$$

4. If you are given an intermediate-value property, use the initial conditions to find the particular solution.

### Example 4    Solving a Separable ODE

$$\text{Solve } \frac{dy}{dt} = t \cdot y$$

$$\frac{dy}{y} = tdt$$

$$\int \frac{dy}{y} = \int tdt$$

$$\ln |y| + C = \frac{t^2}{2} + C$$

$$\ln |y| = \frac{t^2}{2} + C$$

$$e^{\ln|y|} = e^{\frac{t^2}{2}+C}$$

$$y = ce^{\frac{t^2}{2}}$$

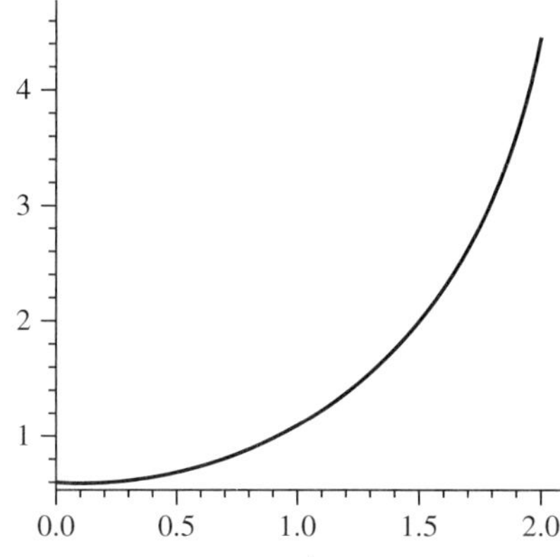

If the ODE $\dfrac{dy}{dt} = t \cdot y$ passes through the point $(1, 1)$, then we use $(1, 1)$ as the initial condition.

$$1 = ce^{\frac{1}{2}}$$

$$c = \frac{1}{e^{\frac{1}{2}}} = e^{-.5} = 0.60653$$

$$y = 0.60653e^{\frac{t^2}{2}}$$

We plot the solution as seen in Figure 12.8.

---

## Example 5        Radioactive Decay as a Separable ODE

Consider the following radioactive-decay model, $\dfrac{dQ}{dt} = -kQ, k > 0$ with a radioactive isotope with a half-life of 16 days. You wish to have 30 grams at the end of 25 days.

$$\frac{dQ}{Q} = -kdt$$

$$ln|Q| = -kt + C$$
$$Q = ce^{-kt}$$

Because the half-life is 16 days,

$$Q_0 = ce^{-kt}$$

$$\frac{Q_0}{2} = Qe^{-16k}$$

$$k = \frac{ln|2|}{16} = .043322$$

$$Q = ce^{-0.043322t}$$

We want 30 grams after 25 days, so

$$30 = ce^{(-0.043322 \cdot 25)}$$

$$c = \frac{30}{e^{-0.043322 \cdot 25}} = 88.61g$$

Initially, we should start with 88.61 grams of the radioactive substance. The model is $Q = 88.61 \cdot e^{-.043322t}$.

## Example 6

## Malthusian Population Model

The population in the United States is provided in the following table.

| Year | Population in Millions |
|------|------------------------|
| 1990 | 249.4 |
| 1991 | 252.2 |
| 1992 | 254.9 |
| 1993 | 257.7 |
| 1994 | 260.2 |
| 1995 | 262.7 |
| 1996 | 265.1 |
| 1997 | 267.7 |
| 1998 | 270.2 |
| 1999 | 272.6 |
| 2000 | 280.5 |

The model that we will initially use is our simple growth model, $\dfrac{dP}{dt} = kP,\ k > 0$.

### Solution

$$\frac{dP}{P} = k\,dt$$

$$\int \frac{dP}{p} = \int k\,dt$$

$$ln|P| = kt + C$$

$$P = ce^{kt}$$

Our initial population in 1990 was 249.4, so $P(0) = 249.4$:

$$P = 249.4e^{kt}$$

We need to find a good estimate for the growth parameter $k$. We find it to be

$$\ln \frac{\left(\dfrac{280.5}{249.4}\right)}{9} = 0.012859$$

$$P = 249.4e^{0.012859t}$$

Predicting the population in 2020 yields

$$P = 249.4e^{0.012859 \cdot 20} = 318.4229 \text{ million people}$$

There are two problems with this model. First, it is not realistic to assume that populations tend to infinity. Second, the predictions of this model always underestimate the real populations. We could use a refinement to this model (to be suggested later) that does not yield the slow exponential growth toward infinity.

## Example 7

## Newton's Law of Cooling

Suppose a cold can of soda is taken from a refrigerator and placed in a warm classroom and the temperature is measured periodically. The temperature of soda is initially 40°F, and the room temperature is a cozy 72°F.

Newton's law of cooling gives

$$\frac{dT}{dt} = -k(T_0 - T), \ T(0) = 40$$

$$\frac{dT}{dt} = -k(72 - T), \ T(0) = 40$$

$$\frac{dT}{(72-T)} = -kdt$$

$$\int \frac{dT}{(72-T)} = \int -kdt$$

$$\ln|72 - T| = -kt + C$$

Assume $k = 0.05$.

$$72 - T = Ce^{-0.05t}$$

When $T(0) = 40$, $C = 32$:

$$T = 72 - 32e^{-0.05t}$$

After 10 minutes, the soda's temperature is

$$T(10) = 72 - 32 \cdot e^{-0.05 \cdot 10} = 52.591$$

After 30 minutes, the soda's temperature is

$$T(30) = 72 - 32 \cdot e^{-0.05 \cdot 30} = 64.860$$

Note that the soda warms and will approximately reach the temperature of the room, a cozy 72°F (but not cozy for drinking a soda).

---

## Example 8      Chemical Mixtures

Initially, 50 pounds of salt is dissolved in a 300-gallon tank of water. A brine solution is pumped into the tank at a rate of 3 gallons per minute, and a well-stirred mixed solution is then pumped out at a rate of 3 gallons per minute. We want to determine the amount of salt present, $A(t)$, at any time $t$, if the concentration of the entering solution is 2 pounds per gallon.

$$\frac{dA}{dt} = R_{in} - R_{out}$$

$$R_{in} = \frac{3 \ gallons}{1 \ minute} \cdot \frac{2 \ pounds}{1 \ gallon} = 6 \ pounds \ per \ minute$$

$$R_{out} = \frac{3 \ gallons}{1 \ minute} \cdot \frac{\frac{A \ pounds}{300 \ gallons}}{1 \ gallon} = \frac{A}{100} \ pounds \ per \ minute$$

$$\frac{dA}{dt} = 6 - \frac{A}{100}, \ A(0) = 50$$

We will rewrite the problem as

$$\frac{dA}{dt} = .01(600 - A), \ A(0) = 50$$

This is separable.

$$\frac{dA}{600 - A} = .01dt$$

$$\int \frac{dA}{600 - A} = \int .01\, dt$$

$$-\ln(600 - A) = .01t + C$$
$$\ln(600 - A) = -.01t + C$$
$$A = 600 + ce^{-0.01t}$$

Using the initial condition, $A(0) = 50$:

$$c = -550$$
$$A = 600 - 550e^{-0.01t}$$

---

## Example 9    Refined Population Model

Consider the logistic growth model defined for blue crabs in Venezuela:

$$\frac{dB}{dt} = .25B(10 - B),\ B(0) = 4\ (B \text{ is in millions of bushels})$$

$$\frac{dB}{B(10 - B)} = .25\, dt$$

$$\int \frac{dB}{B(10 - B)} = \int .25\, dt$$

$$\frac{A}{B} + \frac{D}{10 - B} = 1$$

$$A \cdot (10 - B) + B \cdot D = 1 + 0B$$

$$10A = 1,\ A = \frac{1}{10}$$

$$-A + D = 0,\ A = D,\ D = \frac{1}{10}$$

So,

$$\int \frac{dB}{B(10 - B)} = \int \frac{1}{10B}\, dB + \frac{1}{10}\int \frac{1}{(10 - B)}\, dB = \frac{1}{10}[\ln(B) - \ln(10 - B)]$$

$$\frac{1}{10}[\ln(B) - \ln(10 - B)] = .25t + C$$

Multiply both sides by 10:

$$[\ln(B) - \ln(10 - B)] = 2.5t + C$$

Solve for $C$, using $B(0) = 4$:

$$\ln(4) - \ln(6) = C$$
$$C = -0.405465$$

Using the laws of logarithms and exponents, we obtain

$$\frac{B}{10 - B} = .666666 \cdot e^{2.5t}$$

We simplify to

$$B = \frac{6.66666 \cdot e^{2.5t}}{1 + .66666 \cdot e^{2.5t}} \text{ or } B = \frac{6.6666666}{.666666 + e^{-2.5t}}$$

The result is known as a *logistics curve*. It approaches the maximum sustainable amount of 6.666 million bushels.

### Linear Differential Equations

Let's consider differential equations of the form $a_1(t)\dfrac{dy}{dt} + a_0(t)y = g(t)$, which can be simplified by dividing every term by $a_1(t)$ to obtain $\dfrac{dy}{dt} + P(t)y = f(t)$. We seek a solution to this differential equation. If we put the ODE in the form, $M(t,y)dt + N(t,y)dy = 0$ then we can check to see if $\dfrac{\partial M}{\partial y} = \dfrac{\partial N}{\partial t}$. Using the exact method $dy + P(t)y - f(t)dt = 0$, it is probably not true that $\dfrac{\partial M}{\partial y} = \dfrac{\partial N}{\partial t}$. But if there exists a function $u(t)$ such that $\dfrac{\partial}{ty}u(t) = \dfrac{\partial}{\partial y}[u(t)\cdot(P(t)y - f(t))]$, then we will have an exact equation. $\dfrac{\partial}{ty}u(t) = \dfrac{\partial}{\partial y}[u(t)\cdot(P(t)y - f(t))]$, which reduces to $\dfrac{du}{dt} = u\cdot P(t)$. This is now separable, so

$$\frac{du}{u} = P(t)dt$$

$$ln|u| = \int P(t)dt$$

$$u = e^{\int P(t)dt}$$

Now, this function $u = e^{\int P(t)dt}$ is called the **integrating factor** for the linear equation. We will use this concept of the integrating factor in the linear solution method, also called **variation of parameter**.

**Step 1**   Put the differential equation in the form $\dfrac{dy}{dt} + P(t)y = f(t)$.

**Step 2**   Find the integrating factor: $e^{\int P(t)dt}$.

**Step 3**   Multiply through the differential equation by the integrating factor:

$$e^{\int P(t)dt}\frac{dy}{dt} + e^{\int P(t)dt}\cdot P(t)y = e^{\int P(t)dt}\cdot f(t)$$

**Step 4**   This simplifies to $\dfrac{d}{dt}\left[y\cdot e^{\int P(t)dt}\right] = e^{\int P(t)dt}\cdot f(t)dt$.

**Step 5**   Integrate both sides, including constant of integration:

$$\int\frac{d}{dt}\left[y\cdot e^{\int P(t)dt}\right] = \int e^{\int P(t)dt}\cdot f(t)dt$$

$$y\cdot e^{\int P(t)dt} = \int e^{\int P(t)dt}\cdot f(t)dt + C$$

**Step 6**   Solve for $y$.

---

## Example 10     Solving a Linear ODE by Using the Integrating Factor

Let's take a look at our example from slope fields, $\dfrac{dy}{dt} = \dfrac{t-y}{2}$.

**Step 1**   $\dfrac{dy}{dt} + \dfrac{y}{2} = \dfrac{t}{2}$, $P(t) = \dfrac{1}{2}$

**Step 2**   The integrating factor is $e^{\int\frac{dt}{2}}\cdot u = e^{\frac{t}{2}}$

**Steps 3 and 4**   $e^{\frac{t}{2}}\cdot\dfrac{dy}{dt} + e^{\frac{t}{2}}\dfrac{y}{2} = e^{\frac{t}{2}}\dfrac{t}{2}$ simplifies to $\dfrac{d}{dt}\left[y\cdot e^{\frac{t}{2}}\right] = \dfrac{t}{2}\cdot e^{\frac{t}{2}}$.

**Step 5**   $\int\dfrac{d}{dt}\left[y\cdot e^{\frac{t}{2}}\right] = \int\dfrac{t}{2}\cdot e^{\frac{t}{2}}$ (use integration by parts):

$$y\cdot e^{\frac{t}{2}} = t\cdot e^{\frac{t}{2}} - 2\cdot e^{\frac{t}{2}} + C$$

**Step 6**   $y = t - 2 + C\cdot e^{\frac{-t}{2}}$

## Example 11

## Newton's Law of Cooling

$\dfrac{dT}{dt} = -k\,(T_m - T_s)$, where $k > 0$, $T_s$ = constant temperature of the surroundings; $T_m$ = temperature of the mass being cooled; and $T(0) = \alpha$ (the initial temperature of the mass).

Let's assume that $k = 0.190$, $T_s = 70$, $T(0) = 300$. Now, $\dfrac{dT}{dt} = -.190(T_m - 70)$, $T(0) = 300$.

We notice that this equation is both separable and linear. We will use the linear method.

**Step 1**   Standard form is $\dfrac{dT}{dt} + .190(T_m) = .190 \cdot (70) = 13.3$, $T(0) = 300$.

**Step 2**   $P(t) = .190$

$$u = e^{\int p(t)dt} = e^{.19t}$$

**Steps 3 and 4**   $\dfrac{d}{dt}\left[T_m \cdot e^{.19t}\right] = 13\ 3 \cdot e^{.19t}$

**Step 5**   $\dfrac{d}{dt}\left[T_m \cdot e^{.19t}\right] = 13.3 \cdot e^{.19t}$

$$T_m \cdot e^{.19t} = 70e^{.19t} + C$$

**Step 6**   $T_m = 70 + Ce^{-.19t}$

Use $T(0) = 300$

$300 = 70 + C$

$C = 230$, so $T_m = 70 + 230 \cdot e^{-.19t}$

We note that $\lim\limits_{t \to \infty} 70 + 230 \cdot e^{-.19t} = 70$ (the room temperature) as expected.

### First-Order ODEs and Maple

In Maple, we begin with the command *with(DETools)* to open Maple's differential equation library. We also introduce the *dsolve* command that solves many first-order ordinary differential equations both with and without initial conditions. In *dsolve*, the parameters are *deqns*, which lists the ordinary differential equation in its variables, and *vars*, a variable or set of variables.

```
Command
With(DETools):
dsolve(deqns,vars);
```

We will use Maple to resolve many of our previous examples.

## Example 12

## Revisit Example 4 Using Maple

Solve $\dfrac{dy}{dt} = t \cdot y$

First, we will solve using Maple without initial conditions.

```
> restart;
> with(DETools):
> eqn:=diff(y(t),t)=y(t)*t;
```

$$eqn := \frac{d}{dt}\,y(t) = y(t)t$$

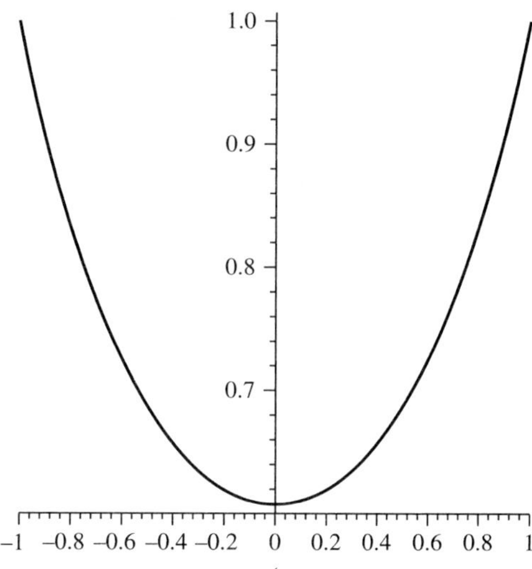

**FIGURE 12.9**
Solution plot to $y' = y\,t$

> *dsolve( eqn,y( t ) );*

$$y(t) = \_C1\; e^{\left(\frac{t^2}{2}\right)}$$

Now, we solve again after we introduce the initial condition $y(1) = 1$.

> *with( DETools ):*
> *eqn:=diff( y( t ),t )=y( t )*t;*

$$eqn := \frac{d}{dt}\,y(t) = y(t)t$$

> *inits:=y( 1 )=1;*

$$inits := y(1) = 1$$

> *dsolve( {eqn,inits},y( t ) );*

$$y(t) = \frac{e^{\left(\frac{t^2}{2}\right)}}{e^{(1/2)}}$$

We plot the following solution shown in Figure 12.9 from the point $[-1,1]$.

> *plot( exp( t^2/2 )/exp( 1/2 ),t=–1..1 );*

---

## Example 13     Refined Population Model Using Maple

Consider the logistic growth model defined for blue crabs in Venezuela:

$$\frac{dB}{dt} = .25B(10 - B),\; B(0) = 4 \;(B \text{ is in millions of bushels})$$

> *eqn:=diff( B( t ), t )=0.25*B( t )*( 10–B( t ) );*

$$eqn := \frac{d}{dt}\,\mathrm{B}(t) = 0.25\,\mathrm{B}(t)\,(10 - \mathrm{B}(t))$$

> *inits:=B( 0 )=4;*

$$inits := \mathrm{B}(0) = 4$$

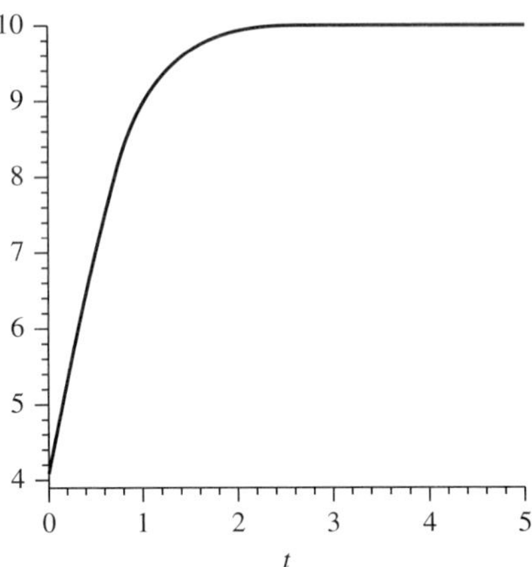

**FIGURE 12.10**
Solution plot for $B' = 0.25B$
$(10 - B)$

```
> dsolve({eqn,inits},B(t));
```

$$B(t) = \frac{20}{2 + 3\,e^{\left(\frac{-5t}{2}\right)}}$$

```
> plot(20/(2+3*exp(-5/2*t)), t=0..5);
```

We view the solution plot in Figure 12.10.

Not all ODEs are, in fact, solvable in closed form. If Maple does not return a solution then there might not be a closed-form analytical solution. We might try a numerical method, as we will discuss later in this chapter.

```
> eqn:=diff(c(t),t)=(c(t)+2*t)/(t^2+c(t)^3);
```

$$eqn := \frac{d}{dt}\,c(t) = \frac{c(t) + 2\,t}{t^2 + c(t)^3}$$

```
> dsolve(eqn,c(t));
```

We note that Maple did not return a closed form. Thus, we need to be able to use numerical solutions.

---

## 12.4  NUMERICAL METHODS FOR SOLUTIONS TO FIRST-ORDER ODEs

As a mentioned in the last section, not all first-order differential equations are solvable in closed form. In this section, we briefly introduce three numerical techniques: Euler's method, improved Euler's method, and Runge–Kutta. We will discuss each technique and its algorithm. We will illustrate the Maple commands with our examples.

### Euler's Method

From the point of view of a mathematician, the *ideal* form of the solution to an initial value problem would be a **formula** for the solution function. After all, if this formula is known, it is usually relatively easy to produce any other form of the solution you may desire, such as a **graphical solution** or a **numerical solution**, in the form of a table of values. You might say that a formulaic solution contains the recipes for these other types of solution within it. Unfortunately, as we have seen in our studies already, obtaining a formulaic solution is not always easy and sometimes is absolutely impossible.

So we often have to "make do" with a numerical solution—that is, a table of values consisting of points that lie along the solution's curve. This can be a perfectly usable form of the answer in many applied problems, but before we go too much further, let's make sure we are aware of the shortcomings of this form of solution.

By its very nature, a numerical solution to an initial value problem consists of a **table of values** that is **finite** in length. On the other hand, the true solution of the initial value problem is most likely a whole continuum of values—that is, it consists of an *infinite* number of points. Obviously, the numerical solution is actually leaving out an infinite number of points. The question might arise, "With this many holes in it, is the solution good for anything at all?" To make this comparison a little clearer, let's look at a very simple specific example starting with Euler's method.

In mathematics and computational science, the **Euler method**, named after Leonhard Euler, is a numerical procedure for solving an ODE with a given initial value. It is the most basic kind of explicit method for numerical integration for ordinary differential equations.

Given the differential equation:

$$\frac{dy}{dt} = g\,(t,y),\ y\,(t_0) = y_0,\ t_0 \ge t \ge b$$

### Euler's Method

**Step 1**    Pick step size $h$ so that the interval $[b - t_0]/n = h$ is divided evenly to make the mathematics using the step sizes easier.

**Step 2**    Start at $y\,(t_0) = y_0$.

**Step 3**    Compute $y_{n+1} = y_n + h{*}g\,(t_n,\ y_n)$

**Step 4**    Let $t_{n+1} = t_n + h,\ n = 0, 1, 2, \ldots$

**Step 5**    Continue until $t_n = b$.

**STOP.**

---

## Example 1

## Euler's Method to Solve: $\dfrac{dy}{dt} = .25 \cdot y \cdot t,\ y(0) = 2$

We want to estimate the solution to $y\,(3)$ using a numerical approach. The interval is $[0,3]$.
A few steps by hand:

Step size of 1.      Point $(t,y)$
When $t_0 = 0$, $y_0 = 2$. This is the initial condition:     $(0,2)$
When $t_1 = 0 + h = 1$, $y_1 = 2 + h{*}\,0.25ty = 2 + (1){*}(0.25)(0)(2) = 2$     $(1,2)$
When $t_2 = 1 + h = 2$, $y_2 = 2 + h{*}0.25{*}t{*}y = 2 + (1){*}(0.25)(1)(2) = 2.5$     $(2,2.5)$
When $t_3 = 3$, $y_3 = 2.5 + h{*}0.25{*}t{*}y = 2.5 + (1){*}(0.25)(2)(2.5) = 3.75$     $(3,3.75)$

Geometrically, we see this in Figure 12.11.
Note that we have underestimated every value between $t = 0$ and $t = 3$.

The exact solution is $y(t) = 2 \cdot e^{\frac{t^2}{8}}$.

**FIGURE 12.11**
Euler's approximation

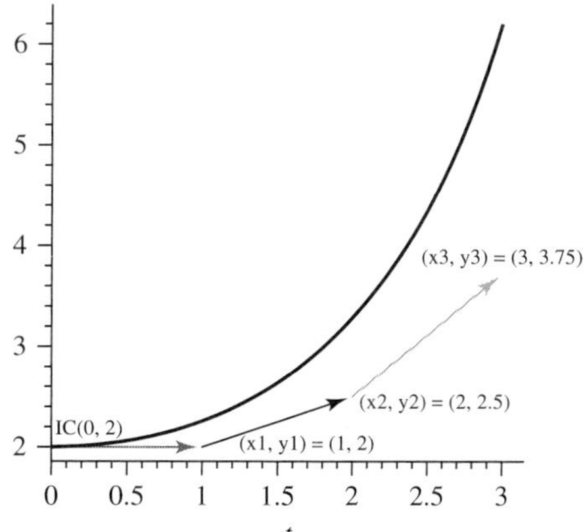

Let's make a table and see how we did:

| $t$ | $Y$ (approximate, $Ya$) | $Y$ (exact, $Ye$) | Error $= \lvert Ye - Ya\rvert$ | Percent Error $= \dfrac{100 \cdot \lvert Ye - Ya\rvert}{Ye}$ (%) |
|---|---|---|---|---|
| 0 | 2 | 2 | 0 | 0 |
| 1 | 2 | 2.266 | 0.266 | 11.73 |
| 2 | 2.5 | 3.297 | 0.797 | 24.17 |
| 3 | 3.75 | 6.16 | 2.41 | 39.12 |

Notice that as we move farther away from the initial condition $y(0) = 2$, the worse our estimate becomes.

Can we do better?

Maybe. We might improve by changing the step size. In this example, let's make the step size 0.5 instead of 1.

We will use Maple to obtain our output.

```
> ans2 := dsolve({eqn, y(0)=2}, numeric,
method=classical[foreuler],
output=array([0,0.5,1,1.5,2,2.5,3]),
stepsize=0.5);
```

$$
ans2 := \begin{bmatrix} [t, & y(t)\,] \\ 0 & 2. \\ 0.5 & 2. \\ 1 & 2.1250 \\ 1.5 & 2.3906250 \\ 2 & 2.83886718750 \\ 2.5 & 3.54858398437500 \\ 3 & 4.657516479492187500 \end{bmatrix}
$$

Let's make a table and see how we did:

| $t$ | $Ya$ | $Ye$ | Error $= \lvert Ye - Ya\rvert$ | Percent Error (%) |
|---|---|---|---|---|
| 0 | 2 | 2 | 0 | 0 |
| 0.5 | 2 | 2.063 | 0.063 | 3.05 |
| 1 | 2.125 | 2.2633 | 0.1383 | 6.11 |
| 1.5 | 2.390625 | 2.6496 | 0.258975 | 9.77 |
| 2 | 2.838867 | 3.2974 | 0.458533 | 13.91 |
| 2.5 | 3.54858 | 4.3684 | 0.81982 | 18.76 |
| 3 | 4.65752 | 6.1604 | 1.503 | 24.84 |

Out plot is shown in Figure 12.12.

**FIGURE 12.12**

Euler's approximation with smaller step size

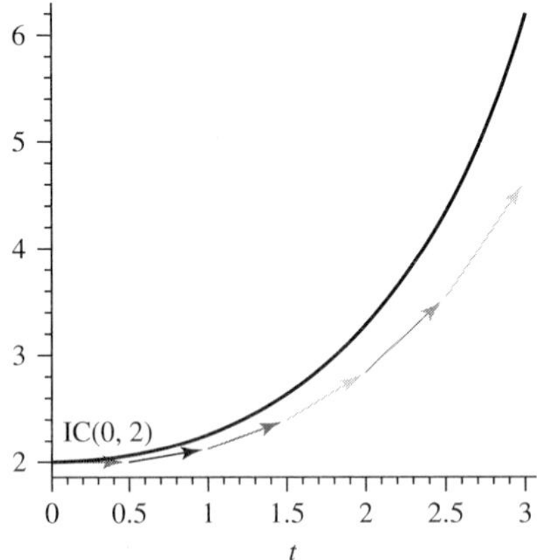

We are still underestimating at all points. We can do better by making this step size small. (*Note*: It is not always true that making the step size smaller improves the estimates.)

| **Example 2** | Euler's Method with Maple |

1. Set up Maple.

   > *restart;*

   > *with(DEtools):with(plots):with(linalg):*

2. Enter your ODE.

   > *eqn:=diff(y(t),t)=.25*t*y(t);*

$$eqn := \frac{d}{dt}\, y\,(t) = 0.25\; ty(t)$$

3. If possible, solve the ODE analytically and plot as we see in Figure 12.13:

   > *dsolve({eqn,y(0)=2},y(t));*

$$y(t) = 2\, e^{\left(\frac{t^2}{8}\right)}$$

   > *plot(2*exp(1/8*t^2),t=0..3, thickness=3);*

   Note you have to copy and paste the output equation from *dsolve* into the plot command because Maple does not recognize the $y(t)$ function directly from the output.

4. Use a numerical method: internal and classical output only.

   You will need to specify both the step size you want and the output array you want to see.

   > *Digits:=20:*

   *ans2:=dsolve({eqn,y(0)=2},numeric,*

   *method=classical[foreuler],*

   *output=array([0,0.5,1,1.5,2,2.5,3]),*

   *stepsize=0.5);*

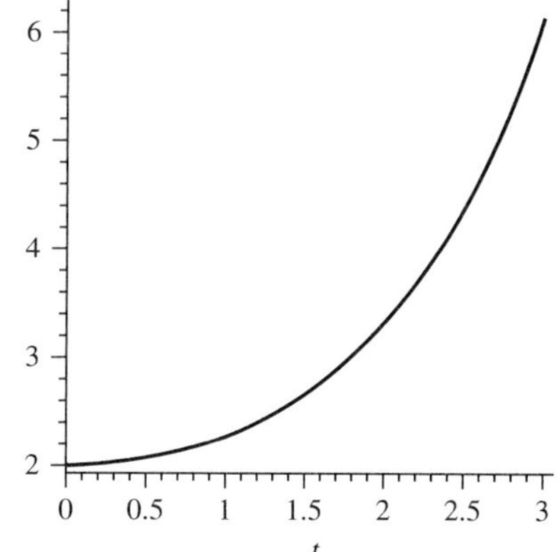

**FIGURE 12.13**

Solution plot from example

$$ans2 := \begin{bmatrix} [t, & y(t)] \\ 0 & 2. \\ 0.5 & 2. \\ 1 & 2.1250 \\ 1.5 & 2.3906250 \\ 2 & 2.83886718750 \\ 2.5 & 3.54858398437500 \\ 3 & 4.657516479492187500 \end{bmatrix}$$

5. Use numerical output for analysis: graphical and percent error.

   First, get the numerical output as a procedure as follows:

   > *dsol2:=dsolve({eqn,y(0)=2},*

   *numeric,method=classical[foreuler],stepsize=1,*
   *output=listprocedure);*

   $$dsol2 := [t = (\text{proc}(t) \ldots \text{end proc }), y(t) = (\text{proc }(t) \ldots \text{end proc})]$$

   Prepare for numerical output.

   > *fy2:=eval(y(t),dsol2);*

   $$fy2 := \text{proc }(t) \ldots \text{end proc}$$

   View output to compare to above output (numerical).

   > *seq(fy2(i),i=0..3);*

   $$2., 2., 2.50, 3.7500$$

   Prepare exact solution for output format:

   > *actual_y2:=evalf(subs(t=i,2*exp(1/8*t^2)));*

   $$actual\_y2 := 2.\ \mathbf{e}^{(0.12500000000000000000\ i^2)}$$

   View actual output.

   > *seq(actual_y2(i),i=0..3);*

   $$2., 2.2662969061336526336, 3.2974425414002562936, 6.1604336978360624900$$

   Putting the solution into an array that lists [*t*, numerical *y(t)*, actual *y(t)*, percent error] as shown below.

   > *array([seq([i,fy2(i),evalf(subs(t=i, 2*exp(1/8*t^2))),*
   *evalf(100*abs(fy2(i)–actual_y2(i))/actual_y2(i))],i=0..3)]);*

   $$\begin{bmatrix} 0 & 2. & 2. & 0. \\ 1 & 2. & 2.2662969061336526336 & 11.750309741540459711 \\ 2 & 2.50 & 3.2974425414002562936 & 24.183667535920822047 \\ 3 & 3.7500 & 6.1604336978360624900 & 39.127662370309425663 \end{bmatrix}$$

6. Make graphical comparisons.

   First, we plot the actual solution (if we have one) such as Figure 12.14 and then the numerical solution. Second, we will overlay them on one plot. Otherwise, compare numerical solutions to either qualitative plots or slope-field plots.

   > *plot(2*exp(1/8*t^2),t=0..3);*

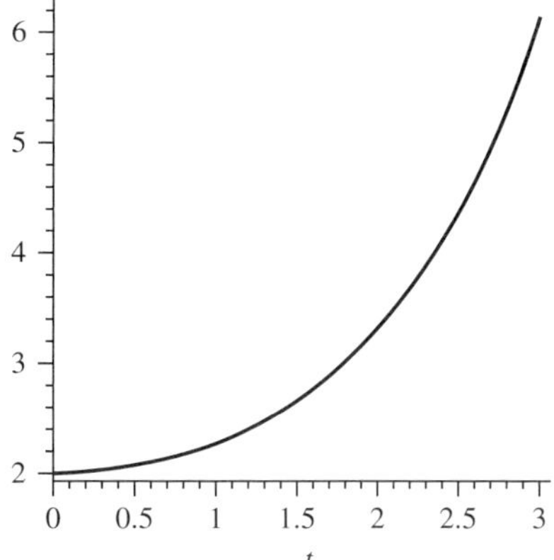

**FIGURE 12.14**
Solution plot from
our example

A numerical plot can be created as shown in Figure 12.15.

```
> data:=({seq([i,fy2(i)],i=0..3)}):
> points:=pointplot(data,symbol=diamond,color=black):
> display(points);
```

We can create overlays when we have an actual solution to see how well our numerical solution works. This is shown in Figure 12.16.

```
> with(plots):
> data:=({seq([i,fy2(i)],i=0..3)}):
> curve:=plot(2*exp(1/8*t^2),t=0..3,color=red):
> points:=pointplot(data, symbol=diamond,color=black):
> display(curve,points);
```

The second method we discuss is the improved Euler's method, which is also known as *Heun's method.*

### Improved Euler's Method (Heun's Method)

The accuracy of Euler's method is improved by using an average of two slopes in the tangent-line approximation. Use the slope obtained at the beginning of the step and the slope obtained at the end of the step to improve our accuracy.

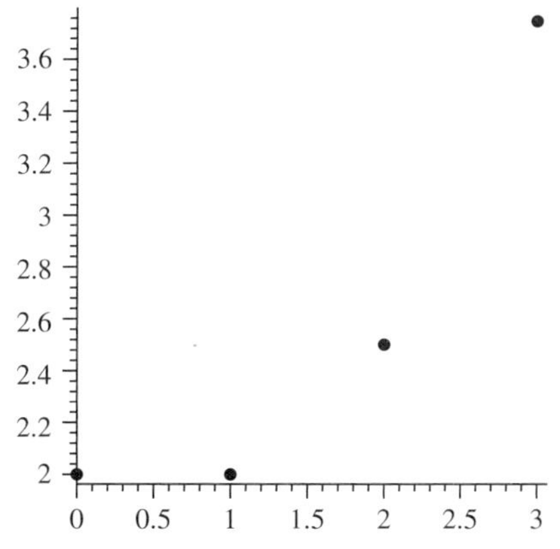

**FIGURE 12.15**

Euler's estimates
for our example

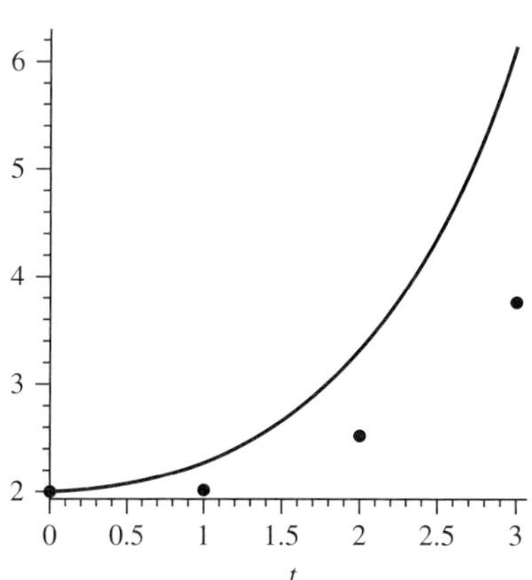

**FIGURE 12.16**

Solution curve and
Euler's estimates
together

**Improved Euler's Method Algorithm**

$$\text{Given } \frac{dy}{dt} = g(t, y), \, y(t_0) = y_0, \, t_0 \geq t \geq b$$

**Step 1**   Pick step size $h$ so that the interval $[b - t_0]/n = h$ is divided evenly.
**Step 2**   Start at $y(t_0) = y_0$.
**Step 3**   Let $t_{n+1} = t_n + h$.
**Step 4**   Compute $y_{n+1} = y_n + (h/2)*[k_1 + k_2]$
**Step 5**   Continue until $t_{n+1} = b$.

**Stop.**

Step 4 uses the formula

$$y_{n+1} = y_n + \frac{h}{2}(k_1 + k_2)$$
$$k_1 = g(t_n, y_n)$$
$$k_2 = g(t_{n+1}, y_n + h*k_1)$$

---

| **Example 3** | # Solving $y' = 0.25ty$, $y(0) = 2$ with Improved Euler's Method |
|---|---|

We want to estimate the solution to $y(3)$ using a numerical approach.

$$k_1 = 0.25 \, t_n y_n$$
$$k_2 = 0.25 \, t_{n+1}(y_n + h*k_1)$$

A few steps of the improved Euler's method by hand:

Step size of h = 1.                                                                Point $(t,y)$

When $t_0 = 0$, $y_0 = 2$ This is the initial condition:                           (0, 2)

When $t_1 = t_0 + h = 1$, $y_1 = 2 + (1/2) \cdot \{[0.25 \cdot (0) \cdot (2)$
$\qquad\qquad + 0.25 \cdot (1) \cdot [2 + 0.25 \cdot (0) \cdot 2]\} = 2.25$         (1, 2.25)

When $t_2 = t_1 + h = 2$, $y_2 = 2.25 + (1/2) \cdot (0.25(1)(2.25)$
$\qquad\qquad + 0.25 \cdot (2)[2.25 + (1)\, 0.25 \cdot (1)(2.25)] = 3.23437$        (2, 3.234375)

When $t_3 = 2 + h = 3$, $y_3 = 3.234375 + (1/2) \cdot \{(0.25) \cdot (2) \cdot (3.234375)$
$+ (0.25) \cdot (3) \cdot [3.234375 + 0.25 \cdot (2) \cdot (3.234375)]\} = 5.862305$   (3, 5.862305)

The exact solution is $y(t) = 2 \cdot e^{\frac{t^2}{8}}$.
Let's make a table and see how we did.

| $t$ | $Y$ (approximate, $Ya$) | $Y$ (exact, $Ye$) | Error $= \|Ye - Ya\|$ | Percent Error $= \dfrac{100 \cdot \|Ye - Ya\|}{Ye}$ (%) |
|---|---|---|---|---|
| 0 | 2 | 2 | 0 | 0 |
| 1 | 2.25 | 2.266 | 0.016 | 0.706 |
| 2 | 3.234375 | 3.297 | 0.062625 | 1.899 |
| 3 | 5.862305 | 6.16 | 0.2977 | 4.83 |

Notice that as we move farther away from the initial condition $y(0) = 2$, the worse our estimate becomes. Also note that these estimates are "better" estimates than Euler's method. Think about it—why?

The plot is shown in Figure 12.17.

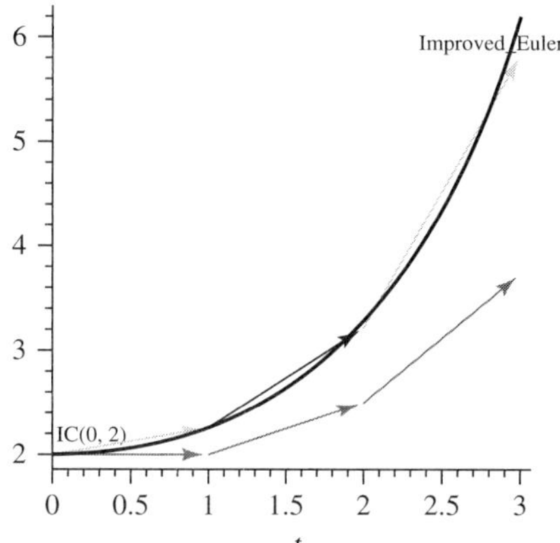



**FIGURE 12.17**
Improved Euler's
approximation

---

| **Example 4** | **Improved Euler's Method Using Maple** |

We note that the only command change is our call within *dsolve*({eqn, $y(0) = 2$}, numeric,method = classical[heunform], output = array([0,0.5,1,1.5,2,2.5,3]), stepsize = 0.5), which calls *heunform* the improved Euler's algorithm.

> *restart;*
> *with( DEtools):with(plots):with(linalg):*

   Enter your ODE.

> *eqn:=diff(y(t),t)=.25*t*y(t);*

$$eqn := \frac{d}{dt}\, y(t) = 0.25\ ty(t)$$

   Solve the ODE analytically and plot as shown in Figure 12.18.

> *dsolve({eqn,y(0)=2},y(t));*

$$y(t) = 2\, e^{\left(\frac{t^2}{8}\right)}$$

> *plot(2*exp(1/8*t^2),t=0..3, thickness=3);*

   *Note*: You must copy and paste the output equation from *dsolve* into the plot command.

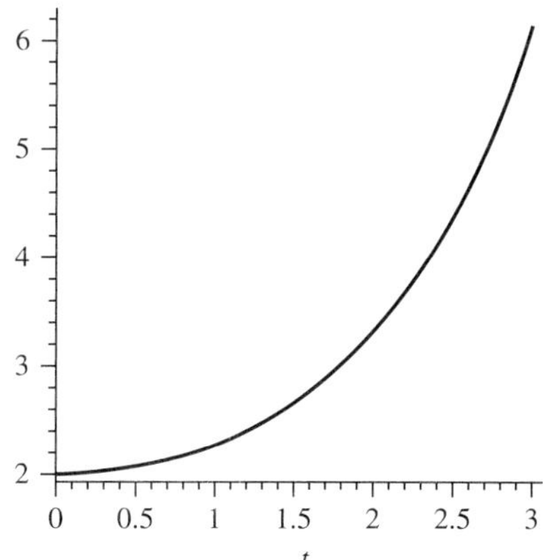

**FIGURE 12.18**
Solution curve from our
example

## Numerical Method: Internal and Classic Output Only

You will need to specify both the step size you want and the output array you want to see.

> *Digits:=20:*
*ans2:=dsolve({eqn,y(0)=2},numeric,*
*method=classical[heunform],*
*output=array([0,0.5,1,1.5,2,2.5,3]),*
*stepsize=0.5);*

$$
ans2 := \begin{bmatrix}
[t, & y(t)\,] \\
0. & 2. \\
0.5 & 2.0625000000000000000 \\
1. & 2.2639160156250000000 \\
1.5 & 2.6441831588745117188 \\
2. & 3.2845712676644325257 \\
2.5 & 4.3366605018381960691 \\
3. & 6.0814887506246577688
\end{bmatrix}
$$

First, we plot the actual solution (if we have one) as seen in Figure 12.19 and then the numerical solution. Second, we will overlay them on one plot. Otherwise, compare numerical solution to either qualitative plots or slope-field plots. We can obtain a numerical plot, as shown in Figure 12.20.

> *plot(2*exp(1/8*t^2),t=0..3);*

## Numerical Plot

> *data:=({seq([i,fy2(i)],i=0..3)}):*
> *points:=pointplot(data, symbol=diamond, color=black):*
> *display(points);*

## Numerical Output for Analysis: Graphical and Percent Error

First, get the numerical output as a procedure as follows:

> *dsol2:=dsolve({eqn,y(0)=2},numeric,method=classical[heunform],stepsize=1,*
*output=listprocedure);*

$$
dsol2 := [t = (\text{proc}(t) \ldots \text{end proc}), y(t) = (\text{proc}(t) \ldots \text{end proc})]
$$

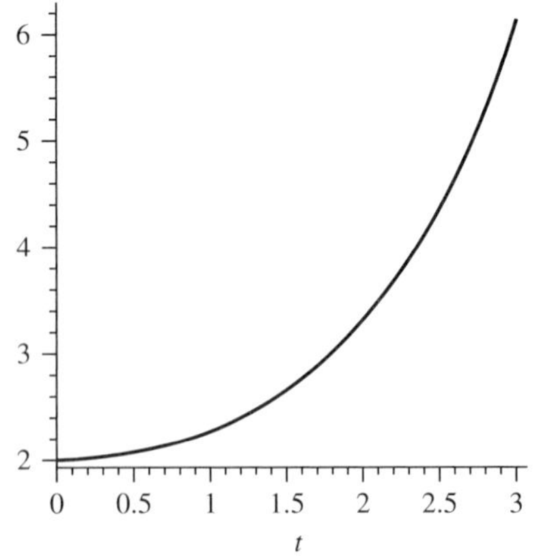

**FIGURE 12.19**
Solution curve from our example

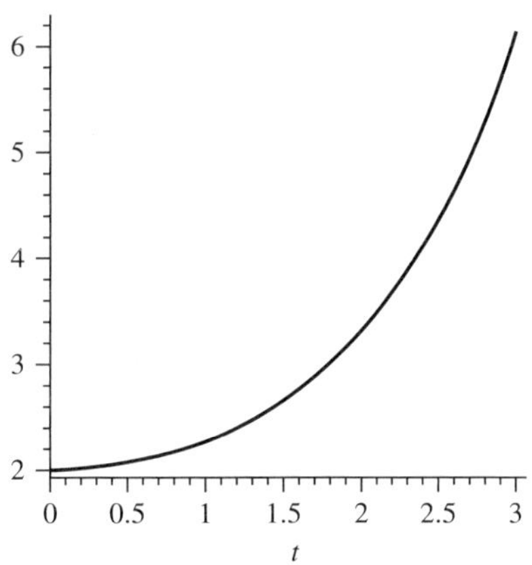

**FIGURE 12.20**
Solution curve from our example

Prepare for numerical output.

> *fy2:=eval(y(t),dsol2);*

$$fy2 := \text{proc}\ (t)\ ...\ \text{end proc}$$

View output to compare to above output (numerical).

> *seq(fy2(i),i=0..3);*

$$2.,\ 2.2500000000000000000,\ 3.2343750000000000000,\ 5.8623046875000000000$$

Prepare exact solution for output format:

> *actual_y2:=evalf(subs(t=i,2*exp(1/8*t^2)));*

$$actual\_y2 := 2.\ e^{(0.12500000000000000000\ i^2)}$$

View Actual Output

> *seq(actual_y2(i),i=0..3);*

$$2.,\ 2.2662969061336526336,\ 3.2974425414002562936,\ 6.1604336978360624900$$

Put into an array that lists [$t$, numerical $y(t)$, actual $y(t)$, percent error].

> *array([seq([i,fy2(i),evalf(subs(t=i,2*exp(1/8*t^2))),* 
*evalf(100*abs(fy2(i)–actual_y2(i))/actual_y2(i))],i=0..3)]);*

$$\begin{bmatrix}
0 & 2. & 2. & 0. \\
1 & 2.2500000000000000000 & 2.2662969061336526336 & 0.71909845923301717516 \\
2 & 3.2343750000000000000 & 3.2974425414002562936 & 1.9126198745975635237 \\
3 & 5.8623046875000000000 & 6.1604336978360624900 & 4.8394159398352818372
\end{bmatrix}$$

## Graphical Comparisons

First, we plot the actual solution (if we have one), and then the numerical solution. Second, we will overlay them on one plot. Otherwise, compare the numerical solution to either qualitative plots or slope-field plots. We illustrate these plots in Figures 12.20 through 12.22.

> *plot(2*exp(1/8*t^2),t=0..3);*

## Numerical Plot

> *data:=({seq([i,fy2(i)],i=0..3)}):* 
> *points:=pointplot(data,symbol=diamond,color=black):* 
> *display(points);*

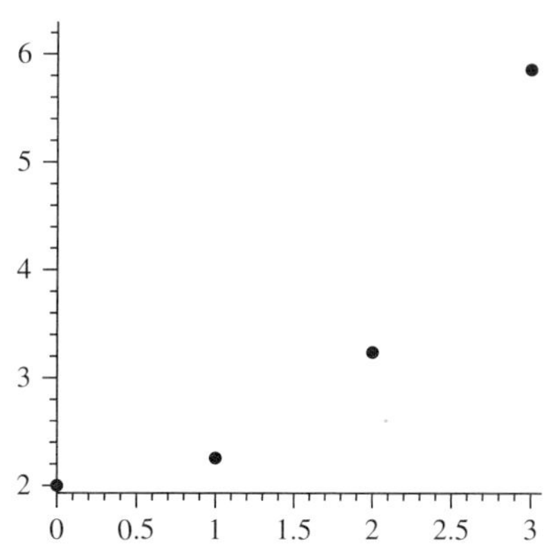

**FIGURE 12.21**

Improved Euler's approximations

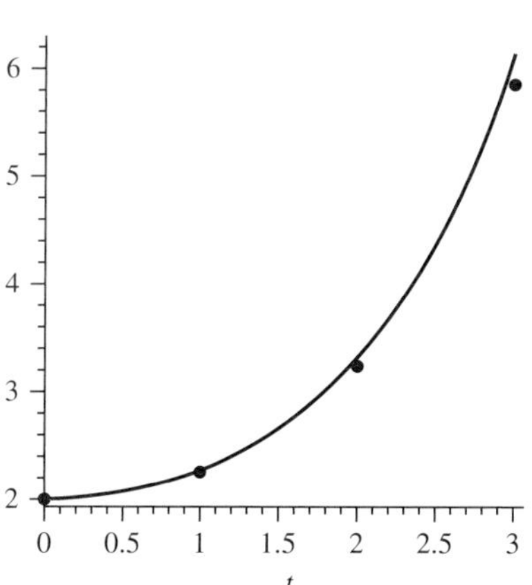

**FIGURE 12.22**

Overlay of Improved Euler's approximations and the solution curve

## Overlay

> *with(plots):*
> *data:=({seq([i,fy2(i)],i=0..3)}):*
> *curve:=plot(2*exp(1/8*t^2),t=0..3, color=red):*
> *points:=pointplot(data,symbol=diamond,color=black):*
> *display(curve,points);*

### Runge–Kutta Methods

When we use the average of the estimates of the derivatives at the end points, we can improve the approximation to the solution. A class of approximation techniques that estimate derivatives at various points within an interval and then computes a *weighted average* are the Runge–Kutta methods, named for two German mathematicians.

The Runge–Kutta methods are classified by order, where the order depends on the number of slope estimates used at each step. A very popular method is the fourth-order Runge–Kutta method.

**Fourth-Order Runge–Kutta Method**  For solving $dy/dt = g(t,y)$, $y(t_0) = y_0$ over an interval.

**Step 1**    First divide the interval $x_0 \leq x \leq b$ into $p$ subintervals using equally spaced points. This yields the step size $h = (b - x_0)/p$.

**Step 2**    For $n = 1, 2, 3, \ldots, p$, obtain the following sequence of approximations.

$$y_{n+1} = y_n + \frac{K_1 + 2K_2 + 2K_3 + K_4}{6}, \text{ where}$$

$$K_1 = g(t_n, y_n)h$$
$$K_2 = g(t_n + h/2, y_n + K_1/2)h$$
$$K_3 = g(t_n + h/2, y_n + K_2/2)h$$
$$K_4 = g(t_n + h, y_n + K_3)h$$

---

**Example 5**    # Runga–Kutta Method by Hand: $y' = 0.25ty$, $y(0) = 2$

We want to estimate the solution to $y(3)$ using a numerical approach.
A few steps of Runge–Kutta method by hand:

Step size of $h = 1$.
When $t = 0$, $y = 2$ This is the initial condition: (0,2)
When $t = 0 + h = 1$,
$K1 = h*g(t(n), y(n)) = 1*.25*0*2 = 0$
$K2 = h*g(t(n) + h/2), y(n) + K1/2) = 1*.25*(0 + 1/2)*(2 + 0) = .25$
$K3 = h*g(t(n) + h/2), y(n) + K2/2) = 1*.25*(0 + 1/2)*(2 + .125) = .265625$
$K4 = h*g(t(n) + h), y(n) + K3) = 1*.25*(0 + 1)*(2 + .265625) = .56640625$
$Y(1) = 2 + (1/6)*(0 + 2*.25 + 2*.265625 + .56640625) = 2.266276$

When $t = 1 + h = 2$
$K1 = 0.566569$      $K2 = 0.956085$      $K3 = 1.029119$      $K4 = 1.647698$
$Y(2) = 2.266276 + (1/6)*(.566569 + 2*.956085 + 2*1.029119 + 1.647698) = 3.297055$

When $t = 2 + h = 3$
$K1 = 1.648528,$      $K2 = 2.575825,$      $K3 = 2.865605,$      $K4 = 4.621995$
$Y(3) = 3.297055 + (1/6)* (1.648528 + 2*2.575825 + 2*2.86560 + 4.621995) = 6.155952$

The exact solution is $y(t) = 2 \cdot e^{\frac{t^2}{8}}$.

Let's make a table and see how we did.

| $t$ | $Y$ (approximate, $Ya$) | $Y$ (exact, $Ye$) | Error = $|Ye - Ya|$ | Percent Error = $\dfrac{100 \cdot |Ye - Ya|}{Ye}$ (%) |
|---|---|---|---|---|
| 0 | 2 | 2 | 0 | 0 |
| 1 | 2.266276 | 2.266297 | 0.00002 | 0.002 |
| 2 | 3.297055 | 3.297443 | 0.000388 | 0.0388 |
| 3 | 6.155952 | 6.160434 | 0.004482 | 0.4482 |

Notice that as we move farther away from the initial condition $y(0) = 2$, the worse our estimate becomes. Also note that these estimates are much better estimates than either Euler's method or the improved Euler's method. Think about it—why? We can visually see this in Figure 12.23.

In Maple, we again change the call from within *dsolve* to RK4 as seen in this sequence of commands.

Set up Maple.

> *restart;*
> *with( DEtools ):with( plots ):with( linalg ):*

Enter your ODE.

> *eqn:=diff( y( t ),t )=.25*t*y( t );*

$$eqn := \frac{d}{dt}\, y(t) = 0.25 \, ty(t)$$

Solve the ODE analytically and plot as shown in Figure 12.24.

> *dsolve( {eqn,y( 0 )=2},y( t ) );*

$$y(t) = 2e^{\left(\frac{t^2}{8}\right)}$$

> *plot( 2*exp( 1/8*t^2 ),t=0..3,thickness=3 );*

Note you have to copy and paste the output equation from *dsolve* into the plot command.

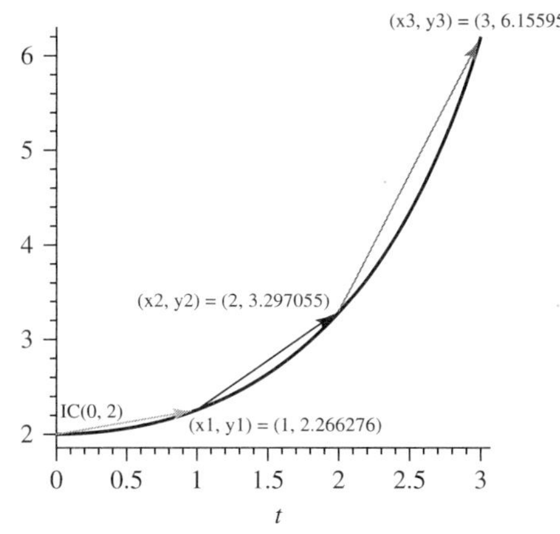

**FIGURE 12.23**

Runga-Kutta approximations: The RK4 plot



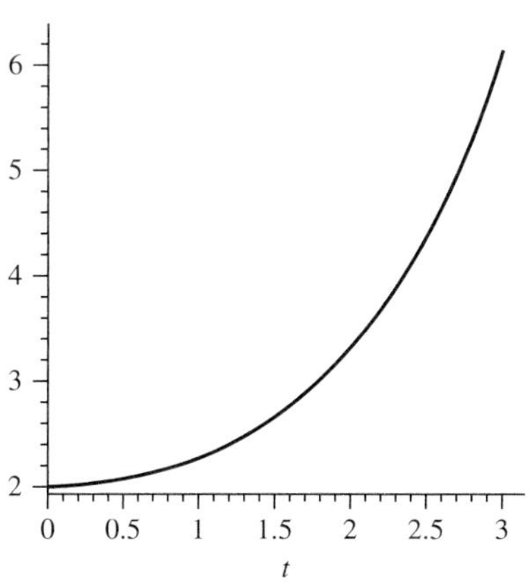

**FIGURE 12.24**

The solution curve for our example

## Numerical Method: Internal and Classic Output Only

You will need to specify both the step size you want and the output array you want to see.

> *Digits:=20:*
*ans2:=dsolve({eqn,y(0)=2},numeric,*
*method=classical[rk4],*
*output=array([0,0.5,1,1.5,2,2.5,3]),*
*stepsize=0.5);*

$$ans2 := \begin{bmatrix} [t, & y(t)] \\ 0. & 2. \\ 0.5 & 2.0634867350260416667 \\ 1. & 2.2662959538380770634 \\ 1.5 & 2.6495643470246741905 \\ 2. & 3.2974194721525799553 \\ 2.5 & 4.3683097289253688453 \\ 3. & 6.1600978602067198076 \end{bmatrix}$$

## Numerical Output for Analysis: Graphical and Percent Error

First, get the numerical output as a procedure as follows.

> *dsol2:=dsolve({eqn,y(0)=2},*
*numeric,method=classical[rk4],stepsize=1, output=listprocedure);*

$$dsol2 := [t = (proc(t) \ldots end\ proc), y(t) = (proc(t) \ldots end\ proc)]$$

Prepare for numerical output.

> *fy2:=eval(y(t),dsol2);*

$$fy2 := proc\ (t) \ldots end\ proc$$

View output to compare to above output (numerical).

> *seq(fy2(i),i=0..3);*

$$2., 2.2662760416666666667, 3.2970554033915201823, 6.1559523209644895461$$

Prepare the exact solution for output format:

> *actual_y2:=evalf(subs(t=i,2*exp(1/8*t^2)));*

$$actual\_y2 := 2.\ e^{(0.12500000000000000000\ i^2)}$$

View the actual output:

> *seq(actual_y2(i),i=0..3);*

$$2., 2.2662969061336526336, 3.2974425414002562936, 6.1604336978360624900$$

Put into an array that lists [$t$, numerical $y(t)$, actual $y(t)$, percent error]

> *array([seq([i,fy2(i),evalf(subs(t=i, 2*exp(1/8*t^2))),*
*evalf(100*abs(fy2(i)–actual_y2(i))/actual_y2(i))],i=0..3)]);*

$$\begin{bmatrix} 0 & 2. & 2. & 0. \\ 1 & 2.2662760416666666667 & 2.2662969061336526336 & 0.00092064137445971691041 \\ 2 & 3.2970554033915201823 & 3.2974425414002562936 & 0.011740553591927441423 \\ 3 & 6.1559523209644895461 & 6.1604336978360624900 & 0.072744502925939929932 \end{bmatrix}$$

**FIGURE 12.25**
The solution curve
for our example

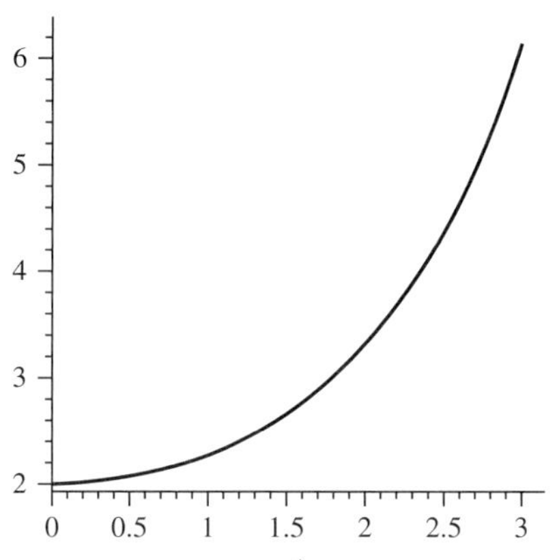

**FIGURE 12.26**
The plot of the
approximations
by RK4

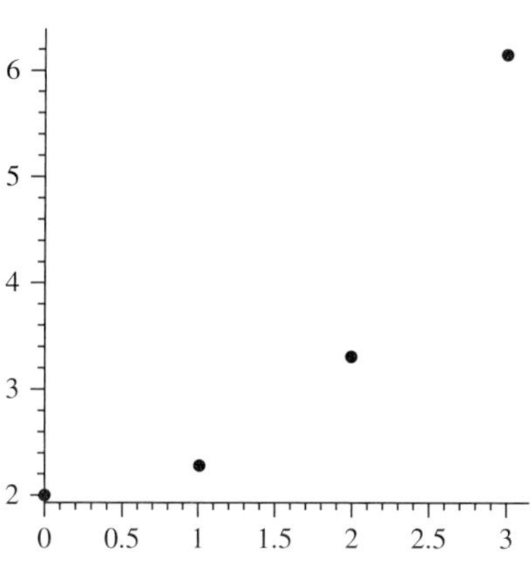

## Graphical Comparisons

We illustrate the plots in Figures 12.25 through 12.27. We plot the actual solution (if we have one) and then the numerical solution. Second, we will overlay them on one plot. Otherwise, compare the numerical solution to either qualitative plots or slope-field plots.

```
> plot(2*exp(1/8*t^2),t=0..3);
```

## Numerical Plot:

```
> data:=( {seq( [i,fy2(i) ],i=0..3)} ):
> points:=pointplot(data, symbol=diamond, color=black):
> display(points);
```

## Overlay

```
> with(plots):
> data:=({seq( [i,fy2(i) ],i=0..3)}):
> curve:=plot(2*exp(1/8*t^2),t=0..3,color=red):
> points:=pointplot(data,symbol=diamond,color=black):
> display(curve,points);
```

**FIGURE 12.27**
The plot of the RK4
approximations with the
solution curve

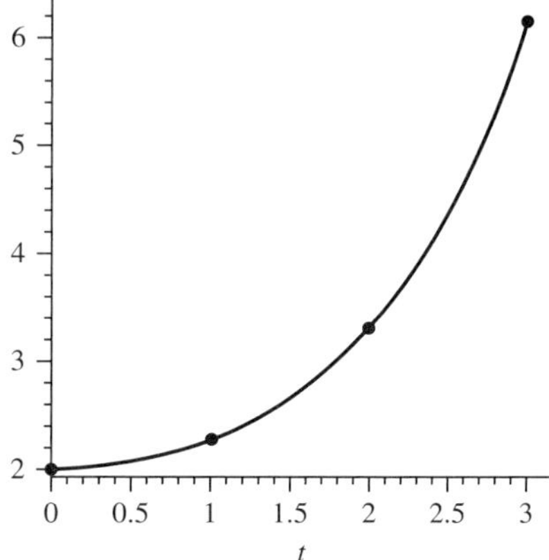

## 12.5 | EXERCISES

1. Get Euler's estimates for $y(3)$ using the DE $y' = y + t$ for step sizes of 0.1, 0.05, and 0.01. Compute the percent error for each. Assume we start at $y(0) = 1$.

2. Use Euler's method by hand for $y' = y + 1$, $y(0) = 1$, with step size of 0.25 to estimate $y(0.5)$. Then do Euler's method using Maple and compare your results to make sure you have the correct solution. Compute the percentage error because you can find this exact solution. Then change the step sizes to 0.1 and 0.05 and use Maple to estimate $y(0.5)$.

3. Consider $v' = 32 - 1.6\,v$, $v(0) = 0$. Using Euler's method in Maple and a step size of 0.05, estimate $v(2)$. What is the terminal velocity? Keep stepping out until you approximate the terminal velocity to three decimal places of accuracy.

4. Try *by hand* $y' = y + 1$, $y(0) = 1$ with step size of 0.25 to estimate $y(0.5)$. Then do the improved Euler's method using Maple and compare your results to make sure you have the correct solution. Compute the percentage error because you can find this exact solution.

5. Consider $v' = 32 - 1.6v$, $v(0) = 0$. Using heunform in Maple and a step size of 0.05, estimate $v(2)$. What is the terminal velocity? Keep stepping out until you approximate the terminal velocity to three decimal places of accuracy.

6. Consider $P' = .P*(15-3*P)$, $P(0) = 2$. Use heunform in Maple and a step size of $h = 1$, $h = 0.5$, $h = 0.1$. Discuss the results as compared to qualitative solution.

7. Try *by hand* $y' = y + 1$, $y(0) = 1$ with step size of 0.25 to estimate $y(0.5)$. Then do the Runge-Kutta (K4) method using Maple and compare your results to make sure you have the correct solution. Compute the percentage error because you can find this exact solution.

8. Consider $v' = 32 - 1.6v$, $v(0) = 0$. Using RK4 in Maple and a step size of 0.05, estimate $v(2)$. What is the terminal velocity? Keep stepping out until you approximate the terminal velocity to three decimal places of accuracy.

9. Consider $P' = .P*(15-3*P)$, $P(0) = 2$. Use RK4 in Maple and a step size of $h = 1$, $h = 0.5$, $h = 0.1$. Discuss the results as compared to qualitative solution.

## 12.5 | PROJECTS

1. The Spread of a Contagious Disease. The following ODE models the spread of a communicable disease as

$$\frac{dN}{dt} = .25\,N\,(10 - N), \quad N(0) = 2$$

where $N$ is in 100s.

Let's completely analyze the behavior of this ODE.

a. Because this is an autonomous ODE, perform a complete graphical analysis:

   (1) Plot $dN/dt$ versus $N$. Find and label all rest points (equilibrium points).

   (2) Find the value where the rate of change of the disease is the fastest. Why did you provide this value?

   (3) Plot $N$ versus $t$ from the following initial conditions:

$$N(0) = 2, \; N(0) = 7, \text{ and } N(0) = 14.$$

   (4) Describe the *stability* of each rest point (equilibrium point).

**b.** Obtain a slope field plot of this ODE from Maple. Briefly discuss how you would analyze the slope field plot. Compare it to the qualitative plot in part (a).

**c.** Solve this ODE using separable variables. (*Hint*: You will need partial fraction decomposition as well.) Ensure that you find the value of arbitrary constant $C$ using the initial condition $N(0) = 2$.

**d.** Simplify your solution in (b) and plot $N$ versus $t$.

**e.** Compare your actual plot with the graphical solution in part (a), item (3), and the slope-field plot in part (b).

**f.** From your graph of the analytical solution in part (d), estimate the solution values of $N(0.5)$ and $N(5)$.

**g.** Find the actual values for these two solutions $N(0.5)$ and $N(5)$ using your analytical solution from part (c) and substitution.

**h.** Compute the time, $t$, when $N$ is changing the fastest using the initial condition $N(0) = 2$. (Recall that you already have found the value of $N$.)

**i.** Use Euler's method with step sizes of $h = 0.1$ and then $h = 1$ to approximate the solutions to the ODE for $N(0.5)$ and $N(5)$. Find the relative error or absolute differences.

**j.** Plot Euler's approximations for $h = 0.1$ and $h = 1.0$ and compare these graphs to your graphical analysis in part (a) and to your other plots. Briefly discuss these plots—you may compare and contrast these plots. What happened with Euler's method? (Any opinions?)

**2.** Chemical Reactions

**Purpose**: To model the changing amounts of salt dissolved in brine as an initial value problem and to use the computer to obtain graphical and numerical solutions.

**Background**: A brine solution is a solution of salt in water. If the brine is in a tank equipped with fill and drain pipes, then the total amount of dissolved salt in the tank varies as the concentration in the inflow stream changes and the inflow and outflow rates are adjusted. The amount of salt in the tank can be modeled by appealing to the following balance law:

$$\text{New rate of change} = \text{rate in} - \text{rate out}$$

The term "rate in" refers to the rate at which salt is added to the brine by means of the inflow stream, and the term "rate out" is the rate at which salt leaves the tank through the outflow pipe. We assume throughout that the inflow stream is instantaneously mixed with the brine in the tank so that the concentration of salt in the tank is uniform at any time. Both terms (rate in and rate out) are products of the appropriate brine flow rate with corresponding salt concentrations.

Consider that a tank with capacity of 4,000 liters holds 2,000 liters of brine that contain 50 kg of dissolved salt. Brine with a salt concentration of 0.2 kg/liter is piped into the tank at a rate of 40 liters/min. Well-mixed brine is drawn off at a rate of $a$ liters/ min.

$$\frac{dx}{dt} = (.2)(40) + \frac{x}{2000 + (40 - a)t}\, a$$

Requirement 1. Show that the equation is the general ODE; find the appropriate initial condition for this brine problem.

Requirement 2. Solve the ODE, leaving $a$ as a parameter.

Requirement 3. Solve the ODE and obtain plots when rate out $a = 30, 40,$ and $50$.

Requirement 4. Bulk salt. Suppose that 200 kg of bulk salt is placed in a tank holding 2,000 liters of brine already containing $x_0$ kg of dissolved salt. Suppose that the saturation level of brine is 300 kg of salt. The model representing this is:

$$\frac{dx}{dt} = .01(200 - x)(300 - x),\ x(0) = x_0$$

Solve this ODE and obtain a plot for various choices of $x_0$ (e.g., 25, 50, 100, 200).

Requirement 5. Bimolecular chemical reactions. In a bimolecular chemical reaction, two species of chemical entities interact and create one or more products. Consider the following reaction and laws:

$$A + B \rightarrow C + D$$

The law of conservation: $\dfrac{dC}{dt} = \dfrac{dD}{dt} = -\dfrac{dA}{dt} = -\dfrac{dB}{dt}$

The law of mass action: The rate of an elementary reaction is proportional to the product of the concentrations of the reactants.

The ODE:

$$\frac{dx}{dt} = k_1(a + c - x)(b + c - x), \, x(0) = c$$

Consider the following reversible chemical reaction:

$$\frac{dx}{dt} = (0.7 - x)(0.4 - x) - 0.1x^2, \, x(0) = c$$

Solve qualitatively (autonomously). Solve this ODE (at least numerically) and show the curve and discuss the equilibrium values. Let $c = 0.2$. This reaction is only valid for $0 \geq c \geq 0.4$.

3. Harvesting a species. Consider the harvesting of blue crabs in South Carolina. There have been many newspaper reports about the declining populations of blue crabs and the difficulty in harvesting them. Let's model the situation and analyze some "what ifs."

The basic balance law for harvesting is:

$$P'(t) = r\left(1 - \frac{P(t)}{k}\right)P(t) - H(t)$$

Where $r$ is the intrinsic rate coefficient (growth rate if $> 0$) and $k$ represents the carrying capacity (or saturation level).

Requirement 1. In the absence of harvesting $[H(t) = 0]$, the ODE is autonomous. Let $r = 0.3$, $k = 12$, and $P(0) = 5$.

a. Perform a qualitative assessment of this situation because it is autonomous. List and classify each equilibrium value.
b. Solve this ODE.
c. Plot the solution to this ODE and compare to your qualitative solution. Briefly discuss any similarities or differences.
d. Solve numerically using Euler's method and obtain a plot of the numerical solution. Compare to the analytical solution.

Requirement 2. Now consider light harvesting in which $H(t) = 0.4$. (Assume $r$ and $k$ are the same values as before.)

a. Obtain a slope-field plot and discuss any equilibrium values observed.
b. Find all equilibrium values of this ODE.
c. Solve this ODE. Assume initial conditions of $P(0) = 19$, $P(0) = 8$, and $P(0) = 0.8$.
d. Solve numerically using Euler's method with $P(0) = 5$.

Requirement 3. Now consider a heavier harvesting in which $H(t) = 1.5$.

a. Obtain a slope-field plot and discuss any equilibrium values observed.
b. Find all equilibrium values of this ODE.
c. Solve this ODE. Assume initial conditions of $P(0) = 19$, $P(0) = 8$, and $P(0) = 0.8$.
d. Solve numerically using Euler's method with $P(0) = 5$.

Requirement 4. Determine which of the models is more likely to represent the current situation for blue crabs in Chesapeake Bay.

4. Rural water supplies. Consider a cylindrical water tank that is discharging water into a pipe located in the middle of its base. The potential energy given by the pressure of the

column of liquid is converted into kinetic energy of the stream of water as it exists in the tank. If the level of the water in the tank is $h$ ft, then it follows that the exit velocity in ft/sec is $v = \sqrt{2gh}$ where $g = 32.2$ ft/sec$^2$ is the acceleration of gravity. Because a convergent flow is set up from all sides to the orifice, inertia from the particles in the jet forces them to overshoot the edge and converge to a smaller cross-section termed the *vena contracta*. The ratio of the diameter of the vena contracta to the diameter of the orifice is called the *contraction coefficient*, $\alpha$, and it ranges form 0.5 to 1.0, depending on the smoothness of the edge of the orifice. The diameter of the vena cross-section and exit velocity of the stream determine the volume rate of the discharge of the stream. Thus, if the tank has cross-section A and is being filled from the top at a rate $K$ while draining from the bottom through an orifice of area $a$, the height in the tank is governed by the differential equation:

$$\frac{dh}{dt} = \frac{(K - a\alpha\sqrt{2gh})}{A}$$

Now consider we are in the mountains outside Stowe, Vermont. The water for a cabin is supplied from an aquifer 125 feet below the surface at a maximum rate of 7 gallons per minute. Because the flow rate is insufficient to meet daily needs, you must design a more complicated system. In your design, water is pumped to a cylindrical tank with diameter 10 feet and height 35 feet; the tank is located on a nearby rise about 75 feet from the cabin, with a base 25 ft above the level of the cabin floor. The water is stored in this gravity tank until needed (water is turned on by opening the spigot).

Assume that the coefficient of contraction at the exit is $\alpha = 0.63$ and that the water main consists of 6-inch-diameter PVC pipe so that the friction losses are less than 2 psi.

### Required

    **a.** Determine the minimum flow rate in the system and the minimum water pressure during the first two hours.

    **b.** The local fire code demands that water supplies maintains a pressure of at least 20 psi at a flow rate of 30 gallons per minute for a minimum of two hours. Will your design pass inspection?

    **c.** At continuous maximum flow, how long would it take for the level of the top of the water in the tank to fall to 10 ft above the base?

**5.** Falling bodies. In bridge jumping, a participant attaches one end of a bungee cord to him- or herself, attaches the other end to a bridge railing, and then drops off the bridge. In this project, the jumper will be dropping off of the Royal Gorge Bridge, a suspension bridge that is 1,053 feet above the floor of the Royal Gorge in Colorado. The jumper will use a 200-foot-long bungee cord. It would be nice if the jumper has a safe jump, meaning that the jumper does not crash into the floor of the gorge or run into the bridge on the rebound. In this project, you will do some analysis of the fall.

Assume the jumper weighs 160 lbs. The jumper will free-fall until the bungee cord begins to exert a force that acts to restore the cord to its natural (equilibrium) position. To determine the spring constant of the bungee cord, you found that that a mass weighing 4 lbs stretches the cord 8 feet. Hopefully, this spring force will help slow down the descent sufficiently so that the jumper does not hit the bottom of the gorge. Throughout this project, we will assume that down is the positive direction.

    Requirement 1. Before the bungee cord begins to retard the fall the jumper, the only forces that act on the jumper are the jumper's weight and wind (air) resistance.

    **a.** If the force from wind resistance is 0.9 times the velocity of the jumper, then use Newton's second law ($\Sigma F = MA$) to write a differential equation that models the fall of the jumper. Be sure to include the initial conditions for the jumper for your differential equation. (*Hint*: Although this problem can be formulated as a second-order DE in position or as a first-order DE in velocity, use the first-order DE model in this project.)

   **b.** Solve this differential equation and find a function that describes the jumper's velocity (as a function of time). Knowing integral calculus, find a function that describes the jumper's position (as a function of time).

   **c.** What is the velocity of the jumper after the jumper has fallen 200 feet?

   **d.** What is the terminal velocity of the jumper, if any?

Requirement 2. After a bit more research, you find that the force from wind resistance is not linear, as assumed earlier. Apparently, the force from wind resistance is more closely modeled by $0.9\,v + 0.0009\,v^2$.

   **a.** Write a new differential equation governing the velocity of the jumper (before the bungee cord comes into effect).

   **b.** You should notice that this DE is no longer easy to solve. Nonetheless, you are determined to find the velocity of the jumper after 4 seconds by using a numerical technique. Use Euler's method with a step size of 0.5 and estimate the velocity of the jumper after 4 seconds.

   **c.** Find the terminal velocity (if it exists).

   **d.** How do your results compare with those found in requirement 1 under the linear assumption?

   **e.** What is the velocity of the jumper after 200 feet? (*Hint*: You can find this from your numerical table.)

# Modeling with Systems of Differential Equations

Interactive situations occur in the study of economics, ecology, electrical engineering, mechanical systems, control systems, systems engineering, and so forth. For example, the dynamics of population growth of various species is an important ecological application of applied mathematics.

Consider a friend who recently retired and bought farm land in South Carolina. He desires to stock a fishing pond with his favorite fish, bass and trout. He finds he has a fair-sized freshwater pond on his land, but it has no fish. He takes a water sample to the local fish and game authority, which analyzes his water and concludes that the water can sustain a fish population. He visits local fish hatchers who provide him the growth rates of bass and trout in isolation—call these

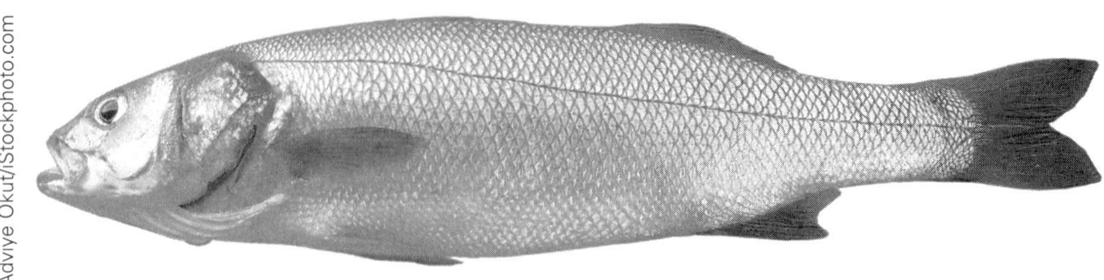


values $r$ and $s$, respectively. The experts tell him that bass and trout have the same food sources in the water and will compete for the oxygen in the water, as well as for the food for survival. The experts estimate the interactions rates between the bass and trout for survival and call these rates $m$ and $n$, respectively. We desire to build a mathematical model to help our friend determine if the pond can sustain both species of fish. This leads to a competitive hunter system of differential equations. We will revisit a similar scenario later in the chapter.

In this chapter, we will consider a variety of models, concepts, and techniques that give the reader some of the basic tools needed in solving and analyzing systems of differential equations.

Section 13.1 provides models that we will solve and analyze. These models come from a variety of disciplines in science and engineering, including chemistry, physics, biology, fluids mechanics, Newtonian mechanics, environmental engineering, and financial mathematics. Modeling techniques are discussed that help determine the necessary coefficients in the models.

Section 13.2 introduces *phase portraits*. Phase portraits provide qualitative viewing of solutions of systems of differential equations.

Section 13.3 covers some analytical methods in Maple for solving systems of differential equations with constant coefficients in both homogeneous and nonhomogeneous cases.

Section 13.4 is devoted to numerical techniques for obtaining numerical tables and plots of solutions to systems of differential equations that do not have closed-form analytical solutions. We use Maple to generate the estimates from Euler's method and the Runge–Kutta method.

## 13.1    APPLIED SYSTEMS OF DIFFERENTIAL EQUATIONS

In this section, we introduce many mathematical models from a variety of disciplines. Our emphasis here is building the mathematical model, or expression, that will be solved later in the chapter. Recall previously that we discussed the modeling process. In this section, we will confine ourselves to the first three steps of the modeling process: (1) identifying the problem, (2) assumptions and variables, and (3) building the model.

| Example 1 | Economics: Basic Supply-and-Demand Models |

Suppose we are interested in variations in the price of a specific product. We observe that a high price for the product attracts more suppliers. However, if we flood the market with the product, the price is driven down. Over time, there is an interaction between price and supply—recall the Tickle Me Elmo doll from Christmas a few years ago. A few years back many children wanted the Tickle Me Elmo doll for Christmas. The number of children desiring the doll and whose parents wanted to buy the doll for Christmas surpassed the capacity to produce the doll. We examine a generic model below.

## Problem Identification

Build a model for price and supply for a specific product.

## Assumptions and Variables

Assume the price is proportional to the quantity supplied. Also assume the change in the quantity supplied is proportional to the price. We define the following variables:

$$P(t) = \text{the price of the product at time } t$$
$$Q(t) = \text{the quantity supplied at time } t$$

We define two proportionality constants as $a$ and $b$. The constant $a$ is negative and represents a decrease in price as quantity increases.

With our limited assumptions, the model could be

$$\frac{dP}{dt} = aQ, \text{ where } a < 0$$

$$\frac{dQ}{dt} = bP, \text{ where } b > 0$$

---

## Example 2          An Electrical Network

Electrical networks with more than one loop give rise to systems of differential equations. Consider the electrical network in Figure 13.1 with two resistors and two inductors. We apply Kirchhoff's law (the sum of the voltage drops in a closed circuit is equal to the impressed voltage) to each loop. We assume that no other factors interact with the flow of electricity in this circuit.

Loop ABEF

$$E(t) = i_1 R_1 + L_1 \frac{di_2}{dt}$$

Loop ABCDEF

$$E(t) = i_1 R_1 + L_2 \frac{di_3}{dt} + i_3 R_2$$

We know that $i_1(t) = i_2(t) + i_3(t)$. We substitute this expression for $i_1$ into the laws to obtain the model:

$$\frac{di_2}{dt} = -\frac{R_1}{L_1} i_2 - \frac{R_1}{L_1} i_3 + \frac{E(t)}{L_1},$$

$$\frac{di_3}{dt} = -\frac{R_1}{L_2} i_2 - \frac{R_1 + R_2}{L_2} i_3 + \frac{E(t)}{L_2}$$

**FIGURE 13.1**
Electrical circuit

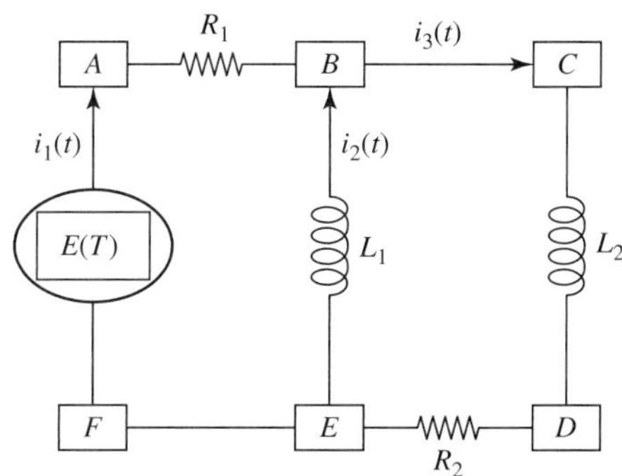

## Example 3

## Competition between Species

Imagine a small fish pond supporting both trout and bass. Let $T(t)$ denote the population of trout at time $t$ and $B(t)$ denote the population of bass at time $t$. We want to know if trout and bass can coexist in the pond. Although population growth depends on many factors, we will limit ourselves to basic isolated growth and the interaction with the other competing species for the scarce life-support resources.

We assume that the species grow in isolation. The population levels of the trout and bass, $T(t)$ and $B(t)$, depend on many variables such as their initial numbers, the amount of competition, the existence of predators, their individual species birthrates and death rates, and so forth. In isolation, we assume the following proportionality models (following the same arguments as the basic population models that we have discussed before) to be true and that the environment can support an unlimited number of trout or bass or both. Later, we might refine this model to incorporate the limited growth assumptions of the logistics model:

$$\frac{dB}{dt} = mB$$

$$\frac{dT}{dt} = aT$$

Next, we modify the proceeding differential equations to take into account the competition of the trout and bass for living space, oxygen, and food supply. The effect is that the interaction decreases the growth of the species. The interaction terms for competition lead to the decay rate, which we call $n$ for bass and $b$ for trout. This leads to the following simplified model:

$$\frac{dB}{dt} = mB - nBT$$

$$\frac{dT}{dt} = aT - bBT$$

If we have the initial stockage level, $B_0$ and $T_0$, we determine how the species coexist over time. If the model is not reasonable, we might try logistic growth instead of isolated growth. Logistic growth in isolation was discussed as a refinement for first-order ODE models.

## Example 4

## Predator–Prey Relationships

We now consider a model of population growth for two species in which one animal is hunted by another animal. An example of this might be wolves and their primary food source, rabbits.

$$\text{Let } R(t) = \text{the population of the rabbits at time } t \text{ and}$$
$$W(t) = \text{the population of the wolves at time } t$$

We assume that rabbits grow in isolation but are killed by interactions with the wolves. We further assume that the constants are proportionality constants:

$$\frac{dR}{dt} = a \cdot R - b \cdot R \cdot W$$

We assume that the wolves will die out without food and grow through their interaction with the rabbits. We further assume that these constants are also proportionality constants:

$$\frac{dW}{dt} = -m \cdot W + n \cdot R \cdot W$$

## Example 5

## Diffusion Models

Diffusion through a membrane leads to a first-order system of ordinary linear differential equations. For example, consider the situation in which two solutions of substance are separated by a membrane of permeability $P$. Assume the amount of substance that passes through the membrane at any particular time is proportional to the difference between the concentrations of the two solutions. Therefore, if we let $x_1$ and $x_2$ represent the two concentrations and $V_1$ and $V_2$ represent their corresponding volumes, then the system of differential equations is given by

$$\frac{dx_1}{dt} = \frac{P}{V_1}(x_2 - x_1)$$

$$\frac{dx_2}{dt} = \frac{P}{V_2}(x_1 - x_2)$$

where the initial amounts of $x_1$ and $x_2$ are given.

If this model does not yield satisfactory results in terms of realism, we might try a refinement of the diffusion model as follows:

Diffusion through a double-walled membrane, where the inner wall has permeability $P_1$ and the outer wall has permeability $P_2$ with $0 < P_1 < P_2$. Suppose the volume of the solution within the inner wall is $V_1$ and $V_2$ between the two walls. We let $x$ represent the concentration of the solution within the inner wall and $y$ the concentration between the two walls. This leads to the following system:

$$\frac{dx}{dt} = \frac{P_1}{V_1}(y - x)$$

$$\frac{dy}{dt} = \frac{1}{V_2}(P_2(C - y) + P_1(x - y))$$

$$x(0) = 2,\ y(0) = 1, C = 10$$

---

| | |
|---|---|
| **Example 6** | ## Insurgency Models |

As we look around the world, we see many conflicts involving insurgencies. We have the political faction (usually the status quo or the new regime) battling insurgents or rebels who are resisting change or the political status quo. We have seen this throughout history—for example, in our own Revolutionary War.

In insurgency operations, we find the following assumptions: They are messy grassroots fights that are confused and brutally contested. The enemy is loosely defined. Positive control of the forces is usually weak. There are few rules of engagement (they are often permissive). There are deep-seated political divisions that leave little room for compromise.

Further, as we consider building a mobilization model, we assume that growth of the insurgency is subject to the same laws as any other natural or human-made population (basic growth or logistical growth as discussed before). In addition, there are three considerations: the pool of potential recruits, the number of recruiters, and the transformation rate.

We assume logistical growth and our systems could look like

$$X(t) = \text{insurgency}$$

$$Y(t) = \text{regime}$$

$$\frac{dX}{dt} = a \cdot (k_1 - X) \cdot X$$

$$\frac{dY}{dt} = b \cdot (k_2 - Y) \cdot Y$$

where

$a$ measures insurgency growth rate,
$b$ measures regime growth rate, and
$k_1$ and $k_2$ are the respective carrying capacities.

## 13.2 PHASE PORTRAITS AND QUALITATIVE ASSESSMENT OF AUTONOMOUS SYSTEMS OF FIRST-ORDER DIFFERENTIAL EQUATIONS

Consider the system of differential equations:

$$\frac{dx}{dt} = f(x, y)$$

$$\frac{dy}{dt} = g(x, y)$$

These are autonomous systems because they do not include $t$.

The solution is a pair of parametric equations, $x = x(t)$, $y = y(t)$. The solution is also a curve that varies over time. We call the solution curve a *trajectory*, *path*, or *orbit*. The $x$–$y$ plane is called the phase plane. We can also obtain plots of $x$ versus $t$ and $y$ versus $t$. *Rest points* or *equilibrium points* are points that satisfy both $f(x, y) = 0$ and $g(x, y) = 0$ *simultaneously*. Once we have the equilibrium values, we desire information about their stability.

### Rules of Stability

We classify equilibrium values as stable, asymptotically stable, or unstable. We define these as follows:

(Stable) If a trajectory starts close to a rest point, it remains close to that point for all future time.

(Asymptotically stable) If a trajectory starts close, then it tends toward the rest point as $t \rightarrow \infty$.

(Unstable) If it does not follow either stable or asymptotically stable rules.

The following results are useful in investigating solutions to autonomous systems. We offer these without proof:

1. There is at most one trajectory through any point in the phase plane.

2. A trajectory that starts at a point other than a rest point cannot reach a rest point in a finite amount of time.

3. No trajectory can cross itself unless it is a closed curve. If it is a closed curve, then it is a periodic solution.

4. The implications and properties of motion from a starting point (not a rest point) are that:

    a. it will move along the same path regardless of starting time,

    b. it cannot return to a starting point unless motion is periodic,

    c. it can *never* cross another trajectory, and

    d. it can only approach and never can reach a rest point.

We illustrate the Maple commands below to obtain phase portraits.

Consider the following autonomous competitive-hunter system of differential equations:

$$\frac{dy}{dt} = .24 \cdot x(t) - .08 \cdot y(t) \cdot x(t)$$

$$\frac{dx}{dt} = y(t) \cdot (4.5 - .9 \cdot x(t))$$

### Qualitative Graphical Assessment

First, we plot $dx/dt$ and $dy/dt$, respectively. We want to see where $dx/dt = 0$ and $dy/dt = 0$, simultaneously. In the equation $dx/dt = 0$, we find that either $y(t) = 0$ (which is the $x$-axis) or $x = 5$ (the horizontal line). In the equation $y(t) = 0$, we find either $x(t) = 0$ (the $y$-axis) or the line $y(t) = 3$. This is shown in Figure 13.2. There are two equilibrium points: the points (0,0), where the $x$-and $y$-axes intersect; and the point (5,3), where the vertical line $x = 5$ intersect the horizontal line $y = 3$.

**FIGURE 13.2**

Lines for *dx/dt* = 0

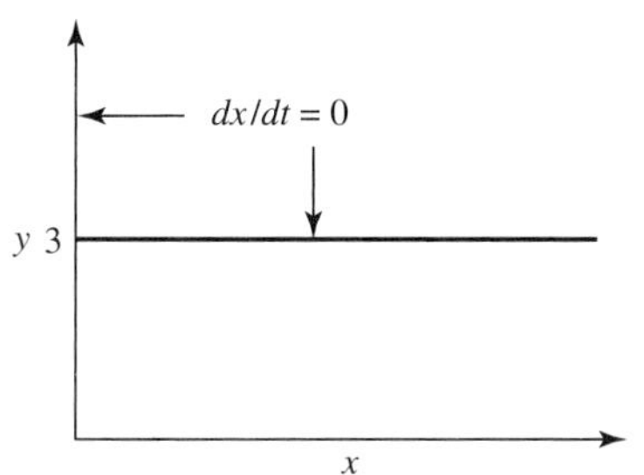


**FIGURE 13.3**

Unstable equilibrium
at (0,0) and (5,3)

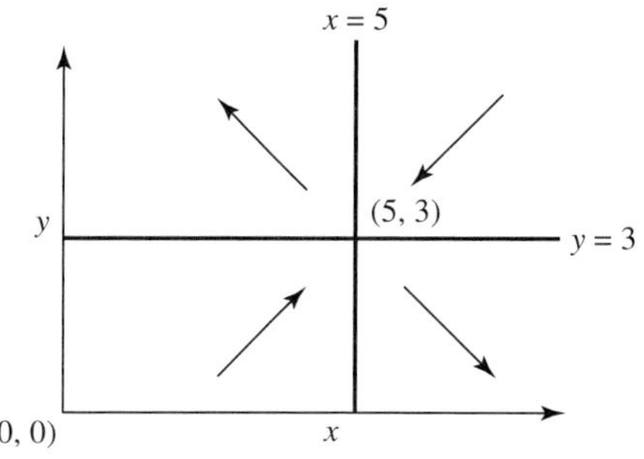


**FIGURE 13.4**

Phase portrait for
competing species

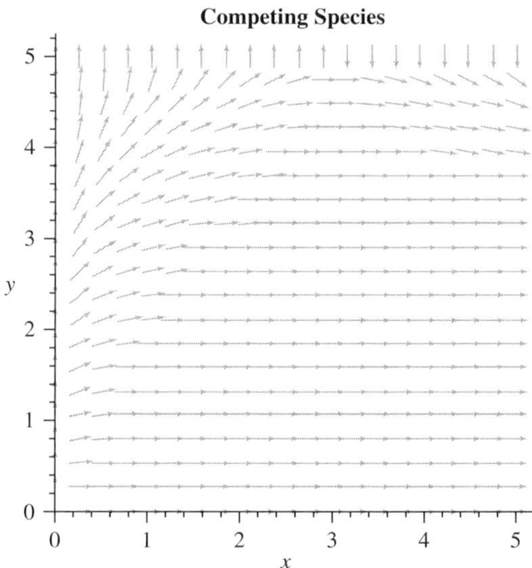


Analysis shows that both (0,0) and (5,3) are unstable equilibrium values as shown in Figure 13.3.

We now examine the phase portrait shown in Figure 13.4 that provided the same information to us but in a slightly different format.

*restart; with(DEtools):with(linalg):*

**Phase Portrait**

> *diffeq1:=diff(y(t),t)=4.5*y(t)−.9*y(t)*x(t);*

$$diffeq1 := \frac{d}{dt}\, y(t) = 4.5\, y(t) - 0.9\, y(t)\, x(t)$$

> *diffeq2:=diff(x(t),t)=.24*x(t)−.08*y(t)*x(t);*

$$diffeq2 := \frac{d}{dt}\, x(t) = 0.24\, x(t) - 0.08\, y(t)\, x(t)$$

> *DEplot({diffeq1,diffeq2},[y(t),x(t)],t=−3..3, y=0..5,x=0..5,title='Competing Species');*

The phase portrait shows who survives from a starting point (the initial condition). The phase portrait traces out a possible solution curve.

---

## Example 1    The Fish Pond

$$B(t) = \text{number of bass fish after time } t$$
$$T(t) = \text{number of trout after } t \text{ time}$$

**FIGURE 13.5**

Phase portrait for fish-pond example for bass and trout

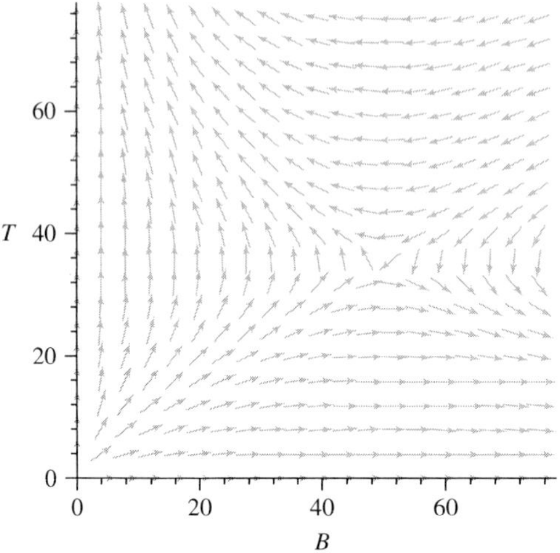



Rate of change of growth = rate in isolation + rate in competition for resources
$dB/dt = 0.7B - 0.02B{*}T$
$dT/dt = 0.5T - 0.01B{*}T$

Solve $dB/dt = 0$ and $dT/dt = 0$:

$$dB/dt = 0 = B(0.7 - 0.02T) = 0, \text{ so } B = 0 \text{ or } T = 35$$
$$dT/dt = 0 = T(0.5 - 0.01B) = 0, \text{ so } T = 0 \text{ or } B = 50$$

The equilibrium values are $(0,0)$ and $(50,35)$.

```
> with(plots):with(DEtools):
> eqn1:=diff(B(t),t)=.7*B(t)-.02*B(t)*T(t);
```

$$eqn1 := \frac{d}{dt} B(t) = 0.7\,B(t) - 0.02\,B(t)\,T(t)$$

```
> eqn2:=diff(T(t),t)=.5*T(t)-.01*B(t)*T(t);
```

$$eqn2 := \frac{d}{dt} T(t) = 0.5\,T(t) - 0.01\,B(t)\,T(t)$$

```
> DEplot([eqn1,eqn2],[B(t),T(t)],t=0..20,B=0..75,T=0..75);
```

Again both equilibrium values are not stable. Depending on the starting value, one species will dominate over time.

Thus, the phase portrait as shown in Figure 13.5 is useful to give us a sense of the possible solutions.

## 13.2 | EXERCISES

Find and classify all the rest points. Then sketch a few trajectories to indicate the motion.

**1.** $dx/dt = x$, $dy/dt = y$

**2.** $dx/dt = -x$, $dy/dt = 2y$

**3.** $dx/dt = y$, $dy/dt = -2x$

**4.** $dx/dt = -x + 1$, $dy/dt = -2y$

## 13.3  SOLVING HOMOGENEOUS AND NONHOMOGENEOUS SYSTEMS OF ODES IN MAPLE

Maple can solve systems of differential equations of the form:

$$\frac{dx}{dt} = ax + by + g(t)$$

$$\frac{dy}{dt} = mx + ny + h(t)$$

where (1) $a$, $b$, $m$, and $n$ are constants and (2) the functions $g(t)$ and $h(t)$ can either be 0 or functions of $t$ with real coefficients.

When $g(t)$ and $h(t)$ are both 0, the system of differential equations is called a *homogeneous system*; otherwise, it is *nonhomogeneous*. We will begin with homogeneous systems.

The method we will use involves eigenvalues and eigenvectors.

---

**Example 1**      Solving Homogeneous Systems

Consider the following homogeneous system with initial conditions:

$$x' = 2x - y + 0$$
$$y' = 3x - 2y + 0$$
$$x(0) = 1, y(0) = 2$$

Basically, if we rewrite the system of differential equation in matrix form

$$X' = AX,$$

*where*

$$A = \begin{bmatrix} 2 & 1 \\ 3 & -2 \end{bmatrix}$$

$$X' = \begin{bmatrix} \dfrac{dx}{dt} \\ \dfrac{dy}{dt} \end{bmatrix}, X = \begin{bmatrix} x \\ y \end{bmatrix}$$

then we can solve $X' = Ax$. This form is highly suggestive of the first-order separable equation that we saw in the previous chapter. We can assume the solution to have a similar form: $\mathbf{X} = \mathbf{K}e^{\lambda t}$, where $\lambda$ is a constant and $\mathbf{X}$ and $\mathbf{K}$ are vectors. The values of $\lambda$ are called **eigenvalues**, and the components of $\mathbf{K}$ are the corresponding **eigenvectors**. We note that a full discussion of the theory and applications of eigenvalues and eigenvectors can be found in linear algebra textbooks as well as in many differential equations textbooks.

Because we have a $2 \times 2$ system, there are two linearly independent solutions that we call $X_1$ and $X_2$. The *complementary solution* or *general solution* is $X = c_1 X_1 + c_2 X_2$ where $c_1$ and $c_2$ are arbitrary constants. We use the initial conditions to find specific values for $c_1$ and $c_2$.

The following steps can be used when we have real distinct eigenvalues.

**Step 1**      Set up the system as a matrix: $X' = AX$, $X(0) = X_{00}$.

**Step 2**      Find the eigenvalues, $\lambda_1$ and $\lambda_2$.

**Step 3**      Find the corresponding eigenvectors, $K_1$ and $K_2$.

**Step 4**  Set up the complementary solution $X_c = c_1 X_1 + c_2 X_2$ where

$$X_1 = K_1 e^{\lambda_1 t}$$
$$X_2 = K_2 e^{\lambda_2 t}$$

**Step 5**  Solve for $c_1$ and $c_2$ and rewrite the solution for $\mathbf{X_c}$.

Maple allows us to find the homogeneous solution in matrix form. The following commands illustrate this:

```
> with(DEtools):with(plots):with(linalg):
> M:=array([[2,-1],[3,-2]]);
> lambda:=eigenvects(M);
> homsol:=matrixDE(M,t);
```

$$M := \begin{bmatrix} 2 & -1 \\ 3 & -2 \end{bmatrix}$$

$$\lambda := [1, 1, \{[1, 1]\}], [-1, 1, \{[1,3]\}]$$

$$homsol := \left[ \begin{bmatrix} e^t & e^{(-t)} \\ e^t & 3\,e^{(-t)} \end{bmatrix}, [0, 0] \right]$$

```
> phi:=homsol[1];
```

$$\phi := \begin{bmatrix} e^t & e^{(-t)} \\ e^t & 3\,e^{(-t)} \end{bmatrix}$$

We find the complementary solution:
The matrix (2,1,[c1,c2]) means that we want a two-row, one-column matrix of $c_1$, $c_2$.

```
> Xc:=multiply(phi, matrix(2,1,[c1,c2]));
```

$$Xc := \begin{bmatrix} e^t\, c1 + e^{(-t)}\, c2 \\ e^t\, c1 + 3\,e^{(-t)}\, c2 \end{bmatrix}$$

Because we only had a homogeneous system, we will solve for $c_1$ and $c_2$ now using the initial conditions $x(0) = 1$, $y(0) = 2$.

```
> eq1:=evalf(subs(t=0,exp(t)*c1+exp(-t)*c2));
```

$$eq1 := 1.\,c1 + 1.\,c2$$

```
> eq2:=evalf(subs(t=0,exp(t)*c1+3*exp(-t)*c2));
```

$$eq2 := 1.\,c1 + 3.\,c2$$

```
> solve({1=eq1,2=eq2},{c1,c2});
```

$$\{c2 = 0.5000000000, c1 = 0.5000000000\}$$

```
> Xg:=multiply(phi, matrix(2,1,[.5,.5]));
```

$$Xg := \begin{bmatrix} 0.5\,e^t + 0.5\,e^{(-t)} \\ 0.5\,e^t + 1.5\,e^{(-t)} \end{bmatrix}$$

**FIGURE 13.6**

Plot of $0.5e^t + 0.5e^{-t}$

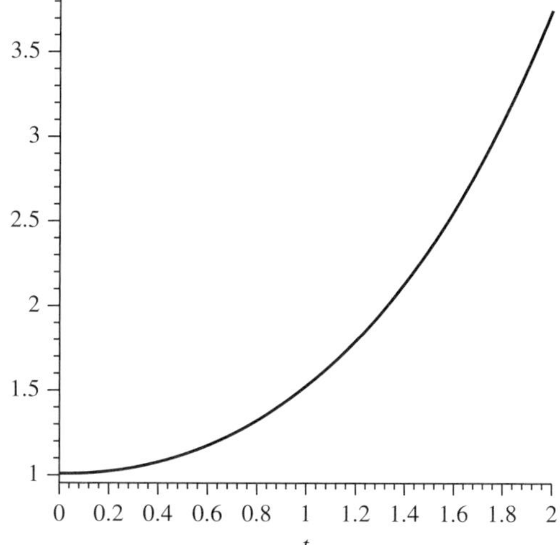

**FIGURE 13.7**

Plot of $0.5e^t + 1.5e^{-t}$

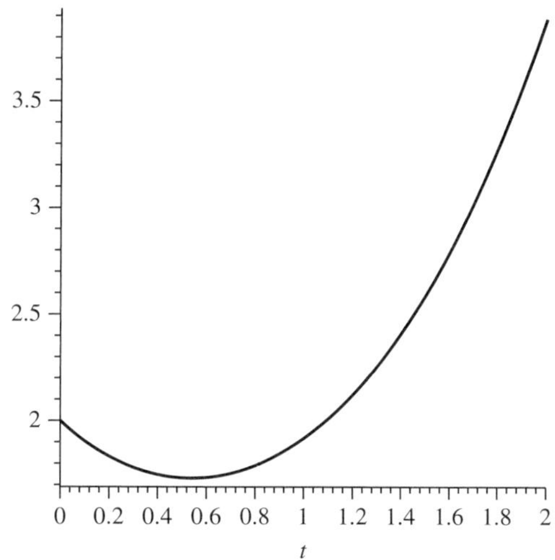

We might plot the solutions to the components $X_1$ and $X_2$, each a function of $t$. We note that both solutions grow without bound as $t \to \infty$; see Figures 13.6 and 13.7.

```
> plot(.5*exp(t)+.5*exp(-t),t=0..2);
> plot(.5*exp(t)+1.5*exp(-t),t=0..2);
```

---

**Example 2**   Complex Eigenvalues (Eigenvalues of the Form $\lambda = a \pm bi$)

We note here that we do not use the form $e^{a \pm bi}$ in the solution as we will see and that complex eigenvalues always appear in conjugate pairs. The key to finding two real linearly independent solutions from complex solutions is Euler's identity:

$$e^{i\theta} = \cos\theta + i\sin\theta$$

We can rewrite the solutions for $X_1$ and $X_2$ using Euler's identity:

$$Ke^{\lambda t} = Ke^{(a+bi)} = Ke^{at}(\cos bt + i\sin bt)$$
$$Ke^{\lambda t} = Ke^{(a-bi)} = Ke^{at}(\cos bt - i\sin bt)$$

Consider the following steps as a summary when we get complex eigenvalues.

**Step 1**    Find the complex eigenvalues: $\lambda = a \pm bi$.

**Step 2**    Find the complex eigenvector, $K$:

$$K = \begin{bmatrix} u_1 + iv_1 \\ u_2 + iv_2 \end{bmatrix}$$

**Step 3**    Form the real vectors:

$$B_1 = \begin{bmatrix} u_1 \\ u_2 \end{bmatrix}$$

$$B_2 = -\begin{bmatrix} v_1 \\ v_2 \end{bmatrix}$$

**Step 4**    Form the linearly independent set of real solutions:

$$X_1 = e^{at}(B_1 \cos bt + B_2 \sin bt)$$
$$X_2 = e^{at}(B_2 \cos bt - B_1 \sin bt)$$

**Step 5**    The solution is $X = c1\ \mathbf{X1} + c2\ \mathbf{X2}$.

We find Maple will allow for an easily manipulation of these steps for us.

```
> with(DEtools):with(plots):with(linalg):
> M:=array([[6,–1],[5,4]]);
> lambda:=eigenvects(M);
> homsol:=matrixDE(M,t);
```

$$M := \begin{bmatrix} 6 & -1 \\ 5 & 4 \end{bmatrix}$$

$$\lambda := [5 + 2\,I,\, 1,\, \{[1,\, 1 - 2\,I]\}],\, [5 - 2\,I,\, 1,\, \{[1,\, 1 + 2\,I]\}]$$

$$homsol := \left[ \begin{bmatrix} e^{(5\,t)}\cos(2\,t) & e^{(5\,t)}\sin(2\,t) \\ e^{(5\,t)}\cos(2\,t) + 2\,e^{(5\,t)}\sin(2\,t) & e^{(5t)}\sin(2\,t) - 2\,e^{(5\,t)}\cos(2\,t) \end{bmatrix},\, [0,0] \right]$$

```
> phi:=homsol[1];
```

$$\phi := \begin{bmatrix} e^{(5\,t)}\cos(2\,t) & e^{(5\,t)}\sin(2\,t) \\ e^{(5\,t)}\cos(2\,t) + 2\,e^{(5\,t)}\sin(2\,t) & e^{(5\,t)}\sin(2\,t) - 2\,e^{(5\,t)}\cos(2t) \end{bmatrix}$$

We find the complementary solution:
The matrix(2,1,[c1,c2]) means that we want a 2 row 1 columns matrix of $c_1$, $c_2$.

```
> Xc:=multiply(phi, matrix(2,1,[c1,c2]));
```

$$Xc := \begin{bmatrix} e^{(5\,t)}\cos(2\,t)\,c1 + e^{(5\,t)}\sin(2\,t)c2 \\ (e^{(5\,t)}\cos(2\,t) + 2e^{(5\,t)}\sin(2\,t))c1 + (e^{(5\,t)}\sin(2\,t) - 2\,e^{(5\,t)}\cos(2\,t))c2 \end{bmatrix}$$

Because we only had a homogeneous system, we will solve for $c_1$ and $c_2$ now using the initial conditions $x(0) = 1$, $y(0) = 2$.

> *eq1:=evalf(subs(t=0,Xc[1,1]));*

$$eq1 := 1.\, c1$$

> *eq2:=evalf(subs(t=0,Xc[2,1]));*

$$eq2 := 1.\, c1 - 2.\, c2$$

> *solve({1=eq1,2=eq2},{c1,c2});*

$$\{c1 = 1.,\ c2 = -0.5000000000\}$$

> *Xg:=multiply(phi,matrix(2,1,[1,-.5]));*

$$Xg := \begin{bmatrix} e^{(5\,t)}\cos(2\,t) - 0.5\,e^{(5\,t)}\sin(2\,t) \\ 2.0\,e^{(5\,t)}\cos(2\,t) + 1.5\,e^{(5\,t)}\sin(2\,t) \end{bmatrix}$$

Again, we obtain plots of $X_1$ and $X_2$ as functions of $t$ as shown in Figures 13.8 and 13.9.

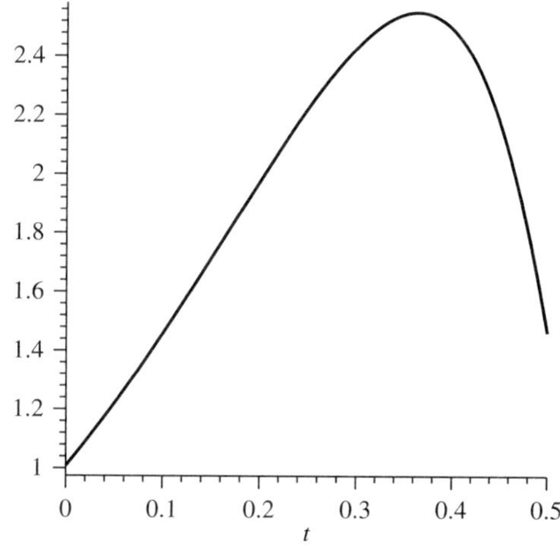

**FIGURE 13.8**

Plot of $e^{5t}\cos(2t) - 0.5e^{5t}\sin(2t)$ from $t = 0$ to $1/2$

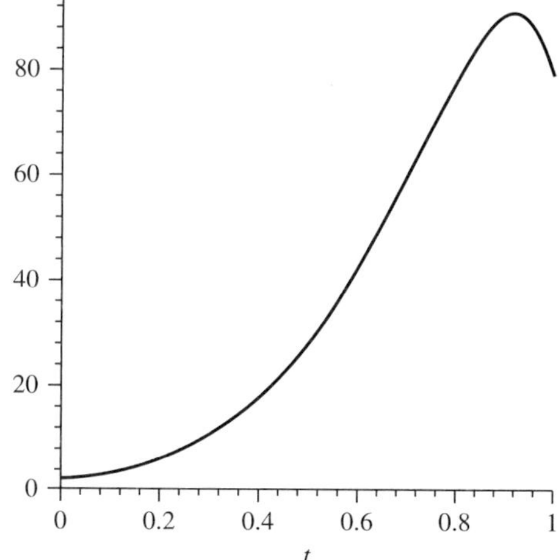

**FIGURE 13.9**

Plot of $2e^{5t}\cos(2t) + 1.5e^{5t}\sin(2t)$ from $t = 0$ to $1$

```
> plot(exp(5*t)*cos(2*t)-.5*exp(5*t)*sin(2*t),t=0..0.5);
> plot(2.0*exp(5*t)*cos(2*t)+1.5*exp(5*t)*sin(2*t),t=0..1);
```

## Example 3    Repeated Eigenvalues Solution

When eigenvalues are repeated, we must find a method to obtain independent solutions. The following is a summary for repeated real eigenvalues.

**Step 1**   Find the repeated eigenvalues $\lambda_1 = \lambda_2 = \lambda$.

**Step 2**   One solution is

$$X_1 = Ke^{\lambda t}$$

and the second linearly independent solution is given by

$$X_2 = Kte^{\lambda t} + Pe^{\lambda t}$$

where the components of $P$ must satisfy the system

$$(a - \lambda)p_1 + bp_2 = k_1$$
$$cp_1 + (d - \lambda)p_2 = k_2$$
$$and$$
$$\begin{bmatrix} a & b \\ c & d \end{bmatrix} = A$$

We now present an example of this method using Maple. Maple recognizes repeated eigenvalues and places the solution in the correct form.

```
> with(DEtools):with(plots):with(linalg):
> M:=array([[3,-18],[2,-9]]);
> lambda:=eigenvects(M);
> homsol:=matrixDE(M,t);
```

$$M := \begin{bmatrix} 3 & -18 \\ 2 & -9 \end{bmatrix}$$

$$\lambda := [-3, 2, \{[3, 1]\}]$$

$$homsol := \left[ \begin{bmatrix} e^{(-3t)} & e^{(-3t)}t \\ \dfrac{1}{3}e^{(-3t)} & \dfrac{1}{3}e^{(-3t)}\left(t - \dfrac{1}{18}\right)e^{(-3t)} \end{bmatrix}, [0, 0] \right]$$

```
> phi:=homsol[1];
```

$$\phi := \begin{bmatrix} e^{(-3t)} & e^{(-3t)}t \\ \dfrac{1}{3}e^{(-3t)} & \dfrac{1}{3}e^{(-3t)}\left(t - \dfrac{1}{18}\right)e^{(-3t)} \end{bmatrix}$$

We find the complementary solution:
The matrix(2,1,[c1,c2]) means that we want a two-row, one-column matrix of $c_1$, $c_2$.

```
> Xc:=multiply(phi, matrix(2,1,[c1,c2]));
```

$$Xc := \begin{bmatrix} e^{(-3t)}c1 + e^{(-3t)}t\,c2 \\ \dfrac{1}{3}e^{(-3t)}c1 + \left(\dfrac{1}{3}e^{(-3t)}t - \dfrac{1}{18}e^{(-3t)}\right)c2 \end{bmatrix}$$

Because we only had a homogeneous system, we will solve for $c_1$ and $c_2$ now using the initial conditions $x(0) = 1$, $y(0) = 2$.

```
> eq1:=evalf(subs(t=0, Xc[1,1]));
```

$$eq1 := 1. \, cl$$

```
> eq2:=evalf(subs(t=0,Xc[2,1]));
```

$$eq2 := 0.3333333333 \, cl - 0.05555555556 \, c2$$

```
> solve({1=eq1,2=eq2},{c1,c2});
```

$$\{cl = 1., c2 = -30.00000000\}$$

```
> Xg:=multiply(phi, matrix(2,1,[1,-30]));
```

$$Xg := \begin{bmatrix} e^{(-3\,t)} - 30 \, e^{\,(-3\,t)}t \\ 2 \, e^{(-3\,t)} - 10 \, e^{\,(-3\,t)}t \end{bmatrix}$$

```
> plot(exp(-3*t)-30*exp(-3*t)*t,t=0..0.5);
> plot(2*exp(-3*t)-10*exp(-3*t)*t,t=0..1);
```

The plots of the two solution are shown in Figures 13.10 and 13.11.

### Maple's Use in Nonhomogeneous Systems of Differential Equations

In the nonhomogeneous form, we need both a complementary solution $X_c$ and a particular solution $X_p$. The complementary solution is found by solving the homogeneous part of the system. The particular solution is found using the variation of parameters. The following summary of the procedure is provided. Students should consult a differential equations textbook for a more detailed explanation and proof of the procedure.

**Step 1**  Find the complementary solution by solving for the homogeneous solution of

$$X' = AX$$

and form the linear independent set of solutions; put them, as columns, into a matrix called $\Phi$.

**Step 2**  Vary the parameters and write the form of the particular solution:

$$X_p = u_1(t)X_1(t) + u_2(t)X_2(t)$$

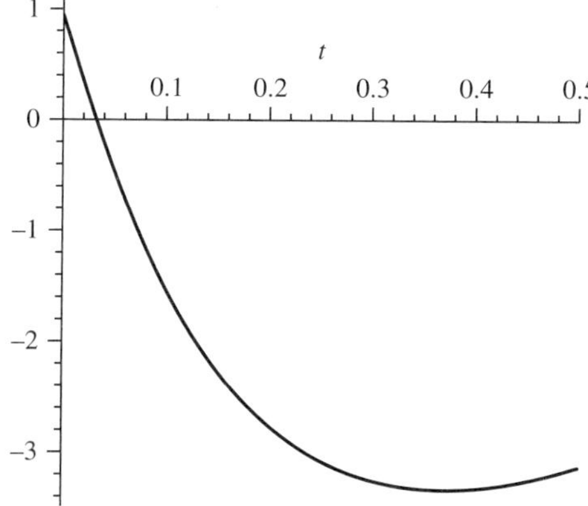

**FIGURE 13.10**
Plot of $e^{-3t} - 30te^{-3t}$ from $t = 0$ to $0.5$

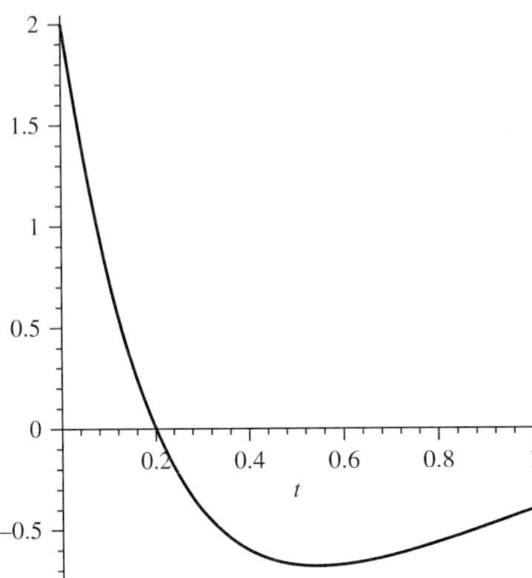

**FIGURE 13.11**
Plot of $2e^{-3t} - 10te^{-3t}$
from $t = 0$ to 1

**Step 3**   Invert the matrix $\Phi$:

$$\Phi^{-1}$$

**Step 4**   Determine the parameters $u_1$ and $u_2$.

$$\begin{bmatrix} u_1 \\ u_2 \end{bmatrix} = \int \Phi^{-1} F(t)dt$$

**Step 5**   Calculate the particular solution $X_p$.

$$X_p = \Phi \cdot U = \Phi \cdot \int \Phi^{-1} F(t)dt$$

**Step 6**   Form the general solution

$$X = X_c + X_p$$

We provide an example using Maple. Given the following system of nonhomogeneous differential equations:

$$x' = 2x - y + 0$$
$$y' = 3x - 2y + 4t$$
$$x(0) = 1, y(0) = 2$$

Note that $g(t) = 0$ and $h(t) = 4t$.

**Part 1: Homogeneous**

```
> with(DEtools):with(plots):with(linalg):
> M:=array([[2,-1],[3,-2]]);
> lambda:=eigenvects(M);
> homsol:=matrixDE(M,t);
```

$$M := \begin{bmatrix} 2 & -1 \\ 3 & -2 \end{bmatrix}$$

$$\lambda := [1, 1, \{[1, 1]\}], [-1, 1, \{[1,3]\}]$$

$$homsol := \left[\left[\begin{array}{cc} e^t & e^{(-t)} \\ e^t & 3\,e^{(-t)} \end{array}\right], [0,0]\right]$$

> *phi:=homsol[1];*

$$\phi := \left[\begin{array}{cc} e^t & e^{(-t)} \\ e^t & 3\,e^{(-t)} \end{array}\right]$$

**The Complementary Solution**  The command matrix (2,1,[c1,c2]) means that we want a two-row, one-column matrix of $c_1$, $c_2$.

> *Xc:=multiply(phi, matrix(2,1,[c1,c2]));*

$$Xc := \left[\begin{array}{c} e^t\,c1 + e^{(-t)}\,c2 \\ e^t\,c1 + 3\,e^{(-t)}\,c2 \end{array}\right]$$

If we only had a homogeneous systems, we would solve for $c_1$ and $c_2$ now. Because this is a nonhomogeneous system, however, we wait until we get $X = X_c + X_p$.

**Finding *Xp***

$$phi*int(phi\text{\textasciicircum}-1*F(t))dt$$

***Nonhomogeneous Systems of Differential Equations***

> *A:=homsol[1];*

$$A := \left[\begin{array}{cc} e^t & e^{(-t)} \\ e^t & 3\,e^{(-t)} \end{array}\right]$$

> *A1:=inverse(A);*

$$A1 := \left[\begin{array}{cc} \dfrac{3}{2}\dfrac{1}{e^t} & -\dfrac{1}{2}\dfrac{1}{e^t} \\ -\dfrac{1}{2}\dfrac{1}{e^{(-t)}} & \dfrac{1}{2}\dfrac{1}{e^{(-t)}} \end{array}\right]$$

> *B:=matrix(2,1,[0,4*t]);*

$$B := \left[\begin{array}{c} 0 \\ 4\,t \end{array}\right]$$

> *B1:=multiply(A1,B);*

$$B1 := \left[\begin{array}{c} -\dfrac{2\,t}{e^t} \\ \dfrac{2t}{e^{(-t)}} \end{array}\right]$$

> *B2:=map(int,B1,t);*

$$B2 := \left[\begin{array}{c} \dfrac{2\,(1+t)}{e^t} \\ \dfrac{2\,(t-1)}{e^{(-t)}} \end{array}\right]$$

> *B3:=multiply( A,B2 );*

$$B3 := \begin{bmatrix} 4\,t \\ 8\,t - 4 \end{bmatrix}$$

> *nB3:=simplify(%);*

$$nB3 := \begin{bmatrix} 4\,t \\ 8\,t - 4 \end{bmatrix}$$

> *x:=evalm(multiply( A,matrix(2,1,[c1,c2]))+nB3);*

$$x := \begin{bmatrix} e^{t}\,c1 + e^{(-t)}\,c2 + 4\,t \\ e^{t}\,c1 + 3\,e^{(-t)}\,c2 + 8\,t - 4 \end{bmatrix}$$

Now use the initial conditions $x_1(0) = 1$, $x_2(0) = 2$ to find $c_1$ and $c_2$.

> *solve({c1+c2=1,c1+3*c2−4=2}, {c1,c2});*

$$\left\{ c2 = \frac{5}{2}, \ c1 = \frac{-3}{2} \right\}$$

> *x1:=subs({c2=2.5,c1=−1.5},x[1,1]);*

$$x1 := -1.5\,e^{t} + 2.5\,e^{(-t)} + 4\,t$$

> *x2:=subs({c2=2.5,c1=−1.5},x[2,1]);*

$$x2 := -1.5\,e^{t} + 7.5\,e^{(-t)} + 8\,t - 4$$

We can plot each versus $t$ as shown in Figures 13.12 and 13.13.

> *plot(x1,t=0..1);*
> *plot(x2,t⁻0..1);*

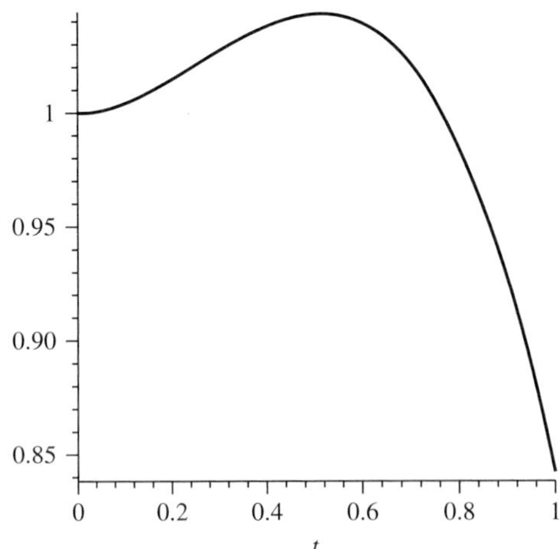

**FIGURE 13.12**
Plot of $-1.5e^{t} + 2.5e^{-t} + 4t$

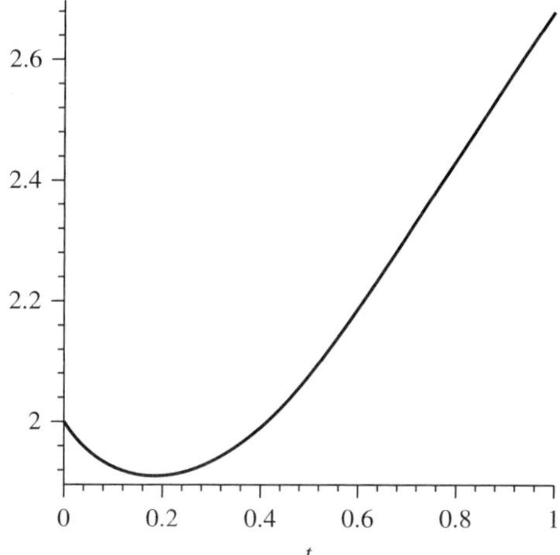

**FIGURE 13.13**
Plot of $-1.5e^t + 7.5e^{-t} + 8t + 4$



## 13.4   APPLIED SYSTEMS OF ODEs WITH MAPLE

Recall the diffusion model from Section 13.1. We will use Maple to assist us in the solution to the diffusion model.

### Example 1

### Diffusion Example

### Part I: Diffusion through a Single Membrane

$$dx_1/dt = (P/V_1)(x_2 - x_1) = -x_1 + x_2$$
$$dx_2/dt = (P/V_2)(x_1 - x_2) = x_1 - x_2$$
$$x_1(0) = 2$$
$$x_2(0) = 10$$

```
> with(linalg):with(DEtools):
> Chem1:=matrix(2,2,[-1,1,-1,01]);
```

$$Chem1 := \begin{bmatrix} -1 & 1 \\ -1 & 1 \end{bmatrix}$$

```
> ch_eigen:=eigenvects(Chem1);
```

$$ch\_eigen := [\,0, 2, \{[1, 1]\}\,]$$

```
> chem1_sol:=matrixDE(Chem1,t);
```

$$chem1\_sol := \left[ \begin{bmatrix} 1 & t \\ 1 & t + 1 \end{bmatrix}, [0, 0] \right]$$

```
> inits:=x1(0)=2,x2(0)=10;
```

$$inits := x1(0) = 2, x2(0) = 10$$

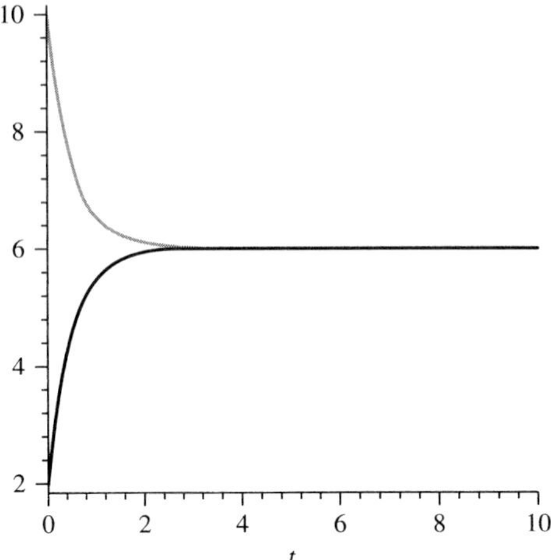

**FIGURE 13.14**
Plot of $6 - 4e^{-2t}$ and $4e^{-2t} + 6$ converging to a stable equilibrium at 6

```
> dsolve({diff(x1(t),t)=-x1(t)+x2(t),diff(x2(t),t)
=x1(t)-x2(t),x1(0)=2,x2(0)=10},{x1(t),x2(t)});
```

$$\{x2(t) = 4\,e^{(-2t)} + 6,\; x1(t) = 6 - 4\,e^{(-2t)}\}$$

```
> plot({6-4*exp(-2*t),4*exp(-2*t)+6},t=0..10,thickness=3);
```

**Interpretation**  The solution shows that we have a stable equilibrium value at six units. This can be seen in Figure 13.14.

---

## Example 2   Diffusion through a Double-Walled Membrane

Consider this system of nonhomogeneous differential equations:

$$dx/dt = 1.5(y - x),\; x(0) = 2$$
$$dy/dt = 0.8\,(10 - y) + 0.3(x - y) = 0.3x - 1.1y + 8,\; y(0) - 1$$

```
> with(linalg):
> M:=array([[-1.5,1.5],[.3,-1.1]]);
> lambda:=eigenvects(M);
```

$$M := \begin{bmatrix} -1.5 & 1.5 \\ 0.3 & -1.1 \end{bmatrix}$$

$$\lambda := [-2.000000000,\ 1,\ \{[-0.9486832981,\ 0.3162277660]\}],$$
$$[-0.6000000000,\ 1,\ \{[-1.129384879,\ -0.6776309273]\}]$$

```
> homsol:=matrixDE(M,t);
```

$$homsol := \left[\begin{bmatrix} 1 \cdot e^{(-2t)} & 1 \cdot e^{\left(-\frac{3t}{5}\right)} \\ -0.333333333\,e^{(-2t)} & 0.6000000000\,e^{\left(-\frac{3t}{5}\right)} \end{bmatrix},\ [0.,\ 0.]\right]$$

```
> phi:=homsol[1];
```

$$\phi := \begin{bmatrix} 1 \cdot e^{(-2t)} & 1 \cdot e^{\left(-\frac{3t}{5}\right)} \\ -0.333333333\,e^{(-2t)} & 0.6000000000\,e^{\left(-\frac{3t}{5}\right)} \end{bmatrix}$$

```
> phi_inv:=inverse(phi);
```

$$phi\_inv := \begin{bmatrix} \dfrac{0.6428571431}{e^{(-2\,t)}} & -\dfrac{1.071428572}{e^{(-2\,t)}} \\[2ex] \dfrac{0.3571428569}{e^{\left(-\frac{3t}{5}\right)}} & \dfrac{1.071428572}{e^{\left(-\frac{3t}{5}\right)}} \end{bmatrix}$$

```
> f:=matrix(2,1,[0,8]);
```

$$f := \begin{bmatrix} 0 \\ 8 \end{bmatrix}$$

```
> p1:=multiply(phi_inv,f);
```

$$p1 := \begin{bmatrix} -\dfrac{8.571428576}{e^{(-2\,t)}} \\[2ex] \dfrac{8.571428576}{e^{\left(-\frac{3t}{5}\right)}} \end{bmatrix}$$

```
> p1_int:=map(int,p1,t);
```

$$p1\_int := \begin{bmatrix} -\dfrac{4.285714288}{e^{(-2.\,t)}} \\[2ex] \dfrac{14.28571429}{e^{(-0.6000000000\,t)}} \end{bmatrix}$$

```
> p2:=multiply(phi,p1_int);
```

$$p2 := \begin{bmatrix} -\dfrac{4.285714288\,e^{(-2\,t)}}{e^{(-2t)}} + \dfrac{14.28571428\,e^{\left(-\frac{3t}{5}\right)}}{e^{(-0.6000000000\,t)}} \\[2ex] \dfrac{1.428571428\,e^{(-2\,t)}}{e^{(-2t)}} + \dfrac{8.571428574\,e^{\left(-\frac{3t}{5}\right)}}{e^{(-0.6000000000\,t)}} \end{bmatrix}$$

```
> xp:=simplify(%);
```

$$xp := \begin{bmatrix} 10. \\ 10. \end{bmatrix}$$

```
> x:=evalm(multiply(phi, matrix(2,1,[c1,c2]))+xp);
```

$$x := \begin{bmatrix} 1.\,e^{(-2\,t)}\,c1 + 1.\,e^{\left(-\frac{3t}{5}\right)}\,c2 + 10. \\[1ex] -0.333333333\,e^{(-2\,t)}\,c1 + 0.6000000000\,e^{\left(-\frac{3t}{5}\right)}\,c2 + 10. \end{bmatrix}$$

Solve for $c_1$, $c_2$.

```
> solve({2=c1+c2+10,10-(1/3)*c1+.6*c2=1},{c1,c2});
```

$$\{c1 = 4.500000000, c2 = -12.50000000\}$$

```
> x1:=subs({c1=4.500000000,c2=-12.50000000},x[1,1]);
```

$$x1 := 4.500000000\,e^{(-2\,t)} - 12.50000000\,e^{\left(-\frac{3t}{5}\right)} + 10.$$

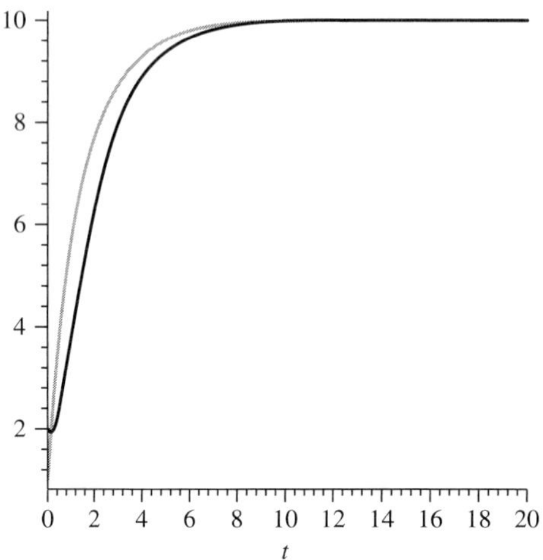



**FIGURE 13.15**

Plot of $4.5e^{-2t} - 12.5e^{-0.6t} + 10$ and $-1.5e^{-2t} - 7.5e^{-0.6t} + 10$ converging to a stable equilibrium at 10

```
> x2:=subs({c1=4.500000000,c2=-12.50000000},x[2,1]);
```

$$x2 := -1.499999998\, e^{(-2\,t)} - 7.50000000\, e^{\left(-\frac{3\,t}{5}\right)} + 10.$$

```
> plot({x1,x2},t=0..20, thickness=3);
```

Interpretation: The solution shows a stable equilibrium value at 10 units as seen in Figure 13.15.

---

## Example 3    Electrical Circuits

### Part 1

```
> ode1:=diff(Q(t),t)=-Q(t)-i2(t);
```

$$ode1 := \frac{d}{dt}\,Q(t) = -\,Q(t) - i2(t)$$

```
> ode2:=diff(i2(t),t)=Q(t)-i2(t);
```

$$ode2 := \frac{d}{dt}\,i2(t) = Q(t) - i2(t)$$

```
> inits:=Q(0)=3,i2(0)=1;
```

$$inits := Q(0) = 3,\ i2(0) = 1;$$

```
> with(linalg):with(DEtools):
> k1:=matrix(2,2,[-1,-1,1,-1]);
```

$$k1 := \begin{bmatrix} -1 & -1 \\ 1 & -1 \end{bmatrix}$$

```
> homogensol:=matrixDE(k1,t);
```

$$homogensol := \left[\left[\begin{matrix} e^{(-t)}\cos(t) & e^{(-t)}\sin(t) \\ e^{(-t)}\sin(t) & -e^{(-t)}\cos(t) \end{matrix}\right], [0,0]\right]$$

> $dsolve(\{ode1,ode2,inits\},\{Q(t),i2(t)\});$

$$\{i2(t) = e^{(-t)}(3\sin(t) + \cos(t)),\ Q(t) = e^{(-t)}(3\cos(t) - \sin(t))\}$$

> $plot(\{exp(-t)*(3*sin(t)+cos(t)),exp(-t)*(3*cos(t)-sin(t))\},t=0..10,\ thickness=3);$
> $dsolve(\{ode1,ode2,inits\},numeric,\ method=classical[rk4],$
  $output=array([0,.5,1,1.5,2,2.5,3,3.5,4]));$

$$
\begin{bmatrix}
\begin{matrix}
[t, & Q(t), & i2(t)] \\
0 & 3. & 1. \\
0.5 & 1.30605590240596148 & 1.40463959485283274 \\
1 & 0.286738455356469190 & 1.12744573728828490 \\
1.5 & -0.175220406714586580 & 0.683497251434332820 \\
2 & -0.292018074783138293 & 0.312860724399885160 \\
2.5 & -0.246411202855808532 & 0.0816148827787278042 \\
3 & -0.154892423813640046 & -0.0282109696518720414 \\
3.5 & -0.0742428892990450868 & -0.0600567485195805245 \\
4 & -0.0220543803444919716 & -0.0535558641615097586
\end{matrix}
\end{bmatrix}
$$

## Part 2: Nonhomogeneous—Finding the Eigenvalues

> $with(linalg):with(DEtools):$
> $M:=array([[-1,-1],[1,-1]]);$
> $lambda:=eigenvects(M);$

$$
M := \begin{bmatrix} -1 & -1 \\ 1 & -1 \end{bmatrix}
$$

$$
\lambda := [-1 + I,\ 1,\ \{[I,\ 1]\}],\ [-1 - I,\ 1,\ \{[-I,\ 1]\}]
$$

> $homsol:=matrixDE(M,t);$

$$
homsol := \left[ \begin{bmatrix} e^{(-t)}\cos(t) & e^{(-t)}\sin(t) \\ e^{(-t)}\sin(t) & -e^{(-t)}\cos(t) \end{bmatrix},\ [0,0] \right]
$$

> $phi:=homsol[1];$

$$
\phi := \begin{bmatrix} e^{(-t)}\cos(t) & e^{(-t)}\sin(t) \\ e^{(-t)}\sin(t) & -e^{(-t)}\cos(t) \end{bmatrix}
$$

> $phi\_inv:=inverse(phi);$

$$
phi\_inv := \begin{bmatrix} \dfrac{\cos(t)}{e^{(-t)}(\cos(t)^2 + \sin(t)^2)} & \dfrac{\sin(t)}{e^{(-t)}(\cos(t)^2 + \sin(t)^2)} \\[3ex] \dfrac{\sin(t)}{e^{(-t)}(\cos(t)^2 + \sin(t)^2)} & -\dfrac{\cos(t)}{e^{(-t)}(\cos(t)^2 + \sin(t)^2)} \end{bmatrix}
$$

> $f:=matrix(2,1,[10*exp(-t),0]);$

$$
f := \begin{bmatrix} 10e^{(-t)} \\ 0 \end{bmatrix}
$$

> *step1:=multiply(phi_inv,f);*

$$step1 := \begin{bmatrix} \dfrac{10\cos(t)}{\cos(t)^2 + \sin(t)^2} \\[2ex] \dfrac{10\sin(t)}{\cos(t)^2 + \sin(t)^2} \end{bmatrix}$$

> *step2:=map(int,step1,t);*

$$step2 := \begin{bmatrix} 10\sin(t) \\ -10\cos(t) \end{bmatrix}$$

> *step3:=multiply(phi,step2);*

$$step3 := \begin{bmatrix} 0 \\ 10\,e^{(-t)}\sin(t)^2 + 10\,e^{(-t)}\cos(t)^2 \end{bmatrix}$$

> *newstep3:=simplify(%);*

$$newstep3 := \begin{bmatrix} 0 \\ 10\,e^{(-t)} \end{bmatrix}$$

> *x:=evalm(multiply(phi, matrix(2,1,[c1,c2]))+newstep3);*

$$x := \begin{bmatrix} e^{(-t)}\cos(t)\,c1 + e^{(-t)}\sin(t)\,c2 \\ e^{(-t)}\sin(t)\,c1 - e^{(-t)}\cos(t)\,c2 + 10\,e^{(-t)} \end{bmatrix}$$

Solve for $c_1$, $c_2$.
> *solve({3=c1,−c2+10=1}, {c1,c2});*

$$\{c1 = 3,\ c2 = 9\}$$

> *x1:=subs({c1 = 3, c2 = 9},x[1,1]);*

$$x1 := 3e^{(-t)}\cos(t) + 9e^{(-t)}\sin(t)$$

> *x2:=subs({c1 = 3, c2 = 9},x[2,1]);*

$$x2 := 3e^{(-t)}\sin(t) - 9e^{(-t)}\cos(t) + 10e^{(-t)}$$

> *plot({x1,x2},t=0..20,thickness=3);*

---

Figures 13.16 and 13.17 show the graphical results of our work.

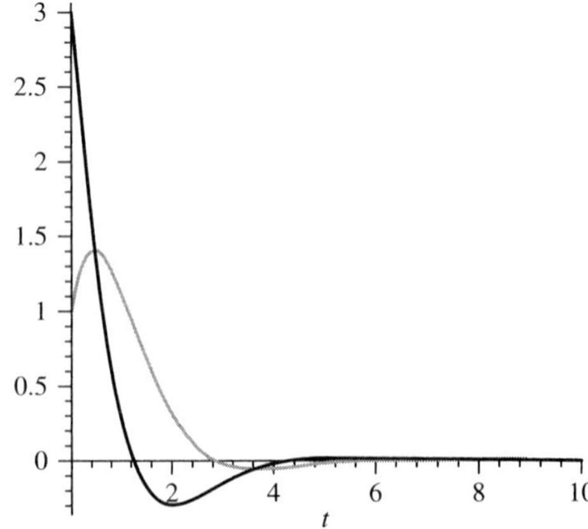

**FIGURE 13.16**

Plot of $e^{-t}\,[3\sin(t) + \cos(t)]$ and $e^{-t}\,[3\cos(t) - \sin(t)]$

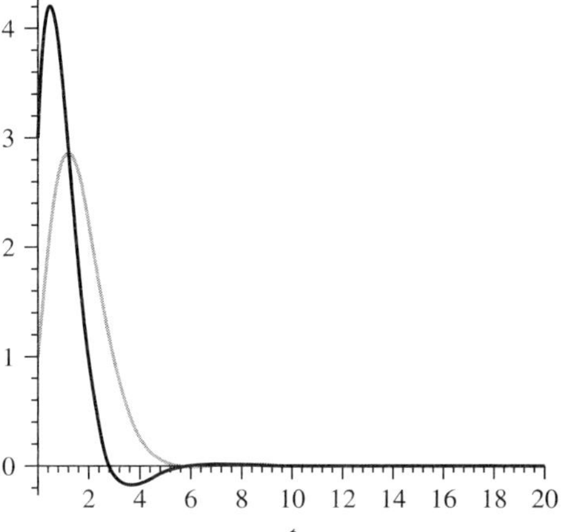

**FIGURE 13.17**

Plot of $3e^{-t}\cos(t) + 9e^{-t}\sin(t)$ and $3e^{-t}\sin(t) - 9e^{-t}\cos(t) + 10e^{-t}$

## 13.4 | FURTHER READING

Abell, M. L., and J. P. Braselton, *Differential Equation with Maple V*, 2nd Ed. San Diego: Academic Press. 2000.

Barrow, D., A. Belmonte, A Boggess, J. Bryant, T. Kiffe, J. Morgan, M. Rahe, K. Smith, and M. Stecher. *Solving Differential Equations with Maple V, Release 4*. Pacific Grove, CA: Brooks-Cole. 1998.

Giordano, F. R., and M. D. Wier. *Differential Equations: A Modeling Approach*. Reading, MA: Addison-Wesley. 1991.

## 13.5 | NUMERICAL SOLUTIONS TO SYSTEMS OF ODE WITH MAPLE

In the previous chapter, we discussed the use of numerical solutions (the Euler, improved Euler, and Runge–Kutta methods) to first-order differential equations. In this chapter, we extend the use of numerical solutions to systems of differential equations. We show only the Euler's and Runge–Kutta methods. Our goal here is to provide a solution method for many models of systems of ODEs that do not have closed-form analytical solutions.

Throughout most of the chapter, we have investigated and modeled autonomous systems of first-order differential equations. A more general form of systems of two ordinary first-order differential equations is given by

$$\frac{dx}{dt} = f(t,x,y)$$

$$\frac{dy}{dt} = g(t,x,y) \tag{13.1}$$

If the variable $t$ appears explicitly in one of the functions $f$ or $g$, then the system is not autonomous. In this section, we present numerical techniques for approximating solutions for $x(t)$ and $y(t)$ subject to initial conditions $x(t_o) = x_o$ and $y(t_o) = y_o$.

We will give the algorithm for each and show the Maple commands to execute a numerical solution. We also show how to obtain both the phase portraits and the plots of approximate numerical solutions.

## Example 1

## Euler's Method

Consider the iterative formula for Euler's method for systems as

$$x(n) = x(n-1) + f[t(n-1), x(n-1), y(n-1)]\Delta t$$
$$y(n) = y(n-1) + g[t(n-1), x(n-1), y(n-1))\Delta t$$

We illustrate a few iterations for the following initial-value problem with a step size of $\Delta t = 0.1$:

$$x' = 3x - 2y, \, x(0) = 3$$
$$y' = 5x - 4y, \, y(0) = 6$$
$$x(0) = 3, \, y(0) = 6$$

given

$$x(1) = 3 + (0.1) \cdot (3 \cdot 3 - 2 \cdot 6) = 2.7$$
$$y(1) = 6 + (0.1) \cdot (5 \cdot 3 - 4 \cdot 6) = 5.1$$

*and*

$$x(2) = 2.7 + (0.1) \cdot (3 \cdot (2.7) - 2 \cdot (5.1)) = 2.49$$
$$y(2) = 5.1 + (0.1) \cdot (5 \cdot (2.7) - 4 \cdot (5.1)) = 4.41$$

and so forth.

In Maple, we enter the system and initial conditions and then use the dsolve with classical numerical methods. Here is the command sequence to obtain the Euler estimates to our example.

> *ode1:=diff(x(t),t)–3·x(t)2·y(t);*

$$ode1 := \frac{d}{dt} x(t) = 3x(t) - 2y(t)$$

> *ode2:=diff(y(t),t)=5·x(t)−4·y(t);*

$$ode2 := \frac{d}{dt} y(t) = 5x(t) - 4y(t)$$

> *inits:=x(0)=3,y(0)=6;*

$$inits := x(0) = 3, \, y(0) = 6$$

> *eulersol:=dsolve({ode1,ode2,inits},numeric,method=classical[foreuler],output=array([0,.1,.2,. 3,.4,.5,.6,.7,.8,.9,1,1.1,1.2,1.3,1.4,1.5,1.6,1.7,1.8,1.9,2,2.1,2.2]),stepsize=0.1);*

$$eulersol := \begin{bmatrix} [[ \, t & x(t) & y(t) \, ]] \\ 0. & 3. & 6. \\ 0.1 & 2.70000000000000016 & 5.09999999999999964 \\ 0.2 & 2.49000000000000022 & 4.41000000000000014 \\ 0.3 & 2.35500000000000042 & 3.89100000000000044 \\ 0.4 & 2.28330000000000056 & 3.51210000000000022 \\ 0.5 & 2.26587000000000050 & 3.24891000000000042 \\ 0.6 & 2.29584900000000046 & 3.08228100000000050 \\ 0.7 & 2.36814750000000052 & 2.99729310000000070 \\ 0.8 & 2.47913313000000058 & 2.98244961000000064 \\ 0.9 & 2.62638314700000075 & 3.02903633100000080 \\ 1. & 2.80849082490000113 & 3.13061337210000090 \\ 1.1 & 3.02491539795000097 & 3.28261343571000142 \\ 1.2 & 3.27586733019300080 & 3.48202576040100098 \\ 1.3 & 3.56222237717070068 & 3.72714912133710108 \\ 1.4 & 3.88545926605448954 & 4.01740066138760987 \\ 1.5 & 4.24761691359331550 & 4.35317002985981106 \end{bmatrix}$$

$$\left|\left|\begin{array}{lll} 1.6 & 4.65126798169934742 & 4.73571047471254403 \\ 1.7 & 5.09950628126664274 & 5.16706027567719950 \\ 1.8 & 5.59594611051119450 & 5.64998930603964044 \\ 1.9 & 6.14473208245662317 & 6.18796663887938080 \\ 2. & 6.75055837941773440 & 6.78514602455594052 \\ 2.1 & 7.41869668833186857 & 7.44636680444243292 \\ 2.2 & 8.15503233394294114 & 8.17716842683139332 \end{array}\right|\right|$$

The power of Euler's method is twofold. First, it is easy to use. Second, as a numerical method, it can be used to estimate a solution to a system of differential equations that does not have a closed-form solution.

---

**Example 2**

## Predator–Prey Model

Assume we have the following predator–prey system that does not have a closed-form analytical solution:

$$\frac{dx}{dt} = 3x - xy$$

$$\frac{dy}{dt} = xy - 2y$$

$$x(0) = 1, y(0) = 2$$

$$t_0 = 1, \Delta t = .1$$

We will obtain an estimate of the solution using Euler's method.

```
> with(linalg):with(DEtools):
```

```
> ode1:=diff(x(t),t)=3*x(t)-x(t)*y(t);
```

$$ode1 := \frac{d}{dt} x(t) = 3\,x(t) - x(t)\,y(t)$$

```
> ode2:=diff(y(t),t)=x(t)*y(t)-2*y(t);
```

$$ode2 := \frac{d}{dt} y(t) = x(t)\,y(t) - 2\,y(t)$$

```
> inits:=x(0)=1,y(0)=2;
```

$$inits := x(0) = 1, y(0) = 2$$

```
> eulersol:=dsolve({ode1,ode2,inits}, numeric,
    method=classical[foreuler],
    output =array([0,.1,.2,.3,.4,.5,.6,.7,.8,.9,1,1.1,1.2,1.3,1.4,1.5,
        1.6,1.7,1.8,1.9,2,2.1,2.2]),stepsize=0.1);
```

$$\left[\begin{array}{lll} [t, & x(t), & y(t)] \\ 0. & 1. & 2. \\ 0.1 & 1.10000000000000008 & 1.80000000000000004 \\ 0.2 & 1.23200000000000021 & 1.63800000000000012 \\ 0.3 & 1.39979840000000010 & 1.51220160000000026 \\ 0.4 & 1.60806018198425638 & 1.42143901801574435 \\ 0.5 & 1.86190228798054114 & 1.36572716301158748 \\ 0.6 & 2.16618792141785876 & 1.34686678336611476 \end{array}\right]$$

$$
eulersol := \begin{bmatrix}
0.7 & 2.52428764205455636 & 1.36925008248155188 \\
0.8 & 2.93593582846188727 & 1.44103817219427799 \\
0.9 & 3.39363701700781206 & 1.57591009774806334 \\
1. & 3.87692143779073328 & 1.79553476251787370 \\
1.1 & 4.34388314781755014 & 2.13254253132470328 \\
1.2 & 4.72069653578025861 & 2.63238558144231760 \\
1.3 & 4.89423614699907182 & 3.34857781466911942 \\
1.4 & 4.72363393293951717 & 4.31773530989456944 \\
1.5 & 4.10118401049446124 & 5.49372835024256556 \\
1.6 & 3.07846012684130698 & 6.64806176699554552 \\
1.7 & 1.95541885784630654 & 7.36502872064382874 \\
1.8 & 1.10187291030753887 & 7.33219458140772317 \\
1.9 & 0.624520125164112150 & 6.67367032336186838 \\
2. & 0.395092020148348210 & 5.75572040125449292 \\
2.1 & 0.286215706118782388 & 4.83198024107766244 \\
2.2 & 0.233781554289212323 & 4.00388305652733401
\end{bmatrix}
$$

> *with(plots):*
> *plot1:=odeplot(eulersol,[t,x(t)],0..10,color=green,title='x(t)'):*
> *plot2:=odeplot(eulersol,[t,y(t)],0..10,color=blue,title='y(t)'):*
> *display(plot1, plot2);*

After viewing the plot in Figure 13.18, we experiment and find that when we plot $x(t)$ versus $y(t)$, we have an approximately closed loop as shown in Figure 13.19.

> *odeplot(eulersol,[x(t),y(t)],0..22,color=green,title='System');*

Another method for numerical estimates is the Runge–Kutta applied to systems. We illustrate with the same predator–prey example.

> *restart;*
> *with(linalg):with(DEtools):*
> *ode1:=diff(x(t),t)=3*x(t)−x(t)*y(t);*

$$
ode1 := \frac{d}{dt}\,x(t) = 3\,x(t) - x(t)\,y(t)
$$

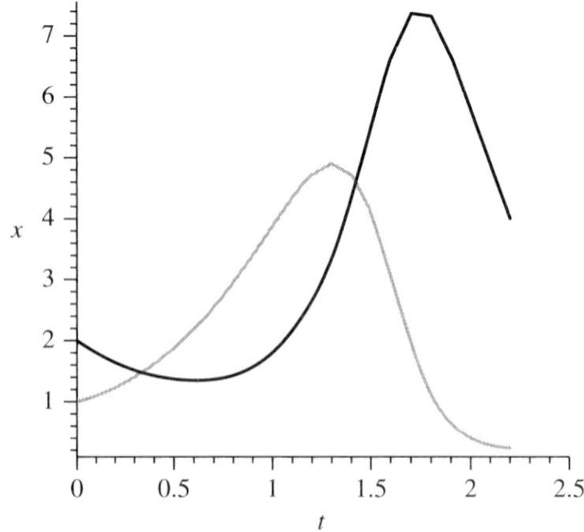

**FIGURE 13.18**
Euler plot over time

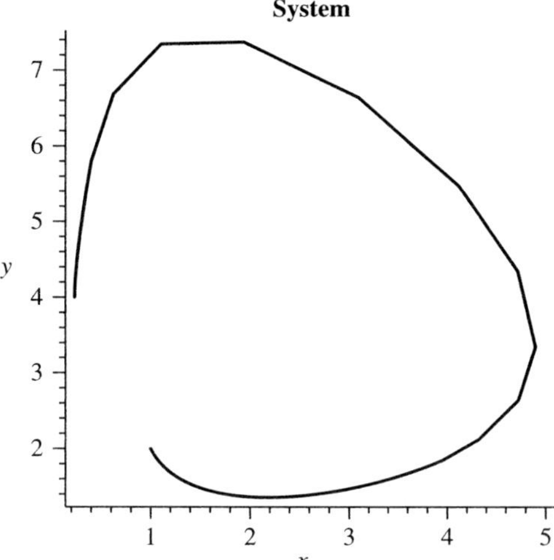



**FIGURE 13.19**
Systems plot with Euler's estimates

> *ode2:=diff(y(t),(t)=x(t)*y(t)−2*y(t);*

$$ode2 := \frac{d}{dt}\, y(t) = x(t)\, y(t) - 2\, y(t)$$

> *inits:=x(0)=1,y(0)=2;*

$$inits := x(0) = 1,\, y(0) = 2$$

> *rk4sol:=dsolve({ode1, ode2, inits},numeric,method=classical[rk4],
output=array( [0,.1,.2,.3,.4,.5,.6,.7,.8,.9,1,1.1,1.2,1.3,1.4,1.5,
1.6, 1.7, 1.8, 1.9, 2, 2.1, 2.2]), stepsize=0.1);*

$$rk4sol :=$$

| [ t, | x(t), | y(t) ] |
|---|---|---|
| 0. | 1. | 2. |
| 0.1 | 1.11554071453956705 | 1.81968188493959970 |
| 0.2 | 1.26463746535620780 | 1.67761981669985926 |
| 0.3 | 1.45146389024689194 | 1.57278931221810914 |
| 0.4 | 1.68031037053259168 | 1.50542579227745321 |
| 0.5 | 1.95456986108324960 | 1.47762563817635396 |
| 0.6 | 2.27500034092209802 | 1.49412980191491540 |
| 0.7 | 2.63687047094120208 | 1.56336399876409616 |
| 0.8 | 3.02561346371586382 | 1.69866644464148696 |
| 0.9 | 3.41107833928229720 | 1.91916110241906601 |
| 1. | 3.74212348522216676 | 2.24852435676575446 |
| 1.1 | 3.94706090545572064 | 2.70772001337807611 |
| 1.2 | 3.95001172191016892 | 3.29658238745704324 |
| 1.3 | 3.70934802583802936 | 3.96638771088000476 |
| 1.4 | 3.25874674267819132 | 4.60727591934640213 |
| 1.5 | 2.70448435236999174 | 5.08419319694337712 |

$$\left[\left[\begin{array}{ccc}
1.6 & 2.16641586882149006 & 5.30768472071020270 \\
1.7 & 1.71962027608752543 & 5.27274016506954890 \\
1.8 & 1.38441109626295588 & 5.03717315896791806 \\
1.9 & 1.14891557150367586 & 4.67751425643222784 \\
2. & 0.991726731312949417 & 4.25984946993614156 \\
2.1 & 0.893335577155144112 & 3.83076393278872685 \\
2.2 & 0.839403883501402824 & 3.41909584046305914
\end{array}\right]\right]$$

```
> with(plots):
> plot1:=odeplot(rk4sol,[t,x(t)],0..10,color=green,title='x(t)'):
> plot2:=odeplot(rk4sol,[t,y(t)],0..10,color=blue, title='y(t)'):
> display(plot1,plot2);
> odeplot(rk4sol,[x(t),y(t)],0..22,color=green,title='System');
```

Figures 13.20 and 13.21 provide visual results to our system.

**FIGURE 13.20**
RK 4 plot over time

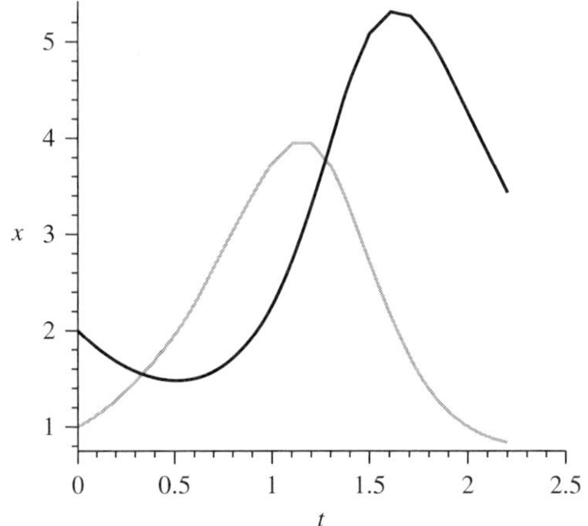

**FIGURE 13.21**
System plot using RK4
estimates

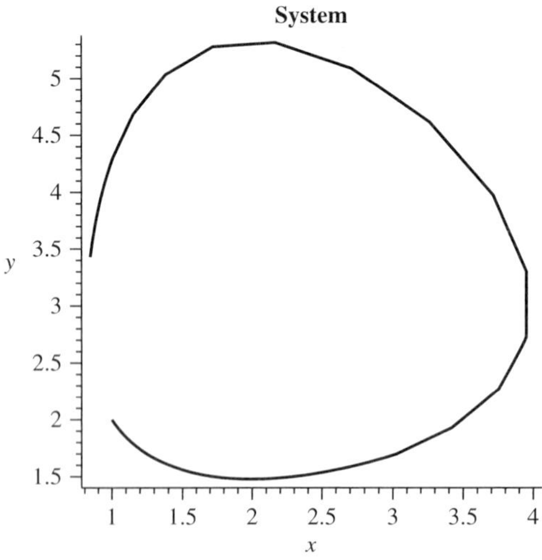


Interpretation: The predator-prey models shows the system is stable.

## 13.5 | EXERCISES

1. Given the following system of linear first-order ODEs of species cooperation (symbiosis):

$$dx_1/dt = -0.5x_1 + x_2,$$
$$dx_2/dt = 0.25x_1 - 0.5x_2,$$
$$x_1(0) = 200, \text{ and}$$
$$x_2(0) = 500$$

   a. Perform Euler's method with step size $h = 0.1$ to obtain graphs of numerical solutions for $x_1(t)$ and $x_2(t)$ versus $t$ and for $x_1$ versus $x_2$. You can put both $x_1(t)$ and $x_2(t)$ versus $t$ on one axis if you want.

   b. From the graphs, discuss the long-term behavior of the system (discuss stability).

   c. Analytically using eigenvalues and eigenvectors, solve the system of DEs to determine the population of each species for $t > 0$.

   d. Determine if there is a steady-state solution for this system.

   e. Obtain real plots of (1) $x_1(t)$ and $x_2(t)$ versus $t$ and (2) for $x_1(t)$ versus $x_2(t)$. Compare to the numerical plots. Briefly discuss.

2. Given a competitive hunter model defined by the system:

$$dx/dt = 15x - x^2 - 2xy = x(15 - x - 2y)$$
$$dy/dt = 12y - y^2 - 1.5xy = y(12 - y - 1.5x)$$

   a. Perform a graphical analysis of this competitive hunter model in the $x$–$y$ plane.

   b. Identify all equilibrium points and classify their stability.

   c. Find the numerical solutions using Euler's method with step size $h = 0.05$. Try it from two separate initial conditions: First, use $x(0) = 5$ and $y(0) = 4$ and then use $x(0) = 3$, $y(0) = 9$. Obtain graphs of $x(t)$, $y(t)$ individually (or on the same axis) and then a plot of $x$ versus $y$ using your numerical approximations. Compare to your phase portrait analysis.

3. Because bass and trout both live in the same lake and eat the same food sources, they are competing for survival. The rate of growth for bass ($dB/dt$) and for trout ($dT/dt$) are estimated by the following equations:

$$dB/dt = (10 - B - T)B$$
$$dT/dt = (15 - B - 3T)T$$

Coefficients and values are in thousands.

   a. Obtain a "qualitative" graphical solution of this system. Find all equilibrium points of the system and classify each as unstable, stable, or asymptotically stable.

   b. If the initial conditions are $B(0) = 5$ and $T(0) = 2$, determine the long-term behavior of the system from your graph in part (a). Sketch it out.

   c. Using Euler's method, $h = 0.1$ and the same initial conditions as above, obtain estimates for $B$ and $T$. Using these estimates, determine a more accurate graph by plotting $B$ versus $T$ for the solution from $t = 0$ to $t = 7$.

   Euler's method:

$$x_{n+1} = x_n + h\,f(x_n, y_n) \text{ and } y_{n+1} = y_n + h\,g(x_n, y_n)$$

   d. Compare the graph in part (c) to the possible solutions found in parts (a) and (b). Briefly comment.

## 13.5 | PROJECTS

1. **Diffusion.** Diffusion through a membrane leads to a first-order system of ordinary linear differential equations. For example, consider the situation in which two solutions of substance are separated by a membrane of permeability $P$. Assume the amount of substance that passes through the membrane at any particular time is proportional to the difference between the concentrations of the two solutions. Therefore, if we let $x_1$ and $x_2$ represent the two concentrations and $V_1$ and $V_2$ their corresponding volumes, then the system of differential equations is given by

$$\frac{dx_1}{dt} = \frac{P}{V_1}(x_2 - x_1)$$

$$\frac{dx_2}{dt} = \frac{P}{V_2}(x_1 - x_2)$$

where the initial amounts of $x_1$ and $x_2$ are given.

Consider two salt concentrations of equal volume $V$ separated by a membrane of permeability $P$. Given that $P = V$, determine the amount of salt in each concentration at time $t$ if $x_1(0) = 2$ and $x_2(0) = 10$.

   a. Write out the system of differential equations that models this behavior.
   b. Using the methods described in Chapter 8, solve this system. Clearly indicate your eigenvalues and eigenvectors.
   c. Plot the solutions for $x_1$ and $x_2$ on the same axis and label each. Comment about the plots.
   d. Use a numerical method (Euler or Runga–Kutta) and iterate a numerical solution to predict $x_1(4)$; use a step size of 0.5. Obtain a plot of your numerical approach. Compare it to the analytical plot. Comment on the plots.

Consider diffusion through a double-walled membrane, where the inner wall has permeability $P_1$ and the outer wall has permeability $P_2$ with $0 < P_1 < P_2$. Suppose the volume of the solution within the inner wall is $V_1$ and between the two walls is $V_2$. Let $x$ represent the concentration of the solution within the inner wall and $y$ the concentration between the two walls. This leads to the following system:

$$\frac{dx}{dt} = \frac{P_1}{V_1}(y - x)$$

$$\frac{dy}{dt} = \frac{1}{V_2}(P_2(C - y) + P_1(x - y))$$

$$x(0) = 2,\ y(0) = 1,\ C = 10$$

Also assume the following:

$$P_1 = 3$$
$$P_2 = 8$$
$$V_1 = 2$$
$$V_2 = 10$$

   a. Set up the system of ODEs with all coefficients.
   b. Use the method of variation of parameter for systems,

$$X = X_c + \phi(t) \int \phi^{-1}(t)\, F(t)\, dt$$

to find both $X_c$ and $X_p$.

   **c.** Use the initial conditions to find the particular solution and the coefficients for $X_c$ in the solution $X_c + X_p$.

   **d.** Plot the solutions for $x(t)$ and $y(t)$ on the same axis. Comment on the solution.

**2.** An electrical network. An electrical network containing more than one loop also gives rise to a system of differential equations. For instance, in the electrical network displayed below, there are two resistors and two inductors. At branch point B in the network, the current $i_1(t)$ splits in two directions. Thus,

$$i_1(t) = i_2(t) + i_3(t)$$

Kirchhoff's law applies to each loop in the network. For loop ABEF, we find that

$$E(t) = i_1 R_1 + L_1\, di_2/dt$$

The sum of the voltage drops across the loop ABCDEF is

$$E(t) = i_1 R_1 + L_2\, di_3/dt + i_3 R_3$$

Substituting, we find the following systems for equations:

$$\frac{di_1}{dt} = -\frac{(R_1 + R_2)}{L_1}\, i_1 + \frac{R_2}{L_2}\, i_2 + 0$$

$$\frac{di_2}{dt} = \left(\frac{R_2}{L_2} - \frac{1}{R_2 C}\right) i_2 - \frac{(R_1 + R_2)}{L_2}\, i_1 + \frac{E(t)}{L_2}$$

$$i_2(0) = 1,\ i_1(0) = 0$$

Initially, let $E(t) = 0$ volts, $L_1 = 1$ henry, $L_2 = 1$ henry, $R_1 = 1$ ohm, $R_2 = 1$ ohm, $C = 3$.

   **a.** Write out the system of differential equation that models this behavior.

   **b.** Using the methods described in Chapter 8, solve this system. Clearly indicate your eigenvalues and eigenvectors.

   **c.** Plot the solutions for $x_1$ and $x_2$ on the same axis and label each. Comment about the plots.

   **d.** Use a numerical method (Euler or Runge–Kutta) and iterate a numerical solution to predict $x_i(4)$; use a step size of 0.5. Obtain a plot of your numerical approach. Compare it to the analytical plot. Comment about the plots.

   Now let $E(t) = 100*\sin(t)$.

   **e.** Set up the system of ODEs with all coefficients.

   **f.** Use the method of variation of parameter for systems,

$$X = X_c + \phi(t) \int \phi^{-1}(t)\, F(t)dt$$

   to find both $X_c$ and $X_p$.

   **g.** Use the initial conditions to find the particular solution and the coefficients for $X_c$ in the solution $X_c + X_p$.

   **h.** Plot the solutions for $x(t)$ and $y(t)$ on the same axis. Comment on the solution.

**3.** Interacting species. Suppose $x(t)$ and $y(t)$ represent respective populations of two species over time, $t$. One model might be

$$X' = R_1 X,\ X(0) = X_0$$
$$Y' = R_2 Y,\ Y(0) = Y_0$$

where $R_1$ and $R_2$ are intrinsic coefficients. Models involving competition between species or predator–prey models most often include interaction terms between the variables. These interactions terms, if included, will preclude any analytical solution attempts, so we will simplify these models for this project.

Let's model bass and trout attempting to coexist in a small pond in South Carolina:

$$B' = -0.5B + T + H$$
$$T' = 0.25B - 0.5T + K$$
$$B(0) = 2{,}000, \; T(0) = 5{,}000$$

Initially, let $H = K = 0$.

**a.** Write out the system of differential equation that models this behavior.
**b.** Using the methods described in Chapter 8, solve this system. Clearly indicate your eigenvalues and eigenvectors.
**c.** Plot the solutions for $x_1$ and $x_2$ on the same axis and label each. Comment on the plots.
**d.** Use a numerical method (Euler or Runge–Kutta) and iterate a numerical solution to predict $x_i(10)$; use a step size of 0.5. Obtain a plot of your numerical approach. Compare it to the analytical plot. Comment on the plots.

Now, let $H = 1{,}500$, $K = 1{,}000$.

**e.** Set up the system of ODEs with all coefficients.
**f.** Use the method of variation of parameter for systems,

$$X = X_c + \phi(t) \int \phi^{-1}(t) \, F(t) \, dt$$

to find both $X_c$ and $X_p$.
**g.** Use the initial conditions to find the particular solution and the coefficients for $X_c$ in the solution $X_c + X_p$.
**h.** Plot the solutions for $x(t)$ and $y(t)$ on the same axis. Comment on the solution.
**i.** Do these species coexist? Briefly explain. If any die out, determine when this happens.

**4.** Trapezoidal method. The trapezoidal method is a more stable numerical method that is shown in numerical analysis textbooks—for example, see Burden and Faires, *Numerical Analysis* (Pacific Grove, CA: Brooks-Cole), pages 344–346. Find the trapezoidal algorithm and modify it for systems of ODEs. Write a Maple program to obtain the trapezoidal estimates and compare these to both Euler and Runge–Kutta estimates.

## 13.6 PREDATOR–PREY, SIR, AND COMBAT MODELS

We examined these type models in Chapter 3 as a dynamical system. In this section, we revisit these as differential equations.

### Example 1    Predator–Prey Revisited

We examined predator–prey models in Chapter 5 as a dynamical system. In this section, we revisit these as differential equations.

We repeat our (admittedly simplistic) assumptions from Chapter 3:

- The predator species is totally dependent on the prey species, its only food supply.
- The prey species has an unlimited food supply and no threat to its growth other than the specific predator.

If there were no predators, the second assumption would imply that the prey species grows exponentially—that is, if $x = x(t)$ is the size of the prey population at time $t$, then we would have $\dfrac{dx}{dt} = ax$. This represents exponential growth when $a > 0$ and decay when $a < 0$.

But there *are* predators, and these must account for a negative component in the prey growth rate. Suppose we write $y = y(t)$ for the size of the predator population at time $t$. Here are the crucial assumptions for completing the model:

- The rate at which predators encounter prey is jointly proportional to the sizes of the two populations.
- A fixed proportion of encounters lead to the death of the prey.

These assumptions lead to the conclusion that the negative component of the prey growth rate is proportional to the product $xy$ of the population sizes—that is,

$$\frac{dx}{dt} = ax - bxy$$

Now we consider the predator population. If there were no food supply, the population would die out at a rate proportional to its size—that is, we would find $\dfrac{dy}{dt} = -cy$. (Keep in mind that the "natural growth rate" is a composite of birthrates and death rates, both presumably proportional to population size. In the absence of food, there is no energy supply to support the birthrate.) But there is a food supply: the prey. And what is bad for hares is good for lynx; that is, the energy to support growth of the predator population is proportional to the deaths of prey, so

$$\frac{dy}{dt} = -cy + pxy$$

This discussion leads to the Lotka–Volterra predator–prey model:

$$\frac{dx}{dt} = ax - bxy$$

$$\frac{dy}{dt} = -cy + pxy$$

where $a$, $b$, $c$, and $p$ are positive constants.

We assume that $\{a,b,c,p\} = \{.1,.005/60,.04\ 0,.00004\}$.

> *with( DETools ):with( plots ):*

**Predator–Prey Model**

> *hare1:=diff(h(t),t)=0.1·h(t)−$\dfrac{0.005}{60}$·h(t)·f(t);*

$$hare1 := \frac{\mathrm{d}}{\mathrm{d}t}\,h(t) = 0.1\,h(t) - 0.00008333333333\,h(t)f(t)$$

> *fox1:=diff(f(t),t)=−0.04·f(t)+0.00004·h(t)·f(t);*

$$fox1 := \frac{\mathrm{d}}{\mathrm{d}t}\,f(t) = -0.04\,f(t) + 0.00004\,h(t)\,f(t)$$

> *Model1:={hare1,fox1};*

$$Model1 := \left[\; \frac{\mathrm{d}}{\mathrm{d}t}\,f(t) = -0.04f(t) + 0.00004\,h(t)\,f(t),\right.$$

$$\left. \frac{\mathrm{d}}{\mathrm{d}t}\,h(t) = 0.1\,h(t) - 0.00008333333333\,h(t)\,f(t)\right]$$

> *vars:={h(t),f(t)};*

$$vars := \{h(t),\, f(t)\}$$

> *init1:=[h(0)=2000,f(t)=600];*

$$init1 := \left[ h(0) = 2000, f(0) = 600 \right]$$

> *domain:=t=0..320;*

$$domain := t = 0\,..320$$

> *L:=DEplot(Model1,vars,domain,[init1],stepsize=0.5,*
> *scene=[t,f],arrows=NONE,linecolor=blue):*

> *H:=DEplot(Model1,vars,domain,[init1],stepsize=0.5,*
> *scene=[t,h],arrows=NONE):*

> *display({L,H},title='PredatorPrey');*
> *init2:=[h(0)=2000,f(0)=1200];*

$$init2 := [h(0) = 2000, f(0) = 1200]$$

> *init2:=[h(0)=2000,f(0)=1200]*

> *init3:=[h(0)=2000,f(0)=3000];*

$$init3 := [h(0) = 2000, f(0) = 3000]$$

> *DEplot(Model1,vars,domain,[init1,init2,init3],stepsize=0.5;*
> *scene=[h,f],title='PredatorPrey',arrows=NONE);*
> *equil:=solve({rhs(hare1),rhs(fox1)},vars);*

$$equil := \{f(t) = 0., h(t) = 0.\}, \{h(t) = 1000., f(t) = 1200.000000\}$$

**Model Interpretation**

From Figures 13.22 and 13.23, we see that our predator–prey model is in equilibrium as we move around the equilibrium value (1,000, 1,200). The point (0,0) is not stable. The ecological system appears stable and does need human intervention at this time. Figure 13.22 specifically shows how the predator–prey relationship operates over time in an oscillatory fashion.

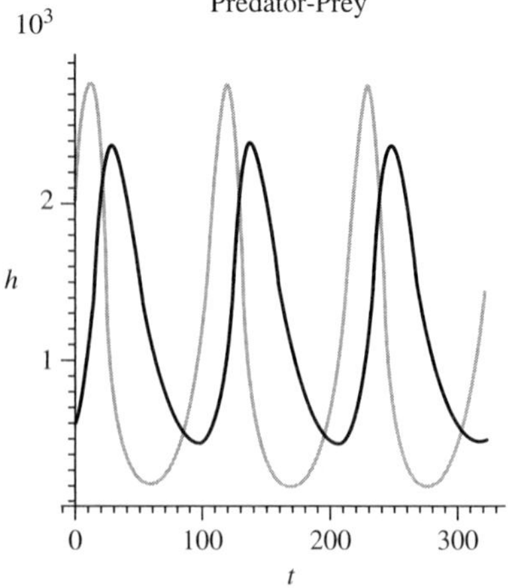


**FIGURE 13.22**
Predator–prey model over time

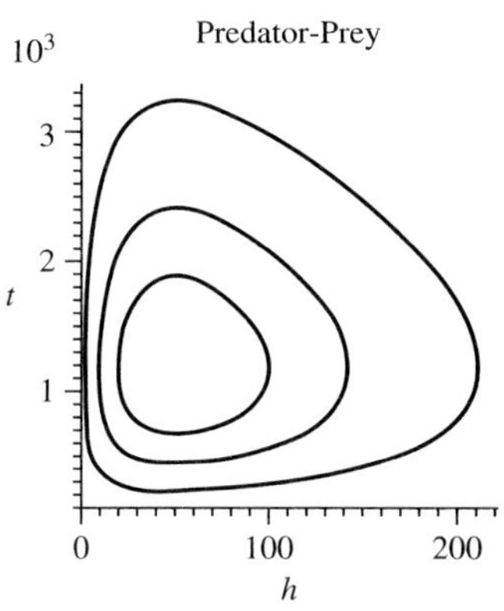


**FIGURE 13.23**
Foxes versus hares

| **Example 2** | ## Continuous SIR Models of Epidemics |
|---|---|

Consider a disease that is spreading throughout the United States, such as the new influenza strain. The Centers for Disease Control and Prevention are interested in knowing and experimenting with a model for this new disease before it actually becomes an actual epidemic. Let's consider the population as being divided into three categories: susceptible, infected, and removed. We make the following assumptions for our model:

- No one enters or leaves the community, and there is no contact outside the community.
- Each person is susceptible, $S$ (able to catch this new flu); infected, $I$ (currently has the flu and can spread the flu); or removed, $R$ (has already had the flu and will not get it again—which includes death).
- Initially, every person is either $S$ or $I$.
- Once people get the flu this year, they cannot get it again.
- The average length of the disease is 2 weeks, over which time the person is deemed infected and can spread the disease.
- Our time period for the model will be per week.
- The model we will consider is the SIR model (Allman, 2004).

Let's assume the following definition for our variables.

$S(n)$ = number in the population susceptible after period $n$
$I(n)$ = number infected after period $n$
$R(n)$ = number removed after period $n$

Let's start our modeling process with $R(n)$. Our assumption for the length of time someone has the flu is 2 weeks. Thus, half the infected people will be removed each week,

$$\frac{dR}{dt} = 0.5* I(t)$$

The value, 0.5, is called the *removal rate per week*. It represents the proportion of the infected persons who are removed from infection each week. If real data are available, then we could do data analysis to obtain the removal rate.

$I(t)$ will have terms that both increase and decrease its amount over time. It is decreased by the number who are removed each week, $0.5*I(t)$. It is increased by the numbers of susceptible who come into contact with infected persons and catch the disease, $aS(t)I(t)$. We define the rate, $a$, as the rate in which the disease is spread or as the *transmission coefficient*. We realize this is a probabilistic coefficient. Initially, we will assume that this rate is a constant value that can be found from initial conditions.

Let's illustrate as follows. Assume we have a population of 1,000 students in the dorms. Our nurse found three students reporting to the infirmary initially. The next week, five students came in to the infirmary with flulike symptoms: $I(0) = 3$, $S(0) = 997$. In week 1, the number of newly infected is 30.

$$5 = aI(n)\, S(n) = a(3)*(997)$$
$$a = 0.00167$$

Let's consider $S(t)$. This number is decreased only by the number who become infected. As before, we may use the same rate, $a$, to obtain the model:

$$\frac{dS}{dt} = -0.00167 \cdot S(t) \cdot I(t)$$

Our coupled SIR model is shown in the following systems of differential equations:

$$\frac{dR}{dt} = 0.5I(t)$$

$$\frac{dI}{dt} = -0.5I(t) + 0.001672\ I(t)S(t) \tag{13.8}$$

$$\frac{dS}{dt} = -0.001672\ S(t)I(t)$$

$$I(0) = 3,\ S(0) = 997,\ R(0) = 0$$

The preceding SIR model can be solved iteratively and viewed graphically. Let's iterate the solution and obtain the graph to observe the behavior and obtain some insights.

> *removed:=diff(r(t),t)=0.5·Inf(t);*

$$removed := \frac{d}{dt}\,r(t) = 0.5\ Inf(t)$$

> *infect:=diff(Inf(t),t)=−0.5·Inf(t)+0.00167·s(t)·Inf(t);*

$$infect := \frac{d}{dt}\,Inf(t) = -0.5\ Inf(t) + 0.00167\ s(t)\ Inf(t)$$

> *susceptible:=diff(s(t),t)=−0.00167·s(t)·Inf(t);*

$$susceptible := \frac{d}{dt}\,s(t) = -0.00167\ s(t)\ Inf(t)$$

> *Model1:={removed,infect,susceptible};*

$$Model1 := \left\{ \frac{d}{dt}\,Inf(t) = -0.5\ Inf(t) + 0.00167\ s(t)\ Inf(t), \right.$$

$$\left. \frac{d}{dt}\,r(t) = 0.5\ Inf(t),\ \frac{d}{dt}\,s(t) = -0.00167\ s(t)\ Inf(t) \right\}$$

> *vars:={r(t),Inf(t),s(t)};*

$$vars := \{ s(t),\ Inf(t),\ r(t) \}$$

> *init1:=[s(0)=997,r(0)=0,Inf(0)=3];*

$$init1 := [s(0) = 997,\ r(0) = 0,\ Inf(0) = 3]$$

> *domain:=t=0..25;*

$$domain := t = 0..25$$

```
> L:=DEplot(Model1,vars,domain,[init1],stepsize=0.5,
  scene=[t,s],arrows=NONE,linecolor=blue):
> H:=DEplot(Model1,vars,domain,[init1],stepsize=0.5,
  scene=[t,Inf],arrows=NONE,linecolor=red):
> G:=DEplot(Model1,vars,domain,[init1],stepsize=0.5,
  scene=[t,r],arrows=NONE):
> display({L,H,G},title}='SIR');
```

In this example and from Figure 13.24, we see that the maximum number of infected persons occurs around day 7. Everyone survives, and not everyone gets the flu. Let's see what happens in another case example.

**FIGURE 13.24**
SIR model over time

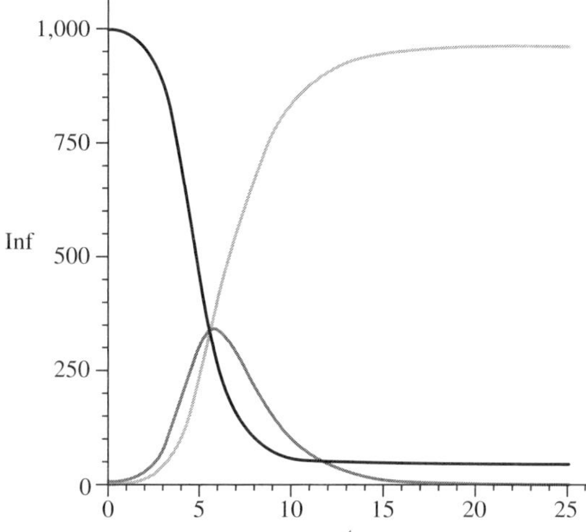

## Example 3     Hong Kong Flu

During the winter of 1968–69, the United States was swept by a virulent new strain of influenza named the *Hong Kong flu* for its place of discovery. At that time, no flu vaccine was available, so many more people were infected than would be the case today. We will study the spread of the disease through a single urban population: New York City. The data displayed in the following table are weekly totals of "excess" pneumonia–influenza deaths—that is, the numbers of such deaths in excess of the average numbers to be expected from other sources. The graph in Figure 13.25 displays the same data (*Source*: Centers for Disease Control and Prevention).

Relatively few of those who contracted the flu died from the disease or its complications, even without a vaccine. However, we may reasonably assume that the number of excess deaths in a week was proportional to the number of new cases of flu in some earlier week, say, three weeks earlier. Thus, the figures in the preceding table reflect (proportionally) the rise and subsequent decline in the number of new cases of Hong Kong flu. We will model the spread of such a disease so that we can predict what might happen with similar epidemics in the future.

At any given time during a flu epidemic, we want to know the number of people who are infected. We also want to know the number who have been infected and have recovered, because these people now have an immunity to the disease. (As a matter of convenience, we include in the recovered group the relative handful who do not recover but die—they, too, can

**FIGURE 13.25**
Hong Kong flu

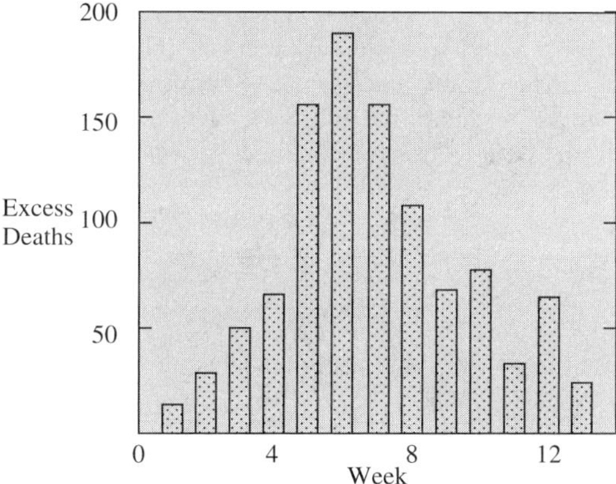

| Week | Flu-Related Deaths |
|---|---|
| 1 | 14 |
| 2 | 28 |
| 3 | 50 |
| 4 | 66 |
| 5 | 156 |
| 6 | 190 |
| 7 | 156 |
| 8 | 108 |
| 9 | 68 |
| 10 | 77 |
| 11 | 33 |
| 12 | 65 |
| 13 | 24 |

no longer contract the disease.) If we ignore movement into and out of the infected area, then the remainder of the population is still susceptible to the disease. Thus, at any time, the fixed total population (approximately 7,900,000 in the case of New York City in the late 1960s) may be divided into three distinct groups: $S$, $I$, and $R$.

## The Model

The first set of dependent variables counts *people* in each of the groups, each as a function of time:

$$S = S(t), \text{ the number of } susceptible \text{ individuals}$$
$$I = I(t), \text{ the number of infected individuals}$$
$$R = R(t), \text{ the number of recovered individuals}$$

The second set of dependent variables represents the *fraction* of the total population in each of the three categories. So, if $N$ is the total population (7,900,000 in our example), we have

$$s(t) = S(t)/N, \text{ the } susceptible \text{ fraction of the population}$$
$$i(t) = I(t)/N, \text{ the } infected \text{ fraction of the population}$$
$$r(t) = R(t)/N, \text{ the } recovered \text{ fraction of the population}$$

It may seem more natural to work with population counts, but some of our calculations will be simpler if we use the fractions instead. The two sets of dependent variables are proportional to each other, so either set will give us the same information about the progress of the epidemic.

Finally, we complete our model by giving each differential equation an initial condition. For this particular virus and incident of infection—Hong Kong flu in New York City in the late 1960s—hardly anyone was immune at the beginning of the epidemic, so almost everyone was susceptible. We will assume that there was a trace level of infection in the population, say, 10 people. Thus, our initial values for the population variables are

$$S(0) = 7,900,000$$
$$I(0) = 10$$
$$R(0) = 0$$

In terms of the scaled variables, these initial conditions are

$$s(0) = 1$$
$$i(0) = 1.27 \times 10^{-6}$$
$$r(0) = 0$$

$$\frac{ds}{dt} = -bs(t)\,i(t), \qquad\qquad s(0) = 1,$$

$$\frac{di}{dt} = bs(t)\,i(t) - ki(t), \qquad i(0) = 1.27 \times 10^{-6},$$

$$\frac{dr}{dt} = ki(t), \qquad\qquad\quad r(0) = 0.$$

This model is identical to the one we just developed. So what is different? We must keep in mind the number lost to deaths from the flu. We have some data, so we can estimate the parameters we need. We don't know values for the parameters **b** and **k** yet, but we can estimate them and then adjust as necessary to fit the excess death data. We have already estimated the average period of infectiousness at three days, so that would suggest **k** = 1/3. If we guess that each infected would make a possibly infecting contact every two days, then **b** would be 1/2. We emphasize that this is just a guess. The following plot shows the solution curves for these choices of **b** and **k**.

> *removed:=diff(r(t),t)=0.33333333·Inf(t);*

$$removed := \frac{\mathrm{d}}{\mathrm{d}t}\,r(t) = 0.33333333\,\mathit{Inf}(t)$$

> *infect:=diff(Inf(t),t)=0.333333·Inf(t)+0.5·s(t)·Inf(t);*

$$infect := \frac{\mathrm{d}}{\mathrm{d}t}\,\mathit{Inf}(t) = -0.333333\,\mathit{Inf}(t) + 0.5\,s(t)\,\mathit{Inf}(t)$$

> *susceptible:=diff(s(t),t)=−0.5·s(t)·Inf(t);*

$$susceptible := \frac{\mathrm{d}}{\mathrm{d}t}\,s(t) = -0.5\,s(t)\,\mathit{Inf}(t)$$

> *Model1:={removed,infect,susceptible};*

$$Model1 := \Bigg\{ \frac{\mathrm{d}}{\mathrm{d}t}\,\mathit{Inf}(t) = -0.333333\,\mathit{Inf}(t) + 0.5\,s(t)\,\mathit{Inf}(t),$$

$$\frac{\mathrm{d}}{\mathrm{d}t}\,s(t) = -0.5\,s(t)\,\mathit{Inf}(t),\; \frac{\mathrm{d}}{\mathrm{d}t}\,r(t) = 0.33333333\,\mathit{Inf}(t)\Bigg\}$$

> *vars:={r(t),Inf(t),s(t)};*

$$vars := \{s(t),\ \mathit{Inf}(t),\ r(t)\}$$

> *init1:=[s(0)=1,r(0)=0,Inf(0)=1.27·10⁻⁶];*

$$init1 := [s(0) = 1,\ r(0) = 0,\ \mathit{Inf}(0) = 0.000001270000000]$$

> *domain:=t=0..150;*

$$domain := t = 0\,..150$$

> *L:=DEplot(Model1,vars,domain,[init1],stepsize=0.5,*
>   *scene=[t,s],arrows=NONE,linecolor=blue):*
> *H:=DEplot(Model1,vars,domain,[init1],stepsize=0.5,*
>   *scene=[t,Inf],arrows=NONE,linecolor=red):*
> *G:=DEplot(Model1,vars,domain,[init1],stepsize=0.5,*
>   *scene=[t,r],arrows=NONE):*
> *display({L,H,G},title='SIR');*

Let's look at these results. The model shows the number of those who were susceptible decreasing as a percentage from 100% to just less than 50%. We also see from Figure 13.26 that the number removed rose from 0% to approximately 60%. Finally, we see the number infected

**FIGURE 13.26**

The SIR model for New York City's Hong Kong flu epidemic

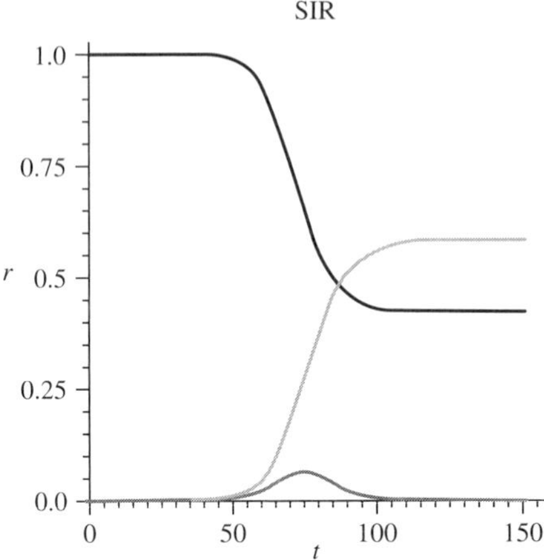


**FIGURE 13.27**

Comparison of the model

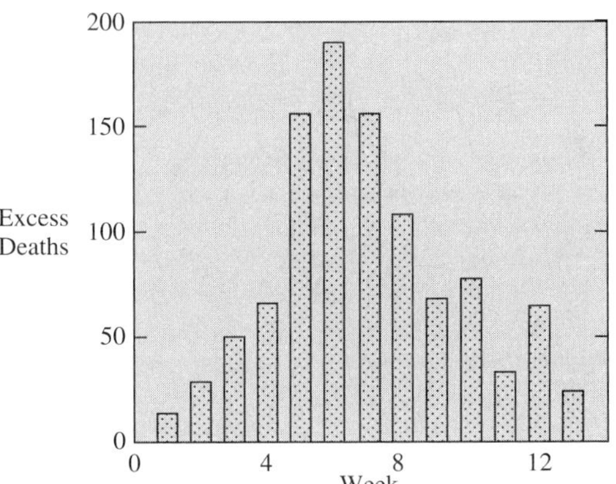


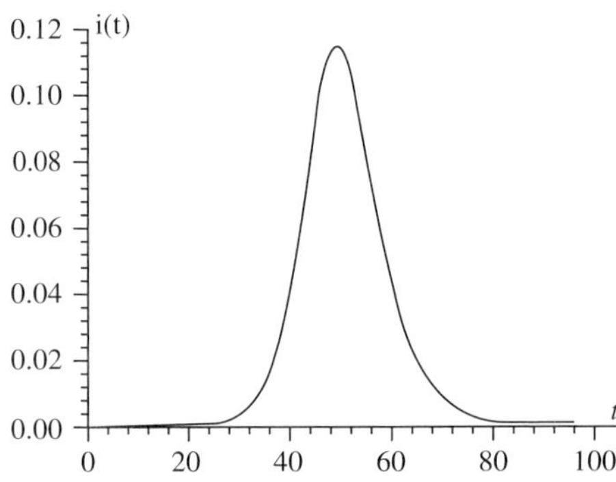


peaking around 75 to a value of about 5%. We started with 7,900,000 people. So about 395,000 have the flu at the maximum time period. The area under the curve would represent the total number who had been infected in the epidemic. So what about our losses from death? The average number of deaths per week was 79.6, or about 0.001% of the population.

Now let's compare our model with the data. Recall that these were the numbers of deaths each week that could be attributed to the flu epidemic. If we assume that the fraction of deaths among infected individuals is constant, then the number of deaths per week should be roughly proportional to the number of infected people in some earlier week. We repeat the graph of the data, along with the graph of $i(t)$ with $k = 1/3$ and $b = 6/10$, and we examine the graphs. The shape of the model appears reasonable (see Figure 13.27).

---

**Example 4**    ## Models of Combat: Iwo Jima

At Iwo Jima, in early 1945 during World War II, the Japanese had 21,500 soldiers and the United States had 73,000 soldiers. The two sides engaged in conventional warfare, but the Japanese were fighting from reinforced entrenchments. The kill rate for the Japanese against the United States was 0.0544, whereas that of the U.S. side against the Japanese was 0.0106 (based on data gathered after the battle). If these rates are correct, which side should win? How many should remain on the winning side when the other side has only 1,500 remaining? Give a brief explanation on the kill rates. (*Historical note*: The battle ended with 1,500 Japanese survivors and 44,314 U.S. survivors and took approximately 34 days.)

Fredrick W. Lanchester developed equations that have been used to model combat for almost 100 years. He developed the following *square law model for modern combat*.

Square Law Modern Combat

$$\frac{dx}{dt} = -a \cdot y(t)$$

$$\frac{dy}{dt} = -b \cdot x(t)$$

where a and b represent the kill rates against the *x* and *y* forces, respectively, by their opponents.

> *b:=0.0544:a:=0.0106 :*
> *forcex:=diff(x(t),t)=−a·y(t);*

$$forcex := \frac{d}{dt} x(t) = -0.0106 y(t)$$

> *forcey:=diff(y(t),t)=−b·x(t);*

$$forcey := \frac{d}{dt} y(t) = -0.0544\, x(t)$$

> *Model1:={forcex,forcey};*

$$Model1 := \left\{ \frac{d}{dt} x(t) = -0.0106\, y(t),\ \frac{d}{dt} y(t) = -0.0544\, x(t) \right\}$$

> *vars:={x(t),y(t)};*

$$vars := \{ x(t), y(t) \}$$

> *init1:=[x(0)=21500,y(0)=73500];*

$$init1 := [x(0) = 21500,\ y(0) = 73500]$$

> *domain:=t=0..35;*

$$domain := t = 0\,..35$$

> *L :=DEplot( Model1,vars,domain,[init1],stepsize=0.5,*
>   *scene=[t,x],arrows=NONE,linecolor=blue,thickness=3):*
> *H :=DEplot( Model1,vars,domain,[init1],stepsize=0.5,*
>   *scene=[t,y],arrows=NONE,linecolor=green,thickness=3):*
> *display({L,H},title='Iwo Jima');*

How did we do? Figure 13.28 shows the Japanese annihilated and United States winning. Actually, we didn't do well. We had a 91% error for the Japanese and a 24% error on the U.S. force. What could account for this?

History shows that two facts were not modeled correctly for the square law. First, the Japanese soldiers were embedded in hillside tunnels and caves and able to operate a sort of guerilla warfare. They saw the U.S. forces attacking, whereas U.S. soldiers probably could not see the Japanese very well. In addition, the U.S. force landed amphibiously over a two-week period, so not all soldiers were on the island at once. You will be asked to consider this in the exercise set to see if you can do better in modeling this historical event.

You might want to consider this model form: Brackney's mixed law (also called the *parabolic law*), which was developed in 1959 and used to represent guerilla warfare:

$$\frac{dx}{dt} = -a \cdot y(t)$$

$$\frac{dy}{dt} = -b \cdot x(t) \cdot y(t)$$

where *a* and *b* are kill rates, *x* represents the conventional force, and *y* represents the guerilla force.

**FIGURE 13.28**
Battle of Iwo Jima

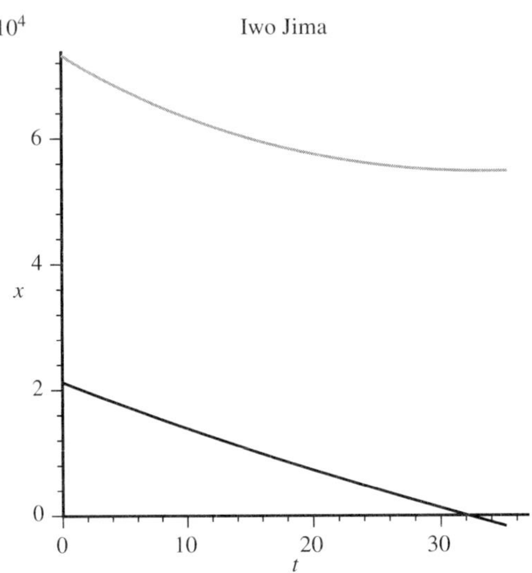


---

## 13.6 | EXERCISES

1. **The Battle of Iwo Jima.** The validity of Lanchester's equation can be demonstrated in an actual situation in which U.S. forces captured the island of Iwo Jima. Information required for the verification and what-if analysis is the number of friendly troops put ashore each day, the number of friendly causalities for each day's engagement, the number of enemy troops at the start of the battle (knowing that enemy troops were neither reinforced nor withdrawn), the number of enemy soldiers at the end of the engagement, and the length of the engagement. The enemy was well entrenched in the rockiest areas of the island, and U.S. forces were attacking prepared defenses. In an idealized situation, the U.S. forces would be considered as following a modified Lanchester's square law, with replacement troops landing each day; the enemy could be considered to follow the standard square law.

   Because the enemy is entrenched and looking down from above on the U.S. forces, it is easier to hit and kill them. We use the following values:

   $P$(hit a U.S. soldier with an enemy weapon) = 0.54
   $P$(kill a U.S. soldier given the shooter already has a hit) = 0.1

   We assume these events are independent and that their product represents the kill coefficient of the aggregated Japanese forces against U.S. forces. Also assume the following values:

   $P$(hit a Japanese soldier with a U.S. weapon) = 0.12
   $P$(kill a Japanese soldier given the shooter already has a hit) = 0.1

   We assume these events are independent and that their product represents the kill coefficient of the aggregated U.S. forces against the Japanese forces.

   **a.** Determine the kill rates for the U.S. and Japanese forces. From your knowledge of probability and statistics, explain why they might be reasonable.
   **b.** Determine who wins a fight to the finish.
   **c.** Parity: Parity is defined to be where both sides are annihilated, there is no winner. Is parity possible in this problem? Can we find it easily? Is it possible or easier to find parity after all the U.S. troops have landed and the battle begins from that point? Under this scenario, at what kill ratio could the enemy have reached parity? Is that value feasible? Explain.

**d.** The real battle ended with 1,500 Japanese survivors and 44,314 U.S. survivors and took approximately 34 days. Relate your result with these real results. If different, why do you think these results are different?

**e.** Reflect on your use of Lanchester equations to adequately explain the results of the Battle of Iwo Jima.

Enemy initial strength was 21,000 troops in fortified positions on the island. Friendly troop strength was modified by landing as shown in the following table.

| Day | Number of Troops |
| --- | --- |
| 1 | 30,000 |
| 2 | 1,200 |
| 3 | 6,735 |
| 4 | 3,626 |
| 5 | 5,158 |
| 6 | 13,227 |
| 7 | 3,054 |
| 8 | 3,359 |
| 9 | 3,180 |
| 10 | 1,456 |
| 11 | 250 |
| Thereafter | 0 |
| Total troops | 71,245 |

**2.** Find the equilibrium values for the predator–prey model presented.

**3.** Find the equilibrium values for the SIR model presented.

**4.** Find the equilibrium values for the combat model presented.

**5.** In the predator–prey model, determine the outcomes with the following sets of parameters.

    **a.** Initially, there are 200 foxes and 400 rabbits.
    **b.** Initially, there are 2,000 foxes and 10,000 rabbits.
    **c.** The birthrate of rabbits increases to 0.1.

**6.** In the SIR model, determine the outcome with the following parameters changed.

    **a.** Initially, five are sick; 10 are sick the next week.
    **b.** The flu lasts 1 week.
    **c.** The flu lasts 4 weeks.
    **d.** There are 4,000 students in the dorm, and five are initially infected. Thirty more are infected the next week.

# 14

# Classical Probability and Discrete Probability Modeling

## Introduction

Why do airlines overbook flights?

Airlines routinely overbook flights to compensate for no-shows: people who reschedule their flights or choose not to fly. An empty seat on a plane means a loss of revenue to an airline.

Routine overbooking leads to *bumping*, or the practice of keeping some passengers from boarding flights even though they have already purchased tickets. In other words, there are more passengers with confirmed reservations who show up for a flight than there are seats on the plane. Passengers who are bumped are offered seats on other flights in compensation.

Airlines overbook because they know from experience that not all passengers who have made reservations show up for their flights. To have any chance of filling the plane, airline computers estimate the number of passengers likely to be no-shows and accept reservations accordingly. (Airlines also ignore the overbook limit when a customer is buying a full-fare ticket because the cost of bribing volunteers with a bump ticket is usually less than the additional income derived from a full-fare ticket.)

When overbooking happens, airline agents will make a gate area announcement asking for volunteers with flexible schedules to give up their seats. Typically, the initial amount offered is based on two factors: the length of the flight, and how long the volunteer must wait in order to be scheduled on a later flight. Usually, the gate agent's first offer is for $250 (this may vary).

If not enough passengers take the first offer, agents will usually only increase the offer once or twice more. For example, Delta Airlines says that it tries to limit increases to just two rounds in order to get flights out on time. If agents are unable to find enough volunteers, they will begin to involuntarily bump a few unlucky passengers, based on a variety of factors, such as the time the passenger arrived for the flight, the amount they paid for their ticket, and their frequent-flyer status.

Overbooking is a standard practice and perfectly legal. Many airlines regularly overbook busy routes by as much as 200 percent. By law, all bumped passengers are entitled to some form of compensation, usually in the form of a free ticket.

Thus, the airlines have to balance the *risk* of a no-show with the compensation they must pay to bumped passengers. They overbook according to a number of variables: whether it's a holiday season, how the airline market is doing in general, and, perhaps most importantly, a specific flight's history of no-shows.

According to several published articles, an average of 50,000 passengers are bumped by the nation's 10 largest airlines every year. If you do lose your seat, you should try to get the most bang from your bump.

Later in this chapter, we will build a mathematical model for airline overbooking.

In this chapter, we introduce the concept of discrete probability and discrete probability distributions. We then use these to build models. We introduce basic reliability models and use probability in the airline-overbooking model.

## 14.1   INTRODUCTION TO CLASSICAL PROBABILITY

Elementary probability theory is required for understanding this chapter in discrete stochastic models. We will provide a quick review of some important concepts in probability.

We define the **probability** that **event** A occurs, $P(A)$, as the number of times $A$ occurs out of the total number of possible outcomes in the **sample space**. An event is any collection of results or outcomes of an experiment. A sample space is a listing of all possible outcomes from an **experiment**. An experiment is any process that allows researchers to obtain observations (data). In a flip of a fair coin two times, the sample space is all the possible outcomes of two flips of a fair coin. If we call a head, $H$, and a tail, $T$, then the possible outcomes are:

$$HH \quad HT \quad TH \quad TT$$

This set constitutes the entire sample space. Let's call the event that exactly one head appeared in the two flips event $A$. That occurred in flip HT and flip TH, or two times. Because there were four possible outcomes, the probability of event $A$, $P(A)$, is 2/4 = 0.5.

Let's consider a tennis match between player A and player B in which the winner is the first to win three sets. The sample space for the winner is:

$$\{AAA, ABAA, ABBAA, AABA, AABBA, ABABA, BBB, BBAB,$$
$$BAABB, BABAB, BAABA, BBAAB, BABB, BAAA, ABBB, ABABB,$$
$$BABAA, BBAAA, AABBB, ABBAB\}$$

If A and B were equally likely to win a set, then we could compute the probability of each event in the sample space. The probability that A wins in three sets is (1/8), in four sets is 3(1/16), and in five sets is 6(1/32). Thus, the probability that A wins is 0.5.

Now let's assume that A is a higher-ranked player with odds to win a set of 3:1. We can recompute the probabilities that A wins from the given sample space in three, four, or five sets.

$P(\text{A wins in 3 sets}) = (0.75^3) = 0.421875$

$P(\text{A wins in 4 sets}) = 3(0.75^3)(0.25) = 0.3164065$

$P(\text{A wins in 5 sets}) = 6(0.75^3)(0.25^2) = 0.158203125$

The probability that A wins the match, $P(\text{A wins the match})$, is 0.89648625.

There are some important rules for probability. Some that we will use include the following:

1. The law of large numbers states that if an experiment is repeated again and again, the relative frequency probability of an event approaches its probability. We saw this in our chapter on simulations.

2. Addition rule: $P(A \text{ or } B) = P(A) + P(B) - P(A \text{ and } B)$
3. For any event $A$ in the sample space, $S$,
   a. $P(A) \geq 0$
   b. $P(A) \leq 1$
   c. $P(S) = 1$
   d. $P(\text{not } A) = 1 - P(A)$
4. For any events $A$ and $B$ in the sample space $S$:
   a. If mutually exclusive, $P(A \text{ and } B) = 0$.
   b. If independent, $P(A \text{ and } B) = P(A) * P(B)$.
   c. Otherwise, $P(A \text{ and } B) = P(A) + P(B) - P(A \text{ or } B)$.
5. Conditional probability: $P(A \text{ given } B \text{ has occurred already}) = P(A \text{ and } B)/P(B)$.

Many models arise that use the concepts of independence and conditional probability. In the following example, we also need our to recall our modeling of discrete dynamical systems (DDSs) to assist in the development of our modeling solution.

---

## Example 1          Fast-Food Selection

### Situation

There are two businesses in our college town that do a majority of the take-out business with students. One business is a pizza parlor, and the other is a Chinese restaurant. You are provided the results of a survey from customers who have used these establishments. You find the probability that you order pizza next time given that you previously ordered pizza is 0.6, and the probability that you order Chinese next time given that you order Chinese this time is 0.7. The managers need to know information about the future of their businesses in order to provide the necessary service.

### Problem Identification

Predict the probabilities of each business in the long term based on the survey data.

### Assumptions

The survey data were randomly collected and are the only viable data currently available. The events of ordering take-out are independent events. The data provided are conditional probabilities based on the manner in which the data were collected. We assume this was a normal take-out day and these data were not collected during rush hour.

### Model Solution

1. We begin by drawing a tree diagram to depict the choices a college student has from the take-out businesses. We label each branch of the tree with its proper notation and fill in the known probabilities. Next, we use our knowledge of DDS to find the initial or steady-state probabilities that a student ordered pizza and that a student ordered Chinese food. We set up this scenario as a **DDS** and solve for the limiting probabilities (see Figure 14.1). We can use our model to find additional probabilities, such as
   a. $P(\text{you order from the same place as last time})$,
   b. $P(\text{you order from a different place})$,
   c. $P(\text{you have ordered pizza, given that you ordered from the same place last time})$.

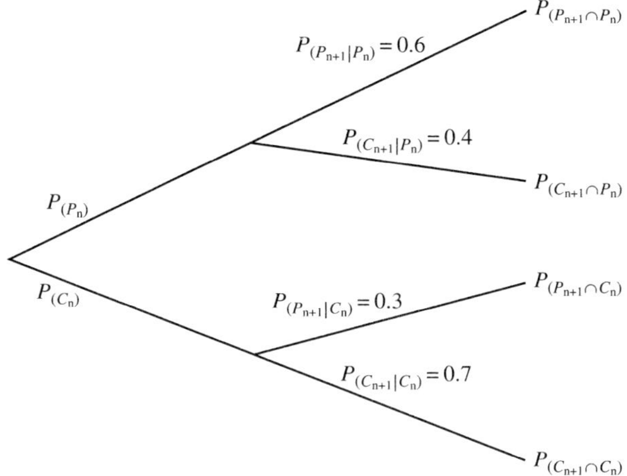


## Solution

Define

$$P_n = \text{order pizza at time } n$$
$$C_n = \text{order Chinese food at time } n$$
$$P(P_n) = \text{probability of ordering pizza at time } n$$
$$P(C_n) = \text{probability of ordering Chinese food at time } n$$

Given

$$P(P_{n+1}|P_n) = 0.6$$
$$P(C_{n+1}|C_n) = 0.7$$
$$\text{Find } P(P_n) \text{ and } P(C_n).$$

Using the law of total probability, we develop the following:

$$P(P_{n+1}) = P(P_{n+1} \cap P_n) + P(P_{n+1} \cap C_n) \tag{14.1}$$
$$P(C_{n+1}) = P(C_{n+1} \cap P_n) + P(C_{n+1} \cap C_n) \tag{14.2}$$

Next, we use Bayes' theorem. We have

$$P(P_{n+1}) = P(P_{n+1}|P_n)P(P_n) + P(P_{n+1}|C_n)P(C_n) \tag{14.3}$$
$$P(C_{n+1}) = P(C_{n+1}|P_n)P(P_n) + P(C_{n+1}|C_n)P(C_n) \tag{14.4}$$

Now we substitute the known conditional probabilities, yielding the discrete dynamical system

$$P(P_{n+1}) = 0.6P(P_n) + 0.3P(C_n) \tag{14.5}$$
$$P(C_{n+1}) = 0.4P(P_n) + 0.7P(C_n) \tag{14.6}$$

Recall, we may solve this using various techniques, including recursive iteration, stability graphs, stability of the A matrix, and back substitution. Each method, if correctly used, yields the correct initial probabilities. Let $x_1 = P(P_n)$ and $(1 - x_1) = P(C_n)$, we substitute and solve.

Because $x_1 + \dfrac{4}{3}x_1 = 1$, $x_1 = \dfrac{3}{7}$.

We find the initial probabilities: $P(P_n) = \dfrac{3}{7}$ and $P(C_n) = \dfrac{4}{7}$.

We make use of Maple to assist in finding the steady-state solution.

**Systems of DDS: The Fast-Food Example**

A fast-food restaurant has the following model as a system of DDS:

$$P(n + 1) = 0.6\ P(n) + 0.3C(n)$$
$$C(n + 1) = 0.4\ P(n) + 0.7C(n)$$
$$P(0) = 1/2 \text{ and } C(0) = 1/2, \text{ respectively.}$$

> *rsolve({P(n+1)=.6*P(n)+.3*C(n),C(n+1)=.4*P(n)+.7*C(n),P(0)=1/2,C(0)=1/2},{P,C});*

$$\left\{ C(n) = \frac{4}{7} - \frac{\left(\frac{3}{10}\right)^n}{14},\ P(n) = \frac{3}{7} + \frac{\left(\frac{3}{10}\right)^n}{14} \right\}$$

> *with(plots):*
> *n1:=pointplot({seq([k,(3000–1000*(3/10)^k)],k=0..20)}):*
> *n2:=pointplot({seq([k,(4000+1000*(3/10)^k)],k=0..20)}):*
> *display(n1,n2);*

Again, you should go back and change the initial conditions, currently $P(0) = 1/2$ and $C(0) = 1/2$, and see what behavior follows (see Figure 14.2):

> *limit(3/7+1/14*(3/10)^n,n=infinity);*

$$\frac{3}{7}$$

> *limit( 4/7–1/14*(3/10)^n,n=infinity);*

$$\frac{4}{7}$$

An alternative method is to take the matrix, A,

$$a := \begin{bmatrix} 0.6 & 0.3 \\ 0.4 & 0.7 \end{bmatrix}$$

and raise it to a sufficient power to see if the matrix becomes steady state. The steady-state solution is:

$$\begin{bmatrix} 0.4285714310 & 0.4285714268 \\ 0.5714285690 & 0.5714285732 \end{bmatrix}$$

We note that $3/7 = 0.428571431$ and $4/7 = 0.571428569$.

Given these results, we can use the law of total probability and Bayes' theorem to answer the scenario questions.

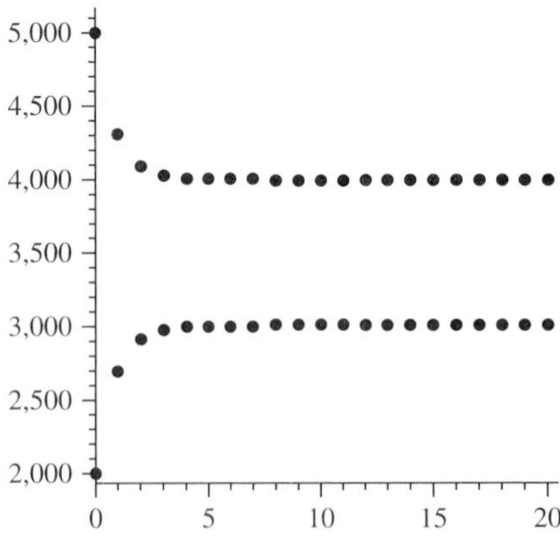

**FIGURE 14.2**

Stable long-term behavior of system

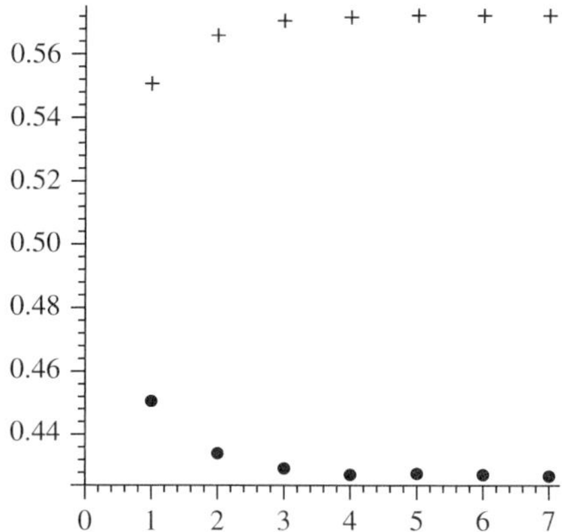

**FIGURE 14.3**

Long-term behavior of graphical solution

Using the completed tree diagram in Figure 14.3 (see also Figure 14.4), we can compute the required probabilities as follows:

$$P(P_{n+1} \cap P_n) = P(P_{n+1} | P_n)P(P_n) = 0.6(3/7) = 0.2571$$
$$P(C_{n+1} \cap P_n) = P(C_{n+1} | P_n)P(P_n) = 0.4(3/7) = 0.1714$$
$$P(P_{n+1} \cap C_n) = P(P_{n+1} | C_n)P(C_n) = 0.3(4/7) = 0.1714$$
$$P(C_{n+1} \cap C_n) = P(C_{n+1} | C_n)P(C_n) = 0.7(4/7) = 0.4000$$

**Scenario Solutions**

1. $P$(order from the same place) $= P(P_{n+1} \cap P_n) + P(C_{n+1} \cap C_n) = 0.2571 + 0.4000 = 0.6571$ (application of the law of total probability).

2. $P$(order from a different place) $= P(C_{n+1} \cap P_n) + P(P_{n+1} \cap C_n) = 0.1714 + 0.1714 = 0.3428$ (application of the law of total probability) $= 1 - 0.6571 = 0.3429$ (concept of the complement).

3. $P$(order pizza | ordered from the same place) $= P$(ordered pizza twice)/
   $P$(ordered from the same place) $= 0.2571/0.6571 = 0.3913$ (application of Bayes' theorem).

### Discrete Distributions in Modeling

We will also use several probability distributions for discrete random variables. A random variable is a rule that assigns a number to every outcome of a sample space. A discrete random

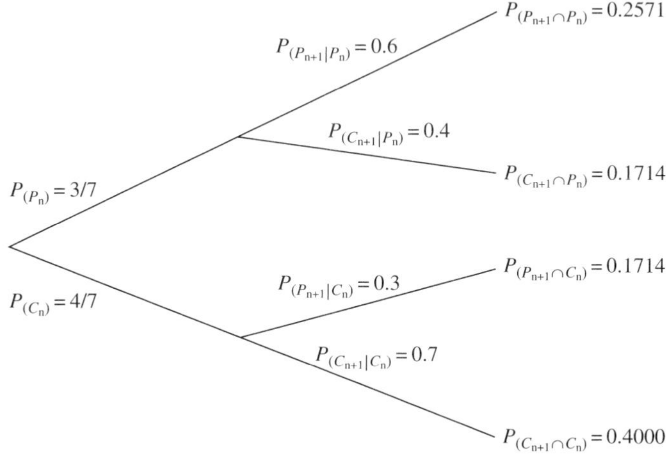

**FIGURE 14.4**

Fast-food example: completed tree diagram

variable takes on counting numbers 0, 1, 2, 3, . . . and so on. These are either finite or countable. Then a probability distribution gives the probability for each value of the random variable.

Let's return to our coin-flipping example. Let the random variable $F$ be the number of heads of the two flips of the coin. The possible values of the random variable $F$ are 0, 1, and 2. We can count the number of outcomes that fall into each category of $F$ as shown in the following *probability mass function* table.

| Random Variable | 0 | 1 | 2 |
| --- | --- | --- | --- |
| Occurrences | 1 | 2 | 1 |
| Corresponding to events | TT | TH, HT | HH |
| $P(F)$ | 1/4 | 2/4 | 1/4 |

Note that the $\Sigma P(F) = 1/4 + 2/4 + 1/4 = 1$. This is a rule for any probability distribution. Let's summarize these rules:

1. $P(\text{each event}) \geq 0$
2. $\Sigma P(\text{events}) = 1$

Thus, the coin flip experiment is a probability distribution.

All probability distributions have means, $\mu$, and variances, $\sigma^2$. We can find the mean and the variance for a random variable $X$ using the following formulas:

$$\mu = E[X] = \Sigma x P(X = x)$$
$$\sigma^2 = E[X^2] - (E[X])^2$$

We define $E[X^2]$ as summation $x^2 P(X = x)$. For our example, we compute the mean and variance as follows:

$$\mu = E[X] = \Sigma x\, P(X = x) = 0(1/4) + 1(2/4) + 2(1/4) = 1$$
$$\sigma^2 = E[X^2] - (E[X])^2 = 0(1/4) + 1(2/4) + 4(1/4) - 1^2 = 0.5$$

There will be several discrete distributions that will arise in our modeling: Bernoulli, binomial, and Poisson.

Consider an experiment made up of a repeated number of independent and identical trials having only two outcomes, such as tossing a fair coin {head, tail} or a stoplight {red, green}. These experiments with only two possible outcomes are called *Bernoulli trials*. Often they are found by assigning either an $S$ (success) or $F$ (failure) or a 0 or a 1 to an outcome. Something either happened (1) or did not happen (0).

A binomial experiment is conducted by counting the number of successes in $n$ trials.

1. A binomial experiment consists of $n$ trials where $n$ is fixed in advance.
2. Trials are identical and can result in either a success or a failure.
3. Trials are independent.
4. The probability of success is constant from trial to trial.

Formula: $b(x; n, p) = p(X = x) = \binom{n}{x} p^x (1 - p)^{n-x}$ for $x = 0, 1, 2, \ldots, n$

Cumulative binomial: $p(X \leq x) = B(x; n, p) = \sum_{y=0}^{x} \binom{n}{y} p^y (1 - p)^{n-y}$ for $x = 0, 1, 2, \ldots, n$

A binomial distribution has the following parameters:

Mean: $\mu = np$

Variance: $\sigma^2 = np(1 - p)$

For example, our coin-flip experiment follows these rules and is a binomial experiment. The probability that we got one head in two flips is:

$$P(X = 1) = \binom{2}{1}.5^1(1 - .5)^{2-1} = 0.50$$

If we wanted five heads in 10 flips of a fair coin, we can compute:

$$P(X = 5) = \binom{10}{5}.5^5(1 - .5)^{10-5} = 0.2461$$

## Example 2    Lightbulbs

Lightbulbs are manufactured in a small local plant. Before the lightbulbs are packaged and shipped, they are tested and either work, $S$, or fail to work, $F$. The company cannot test all the lightbulbs, but it does test a random batch of 100 lightbulbs per hour. In this batch, it finds that 0.2 percent did not work, but all batches were shipped to distributors.

As a distributor, you are worried about past performance of these lightbulbs, which you sell as individual units. If a customer buys 20 lightbulbs, what is the probability that all of them work?

PID: Predict the probability that $x$ lights bulbs out of $N$ work.

Assumptions: The lightbulbs follow the binomial distribution rules stated earlier.

Model: The formula is $b(x; n, p) = p(X = x) = \binom{n}{x} p^x(1 - p)^{n-x}$ for $x = 0, 1, 2, ..., n$.

```
> Binomial:=proc(x,n,p)
> binomial(n,x)*(p^x)*((1-p)^(n-x));
> evalf(%);
> end;
```

$Binomial := \text{proc}(x, n, p) \; binomial(n, x) \times p\text{^}x \times (1 - p)\text{^}(n - x); evalf(\%) \text{ end proc}$

```
> Binomial(20,20,.998);
```

$$0.9607509570$$

### Poisson Distribution

A random variable is said to have a Poisson distribution if the probability distribution function of $X$ is

$$p(x;\lambda) = \frac{e^{-\lambda} \lambda^x}{x!} 0, \text{ for } x = 0, 1, 2, 3, \ldots, \text{ for some } \lambda > 0.$$

We consider $\lambda$ as a *rate per unit time or per unit area*.

## Example 3    Flaws on a Glass Surface

In the following example, let $X$ represent the number of flaws on the surface of a randomly selected crystal glass. On average, five flaws are found per glass surface. Find the probability that a randomly selected glass has exactly two flaws.

```
> Poisson:=proc(x,lam)
> (exp(-lam)*lam^x)/x!;
> evalf(%);
> end;
```

$Poisson := \text{proc}(x, lam) \; exp(-lam) \times lam\text{^}x/x!; evalf(\%) \text{ end proc}$

> *Poisson(2,5);*

$$0.08422433749$$

$$p(X = 2) = \frac{e^{-5}5^2}{2!} = .084$$

A Poisson distribution has a mean, $\mu$, of $\lambda$ and variance $\sigma^2$ of $\lambda$.

A *Poisson process* is a Poisson distribution that varies over time (generally, its time). There exists a rate called $\alpha$ for a short time period. Over a longer period of time, $\lambda$ becomes $\alpha t$.

## Example 4    A Poisson Process

Suppose your pulse is read by an electronic machine at a rate of five times per minute. Find the probability that your pulse is read 15 times in a 4-minute interval.

$$\lambda = \alpha t = 5 \times 4 \text{ minutes} = 20 \text{ pulses in a 4-minute period}$$

$$p(X = 15) = \frac{e^{-20}20^{15}}{15!} = .052$$

> *Poisson(15,20);*

$$0.05164885353$$

## 14.1 | EXERCISES

1. If 75 percent of all purchases at Walmart are made with a credit card, and $X$ is the number among 10 randomly selected purchases made with a credit card, find the following:
   a. $p(X = 5)$
   b. $p(X \leq 5)$
   c. $\mu$ and $\sigma^2$

2. Russell Stover produces fine chocolates, but from experience it is known that 10 percent of its chocolate boxes have flaws and must be classified as *seconds*.
   a. Among six randomly selected chocolate boxes, how likely is it that one is a second?
   b. Among the six randomly selected boxes, what is the probability that at least two are seconds?
   c. What is the mean and variance for seconds?

3. Consider the following TV ad for an exercise program: 17 percent of the participants lose 3 pounds, 34 percent lose 5 pounds, 28 percent lose 6 pounds, 12 percent lose 8 pounds, and 9 percent lose 10 pounds. Let $X$ = the number of pounds lost on the program.
   a. Give the probability mass function of X in a table.
   b. What is the probability that the number of pounds lost is at most 6? At least 6?
   c. What is the probability that the number of pounds lost is between 6 and 10?
   d. What are the values of $\mu$ and $\sigma^2$?

4. A machine fails on average 0.4 times a month (defined as 30 consecutive days). Determine the probability that there are 10 failures in the next year.

## 14.2  RELIABILITY MODELS IN ENGINEERING AND SCIENCE

Consider your cellular telephone, graphing calculator, or automobile. How often do you replace them? Probably not that often because they perform well over a reasonably long period of time. If they do last for a long period of time, then we say that these systems are *reliable*. The reliability of a system or its components is the probability that it will not fail over a specific time period $t$. Let's define $f(t)$ to be the failure rate of an item, component, or system over time $t$ and $f(t)$ be a probability distribution. Let $F(t)$ be the cumulative distribution function corresponding to $f(t)$ as discussed in the last section. We define reliability of an item, component, or system by

$$R(t) = 1 - F(t)$$

Human–machine systems, whether electrical, mechanical, analog, or digital, consist of components that interact to form systems. Consider your personal computer, stereo, PlayStation (or Nintendo game systems), or automobile. We want to build simple models to examine the reliability of complex systems. We consider relationships in series, parallel, and combinations of these. Although individual item failure rates can follow a wide variety of distributions, we consider only a few elementary examples, using distributions we have already reviewed.

### Modeling Series Systems

Now we consider a system with $n$ components $C_1, C_2, \ldots, C_n$, where each of the individual components must work for the system to function. A model of this type of system is shown in Figure 14.5.

If we assume these components are mutually independent, the reliability of this type of system is easy to compute. We denote the reliability of component $i$ at time $t$ by $R_i(t)$. In other words, $R_i(t)$ is simply the probability that component $i$ will function continuously from time 0 until time $t$. We are interested in the reliability of the entire system of $n$ components, but because these components are mutually independent, the system reliability is

$$R(t) = R_i(t) \cdot R_2(t) \ldots R_n(t)$$

---

**Example 1**    Systems in a Series Configuration

A **series** system is one that performs well as long as every item or component is performing well. Consider a video game with four components: the game, the controller, the PlayStation unit, and the television as displayed in Figure 14.6. This is an example of a series system because failure of any one of the independent items will result in total system failure. If the component reliabilities are given by $R_1(t) = 0.90$, $R_2(t) = 0.95$, $R_3(t) = 0.96$, and $R_4(t) = 0.99$, then the system reliability is defined to be their product:

$$R_s(t) = R_1(t)\, R_2(t)\, R_3(t)\, R_4(t) = (0.90)(0.95)(0.96)(0.96) = 0.812592$$

Thus, we assume that the system will work correctly 81.2 percent of the time. Note that the reliability of the system is less than any single component because each has a reliability that is less than 1.0 (see Figure 14.6).

---

**FIGURE 14.5**
Series system

**FIGURE 14.6**
Series setup for game

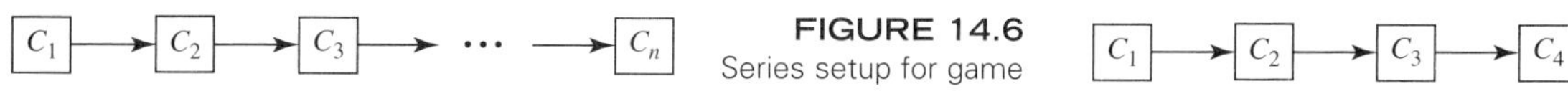

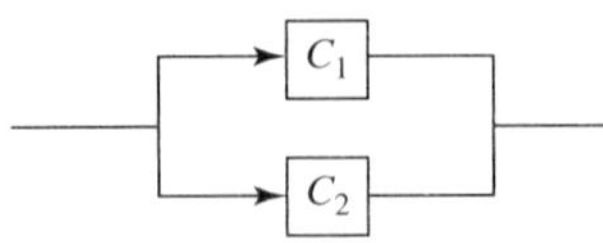

### Modeling Parallel Systems (Two Components)

Now we consider a system with two components in which only one of the components must work for the system to function. A system of this type is depicted in Figure 14.7.

Notice that in this situation the two components are *both* put in operation at time 0; they are both subject to failure throughout the period of interest. Only when *both* components fail before time $t$ does the system fail. Again, we also assume that the components are independent. The reliability of this type of system can be found using the following well-known model:

$$P(A \cup B) = P(A) + P(B) - P(A \cap B)$$

In this case, $A$ is the event that the first component functions for longer than some time $t$, and $B$ is the event that the second component functions longer than the same time $t$. Because reliabilities *are* probabilities, we can translate the above formula into the following:

$$R(t) = R_1(t) \cdot R_2(t) - R_1(t)\, R_2(t)$$

---

## Example 2    Parallel Systems

A **parallel** system is one that performs as long as any one of its components remains operational. Consider your home communications system of cordless telephone and conventional telephone as shown in Figure 14.7. Note that there are two separate and independent routes to transverse the network (input to output), either of which can be used to complete the call. If either is working, the call can be completed. Let $C_1$ and $C_2$ be as in Figure 14.7 with reliability values of 0.95 and 0.96, respectively.

We apply the definition of the system reliability for parallel components as

$$R_s(t) = R_1(t) + R_2(t) - R_1(t)\, R_2(t) = 0.95 + 0.96 - (0.95)(0.96) = 0.998$$

Note that in parallel relationships the system reliability is higher than any single component's reliability.

---

## Example 3    Series and Parallel Systems Combined

Let's consider systems a modified setup in combination of series and parallel. Consider the PlayStation in Example 1, now with two controllers in parallel as shown in Figure 14.8.



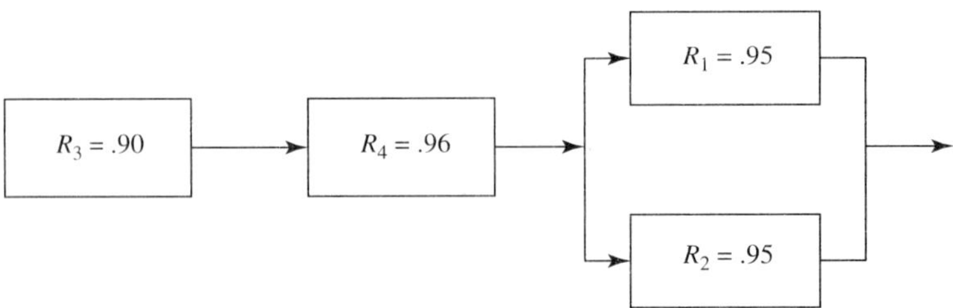

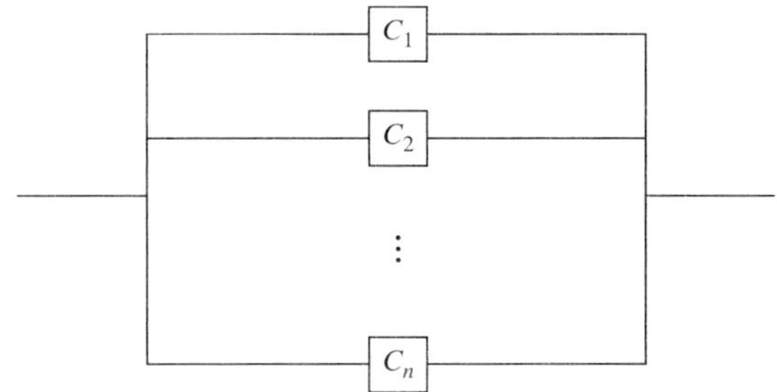

**FIGURE 14.9**
Active redundant system



The reliability can be seen as two subsystems in series. Subsystem A is the two controllers in parallel, and subsystem B is the PlayStation and television in series:

$$R_a(t) = R_1(t) + R_2(t) - R_1(t)\,R_2(t) = 0.95 + 0.95 - (0.95)(0.95) = 0.9975$$
$$R_b(t) = R_3(t)\,R_4(t) = (0.90)(0.96) = 0.864$$
$$R_s(t) = R_a(t)\,R_b(t) = (0.9975)(0.864) = 0.86184$$

### Modeling Active Redundant Systems

Consider the situation in which a system has $n$ components, all of which begin operating (are active) at time $t = 0$. The system continues to function properly as long as at least $k$ of the components do *not* fail. In other words, if $n - k + 1$ components fail, the system fails. This type of component system is called an **active redundant system**. The active redundant system can be modeled as a **parallel** system of components as shown in Figure 14.9.

We assume that all $n$ components are identical and will fail independently. If we let $T_i$ be the time to failure of the $i$th component, then the $T_i$ terms are independent and identically distributed for $i = 1, 2, 3, \ldots, n$. Thus $R_i(t)$, the reliability at time $t$ for component $i$, is identical for all components.

Recall that our system operates if at least $k$ components function properly. Now we define the random variables $X$ and $T$ as follow:

$$X = \text{number of components functioning at time } t, \text{ and}$$
$$T = \text{time to failure of the entire system.}$$

Then we have

$$R(t) = P(T > t) = P(X \geq k)$$

It is easy to see that we now have $n$ identical and independent components with the same probability of failure by time $t$. This situation corresponds to a binomial experiment, and we can solve for the system reliability using the binomial distribution with parameters $n$ and $R_i(t)$.

## Example 4

## Listening Devices for an Undercover Operation as an Active Redundant System

Three undercover police persons on a stakeout have been instructed to put out 15 listening devices to detect and listen to plans for a robbery. The experts estimate that all communications can be heard as long as at least 12 of the devices are operating. These devices are assumed to be in parallel and that if any of them fail, they will fail independently. Each device has a 0.75 probability of working properly for 24 hours. The reliability of the systems for 24 hours is computed as:

$$R(24) = P(T > 24) = P(X \geq 12) = 1 - P(X \leq 11) = 1 - .5387 = 0.4613.$$

FIGURE 14.10
Standby redundant system

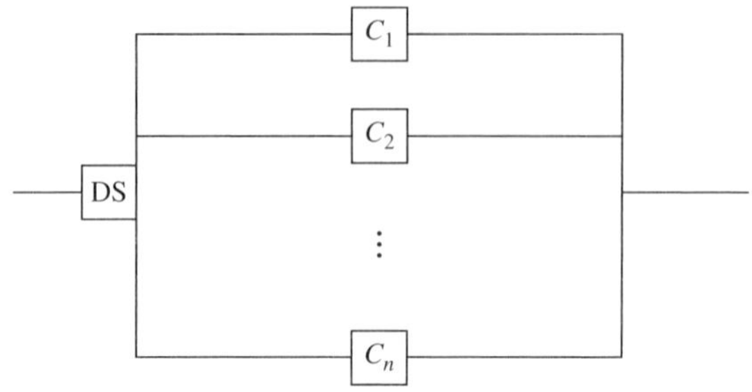


Thus, the reliability of the system for 24 hours is 0.4613.

If only ten had to work, then R(24) = 1 − P(X ≤ 9) = 0.8517.

### Modeling Standby Redundant Systems

Active redundant systems can sometimes be inefficient. These systems require only $k$ of the $n$ components to be operational, but all $n$ components are initially in operation and thus subject to failure. An alternative is the use of spare components. Such systems have only $k$ components initially in operation—exactly what we need for the whole system to be operational. When a component fails, we have a spare standing by that is immediately put in to operation. For this reason, we call these **standby redundant systems**. Suppose our system requires $k$ operational components, and we initially have $n - k$ spares available. When a component in operation fails, a decision switch causes a spare or standby component to activate (becoming an operational component). The system will continue to function until there are less than $k$ operational components remaining. In other words, the system works until $n - k + 1$ components have failed. We will consider only the case where one operational component is required (the special case where $k = 1$) and there are $n - 1$ standby (spare) components available. We will assume that a decision switch (DS) controls the activation of the standby components instantaneously and 100 percent reliably. We use the model in Figure 14.10 to represent this situation.

If we let $T_i$ be the time to failure of the $i$th component, then the $T_i$'s are independent and identically distributed for $i = 1, 2, 3, \ldots, n$. Thus $R_i(t)$ is identical for all components. Let $T =$ time to failure of the entire system. Because the system fails only when all $n$ components have failed, and component $i + 1$ is put into operation only when component $i$ fails, it is easy to see that

$$T = T_1 + T_2 + \ldots T_n$$

In other words, we can compute the system failure time easily if we know the failure times of the individual components.

We further assume the DS is one-hundred-percent reliable and instantaneously switches to a standby component. We let $T_i$ be the time to failure of the $i$th component; then all $T_i$'s are independent and identically distributed for $I = 1, 2, \ldots, N$. Thus, $R_i(t)$ is identical for all components. The reliability of the system is equal to the probability, $P(X < N)$, which is the probability that fewer than $N$ components fail during the interval $(0, t)$. Since this is a property of a Poisson experiment, we can use the Poisson to solve. In fact, we can show that the random variable $X$ follows a Poisson distribution with parameter $\lambda = \alpha t$, where $\alpha$ is the component failure rate. The reliability for some specific time, $t$, becomes

$$R(t) = P(X < N) = \text{Poisson} (\lambda = \alpha t, N - 1)$$

## Example 5    Playing Tetris on Game Boy as a Standby Redundant System

Game Boy runs on AA batteries. For 24 hours, the reliability of the batteries with constant use is 0.80. You hate it when the game ends because the batteries run low, so you carry spare battery packs. The addition of a spare should increase the system's reliability. In this case, $N = 3$ total battery packs and $\alpha = 1/30$ per hour. The reliability is found by

$$R(24) = P[X < 3\lambda = (1/30)(24)] = P(0 \le X \le 2) = 0.9526$$

The system reliability with the spare battery packs for 24 hours is now 0.9526.

If we want to keep the reliability at 95 percent for 48 hours, how many spare battery packs are required? We use trial and error to solve this problem. Using the Poisson, we find the reliability for 48 hours and vary the number of spare until the reliability is above 95 percent.

| Number of Battery Packs as Spares | System Reliability ( Using Poisson) |
| --- | --- |
| 2 | 0.7834 |
| 3 | 0.9212 |
| 4 | 0.9763 |

This indicates that four spare battery packs are required for the 48-hour period to ensure at least a 95-percent reliability of the Game Boy based only on the batteries.

## 14.2 | EXERCISES

1. Your college football stadium is to be the site of a televised night football game. The TV lighting experts have told you that a minimum level of lighting must be maintained throughout the game. If more than three of the stadium's 12 light grids go out, the field will not be illuminated enough for the TV coverage. The maintenance crew services all the light grids before the game, and this alone yields a probability of 0.7596 of working for the required 5.5 hours. What is the reliability of the lighting system for providing minimum TV illumination throughout the game? What assumptions are made to model this problem?

2. You are the emergency preparedness (EP) coordinator for an East Coast beach resort. You fear hurricane season and know that you must maintain power to your EP center during weather emergencies. You use generators for both main and backup power. Their failure rate has been about once for every 7.5 hours of operation. Assuming all generators have the same time to failure and that they fail independently, find the reliability of the power system for 10 hours if two spares are available.

3. In problem 2, how many spares are needed so the system has at least a 99-percent reliability for the 10-hour period?

4. Consider a car stereo system with a CD player, an AM-FM radio turner, dual speakers, and a power supply, as displayed with their reliabilities in the following figure. What assumptions are required by your model? Determine the reliability of the system.

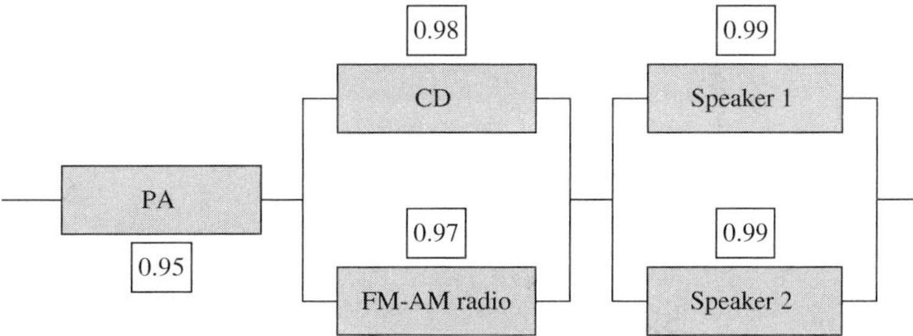


5. Consider your personal computer with each component's reliability as shown in the following figure. What assumptions are required? Determine the system's reliability.

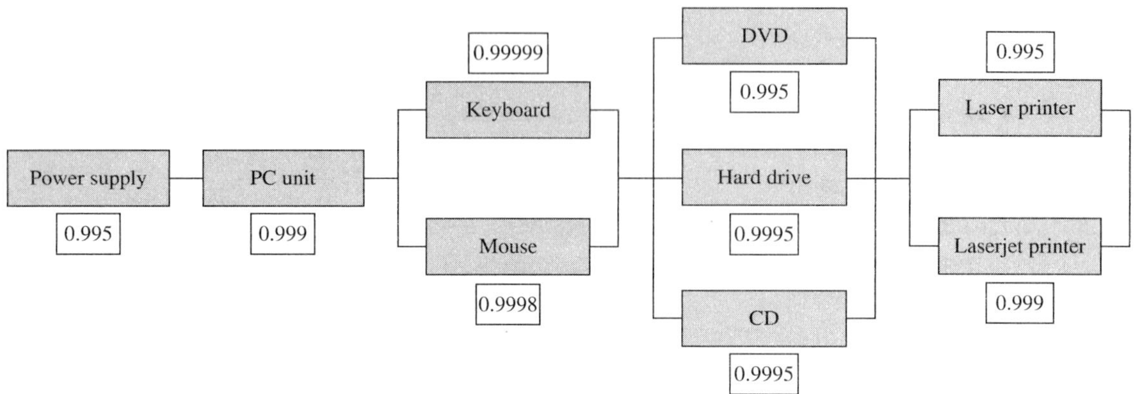



## 14.2 | PROJECTS

1. Two alternate designs are submitted for the Mars landing probe. The mission is to land safely on Mars, collect soil and other samples, and then return to its orbiting shuttle. Determine which design you would recommend to NASA for the mission. What assumptions are required? What assumptions are reasonable?

   Power, communications, and storage are the same for each alternative. The landing and rocket modules are the different for each alternative. Power is the main power supply and has a reliability of 0.998. Communications is a parallel system of two radios, each having a reliability of 0.995. Storage is a storage arm component with a reliability of 0.998. Each of the five major modules is in series.

**Alternative designs for the Mars module**

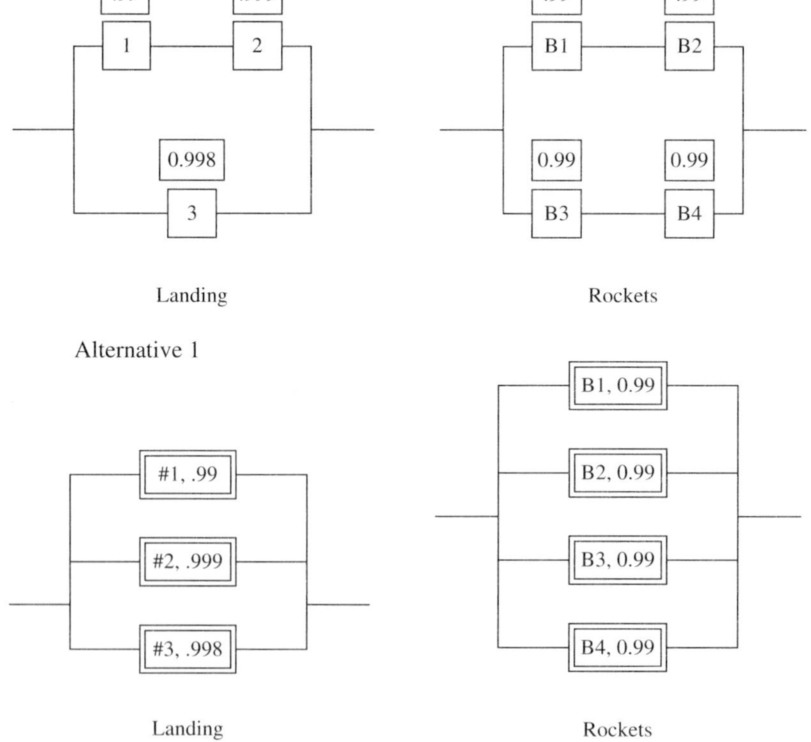

2. You are assigned as the transportation coordinator for emergency evacuations. It is your job to analyze the maps with bridges (annotated as A, B, C, D, E, F, G, H, and J) in the figure below to determine the reliability of the bridge system that must be used to travel safely out of the area. In addition, you must determine the best route over which to travel. The bridges operate independently of each other, and the lifetime of each bridge is distributed exponentially with reliability $1 - e^{at}$. You find historical records to indicate the failure rates per year (52-week period) for the bridges are:

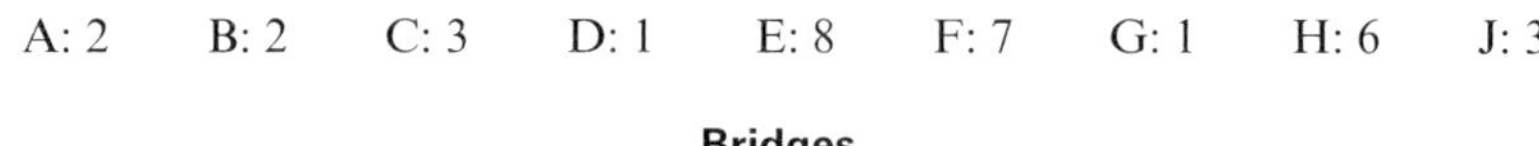


**Bridges**

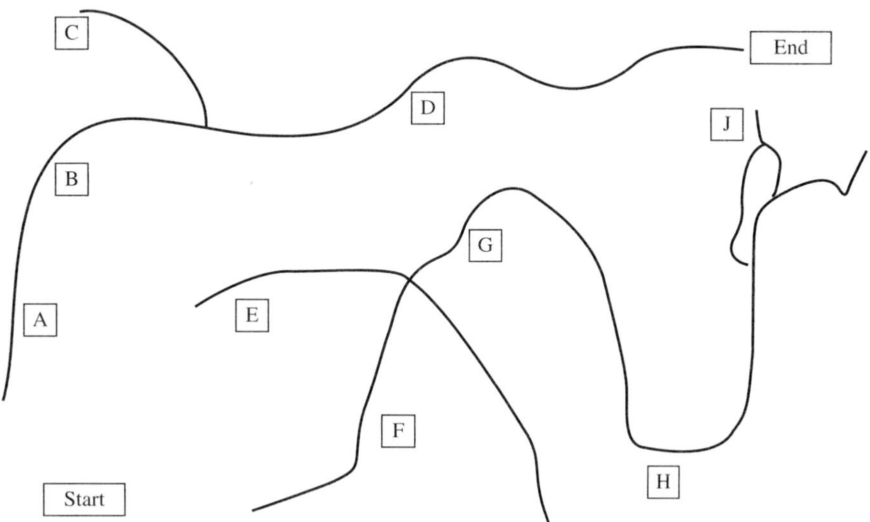


At bridge position E, the Army Corps of Engineers has two spare bridges to use in case one fails. At bridge position J, it is determined that at least two of the four bridges must operate for transportation to be effective.

**a.** Determine the reliability of the bridge system for any two-week period.

**b.** Determine the best route over which to travel.

**c.** Consider points 1 and 2 as decision points for your routes. Compute the reliabilities of each route. Which has the highest reliability?

**d.** If you need a three-week window, will the reliabilities increase or decrease? Why?

**e.** Make a recommendation to the governor for an evacuation plan.



# OVERBOOKING AIRLINES MODEL

### Introduction

Recently, on an airplane, I heard several passengers complain that they were bumped off their flight from Rome to New York City because the plane was overbooked. The passengers were told that the airlines sold more than 100 seats than the plane held. Because of their delay at the Rome airport, these passengers found that their seats were occupied by other passengers. Of course, the bumped passengers were compensated, but they were extremely upset and threatened to sue the airlines. Is overbooking common? Let's build a model from the standpoint of the airlines and examine the overbooking phenomenon.

Airlines know that only a certain percentage of passengers who have made reservations on a particular flight will actually take that flight, which is why they overbook. Airlines deal with bumped passengers in various ways to try to compensate them. Passengers will be allowed to fly on a later flight and may be given some kind of cash settlement or free plane tickets.

We will model the effects of the various types of overbooking strategies that an airline company might employ. In particular, we'll examine the effect that different schemes have on the revenue received by the airline company. Given the probability that a person who purchased a ticket will show up for his or her flight, we will attempt to discover an optimal overbooking strategy—that is, the number of people by which an airline should overbook on a particular flight so that the company's revenue is maximized. As you might imagine, there are many variables to consider in analyzing these types of situations. We will simplify as many as necessary.

---

## Example 1    Overbooking Airline Flights

### PID

Predict the number of overbooked reservations to be made on an airline in order to maximize the airline's revenue.

### Assumptions

For the purposes of this model, we are going to assume the following:

1. Only one passenger class is available, and all $n$ passengers pay the same price for a ticket. No first class is available, and there is no business class.

2. Bumped passengers are given cash settlements of exactly the same amount. This settlement is over and above the price of the ticket, which is also refunded to the passenger.

3. The probability that a passenger will actually take the flight is the same from passenger to passenger—that is, it may be regarded as a constant. This is provided from historical data.

4. The decision to take the flight is independent among the passengers; that is, passenger A's decision to take (or not take) the flight has no effect on the decision of passenger B.

5. The airlines have historical files that provide information about passenger no-shows and revenues for flights.

The assumptions that have been made are essentially the assumptions of the *binomial probability* function. Suppose there are $n$ independent trials, the outcome of each of which may be regarded as a success or failure. Let $p$ be the probability of success on a single trial and $1 - p$ be the probability of failure. Then the probability of $x$ successes in $n$ trials is given by

$$\binom{n}{x} p^x (1 - p)^{n-x}$$

The analysis will be performed on the basis of *expected value* of the revenue. If $X$ is a random variable that represents the possible outcomes in a random experiment, then the expected value of $X$ is defined by $E[X] = \Sigma x\, p(X = x)$.

In the special case of a binomial variable, the expected value of the random variable $X$ is given by

$$E(X) = np$$

Let us assume that the price in dollars of an airline ticket is $P$ and the refund in dollars to bumped passengers is $R$, where $R > P$. Assume that $n$ tickets are sold and $c \leq n$ is the capacity of the aircraft.

The airline makes $P$ dollars if one passenger shows up, $2P$ dollars if two take the flight, and so on up to $cP$ dollars. If one passenger gets bumped, the revenue is reduced to $cP - R$; if two get bumped, $cP - 2R$; and so on. Thus, if $M$ is a random variable that represents the total revenue, the following equations represent the model:

$$P(\text{at least one passenger gets bumped}) = \sum_{k=1}^{n-c} p^{c+k} (1 - p)^{n-c-k}$$

and

$$E[m] = \sum_{k=1}^{c} kP\binom{n}{k} p^k(1 - p)^{n-k} + \sum_{k=1}^{n-c} (cP - kR)\binom{n}{c+k} p^{c+k}(1 - p)^{n-c-k}$$

To present a specific example, suppose that an airplane has room for five passengers. If it charges all fliers \$100 per ticket, refunds \$200 to a bumped passenger, and overbooks by one, and the probability that a passenger shows up is 0.6, then the expected revenue reduces to:

$$E(M) = 100(0.036864) + 200(0.13824) + 300(0.27648) + 400(0.31104)$$
$$+ 500(0.186624) + 300(0.046656) = \$346.00$$

Note that the \$300 in the last term of the above computation is $5(100) - 200$ as stated in the description of the model. The numbers in parentheses are the binomial probabilities with $n = 6$ and $p = 0.6$.

If seven passengers are booked, a similar computation yields

$$E(M) = 100(0.0172032) + 200(0.0774144) + 300(0.193536) + 400(0.290304)$$
$$+ 500 (0.2612736) + 300(0.1306368) + 100(0.0279936) = \$364.01$$

Note that the expected revenue, if there are no airline overbookings, is $(0.6)(5)(500) = \$300$. Thus, it is clear that the airlines make money by overbooking their flights. Under these conditions, the best of the three strategies is, of course, the second: overbook by two passengers. On the other hand, if $p = 0.8$, these same calculations for overbooking by 0, 1, and 2, respectively, yield \$401.35, \$400.00, and \$324.07. In other words, accepting zero or one overbookings are approximately equivalent, but accepting two overbookings is a bad strategy. If $p = 0.9$, the expected revenues are \$450.00, \$380.57, and \$231.42, respectively.

Indeed, the results are intuitive. An airline should overbook if the "show up" probability is relatively low, and should not (or not much) if the probability is high.

In what follows, expected revenues have been compiled for various values of $p$, using $P = 100$ and $R = 200$, with $c$ (the airplane capacity) equal to 25. Note that for $p = 0.4$, the optimal strategy is to overbook by 33 people! This number drops to 14 for $p = 0.6$ and to 5 for $p = 0.8$.

| Number Overbooked ($p = 0.4$) | Expected Revenue (\$) |
|:---:|:---:|
| 0 | 1,000.00 |
| 2 | 1,079.84 |
| 4 | 1,159.79 |
| 6 | 1,239.94 |
| 8 | 1,319.92 |
| 10 | 1,399.89 |
| 12 | 1,479.82 |
| 14 | 1,559.60 |
| 16 | 1,638.95 |
| 18 | 1,717.32 |
| 20 | 1,793.72 |
| 22 | 1,866.61 |
| 24 | 1,933.85 |
| 26 | 1,992.87 |
| 28 | 2,040.63 |
| 30 | 2,074.26 |
| 32 | 2,091.16 |
| 33 | 2,092.71 (optimal) |
| 34 | 2,089.40 |

| Number Overbooked ($p = 0.6$) | Expected Revenue (\$) |
| --- | --- |
| 0 | 1,500.00 |
| 2 | 1,619.95 |
| 4 | 1,739.68 |
| 6 | 1,858.37 |
| 8 | 1,971.56 |
| 10 | 2,070.48 |
| 12 | 2,141.07 |
| 13 | 2,161.21 |
| 14 | 2,169.09 (optimal) |
| 15 | 2,163.67 |
| 16 | 2,144.42 |

| Number Overbooked ($p = 0.8$) | Expected Revenue (\$) |
| --- | --- |
| 0 | 2,000.00 |
| 2 | 2,153.65 |
| 4 | 2,257.98 |
| 5 | 2,269.83 (optimal) |
| 6 | 1,971.56 |

| Number Overbooked ($p = 0.9$) | Expected Revenue (\$) |
| --- | --- |
| 0 | 2,250.00 |
| 1 | 2,320.63 |
| 2 | 2,342.78 (optimal) |
| 3 | 2,301.94 |
| 4 | 2,204.41 |

Back to our Rome to New York flight: Let's assume that for the overseas flights the probability of showing up is 0.75. Let's assume the cost per ticket is \$1,000, and a refund for a bumped passenger costs the airlines \$2,000. Let's also assume the flight holds a total of 200 passengers.

The model, written in Maple code, shows that 62 additional tickets will be sold to maximize the airline profits.

## Maple Procedure to Compute Overbooking Values

```
> overbook := proc(c,n,p,P,R)
> f := sum(P*k*binomial(n,k)*(p^k)*(1 − p)^(n − k),k = 1..c) + sum
((c*P − k*R)*binomial(n,c + k)*p^(c + k)*(1 − p)^(n − c − k), k = 1..(n − c));
> print(f);
> end:
> overbook(200, 260,.75,1000,2000);

                                   192154.2097
> overbook(200,261,.75,1000,2000);

                                   192314.8455
> overbook(200,262,.75,1000,2000);

                                   192392.1834
> overbook(200,263,.75,1000,2000);

                                   192381.8571
```

The moral of the story is to arrive to the airport early for your flight because the chances are very high that your flight is overbooked.

## 14.3 | EXERCISES

1. Rework the calculation for $E(M)$ given in this section using $R$ equals:
   a. $150.
   b. $250.
      Recall that $c = 5$, $n = 6$ or $7$, and $p = 0.6, 0.8$, or $0.9$. For $n = 6$ and $p = 0.6$, use the numbers in the previous section with $350 ($350 = 500 - 150$) and $250 ($250 = 500 - 250$) instead of $300 for the last term in each sum.
      For $n = 7$, add an additional term of either $200$ ($200 = 500 - 2 \cdot 150$) or $0$ ($0 = 500 - 2 \cdot 250$). For $p = 0.8$ and $p = 0.9$, do the same kind of calculations using a table of binomial probabilities.

2. Find the optimal overbooking for $p = 0.6$ and $p = 0.8$ with $c = 5$, $P = 100$, and $R = 200$. (*Hint*: Use the binomial tables or Excel.)

3. Do Exercise 2 with $R = $250$.

4. Do Exercise 2 with $R = $150$.

## 14.3 | PROJECTS

1. Redo the overbooking analysis if the distribution were Poisson and not binomial.

2. Do sensitivity analysis by varying the mean and variance of each distribution and measure the impact on the results.

## 14.4   MARKOV CHAINS AS A DDS

In this section, we revisit the systems of difference equations studied at the beginning of this chapter. A special case called a *Markov chain* is a process in which there is the same finite number of states or outcomes than can be occupied at any given time. In a Markov process, the system may move from one step to another one at each time step, and there is a probability associated with this transition for each possible outcome. The sum of all the transitioning probabilities from the current state to the next state is equal to 1 for each state at each time step. A Markov chain with two states is illustrated in Figure 14.11.

### Example 1     Downtown and the Mall Revisited

Let's consider the attempt to revitalize the downtown section of a small city with merchants. There are merchants downtown and others in the large mall. Suppose historical records

**FIGURE 14.11**
A Markov chain with two states, 1 and 2, with probabilities $p$ and $q$

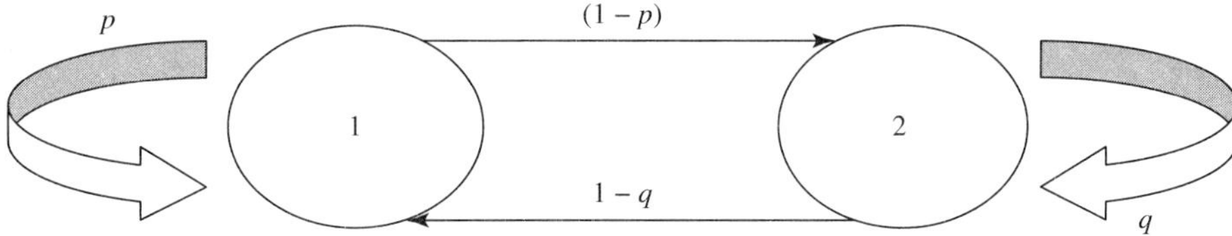

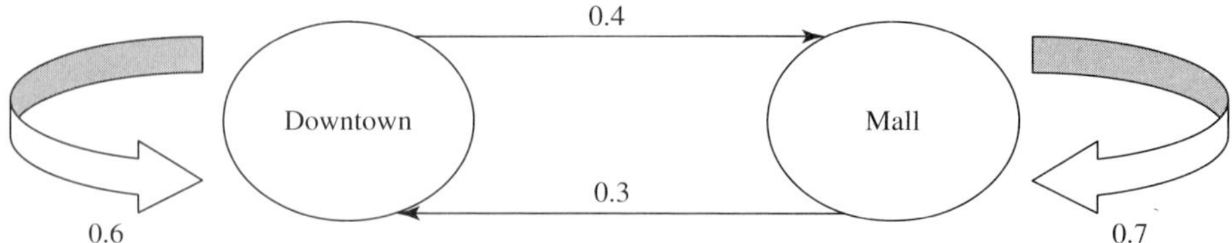


**FIGURE 14.12**

Markov chain stage diagram for merchants

determined that 60 percent of the downtown merchants remain downtown, whereas 40 percent move to the mall. We find that 70 percent of the mall merchants want to remain in the mall, but 30 percent want to move to downtown. Build a model to determine the long-term behavior of these merchants based on these historical data.

|  |  | Next State | |
| --- | --- | --- | --- |
|  |  | **Downtown** | **Mall** |
| **Current State** | **Downtown** | 0.60 | 0.40 |
|  | **Mall** | 0.30 | 0.70 |

The recorded data form the transition matrix show the probability for transitioning states: Staying downtown when originally downtown is 0.60, whereas moving to the mall is 0.40. Likewise, if a store is originally in the mall, it will stay there with a 70-percent likelihood; it has a 30-percent likelihood of moving downtown. This represents a Markov process with two states: downtown and the mall. Notice that the sum of the probabilities for transitioning from a current state to the next state, which is the sum of the probabilities in each row, equals 1. The process is illustrated in Figure 14.12.

## Model Information

$D(n)$ = the number of merchants operating downtown at the end of $n$ months
$M(n)$ = the number of merchants operating at the mall at the end of $n$ months

The number of merchants downtown in any time period is equal to the number of downtown merchants that stay downtown plus the number of mall merchants that relocate downtown. The same is true for the number of mall merchants in any time period: It is equal to the number that remain in the mall plus the number of downtown merchants that move to the mall. Mathematically, this is written as:

$$D(n + 1) = 0.60\ D(n) + 0.30\ M(n)$$
$$M(n + 1) = 0.40\ D(n) + 0.70\ M(n)$$

```
> rsolve({DD(n+1)=.6*DD(n)+.3*M(n),M(n+1)=.4*DD(n)+.7*M(n),
DD(0)=1/2,M(0)=1/2},{DD, M});
> with(plots):
> n1:=pointplot({seq([k,(4/7-(1/14)*(3/10)^k)],k=0..7)},symbol=cross):
> n2:=pointplot({seq([k,(3/7+(1/14)*(3/10)^k)],k=0..7)}):
> display({n1,n2});
```

Again, go back and change the initial conditions and see what behavior follows, as shown in Figure 14.13.

$$\left[ DD(n) = \frac{3}{7} + \frac{\left(\dfrac{3}{10}\right)^{n}}{14},\ M(n) = \frac{4}{7} - \frac{\left(\dfrac{3}{10}\right)^{n}}{14} \right]$$

**FIGURE 14.13**
Long-term behavior

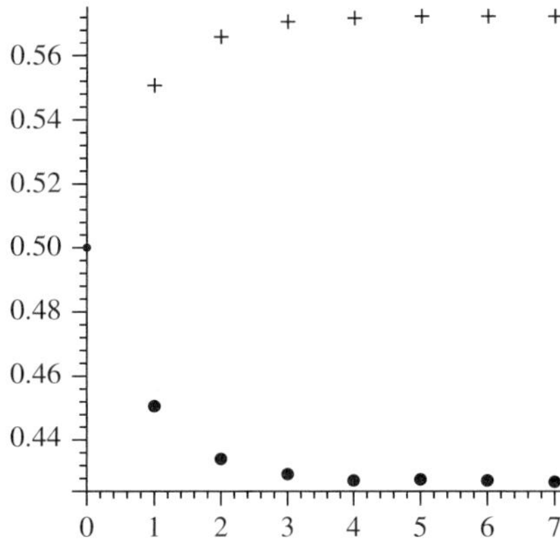

> $limit(3/7+1/14*(3/10)^n,n=infinity)$;

$$\frac{3}{7}$$

> $limit(4/7-1/14*(3/10)^n,n=infinity)$;

$$\frac{4}{7}$$

## Model Solution

Assuming all store owners are originally downtown, numerical solutions to the system give the long-term behavior of the percentages of store owners at each location. The sum of these long-term percentages or probabilities also equals 1.

As before, notice that

$$D(k) \rightarrow 3/7 = 0.428571$$
$$M(k \rightarrow 4/7 = 0.571429$$

## Model Interpretation

If the two branches begin the year with a total of $n$ stores, after 14 time periods approximately 57 percent will be in the mall and 43 percent will be downtown. Starting with 100 stores in each location, we will end up with 114 stores in the mall and 86 stores downtown. If better transitioning to downtown is required, recommend additional incentives to move.

---

## Example 2          Fast-Food Tendencies

Your student union center desires to have three fast-food centers—serving burgers, tacos, and pizza—available to students. These chains run a survey of students and find the following information concerning lunch: 75 percent who ate burgers will eat burgers again at the next lunch, 5 percent will eat tacos next, and 20 percent will eat pizza next. Of those who ate tacos last, 20 percent will eat burgers next, 60 percent will stay with tacos, and 35 percent will eat pizza next. Of those who ate pizza, 40 percent will eat burgers next, 20 percent tacos, and 40 percent pizza again.

### Problem Identification

Can we find the long-term behavior of the students regarding fast food?

## Assumptions

Over the last few years, data were collected on eating tendencies at the student union. The data are presented in the following transition matrix.

|  |  | Next State |  |  |
| --- | --- | --- | --- | --- |
|  |  | **Burgers** | **Tacos** | **Pizza** |
|  | **Burgers** | 0.75 | 0.05 | 0.20 |
| **Current State** | **Tacos** | 0.20 | 0.60 | 0.20 |
|  | **Pizza** | 0.40 | 0.20 | 0.40 |

We formulate the problem as follows:

Let $n$ represent the $n$th days lunch and define
$B(n)$ = the number of burger eaters in the $n$th lunch
$T(n)$ = the number of taco eaters in the $n$th lunch
$P(n)$ = the number of pizza eaters in the $n$th lunch

Formulating the system, we have the following dynamical system:

$$B(n + 1) = 0.75\ B(n) + 0.20\ T(n) + 0.40\ P(n)$$
$$T(n + 1) = 0.05\ B(n) + 0.60\ T(n) + 0.20\ P(n)$$
$$P(n + 1) = 0.20\ B(n) + 0.20\ T(n) + 40\ P(n)$$

> *rsolve({B(n+1)=.75*B(n)+.2*T(n)+.4*P(n),*
*T(n+1)=.05*B(n)+.6*T(n)+.2*P(n),P(n+1)=.2*B(n)+.2*T(n)+.4*P(n),*
*B(0)=1/3,T(0)=1/3,P(0)=1/3}, {B,T,P});*

$$\left\{ B(n) = -\frac{\left(\frac{1}{5}\right)^n}{21} - \frac{11\left(\frac{11}{20}\right)^n}{63} + \frac{5}{9},\ T(n) = -\frac{\left(\frac{1}{5}\right)^n}{28} + \frac{11\left(\frac{11}{20}\right)^n}{63} + \frac{7}{36},\ P(n) = \frac{\left(\frac{1}{5}\right)^n}{12} + \frac{1}{4} \right\}$$

## Model Construction

Assume initially that one-third of students ate at each of the burger, taco, and pizza establishments at the student union.

## Model Interpretation

The table shows that in the long run, as $n$ approaches infinity, about 56 percent (5/9) prefer burgers, 19 percent (7/36) prefer tacos, and 25 percent (1/4) prefer pizza.

---

## 14.4 | EXERCISES

1. Consider a model for long-term dining behavior of the students at University College. It is found that 25 percent of the students who eat at the dining hall return to eat their next meal at the dining hall, whereas those who eat at the deli bar have a 93-percent return rate. These are the only available eating establishments around or on the campus. Determine the long-term behavior of the system.

**2.** Consider adding a pizza delivery service as an option for University College. The following table gives the transition matrix based on a student survey. Determine the long-term behavior of the system.

|  |  | Next State | | |
| --- | --- | --- | --- | --- |
|  |  | **Dining Hall** | **Deli** | **Pizza** |
| **Current State** | **Dining Hall** | 0.25 | 0.25 | 0.50 |
|  | **Deli** | 0.10 | 0.30 | 0.60 |
|  | **Pizza** | 0.05 | 0.15 | 0.80 |

# 15

# Continuous Probability Models

## Introduction

Consider a situation in which a lumber company is looking at a forest of ponderosa pines that appears ripe for cutting. We can build a mathematical model that could predict approximately how many board feet of lumber the company can cut from those trees.

Consider another situation in which a company claims that only 0.5 percent of the items it produces are defective and not usable. We can build a mathematical model to support or deny the claim. (We will use hypothesis testing.)

Some random variables do not have a discrete range of values. In Chapter 14, we saw examples of discrete random variables and discrete distributions. What if we were looking at time as a random event? Time has a continuous range of values, so, as a continuous random variable, it can be a continuous probability distribution. We define a continuous random variable as any random variable measured on a continuous scale. Other examples include altitude of a plane, the percentage of alcohol in a person's blood, the net weight of a package of frozen chicken wings, or the time to failure of an electric lightbulb. We cannot list the sample space because the sample space is infinite. We need to be able to define a distribution as well as its domain and range.

For any continuous random variable, we can define the cumulative distribution function (CDF) as $F(b) = P(X \le b)$.

The probability density function (PDF) of $f(x)$ is defined to be $P(a \le x \le b) = \int_a^b f(x)\,dx$. To be a valid PDF:

**1.** $f(x)$ must be greater than or equal to zero for all $x$ in its domain, and

**2.** the integral $\int_{-\infty}^{\infty} f(x)\,dx = 1 =$ the area under the entire graph of $f(x)$.

Expected value or average value of a random variable $x$, with PDF defined as above, is defined as

$$E[X] = \int_{-\infty}^{\infty} x \cdot f(x)\,dx$$

In this chapter, we will see some modeling applications using many continuous distributions such as the exponential distribution and the normal distribution.

# 15.1   RELIABILITY REVISITED

You are an undercover police investigator on a stakeout. Your mission is to cover the stakeout with three police officers for at least 24 hours. All necessary meals, equipment, and supplies for the 24-hour period must be carried with your team. The stakeout is ineffective unless your team can communicate with headquarters in a timely manner. Therefore, radio communications must be reliable. The radio has several components that affect its reliability, the most essential being the battery. Batteries have a useful life that is not deterministic (we do not know exactly how long a battery will last when we install it). Its lifetime is a variable that may depend on previous use, manufacturing defects, weather, and so on. The battery that is installed in the radio before it is used on the stakeout could last only a few minutes or for the entire 24 hours. Because communications are so important to this mission, we are interested in modeling and analyzing the battery's reliability.

We will use the following definition from Chapter 14 as indicated below:

**Definition:** If $T$ is the time to failure of a component of a system, and $f(t)$ is the probability distribution function of $T$, then the components' reliability at time $t$ is $R(t) = P(T > t) = 1 - F(t)$, where $R(t)$ is called the *reliability function* and $F(t)$ is the cumulative distribution function of $f(t)$.

A measure of this reliability is the probability that a given battery will last more than 24 hours, R(24). If we know the probability distribution for the battery life, we can use our knowledge of probability theory to determine the reliability. If the battery reliability is below acceptable standards, one solution is to have police officers carry extra batteries. Clearly, the more extra batteries they carry, the less likely there is to be a failure in communications because of batteries. Of course, the battery is not the only component of the radio system. Others include the antenna and the handset. Failure of any one of the essential components causes the system to fail.

This is a relatively simple example of one of many applications of reliability. This chapter will show we can use elementary continuous probability to generate models that can be used to determine the reliability of police equipment.

### Component Reliability

In this section, we will discuss how to model component reliability. Recall that the reliability function, $R(t)$, is defined as $R(t) = P(T > t) = P$ (component fails after time $t$).

This can also be stated, using $T$ as the component failure time, as

$$R(t) = P(T > t) = 1 - P(T \le t) = 1 - \int_{-\infty}^{t} f(x)\,dx = 1 - F(t)$$

Thus, if we know the probability density function, $f(t)$, of the time to failure $T$, we can use probability theory to determine the reliability function, $R(t)$. We normally think of these functions as being time dependent; however, this is not always the case. The function might be discrete—for example, the lifetime of a cell phone because it is dependent on the number of calls through it (a discrete random variable).

A useful probability distribution in reliability is the exponential distribution. Its probability density function is given by:

$$f(t) = \begin{cases} \lambda e^{-\lambda t} & t > 0 \\ 0 & otherwise \end{cases}$$

where the parameter $\lambda$ is such that $\frac{1}{\lambda}$ equals the mean of the random variable $T$. If $T$ denotes the time to failure of a piece of equipment or a system, then $\frac{1}{\lambda}$ is the mean time to failure, which is expressed in units of time. For applications of reliability, we will use the parameter $\lambda$.

Because $\frac{1}{\lambda}$ is the mean time to failure, $\lambda$ is the average number of failures per unit time or the failure rate. For example, if a lightbulb has a time to failure that follows an exponential distribution with a mean time to failure of 50 hours, then its failure rate is 1 lightbulb per 50 hours or 1/50 per hour, so in this case $\lambda = 0.02$ per hour. Note that the mean of the continuous variable, $T$, is the mean time to failure of the component. It is equal to $\frac{1}{\lambda}$.

---

**Example 1**     ## Battery Reliability

Let's consider the example presented in the introduction. Let the random variable $T$ be defined as follows: $T$ = time until a randomly selected battery fails. Suppose radio batteries have a time to failure that is exponentially distributed with a mean of 30 hours. In this case, we could write

$T$ is approximated an exponential distribution with $\lambda = \frac{1}{30}$.

Therefore, $\lambda = \frac{1}{30}$ per hour, so $f(t) = \frac{1}{30} e^{-\frac{t}{30}}, t > 0$ and $F(t) = \int_0^t \frac{1}{30} e^{-\frac{x}{30}} dx$.

$F(t)$, the CDF of the exponential distribution, can be evaluated to obtain $1 - e^{-\frac{t}{30}}, t > 0$. Now we can compute the reliability function for a battery, $R(t) = 1 - F(t) = 1 - (1 - e^{-\frac{t}{30}}) = e^{-\frac{t}{30}}, t > 0$.

Recall that in the earlier example, the police must occupy the stakeout for 24 hours. The reliability of the battery for 24 hours is $R(24) = e^{-\frac{24}{30}} = 0.4493$. Thus, the probability that the battery lasts more than 24 hours is 0.4493. This is less than 50 percent, so we want to take lots of extra batteries.

We can model this easily in Maple without any new commands:

```
> f:=(1/30)·exp(-x/30);
```

$$f := \frac{1}{30} e^{-\frac{1}{30}x}$$

```
> F:=int(f,x=0..t);
```

$$F := 1 - e^{-\frac{1}{30}t}$$

```
> r:=1-F;
```

$$r := e^{-\frac{1}{30}t}$$

```
> evalf(subs(t=24,r));
```

$$0.4493289641$$

---

**Example 2**     ## Reliability of New Rechargeable Batteries

We have the option to purchase a new nickel metal hydride (NiMh) battery for our police stakeout operation. Testing has shown that the distribution of the time to failure can be modeled using a parabolic function:

$$f(x) = \begin{cases} \left(\frac{x}{384}\right) \cdot \left(1 - \frac{x}{48}\right) & 0 \le x \le 48 \\ 0 & otherwise \end{cases}$$

Let the random variable $T$ be defined as follows: $T$ = time until a randomly selected battery fails. In this case, we could write $f(t) = \left(\frac{t}{384}\right) \cdot \left(1 - \frac{t}{48}\right)$, $0 \le t \le 48$ and $F(t) = \int_0^t \left(\frac{x}{384}\right) \cdot \left(1 - \frac{x}{48}\right) dx$.

Recall that in the earlier example, the police must cover the stakeout for 24 hours. The reliability of the battery for 24 hours is therefore

$$R(24) = 1 - F(24) = 1 - \int_0^{24} \left(\frac{t}{384}\right) \cdot \left(1 - \frac{t}{48}\right) dt = 0.500,$$

This is an improvement over the batteries from Example 1. Thus, we would recommend the new battery.

$> f := \left(\frac{x}{384}\right) \cdot \left(1 - \frac{x}{48}\right);$

$$f := \frac{1}{384}\, x \left(1 - \frac{1}{48}\, x\right)$$

$> F := int(f, x = 0..24);$

$$F := \frac{1}{2}$$

$> r := 1 - F;$

$$r := \frac{1}{2}$$

---

## Models of Large-Scale Systems

In our discussion of reliability in Chapter 14, we discussed series systems, active redundant systems, and standby redundant systems. Unfortunately, things are not always this simple. The system types just listed often appear as subsystems in larger arrangements of components that we shall call *large-scale systems*. Fortunately, if you know how to deal with series systems, active redundant systems, and standby redundant systems, finding system reliabilities for large-scale systems is easy. Consider the following example.

The first and most important step in developing a model to analyze a large-scale system is to draw a picture. Consider the network in Figure 15.1. Subsystem A is the standby redundant system of three components (each with a failure rate of five per year) with the decision switch on the left of the figure. Subsystem $B_1$ is the active redundant system of three components (each with a failure rate of three per year) in which at least two of the three components must be working for the subsystem to work. Subsystem $B_2$ is the two-component serial system in the lower right. We define subsystem B as being subsystems $B_1$ and $B_2$ together. We assume all components have exponentially distributed times to failure with failure rates as shown in Figure 15.1.

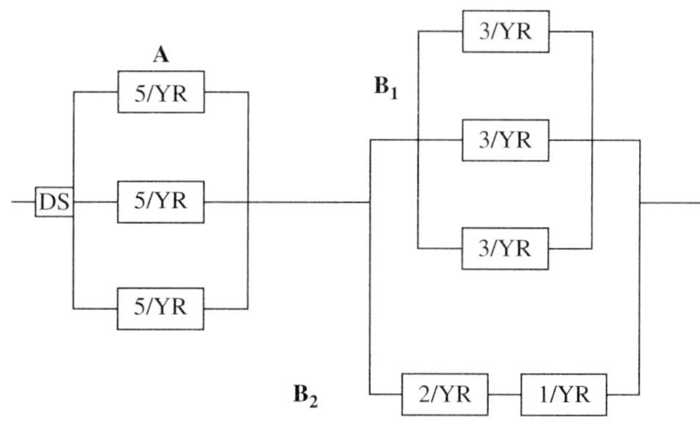


**FIGURE 15.1**
System B

## Example 3     A System Network

Suppose we want to know the reliability of the whole system for six months. Observe that you already know how to compute the reliabilities for subsystems A, $B_1$, and $B_2$. Let's review these computations and then see how we can use them to simplify our problem.

Subsystem A is a standby redundant system, so we will use the Poisson model (see Chapter 14). We let $X$ = the number of components that fail in one year.

Because six months is 0.5 years, we seek $R_A(0.5) = p(x < 3)$, where $X$ follows a Poisson distribution with parameter $\lambda = at = (5)(0.5) = 2.5$. Then,

$$R_A(0.5) = p(X < 3) = p(0 \le x \le 2) = 0.5438$$

Now we consider subsystem $B_1$. In our previous sections, we learned how to find individual component reliabilities when the time to failure followed an exponential distribution. For subsystem $B_1$, the failure rate is three per year, so our individual component reliability is

$$R(0.5) = 1 - F(0.5) = 1 - (1 - e^{-(3)(0.5)}) = e^{-(3)(0.5)} = 0.2231$$

Now recall that subsystem $B_1$ is an active redundant system in which two of three components must work for the subsystem to work. If we let $Y$ = the number of components that function for six months and recognize that Y follows a binomial distribution with $n = 3$ and $p(\text{success}) = 0.2231$, we can quickly compute the reliability of the subsystem $B_1$ as follows:

$$R_{B_1}(0.5) = p(Y \ge 2) = 1 - p(Y < 2) = 1 - p(Y \le 1) = 1 - 0.8729 = 0.1271$$

Finally, we can look at subsystem $B_2$. Again we use the fact the failure times follow an exponential distribution. The subsystem consists of two components; obviously, they both need to work for the subsystem to work. The first component's reliability is

$$R(0.5) = 1 - F(0.5) = e^{-(2)(0.5)} = 0.3679$$

and for the other component the reliability is $R(0.5) = 1 - F(0.5) = e^{-(1)(0.5)} = 0.6065$. Therefore, the reliability of the subsystem is $R_{B_2}(0.5) = (0.3679) \cdot (0.6065) = 0.2231$.

Our overall system can now be drawn as shown in Figure 15.2.

## Example 4     A Simplified Network

From here we determine the reliability of subsystem B by treating it as a system of two independent components in parallel in which only one component must work. Therefore,

$$R_B = 0.1271 + 0.2231 - (.1271) \cdot (0.2231) = 0.3218$$

Finally, because subsystems A and B are in series, we can find the overall system reliability for six months by taking the product of the two subsystem reliabilities,

$$R_{system}(0.5) = (0.5438) \cdot (0.3218) = 0.1750$$

**FIGURE 15.2**
Simplified system after reduction

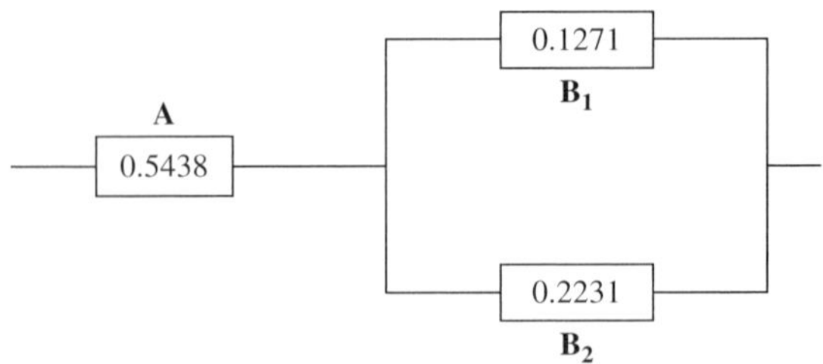

We have used a network-reduction approach to determine the reliability for a large-scale system for a given time period. Starting with those subsystems, which consist of components independent of other subsystems, we reduced the size of our network by individually evaluating each subsystem's reliability. This approach works for any large-scale network consisting of basic subsystems of the types we have studied (series, active redundant, and standby redundant).

We have seen how methods from elementary probability can be used to model reliability problems. The modeling approach presented here is useful in helping students simultaneously improve their understanding of both the modeling problems addressed and the mathematics behind these problems.

## 15.1 | EXERCISES

1. A continuous random variable, $Y$, representing the time to failure of a soda-bottling machine, has a probability density function given by

$$f(y) = \begin{cases} \dfrac{1}{3} e^{-\frac{y}{3}} & y > 0 \\ 0 & otherwise \end{cases}.$$

   **a.** Find the reliability function for $Y$.
   **b.** Find the reliability for 1.2 time periods, $R(1.2)$.

2. The lifetime of a generator engine (measured in time of operation) is exponentially distributed with a mean time to failure (MTTF) of 400 hours. You have received a mission that requires 12 hours of continuous operation after a major storm. Your maintenance book indicates that the generator engine has been operating for 158 hours.
   **a.** Find the reliability of your engine for this mission.
   **b.** If your vehicle's engine had operated for 250 hours before the mission, what is the reliability for the mission?

3. You are a project manager for a new system being developed in Huntsville, Alabama. A critical subsystem has two components arranged in a parallel configuration. You have told the contractor that you require this subsystem to be at least 0.995 percent reliable. One subsystem came from an older system and has a known reliability of 0.95. What is the minimum reliability of the other component so that we meet our specifications?

4. You are in a national guard reserve battalion that has a mission to observe several planned night operations. There is some concern about the reliability of the lighting system for the FEMA operations center. The lights are powered by a 1.5-kW generator that has a MTTF of 7.5 hours.
   **a.** Find the reliability of the generator for 10 hours if the generator's reliability is exponential.
   **b.** Find the reliability of the power system if two other identical 1.5-kW generators are available. First consider the system as active redundant and then as standby redundant. Which would improve reliability the most?
   **c.** How many generators would be necessary to ensure 99-percent reliability?

5. Consider the air-traffic-control system for a small airport in the following figure; the reliability for each component is indicated. Assume all components are independent and the radars are active redundant.

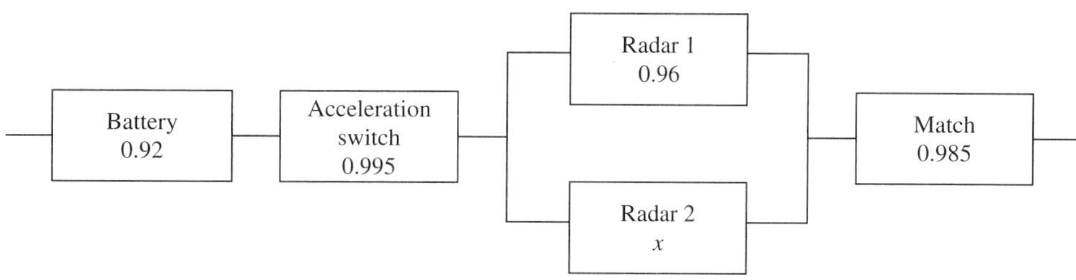

**a.** Find the system reliability for six months when $x = 0.96$.
**b.** Find the system reliability for six months when $x = 0.939$.

# 15.1 | PROJECTS

1. Math in space: The U.S. space program is very concerned about the reliability of its space shuttle. By system "reliability," we mean the probability that the system will meet a set of specifications for a given period of time. The failure time of a system is the length of time that the system works according to specification. The failure time distribution for a system is the density function for the failure time $t$. If we denote the failure time density by the term $f(t)$, then the probability that the system will fail before time $t_0$ is defined by the integral:

$$F(t_0) = \int_0^{t_0} f(t)\, dt$$

Reliability is then defined as $R(t_0) = 1 - F(t_0)$, where $F(t_0)$ is the probability that the system fails within time $t$ as previously defined.

Several alternative shuttle crafts are being purposed for building by Boeing Corporation under a Department of Defense contract. Your consultant team has been hired to examine these alternatives and make recommendations to NASA.

Your math skills are critical because you need to verify the distribution assumptions as well as find reliabilities to make your recommendations to NASA.

Alternative 1 was built using a Raleigh failure rate. The Raleigh distribution is given by

$$f(x) = \begin{cases} 2\alpha x e^{-\alpha x^2} & x > 0 \\ 0 & otherwise \end{cases}$$

where $\alpha > 0$.
Alternative 2 was built using an exponential failure rate defined by

$$f(t) = \begin{cases} c e^{-ct} & t > 0 \\ 0 & otherwise \end{cases}$$

where $c > 0$.
Alternative 3 was built using a Pareto distribution given by

$$f(x) = \begin{cases} \dfrac{\delta}{x^{\delta+1}} & x > 1 \\ 0 & otherwise \end{cases}$$

where $\delta > 0$.
Requirements

**a.** To validate the assumptions that each distribution is indeed a valid probability function, show that each failure rate follows the definition of a probability distribution, $\int_{-\infty}^{\infty} f(x)dx = 1$. Use improper integrals. Also show the values for the constants $\{\alpha, c, \delta\}$ for which the distributions converge.

**b.** Given the following estimated constants by Boeing, recommend the alternative that should be selected for the new space shuttle based on it lasting 250 hours:

$$\text{Alternative 1} \qquad \alpha = 0.000016$$
$$\text{Alternative 2} \qquad c = 0.0001$$
$$\text{Alternative 3} \qquad \delta = 0.0005$$

As a minimum, include the following in your report: (1) plots of the failure rates with the estimated constants, and (2) the expected MTTF value of alternative 2. Show all calculations that support your recommendations, and make a short list of possible recommendations for NASA in future operations.

## 15.2 MODELING USING THE NORMAL DISTRIBUTION

A continuous random variable $X$ is said to have a normal distribution with parameters $\mu$ and $\sigma$ (or $\mu$ and $\sigma^2$), where $-\infty < \mu < \infty$ and $\sigma > 0$, if the probability density function (PDF) of $X$ is

$$f(x; \mu, \sigma) = \frac{1}{\sqrt{2\pi}\sigma} e^{\frac{-(x-\mu)^2}{(2\sigma^2)}}, \quad -\infty \le x \le \infty$$

The plot of the normal distribution is our bell-shaped curve (see Figure 15.3).

To compute $P(a < x < b)$ when $X$ is a normal random variable, with parameters $\mu$ and $\sigma$, we must evaluate $\displaystyle\int_a^b \frac{1}{\sqrt{2\pi}\sigma} e^{\frac{-(x-\mu)^2}{(2\sigma^2)}}\, dx.$

Because none of the standard integration techniques can be used to evaluate this integral, the standard normal random variable $Z$ with parameters $\mu = 0$ and $\sigma = 1$ has been numerically evaluated and tabulated for certain values. Because most applied problems do not have parameters of $\mu = 0$ and $\sigma = 1$, a "standardizing" transformation can be used: $Z = \dfrac{x - \mu}{\sigma}$. Many statistics textbooks illustrate how to convert the random variable $X$ to the standard normal $Z$. In Maple, we can numerically evaluate $\displaystyle\int_a^b \frac{1}{\sqrt{2\pi}\sigma} e^{\frac{-(x-\mu)^2}{(2\sigma^2)}}\, dx.$

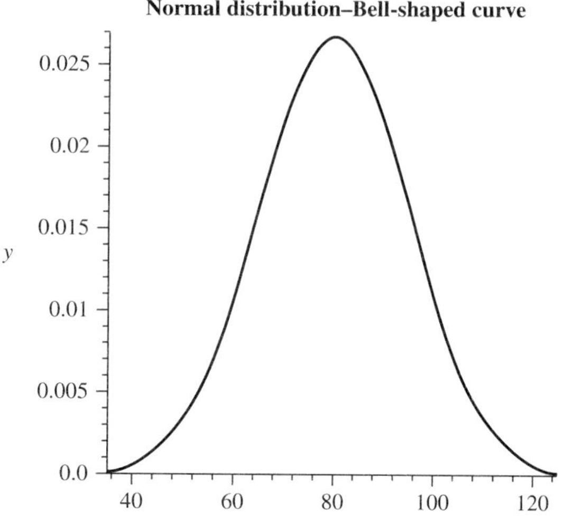


**FIGURE 15.3**
Bell-shaped curve of the normal distribution

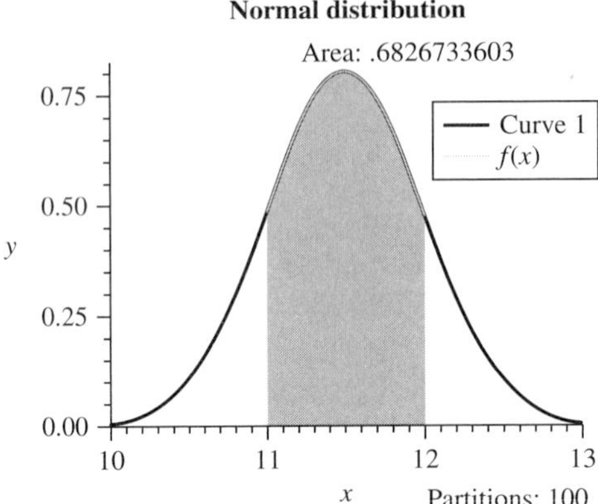


**FIGURE 15.4**
Normal distribution area from 11 to 12

For example, the amount of fluid dispensed into a can of diet soda by a filling machine is approximately a normal random variable with mean 11.5 fluid ounces and a standard deviation of 0.5 fluid ounces. We want to determine the probability that between 11 and 12 fluid ounces, $p(11 < x < 12)$, are dispensed:

$$Z_1 = 11 - 11.5/.5 = -1$$
$$Z_2 = 12 - 11.5/.5 = 1$$

This probability statement $p(11 < x < 12)$ is equivalent to $p(-1 < Z < 1)$. If we used the standard normal tables from a statistics book, we can compute this to be $0.8413 - 0.1587 = 0.6826$. However, we can use Maple to compute the area between 11 and 12. Because Maple does not have a stand-alone normal distribution command, the following program was created to calculate the area and produce a partially shaded bell curve (see Figure 15.4).

```
> normprob:=proc(mu,sigma,a,b)
> u:=mu:s:=sigma:aa:=a:bb:=b:
> #produce normal distribution
> nd:=exp(-(x-u)^2/(2*s^2))/(s*sqrt(2*Pi));
> p3d:=u+3*s:n3d:=u-3*s:
> with(Student[Calculus1]):
> prob:=int(nd,x=aa..bb);
> print('Probability is',prob);
> curve:=plot(nd,x=n3d..p3d,thickness=3,title='Normal_Distribution'):
> nc:=RiemannSum(nd,x=aa..bb,method=left,output=plot,partition=100):
> display(curve,nc);
> end;
> normprob(11.5,.5,11,12);
```

*Probability is*, 0.6826894921

Therefore, 68.267 percent of the time, the cans are filled with between 11 and 12 fluid ounces of soda.

In Maple, we did not need to use the standard normal probability tables. Maple evaluates the integral of the normal function from $a$ to $b$.

The central limit theorem is one of the most important theorems in probability. It states that if $X_1, X_2, \ldots X_n$ are a random sample from a distribution with a mean $\mu$ and a standard deviation $\sigma$ and $n$ is sufficiently large ($n > 30$) then the distribution of the average $\overline{X}$ or the total (T) have normal distributions with the following parameters:

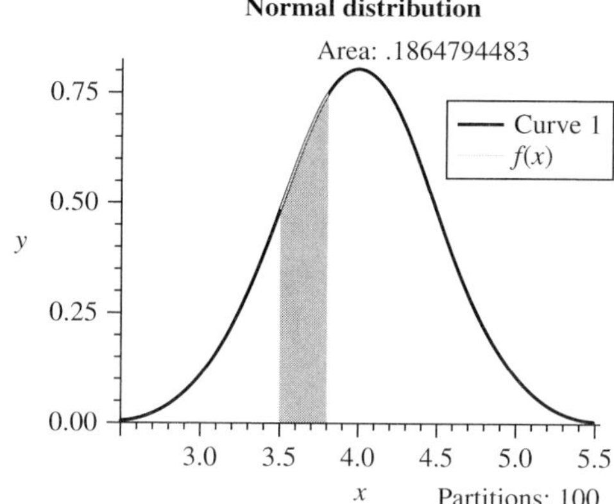


$$\mu_{\overline{X}} = \mu$$
$$\sigma_{\overline{X}}^{2} = \frac{\sigma^{2}}{n}$$
$$\mu_{T} = n \cdot \mu$$
$$\sigma_{T}^{2} = n \cdot \sigma$$

For example, when a batch of a certain pharmaceutical is prepared, the amount of aloe, a natural substance, is a random variable with mean value 4.0 grams and standard deviation 0.35 grams. If 50 batches are prepared, what is the probability that the sample average of aloe is between 3.5 and 3.8 grams?

Because $n = 50$ ($n > 30$), the aloe sample average random variable, $\overline{X}$, follows a normal distribution with mean 4.0 and standard deviation $\frac{0.35}{\sqrt{50}} = 0.4950$ (see Figure 15.5).

$$P(3.5 < \overline{X} < 3.8) = P\left(\frac{(3.5 - 4.0)}{0.4950} < Z < \frac{(3.8 - 4)}{0.4950}\right) = 0.1869$$

Using our Maple commands,

```
> normprob(4, .4950, 3.5, 3.8);
```

*Probability is,* 0.1868680548

The normal distribution and the central limit (when applicable) are used in many applications of confidence intervals and hypothesis testing.

## 15.2 | EXERCISES

Find the following probabilities:

1. $X \sim N(\mu = 10, \sigma = 2)$, $P(X > 6)$

2. $X \sim N(\mu = 10, \sigma = 2)$, $P(6 < x < 14)$

3. Determine the probability that lies within one standard deviation of the mean, two standard deviations of the mean, and three standard deviations of the mean. Draw a sketch of each region.

4. A tire manufacturer thinks that the amount of wear per normal driving year of the rubber used in its tires follows a normal distribution with mean 0.05 inches and standard deviation 0.05 inches. If 0.10 inches is considered dangerous, then determine the probability that $P(X > 0.10)$.

## 15.3 CONFIDENCE INTERVALS AND HYPOTHESIS TESTING

The basic concepts and properties of confidence intervals involve initially understanding and using two assumptions:

1. The population distribution is normal.

2. The standard deviation $\sigma$ is known or can be easily estimated.

In its simplest form, we are trying to find a region for $\mu$ (and thus a confidence interval) that will contain the value of the true parameter of interest. The formula for finding the confidence interval for an unknown population mean from a sample is $\overline{X} \pm Z_{\frac{\alpha}{2}} \frac{\sigma}{\sqrt{n}}$.

The value of $Z_{\frac{\alpha}{2}}$ is computed from the normality assumption and the level of confidence, $1 - \alpha$, desired.

Let's consider a variation of the diet soda example in the previous section. For example, the amount of fluid dispensed into a can of diet soda is approximately a normal random variable with unknown mean fluid ounces and a standard deviation of 0.5 fluid ounces. We want to determine a 95-percent confidence interval for the true mean. A sample of 36 diet sodas was taken, and we found a sample mean of $\overline{x} = 11.35$.

Now, we want a 95-percent confidence interval, so $1 - \alpha = 0.95$. Therefore, $\alpha = 0.05$, and because there are two tails, we need each tail to contain only 0.025 probability. We need $\frac{\alpha}{2} = 0.025$. We use

$$\overline{X} \pm Z_{\frac{\alpha}{2}} \frac{\sigma}{\sqrt{n}}$$

with $\overline{x} = 11.35$, $\sigma = .5$, $Z_{\frac{\alpha}{2}} = 1.96$, and $n = 36$.

> CI:=**proc**(*xbar*,sigma *n*,alpha)*z:=evalf*

$$\left( solve\left( int\left( \frac{1}{\text{sqrt}(2 . \pi)} \cdot \exp\left( -\frac{z^2}{2} \right), z = -5..ep \right) = 1 - \frac{\alpha}{2}, ep \right) \right);$$

$$leftx:=xbar - \frac{z \cdot sigma}{\text{sqrt}(n)}; xright:=xbar + \frac{z \cdot sigma}{\text{sqrt}(n)}; printf$$

("Z=%axbar=%a sigma=%a N=%a CI=%q\n,"*z,xbar,*sigma *n leftx,xright*)**end proc;**

```
CI := proc(xbar, sigma n, alpha)
    local z, leftx, xright;
    z := evalf(solve(int(exp(-1/2*z^2)/sqrt(2*Pi), z = -5..ep)
        = 1 - 1/2* alpha, ep));
    leftx := xbar - z* sigma/sqrt(n);
    xright := xbar + z*sigma/sqrt(n);
    printf ("Z = %a xbar = %a sigma = %a N = %a CI = %q ", z, xbar,
    sigma, n, leftx, xright) end proc
```

> *CI(11.35,.5,36,.05);*

Z = 1.959968889 xbar = 11.35 sigma = .5 N = 36 CI = 11.18666926, 11.51333074

This is seen in Figure 15.6.

Our confidence interval for the parameter, $\mu$, is $11.35 \pm 1.96 \cdot \frac{0.5}{\sqrt{36}}$ or from (11.18666, 11.51333).

Let's interpret this confidence interval. If we took 100 experiments of 36 random samples each and calculated the 100 confidence intervals in the same manner: $\overline{X} \pm Z_{\frac{\alpha}{2}} \frac{\sigma}{\sqrt{n}}$.

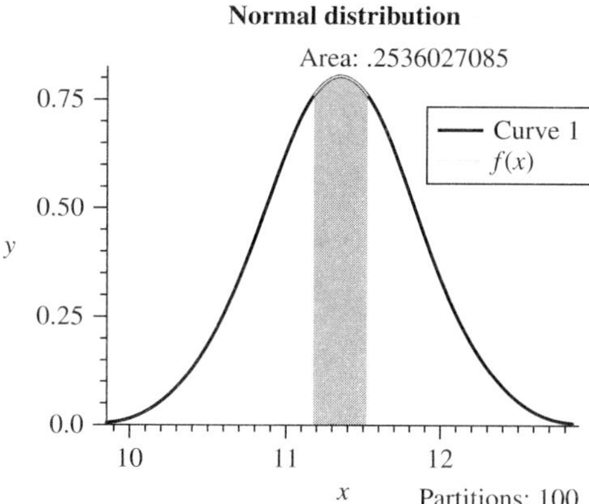



Thus, 95 of the 100 confidence intervals would contain the true mean, $\mu$. We do not know which of the 95 confidence intervals contains the true mean. Thus, to a modeler, each confidence interval built will either contain or not contain the true mean.

**Simple Hypothesis Testing**

A more powerful technique for inferring information about a parameter is a *hypothesis test*. A statistical hypothesis test is a claim about a single population characteristic or about values of several population characteristics. There is a *null hypothesis* (which is the claim initially favored or believed to be true), which is denoted as $H_0$. The other hypothesis, the alternate hypothesis, is denoted as $H_a$. We will always keep equality with the null hypothesis, see the Cases 1, 2, and 3 below. The objective is to decide, based on sample information, which of the two claims is correct. Typical hypothesis tests can be categorized by three cases listed below and seen visually in Figure 15.7.

$$
\begin{array}{llllll}
\text{Case 1} & H_0: & \mu = \mu_0 & \text{versus} & H_a: & \mu \neq \mu_0 \\
\text{Case 2} & H_0: & \mu \leq \mu_0 & \text{versus} & H_a: & \mu > \mu_0 \\
\text{Case 3} & H_0 & \mu \geq \mu_0 & \text{versus} & H_a: & \mu < \mu_0
\end{array}
$$

Two types of errors can be made in hypothesis testing: type 1 errors (called $\alpha$ errors) and Type II errors (called $\beta$ errors). It is important to understand these. Consider the information in the following table.



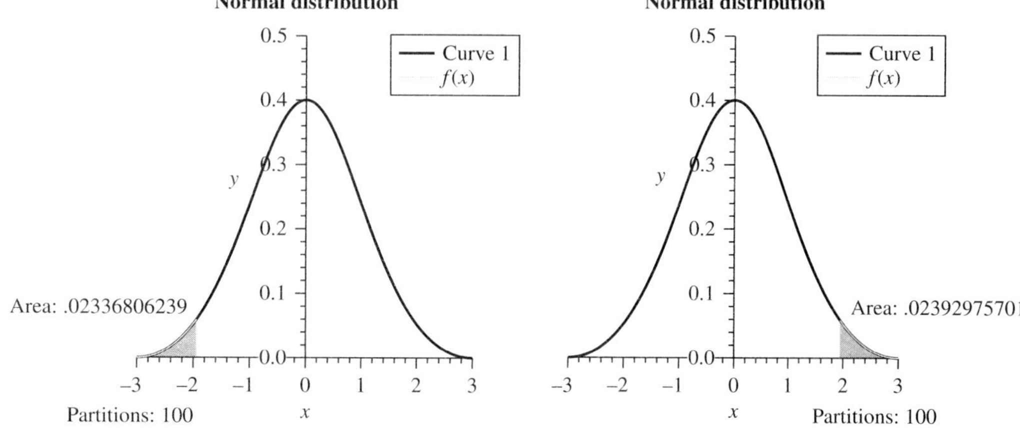

**FIGURE 15.7**
(*continued*)

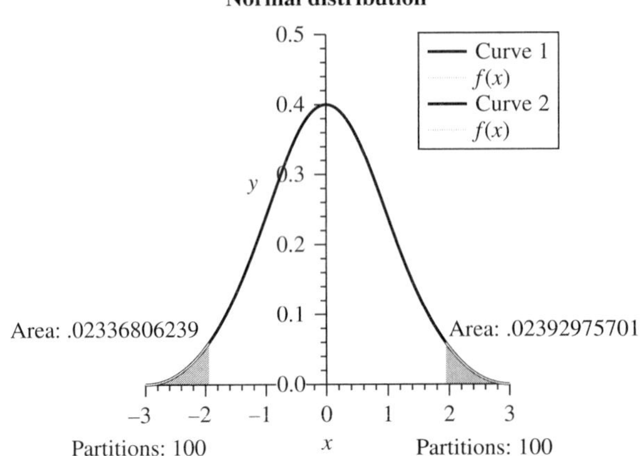


State of Nature

| | | $H_0$ **True** | $H_a$ **True** |
|---|---|---|---|
| **Test conclusion** | Fail to reject $H_0$ | $1 - \alpha$ | $\beta$ |
| | Reject $H_0$ | $\alpha$ | $1 - \beta$ |

Some important facts about both $\alpha$ and $\beta$:

1. $\alpha = P(\text{reject } H_0 \mid H_0 \text{ is true}) = P(\text{Type I error})$.
2. $\beta = P(\text{fail to reject } H_0 \mid H_0 \text{ is false}) = P(\text{Type II error})$.
3. $\alpha$ is the *level of significance* of the test.
4. $1 - \beta$ is the *power* of the test.

Thus, referring to the table, we would like $\alpha$ to be small because it is the probability that we reject $H_0$ when $H_0$ is true. We would also want $1 - \beta$ to be large because it represents the probability that we reject $H_0$ when $H_0$ is false. Part of the modeling process is to determine which of these errors is the most costly and then work to control that error as the primary error of interest.

The following template is provided for hypothesis testing:

**Step 1**    Identify the parameter of interest.

**Step 2**    Determine the null hypothesis, $H_0$.

**Step 3**    State the alternative hypothesis, $H_a$.

**Step 4**    Give the formula for the test statistic based on the assumptions that are satisfied.

**Step 5**    State the rejection criteria based on the value of $\alpha$.

**Step 6**    Obtain your sample data and substitute into your test statistic.

**Step 7**    Determine the region in which your test statistics lies (rejection region or fail-to-reject region).

**Step 8**    Make your statistical conclusion. Your choices are to either reject the null hypothesis or fail to reject the null hypothesis. Ensure that the conclusion is scenario oriented.

| **Example 1** | ## Hypothesis Testing for a Small Aviation Company's Crews |
|---|---|

You run a small aviation transport company for a major corporation. You are tired of hearing management complain that your flight crews rest too much during the day. Aviation rules require a crew to get about nine hours of rest each day. You collect a sample of 37 crew members and determine that their sample average, $\bar{x}$, is 8.94 hours, with a sample deviation of 0.2 hours.

The parameter of interest is the true population mean, $\mu$.

$$H_0: \mu \geq 9$$
$$H_a: \mu < 9$$

The test statistic is $Z = \dfrac{\bar{x} - \mu}{s/\sqrt{n}}$. This is Case 3, and it is a one-tailed test.

We select $\alpha$ to be 0.05.
We reject $H_0$ at $\alpha = 0.05$ if $Z < -1.645$.

From our sample of 36 aviators, we find

$$Z = \frac{\bar{x} - \mu}{s/\sqrt{n}} = \frac{8.94 - 9}{.2/\sqrt{36}} = -0.06(6)/.2 = -1.8. \ Z = -1.8$$

Because $-1.8 < -1.645$, we reject the null hypothesis that aviators rest nine or more hours per day and conclude the alternate hypothesis is true: Your aviators rest fewer than nine hours per day. Rejecting the null hypothesis is the better strategy. Our conclusion is that we reject the null hypothesis that the aviator crews rest nine or more hours a day, so the crews do not rest nine or more hours per day.

```
> hypothesis_test:=proc(test,realmean,xbar,n,sigma,alpha)
> if test=1 then
> z:=evalf((xbar-realmean)/(sigma/sqrt(n)));
> z2side:=evalf(solve((1/sqrt(2*Pi))*exp(-(z1^2)/2)=alpha,z1));;
> if abs(z)>abs(z2side[1])then
> print('Reject H0');
> else print('Fail to Reject')
> end if;end if;
> if test=2 then
> z:=evalf((xbar-realmean)/(sigma/sqrt(n)));
> z2side:=evalf(solve((1/sqrt(2*Pi))*exp(-(z1^2)/2)=alpha,z1));;
> if abs(z)>abs(z2side[1])then
> print('Reject H0');
> else print('Fail to Reject')
> end if;end if;
> if test=3 then
> z:=evalf((xbar-realmean)/(sigma/sqrt(n)));
> z2side:=evalf(solve((1/sqrt(2*Pi))*exp(-(z1^2)/2)=.01,z1));
> if(z)<(z2side[1])then
> print('Reject H0');
> else print('Fail to Reject')
> end if; end if;
> print('Z statistic is',z,'Z_alpha is',z2side);
> end;

> hypothesis_test(2,9,8.94,36,.2,.05);
```

*Fail to Reject*

*Z statistic is, $-1.800000000$, Z_alpha is, $-2.038035201$, $2.038035201$*

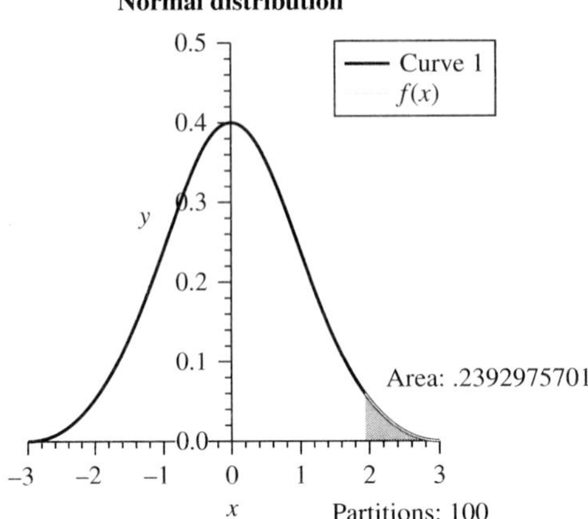

**FIGURE 15.8**
Type I error



Given our hypothesis test above, the probability of a Type I error, $\alpha$, is the area under the normal bell-shaped curve centered at $\mu_0$ and corresponding to the rejection region (see Figure 15.8). This value is 0.05.

You must know about or suspect another choice of $\mu$ to find this Type II error. Let us continue our example. Assume the new true mean is 8.87 hours of crew rest, and let's find the probability of Type II error.

$$\beta = p(\text{Fail to Reject } H_0 \mid H_0 \text{ is false})$$

$$\beta = p(\overline{X} > \overline{X}_{cr} \mid \mu = 8.87) = p\left(z > \frac{\overline{x} - \mu}{s/\sqrt{n}}\right) = p\left(z > \frac{8.94 - 8.87}{.2/6}\right)$$

$$= 1 - p(z < 2.1) = 1 - 0.9821 = 0.01786$$

This means that more than 1 percent of the time we will fail to reject $H_0$ when $\mu$ really equals 8.87 and not 9.0.

---

## 15.3 │ EXERCISES

Discuss how to set up each of the following as a hypothesis test.

1. Does drinking coffee increase the risk of getting cancer?

2. Does taking aspirin every day reduce the chance of a heart attack?

3. Which of two gauges is more accurate?

4. Why is a person "innocent until proven guilty"?

5. Is the drinking water safe to drink?

6. Set up a fake trial for a suspected felon. Build a matrix for his or her innocence or guilt with an appropriate null hypothesis. Which error, Type I or Type II, is the worst error?

## 15.3 │ PROJECTS

1. Numerous complaints have been made that a certain hot coffee machine is not dispensing enough coffee into each cup. The vendor claims that on average the machine dispenses at least 8 ounces of coffee per cup. You take a random sample of 36 hot drinks and calculate the mean to be 7.65 ounces, with a standard deviation of 1.05 ounces. Find a 95-percent confidence interval for the true mean. Next, find some statistical software (or a calculator) and take 30 random samples of size 36 from a normal

distribution whose mean is 8 ounces and standard deviation is $\dfrac{1.05}{\sqrt{30}}$. Plot these intervals and determine how many would contain the true mean, $\mu = 8$ ounces.

2. Use the information provided in problem 1 above and set up and conduct a hypothesis test to determine if the vendor's claim is correct. Use an $\alpha = .05$ level of significance. Determine the Type II error if the true mean were 7.65 ounces.

## 15.4  REGRESSION: LINEAR, TRANSFORMED, AND NONLINEAR

In several chapters, we have discussed the criteria for fitting models to collected data. We have considered a single observation $y_i$ for each value of the independent variable $x_i$. Now, we examine several other regression models. In this section, we explore a statistical methodology for minimizing the sum of squared deviations: linear regression. Our objectives in this section are to:

1. illustrate the basic linear regression model and its assumptions,

2. illustrate the statistics $r$ and $R^2$ and provide some rules for interpretation,

3. illustrate a graphical interpretation of the fit of the linear regression model through examining and interpreting residual plots,

4. revisit the transformed regression model, and

5. introduce nonlinear regression models.

We plan to introduce only basic concepts and interpretations. More in-depth studies of linear regression are gained in advanced statistics courses. In this section, we present both the basic Maple commands to get the model and graphs and more sophisticated output for the advanced student. Only those versed in advanced statistics should study the detailed output of the statistical modeling  programs.

The basic linear regression model is $y_i = \beta_0 + \beta_1 x_i$ for $I = 1, 2, \ldots , m$ data points.

In Chapter 4, we provided Equations (4.4), (4.5), and (4.6) to build the normal equations and find the coefficient of the slope $\beta_1$ and the intercept of $\beta_0$. We will now add a few more equations that will enable us to obtain a few basic statistics to aid in the analysis. Additionally, we define DF as degrees of freedom and a use an $F$ statistic, both these are more advanced statistical terms. These equations are provided in Table 15.1.

A measure of fit is $R^2$, the coefficient of determination. $R^2$ is found by $R^2 = 1 - \dfrac{SSE}{SST}$.

The formulas for sum of squared error (SSE) and sum of squares total (SST) are provided in Table 15.1. The square root of $R^2$ is called $r$, which measures the linear relationship between the variables $y$ and $x$. The following are properties of $r$:

1. The value of $r$ does not depend on which of the two variables is labeled $x$ or $y$.

2. The value of $r$ is independent of the units of $x$ and $y$.

**Table 15.1**  Analysis of Variance

| Source | DF (degrees of freedom) | Sum of Squares | Mean Square Error | $F$ Ratio |
|---|---|---|---|---|
| Regression | 1 | SSR = SST − SSE | MSR = SSR/1 | $F$ = MSR/MSE |
| Error | $n - 2$ | $SSE = \sum_{i=1}^{m} [y_i - (\beta_0 + \beta_1 x_i)]^2$ | MSE = SSE/$(n - 2)$ | |
| Total | $n - 1$ | $SST = \sum_{i=1}^{m} (y_i - \bar{y})^2$ | | |

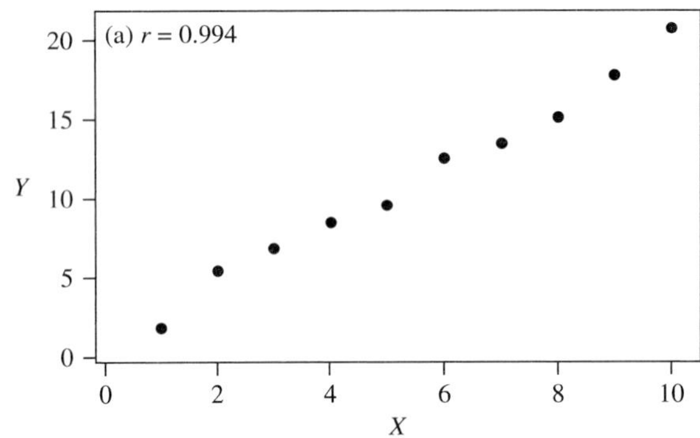





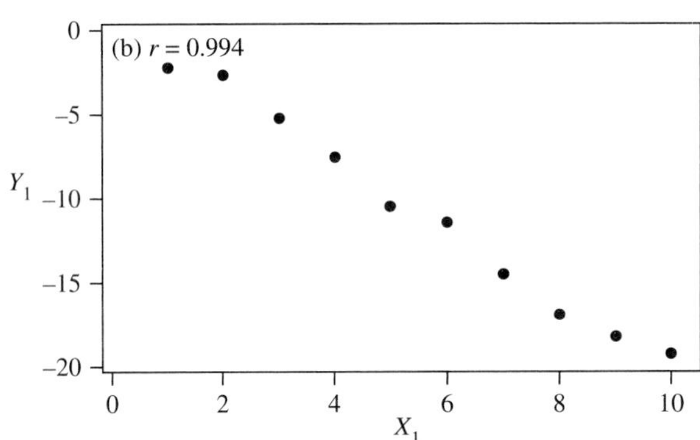



3. $-1 \leq r \leq 1$. A value of $r = -1$ corresponds to data that has a graph that is "exactly" a line with negative slope. Similarly, $r = 1$ represents a "perfectly" linear graph with positive slope. These are illustrated in Figures 15.9 and 15.10.

4. $R^2 = (r)^2$

    **a.** $r = 0.994$ (see Figure 15.10).

    **b.** $r = -0.994$

Another indicator of the reasonableness of the fit is the plot of the residuals versus the independent variable. Recall that residuals are the errors and are defined using the data and Equation (15.1) as follows: $e = (y_i - f(x_i))$.

If we plot the errors versus the independent variable, the plot will provide us with some valuable information.

1. The residuals should be randomly distributed and contained in a reasonable small band that is commensurate with the accuracy of the data.

2. An extremely large residual warrants further investigation of that data point in question to discover the cause of the large residual.

3. A pattern or trend in the residuals indicates that a predicable effect remains to be modeled, and the nature of the pattern gives clues on how to refine the model, if refinement is warranted. These are illustrated in Figure 15.11.

We can also have a linear pattern that appears as a band that is either increasing or decreasing.

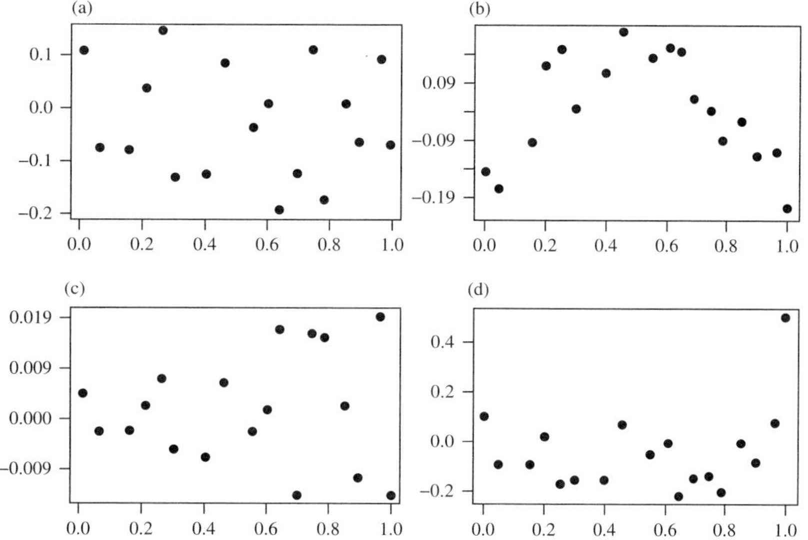



**FIGURE 15.11**
Residual patterns: (a) no pattern, (b) curved pattern, (c) Fanning pattern, (d) possible outlier

## Example 1

## Ponderosa Pines

The ponderosa pine data are displayed in the following Maple output. Figure 15.12 illustrates the graphical trends in the data. The plot is concave upward and increasing, suggesting that it is a power function model.

$$diameter := [36, 28, 28, 41, 19, 32, 22, 38, 25, 17, 31, 20, 25, 19, 39, 33, 17, 37, 23, 39]$$

$$board\_feet := [192, 113, 88, 294, 28, 123, 51, 252, 56, 20, 141, 32, 86, 21, 231, 187, 22, 205, 57, 265]$$

### Problem Identification

Predict the number of board feet as a function of the diameter of the ponderosa pine.

### Assumptions

We assume that the ponderosa pines are geometrically similar in the shape of a right circular cylinder. This allows us to use a *characteristic dimension*—diameter—to predict the volume. (Board feet are proportional to volume.) We can reasonably assume the height is proportional to diameter.

**FIGURE 15.12**
Scatterplot of ponderosa pine data

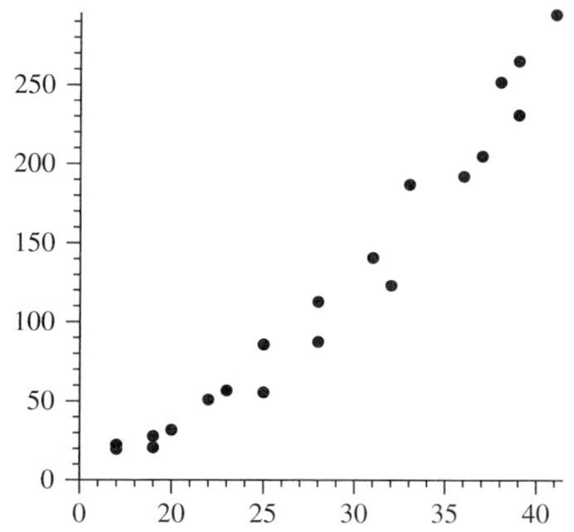

## Model Formulation

We can immediately obtain the model $V \propto d^3$.

Instead if we assume that the ponderosa pines have a constant height rather than the height being proportional to the diameter, then we can obtain that $V \propto d^2$.

Using these as basic model forms, we expand the use of the definition to a complete quadratic and a complete cubic model as a comparison:

$$V = \beta_0 + \beta_1 d_i + \beta_2 d_i^2 + \beta_3 d_i^3 \text{ and } V = A_0 + A_1 d_i + A_2 d_i^2$$

## Model Solution

The following solutions were obtained by using *Fit ('equation', diameter, volume, x)* command in Maple.

$> Model1:=Fit(a*x^2, diameter, board\_feet, x);$

$$Model1 := 0.15238186x^2.$$

$> Model2:=Fit(a*x^2+b*x+c, diameter, board\_feet, x);$

$$Model2 := 0.2970327x^2 - 6.034394x + 35.859265$$

$> Model3:=Fit(a*x^3+b*x^2+c*x+d, diameter, board\_feet, x);$

$$Model3 := 0.002908x^3 + 0.04743314x^2 + 0.81522984x - 23.9486297$$

The residuals are calculated using the difference between the given data and the model. We examine the residual plots in Figure 15.12. Recall we are looking for a random distribution of the residuals with no apparent pattern. Note the apparent trend in the errors corresponding to the model $V = 0.152d^2$. We would probably refine this model based on this plot shown in figure 15.13. The other models appear reasonable.

**FIGURE 15.13**

Residual plot for $y = 0.152\, x^2$ shows trend

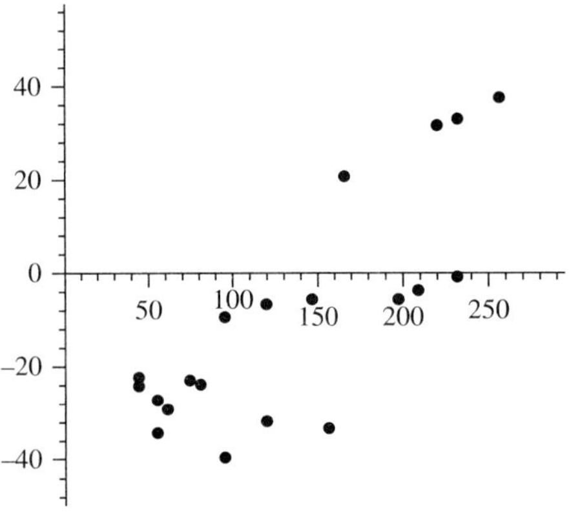

**FIGURE 15.14**

Residual plot shows no trend for the complete quadratic model

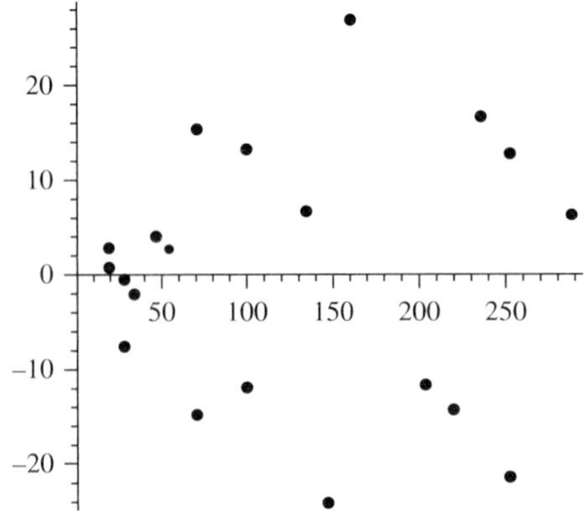

**FIGURE 15.15**
Residual plot shows no trend
for the complete cubic model

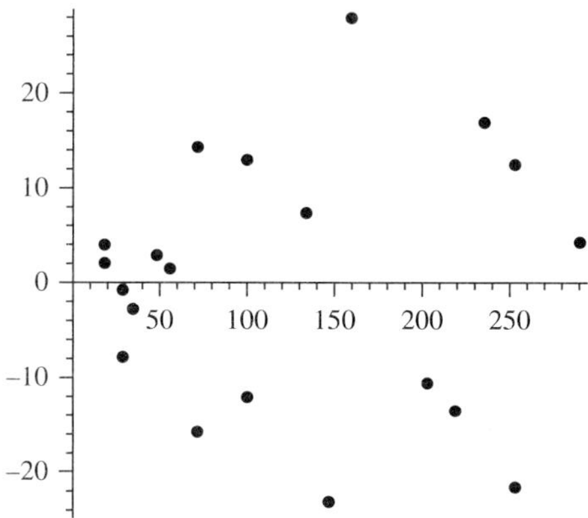

1.  Model $V = 0.152x^2$. Note that a model that indicates a linear trend is not adequate, and this is seen in Figure 15.13.
2.  Model $V = 0.2970327\, x^2 - 6.03434937\, x + 35.8592654$ has no trend in the residual plot as shown in figure 15.14.
3.  Model $V = 0.002908\, x^3 + 0.04743314\, x^2 + 0.81522984\, x - 23.9486202$ has no trends in the residuals plot as seen in Figure 15.15.

---

## Example 2     Telemetry Model

We have data for a missile being fired for time and speed. These data are displayed in Figure 15.16.

> *Tdata*
*:=  [0.00,0.25,0.50,0.75,1.00,1.25,1.50,1.75,2.00,2.25,2.75,3.00,3.25,3.57,3.70,3.86,4.16,4.40,4.73,4.93,5.13,5.43,5.65,6.20,6.43];*

> *Tdata:= [0., 0.25, 0.50, 0.75, 1.00, 1.25, 1.50, 1.75, 2.00, 2.25, 2.75, 3.00, 3.25, 3.57, 3.70, 3.86, 4.16, 4.40, 4.73, 4.93, 5.13, 5.43, 5.65, 6.20, 6.43]*

> *Vdata:=[0.00,1.29,1.96,3.29,5.70,9.58,15.34,23.37,34.05,47.83,86.20,111.70,141.60,187.40,208.60,240.80,296.70,351.40,436.70,495.00,557.10,662.10,746.40,987.80,1102.10];*

> *Vdata := [0., 1.29, 1.96, 3.29, 5.70, 9.58, 15.34, 23.37, 34.05, 47.83, 86.20, 111.70, 141.60, 187.40, 208.60, 240.80, 296.70, 351.40, 436.70, 495.00, 557.10, 662.10, 746.40, 987.80, 1102.10]*

> *pointplot(zip((x,y)→[x,y],Tdata,Vdata));*

**FIGURE 15.16**
Scatterplot of telemetry
data shows increasing and
concave up

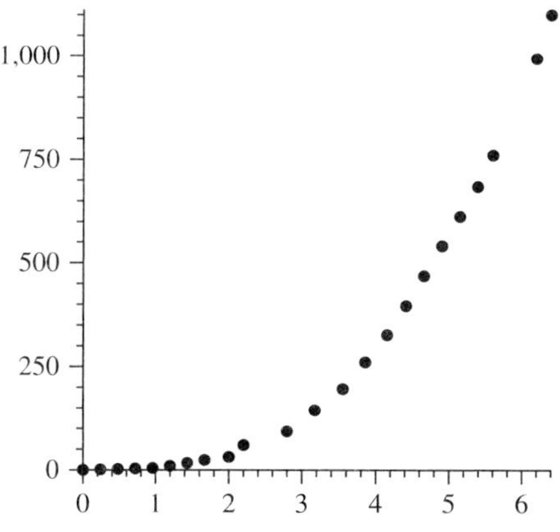

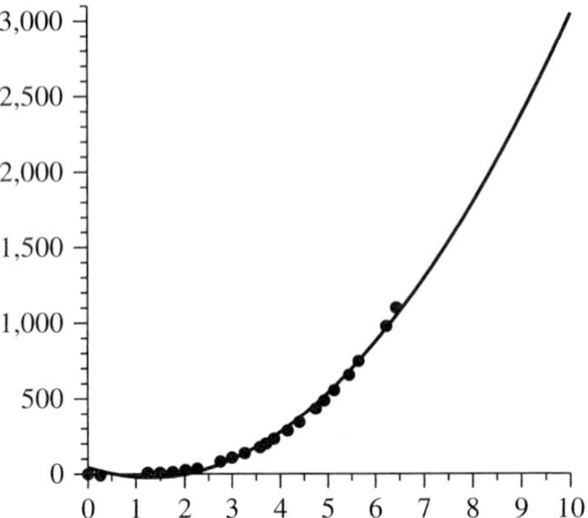

**FIGURE 15.17**
Graph of quadratic model with telemetry data. Visually, this graph looks good.

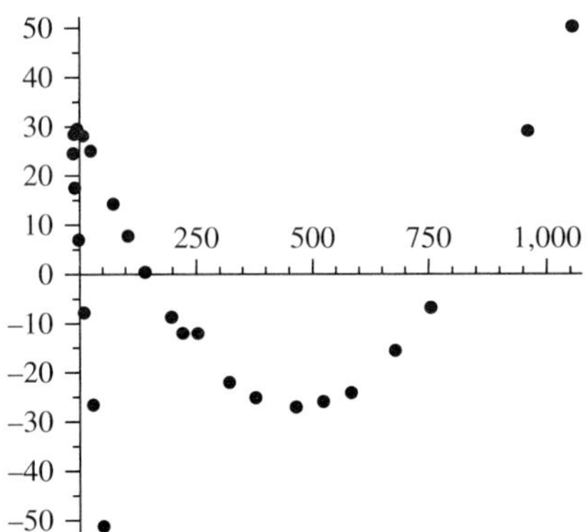

**FIGURE 15.18**
Residual plot clearly shows a curved trend

Based on the empirical model building, we decide to try a quadratic model as shown visually in Figure 15.17.

> $quad1 := Fit\ (ax^2+bx+c,\ Tdata,\ Vdata,\ x)$;

$$quad1 := 40.23500874x^2 - 103.0647504x + 50.96791171$$

We see an apparent trend in Figure 15.18 and hence should look to refine our model.

### Advanced Regression Models (Optional)

We consider the empirical model building alternative of looking at using an ln−ln transformation model. We use the ln−ln method to transform the equation wanted from a nonlinear equation to a linear equation. This method can be used with two or multiple variables. Let's look at a few examples.

## Example 3    Tire Tread Wear

In our first example, the data in the following table show average tread on a radial racing tire over time. The variable $x$ will be the hours in heavy racing use, and the variable $y$ will be the average tread thickness in centimeters (cm).

| Number | Hours | Tread (cm) |
|---|---|---|
| 1 | 2 | 5.4 |
| 2 | 5 | 5.0 |
| 3 | 7 | 4.5 |
| 4 | 10 | 3.7 |
| 5 | 14 | 3.5 |
| 6 | 19 | 2.5 |
| 7 | 26 | 2.0 |
| 8 | 31 | 1.6 |
| 9 | 34 | 1.8 |
| 10 | 38 | 1.3 |
| 11 | 45 | 0.8 |
| 12 | 52 | 1.1 |
| 13 | 53 | 0.8 |
| 14 | 60 | 0.4 |
| 15 | 65 | 0.6 |

*> hours:=[2,5,7,10,14,19,26,31,34,38,45,52,53,60,65];*

$$hours := [2, 5, 7, 10, 14, 19, 26, 31, 34, 38, 45, 52, 53, 60, 65]$$

*> tread:=[5.4,5.0,4.5,3.7,3.5,2.5,2.0,1.6,1.8,1.3,.8,1.1,.8,.4,.6];*

$$tread := [5.4, 5.0, 4.5, 3.7, 3.5, 2.5, 2.0, 1.6, 1.8, 1.3, 0.8, 1.1, 0.8, 0.4, 0.6]$$

*> pointplot(zip((x,y)→[x,y],hours, tread));*

---

The graph in Figure 15.19 shows a decreasing trend that is concave upward. This suggests a model of the form $y = ax^b$.

Using an $\ln - \ln$ transformation to the model yields the model form $\ln y = \ln a + b \cdot \ln x$.

After taking the natural log of each data point, we look at a scatterplot to see if it appears linear. We see the plot of this in Figure 15.20, which is approximately linear. We fit the model, $\ln y = \ln a + b \cdot \ln x$, to find the parameters, $\ln a$ and $\ln b$. The results were

$$\ln y = 2.7721 - 0.07191 \cdot \ln x$$

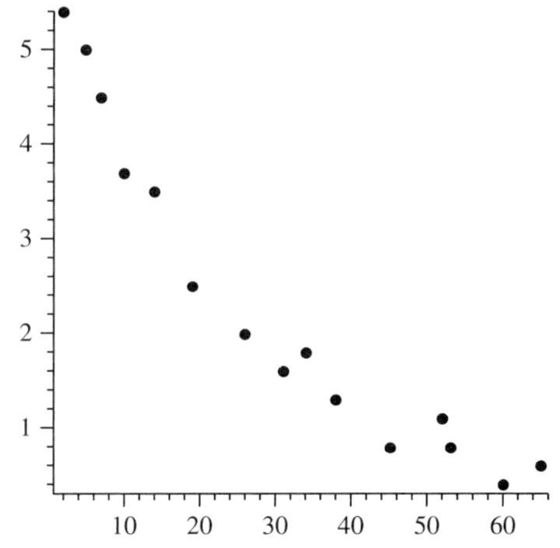

**FIGURE 15.19**

Scatterplot of hours versus tire tread wear

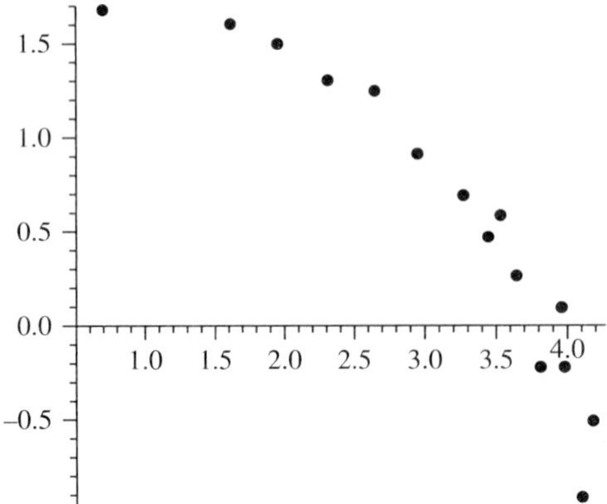

**FIGURE 15.20**
The ln–ln plot for tire tread wear

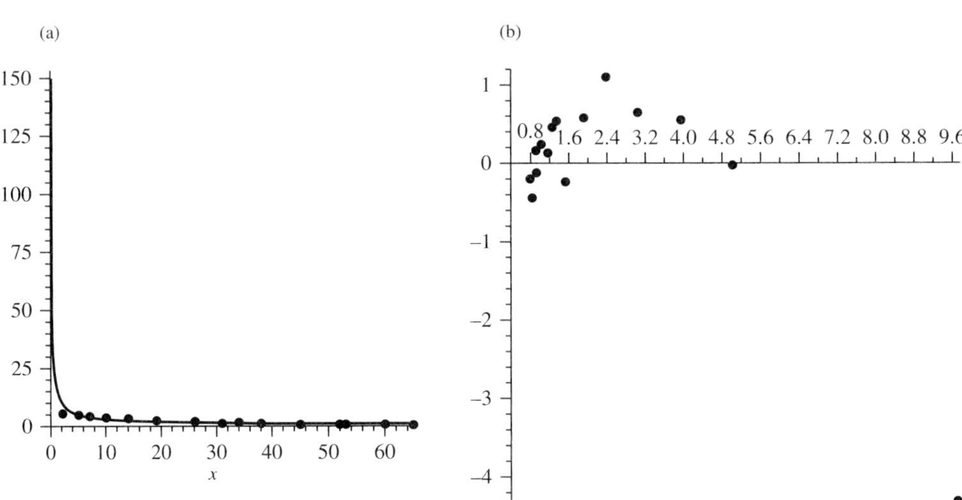

**FIGURE 15.21**
Plots of (a) data with curve and (b) residual with trend

We put this back into its original space by raising $e$ to each side of the expression: $e^{\ln y} = e^{2.771 - 0.7191 \cdot \ln x}$ $y = 15.997 \cdot x^{-0.7191}$.

We plot the curve with the data [see Figure 15.21(a)] and then the residuals in their original space [see Figure 15.21(b)] and note the nonlinear trend, which indicates the model is not adequate.

We will revisit this problem later in the chapter.

---

**Example 4**　　Fitting a Model Form: $Z = a \dfrac{x^b}{y^c}$

Let's assume our suspected nonlinear model form is $Z = a \dfrac{x^b}{y^c}$ for the following data. If we use our $\ln - \ln$ transformation, we obtain: $\ln Z = \ln a + b \ln x - c \ln y$.

We use regression techniques to estimate the parameters $a$, $b$, and $c$.

| Row | $x$ | $y$ | $Z$ |
|-----|-----|-----|-----|
| 1 | 101 | 15 | 0.788 |
| 2 | 73 | 3 | 304.149 |
| 3 | 122 | 5 | 98.245 |
| 4 | 56 | 20 | 0.051 |
| 5 | 107 | 20 | 0.270 |
| 6 | 77 | 5 | 30.485 |
| 7 | 140 | 15 | 1.653 |
| 8 | 66 | 16 | 0.192 |
| 9 | 109 | 5 | 159.918 |
| 10 | 103 | 14 | 1.109 |
| 11 | 93 | 3 | 699.447 |
| 12 | 98 | 4 | 281.184 |
| 13 | 76 | 14 | 0.476 |
| 14 | 83 | 5 | 54.468 |
| 15 | 113 | 12 | 2.810 |
| 16 | 167 | 6 | 144.923 |
| 17 | 82 | 5 | 79.733 |
| 18 | 85 | 6 | 21.821 |
| 19 | 103 | 20 | 0.223 |
| 20 | 86 | 11 | 1.899 |
| 21 | 67 | 8 | 5.180 |
| 22 | 104 | 13 | 1.334 |
| 23 | 114 | 5 | 110.378 |
| 24 | 118 | 21 | 0.274 |
| 25 | 94 | 5 | 81.304 |

The regression equation is $\ln Z = -0.046 + 2.50 \ln x - 4.34 \ln y$.

| Predictor | Coefficient | Standard Deviation | $t$-ratio | $p$ |
|-----------|-------------|--------------------|-----------|-----|
| Constant | $-0.0457$ | 0.7535 | $-0.06$ | 0.952 |
| $\ln x$ | 2.4972 | 0.1630 | 15.32 | 0.0000 |
| $\ln y$ | $-4.33823$ | 0.06268 | $-69.21$ | 0.0000 |
| $s = 0.1970$ | $R - \text{sq} = 99.6\%$ | $R - \text{sq (adj)} = 99.5\%$ | | |

| Source | DF | SS | MS | $F$ | $p$ |
|--------|-----|---------|--------|---------|--------|
| Regression | 2 | 194.113 | 97.057 | 2500.94 | 0.0000 |
| Error | 22 | 0.854 | 0.039 | | |
| Total | 24 | 194.967 | | | |

| Source | DF | SEQ | SS |
|--------|-----|-------|-----|
| $\ln x$ | 1 | 8.222 | |
| $\ln y$ | 1 | 185.891 | |

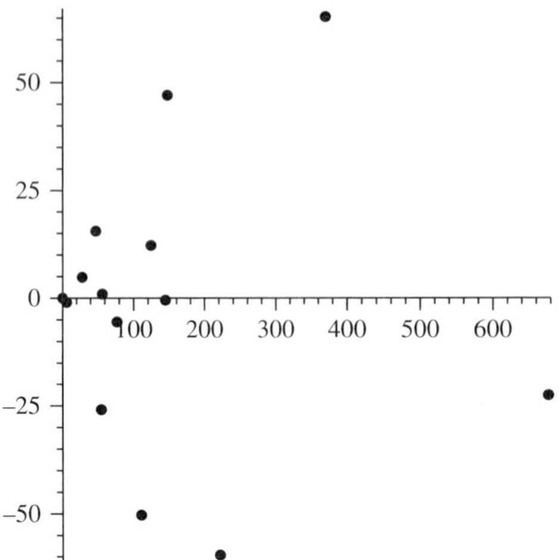

**FIGURE 15.22**

Residual plot for model; trend of residuals is fanning out

Unusual Observations

| Observations | ln $x$ | ln $Z$ | Fit | St. Dev. | Fit Residual | St. Resid. |
|---|---|---|---|---|---|---|
| 3 | 4.80 | 4.5875 | 4.9691 | 0.0663 | −0.3816 | −2.06R |
| 6 | 4.34 | 3.4172 | 3.8198 | 0.0613 | −0.4025 | −2.15R |
| 9 | 4.69 | 5.0747 | 4.6877 | 0.0565 | 0.3870 | 2.05R |
| 17 | 4.41 | 4.3787 | 3.9769 | 0.0564 | 0.4018 | 2.13R |

*$R$ denotes an observation with a large standard residual.

Bringing the model back into the original space yields

$$Z = \exp(-0.046)\!*\!x^{\wedge}2.50/y^{\wedge}4.34 = 0.95504 \cdot \frac{x^{2.50}}{y^{4.34}}$$

The residual plot is shown in Figure 15.22. The trend is fanning out.

### Nonlinear Regression Models with Technology: Maple

The intrinsic, built-in features in Maple allow for a direct calling of least squares for nonlinear models, provided that the parameters are linear. The following examples can be used for exponential and logarithmic nonlinear functions, respectively, that have linear parameters.

```
> with(Statistics):
> eq:=Fit(a+b*exp(x),Xvals,Yvals,x);
or
> eq:=Fit(a*ln(x)+b,Xvals,Yvals,x);
```

These two previous examples illustrate Maple's built-in functional capabilities. However, it is important to note that Maple is a programming language and can be used to implement different algorithms or methods. Thus, we have written a program to perform a general nonlinear regression using Maple where the parameters are also nonlinear. Examples of some of these nonlinear forms are:

**1.** $y = a \cdot e^{bx}$

**2.** $y = a \cdot x^{b}$

**3.** $y = \dfrac{a_1}{a_2 + (e^{a}\,3^{x})}$

The method chosen for this generalized nonlinear least-squares program is Gauss–Newton. The gradient of numerical partial derivatives is used in the solution process. The least-squares estimates are found by iterating the matrix algebra equations:

$$b = (D'D)^{-1} DY$$
$$g_1 = g_0 + b$$

where
  $g$ is the matrix of refined estimates,
  $b$ is the change in the estimates,
  $Y$ is the matrix of errors, $[y - f(x)]$, and
  $D$ is the matrix of the partial derivatives with respect to the unknown parameters.

The procedure iterates until the absolute difference between the sum of squared errors of successive iterations is less than some user-defined tolerance. The program is provided at the end of this chapter.

---

**Example 5**

## Tire Tread Wear Revisited

> *Xpts:=[2,5,7,10,14,19,26,31,34,38,45,52,53,60,65];*
> *Ypts:=[5.4,5,4.5,3.7,3.5,2.5,2,1.6,1.8,1.3,.8,1.1,.8,.4,];*
> *nda:=15:*

*xx:=matrix(nda,1,Xpts):y:=vector(nda,Ypts):p2:=vector(2,[15.99,–.719]):f:=a[1]*t[1]^a[2];*

$$f := a_1 \, t_1^{a_2}$$

> *gnfit(f,t,xx,y,a,p2,10^(−7),50);*

```
            iteration No. 1
initial/old SSE =    21.791
      new SSE =    11.269
            iteration No. 2
initial/old SSE =    11.269
      new SSE =    6.3297
            iteration No. 3
initial/old SSE =    6.3297
      new SSE =    6.2590
            iteration No. 4
initial/old SSE =    6.2590
      new SSE =    6.2572
            iteration No. 5
initial/old SSE =    6.2572
      new SSE =    6.2572
            iteration No. 6
initial/old SSE =    6.2572
      new SSE =    6.2572
            iteration No. 7
initial/old SSE =    6.2572
      new SSE =    6.2572
      vector of corrections
   −0.00049233,      0.000029342
      final values of parameters
      No.            value            error
      1,             8.879,           1.27
      2,             −.46622,         0.00334
```

*standard deviation of fit* = 0.4813
*F statistic* = 58.58
*matrix of correlation coefficients*
1.0000
−.86022,      1.0000
*table of residuals*

| *x obs,* | *y obs,* | *y calc,* | *difference* |
|---|---|---|---|
| [2], | 5.4, | 6.4272175677012572018, | −1.0272 |
| [5], | 5, | 4.1926159615996923019, | 0.8074 |
| [7], | 4.5, | 3.5838801542210871160, | 0.9161 |
| [10], | 3.7, | 3.0348069262443679093, | 0.6652 |
| [14], | 3.5, | 2.5941761455084488600, | 0.9058 |
| [19], | 2.5, | 2.2499005848633129059, | 0.2501 |
| [26], | 2, | 1.9437993072452938905, | 0.0562 |
| [31], | 1.6, | 1.7907529117624356097, | −.1908 |
| [34], | 1.8, | 1.7152647912934610163, | 0.0847 |
| [38], | 1.3, | 1.6285808051208220118, | −.3286 |
| [45], | 0.8, | 1.5051277968491743577, | −.7051 |
| [52], | 1.1, | 1.4070107195332619580, | −.3070 |
| [53], | 0.8, | 1.3945701798964234756, | −.5946 |
| [60], | 0.4, | 1.3161979082246598795, | −.9162 |
| [65 ], | 0.6, | 1.2679832160431766548, | −.6680 |

*Covariance Matrix*
1.2708
−0.056021,      0.0033374

*ANOVA TABLE*

| | *SS,* | *deg of free,* | *ms* |
|---|---|---|---|
| *Regression,* | 28.193613150819349454, | 1, | 28.193613150819349454 |
| *Error,* | 6.2571541537485045139, | 13, | 0.48131955028834650107 |
| *Total,* | 39.507097466863957422, | 14 | |

The output has lots of statistical results that can be used by those familiar with these statistical measures. The novice modeler might only be interested in the coefficient, plots, and residual plots for their analysis. The generalized nonlinear model fit is $(8.879)*x^{-.46622}$. We plot both its fits with the data [see Figure 15.23(a)] and the residual plot [see Figure 15.23(b)].

**FIGURE 15.23**

Plot of curve with data (a) and residual plot (b) without a trend

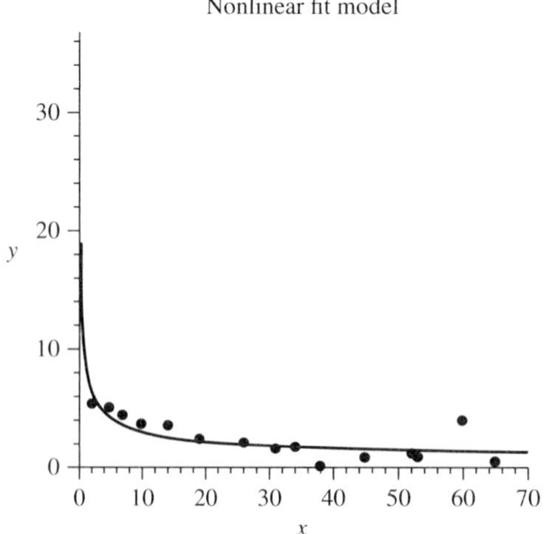


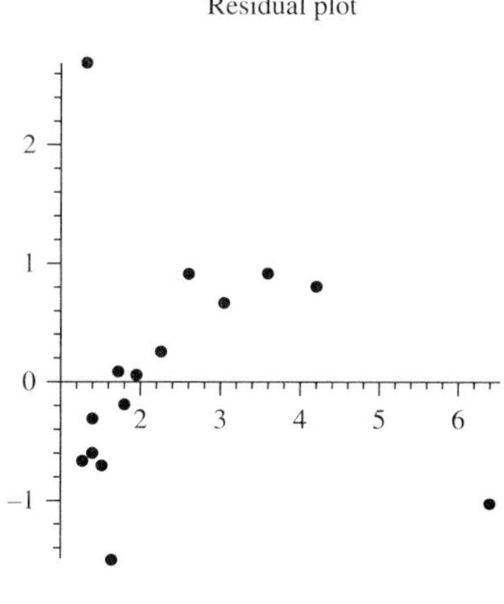

**FIGURE 15.24**
Scatterplot of data
(exponential decay)

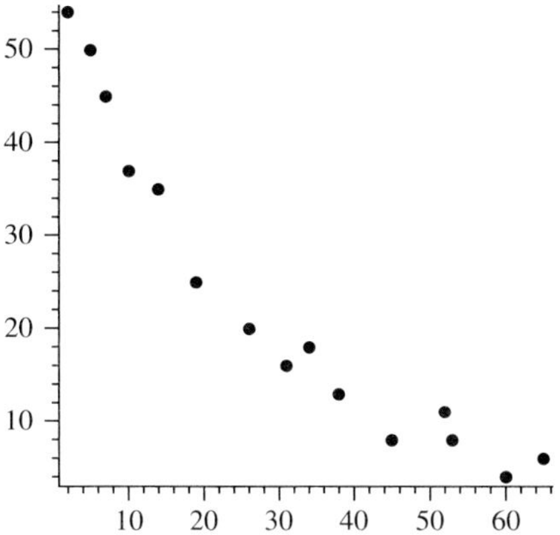

## More Advanced Models with Regression and Other Forecasts (Optional)

Newer versions of Maple and their commands allow for models to used in more advanced regression techniques without having to write programs. Let's start by looking at data that could be modeled with exponential decay. We also illustrate the use of the *zip* command in obtaining the point plot.

Consider data for patients over time. We think it should follow an exponential decay (see Figure 15.24).

> *days:=[2,5,7,10,14,19,26,31,34,38,45,52,53,60,65];*

$$days := [2, 5, 7, 10, 14, 19, 26, 31, 34, 38, 45, 52, 53, 60, 65]$$

> *PI:=[54,50,45,37,35,25,20,16,18,13,8,11,8,4,6];*

$$\Pi := [54, 50, 45, 37, 35, 25, 20, 16, 18, 13, 8, 11, 8, 4, 6]$$

> *pointplot(zip'[ ]'days,PI),symbol=circle);*
> *f5:=Fit(a·exp(−b·t),days,PI,t);*

$$f5 := 58.6065630985571248 \, e^{-0.0395864490777913386 \, t}$$

> *display(pointplot(zip('[ ]',days,PI),symbol=circle),plot(f5,t=0..70))*

Visually, we have a good fit as seen in Figure 15.25. The sum of squared error for this model is 49.4593. Because $R^2$ is not meaningful in nonlinear regression, we omit it from our analysis.

**FIGURE 15.25**
Plot of our decay model and
the original data

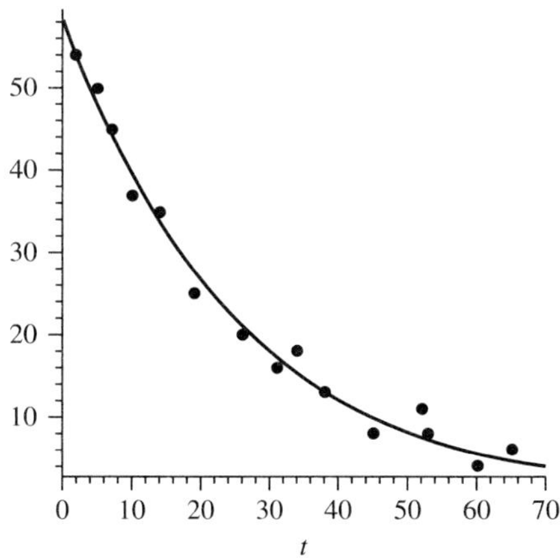

If we used our previous gnfit, then we obtain the same model. However, note that the use of gnfit is very sensitive to the input vector. A poor choice might not allow the program to converge to the result we seek.

> *gnfit( f,t,xx,y,a,p2,10^(−3),50 );*

```
              Iteration Number: 1
Initial/old SSE = 1312000.00
      New SSE  =   65597.00

              Iteration Number: 2
Initial/old SSE =   65597.00
      New SSE  =   10354.00

              Iteration Number: 3
Initial/old SSE =   10354.00
      New SSE  =    1210.80

              Iteration Number: 4
Initial/old SSE =    1210.80
      New SSE  =      83.12

              Iteration Number: 5
Initial/old SSE =      83.12
      New SSE  =      49.51

              Iteration Number: 6
Initial/old SSE =      49.51
      New SSE  =      49.46

              Iteration Number: 7
Initial/old SSE =      49.46
      New SSE  =      49.46
```

Vector of Corrections:
    0.002964953276     4.673948081e-06

Final Values of Parameters:

| No. | Value | Error |
|---|---|---|
| 1 | 58.6100000000 | 2.17 |
| 2 | 0.0395863000 | 2.93e-06 |

Standard deviation of fit = 3.80

F statistic = 1045.00

Matrix of correlation coefficients:
    1.00000

    0.70715          1.00000

Table of residuals:

| [x obs] | [y obs] | y calc | diff] |
|---|---|---|---|
| 2.00 | 54.00 | 54.14 | −0.14 |
| 5.00 | 50.00 | 48.08 | 1.92 |
| 7.00 | 45.00 | 44.42 | 0.58 |
| 10.00 | 37.00 | 39.45 | −2.45 |
| 14.00 | 35.00 | 33.67 | 1.33 |
| 19.00 | 25.00 | 27.63 | −2.63 |
| 26.00 | 20.00 | 20.94 | −0.94 |
| 31.00 | 16.00 | 17.18 | −1.18 |
| 34.00 | 18.00 | 15.26 | 2.74 |
| 38.00 | 13.00 | 13.02 | −0.02 |

| 45.00 | 8.00 | 9.87 | $-1.87$ |
| 52.00 | 11.00 | 7.48 | 3.52 |
| 53.00 | 8.00 | 7.19 | 0.81 |
| 60.00 | 4.00 | 5.45 | $-1.45$ |
| 65.00 | 6.00 | 4.47 | 1.53 |

Covariance matrix:
2.1670296398
0.0017812355     0.0000029279

ANOVA Table<br>(F* = 1045.0000)

|  | df | SS | MS |
|---|---|---|---|
| Regression | 1 | 3974.8106 | 3974.8106 |
| Error | 13 | 49.4593 | 3.8046 |
| Total | 14 | 3943.5351 | |

Let's look at the carbon dioxide emissions from 1960 to 2004 to predict emissions from 2005 to 2010. The following data and plot are provided below (see also Figure 15.26).

```
> T:=[0,1,2,3,4,5,6,7,8,9,10,11,12,13,14,15,16,17,18,19,20,21,22,23,24,25,26,27,28,29,30,31,32
,33,34,35,36,37,38,39,40,41,42,43,44];

> T:=[seq(i,i=0..44)];
```

$$T := [0, 1, 2, 3, 4, 5, 6, 7, 8, 9, 10, 11, 12, 13, 14, 15, 16, 17, 18, 19, 20, 21, 22, 23, 24, 25,$$
$$26, 27, 28, 29, 30, 31, 32, 33, 34, 35, 36, 37, 38, 39, 40, 41, 42, 43, 44]$$

```
> CO2:=[16.2,16.3,16.7,17.2,17.7,18.4,18.9,19.5,20.2,20.4,20.6,20.4,21.1,21.7,20.8,19.7,20.8,
20.8,21.4,21.2,20.3,19.3,18.2,18.2,18.6,18.4,19,19.6,19.8,19.2,19,18.8,19.7,19.8,19.5,19.7,20,
19.5,19.6,20,19.6,19.6,19.5,19.5];
```

$$CO2 := [16.2, 16.3, 16.7, 17.2, 17.7, 18.4, 18.9, 19.5, 20.2, 20.4, 20.6, 20.4, 21.1, 21.7, 20.8,$$
$$19.7, 20.8, 20.8, 21.4, 21.2, 20.3, 19.3, 18.2, 18.2, 18.6, 18.4, 19, 19.6, 19.8, 19.2, 19,$$
$$18.8, 19.7, 19.8, 19.5, 19.7, 20, 19.5, 19.6, 20, 19.6, 19.6, 19.4, 19.5, 19.5]$$

```
> pointplot(zip('[]', T, CO2), symbol = circle);
```

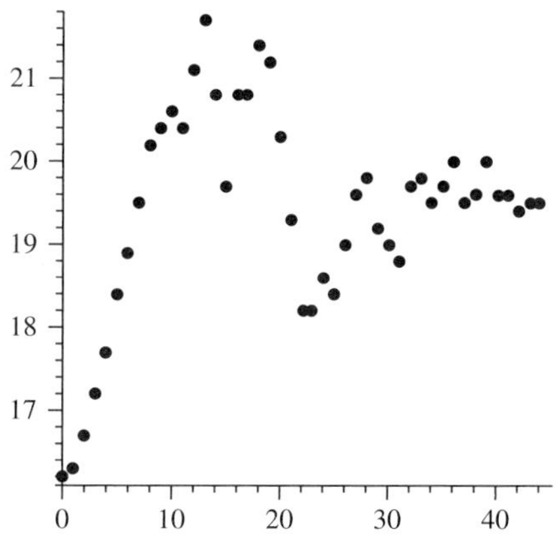



**FIGURE 15.26**
Scatterplot of emissions data

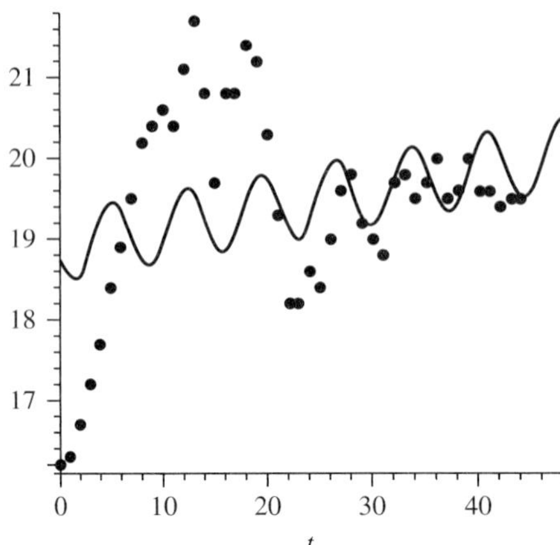

**FIGURE 15.27**
Poor model for emissions data

The plot is a slightly oscillating data that appears to be damping out. If we assume that conditions remain the same over the next years, can we predict the amount of emission (in metric tons) from 2005 to 2010?

Let's look at several models that can be developed. We can try some transcendental functions. First, we try a sine function.

```
> with(Statistics) :
> f1:=Fit(a·sin(b·t+c)+d·t+e,T,C02,t);
```

$$f1 := 0.443601279730423847 \sin(0.879172065571240944\, t + 3.54720898490686932)$$
$$+ 0.0236924559319273386t + 18.9111905209491874$$

```
> with(plots):
> display(pointplot(zip('[ ]',T,CO2),symbol=circle),plot(f1,t=0..48));
```

We have some oscillations but no damping. The visual fit in Figure 15.27 is not that appealing. Let's try something from logistic.

```
> f2:=Fit(a+b·exp(−c·t),T,CO2,t);
```

$$f2 := 19.79348331367 96122 − 4.45955870262544352\, e^{-0.306154787566737474t}$$

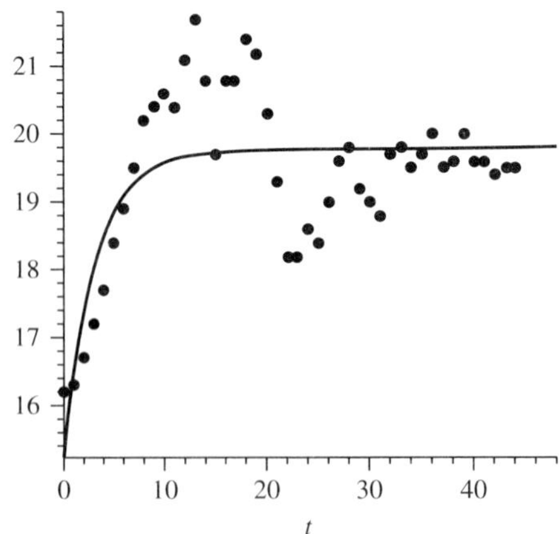

**FIGURE 15.28**
Better model for emissions data

> *with (plots) :*
> *display(pointplot(zip('[ ]',T,CO2),symbol=circle),plot(f2,t=0..48));*

We have captured the essence of the data as seen in figure 15.28. This models tells us the prediction of our model is that we are achieving steady state at about 19.79 metric tons.

## 15.4 | EXERCISES

Using the basic linear model $y_i = \beta_0 + \beta_1 x_i$, fit the following data sets. Provide the model, the analysis of variance information, the value of $R^2$, and a residual plot.

**1.**

| $x$ | $y$ |
| --- | --- |
| 100 | 150 |
| 125 | 140 |
| 125 | 180 |
| 150 | 210 |
| 150 | 190 |
| 200 | 320 |
| 200 | 280 |
| 250 | 400 |
| 250 | 430 |
| 300 | 440 |
| 300 | 390 |
| 350 | 600 |
| 400 | 610 |
| 400 | 670 |

**2.** The following data represent changes in growth where $x$ = body weight and $y$ = normalized metabolic rates for 13 animals.

| $Row$ | $x$ | $y$ |
| --- | --- | --- |
| 1 | 110 | 198 |
| 2 | 115 | 173 |
| 3 | 120 | 174 |
| 4 | 230 | 149 |
| 5 | 235 | 124 |
| 6 | 240 | 115 |
| 7 | 360 | 130 |
| 8 | 362 | 102 |
| 9 | 363 | 95 |
| 10 | 500 | 122 |
| 11 | 505 | 112 |
| 12 | 510 | 98 |
| 13 | 515 | 96 |

**3.** Discuss reasons why the sine model is so poor. Then use the *gnfit* commands and better estimate the parameters; separately use trigonometry and see if you can obtain a better model.

## 15.4 | PROJECTS

**1–5.** Use linear regression to rebuild and analyze projects 1–5 in Section 4.5.

**6.** Fit the following nonlinear model with the provided data:

$$\text{Model: } y = ax^b$$

| $t$ | 7 | 14 | 21 | 28 | 35 | 42 |
|---|---|---|---|---|---|---|
| $y$ | 8 | 41 | 133 | 250 | 280 | 297 |

**7.** Fit the following model, $y = ax^b$, with the provided data.

| Year | 0 | 1 | 2 | 3 | 4 | 5 | 6 | 7 | 8 | 9 | 10 |
|---|---|---|---|---|---|---|---|---|---|---|---|
| Quantity | 15 | 150 | 250 | 275 | 270 | 280 | 290 | 650 | 1,200 | 1,550 | 2,750 |

## 15.4 | FURTHER READING

Mendenhall, W., and T. Sincich. A Second Course in Statistics: Regression Analysis, 5th Ed. Upper Saddle River, NJ: Prentice Hall. 1996.

Neter, J., W. Wasserman, and J. Kutner. Applied Statistical Models. 4th Ed. Boston: McGraw-Hill. 1996.

# 16

# Simulation Modeling

## Modeling Deterministic and Probabilistic Behavior Using Monte Carlo Simulation in Maple

## Introduction

Consider an engineering company that conducts vehicle inspections for a specific state. We have data for times of vehicle arrivals and departures, service times for inspectors under various conditions, numbers of inspection stations, and penalties levied for failure to meet state inspection standards in terms of waiting time for customers. The company wants to know how it can improve its inspection process throughout the state in order to both maximize its profit and minimize the penalties it receives. This type of analysis for a complex system has many variables, and we could use a computer simulation to model this operation.

A modeler may encounter situations where the construction of an analytic model is infeasible because of the complexity of the situation. In instances where the behavior cannot be modeled analytically or where data are collected directly, the modeler might simulate the behavior indirectly and then test various alternatives to estimate how each affects the behavior. Data can then be collected to determine which alternative is best. Monte Carlo simulation is a common simulation method that a modeler can use, usually with the aid of a computer. The proliferation of today's computers in the academic and business worlds makes Monte Carlo simulation very attractive. It is imperative that students have at least a basic understanding of how to use and interpret Monte Carlo simulations as a modeling tool.

There are many forms of simulation ranging from building scale models such as those used by scientists or designers in experimentation to various types of computer simulations. One preferred type of simulation is the Monte Carlo simulation. Monte Carlo simulation deals with the use of random numbers. There are many serious mathematical concerns associated with the construction and interpretation of Monte Carlo simulations. Here we are concerned only with reinforcing the techniques of simulations with these random variates.

A principal advantage of Monte Carlo simulation is the ease with which it can be used to approximate the behavior of very complex systems. Often, simplifying assumptions must be made to reduce this complex system into a manageable model. In the environment forced on

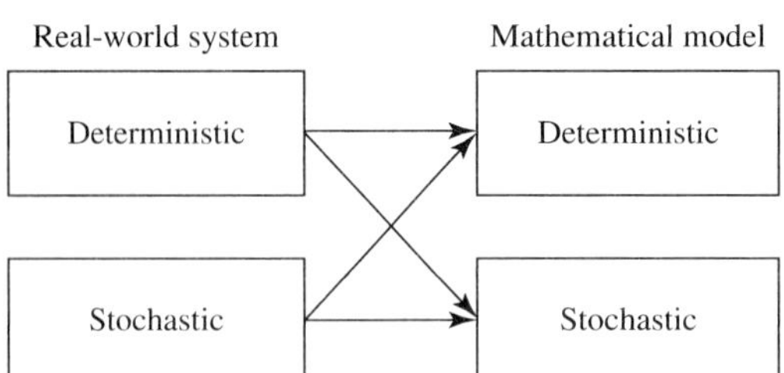

**FIGURE 16.1**

System versus model relationship

the system, the modeler attempts to represent the real system as closely as possible. Complex systems probably are stochastic systems; however, simulation can allow either a deterministic or stochastic approach (see Figure 16.1). Processes with an element of chance involved are called stochastic, as opposed to deterministic, processes. The elements of chance are what make Monte Carlo simulation a stochastic process. We will concentrate on the stochastic modeling approach to deterministic behavior.

Many undergraduate mathematical science, engineering, and operations research programs currently require or offer a course involving simulation. Typically, such a course will use a high-level simulation language such as C++, Java, FORTRAN, SLAM, Prolog, STELLA, Siman, or GPSS as the tool to teach simulation. Here we will use Maple to simulate some simple modeling scenarios.

Our emphasis is twofold. First, we want you (the student) to think in terms of an algorithm, not a specific language. Second, we want you to understand that *more* is better in Monte Carlo simulations. The "More Is Better" rule is based on the law of large numbers where probabilities are assigned to events in accordance with their limiting relative frequencies.

## 16.1 MONTE CARLO SIMULATION

A Monte Carlo simulation model is a model that uses random numbers to simulate behavior of a situation. Using a known probability distribution (such as uniform, exponential, or normal) or an empirical probability distribution, a modeler assigns a behavior to a specific range of random numbers. The behavior returned from the random number generated is then used in analyzing the problem. For example, if a modeler is simulating the tossing of a fair coin using a uniform random number generator that gives numbers in the range $0 \leq x < 1$, then he or she may assign all numbers less than 0.5 to be a head while numbers from 0.5 to 1 are tails. We simplify the important part in Figure 16.2.

A Monte Carlo simulation can be used to model either stochastic or deterministic behavior. It is possible to use a Monte Carlo simulation to determine the area under a curve (a deterministic problem) or stochastic behavior like the probability of winning in craps (a stochastic problem).

**FIGURE 16.2**

Relationship between random numbers and outcomes

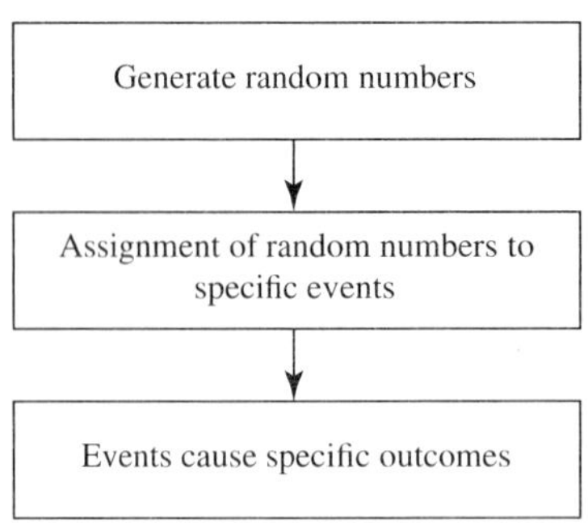

In this chapter, we will introduce both a deterministic problem and a stochastic problem. We discuss how to create algorithms to solve both. We will start with the deterministic simulation modeling.

## Random-Number Generators in Maple

Using random numbers is of paramount importance in running Monte Carlo simulations, so a good random-number generator is critical. In particular, a modeler must have a method of generating uniform, $U(0,1)$, random numbers—that is, numbers that are uniformly distributed between 0 and 1. All other distributions, known and empirical, can be derived from the $U(0,1)$ distribution. At the graduate level, a lot of class time is spent on the theory behind good and bad random-number generators, and the tests that can be made on them. More and more is being learned about what does and does not make up a true random-number generator. At the undergraduate level, this is not necessary, provided the students have access to either random numbers or a good algorithm for generating pseudo-random numbers.

In addition, most computer languages now use good pseudo–random-number generators (although this has not always been the case—the old RANDU generator distributed by IBM was statistically unsound). These good generators use the recursive sequence $X_i = (aX_{i-1} + c)$ mod $m$ where $a$, $c$, and $m$ determine the statistical quality of the generator. Because we do not discuss the testing of random-number generators in our course, we trust the generators provided by our software packages. Serious study of simulation must, of course, include a study of random-number generators because a bad generator will provide output from which a modeler may make poor conclusions.

Maple has several choices to generate random numbers. We now present a few of Maple's commands. Note that the command *rand(r)* returns a 12-digit nonnegative number.

rand: **random-number generator**
Calling sequence
rand(**r**)
Parameters
r:(optional) integer range or integer

**Description**

- With no arguments, the call **rand()** returns a random 12-digit nonnegative integer.

- With an integer range as an argument, the call **rand(a..b)** returns a procedure that, when called, generates random integers in the range **a..b**.

- With a single integer as an argument, the call **rand(n)** is the abbreviated form of **rand(0..n-1)**.

- More than one random-number generator may be used at the same time, because **rand(a..b)** returns a Maple procedure, However, because all random-number generators use the same underlying random-number sequence, calls to one random-number generator will affect the random numbers returned from another.

- **rand** calls *RandomTools[MersenneTwister][GenerateInteger]* or *RandomTools[Mersenne Twister][NewGenerator]* depending on whether or not a number or procedure is to be returned. It is more efficient to make these calls directly than to call **rand**.

- The random number generator used by **rand** can be seeded by using the *randomize* or *RandomTools[MersenneTwister][SetState]* functions.

- The use of the **_seed** global variable to seed the generator is deprecated.

- The algorithm used by **rand** in Maple versions up to and including 9.5 has been moved into the *RandomTools* package as *RandomTools[LinearCongruence]*.

- To generate more complex Maple objects, the *RandomTools[Generate]* function can be used.

**Examples**

> *rand( );*

$$395718860534$$

> *rand( );*

$$193139816415$$

> *roll:=rand(1..6):roll( );*

$$6$$

> *roll( );*

$$2$$

> *RandomTools[Generate](integer(range=1..6));*

$$3$$

The seed of the default generator can be modified by using the function **randomize()**.

**Examples to Generate Random Numbers**

> *with(Statistics):*

Generate 20 numbers with the standard normal distribution.

> *Generate(list(distribution(Normal(0,1)),20));*

1.275676034, 0.4761860895, −2.620054215, 0.8547207946, −1.089001803,
0.7165427609, −0.02996436194, 1.330855643, −1.296345120, −0.4703304005,
−0.7122630822, −0.08717900722, 1.282253011, −1.683720684, 0.3265473243,
0.08179818206, 0.2127564698, −0.7279468705, 0.6855773713, −0.3760280547

This time, we want the normal inverse. Since we use the formula $z = (x - \mu)/\sigma$ then we want $z$. When $\mu = 0$ and $\sigma = 1$ then $z$ is equal to our random numbers. The inverse method.

> *z:=Generate(list(distribution(Normal(0,1)),20));*

$z$:=0.2942905914, 1.023730745, −0.07256069586, −0.2726670238, 0.9797884030,
−0.06456731317, −0.7584692928, −1.414509920, 0.5848741507, 0.1502221444,
0.7052134613, −0.1760249535, 1.114921264, 1.111477077, 1.837138817,
−0.007014805706, 0.1412349839, −0.3487795280, 2.269571409, 0.6290512530

Generate 5 numbers with the Poisson$[\lambda = 3]$ distribution with a user-specified generator.

> *seed:=randomize( );*

$$seed := 1295554922$$

> *Generate(list(distribution(Poisson(3)),5));*

$$[3,2,3,2,2]$$

### Generating Uniform Random Numbers from [0,1] in Maple

In the next example, we generate 20 uniform random number from [0,1].

$> random:=$**proc**$(n)$ **option** $remember$ **for** $i$ **from** $1$ **to** $n$ **do** $x(i):=evalf\left(\dfrac{rand()}{999999999999}\right)$ **end do**
**end;**

Warning, $i$ is implicitly declared local to procedure $random$.

```
random := proc(n)
    option remember;
    local i;
    for i to n do
        x(i):= evalf (1/999999999999*rand( ) )
    end do
end proc
```

$> random\ (20);$

$$0.8950382012$$

$> seq(x(i),i=1..20);$

> 0.2240171515, 0.2008401063, 0.8685719066, 0.5704134665, 0.9920881460,
> 0.04437752746, 0.4780291372, 0.01294305199, 0.6408831565, 0.2487102377,
> 0.9700117830, 0.04705144220, 0.5745759906, 0.1717448313, 0.6488303816,
> 0.4114303688, 0.09704265386, 0.7719464925, 0.2477991384, 0.8950382012

We can generate many distributions that we might need in modeling from the uniform [0,1] random numbers. For example,

1. Uniform [a,b]
   a. Generate a random uniform number $U$ from [0,1]
   b. Return $X = a + (b - a)*U$

2. Exponential with mean $\beta$
   a. Generate a random uniform number $U$ from [0,1]
   b. Return $X = -\beta\ln(U)$

3. Normal(0,1)
   a. Generate $U_1$ and $U_2$ from uniform [0,1].
   b. Let $V_i = 2U_i - 1$ for $i = 1,2$.
   c. Let $W = V_1^2 + V_2^2$
   d. If $W > 1$, go back to step a. Otherwise, let $Y = \sqrt{(-2\ln(W)/W}$, $X_1 = V_1 Y$, $X_2 = V_2 Y$.
   e. $X_1$ and $X_2$ are normal (0,1).

## **16.1** | EXERCISES

For each problem, generate 20 random numbers.

1. Uniform (0,1)

2. Uniform (−10,10)

3. Exponential ($\lambda = 0.5$)

4. Normal(0,1)

5. Normal (5,0.5)

## 16.2  PROBABILITY AND MONTE CARLO SIMULATION USING DETERMINISTIC BEHAVIOR

One key to good Monte Carlo simulation is an understanding of the axioms of probability. *Probability* is a long-term average. For example, if the probability of an event occurring is 1/5, this means that "in the long term, the chance of the event happening is $1/5 = 0.2$." not that it will occur exactly once out of every five trials.

### Deterministic Simulation Examples

Let's consider the following deterministic examples. We want to compute the area under a non-negative curve in examples 1 and 2 and compute a volume in example 3.

1.  $y = x^3$ from $0 \le x \le 2$.

2.  $y = e^{x^2} \cdot \cos(x) \cdot \sqrt{x}$ from $[0,1.4]$. (We note that there is no closed-form solution to

$$\int_{x=0}^{1.4} \cos(x^2) \cdot \sqrt{x} \cdot e^{x^2}\, dx \text{ from } [0,1.4].)$$

3.  Compute the volume in the first octant of $x^2 + y^2 + z^2 \le 1$.

We will present algorithms for their models as well as produce output of the Monte Carlo simulation to analyze. These algorithms are important to the understanding of simulation as a mathematical modeling tool.

Here is a generic framework for an algorithm. This framework includes inputs, outputs, and the steps required to achieve the desired output.

---

## Example 1   A Deterministic Example

Consider the area under the curve $y = x^3$ from $0 \le x \le 2$. A graphical illustration of the region is provided in Figure 16.3.

The algorithm for determining the area under a nonnegative curve between $[a, b]$ is described in Figure 16.4.

We begin with an easy function such as $y = x^3$ over the interval $[0, 2]$. We can easily integrate the function and find the following answer:

$$\int_0^2 x^3 dx = 4$$

**FIGURE 16.3**
Graph of $y = x^3$ from $0 \le x \le 2$

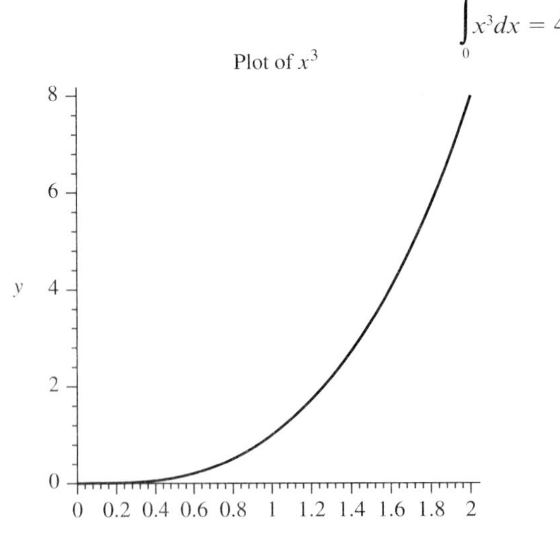

Algorithm for determining the area under a nonnegative curve

| | |
|---|---|
| **INPUT** | The total number of random points, N. The nonnegative function, f(x), the interval for x [a,b] and an interval for y [0,M] where M > max f(x), a ≤ x ≤ b. |
| **OUTPUT** | The approximate area under the curve, f(x) over the interval [a,b] |

**Step 1.** Specify the function, f(x) and set all counters at 0
**Step 2.** For i from 1 to N do steps 3-5
**Step 3.** Calculate random coordinates in the rectangular region:

$$a \le x_i \le b, \ 0 \le y_i \le M$$

**Step 4.** Calculate $f(x_i)$
**Step 5.** Compare $f(x_i)$ and $y_i$. If $y_i \le f(x_i)$ then increment counter by 1. Otherwise, do not increment counter.
**Step 6.** Estimate the area by $A = M \cdot (b - a) \cdot \dfrac{counter}{N}$
**Stop**

Now we are ready to *approximate* the area by using Monte Carlo simulation. The simulation only approximates the solution. We increase the number of trials attempting to get closer to the value. We present the results in Table 16.1. Recall that we introduced randomness into the procedure with the Monte Carlo simulation area algorithm. In our Maple program, we provide graphical output as well so that the algorithm may be seen as a process. In our graphical output, each generated coordinate $(x_i, y_i)$ is a point on the graph. Points are randomly generated in our intervals [a, b] for x and [0, M] for y. The curve for the function f(x) is overlaid with the points. The output also includes the approximate area.

Calculate the area under a curve.

First, let's view the graphical outputs from our simulation runs when $N = 100$ (Figure 16.5) and $N = 5,000$ (Figure 16.6).

**Table 16.1**  Summary of output for the area under $x^3$ from 0 to 2

| Number of Trials | Approximate Area | Percent Error (%) |
|---|---|---|
| 100 | 3.36 | 16 |
| 500 | 3.872 | 3.2 |
| 1000 | 4.32 | 8 |
| 5000 | 4.1056 | 2.64 |
| 10,000 | 4.136 | 3.4 |


Graphical output with $N = 100$; area estimate is 3.36

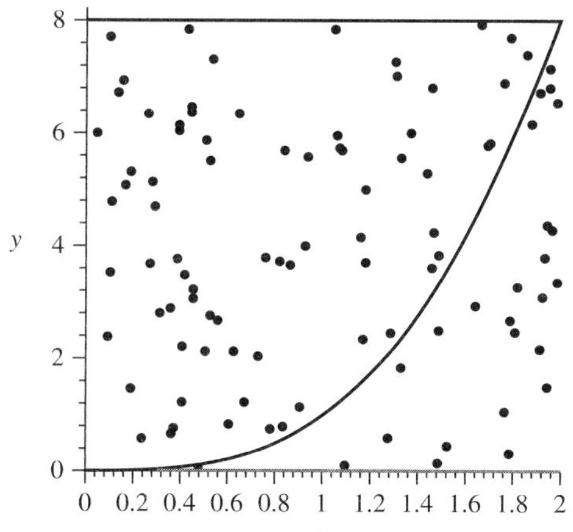

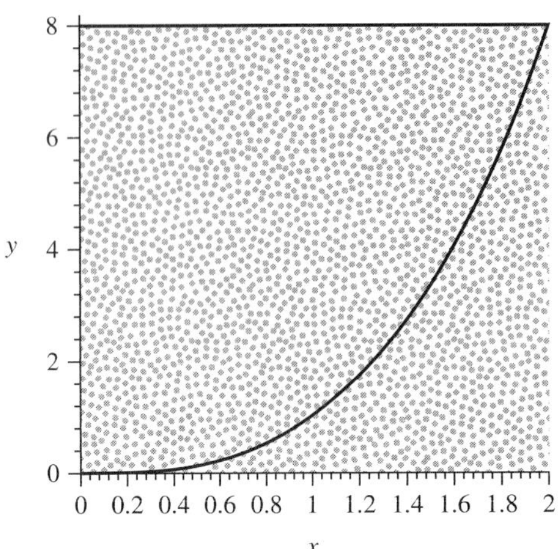

**FIGURE 16.6**
Graphical output with
$N = 5{,}000$; area estimate is
4.1056

We need to stress that in modeling deterministic behavior with stochastic features, we not nature have introduced the randomness into the problem. Although more runs is better, it is not true that the solution uniformly becomes closer to reality as we increase the number of trials, $N$, $\to \infty$. It is generally true that more runs is better than a small number of runs (16 percent error was the worst by almost an order of magnitude, and that occurred at $N = 100$). In general, more trials are better.

## Example 2

## Area Under a Nonnegative Curve

Find the area under the curve for a nonnegative function where we cannot integrate to find a closed-form answer, $e^{x^2} \cdot \cos(x) \cdot \sqrt{x}$ from $[0, 1.4]$.

We modify the algorithm to compute the area under the curve for the function $e^{x^2} \cdot \cos(x) \cdot \sqrt{x}$ over the interval $[0, 1.4]$ (Figure 16.7). We use the algorithm presented in Figure 16.4 with our equations, intervals, and the final output for runs of 50, 100, 500, and 1,000 trials.

We provide one graphical output from our program in Figure 16.8 with $N = 2{,}000$. Because this function has no closed-form solution, we used Simpson's method to obtain a numerical solution to

$$\int_0^{1.4} e^{(x^2)} \cos(x) \sqrt{x}\, dx$$

**FIGURE 16.7**
Plot of $f(x)$ from $[0, 1.40]$

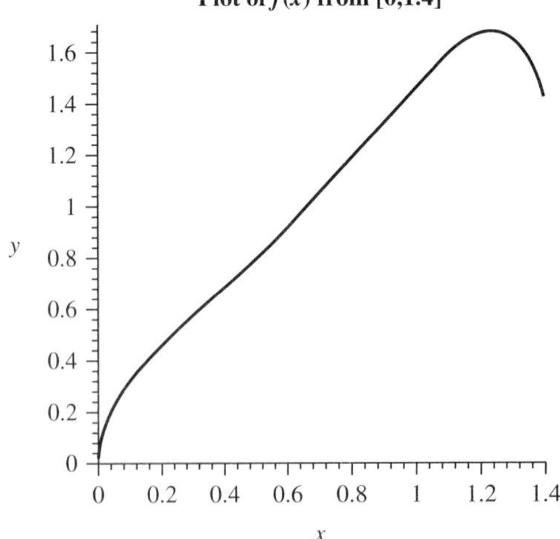

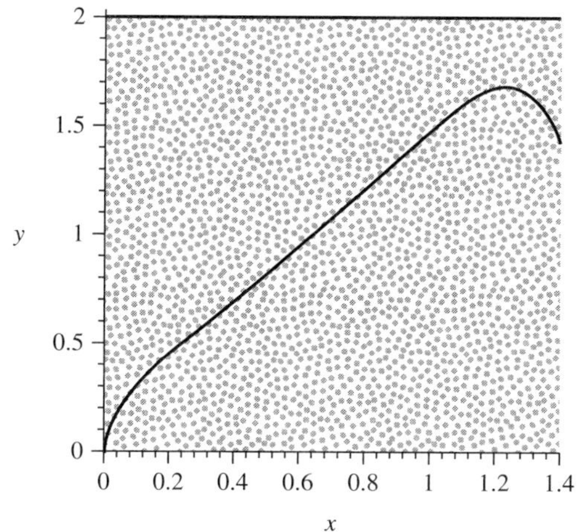

The numerical integration solution is 1.448293361.
The percent error for this run when $N = 2,000$ is only 0 .927 percent.

## Example 3

## Finding the Volume in the First Octant

We can also extend this concept to multiple dimensions. We develop an algorithm for the volume under a surface in the first octant (Figure 16.9).

Table 16.2 provides the numerical output. The actual volume in the first octant is $\pi/6$ (with radius as 1). We take $\pi/6$ to four decimals as 0.5236 cubic units. Figure 16.10 graphically displays the algorithm.

Generally, though not uniformly, the percentage errors become smaller as the number of points $N$ is increased.

FIGURE 16.9

Algorithm for volume under a nonnegative surface in the first octant

Monte Carlo Volume Algorithm

**INPUT**   The total number of random points, N. The nonnegative function, f(x), the interval for x [a,b], interval for y [c,d] and an interval for z [0,M] where M > Max f(x,y), $a < x < b$, $c < y < d$

**OUTPUT**   The approximate volume enclosed for the function f(x,y) in the first octant, $x > 0, y > 0,$ and $z > 0.$

**Step 1.**   Set all counters at 0
**Step 2.**   For i from 1 to N do step 3- 5
**Step 3.**   Calculate random coordinates in the rectangular region:

$$a < xi < b, c < yi < d, 0 < zi < M$$

**Step 4.**   Calculate f(xi, yi)
**Step 5.**   Compare f(xi, yi) and zi . If zi < f(xi, yi) then increment counter by 1. Otherwise, do not increment counter.

**Step 6.**   Estimate the Volume by $V = (M - 0) \cdot (c - d) \cdot (b - a) \cdot \dfrac{counter}{N}$

**Stop**

**Table 16.2** Volume in first octant

| Number of Points | Approximate Volume | Percent Error (%) |
|---|---|---|
| 100 | 0.47 | 10.24 |
| 200 | 0.595 | 13.64 |
| 300 | 0.5030 | 3.93 |
| 500 | 0.514 | 1.833 |
| 1,000 | 0.518 | 1.069 |
| 2,000 | 0.512 | 2.21 |
| 5,000 | 0.518 | 1.069 |
| 10,000 | 0.5234 | 0.13368 |
| 20,000 | 0.5242 | 0.11459 |

**FIGURE 16.10**

A 3-D view of the graphical output from our algorithm $N = 1,000$, $V = 518$

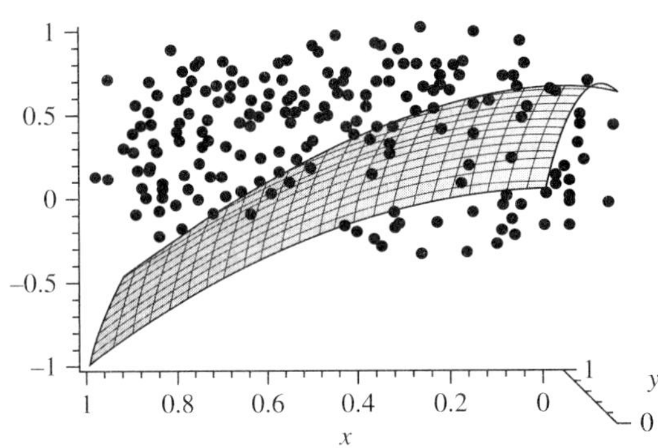

## 16.2 | EXERCISES

1. Use Monte Carlo simulation to approximate the area under the curve $f(x) = 1 + \sin x$ over the interval $\frac{-\pi}{2} \leq x \leq \frac{\pi}{2}$.

2. Use Monte Carlo simulation to approximate the area under the curve $f(x) = x^{0.5}$ over the interval $\frac{1}{2} \leq x \leq \frac{3}{2}$.

3. Use Monte Carlo simulation to approximate the area under the curve $f(x) = \sqrt{1 - x^2}$ over the interval $0 \leq x \leq 1$.

4. How would you modify question 3 to obtain an approximation to $\pi$?

5. Use Monte Carlo simulation to approximate the volume under the surface $f(z) = x^2 + y^2$, in the first octant.

# 16.3  PROBABILITY AND MONTE CARLO SIMULATION USING PROBABILISTIC BEHAVIOR

Let's consider the following probabilistic examples.

1. Compute the probability of getting a head or a tail if you flip a fair coin.
2. Compute the probability of rolling a number from 1 to 6 using a fair die.

## Example 1     Flip a Fair Coin

Algorithm

**Input:**   The number of trials, $N$
**Output:**   The probability of a head or a tail

**Step 1**   Initialize counters to 0.

**Step 2**   For $i = 1, 2, \ldots, N$ do.

**Step 3**   Generate a random number, $x$, $U(0,1)$.

**Step 4**   If $0 \leq x < 0.5$ increment heads, $H = H + 1$, otherwise $T = T + 1$.

**Step 5**   Output $H/N$ and $T/N$, the probabilities for heads and tails.

```
> restart;
> flip:=proc(n)option remember;
> head:=0:tail:=0;
> i:=0;
> while i<n do
> seed:=n:
> x[i]:=(rand( )/10^12);
> if x[i]<=0.500 then
> head:=head+1;
> i:=i+1;
> else;
> tail:=tail+1:i:=i+1;
> end if;
> end do;
> prob_head:=evalf(head/n);
> prob_tail:=evalf(tail/n);
> print (prob_head,prob_tail);
> end;
```

$$flip := \mathbf{proc}\ (n)$$
$$\mathbf{local}\ head,\ tail,\ i,\ seed,\ x,\ prob\_head,\ prob\_tail;$$
$$\mathbf{option}\ remember;$$
$$head := 0;$$
$$tail := 0;$$
$$i := 0;$$
$$\mathbf{while}\ i < n\ \mathbf{do}$$
$$seed := n;$$
$$x[i] := 1/1000000000000 \times \mathrm{rand}(\ );$$
$$\mathbf{if}\ x[i] \leq 0.500\ \mathbf{then}\ head := head + 1\ ;\ i := i + 1$$
$$\mathbf{else}\ tail := tail + 1\ ;\ i := i + 1$$
$$\mathbf{end\ if}$$
$$\mathbf{end\ do};$$
$$prob\_head := \mathrm{evalf}(head/n); \qquad \mathrm{print}(prob\_head,\ prob\_tail)$$
$$prob\_tail := \mathrm{evalf}(tail/n); \qquad \mathbf{end\ proc}$$

```
> flip(10);
```

$$0.5000000000,\ 0.5000000000$$

```
> flip(1000);
```

$$0.5310000000,\ 0.4690000000$$

*> flip(2000);*

$$0.5175000000, 0.4825000000$$

*> flip(5000);*

$$0.5058000000, 0.4942000000$$

*> flip(10000);*

$$0.4968000000, 0.5032000000$$

*> flip(20000);*

$$0.5016000000, 0.4984000000$$

---

## Example 2    Roll of a Fair Die

Rolling a fair die adds the additional process of multiple assignments (six for a six-sided die). The probability will be the number of occurrences of each number divided by the total number of trials.

**Input:**    Number of rolls
**Output:**    Probability of getting a $\{1,2,3,4,5,6\}$

**Step 1**    Initialize all counters (counter 1 through counter 6) to 0.

**Step 2**    For $i = 1, 2, \ldots , n$, do steps 3 and 4.

**Step 3**    Obtain a random number $j$ from integers $(1,6)$.

**Step 4**    Increment the counter for the value of $j$ so that

$$\text{Counter } j = \text{counter } j + 1$$

**Step 5**    Calculate the probability of each roll $\{1,2,3,4,5,6\}$ by

$$\text{Counter } j/n$$

**Step 6**    Output probabilities.

**Step 7**    Stop.

**Roll-a-Fair-Die Program**

```
> rolldie:=proc(n)
> c1:=0:c2:=0:c3:=0:c4:=0:c5:=0:c6:=0;
> # Loop
> for i from 1 to n do
> roll:=rand(1..6);
> x:=roll( );
> if x=1 then
> c1:=c1+1;
> else
> if x=2 then
> c2:=c2+1;
> else
> if x=3 then
> c3:=c3+1;
> else
> if x=4 then
> c4:=c4+1;
```

```
> else
> if x=5 then
> c5:=c5+1;
> else
> c6:=c6+1;
> end if;
> end if;
> end if;
> end if;
> end if;
> end do;
>r1:=evalf(c1/n):r2:=evalf(c2/n):r3:=evalf(c3/n):r4:=evalf(c4/n):r5:=evalf
(c5/n):r6:=evalf(c6/n);
> print(r1,r2,r3,r4,r5,r6);
> end;
```

```
rolldie := proc (n)
local c, roll, i, x, r;
    c[1] := 0; c[2] := 0; c[3] := 0; c[4] := 0; c[5] := 0; c[6] := 0;
    roll := rand(1 .. 6);
    for i to n do x := roll();
    c[x] := c[x] + 1
end do;
    for i to 6 do
    r[i] := evalf(c[i]/n)
end do;
    print (r[1], r[2], r[3], r[4], r[5], r[6])
end proc
```

```
> rolldie(10);
```

$$0.3000000000, 0.1000000000, 0.4000000000, 0.1000000000, 0.1000000000, 0.$$

```
> rolldie(100);
```

$$0.1500000000, 0.1900000000, 0.1800000000, 0.1200000000, 0.1400000000, 0.2200000000$$

```
> rolldie(1000);
```

$$0.1800000000, 0.1770000000, 0.1470000000, 0.1550000000, 0.1690000000, 0.1720000000$$

```
> rolldie(10000);
```

$$0.1649000000, 0.1709000000, 0.1615000000, 0.1679000000, 0.1643000000, 0.1705000000$$

The expected probability is 1/6 or 0.1667. We note that as the number of trials increases, the closer our probabilities are to the expected long-run values. We offer the following concluding remark: when you have to run simulations, run them for a very large number of trials.

## 16.3  EXERCISES

Use Monte Carlo Simulation in problems 1-4.

**1.** Obtain 20 flips of a fair coin and approximate the probabilities of a head or a tail.

**2.** Obtain 20 rolls of a fair die.

3. Obtain 2000 flips of a fair coin and approximate the probabilities of a head or a tail.

4. Obtain 2000 rolls of a fair die.

## 16.3 | PROJECTS

1. Blackjack: Construct and Perform a Monte Carlo Simulation

   Also called "21," blackjack is played at most casinos with six or eight decks of cards to inhibit card counters, or people who are able to keep mental track of the cards and their values and thus understand when the odds of various hands are in their favor.

   You will use two decks of cards in your simulation (104 cards total). There are only two players: you and the dealer. Each player receives two cards to begin play. Numbered cards are worth their face value (2–10) and face cards (jack, queen, and king), are worth 10, with aces being worth either 1 point or 11 points. The object of the game is to obtain a total as close to 21 *without going over* (which is called "going bust") so that your total is more than the dealer's. If the first two cards total 21 (ace plus 10 or ace plus a face card), this is called *blackjack* and automatically wins the hand—unless both you and the dealer have blackjack, in which case it is a tie (or "push") and your bet remains on the table. Winning via blackjack pays 3:2 or 1.5:1 (a one-dollar bet reaps $1.50, and you don't lose the dollar you bet).

   If neither you nor the dealer has blackjack, you (the player) can take as many cards as you want, one at a time, to try to get as close to 21 as possible. If you go over 21, you lose, and the game ends. Once you are satisfied with your score, you "stand." The dealer then draws cards according to the following rules:

   The dealer stands on 17, 18, 19, 20, or 21. The dealer must draw a card if the total is 16 or less. The dealer always counts aces as 11 unless it causes him or her to bust, in which case it is counted as a 1. For example, an ace–six combo for the dealer is 17, not 7 (the dealer has no option), and the dealer must stand on 17. However, if the dealer has an ace–four (for 15) and draws a king, then the new total is 15 because the ace reverts to its value of 1 (so as not to go over 21.) The dealer would then draw another card.

   If the dealer goes over 21, you win (even your bet money—you gain $1 for every $1 you bet). If the dealer's total exceeds your total, you lose all the money you bet. If the dealer's total equals your total, it is a "push" (no money exchanges hands—you don't lose your bet but you don't gain any money).

   What makes the game exciting in a casino is that the dealer's original two cards are one up and then one down, so you don't know the dealer's total and must play the odds based on the one card showing. You do not need to incorporate this twist into your simulation for this project.

   Here's what you are required to do. Run through 12 sets of two decks playing the game. You have an unlimited bankroll (don't you wish!) and bet $2 on each hand. Each time the two decks run out, the hand in play continues with two fresh decks (104 cards). At that point, record your standing (plus or minus $X$ dollars). Then start again at 0 for the next deck. So your output will be the 12 results from playing each of the 12 decks, which you can then average or total to determine your overall performance.

   What about *your* strategy? That's up to you! But here's the catch: You will assume that you can see *neither* of the dealer's cards (so you have no idea what cards the dealer has). Choose a strategy to play and then play it throughout the entire simulation. (Blackjack enthusiasts can consider doubling down and splitting pairs into their simulation, but this is not necessary.)

   Provide your instructor with the simulation algorithm, computer code, and output results from each of the 12 decks.

2. Darts: Construct and Perform a Monte Carlo Simulation of a Darts Game

The ring colors and points are shown in the following table.

| Dart Board Area | Points |
|---|---|
| Bull's-eye | 50 |
| Yellow ring | 25 |
| Blue ring | 16 |
| Red ring | 10 |
| White ring | 5 |

The origin (the center of the bull's-eye), the radius of each ring, and the distance from the center are given in the next table.

| Ring | Thickness (in) | Distance to Outer Ring Edge from the Origin (in) |
|---|---|---|
| Bull's-eye | 1 | 1 |
| Yellow | 1.5 | 2.5 |
| Blue | 2.5 | 5 |
| Red | 3 | 8 |
| White | 4 | 12 |

The board has a radius of 1 foot (12″).

Make an assumption about the distribution of how the darts hit on the board. Then compare your assumption about using appropriate areas. Write an algorithm and code it in the computer language of your choice. Run 1,000 simulations to determine the mean score for throwing five darts. Also determine which ring has the highest expected value (point value times the probability of hitting that ring).

3. Craps: Construct and Perform a Monte Carlo Simulation

The rules for this popular casino game are as follow. There are two basic bets in craps: Pass and Don't Pass. In the Pass bet, you wager that the person throwing the two dice (the shooter) will win. In the Don't Pass bet, you wager that the shooter will lose. We will play by the rule that on an initial roll of two sixes for a total of 12 ("boxcars"), both Pass and Don't Pass bets are losers. Both are so-called even-money bets.

Here's how the game is conducted.

Roll a 7 or 11 on the first roll: The shooter wins, Pass bets win, and Don't Pass bets lose.
Roll a 12 (boxcars) on the first roll: The shooter loses and both Pass and Don't Pass bets lose.
Roll a 2 or 3 on the first roll: Shooter loses, Pass bets lose, and Don't Pass bets win.
Roll 4, 5, 6, 8, 9, or 10 on the first roll: The number rolled becomes the "point." The object then becomes to roll the point again before rolling a 7. The shooter continues to roll the dice until the point or a 7 appears.
Pass betters win if the shooter rolls the point again before rolling a 7.
Don't Pass betters win if the shooter rolls a 7 before rolling the point again.

Write an algorithm and code it in the computer language of your choice. Run the simulation to estimate the probability of winning a Pass bet and the probability of winning a Don't Pass bet. Which is the better bet? As the number of trials increases, to what do the probabilities converge?

4. Horse Race: Construct and Perform a Monte Carlo Simulation

You can be creative here and use odds from the newspaper or simulate a "mathematical derby" with entries and odds from the following table.

| Horse Entry | Odds |
| --- | --- |
| Euler's Folly | 7−1 |
| Leapin' Leibnitz | 5−1 |
| Newton Lobell | 9−1 |
| Count Cauchy | 12−1 |
| Pumped Up Poisson | 4−1 |
| Loping L'Hopital | 35−1 |
| Steamin' Stokes | 15−1 |
| Dancing Danzig | 4−1 |

Construct and perform a Monte Carlo simulation of 1,000 horse races. Which horse won the most races? Which horse won the least number of races? Do these results surprise you? Provide the tallies of how many races each horse won with your output.

5. Roulette. An American roulette wheel has 38 spaces numbered 0, 00, and 1 through 36. Half the spaces numbered 1–36 are red, and half are black. The two spaces 0 and 00 are green.

   Simulate the playing of 1,000 games betting either red or black (which pay even money, 1:1). Bet $1 on each game and keep track of your earnings. What are the earnings per game betting red or black according to your simulation? What was your longest winning streak? Longest losing streak?

   Now simulate 1,000 games betting green (pays 17:1, so if you win, you add $17 to your kitty, and if you lose, you lose $1). What are your earnings per game betting green according to your simulation? How does it differ from your earnings betting red or black? What was your longest winning streak betting green? Longest losing streak? Which strategy do you recommend using and why?

6. *The Price Is Right.* On the popular TV game show *The Price Is Right*, at the end of each half hour, the three winning contestants face off in what is called the "Showcase Showdown." The game consists of spinning a large wheel with 20 spaces on which the pointer can land; the spaces are numbered from $0.05 to $1.00 in 5-cent increments. The contestant who has won the least amount of money at this point in the show spins first, followed by the one who has won the next most, and then by the biggest winner for that half hour.

   The objective of the game is to obtain as close to $1.00 as possible without going over that amount with an allowed maximum of two spins. Naturally, if the first player does not go over, the other two will use one or both spins in their attempts to overtake the leader.

   But what of the person spinning first? If he or she is an expected-value decision maker, how high a value on the first spin does he or she need to not want to take a second spin? Remember, the person can lose if
   a. either of the other two players surpasses the player's total, or
   b. the player spins again and goes over.

7. *Let's Make a Deal.* You are "dressed to kill" in your favorite costume, and host Monte Hall picks you out of the audience. You are offered the choice of three wallets. Two wallets contain a single $50 bill, and the third contains a $1,000 bill. You choose one of the three wallets. Monte knows which wallet contains the $1,000, so he shows you one of the other two wallets—one with one of the two $50 bills inside. Monte does this on purpose because he must have at least one wallet with $50 inside. If he holds the $1,000 wallet, he shows you the other wallet, the one with $50. Otherwise, he just shows you one of his two $50 wallets. Monte then asks you if you want to trade your choice for the one he's still holding. Should you trade?

   Develop an algorithm and construct a computer simulation to support your answer.

# 16.4  APPLIED SIMULATION MODELS

In this section, we present algorithm and Maple code for the following simulations.

1. an aircraft missile attack, and
2. the amount of gas that a series of gas stations will need.

## Example 1

## Missile Attack

An analyst plans a missile strike using F-15 aircraft. The F-15 must fly through air-defense sites that hold a maximum of eight missiles. It is vital to ensure success early in the attack. Each aircraft has a probability of 0.5 of destroying the target, assuming it can get to the target through the air-defense systems and then acquire and attack its target. The probability that a single F-15 will acquire a target is approximately 0.9. The target is protected by air-defense equipment with a 0.40 probability of stopping the F-15 from either arriving at or acquiring the target. How many F-15 are needed to have a successful mission assuming we need a 99-percent success rate?

**Algorithm:**  Missiles
**Input:**  $N$ = number of F-15s
$M$ = number of missiles fired
$P$ = probability that one F-15 can destroy the target
$Q$ = probability that air defense can disable an F-15
**Output:**  $S$ = probability of mission success

**Step 1**  Initialize $S = 0$
**Step 2**  For $I = 0$ to $M$ do
**Step 3**  $P(i) = [1 - (1 - P)^{N-I}]$
**Step 4**  $B(i) = $ binomial distribution for $(m,i,q)$
**Step 5**  Compute $S = S + P(i) * B(i)$
**Step 6**  Output $S$.
**Step 7**  Stop.

```
> restart;
> with(Statistics):
> bombsaway:=proc(n,m,p,q)
> s:=0:qn:=q:nn:=n:pn:=p;
> for i from 0 to m do
> pn:=1-(1-pn)^(nn-i);
> x:=RandomVariable(Binomial(m,qn));
> b:=ProbabilityFunction(x, i);
> s:=s+pn*b;
> end do;
> print(s);
> end;
```

$$bombsaway := \text{proc}(n, m, p, q)$$
$$\text{local } s, qn, nn, pn, i, x, b;$$
$$s := 0;$$
$$qn := q;$$
$$nn := n;$$
$$pn := p;$$
$$\text{for } i \text{ from } 0 \text{ to } m \text{ do} \quad s := s + pn \times b$$
$$pn := 1 - (1 - pn)^{\wedge}(nn - i); \quad \text{end do;}$$
$$x := Statistics:-RandomVariable (Binomial(m, qn)); \quad print(s)$$
$$b := Statistics:-ProbabilityFunction (x, i); \quad \text{end proc}$$

```
>
```

We run the simulation letting the number of F-15s vary and calculate the probability of success.

```
> for i from 1 to 10 do
> bombsaway(i,8,.45,.4);
> end do;
```

$$
\begin{aligned}
&0.0075582720 \\
&0.07419703680 \\
&0.3043347039 \\
&0.5924805227 \\
&0.8254837008 \\
&0.9497277101 \\
&0.9912246093 \\
&0.9992039991 \\
&0.9999226475 \\
&0.9999574561
\end{aligned}
$$

We find that seven F-15s gives us $P(s) = 0.99122$.

Actually, any number of F-15 greater than seven provides a result with the probability of success we desire. Ten F-15s yielding a $P(s) = 0.999957$ would suffice. Any more would be overkill.

---

| Example 2 | Gasoline-Inventory Simulation |
|---|---|

## Background

You are a consultant to an owner of a chain of gasoline stations along a freeway. The owner wants to maximize profits and meet consumer demand for gasoline. You decide to look at the following problem.

## Problem Identification Statement

Minimize the average daily cost of delivering and storing sufficient gasoline at each station to meet consumer demand.

## Assumptions

For an initial model, consider that, in the short run, the average daily cost is a function of demand rate, storage costs, and delivery costs. You also assume that you need a model for the demand rate. You decide that historical data will assist you.

| Demand: Number of Gallons | Number of Occurrences (Days) |
|---|:---:|
| 1,000–1,099 | 10 |
| 1,100–1,199 | 20 |
| 1,200–1,299 | 50 |
| 1,300–1,399 | 120 |
| 1,400–1,499 | 200 |
| 1,500–1,599 | 270 |
| 1,600–1,699 | 180 |
| 1,700–1,799 | 80 |
| 1,800–1,899 | 40 |
| 1,900–1,999 | 30 |
| Total number of days = | 1000 |

## Model Formulation

We convert the number of days into probabilities by dividing by the total and we use the midpoint of the interval of demand for simplification.

| Demand: Number of Gallons | Probabilities |
|---|---|
| 1,000 | 0.010 |
| 1,150 | 0.020 |
| 1,250 | 0.050 |
| 1,350 | 0.120 |
| 1,450 | 0.200 |
| 1,550 | 0.270 |
| 1,650 | 0.180 |
| 1,750 | 0.080 |
| 1,850 | 0.040 |
| 2,000 | 0.030 |
| Total number of days = | 1.000 |

Because cumulative probabilities will be more useful we convert to a CDF.

| Demand: Number of Gallons | Probabilities |
|---|---|
| 1,000 | 0.010 |
| 1,150 | 0.030 |
| 1,250 | 0.080 |
| 1,350 | 0.20 |
| 1,450 | 0.4 |
| 1,550 | 0.670 |
| 1,650 | 0.850 |
| 1,750 | 0.93 |
| 1,850 | 0.97 |
| 2,000 | 1.0 |

We will use cubic splines to model the function for demand (see Chapter 6 for a discussion of cubic splines).

### Inventory Algorithm

**Input:**  $Q$ = delivery quantity in gallons
$T$ = time between deliveries in days
$D$ = delivery cost in dollars per delivery
$S$ = storage costs in dollars per gallons
$N$ = number of days in the simulation

**Output:**  $C$ = average daily cost

**Step 1**  Initialize: Inventory $\rightarrow I = 0$ and $C = 0$.

**Step 2**  Begin the next cycle with a delivery:

$$I = I + Q$$
$$C = C + D$$

**Step 3**  Simulate each day of the cycle.

For $i = 1, 2, \ldots, T$, do steps 4–6.

**Step 4**    Generate a demand, $q_i$. Use cubic splines to generate a demand based on a random CDF value, $x_i$.

**Step 5**    Update the inventory: $I = I - q^i$.

**Step 6**    Calculate the updated cost: $C = C + s * I$ if the inventory is positive.

If the inventory is $\leq 0$, then set $I = 0$ and go to step 7.

**Step 7**    Return to step 2 until the simulation cycle is completed.

**Step 8**    Compute the average daily cost: $C = C/n$.

**Step 9**    Output $C$.
Stop.

Maple program

```
> inventorygas:=proc(q,t,d,s,n,xdat,qdat)
> #print(xdat,qdat);
> k:=n:i:=0:c:=0:Flag:=0:nq:=q: nt:=t:nd:=d:ns:=s;
> label_2:i:=i+nq;#print(i,nt);
> c:=c+nd:#print(c);
> if(nt>=k) then
> nt:=k: Flag:=1;
> readlib(spline):
> spline(xdat,qdat,x,cubic);
> nfunc:=unapply(%,x):
> end if;
> for j from 1 to nt do
> readlib(spline):
> spline(xdat,qdat,x,cubic);
> nfunc:=unapply(%,x):
> nx:=evalf(rand()/(1.0*10^12));
> newq:=evalf(nfunc(nx));
> i:=i-newq:#print(i);
> if (i<=0) then
> i:=0;
> goto(label_9);
> else
> c:=c+i*ns:#print(c);
> end if;
> label_9;
> k:=k-1: #print(k);
> if k>0 then
> goto(label_2);
> else
> newc:=c/n:print(newc);
> goto(label_3);
> end if;
> print(newc);
> end do;
> label_3;
> print(i,newc);
> end;
```

$$inventorygas := \mathbf{proc}(q,\ t,\ d,\ s,\ n,\ xdat,\ qdat)$$
$$\mathbf{local}\ k,\ i,\ c,\ Flag,\ nq,\ nt,\ nd,\ ns,\ nfunc,\ j,\ nx,\ newq,\ newc;$$
$$k := n;$$
$$i := 0;$$
$$c := 0;$$
$$Flag := 0;$$

```
               nq :=q;
               nt :=t;
               nd :=d;
               ns :=s;
               label_2;
               i :=i + nq;
               c :=c + nd;
               if k <= nt then
                   nt := k;
                   Flag := 1;
                   readlib(spline);
                   spline(xdat, qdat, x, cubic);
                   nfunc :=unapply('%', x)
               end if;
               for j to nt do
                   readlib(spline);
                   spline(xdat, qdat, x, cubic);
                   nfunc :=unapply('%', x);
                   nx :=evalf(rand( )/(1.0*1000000000000));
                   newq :=evalf(nfunc (nx));
                   i :=i − newq;
                   if i <= 0 then
                       i :=0;
                       goto (label_9)
                   else
                       c :=c + i*ns
                   end if;
                   label_9;
                   k :=k − 1;
                   if 0 < k then
                       goto (label_2)
                   else
                       newc :=c/n;
                       print(newc);
                       goto (label_3)
                   end if;
                   print(newc)
               end do;
               label_3;
               print(i, newc)
           endproc
```

> *xdat:=[0,.01,.03,.08,.2,.4,.67,.85,.93,.97,1];*

$$xdat := [0, 0.01, 0.03, 0.08, 0.2, 0.4, 0.67, 0.85, 0.93, 0.97, 1]$$

> *qdat:=[1000,1050,1150,1250,1350,1450,1550,1650,1750,1850,2000];*

$$qdat := [1000, 1050, 1150, 1250, 1350, 1450, 1550, 1650, 1750, 1850, 2000]$$

> *inventorygas( 11500,7,500,.05,20,xdat,qdat );*

$$5753.039330$$
$$199862.4518,\ 5753.039330$$

The average cost is \$5,753.04, and the inventory on hand is 199,862.4518 gallons.

## 16.4 | EXERCISES

1. Modify the missile strike problem if the probability of $S$ were only 0.95 and the probability of an F-15 being deterred by air defense were 0.3. Determine the number of F-15s needed to complete the mission.

2. What if in the missile attack problem the air-defense units were modified to carry 10 missiles each? What effect does that have on the number of F-15s needed?

3. Perform sensitivity analysis on the gasoline-inventory problem by modifying the delivery to 11,450 gallons per week. What effect does this have on the average daily cost?

## 16.4 | PROJECTS

1. Tollbooths. Heavily traveled toll roads such as the Garden State Parkway, Interstate 95, and so forth, are multilane divided highways that are interrupted at intervals by toll plazas. Because collecting tolls is usually unpopular, it is desirable to minimize motorist annoyance by limiting the amount of traffic disruption caused by the toll plazas. Commonly, a much larger number of tollbooths are provided than the number of travel lanes entering the toll plaza. On entering the toll plaza, the flow of vehicles fans out to the larger number of tollbooths; when leaving the toll plaza, the flow of vehicles is forced to squeeze down to a number of travel lanes equal to the number of travel lanes before the toll plaza. Consequently, when traffic is heavy, congestion increases when vehicles leave the toll plaza. When traffic is very heavy, congestion also builds at the entry to the toll plaza because of the time required for each vehicle to pay the toll.

   Construct a mathematical model to help you determine the **optimal number** of tollbooths to deploy in a barrier-toll plaza. Explicitly, first consider the scenario in which there is exactly one tollbooth per incoming travel lane. Then consider multiple tollbooths per incoming lane. Under what conditions is one tollbooth per lane more or less effective than the current practice? Note that the definition of *optimal* is up to you to determine.

2. Major League Baseball. Build a simulation to model a baseball game. Use your two favorite teams or favorite all-star players to play a regulation game.

3. NBA Basketball. Build a simulation to model the NBA basketball playoffs.

4. Hospital Facilities. Build a simulation to model surgical and recovery rooms for the hospital.

5. Class Schedules. Build a simulation to model the registrar's scheduling changes for students or final exam schedules.

6. Automobile Emissions. Consider a large engineering company that performs emissions control inspections on automobiles for the state. During the peak period, cars arrive at a single location that has four lanes for inspections following exponential arrivals with a mean of 15 minutes. Service times during the same period are uniform: between [15,30] minutes. Build a simulation for the length of the queue. If cars wait more than 1 hour, the company pays a penalty of $200 per car. How much money, if any, does the company pay in penalties? Would more inspection lanes help? What costs associated with the inspection lanes need to be considered?

## 16.4 | FURTHER READING

F. R. Giordano, M. D. Weir, and W. P. Fox. *A First Course in Mathematical Modeling.* 3rd Ed. Pacific Grove, CA: Brooks-Cole. 2003.

Law, A., and D. Kelton. *Simulation Modeling and Analysis.* 4th Ed. New York: McGraw Hill. 2007.

Meerschaert, M. M. *Mathematical Modeling.* San Diego: Academic Press. 1993.

Winston, W. *Operations Research: Applications and Algorithms.* 3rd Ed. Belmont, CA: Duxbury Press. 1994.

# PROGRAMS IN MAPLE

Programs have been written in Maple and are available from the accompanying web site for this text.

### Area Under Curve with Points for Any Nonnegative Function, $f(x)$, Over a Domain $a < x < b$

```
> Area:=proc(f::procedure,f1::algebraic,an::numeric, bn::numeric, ymin::numeric,
ymax::numeric,N::posint)option remember;
> local A,B,y1,y2,areaR,count,i,area,xpt,ypt,xrpt,yrpt,gen_x,gen_y,c1,c2,c3,c4,c5;
> y1:=ymin;
> y2:=ymax;
> A:=an;
> B:=bn;
> areaR:=(B–A)*(y2–y1);
> count:=0;
> i:=0;
> gen_x:=evalf(A+(B–A)*rand(0..10^5)/10^5):
> gen_y:=evalf(y1+(y2–y1)*rand(0..10^5)/10^5):
> xrpt:=[ ]:
> yrpt:=[ ]:
> while i<N do
Step 1
> xpt[i]:=A+(B–A)*(rand( )/10^12);
> ypt[i]:=y1+(y2–y1)*(rand( )/10^12);
Step 2
> xrpt:=[op(xrpt),gen_x( )]:
> yrpt:=[op(yrpt),gen_y( )]:
> if ypt[i]<evalf(f(xpt[i])) then
> count:=count+1;
> i:=i+1;
> else
> i:=i+1;
> fi;
> od;
> area:=areaR*count/N;
> print(area);
> with(plots):
> c1:=pointplot(zip((x,y)–>[x,y],xrpt,yrpt)):
> c2:=plot(f1,x=A..B,y=y1..y2):
> c3:=area:
> c4:=plot(y2,x=A..B,thickness=3,colour=red):
> c5:=plot(y1,x=A..B,thickness=3,colour=blue):
> display(c1,c2,c4,c5);
> end:
> f:=x–>x^2;;
```

$$f := x \rightarrow x^2$$

> *f1:=x^2;*

$$f1 := x^2$$

> *Area(f,f1,−1.57,1.57,0,2,200);*

$$2.857400000$$

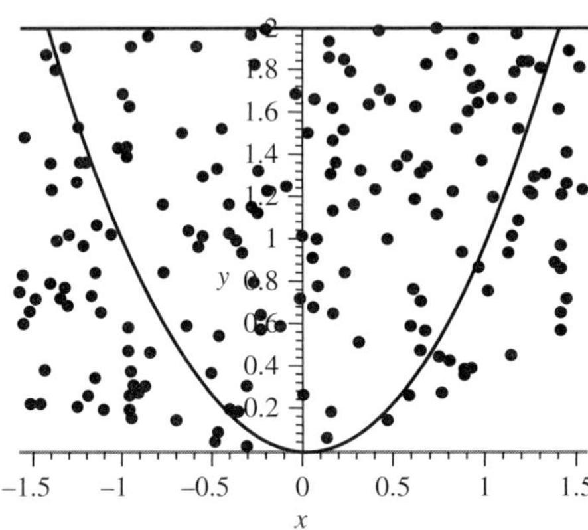

> *int(x^2,x=−1.57..1.57);*

$$2.579928667$$

**Simulation**

**Deterministic Modeling: Volume Under a Surface** Start with the curve $f(x,y,z) = x$^2$ + y$^2$ + z$^2$ − 1 with $x > 0$, $y > 0$, $z > 0$.

Use the algorithm from the text, pages 221−222.

This program will take any given function and approximate the area for *n* trials.

> *restart;*
> *f:=(x,y)−>(1−(x^2+y^2));*

$$f := (x, y) \rightarrow 1 - x^2 - y^2$$

> *Volume:=proc(f::procedure,f1::algebraic,an::numeric, bn::numeric, ymin::numeric,*
*ymax::numeric,zlow::numeric, zhigh::numeric,N::posint)option remember;*
> *local A,B, y1,y2,zz1,zz2,c1,c2,c3,c4,c5, volumeR, count,xrpt,yrpt,zrpt,gen_x,gen_y,gen_z,xpt,*
*ypt,zpt,volume;*
> *y1:=ymin;*
> *y2:=ymax;*
> *A:=an;*
> *B:=bn;*
> *zz1:=zlow;*
> *zz2:=zhigh;*
> *volumeR:=(B−A)*(y2−y1)*(zz2−zz1);*
> *count:=0;*
> *i:=0;*
> *gen_x:=evalf(A+(B−A)*rand(0..10^5)/10^5):*
> *gen_y:=evalf(y1+(y2−y1)*rand(0..10^5)/10^5):*
> *gen_z:=evalf(zz1+(zz2−zz1)*rand(0..10^5)/10^5):*
> *zrpt:=[]:*
> *xrpt:=[]:*
> *yrpt:=[]:*
> *zrpt:=[]:*
> *while i < N do*

```
> xpt[i]:=A+(B-A)*(rand( )/10^12);
> ypt[i]:=y1+(y2-y1)*(rand( )/10^12);
> zpt[i]:=zz1+(zz2-zz1)*(rand( )/10^12);
> xrpt:=[op(xrpt),gen_x( )]:
> yrpt:=[op(yrpt),gen_y( )]:
> zrpt:=[op(zrpt),gen_z( )]:
> if evalf((zpt[i])^2)<evalf(f(xpt[i],ypt[i])) then
> count:=count+1;
> i:=i+1;
> else
> i:=i+1;
> fi;
> od;
> volume:=volumeR*count/N;
> with(plots):
> print(volume);
> c1:=pointplot3d( {seq([xrpt[i],yrpt[i],zrpt[i]], i=1..nops(zrpt)) }, symbol=circle,
color=blue):
> c2:=plot3d(f1,x=A..B,y=y1..y2):
> #c4:=plot3d(y2,x=A..B,y=y1..y1,thickness=3, colour=red):
> #c5:=plot3d(y1,x=A..B,y=y1..y2,thickness=3,colour=blue):
> display(c1,c2);
> end:
> f:=(x,y)->1-x^2-y^2;
```

$$f := (x, y) \rightarrow 1 - x^2 - y^2$$

```
> f1:=1-x^2-y^2;
```

$$f1 := 1 - x^2 - y^2$$

```
> Volume(f,f1,0,1,0,1,0,1,300);
```

$$\frac{11}{20}$$

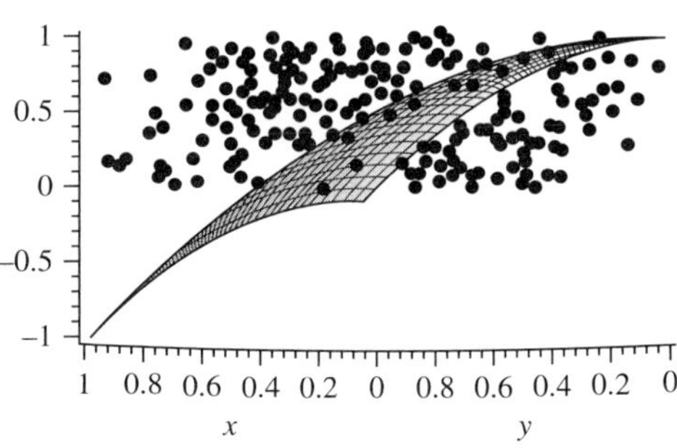

```
> convert(%,float);
```

$$.5180000000$$

Check simulation (when possible).

```
> convert(Pi/6,float);
```

$$.5235987758$$

The exact answer is pi/6 or about 0.5235987758. Notice that our volume is, in fact, an approximation.

# 17

# Modeling with Game Theory

## Introduction

Conflict has been a central concern in human history. It arises when two or more individuals with different views, goals, or aspirations compete to control the course of future events. Game theory uses mathematics and mathematical tools to study situations in which rational players are involved in conflict and cooperation. Game theory is a branch of applied mathematics that is used in the social sciences (most notably in economics), biology, decision sciences, engineering, political science, international relations, operations research, applied mathematics, computer science, and philosophy. Game theory attempts to mathematically capture behavior in strategic situations in which an individual's success in making choices depends on the choices of others. Although initially developed to analyze competitions in which one individual does better at another's expense (zero-sum games), it has been expanded to treat a wide class of interactions that are classified according to several criteria.

Let's first consider zero-sum games. The players in a game may be individuals, groups of people, organizations, or countries. These players have strategies or courses of action that they could follow. These strategies lead to outcomes that describe the results of the decisions made. We assume that all players have certain preferred outcomes. We further assume that all players are rational in their decisions.

Outcomes or payoffs for a game may come from calculated values or expected values, ordinal ranking, or cardinal values developed from a lottery system (see Von Neumann and Morgenstern). In this chapter, we will assume that we have these outcomes or payoffs for our games.

## Example 1    Coin Matching

We have two players who each have a coin that they choose to show a head or a tail. The bet is always \$1. If both coins match (both heads or both tails), then player A wins the dollar; if the coins are different (one head and one tail), then player B wins the dollar. The payoff matrix for the game would like the following for player A:

|  |  | Player B | |
|---|---|---|---|
|  |  | **Heads** | **Tails** |
| **Player A** | **Heads** | $1 | −$1 |
|  | **Tails** | −$1 | $1 |

In Section 17.1 we examine zero-sum games. We will illustrate the movement diagrams to look for dominance and saddle-point solutions, if they exist. We illustrate the linear programming formulation and the primal–dual solution method to obtain the game solutions. All other methods of solutions can be found in a game theory course but will not be covered here.

In Section 17.2, we examine the non–zero-sum game. We will illustrate the movement diagrams for non-zero sum games to look for dominance and for dominant strategies leading to solutions, if they exist. We illustrate the linear programming formulation and the primal–primal solution method to obtain the game solutions as all players now seek to maximize their preferred outcome. We use the concept that every non–zero-sum game has a Nash's equalizing mixed-strategy solution even if it has a pure-strategy equilibrium. According to Gillman and Housman (2009, p. 189) non–zero-sum games with pure-strategy equilibrium also have non–pure-strategy equilibriums using Nash's equalizing strategy.

In Section 17.3, we show how to apply nonlinear optimization to the Nash arbitration scheme. Both Nash equilibrium and Nash arbitration solutions to non–zero-sum games are explored. The Nash equilibrium is found using linear programming formulations, but now all players are maximizing. A separate linear programming formulation is required for each player, which often is easier for solving Nash equilibrium problems than the alternative methods suggested in other textbooks. The reduced costs will be conjectured.

In Section 17.4, we present an illustrative modeling scenario using the techniques from Sections 17.1–17.3. We also look at the Nash arbitration as an example of constrained nonlinear optimization. We finish will several modeling scenarios.

Throughout this chapter, we provide examples from game theory. We will illustrate a few basic procedures but will concentrate on using linear programming for the zero-sum game using Equation (17.2) (in Section 17.2), and we conclude that often the linear programming model is best when games are lager than $2 \times 2$.

## 17.1  TWO-PERSON ZERO-SUM GAMES

In this section, we examine zero-sum games such as the previous coin-matching example. We will illustrate the movement diagrams to look for dominance and saddle-point solutions if they exist. We illustrate the linear programming formulation and the primal–dual solution method to obtain the game solutions. We must begin with some basic explanations concerning game theory. Characteristics of the two-person zero-sum game include the following.

1. There are two persons, one called the *row player* (who we will refer to as Rose) and the other called the *column player* (who we will refer to as Colin).

2. Rose must choose 1 of $m$ strategies, and Colin must choose one of $n$ strategies.

3. If Rose chooses the $i$th strategy and Colin the $j$th strategy, then Rose receives a payoff of $a_{ij}$ and Colin loses an amount $a_{ij}$.

In general the payoff matrix look as shown in Table 17.1.

For two-player, finite zero-sum games, the different game-theoretic solution concepts of Nash equilibrium, minimax, and maximin all give the same solution. We begin examining the payoff matrix. We look for dominance first, and the movement diagram is a useful technique. We illustrate how linear programming works well with both mixed-strategy and pure-strategy solutions.

**Table 17.1** Payoff matrix of a two-person zero-sum game

| | | Colin's Strategies | | | |
|---|---|---|---|---|---|
| | | **Column 1** | **Column 2** | ... | **Column $n$** |
| | **Row 1** | $a_{11}$ | $a_{12}$ | ... | $a_{1n}$ |
| | **Row 2** | $a_{21}$ | $a_{22}$ | ... | $a_{2n}$ |
| **Rose's Strategies** | . | . | . | . | . |
| | . | . | . | . | . |
| | . | . | . | . | . |
| | **Row $m$** | $a_{m1}$ | $a_{m2}$ | ... | $a_{mn}$ |

Traditional applications of game theory provide techniques to find an equilibrium value in zero-sum games. John F. Nash, Jr., proved that every game has an equilibrium at which point each player has adopted a strategy that is unlikely to change; this solution is thus called the *Nash equilibrium*. This methodology is not without criticism, and debates continue over the appropriateness of particular equilibrium concepts, the appropriateness of equilibrium altogether, and the usefulness of mathematical models. However, the Nash equilibrium is still widely used.

**Finding the Nash Equilibrium**

We will present only two methods in looking for the Nash equilibrium: (1) pure-strategy equilibrium through movement diagram and linear programming and (2) mixed-strategy equilibrium through linear programming. We begin with movement diagrams.

> Movement Diagrams (Zero-Sum Games)
>
> In each row, draw an arrow from an entry to the smallest entry in that row. In each column, draw arrows to the largest entry in each column. When all arrows point to a point (or points), these points describe the pure Nash equilibrium.

Let's begin with an example.

---

## Example 1     Predator–Prey Game

Suppose we have a predator and a prey. The predator has two strategies for catching the prey: (1) lie in wait and ambush and (2) fast pursuit. The prey has two strategies for escaping: (1) hide and (2) run faster than the predator. We provide a payoff matrix in Table 17.2. The value indicates the probability that prey is not eaten (escapes). We want to calculate what strategies the prey and predator should use to maximize their respective benefits. The prey is a pessimist: It tries to make the best of the worst of all possible outcomes.

We start with the movement diagram. In the row "Hide," we draw an arrow from 0.4 to 0.2; in row "Run," we draw an arrow from 0.8 to 0.6. In column "Ambush," we draw an arrow from 0.2 to 0.8; and in column "Pursue," we draw an arrow from 0.4 to 0.6.

**Table 17.2** Payoff matrix for predator–prey game

| | | Predator | |
|---|---|---|---|
| | | **Ambush** | **Pursue** |
| **Prey** | **Hide** | 0.2 ⟵ | 0.4 |
| | **Run** | 0.8 ⟶ | 0.6 |

The arrows point to 0.6. This represents a strategy by the prey to run and for the predator to pursue. This value is a pure-strategy equilibrium.

## Linear Programming of Zero-Sum Games

Consider a constant-sum two person game in which maximizing player X has $m$ strategies and minimizing player Y has $n$ strategies. The entry $a_{ij}$ from the $i$th row and $j$th column of the payoff matrix represents the value of the game. Without loss of generality, we can assume that every element of the matrix is greater than or equal to zero. If this is not true, then a constant can be added to every element in the payoff matrix to make all the entries positive.

Suppose that player X plays a weighted mixed strategy defined by assigning a weight $x_i$ to the $i$th strategy in which $\Sigma x_i = 1$. Then, according to Dorfman (1951) and Danzig (1951), the value of the game will be

$$v = \sum_{j=1}^{n} y_j \, (a_1 x_1 + a_2 x_2 + \ldots + a_m x_m)$$

where $y_j$ is the $j$th strategy for player Y. We use this concept to construct the linear programming equivalent.

The Nash equilibrium for a two-player, zero-sum game can be found by solving a linear programming problem. Suppose a zero-sum game has a payoff matrix $M$ in which element $M_{i,j}$ is the payoff obtained when the minimizing player chooses pure strategy $i$ and the maximizing player chooses pure strategy $j$ (i.e., the player trying to minimize the payoff chooses the row, and the player trying to maximize the payoff chooses the column). In their work, Danzig and Dorman assume that every element of $M$ is positive. A general payoff matrix would be

$$M = \begin{bmatrix} M_{1,1} & M_{1,2} & \cdots & M_{1,n} \\ M_{2,1} & M_{2,2} & \cdots & M_{2,n} \\ \cdot & \cdot & \cdots & \cdot \\ \cdot & \cdot & \cdots & \cdot \\ \cdot & \cdot & \cdots & \cdot \\ M_{m,1} & M_{m,2} & \cdots & M_{m,n} \end{bmatrix}$$

The game will have at least one Nash equilibrium. The Nash equilibrium can be found by solving the following linear program to find a vector $u$ as shown by Dorfman (1951) to solve for the solution to the column player's game:

$$\text{Minimize } u_1 + u_2 + u_3 + \ldots + u_n \tag{17.1}$$

Subject to:
$$u_1 \geq 0, u_2 \geq 0, u_3 \geq 0, \ldots, u_n \geq 0$$
$$M_{1,1} u_1 + M_{1,2} u_2 + \ldots + M_{1,n} u_n \geq 1$$
$$M_{2,1} u_1 + M_{2,2} u_2 + \ldots + M_{2,n} u_n \geq 1$$
$$\cdot$$
$$\cdot$$
$$\cdot$$
$$M_{m,1} u_1 + M_{m,2} u_2 + \ldots + M_{m,n} u_n \geq 1$$

The first constraint says that each element of the $u$ vector must be nonnegative, and the remaining constraints says that each element of the $Mu$ vector must be at least 1. For the resulting $u$ vector, the inverse of the sum of its elements is the value of the game. Multiplying $u$ by that value gives a probability vector that shows the probability that the maximizing player will choose each possible pure strategy.

Nonnegativity is an important concept in linear programming. If the game matrix does not have all positive elements, then a constant is simply added to every element that is large enough to make them all positive. That will increase the value of the game by that constant but will have no effect on the equilibrium mixed strategies for the equilibrium.

The equilibrium mixed strategy for the minimizing player can be found by solving the dual of the given linear program. Or it can be found by using the above procedure to solve a modified payoff matrix, which is the transpose and negation of $M$ (adding a constant so it is positive), and then solving the resulting game.

If all the solutions to the linear program are found, they will constitute all of the Nash equilibria for the game. Conversely, any linear program can be converted into a two-player, zero-sum game by using a change of variables that puts it in the form of the above equations. So, in general, such games are equivalent to linear programs.

We can also consider the following formulation for the maximizing player that provides results for the value of the game and the probabilities $x_i$ as illustrated by Winston (1995, p. 636). This formulation is a simpler modification to Equation (17.1) and avoids any mathematics being used after the solution is found.

Again, we use the same format for the elements of the payoff matrix $M$, but we no longer restrict the elements to be positive:

$$\text{Maximize } V \tag{17.2}$$

Subject to:
$$M_{1,1}x_1 + M_{1,2}x_2 + \ldots + M_{1,n}x_n - V \geq 0$$
$$M_{2,1}x_1 + M_{2,2}x_2 + \ldots + M_{2,n}x_n - V \geq 0$$
$$\vdots$$
$$M_{m,1}x_1 + M_{m,2}x_2 + \ldots + M_{m,n}x_n - V \geq 0$$
$$x_1 + x_2 + \ldots + x_n = 1$$
$$V, x_i \geq 0$$

where $V$ is the value of the game, $M_{m,n}$ are payoff-matrix entries, and $x$'s are the weights (probabilities to play the strategies).

Consider our example of the predator–prey game. The prey decisions are given in rows, and the predator information is given in the columns. For convenience, we will refer to the row player as Rose and the column player as Colin. For the zero-sum games only, we can take advantage of the fact the Rose is maximizing and Colin is minimizing by using the *primal* (maximizing linear program) and the *dual* (minimizing linear program). If we solve for Rose's maximizing solution in our primal linear program, then we find Colin's solution in the dual.

If we let our primal linear program for Rose be Equation (17.2), then the dual linear program for Colin would be:

Minimize $V$ (The Dual Model)
Subject to:
$$M_{1,1}y_1 + M_{2,1}y_2 + \ldots + M_{n,1}y_n - V \leq 0$$
$$M_{1,2}y_1 + M_{2,2}y_2 + \ldots + M_{n,2}y_n - V \leq 0$$
$$\vdots$$
$$M_{1,n}y_1 + M_{2,n}y_2 + \ldots + M_{n,n}y_n - V \leq 0$$
$$y_1 + y_2 + \ldots + y_n = 1$$
$$V, y_i \geq 0$$

This formulation yields a solution to the probabilities $y_1, y_2, \ldots, y_m$ and $V$. We may read the dual solution directly from the primal output. Each dual variable $y_i = (-$ coefficient of the slack variable) (shown as dual prices in many software packages). We will illustrate the possible methods to get these values from Maple.

We illustrate several examples of the applications of linear programming to zero-sum games in order to establish a working procedure. In the zero-sum games, because we only have Rose's game presented, the solution to Colin is found in the solution to the linear programming known as the *dual problem*. We use Equation (17.2) to illustrate the procedure. We start by showing both formulations in the following simple example.

$$\text{Maximize } V_r$$
$$\text{Subject to}$$
$$0.2x_1 + 0.08x_2 - V_r \geq 0$$
$$0.4x_1 + 0.6x_2 - V_r \geq 0$$
$$x_1 + x_2 = 1$$
$$x_1, x_2, V_r \geq 0$$

We use our Maple commands from Chapter 7 to obtain the solution as follows:

> *with(plots):with(Optimization):with(simplex):*
> *obj:=Vr;*

$$obj := Vr$$

> *const:={.2·x1+.8·x2−Vr≥0,·4·x1+.6·x2−Vr≥0,x1+x2=1}*

$$const := \{x1 + x2 = 1, 0 \leq 0.4\,x1 + 0.6\,x2 - Vr, 0 \leq 0.2\,x1 + 0.8\,x2 - Vr\}$$

> *LPSolve(obj,const,assume=nonnegative,maximize);*

$$[0.60000000000000, [x1 = 0., x2 = 1., Vr = 0.599999999999999976]]$$

We interpret the solution as the row player chooses strategy 2 and receives a value of 0.6.
   The dual formulation is found by associating the multipliers $z_1$, $z_2$, $z_3$, and $z_4$ with the original formulation in standard form.

$$\text{Minimize } z_3 - z_4$$
$$0.2z_1 + 0.4z_2 - z_3 + z_4 \leq 0$$
$$0.8z_1 + 0.6z_2 - z_3 + z_4 \leq 0$$
$$z_1 + z_2 = 1$$

> *LPSolve(z3−z4,{.2·z1+.4·z2−z3+z4≤0,.8·z1+.6·z2−z+z4≤0,z1+z2=1},assume=nonnegative);*

$$[0.60000000000000, [z1 = 0., z2 = 1., z3 = 0.599999999999999866, z4 = 0.]]$$

From these formulations, we conclude that the answer is 0.6 when the prey plays strategy $x_2$, which represents run, and the predator plays $z_2$, which represents pursue. The prey selects strategy 2 (run) because it gives the highest minimum survival value (0.6). The predator wishes to minimize the survival probability of the prey given that it doesn't know what the prey will do. The predator compares the maximum of each column and selects to pursue because this gives the lowest of the best survival probabilities for each of the prey's strategies. Both predator and prey select the strategy that gives the prey a 0.6 probability of survival. This point (the solution of the game) is called a *saddle point* and is an equilibrium point. Any deviation from this leads to lower returns for either the predator or prey.

---

| **Example 2** | ## Baseball's Hitter–Pitcher Duel |

Roy Halladay is facing Derek Jeter. Roy is known for his fierce fastball and his devastating split-finger pitch. Derek is a tremendous competitor and a great fastball hitter. We have the following historical statistics on these players.

| | | Roy Halladay | |
| --- | --- | --- | --- |
| | **His Guess** | **Fastball** | **Split-Finger Fastball** |
| **Derek Jeter** | **Fastball** | 0.475 | 0.100 |
| | **Split-Finger Fastball** | 0.125 | 0.250 |

Derek wants to maximize his batting average by doing a better job at guessing which pitch Roy will throw during an at bat. Halladay wants to limit Derek to his minimal batting average by throwing the most unexpected pitch during an at bat. We define the decision variables as follows:

$BA$ = Derek Jeter's batting average,
$x_1$ = the percentage of time Derek Jeter will be looking for the fastball, and
$(1 - x_1)$ = the percentage of time Derek Jeter is guessing that the split-fingered fastball is coming

We set up the linear programming problem for Derek Jeter using Equation (17.2).

$$\text{Maximize } BA$$
$$\text{Subject to}$$
$$0.475x_1 + 0.125x_2 - BA \geq 0$$
$$0.1x_1 + 0.250x_2 - BA \geq 0$$
$$x_1 + x_2 = 1$$
$$x_1, x_2, BA \geq 0 \text{ (nonnegativity)}$$

The solution via LPSolve in the *with(Optimization)* package in Maple is as follows:

*>with(Optimization):*
*> obj1:=BA;*

$$obj1 := BA$$

*> const1:={−.475·x1−.125·x2+BA≤0,−.1·x1−.25·x2+BA≤0,x1+x2=1,x1≥0,x2≥0,BA≥0};*

$$const1 := \{-0.475\,x1 - 0.125\,x2 + BA \leq 0, -0.1\,x1 - 0.25\,x2 + BA \leq 0,$$
$$0 \leq x2, 0 \leq x1, 0 \leq BA, x1 + x2 = 1\}$$

*LPSolve(obj1,const1,maximize);*

$$[0.21250000000000, [x2 = 0.750000000000000000, x1 = 0.249999999999999832,$$
$$BA = 0.212499999999999967]$$

We interpret the solution as the Derek's batting average will be 0.212 if he guesses fastball 25 percent of the time and split-finger 75 percent of the time.

To complete the analysis, we will need the dual. We can formulate the dual ourselves, as in Example 1, or we can make use of the dual command in Maple. The following definition is in Maple, but Maple does not have a separate command to give the dual solution. We should be proficient at doing this so you can interpret the variables in Maple.

---

### Duality Theory of Linear Programming

The assertion that a dual pair of *linear programs* is in *strong duality*, if both are feasible. When the *primal linear program* is written

$$p = \max \{<c, x>: Ax \leq b, x \geq 0\}$$

with inequalities taken coordinate-wise, then the dual linear program is

$$d = \min \{<b, y>, Ay \geq c, y \geq 0\}$$

where $A^*$ is the transpose of the original matrix. The strong duality assertion is that the two optimal values ($p$ and $d$) agree and are attained. These terms are used somewhat analogously to relate nonlinear programs.

Reformulate the dual linear programming program and enter in Maple as follows:

*> obj3:=ba;*

$$obj3 := ba$$

> *const3:={.475·y1+·1·y2−ba≤0,.125·y1+.25·y2−ba≤0,y1+y2=1,y1≥0,y2≥0,ba≥0};*

$$const3 := \{0.475\,y1 + 0.1\,y2 - ba \le 0, 0.125\,y1 + 0.25\,y2 - ba \le 0,\, y1 + y2 = 1,\, 0 \le y1,$$
$$0 \le y2,\, 0 \le b\}$$

> *LPSolve(obj3,const3);*

$$[0.21250000000000, [ba = 0.212499999999999939,\, y2 = 0.699999999999999844$$
$$y1 = 0.299999999999999654]]$$

The 0.212 batting average for Derek Jeter is obtained if Roy throws the fastball 70 percent of the time and the split-finger 30 percent of the time.

Maple can create the dual formulation, so we may try the following Maple commands:

> *with(simplex):*
> *dual(obj1,(−.475·x1−.125·x2+BA≤0,−.1·x1−.25·x2+BA≤0,x1+x2≤1,x1+x2≥1},y);*

$$y3 - y4,\, \{0 \le\, -0.125\,y1 - 0.25\,y2 + y3 - y4, 0 \le\, -0.475\,y1 - 0.1\,y2 + y3 - y4,\, 1 \le y1 + y2\}$$

> *LPSolve(y3−y4, {0≤−0.125y1−0.25y2+y3−y4,0≤−0.475y1−0.1y2+y3−y4,1≤y1+y2},*
> *assume=nonnegative);*

$$[0.21249999974185, [y1 = 0.300000002065195281,\, y2 = 0.699999997934804496,$$
$$y3 = 0.212499999741850465,\, y4 = 0.]]$$

---

|                  | Dorfman's Original Example (1951) |
| :--------------- | :-------------------------------- |
| **Example 3**    |                                   |

## Example 3 — Dorfman's Original Example (1951)

|                      |       | Player A Strategy | | | | |
| -------------------- | ----- | ----- | ----- | ----- | ----- | ----- |
|                      |       | **1** | **2** | **3** | **4** | **5** |
|                      | **1** | 5.31  | 8.52  | 12.05 | 16    | 20.00 |
|                      | **2** | 2.70  | 3.77  | 6.30  | 9.7   | 13.40 |
| **Player B Strategy**| **3** | 3.64  | 2.70  | 3.60  | 5.91  | 8.99  |
|                      | **4** | 5.91  | 3.60  | 2.70  | 3.64  | 6.02  |
|                      | **5** | 9.70  | 6.30  | 3.77  | 2.70  | 4.04  |
|                      | **6** | 16    | 12.05 | 8.52  | 5.31  | 2.70  |

Using Equation (17.2), we set up to obtain the following linear program and solve.

Maximize $V$
Subject to
$$5.31x_1 + 8.52x_2 + 12.05x_3 + 16x_4 + 20x_5 - V \ge 0$$
$$2.7x_1 + 3.77x_2 + 6.3x_3 + 9.7x_4 + 13.4x_5 - V \ge 0$$
$$3.64x_1 + 2.7x_2 + 3.6x_3 + 5.91x_4 + 8.99x_5 - V \ge 0$$
$$5.91x_1 + 3.6x_2 + 2.7x_3 + 3.64x_4 + 6.02x_5 - V \ge 0$$
$$9.7x_1 + 6.3x_2 + 3.77x_3 + 2.7x_4 + 4.04x_5 - V \ge 0$$
$$16x_1 + 12.05x_2 + 8.52x_3 + 5.31x_4 + 2.7x_5 - V \ge 0$$
$$x_1 + x_2 + x_3 + x_4 + x_5 = 1$$
$$V, x_i \ge 0$$

> *obj4:=V;*

$$obj4 := V$$

> *const4:={5.31·x1+8.58·x2+12.05·x3+16·x4*
> *+20·x5−V≥0,2.7·x1+3.77·x2+6.3·x3+9.7·x4*
> *+13·4.x5−V≥0,3.64·x1+2.7·x2+3.6·x3+5.91·x4*
> *+8.99·x5−V≥0,5.91·x1+3.6·x2+2.7·x3+3.64·x4*

$$+6.02 \cdot x5 - V \geq 0, 9.7 \cdot x1 + 6.3 \cdot x2 + 3.77 \cdot x3 + 2.7 \cdot x4$$
$$+4.04 \cdot x5 - V \geq 0, 16 \cdot x1 + 12.05 \cdot x2 + 8.52 \cdot x3 + 5.31 \cdot x4$$
$$+2.7 \cdot x5 - V \geq 0, x1 + x2 + x3 + x4 + x5 = 1\};$$

$$const4 := \{0 \leq 5.31\ x1 + 8.58\ x2 + 12.05\ x3 + 16\ x4 + 20\ x5 - V,$$
$$0 \leq 2.7\ x1 + 3.77\ x2 + 6.3\ x3 + 9.7\ x4 + 13.4\ x5 - V,$$
$$0 \leq 3.64\ x1 + 2.7\ x2 + 3.6\ x3 + 5.91\ x4 + 8.99\ x5 - V,$$
$$0 \leq 5.91\ x1 + 3.6\ x2 + 2.7\ x3 + 3.64\ x4 + 6.02\ x5 - V,$$
$$0 \leq 9.7\ x1 + 6.3\ x2 + 3.77\ x3 + 2.7\ x4 + 4.04\ x5 - V,$$
$$0 \leq 16\ x1 + 12.05\ x2 + 8.52\ x3 + 5.31\ x4 + 2.7\ x5 - V,$$
$$x1 + x2 + x3 + x4 + x5 = 1\}$$

$> LPSolve(obj4, const4, maximize, assume=nonnegative);$

The solution is directly found as $V = 5.982253$ when

$$x_1 = 0.343154,\ x_2 = x_3 = x_4 = 0,\ \text{and}\ x_5 = 0.656846$$

The dual solution (for the minimizing player) is found to be $V = 5.982253$, $y_4 = 0.980936$, and $y_5 = 0.019064$.

---

| Example 4 | The 2 × 3 Games |
| --- | --- |

Consider the following example, which is used in the literature to show a shortcut graphical method introduced by Williams (1986) to reduce the 2 × 3 game to a 2 × 2 game. Rather than use his shortcut method, we can use linear programming directly to obtain the solution.

|  |  | Colin | | |
| --- | --- | --- | --- | --- |
|  |  | **C** | **D** | **E** |
| **Rose** | **A** | 1 | −1 | 3 |
|  | **B** | −2 | 1 | −1 |

We note that there are negative values in the payoff matrix as we illustrate. This procedure is valid for any values in the payoff matrix, $M$. Our best option is to use the method suggested by Winston (1995, pp. 172–178) to replace any variable that could take on negative values with the difference in two positive variables, $x - y$, where both $x \geq 0$ and $y \geq 0$. We assume that the value of the game could be positive or negative. The other values we are looking for are probabilities that are always between 0 and 1.

$$\text{Maximize } V = v_1 - v_2$$
$$\text{Subject to}$$
$$x_1 - 2x_2 - v_1 + v_2 \geq 0$$
$$-x_1 + x_2 - v_1 + v_2 \geq 0$$
$$3x_1 - x_2 - v_1 + v_2 \geq 0$$
$$x_1 + x_2 = 1$$
$$v_1, v_2, x_i \geq 0$$

*Note*: Since there are negative values in the payoff matrix, there is a chance that the value of $V$ could be negative. Thus, we need to replace $V$ with the difference of positive variables, $v_1 - v_2$. We needed to replace $V$ with $v_1 - v_2$ because the solution to $V$ can be positive or negative.

Solving this linear program using Maple, we find

$> obj3:=v1-v2;$

$$obj3 := v1 - v2$$

$> const3:=\{x1-2 \cdot x2-v1+v2 \geq 0,-x1+x2-v1+v2 \geq 0,3 \cdot x1-x2-v1+v2 \geq 0,x1+x2 \leq 1,x1+x2 \geq 1\};$

$$const3 := \{0 \leq x1 - 2\,x2 - v1 + v2,\ x1 + x2 \leq 1,\ 1 \leq x1 + x2,$$
$$0 \leq -x1 + x2 - v1 + v2,\ 0 \leq 3\,x1 - x2 - v1 + v2\}$$

> $LPSolve(obj3, const3, assume{=}nonnegative, maximize);$

$$[-19999999927718, [v2 = 0.199999999277181538,\ x2 = 0.400000000103259756,$$
$$x1 = 0.599999999896740242,\ v1 = 0.]]$$

$$V = v_1 - v_2 = -0.2$$
$$x_1 = 0.60,\ x_2 = 0.40,\ v_1 = 0.00,\ v_2 = 0.20$$

and the dual variables are found to be

$$y_1 = 0.40,\ y_2 = 0.60,\ y_3 = 0.00$$

Thus, Rose plays 60 percent A and 40 percent B to get a value of the game of $-0.2$, whereas Colin plays 40 percent C, 60 percent D, and 0 percent E to obtain the value of $-0.2$.

The interpretation of the solution to the game is found as Colin plays strategy C 40 percent of the time and strategy E 60 percent of the time, but never plays strategy F. Colin's game is worth 0.2. Rose plays strategy A 40 percent of the time and strategy B 60 percent of the time. The value of the game to Rose is $-0.2$.

In the previous two games, it might have been easier to use the mixed-strategy methods to find the solutions. However, these examples are nice because they provide a vehicle to set up and solve games as linear programming problems in which we can obtain solutions via other methods. These examples show the direct application of linear programming to zero-sum games.

---

| **Example 5** | **A 3 × 3 Game in Which Traditional Equalizing Strategies Do Not Work** |
| --- | --- |

In this $3 \times 3$ game, we first checked the movement diagram and dominance, and then we attempted to employ equaling strategies. The method fails. Textbooks in game theory, such as Straffin's text (2004), suggest trying every subgame to find the solution. The literature is hazy on which subgame solution we want to use and also states these methods can be very "tedious." Straffin (2004, p.19) confesses that linear programming is the most efficient method to solve larger games, but his text contains no examples of the use of linear programming. Linear programming is the better choice to find the solutions to large game theory problems where there is no dominant solution or solutions found through the movement diagrams. In Winston's text (1995), he provides a three-step method to consider.

**Step 1**    Check for a saddle point. If the game has a saddle point, the saddle point is the solution. If Max{Row{Min}=Min[Column{Max} then we have a saddle point. In the example below.

| Row | Row[Min} | Max |
| --- | --- | --- |
| A | 2 | |
| B | 3 | 3 |
| C | 1 | |

| Column | Column[Max} | Min |
| --- | --- | --- |
| D | 9 | |
| E | 6 | 6 |
| F | 7 | |

Since 6 does not equal 3, there is not a saddle point. If the game does not have a saddle point, go to step 2.

**Step 2** Eliminate any of the row player's dominated strategies. Looking at the reduced payoff matrix, eliminate any column player's dominated strategies. Continue until all dominated strategies are removed. Go to step 3.

**Step 3** If the game has been reduced greater than to 2 × 2, then solve by linear programming.

We illustrate below.

Payoff matrix

|  |  | Colin | | | |
|---|---|---|---|---|---|
|  |  | **D** | **E** | **F** |  |
|  | **A** | 9 | 2 | 7 | $x$ |
| **Rose** | **B** | 3 | 6 | 4 | $y$ |
|  | **C** | 5 | 3 | 1 | $z$ |
|  |  | $s$ | $t$ | $u$ |  |

For Rose, the decision variables are

$$v = \text{expected value of the game}$$
$$x = \text{probability for playing strategy A}$$
$$y = \text{probability for playing strategy B}$$
$$z = \text{probability for playing strategy C}$$

We formulate the problem as follows:

$$\text{Maximize } v$$
$$\text{Subject to}$$
$$9x + 3y + 5z - v \geq 0$$
$$2x + 6y + 3z - v \geq 0$$
$$7x + 4y + 1z - v \geq 0$$
$$x + y + z = 1$$
$$\text{nonnegativity } x, y, z, v \geq 0$$

> *obj5:=v;*

$$obj5 := v$$

> *const5:= {9·x+3·y+5·z−v≥0,2·x+6·y+3·z−v≥0,7·x+4·y+z−v≥0,x+y+z≥1, x+y+z≤1};*

$$const5 := \{0 \leq 9\,x + 3\,y + 5\,z - v, 0 \leq 2\,x + 6\,y + 3\,z - v,$$
$$0 \leq 7\,x + 4\,y + z - v, 1 \leq x + y + z, x + y + z \leq 1\}$$

> *LPSolve(obj5,const5,assume=nonnegative,maximize);*

$$[4.8000000049565, [v = 4.80000000495646884, z = 0.,$$
$$x = 0.300000000309779413, y = 0.700000000722818205]]$$

Our solution is $v = 4.8$ when $x = 0.03$, $y = 0.70$, and $z = 0.0$, and the dual variables are 0.40, 0.60, and 0.00, respectively.

Colin will play his strategies with probabilities 0.40, 0.60, 0.0 with the game yielding the same results ($V_c = -4.8$). The use of linear program is quick, concise, and direct.

## Example 6

## A 3 × 3 Game with a Saddle-Point Solution

Consider the game with payoff matrix, $M$, and saddle-point solution at pure strategy CE with value of the game as 5 to Rose.

Payoff matrix

|  |  | Colin | | | |
| --- | --- | --- | --- | --- | --- |
|  |  | **D** | **E** | **F** | **Max{RowMin}** |
|  | **A** | 4 | 4 | 10 |  |
| **Rose** | **B** | 2 | 3 | 1 |  |
|  | **C** | 6 | 5 | 7 | 5 |
|  | **Min{ColMax}** |  | 5 | 5 | Saddle point |

Our formulation would be:

$$\text{Maximize } V$$
$$\text{Subject to}$$
$$4x_1 + 2x_2 + 6x_3 - V > 0$$
$$4x_1 + 3x_2 + 5x_3 - V > 0$$
$$10x_1 + x_2 + 7x_3 - V > 0$$
$$x_1 + x_2 + x_3 = 1$$
$$x_i, V \geq 0$$

Solving the linear program yields $x_1 = x_2 = 0$, $x_3 = 1$, and $V = 5$ for Rose with a dual solution as $y_1 = y_3 = 0$, $y_2 = 1$, and $V = -5$ for Colin.

## Example 7

## Multiple Saddle Points

Consider the game with multiple saddle-point solutions at strategies AF, AH, BE, and CF with value of 2.

|  |  | Colin | | | |
| --- | --- | --- | --- | --- | --- |
|  |  | **E** | **F** | **G** | **H** |
|  | **A** | 4 | 2 | 5 | 2 |
| **Rose** | **B** | 2 | 1 | -1 | -20 |
|  | **C** | 3 | 2 | 4 | 2 |
|  | **D** | -16 | 0 | 16 | 1 |

We formulate the problem as

$$\text{Maximize } V$$
$$\text{Subject to}$$
$$4x_1 + 2x_2 + 3x_3 - 16x_4 - V > 0$$
$$2x_1 + x_2 + 2x_3 - V > 0$$
$$5x_1 - x_2 + 4x_3 + 16x_4 - V > 0$$
$$2x_1 - 20x_2 + 2x_3 + x_4 - V > 0$$
$$x_1 + x_2 + x_3 + x_4 = 1$$
$$x_i, V \geq 0$$

> *obj6:=V;*

$$obj6 := V$$

> *const6:={48x1+2·x2+3·x3−16·x4−V≥0,2·x1+x2+2·x3−V≤0,5·x1−x2+4·x3+16·x4−V≥0, 2·x1−20·x2+2·x3+x4−V≥0,x1+x2+x3+x4≥1x1+x2+x3+x4≤1};*

$$const6 := \{0 \leq 48\,x1 + 2\,x2 + 3\,x3 - 16\,x4 - V,$$
$$0 \leq 2\,x1 + x2 + 2\,x3 - V, 0 \leq 5\,x1 - x2 + 4\,x3 + 16\,x4 - V,$$
$$0 \leq 2\,x1 - 20\,x2 + 2\,x3 + x4 - V, 1 \leq x1 + x2 + x3 + x4,$$
$$x1 + x2 + x3 + x4 \leq 1\}$$

> *LPSolve(obj6,const6,assume=nonnegative,maximize);*

$$[2.0000000030486, [V = 2.00000000304862180, x2 = 0.,$$
$$x1 = 1.00000000201602424, x3 = 0., x4 = 0.]]$$

We find a solution of $V = 2$ when $x_1 = 1$ and the corresponding dual $y_4 = 1$. This is our solution for pure strategy, AH, at $V = 2$. How do we know that there are alternate optimal solutions? We must use the tableau form (as shown in Chapter 7). If a nonbasic variable has a value of 0 in the objective function row that indicates that we might pivot (if a pivot is possible) and not alter the value of the objective function. When that occurs, we have alternate optimal solutions. The other solutions are found by alternate optimal techniques in linear programming. We suggest examining a solid discussion of this technique such as in Winston (1995, pp.142–144). We will find all the alternate optimal solutions to this game-theory problem by applying the techniques to find such solutions in the optimal tableau.

So far in our examples and discussions, we have reviewed the concepts of applying linear programming to the zero-sum games. We showed that both pure-strategy and mixed-strategy solutions can be found using linear programming. We make a strong recommendation that using linear programming more easily provides a solution to zero-sum game-theory problems that are larger than $2 \times 2$. We have shown the linear programming can be used to find solutions to pure-strategy games as well, although we believe that the application of movement diagrams and dominance to find pure-strategy solutions might be easier. Now we present some new ground and introduce the concept in applying the use of linear programming in game theory to non–zero-sum games.

## 17.1 | EXERCISES

In Problems 1–12, first use a movement diagram to check for saddle-point solutions and then formulate and solve the LP for each game.

**1.** Consider the following game and find the solution.

| Rose/Colin | $x$ | $y$ | $z$ |
|---|---|---|---|
| No. 1 | 80 | 40 | 75 |
| No. 2 | 70 | 35 | 30 |

**2.** Consider the following game and find the solution:

| Rose/Colin | W | X | Y | Z |
|---|---|---|---|---|
| No. 1 | 40 | 80 | 35 | 60 |
| No. 2 | 65 | 90 | 55 | 70 |
| No. 3 | 55 | 40 | 45 | 75 |
| No. 4 | 45 | 25 | 50 | 50 |

3. The following represents a game between a professional athlete (Rose) and management (Colin). The values are in (1,000s). What decision should each make?

|        | $x$ | $y$ | $z$ |
|--------|-----|-----|-----|
| No. 1  | 490 | 220 | 195 |
| No. 2  | 425 | 350 | 150 |

4. $\begin{pmatrix} 6 & 4 \\ 4 & 2 \end{pmatrix}$

5. $\begin{pmatrix} -2 & 3 \\ 2 & -2 \end{pmatrix}$

6. $\begin{pmatrix} .30 & .20 \\ .10 & .40 \end{pmatrix}$

7. $\begin{pmatrix} 3 & 7 & 2 \\ 8 & 5 & 1 \\ 6 & 9 & 4 \end{pmatrix}$

8. $\begin{pmatrix} .5 & .9 & .9 \\ .1 & 0 & .1 \\ .9 & .9 & .5 \end{pmatrix}$

9. $\begin{pmatrix} 6 & 5 \\ 1 & 4 \\ 8 & 5 \end{pmatrix}$

10. $\begin{pmatrix} 2 & -1 \\ 1 & 4 \\ 6 & 2 \end{pmatrix}$

11. $\begin{pmatrix} 1 & -1 & 2 & 3 \\ 2 & 4 & 0 & 5 \end{pmatrix}$

12.

| Team A/Team B | 1 | 2 | 3 |
|---------------|----|-----|-----|
| 1 | 0 | -1 | 5 |
| 2 | 7 | 5 | 10 |
| 3 | 15 | -4 | -5 |
| 4 | 5 | 0 | 10 |
| 5 | -5 | -10 | 10 |

## 17.1  PROJECTS

1. Now suppose the payoff matrix of prey survival probabilities is as shown in the following table.

|      |        | Predator | |
|------|--------|----------|--------|
|      |        | **Ambush** | **Pursue** |
| **Prey** | **Hide** | 0.3 | 0.6 |
|      | **Run** | 0.5 | 0.4 |

Analyze this game and present the results.

2. Take your favorite baseball batter and pitcher from the same league and era. Find batting and pitching statistics for each player. Determine the best strategy for each in a head-to-head competition.

3. In tennis, statistics are kept for percentages of successful return of first serves by forehand and backhand. Choose two tennis players and determine the best strategy for each player.

## 17.2 THE NON–ZERO-SUM GAME AND OPTIMIZATION

The literature for noncooperative game solutions that are non–zero-sum games is to find the Nash equilibrium. Recall, in a simple zero-sum game such as flipping and matching coins where, if we match, the Rose player loses $1 and the Colin player wins $1, and if we do not match, the Rose player wins a dollar and the Colin player loses a dollar, the total of the players, entries always equal zero.

Zero-sum game

| | | Colin | |
|---|---|---|---|
| | | **Heads** | **Tails** |
| **Rose** | **Heads** | (–1,1) | (1,–1) |
| | **Tails** | (1,–1) | (–1,1) |

Non–zero-sum games are games in which one player wins but the other player does not have to lose. Both players could win something or lose something. The following payoff matrix is a simple example of a non–zero-sum game between two players.

| | | Colin | |
|---|---|---|---|
| | | **C** | **D** |
| **Rose** | **A** | (1,1) | (–1,–1) |
| | **B** | (–1,–1) | (1,1) |

Solution methods for non–zero-sum games include looking for dominance, movement diagrams, and equalizing strategies. Here we present an extension to the application of linear programming from the zero-sum game to the non–zero-sum game. The primal–dual relationship from zero-sum games does not hold for non–zero-sum games because both players' strategy is to maximize their game. However, the theory of Danzig, Kuhn, and Tucker (1951) that allowed for linear programming to be used in setting up a zero-sum game to maximize $V$ subject to the constraints still holds for each individual player, as we will show. Because of the nature of non–zero-sum games, where both players are trying to maximize their outcomes, we can model all players' strategies as their own *maximizing* linear programs. Therefore, the theory accompanying the application to the maximizing player in a zero-sum game holds for each player maximizing in a non–zero-sum game. We treat each player as separate linear programming problems.

Again, let's define the following payoff matrix that has components for both Rose and Colin:

$$(M, N) = \begin{bmatrix} (M_{1,1}, N_{1,1}) & (M_{1,2}, N_{1,2}) & \cdots & (M_{1,n}, N_{1,n}) \\ (M_{2,1}, N_{2,1}) & (M_{2,2}, N_{2,2}) & \cdots & (M_{2,n}, N_{2,n}) \\ \cdot & \cdot & \cdots & \cdot \\ \cdot & \cdot & \cdots & \cdot \\ \cdot & \cdot & \cdots & \cdot \\ (M_{m,1}, N_{m,1}) & (M_{m,2}, N_{m,2}) & \cdots & (M_{m,n}, N_{m,n}) \end{bmatrix}$$

In noncooperative non–zero-sum games, we use similar concepts of pure strategies and equalizing strategies (mixed strategy) to solve the game. We look for a pure-strategy solution using the *movement diagram*.

Rose would maximize payoffs, so she would prefer the highest payoff at each *column*. Arrows are in columns, but the values are Rose's. Similarly for Colin, he wants to maximize his payoffs, so he would prefer the highest payoff in each row. We draw an arrow to the highest payoff in that row. Arrows are in rows, but the values are Colin's. If all arrows point in from every direction, then that or those points will be pure Nash equilibrium.

If all the arrows do not point at a value or values, then we must use equalizing strategies to find the weights (probabilities) for each player. Basically, we would proceed as follows:

- Rose's game: Rose maximizing, Colin "equalizing" is a zero-sum game that yields Colin's equalizing strategy.

- Colin's game: Colin maximizing, Rose "equalizing" is a zero-sum game that yields Rose's equalizing strategy.

- *Note*: If either side plays its equalizing strategy, then the other side "unilaterally" cannot improve its own situation (it stymies the other player).

This translates into two additional linear programming formulations, one for each maximizing player. We create two more formulations in the form of Equation (17.2) that we call Equations (17.3) and (17.4), and that provide the values of the game and the probabilities that the players should play their strategies in the equalizing strategy concept just presented.

$$\text{Maximize } V \tag{17.3}$$

Subject to:
$$N_{1,1}x_1 + N_{2,1}x_2 + \ldots + N_{m,1}x_n - V \geq 0$$
$$N_{2,1}x_1 + N_{2,2}x_2 + \ldots + N_{m,2}x_n - V \geq 0$$
$$\ldots$$
$$N_{m,1}x_1 + N_{m,2}x_2 + \ldots + N_{m,n}x_n - V \geq 0$$
$$x_1 + x_2 + \ldots + x_n = 1$$
*Nonnegativity*

where the weights $x_i$ yield Rose's strategy and the value of $V$ is the value of the game to Colin.

$$\text{Maximize } v \tag{17.4}$$

Subject to:
$$M_{1,1}y_1 + M_{2,1}y_2 + \ldots + M_{m,1}y_n - v \geq 0$$
$$M_{2,1}y_1 + M_{2,2}y_2 + \ldots + M_{m,2}y_n - v \geq 0$$
$$\ldots$$
$$M_{m,1}y_1 + M_{m,2}y_2 + \ldots + M_{m,n}y_n - v \geq 0$$
$$y_1 + y_2 + \ldots + y_n = 1$$
*Nonnegativity*

where the weights $y_i$ yield Colin's strategy and the value of $v$ is the value of the game to Rose.

We also point out that looking for dominance and the movement diagrams are still critical as initial steps. If you find total dominance and pure-strategy solutions by the movement diagram, you have found a pure-strategy Nash equilibrium. According to Gillman and Housman (2009, p. 189) non–zero-sum games with pure-strategy equilibrium also have non–pure-strategy equilibriums using Nash's equalizing strategy. Linear programming provides a solution methodology for these non–pure-strategy equilibriums.

## Example 8

## Partial Conflict Mixed-Strategy Game Solution

Consider the following partial conflict mixed-strategy game.

|  |  | Colin | |
|---|---|---|---|
|  |  | **C** | **D** |
| **Rose** | **A** | (2,4) | (1,0) |
|  | **B** | (3,1) | (0,4) |

The linear programming formulations for each of our two players to find the Nash equilibrium values for both Rose and Colin are found as follows:

**1.** Maximize $V_c$

Subject to
$$4x_1 + x_2 - V_c \geq 0$$
$$0x_1 + 4x_2 - V_c \geq 0$$
$$x_1 + x_2 = 1$$
$$x_i, V_c \geq 0$$

**2.** Maximize $V_r$

Subject to
$$2y_1 + y_2 - V_r \geq 0$$
$$3y_1 - V_r \geq 0$$
$$y_1 + y_2 = 1$$
$$y_i, V_r \geq 0$$

> *obj7:=Vc;*

$$obj7 := Vc$$

> *obj8:=Vr;*

$$obj8 := Vr$$

> *constr7:={4·x1+x2−Vc≥0,4·x2−Vc≥0,x1+x2≤1,x1+x2≥1};*

$$constr7 := \{0 \leq 4\,x1 + x2 - Vc, 0 \leq 4\,x2 - Vc, x1 + x2 \leq 1, 1 \leq x1 + x2\}$$

> *const8:={2·y1+y2−Vr≥0,3·y1−Vr≥0,y1+y2≤1,y1+y2≥1};*

$$const8 := \{0 \leq 2\,y1 + y2 - Vr, 0 \leq 3\,y1 - Vr, y1 + y2 \leq 1, 1 \leq y1 + y2\}$$

> *LPSolve(obj7,constr7,assume=nonnegative,maximize);*

$$[2.2857142880745, [x2 = 0.571428572018627290, x1 = 0.428571429013970494,$$
$$Vc = 2.28571428807450916]]$$

> *LPSolve(obj8,const8,assume=nonnegative, maximize);*

$$[2.0000000020652, [Vr = 2.00000000206519556, y1 = 1.00000000103259778, y2 = 0.]]$$

The solutions are (a) $V_c = 2.285714286$ when $x_1 = 0.5714285714$ or 4/7 and $x_2 = 0.4285714286$ or 3/7 and (b) $V_r = 2.0000$ when $y_1 = 1.0000$ and $y_2 = 0.0000$. This game results in Colin playing

C and ensuring a value of the game of 2.000 for Rose, while Rose plays 4/7 A, 3/7 B, which yields a value of the game of 2.285714286 for Colin. The solution is (2.000, 2.285714286).

---

| **Example 9** | ## Two-Player Games with More Than Two Strategies Each |

Consider the non–zero-sum game with more than two strategies per player, where there is no pure-strategy equilibrium.

|  |  | Colin | | |
|---|---|---|---|---|
|  |  | **D** | **E** | **F** |
|  | **A** | (9,1) | (2,2) | (7,2) |
| **Rose** | **B** | (3,2) | (6,1) | (4,2) |
|  | **C** | (5,2) | (3,2) | (5,0) |

The movement arrows reveal no pure strategy, so we turn to linear programming to find our solutions using the equalizing strategies.

$$\text{Maximize } V_c$$
$$\text{Subject to}$$
$$x_1 + 2x_2 + 2x_3 - Vc \geq 0$$
$$2x_1 + x_2 + 2x_3 - Vc \geq 0$$
$$2x_1 + 2x_2 - Vc \geq 0$$
$$x_1 + x_2 + x_3 = 1$$
$$Vc, x_i \geq 0$$

$$\text{Maximize } V_r$$
$$\text{Subject to}$$
$$9y_1 + 2y_2 + 7y_3 - Vr \geq 0$$
$$3y_1 + 6y_2 + 4y_3 - Vr \geq 0$$
$$5y_1 + 3y_2 + 5y_3 - Vr \geq 0$$
$$y_1 + y_2 + y_3 = 1$$
$$Vr, y_i \geq 0$$

Solving these two linear programming problems yields the following results:

$$V_r = 4.5, \ V_c = 1.6 \text{ when Rose plays 0.4A, 0.4B, 0.2C, and}$$
Colin plays 0.0D, 0.25E, 0.75F

Thus, the Nash equilibrium is (4.5,1.6), and the probabilities to play strategies are $x_1 = 0.4$, $x_2 = 0.4$, $x_3 = 0.2$, $y_1 = 0$, $y_2 = 0.25$, and $y_3 = 0.75$.

We have extended the application of linear programming in zero-sum games to finding the Nash equilibrium by finding the equalizing strategies in non–zero-sum games. The theory for linear programming use holds as each player is maximizing, so we treat each problem as a *primal* maximizing problem. The dual solution (the reduced costs) only provides us information about the utilities as resources. For example, assume we have a reduced cost of 0.5. Then an increase in the utility value of one unit of that constraint increases the objective function value by approximately 0.5 utility units.

Let's extend our optimization applications to the Nash arbitration scheme as a nonlinear programming problem using Kuhn–Tucker conditions (Kuhn, 1951) in the next section.

## 17.2 | EXERCISES

1. Find all the Nash equilibria for the following strategies:

|  |  | State Sponsor | |
| --- | --- | --- | --- |
|  |  | **Sponsor Terrorism** | **Stop Sponsoring Terrorism** |
| **United States** | **Strike Militarily** | (2, 3) | (3, 1) |
|  | **Don't Strike Militarily** | (1, 4) | (4, 2) |

2. Find all the Nash equilibria for the following strategies:

|  |  | State Sponsor | |
| --- | --- | --- | --- |
|  |  | **Sponsor Terrorism** | **Stop Sponsoring Terrorism** |
| **United States** | **Strike Militarily** | (1, 2) | (3, 1) |
|  | **Don't Strike Militarily** | (2, 4) | (4, 3) |

3. Find all the Nash equilibria for the following strategies:

|  |  | Colin | |
| --- | --- | --- | --- |
|  |  | **Arm** | **Disarm** |
| **Rose** | **Arm** | (2, 2) | (4, 1) |
|  | **Disarm** | (1, 4) | (3, 3) |

## 17.3 | NASH ARBITRATION AND NONLINEAR PROGRAMMING FORMULATION

We utilize the Kuhn–Tucker condition (KTC) to find the optimal solution to a nonlinear optimization problem as listed in Equation (17.5):

$$\text{Max (or min) } f(x_1, x_2, \dots, x_n) \tag{17.5}$$

Subject to
$$g_1[(x_1, x_2, \dots, x_n)] \le b_1$$
$$g_2(x_1, x_2, \dots, x_n) \le b_2$$
$$.$$
$$.$$
$$.$$
$$g_m(x_1, x_2, \dots, x_n) \le b_m$$

Because we want to find the Nash arbitration point, we desire the maximization of the function. We want to find the values of $(x_1, x_2, \dots, x_n)$ and multiplier $(\lambda_1, \lambda_2, \dots, \lambda_m)$ that satisfy the following KTC conditions in Equation (17.6):

$$\frac{\partial f(x)}{\partial x_j} - \sum_{i=1}^{m} \lambda_i \frac{\partial g_i(x)}{\partial x_j} = 0$$
$$\lambda_i [b_i - g_i(x)] = 0 \tag{17.6}$$
$$\lambda_i \ge 0$$

### Nash Arbitration as a Nonlinear Programming Problem

In the bargaining problem, Nash (1950) developed a scheme for producing a single fair outcome. The goals for the Nash arbitrations scheme are that the result will be at or above the stats quo point for each player and that the result must be "fair."

Nash introduced the following terminology:

*status quo point* (we will typically use the intersection of Rose's security level and Colin's security level; the threat positions may also be used), and

*negotiation set* (those points in the Pareto optimal set that are at or above the status quo of both players).

We use Nash's four axioms. He believed that a reasonable arbitration scheme should satisfy rationality, linear invariance, symmetry, and invariance. A good discussion of these axioms can be found in Straffin (2004, pp. 104–105). Simply put, the Nash arbitration point is the point that follows all four axioms. This leads to Nash's theorem:

| | |
|---|---|
| **Nash's theorem (1950)** | There is one and only one arbitration scheme that satisfies axioms 1 through 4. It is this: If (status quo) $SQ = (x_0, y_0)$ then the arbitrated solution point $N$ is the point $(x, y)$ in the polygon with $x \geq x_0$ and $y \geq y_0$ which *maximizes the product:* $(x - x_0)(y - y_0)$.  ∎ |

Let's examine this geometrically first because it provides insights into using nonlinear optimization methods. We produce the contour plot of our nonlinear function: $(x - x_0)(y - y_0)$ when our status quo point is assumed to be (0,0). Using Nash's theorem, it is obvious that the northeast corner of quadrant 1 is where this function is maximized. This is illustrated in Figure 17.1.

In his theory for the arbitration and cooperative solutions, Nash (1950) stated the "reasonable" solution should be Pareto optimal and will be at or above the security level. The set of outcomes that satisfy these two conditions is called the *negotiation set*. The line segments that join the negotiation set must form a convex region as shown in Nash's proof. Methodologies for solving for this point use basic calculus, algebra, and geometry.

We present another methodology. Because we earlier used linear programming to solve for Nash equilibrium values, we show that we can use nonlinear programming to solve for the arbitration point. It is nonlinear because of the choice of the Nash function, $(x - x^*)(y - y^*)$, which we want to maximize. The constraints are linear and must form a convex set.

For any game-theory problem, we next overlay the convex polygon onto our contour plot (Figure 17.1). The most northeastern point in the feasible region is our optimal point and the Nash arbitration point. This will be where the feasible region is tangent to the hyperbola.



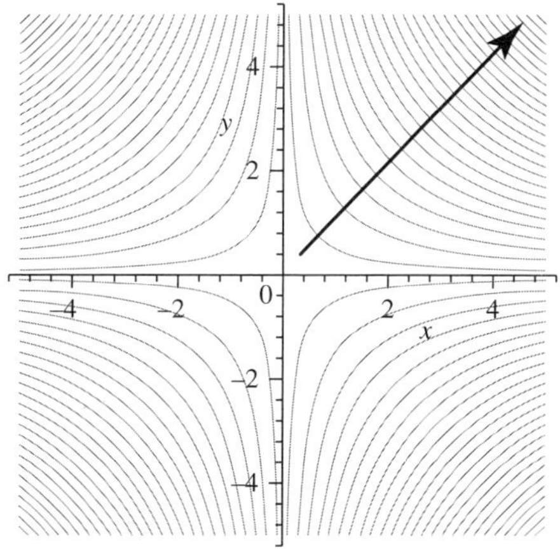

Without generalization, maximizing a nonlinear concave function over a linear convex region produces a maximum directly from the Kuhn–Tucker conditions (1951).

In our examples, we will use the security value as the status quo point to use in the Nash arbitration procedure. We additionally define the procedure to find the security value as follows:

> **Definition** | In a non–zero-sum game, Rose's optimal strategy in Rose's game is called Rose's **prudential strategy**, the value is called Rose's **security level**. Colin's optimal strategy in Colin's game is called Colin's **security level**. We will illustrate this during the solution to find the Nash arbitration point in Example 9.

## Example 10  Nash Arbitration Example from a Non–Zero-Sum Game

|  |  | Colin | |
|---|---|---|---|
|  |  | **C** | **D** |
| **Rose** | **A** | (2,6) | (10,5) |
|  | **B** | (4,8) | (0,0) |

To find the security level (status quo point), we look at the following two separate games extracted from the original game and use movement diagrams, dominance, or our linear programming method to solve each game for those players' values.

In a prudential strategy, we allow players to find their optimal strategies in their own games. Rose would need to find her optimal solution in her own game. Rose's game, which follows, has a mixed-strategy solution: $V = 10/3$.

|  |  | Colin | |
|---|---|---|---|
|  |  | **C** | **D** |
| **Rose** | **A** | 2 | 10 |
|  | **B** | 4 | 0 |

Colin would need to find his optimal solution in his own game. Colin's game, which follows, has a pure-strategy solution: $V = 6$.

|  |  | Colin | |
|---|---|---|---|
|  |  | **C** | **D** |
| **Rose** | **A** | 6 | 5 |
|  | **B** | 8 | 0 |

The status quo point or security level from the prudential strategy is found to be (10/3,6). We will use this point in the formulation of the nonlinear program.

We set up the convex polygon (constraints) for the function that we want to maximize, which is $(x - \frac{10}{3}) \cdot (y - 6)$. The convex polygon is the convex set from the values in the payoff matrix.

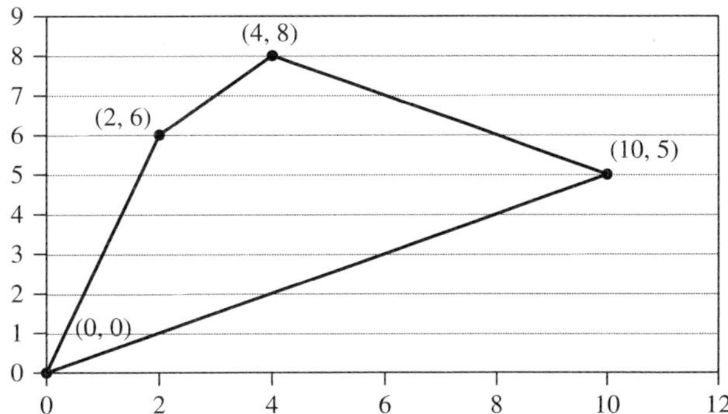

**FIGURE 17.2**

Payoff polygon, Example 9



Its boundary and interior points represent all possible combinations of strategies. Corner points represent **pure** strategies. All other points are mixed strategies. Occasionally, a pure strategy is an interior point. Thus, we start by plotting the strategies from our payoff matrix set of values {(2,6), (4,8), (10,5), (0,0)} (see Figure 17.2).

We note that our convex region has four sides whose coordinates are our pure strategies. We use the point–slope formula to find the equations of the line and then test points to transform the equations to inequalities. For example, the line form (4,8) to (10,5) is $y = -0.5x + 10$. We rewrite as $y + 0.5x = 10$. Our test point (0,0) shows that our inequality is $0.5x + y \leq 10$. We use this technique to find all boundary lines as well as add our security levels as lines that we need to be above.

The convex polygon is bounded by the following in equalities:

$$.5x + y \leq 10$$
$$-3x + y \leq 0$$
$$0.5x - y \leq 0$$
$$-x + y \leq 4$$
$$x \geq x^*$$
$$y \geq y^*$$

where $x^*$ and $y^*$ are the security levels (10/3,6).

The NLP formulation to find the Nash arbitration value following the format of Equation (17.5) is as follows:

$$\text{Maximize } Z = \left(x - \frac{10}{3}\right) \cdot (y - 6) \tag{17.7}$$

Subject to
$$0.5x + y \leq 10$$
$$-3x + y \leq 0$$
$$0.5x - y \leq 0$$
$$-x + y \leq 4$$
$$x \geq \frac{10}{3}$$
$$y \geq 6$$

We use Maple as our software to provide both the graphs and the solution outputs using programs previously written (Fox, 2000). We display the feasible region graphically in Figure 17.2. The feasible region is the solid region. From the figure, we can approximate the solution as the point of tangency between the feasible region and the hyperbolic contours in the northeastern region.

We use the conditions of Equation (17.6) to solve our NLP as shown in Equation (17.7) to find the point indicated by the arrow in Figure 17.3.

Our optimal solution, the Nash arbitration point, is found to be $x = 5.667$ and $y = 7.167$, and the value of the objective function is 2.72 with shadow prices: $\lambda_2 = 2.3333$ and all other $\lambda_i = 0$.

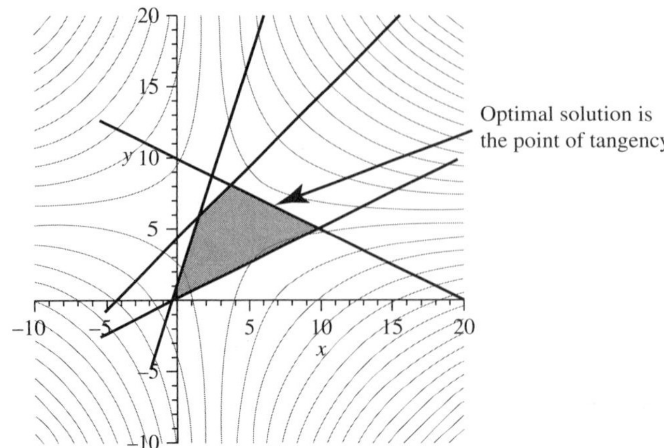

**FIGURE 17.3**
Convex polygon and function contour plot



The value is 2.72. The interpretation of the Lagrange multiplier, $\lambda_p$, would be that an increase in the utility associated with strategy AD of one unit would yield an increase in the value of the game solution of 2.3333 units.

We choose a more complicated example next: the labor–management game from Straffin (2004, pp. 115–117). This game has two players, and the Rose player has four strategies and the Colin player four.

## Example 11    Management–Labor Arbitration

| | | Labor Concedes | | | |
|---|---|---|---|---|---|
| | | **Nothing** | **C** | **A** | **CA** |
| | **Nothing** | (0,0) | (4,−1) | (4,−2) | (8,−3) |
| **Management Concedes** | **P** | (−2,2) | (2,1) | (2,0) | (6,−1) |
| | **R** | (−3,3) | (1,2) | (1,1) | (5,0) |
| | **PR** | (−5,5) | (−1,4) | (−1,3) | (3,2) |

The convex polygon is graphed from the following constraints (see the plots in Figures 17.3 and 17.4):

$$x + y \geq 0$$
$$0.5x + y \geq 0$$
$$0.25x + y \geq -1$$
$$x + y \geq 5$$
$$0.5x + y \leq 3.5$$
$$0.25x + y \leq \frac{15}{4}$$

The status quo point (our security level) is (0,0), making the function to maximize simply $x \cdot y$. Our formulation is:

$$\text{Maximize } x \cdot y$$
$$\text{Subject to}$$
$$x + y \geq 0$$
$$0.5x + y \geq 0$$
$$0.25x + y \geq -1$$
$$x + y \geq 5$$
$$0.5x + y \leq 3.5$$
$$0.25x + y \leq \frac{15}{4}$$

**FIGURE 17.4**

The graphical NLP problem for management–labor arbitration

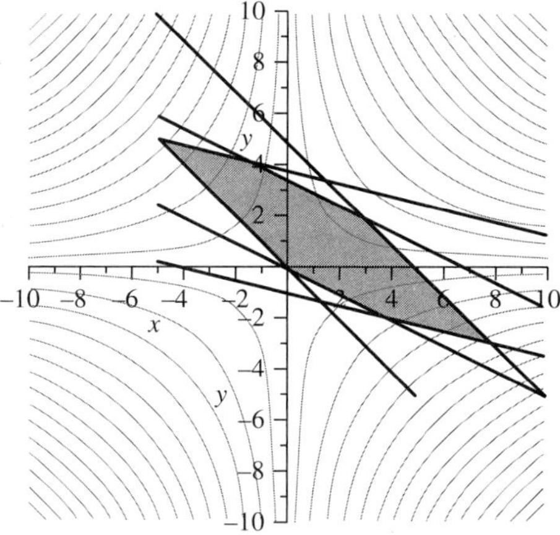

**FIGURE 17.5**

The Nash function's contours with status quo (0,0) and the northeastern boundary line

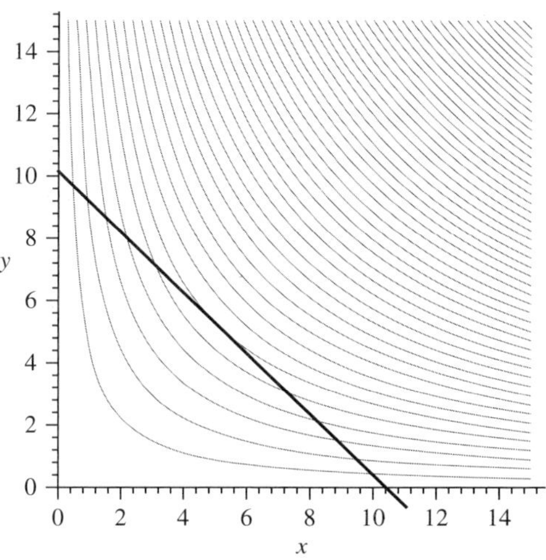

We use Maple to assist in finding the results.

> *sol:=NLPSolve(objective,constr,maximize=true);*

$$sol := [5.99999999999999912, [x = 3.00000000000000088, y = 1.99999999999999912]]$$

The product is taken as $xy = 6.0$, and the values are taken as $x = 3$ and $y = 2$.

With our function, we are concerned with a point of the boundary of the constraint. The constraint region is the Pareto optimal region (the northeast boundary).

In particular, we are looking for the optimal point—that is, the point on the line that is tangent to the contours in the direction of the northeastern increase (Figure 17.5).

## **17.3** | EXERCISES

Find the Nash arbitration solution to problems 1–3 from Section 17.2.

## 17.4 ILLUSTRATIVE EXAMPLE: THE 2007–2008 WRITERS GUILD STRIKE

Models of conflict in game theory assume rational decision makers as players in the game trying to maximize some payoffs. In game theory, the term *rational* has a different meaning than most people think. Rational does not mean what we think is best or wise; to be rational, actors have to be able to (1) define their objectives, however foolish they appear to others, (2) formulate sufficiently different alternative strategies, and (3) choose strategies that maximize their objectives. So the main question we study is, "What should or will the rational-value-maximizing player do?"

### Writers Guild Strike Background

The 2007–2008 writers strike was a strike by the Writers Guild of America, East (WGAE), and the Writers Guild of America, West (WGAW), which started on November 5, 2007. The WGAE and WGAW were two labor unions representing film, television, and radio writers working in the United States. More than 12,000 writers joined the strike. We refer to these entities in the model as the Writers Guild.

The strike was against the Alliance of Motion Picture and Television Producers (AMPTP), a trade organization representing the interests of 397 U.S. film and television producers. The most influential of these are eight corporations: CBS Corporation, Metro-Goldwyn-Mayer, NBC Universal, News Corp/Fox, Paramount Pictures, Sony Pictures Entertainment, the Walt Disney Company, and Warner Brothers. We will refer to this group as management.

The Writers Guild said that its industrial action would be a "marathon." AMPTP negotiator Nick Counter indicated that negotiations would not resume as long as strike action continued: "We're not going to negotiate with a gun to our heads—that's just stupid."

The previous strike in 1988 lasted 21 weeks and 6 days, costing the U.S. entertainment industry an estimated $500 million ($870 million in 2007 dollars).

The strike was projected to cost the industry approximately $1 billion in everything from lost wages to cast and crew members to payments for services provided by janitors, caterers, prop- and costume-rental companies, and the like.

The TV and movie companies stockpiled TV shows and movies so that they could possibly outlast the strike rather than work to meet the demands of the writers and avoid the strike.

Now, let's examine this strike as an example of game theory.

### Game-Theory Approach

Let us begin by stating each side's strategies. Our two rational players will be the writer's guild and the management. We develop strategies for each player.

#### Strategies
- Writers Guild: Its strategies are to strike (S) or not strike (NS).
- Management: Its strategy is give salary increases and revenue sharing (IN) or keep status quo (do not pay the increases) (SQ).

We used the lottery method by Von Neumann and Morgenstern (see the reading references at the end of the chapter) in order to create cardinal utilities for the payoff matrix.

First, we rank order the outcomes for each side in order of preference. (These rank orderings are ordinal utilities.)

#### Writers Guild's Alternatives and Rankings
- Strike–status quo (S–SQ), the Writers Guild's worst case (1)
- No strike–status quo (NS–SQ), the Writers Guild's next-to-worst case (2)
- Strike–salary increase and revenue sharing (S–IN), the Writers Guild's next-to-best case (3)
- No strike–salary increase and revenue sharing (NS–IN), the Writers Guild's best case (4)

#### Management's Alternatives and Rankings
- Strike–status quo, management's next-to-best case (3)
- No strike–status quo, management's best case (4)

**FIGURE 17.6**
Payoff matrix for Writers Guild strike

|  |  | Management (Colin) | |
|---|---|---|---|
|  |  | *SQ* | *IN* |
| | *S* | (0,5) | (6,2) |
| **Writers Guild (Rose)** | | | |
| | *NS* | (4,10) | (10,0) |

- Strike–salary increase and revenue sharing, management's next-to-worst case (2)
- No strike–salary increase and revenue sharing, management's worst case (1)

Then we provide a lottery range using the Method of Von Neumann and Morgenstern to enable us to create the cardinal utilities for each outcome.

- Writers Guild's range of values for [*S–SQ, NS–IN*] are from [0,10]
- Lottery method of Von Neumann and Morgenstern provides *S–IN* as 6 and *NS-SQ* as 4.
- Management's range of values for their choices [*NS–IN, NS–SQ*] are from [0, 10]
- Lottery method of Von Neumann and Morgenstern provides *S–IN* as 2 and *S–SQ* as 5.

This provides us with a payoff matrix consisting of cardinal utilities (see Figure 17.6). This use of cardinal utilities is important because we can then employ cardinal utilities in a Nash arbitration scheme. We will refer to the Writers Guild as Rose and the management as Colin.

John F. Nash proved that every two-person game has at least one equilibrium value either in pure or in mixed strategies (see the readings listed at the end of the chapter). The equilibriums are also called Nash equilibriums. We can use a movement diagram to determine if a pure Nash equilibrium exists or does not exist:

> Movement diagrams in non–zero-sum games will be as follows: Rose would maximize payoffs, so she would prefer the highest payoff at each column. Similarly, for Colin, he wants to maximize his payoffs, so he would prefer the high payoff at each row. We draw an arrow to the highest payoff in that row. If all arrows point in from every direction, then that point or those points will be pure Nash equilibrium.

Examining the movement diagram in Figure 17.7, we find that (4,10) is the Nash equilibrium.

We notice that (4,10) as the pure Nash equilibrium is not satisfying to the writer's guild and that it would like to have a better outcome. Both (6,2) and (10,0) within the payoff matrix provide a better outcome to the writers. As we showed earlier, we could write two linear programming formulations for the problem to find the mixed-strategy solutions because every game has a mixed-strategy solution.

**1.** Maximize $V$

$$\text{Subject to}$$
$$5x_1 + 2x_2 - V \geq 0$$
$$10x_1 - V \geq 0$$
$$x_1 + x_2 = 1$$
$$x_1, x_2, V \geq 0$$

and

**FIGURE 17.7**
Movement diagram for Writers Guild strike

|  |  | Management | |
|---|---|---|---|
|  |  | *SQ* | *IN* |
| | *S* | (0,5) ⟵ | (6,2) |
| **Writers Guild** | | ↓ | ↓ |
| | *NS* | (4,10) ⟵ | (10,0) |

**2.** Maximize $v$

Subject to
$$0y_1 + 6y_2 - v \geq 0$$
$$4y_1 + 10y_2 - v \geq 0$$
$$y_1 + y_2 = 1$$
$$y_1, y_2, v \geq 0$$

> *obj9:=V;*

$$obj9 := V$$

> *const9:={5·x1+2·x2−V≥0,10·x1−V≥0,x1+x2≤1,x1+x2≥1};*

$$const9 := \{x1 + x2 \leq 1, 1 \leq x1 + x2, 0 \leq 5\,x1 + 2\,x2 - V, 0 \leq 10\,x1 - V\}$$

> *LPSolve( V,const9,assume=nonnegative,maximize );*

$$[5.0000000051630, [V = 5.00000000516298915, x2 = 0., x1 = 1.00000000103259778]]$$

> *LPSolve( v, {6·y2−v≥0,4·y1+10·y2−v≥0,y1+y2≤1,y1+y2≥1},assume=nonnegative,maximize );*

$$[6.0000000072282, [v = 6.00000000722818516, y1 = 0., y2 = 1.00000000103259778]]$$

We define the following terms used in game-theory analysis to find an acceptable solution.

**Pareto optimal:**  The outcome where neither player can improve its payoff without hurting (decreasing the payoff) of the other player.

**Pareto principle:**  To be acceptable as a solution of the game, an outcome should be Pareto optimal.

As in this case, group rationality (Pareto) is sometimes in conflict with the individual rationality (dominant). The eventual outcome depends on the players. Obtaining a Pareto optimal outcome usually requires some sort of communication and cooperation among the players.

With the assumption that the outcome should be Pareto optimal, the next question is, "What is Pareto optimal, and what is it not (Pareto inferior)?" The simplest way for this to be understood is to draw a **payoff polygon** of the game. On the chart, the $X$-axis depicts the payoffs of Rose, and the $Y$-axis depicts the payoffs of Colin. By plotting the pure-strategy solutions on the chart, one can see that the convex (everything inside) polygon enclosing the pure-strategy solutions is then the **payoff polygon** or the **feasible region**. Therefore, the points inside the polygon are the possible solutions of the game.

We plot these coordinates from the payoff matrix to determine if any points are Pareto optimal (see Figure 17.8).

The Nash equilibrium value (4,10) lies along the Pareto optimal line segment. But the writers can do better by going on strike and forcing arbitration, which is what they did.

We can employ several options to try to secure a better outcome for the writers. We can first try strategic moves; if those fail to produce a better outcome, then we can move on to Nash

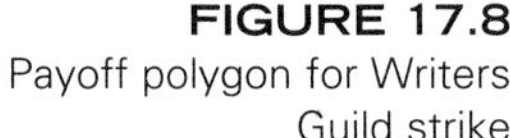

**FIGURE 17.8**
Payoff polygon for Writers Guild strike

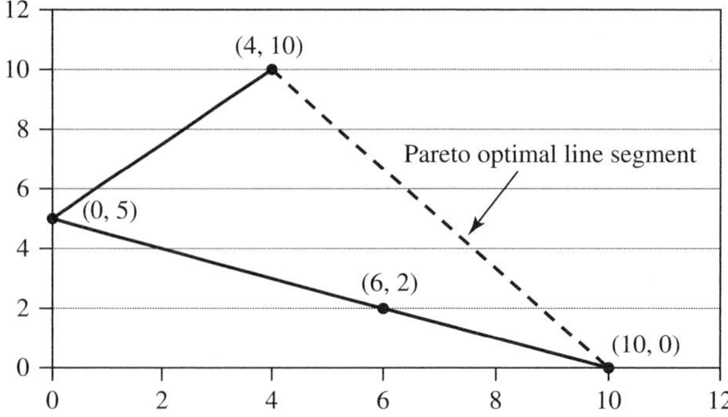

arbitration. Both methods employ communications in the game. We mention the results of the strategic moves, but our interest is in the Nash arbitration method. The result of strategic moves is that none of them—moving first, promises, threats, combination of threats and promises—improved the outcome from the Nash equilibrium value of (4,10). We move on to Nash arbitration.

### Nash Arbitration Scheme

Because our strategic moves did not improve our outcomes for the writers, we consider the Nash arbitration method. In this method, we consider *binding arbitration* in which a third party works out the outcomes that best meet both parties' desires and are acceptable to all players. John Nash found that this outcome can be obtained as follows:

> If SQ (status quo) = $(x_0, y_0)$, then the arbitrated solution point $N$ is the point $(x,y)$ in the polygon with $x \geq x_0$ and $y \geq y_0$, which maximizes the product $(x - x_0)*(y - y_0)$.

The status quo point in the definition is the likely outcome of the game when the negotiation fails. An arbitrated solution should be better for both players than the status quo; this is incorporated in the definition by $x \geq x_0$ and $y \geq y_0$. Status quo is the minimum the players can get. Everything above is an improvement of their gain. The solution has to maximize their joint utility. The objective function $-(x - x_0)*(y - y_0)$ maximizes these "above security level" utilities. In other words, it has to maximize the area of the rectangle.

The status quo point is the security level of each side. We find these values using prudential strategies. Again, the software can assist us in finding these values as (4,5). We start by entering the outcomes into the payoff matrix and requesting the prudential strategies. The value (4,5) is the security levels for our two players. The security level is an important component in the Nash arbitration scheme.

The Nash equilibrium value, (4,10), lies along the Pareto optimal line segment. But the writers can do better by going on strike and forcing arbitration, which is what they did.

The status quo point is the security levels of each side. We find these values using prudential strategies as (4,5). The function for the Nash arbitration scheme is maximize $(x - 4)(y - 5)$.

Our formulation is

$$\text{Maximize } (x - 4)(y - 5)$$
$$\text{Subject to}$$
$$\frac{5}{3}x + y \leq \frac{50}{3}$$
$$\frac{-5}{4}x + y \leq 5$$
$$\frac{1}{2}x + y \geq 5$$

We can find the convex polygon from the payoff values and plot in Figure 17.8. We graph the convex polygon in Figure 17.9 with the contours of our function, $(x - 4)\,(y - 5)$.

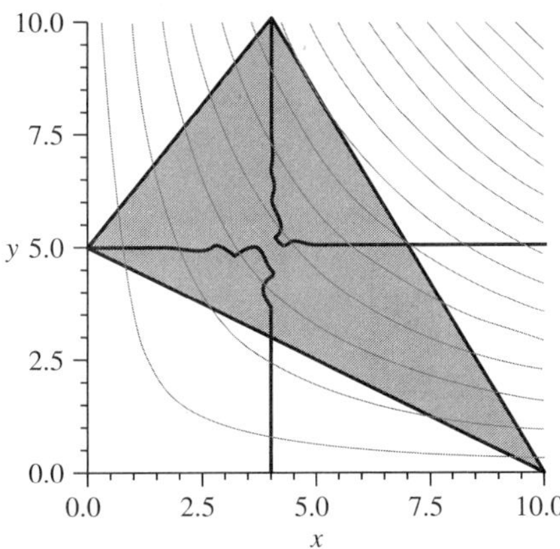

**FIGURE 17.9**
Convex polygon and Nash arbitration contours

1. Find the Nash equilibrium.

2. Draw the payoff polygon and state whether or not the Nash equilibrium is Pareto optimal.

3. Assume both sides agree to arbitration. Find the Nash arbitration point (use any of the methods) and also find how the negotiators get there in terms of the nearest Pareto optimal points. Provide a graph showing the payoff polygon, the negotiation set, and the Nash arbitration point.

Using Maple, we find the desired solution to our NLP as

$$x = 5.5$$
$$y = 7.5$$
$$l_1 = 1.5$$
$$l_2 = 0$$
$$l_3 = 0$$
$$u_1 = 0$$
$$u_2 = 2.09165$$
$$u_3 = 2.29128$$

We have the $x$ and $y$ coordinate $(5.5, 7.5)$ as our arbitrated solution. We also have obtained some information about $\lambda_1$. Its value of 1.5 means that an increase in utility of 1 unit for this constraint provides an increase in value of approximately 1.5 units.

**Results**

One of the strongest elements of the Nash arbitration is that by using the security levels as the basis of the arbitration our players can do no worse than those values $(4,5)$. For the Writers Guild, that value of 4 is equal to their value under the Nash equilibrium. That means that, in the worst case, their outcome is at least the equilibrium value. The Excel programs enabled our students to find the Nash arbitrated solution value of $(5.5, 7.5)$. The Writers Guild should be able to improve its outcome from 4 to 5.5 through an arbitrated solution.

How should the negotiators achieve this result?

We find that the negotiators should consider the following strategy: Writers should not strike, and management should offer status quo at 75 percent and salary increasing and revenue sharing at 25 percent.

The Nash arbitrated solution was found to be $(5.5, 7.5)$. The Writers Guild was able to improve its outcome from 4 to 5.5 through an arbitrated solution. This is a substantial increase for the writers and would effectively satisfy both players in the game.

1. Find the Nash equilibrium.

2. Draw the payoff polygon and state whether or not the Nash equilibrium is Pareto optimal.

3. Assume both sides agree to arbitration. Find the Nash arbitration point (use any of the methods) and also find how the negotiators get there in terms of the nearest Pareto optimal points. Provide a graph showing the payoff polygon, the negotiation set, and the Nash arbitration point.

## 17.4 | READINGS

Nash, J. F. "The Bargaining Problem," *Econometrica*, 18(2) (April 1950).

Nash, J. F. "Equilibrium Points in *n*-Person Games," *Proceedings of the National Academy of Sciences*, 36(1) (January 15, 1950).

Straffin, P. D. *Game Theory and Strategy* Vol. 25, New Mathematical Library. Washington, DC: Mathematical Association of America, 1993.

Von Neumann, J., and O. Morgenstern, *Theory of Games and Economic Behavior* (60th anniversary ed.). Princeton, NJ; Princeton University Press, 2004.

Wikipedia, Writers_Guild_of_America_strike, The Writer's Guild Strike, 2007 (http://en.wikipedia.org/wiki/2007).

**Additional Readings**

Aumann, R. J. (1987). "Game Theory." *The New Palgrave: A Dictionary of Economics*, Vol. 2, pp. 460–482 (London: Palgrave Macmillan, 1987).

Bazarra, M. S., H. D. Sherali, and C. M . Shetty. (1993). *Nonlinear Programming*. New York: Wiley.

Crawford, V. (1974). "Learning the Optimal Strategy in a Zero-Sum Game." *Econometrica* 42(5), pp. 885–891.

Danzig, G. (2002). "Linear Programming," *Operations Research*, 50(1), pp. 42–47.

Danzig, G. (1951). "Maximization of a Linear Function of Variables Subject to Linear Inequalities." Chapter 21 from *Activity Analysis of Production and Allocation Conference Proceeding* (pp. 339–347), T. Koopman (Ed.). New York: John Wiley.

Daskalakis, C., P. W. Goldberg, and C. H. Papadimirtriou. (In press.) "The Complexity of Computing a Nash Equilibrium." To appear in SICOMP.

Dorfman, R. (1951). "Application of the Simplex Method to a Game Theory Problem." Chapter 22 from *Activity Analysis of Production and Allocation Conference Proceeding* (pp. 348–358), T. Koopman (Ed.). New York: John Wiley.

Fox, W. P. (2008). "Mathematical Modeling of Conflict and Decision Making: 'The Writers Guild Strike 2007–2008,'" *Computers in Education Journal*, 18(3), pp. 2–11.

Gale, D., H. Kuhn, and A. Tucker. (1951). "Linear Programming and the Theory of Games." Chapter 19 from *Activity Analysis of Production and Allocation Conference Proceeding* (pp. 317–329), T. Koopman (Ed.). New York: John Wiley.

Gillman, R., and D. Housman. (2009). *Models of Conflict and Cooperation*, pp.189–195. Providence, RI: American Mathematical Society.

Klarrich, E. (2009). "The Mathematics of Strategy." *Classics of the Scientific Literature.* Oct 2009 (www.pnas.org/site/misc/classics5.shtml).

Kuhn, H. W., and A.W. Tucker. (1951). "Nonlinear Programming." *Proceedings of the Second Berkley Symposium on Mathematical Statistics and Probability*, J. Newman (Ed.). Berkeley: University of California Press.

Maple (2006). Product of @Maplesoft, Waterloo, Canada.

Nash, J. (1950). "The Bargaining Problem," *Econometrica*, 18: pp. 155–162.

Nash, J. (1951). "Non-Cooperative Games," *Annals of Mathematics*, 54: pp. 289–295.

Nash, J. (2009). Lecture at NPS. Feb 19, 2009.

Straffin, P. D. (2004). *Game Theory and Strategy.* Washington, DC: Mathematical Association of America.

Williams, J. D. (1986). *The Compleat Strategyst.* New York: Dover Press. (original edition by RAND Corporation, 1954).

Winston, W. L. (1995). *Introduction to Mathematical Programming*, 2nd Ed. Belmont, CA: Duxbury Press.

# Chapter 1

**1.** $sqrt(2.3^3 \cdot 4.5)$;

$$7.399425653$$

**2.** $11.3^3 + 5.1^2$;

$$1468.907$$

**3.** $21.6^{\left(\frac{1}{3}\right)}$;

$$2.784953300$$

**4.** $a := 8 : b := 7 : (a^2 - b^2)$;

$$15$$

**5.** $2 \cdot (11.5) + 6.2^2 \cdot (.7)$;

$$49.908$$

**6.** $f6 := -x^2 + 3 \cdot x + 3$;

$$f6 := -x^2 + 3x + 3$$

$plot(f6, x=-5..5, thickness=3)$;

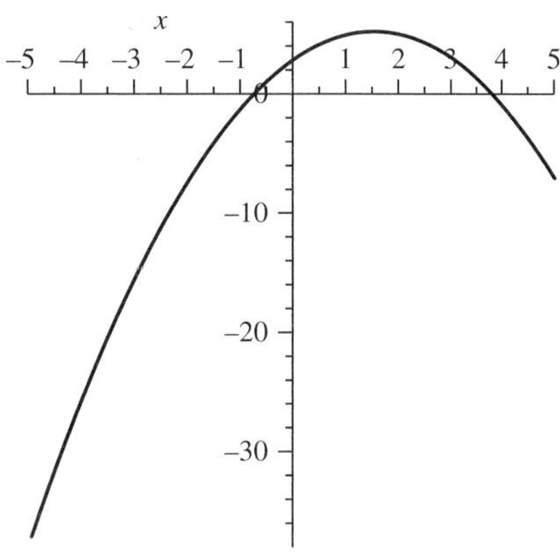

$fsolve(f6=0, x)$;

$$-.7912878475, \; 3.791287847$$

**7.** $f7 := x^2 - 3 \cdot x - 1$;

$$f7 := x^2 - 3x - 1$$

$plot(f7, x=-10..10, thickness=3)$;

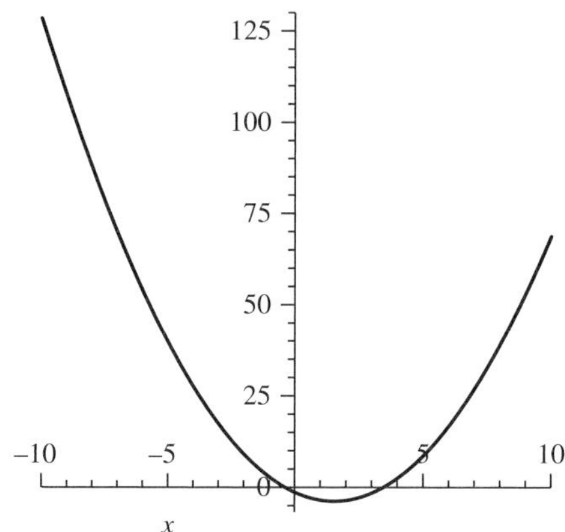

$fsolve(f7=0, x)$;

$$-.3027756377, \; 3.302775638$$

**8.** $> f8 := -0.1213 \cdot x^4 + 3.462 \cdot x^3 - 29.22 \cdot x^2 + 64.68 \cdot x + 97.69$;

$$f8 := -0.1213\,x^4 + 3.462\,x^3 - 29.22\,x^2 + 64.68\,x + 97.69$$

$> plot(f8, x=-1..1, thickness=3)$;

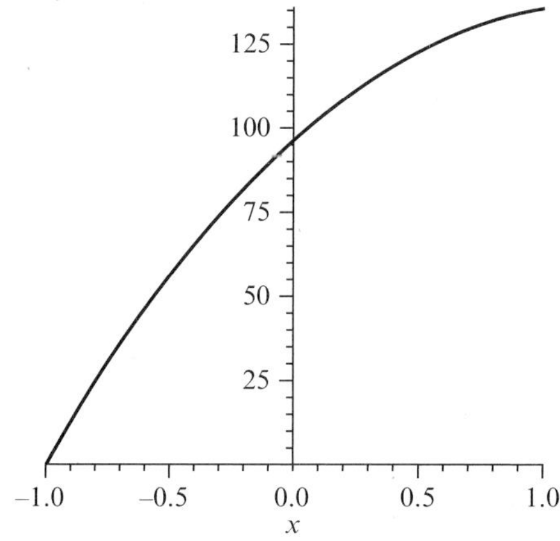

$> fsolve(f8=0, x)$;

$$-.4721368076, \; 0.5672715705$$

> *solve(f8=0,x);*

$$0.5672715705,\ -0.04614034104 + 0.5464146718\ \mathrm{I},$$
$$-.4721368076,\ -0.04614034104 - 0.5464146718\ \mathrm{I}$$

**9.** > *f9:=x³−2·x²−5·x+6;*

$$f9 := x^3 - 2\,x^2 - 5\,x + 6$$

> *plot(f9,x=−5..5,thickness=3);*

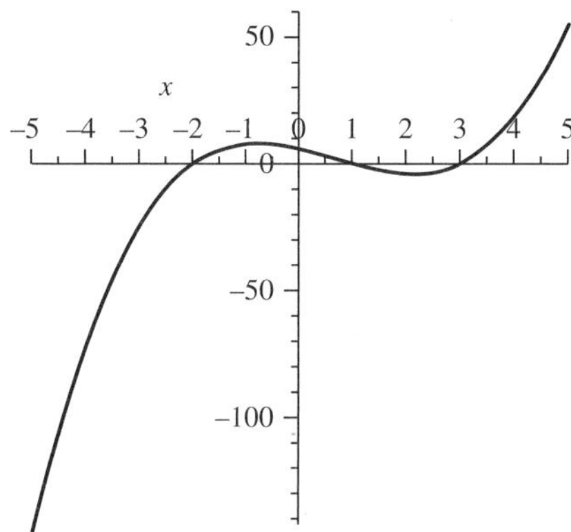

> *solve(f9=0,x);*

$$1,\ -2,\ 3$$

**10.** > *f10:=2·x³−3·x²−11·x+7;*

$$f10 := 2\,x^3 - 3\,x^2 - 11\,x + 7$$

> *plot(f10,x=−3..2,thickness=3);*

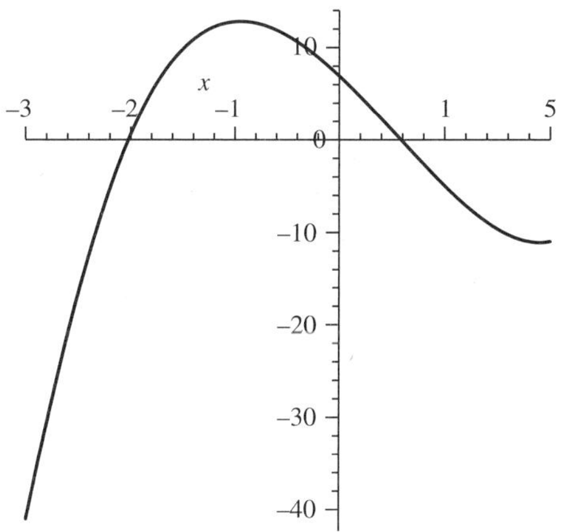

> *solve(f10=0,x);*

$$\frac{1}{6}\left(-54 + 3\,\mathrm{I}\,\sqrt{46551}\,\right)^{1/3} + \frac{25}{2\left(-54 + 3\mathrm{I}\,\sqrt{46551}\,\right)^{1/3}}$$

$$+\frac{1}{2},\ -\frac{1}{12}\left(-54 + 3\,\mathrm{I}\,\sqrt{46551}\,\right)^{1/3}$$

$$-\frac{25}{4\left(-54 + 3\,\mathrm{I}\,\sqrt{46551}\,\right)^{1/3}} + \frac{1}{2}$$

$$+\frac{1}{2}\mathrm{I}\,\sqrt{3}\left(\frac{1}{6}\left(-54 + 3\,\mathrm{I}\,\sqrt{46551}\,\right)^{1/3}\right.$$

$$\left.-\frac{25}{2\left(-54 + 3\,\mathrm{I}\,\sqrt{46551}\,\right)^{1/3}}\right),\ -\frac{1}{12}\left(-54 + 3\,\mathrm{I}\,\sqrt{46551}\,\right)^{1/3}$$

$$-\frac{25}{4\left(-54 + 3\,\mathrm{I}\,\sqrt{46551}\,\right)^{1/3}}$$

$$+\frac{1}{2} - \frac{1}{2}\mathrm{I}\,\sqrt{3}\left(\frac{1}{6}\left(-54 + 3\,\mathrm{I}\,\sqrt{46551}\,\right)^{1/3}\right.$$

$$\left.-\frac{25}{2\left(-54 + 3\,\mathrm{I}\,\sqrt{46551}\,\right)^{1/3}}\right)$$

> *fsolve(f10=0,x);*

$$-2.039078928,\ 0.5800821727,\ 2.958996755$$

**11.** > *xdat:=[1,3,8,10];ydat:=[.7,5,15.2,36];*

$$xdat := [1,\ 3,\ 8,\ 10]$$

$$ydat := [0.7,\ 5,\ 15.2,\ 36]$$

> *with (plots):*
> *pointplot({seq{[xdat[i],ydat[i]],i=1..4}},symbol= circle,thickness=3,title='Scatter Plot of Raw Data');*

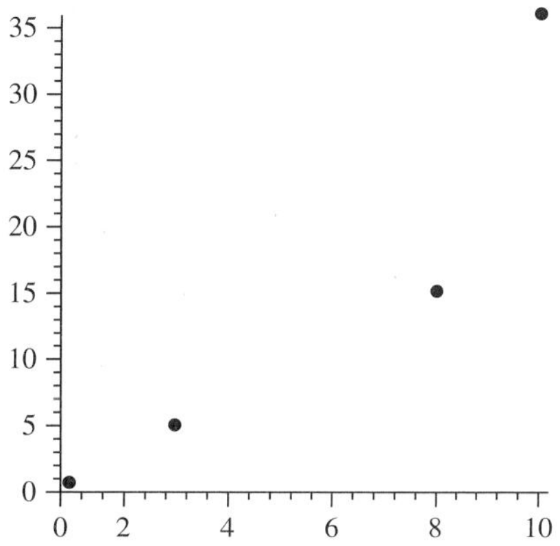

**Scatter Plot of Raw Data**

**12.** > *xdat2:=[7,14,21,28,35,42];ydat2:=[8,41,133,250,280,297];*

$$xdat2 := [7,\ 14,\ 21,\ 28,\ 35,\ 42]$$

$$ydat2 := [8,\ 41,\ 133,\ 250,\ 280,\ 297]$$

> *with (plots):*
> *pointplot({seq([xdat2[i],ydat2[i]],i=1..6)}, symbol=circle, thickness=3,title='Scatter Plot of Raw Data');*

**Scatter Plot of Raw Data**

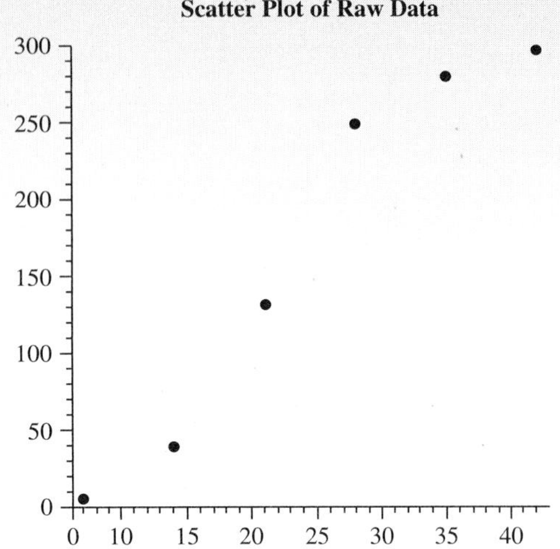

**13.** > *xdati3:=[29,48,72.7,92,118,140,165,199];*

$$xdat3 := [29, 48, 72.7, 92, 118, 140, 165, 199]$$

> *ydat3:=[.49,.82,1.23,1.54,1.97,2.34,2.74,3.3];*

$$ydat3 := [0.49, 0.82, 1.23, 1.54, 1.97, 2.34, 2.74, 3.3]$$

> *with(plots):*
> *pointplot({seq([xdat3[i],ydat3[i]],i=1..7)},symbol= circle,thickness=3,title='Scatter Plot of Raw Data');*

**Scatter Plot of Raw Data**

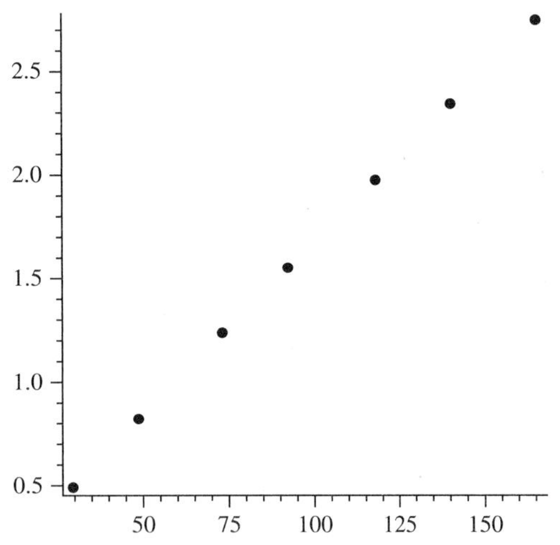

**14.** *diff(1.104·x−0.542·x²,x);*

$$1.104 - 1.084\,x$$

**15.** *int(1.104·x−0.542·x²,x);*

$$0.5520000000\,x^2 - 0.1806666667\,x^3$$

**16.** *f14:=1.104·x−0.542·x²;*

$$f14 := 1.104\,x - 0.542\,x^2$$

> *int(f14,x=1..5);*

$$-9.154666667$$

# Chapter 3

### Section 3.1

**1.** > *p1:=rsolve({a(n+1)=.5*a(n)+.10,a(0)=0.1}, a(n));*

$$-\frac{1}{10}\left(\frac{1}{2}\right)^n + \frac{1}{5}$$

0.1, 0.15, 0.175, 0.1875, 0.19375, 0.196875, 0.1984375,
0.19921875, 0.199609375, 0.1998046875, 0.1999023438,
0.1999511719, 0.1999755860, 0.1999877930, 0.1999938965,
0.1999969482, 0.1999984741, 0.1999992370, 0.1999996185,
0.1999998092, 0.1999999046, 0.1999999523, 0.1999999762,
0.1999999881, 0.1999999940, 0.1999999970, 0.1999999985,
0.1999999992, 0.1999999996, 0.1999999998, 0.1999999999,
0.2000000000, 0.2000000000, 0.2000000000, 0.2000000000,
0.2000000000, 0.2000000000, 0.2000000000, 0.2000000000,
0.2000000000, 0.2000000000, 0.2000000000, 0.2000000000,
0.2000000000, 0.2000000000, 0.2000000000, 0.2000000000,
0.2000000000, 0.2000000000, 0.2000000000

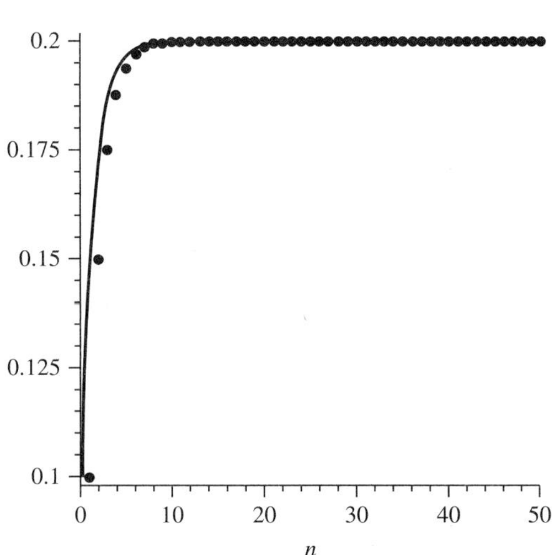

**2.** > *p1:=rsolve({a(n+1)=.5*a(n)+.10,a(0)=0.2}, a(n));*

$$\frac{1}{5}$$

0.2, 0.20, 0.200, 0.2000, 0.20000, 0.200000,
0.2000000, 0.20000000, 0.200000000,
0.2000000000, 0.2000000000, 0.2000000000,
0.2000000000, 0.2000000000, 0.2000000000,
0.2000000000, 0.2000000000, 0.2000000000,
0.2000000000, 0.2000000000, 0.2000000000,
0.2000000000, 0.2000000000, 0.2000000000,
0.2000000000, 0.2000000000, 0.2000000000,
0.2000000000, 0.2000000000, 0.2000000000,
0.2000000000, 0.2000000000, 0.2000000000,
0.2000000000, 0.2000000000, 0.2000000000,
0.2000000000, 0.2000000000, 0.2000000000,
0.2000000000, 0.2000000000, 0.2000000000,
0.2000000000, 0.2000000000, 0.2000000000,
0.2000000000, 0.2000000000, 0.2000000000,
0.2000000000, 0.2000000000

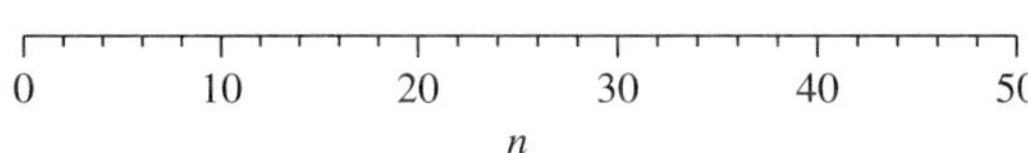

**3.**
$$\frac{1}{10}\left(\frac{1}{2}\right)^n + \frac{1}{5}$$

0.3, 0.25, 0.225, 0.2125, 0.20625, 0.203125, 0.2015625,
0.20078125, 0.200390625, 0.2001953125, 0.2000976562,
0.2000488281, 0.2000244140, 0.2000122070,
0.2000061035, 0.2000030518, 0.2000015259,
0.2000007630, 0.2000003815, 0.2000001908,
0.2000000954, 0.2000000477, 0.2000000238,
0.2000000119, 0.2000000060, 0.2000000030,
0.2000000015, 0.2000000008, 0.2000000004,
0.2000000002, 0.2000000001, 0.2000000000,
0.2000000000, 0.2000000000, 0.2000000000,
0.2000000000, 0.2000000000, 0.2000000000,
0.2000000000, 0.2000000000, 0.2000000000,
0.2000000000, 0.2000000000, 0.2000000000,
0.2000000000, 0.2000000000, 0.2000000000,
0.2000000000, 0.2000000000, 0.2000000000

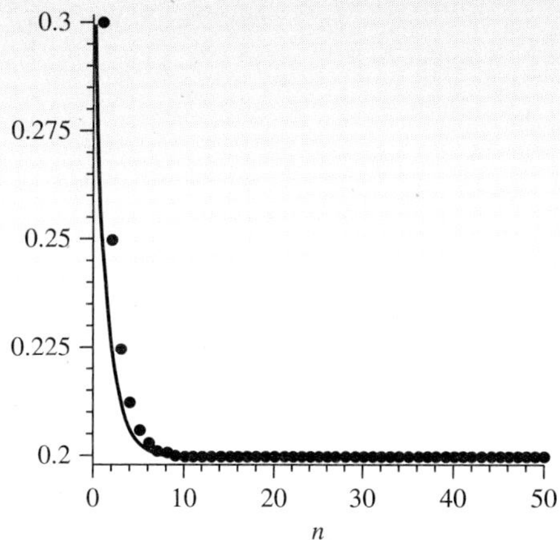

**4.**
$$-10000\left(\frac{101}{100}\right)^n + 100000$$

90000, 89900.00, 89799.0000, 89696.99000, 89593.95990,
89489.89950, 89384.79850, 89278.64648, 89171.43294,
89063.14727, 88953.77874, 88843.31653, 88731.74970,
88619.06720, 88505.25787, 88390.31045, 88274.21355,
88156.95569, 88038.52525, 87918.91050, 87798.09960,
87676.08060, 87552.84141, 87428.36982, 87302.65352,
87175.68006, 87047.43686, 86917.91123, 86787.09034,
86654.96124, 86521.51085, 86386.72596, 86250.59322,
86113.09915, 85974.23014, 85833.97244, 85692.31216,
85549.23528, 85404.72763, 85258.77491, 85111.36266,
84962.47629, 84812.10105, 84660.22206, 84506.82428,
84351.89252, 84195.41145, 84037.36556, 83877.73922,
83716.51661

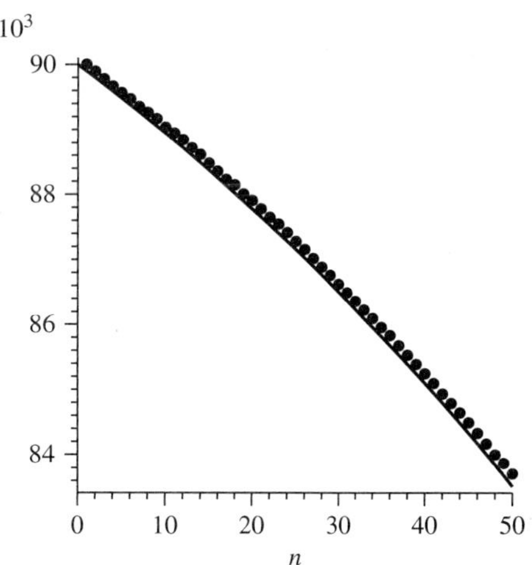

**5.**

$$100000$$

$100000$, $1.0000000 \ 10^5$, $1.000000000 \ 10^5$, $1.000000000 \ 10^5$,
$1.000000000 \ 10^5$, $1.000000000 \ 10^5$, $1.000000000 \ 10^5$,
$1.000000000 \ 10^5$, $1.000000000 \ 10^5$, $1.000000000 \ 10^5$,
$1.000000000 \ 10^5$, $1.000000000 \ 10^5$, $1.000000000 \ 10^5$,
$1.000000000 \ 10^5$, $1.000000000 \ 10^5$, $1.000000000 \ 10^5$,
$1.000000000 \ 10^5$, $1.000000000 \ 10^5$, $1.000000000 \ 10^5$,
$1.000000000 \ 10^5$, $1.000000000 \ 10^5$, $1.000000000 \ 10^5$,
$1.000000000 \ 10^5$, $1.000000000 \ 10^5$, $1.000000000 \ 10^5$,
$1.000000000 \ 10^5$, $1.000000000 \ 10^5$, $1.000000000 \ 10^5$,
$1.000000000 \ 10^5$, $1.000000000 \ 10^5$, $1.000000000 \ 10^5$,
$1.000000000 \ 10^5$, $1.000000000 \ 10^5$, $1.000000000 \ 10^5$,
$1.000000000 \ 10^5$, $1.000000000 \ 10^5$, $1.000000000 \ 10^5$,
$1.000000000 \ 10^5$, $1.000000000 \ 10^5$, $1.000000000 \ 10^5$,
$1.000000000 \ 10^5$, $1.000000000 \ 10^5$, $1.000000000 \ 10^5$,
$1.000000000 \ 10^5$,

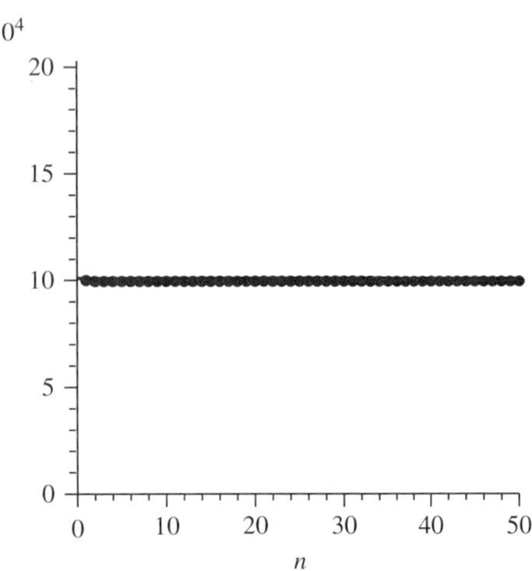

**6.**

$$10000 \left(\frac{101}{100}\right)^n + 100000$$

$110000$, $1.1010000 \ 10^5$, $1.102010000 \ 10^5$, $1.103030100 \ 10^5$,
$1.104060401 \ 10^5$, $1.105101005 \ 10^5$, $1.106152015 \ 10^5$,
$1.107213535 \ 10^5$, $1.108285670 \ 10^5$, $1.109368527 \ 10^5$,
$1.110462212 \ 10^5$, $1.111566834 \ 10^5$, $1.112682502 \ 10^5$,
$1.113809327 \ 10^5$, $1.114947420 \ 10^5$, $1.116096894 \ 10^5$,
$1.117257863 \ 10^5$, $1.118430442 \ 10^5$, $1.119614746 \ 10^5$,
$1.120810893 \ 10^5$, $1.122019002 \ 10^5$, $1.123239192 \ 10^5$,
$1.124471584 \ 10^5$, $1.125716300 \ 10^5$, $1.126973463 \ 10^5$,
$1.128243198 \ 10^5$, $1.129525630 \ 10^5$, $1.130820886 \ 10^5$,
$1.132129095 \ 10^5$, $1.133450386 \ 10^5$, $1.134784890 \ 10^5$,
$1.136132739 \ 10^5$, $1.137494066 \ 10^5$, $1.138869007 \ 10^5$,
$1.140257697 \ 10^5$, $1.141660274 \ 10^5$, $1.143076877 \ 10^5$,
$1.144507646 \ 10^5$, $1.145952722 \ 10^5$, $1.147412249 \ 10^5$,
$1.148886371 \ 10^5$, $1.150375235 \ 10^5$, $1.151878987 \ 10^5$,

$1.153397777 \ 10^5$, $1.154931755 \ 10^5$, $1.156481073 \ 10^5$,
$1.158045884 \ 10^5$, $1.159626343 \ 10^5$, $1.161222606 \ 10^5$,
$1.162834832 \ 10^5$

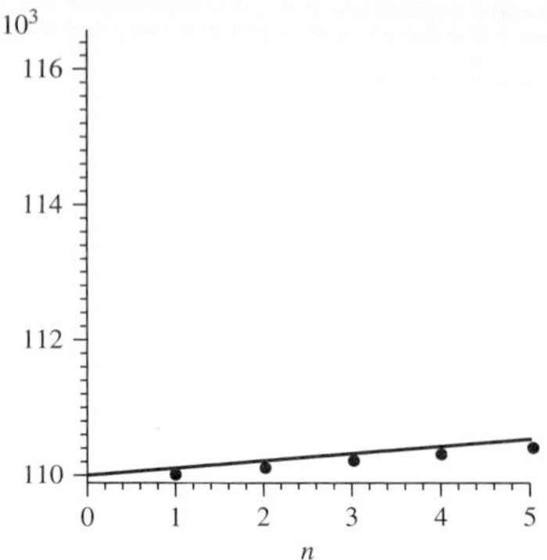

**7.**

$$\frac{7}{23}\left(-\frac{13}{10}\right)^n + \frac{200}{23}$$

$9$, $8.3$, $9.21$, $8.027$, $9.5649$, $7.56563$, $10.164681$,
$6.7859147$, $11.17831089$, $5.46819584$, $12.89134541$,
$3.24125097$, $15.78637374$, $-.52228586$, $20.67897162$,
$-6.88266311$, $28.94746204$, $-17.63170065$,
$42.92121084$, $-35.79757409$, $66.53684632$,
$-66.49790022$, $106.4472703$, $-118.3814514$,
$173.8958868$, $-206.0646528$, $287.8840486$,
$-354.2492632$, $480.5240422$, $-604.6812549$,
$806.0856314$, $-1027.911321$, $1356.284717$,
$-1743.170132$, $2286.121172$, $-2951.957524$,
$3857.544781$, $-4994.808215$, $6513.250680$,
$-8447.225884$, $11001.39365$, $-14281.81174$,
$18586.35526$, $-24142.26184$, $31404.94039$,
$-40806.42251$, $53068.34926$, $-68968.85404$,
$89679.51025$, $-1.165633633 \ 10^5$

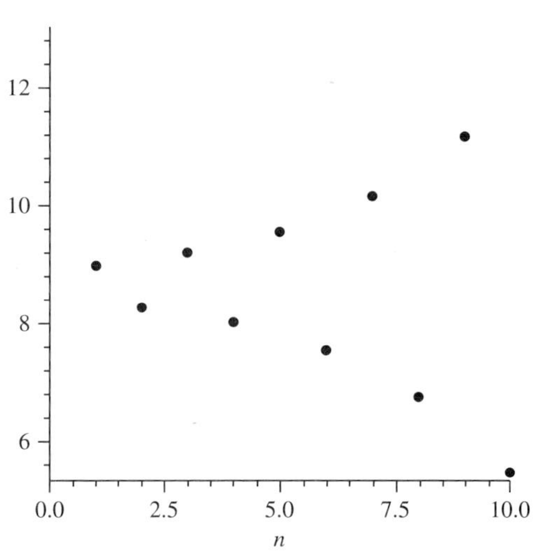

$$-(-4)^n + 10$$

9, 14, −6, 74, −246, 1034, −4086, 16394, −65526, 262154, −1048566, 4194314, −16777206, 67108874, −268435446, 1073741834, −4294967286, 17179869194, −68719476726, 274877906954, −1099511627766, 4398046511114, −17592186044406, 70368744177674, −281474976710646, 1125899906842634, −4503599627370486, 18014398509481994, −72057594037927926, 288230376151711754, −1152921504606846966, 4611686018427387914, −18446744073709551606, 73786976294838206474, −295147905179352825846, 1180591620717411303434, −4722366482869645213686, 18889465931478580854794, −75557863725914323419126, 302231454903657293676554, −1208925819614629174706166, 4835703278458516698824714, −19342813113834066795298806, 77371252455336267181195274, −309485009821345068724781046, 1237940039285380274899124234, −4951760157141521099596496886, 19807040628566084398385987594, −79228162514264337593543950326, 316912650057057350374175801354

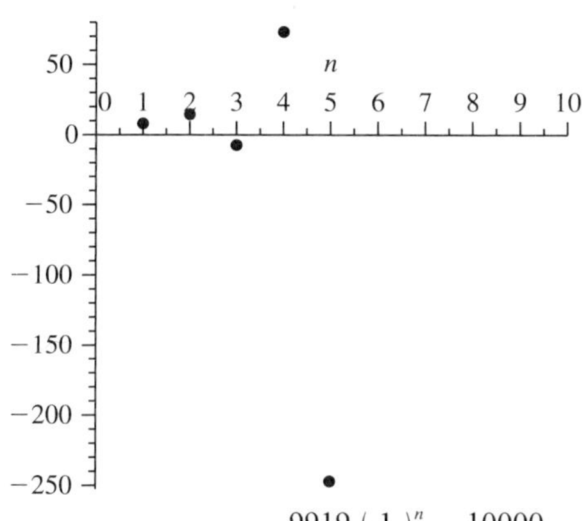

**9.**
$$-\frac{9919}{9}\left(\frac{1}{10}\right)^n + \frac{10000}{9}$$

9, 1000.9, 1100.09, 1110.009, 1111.0009, 1111.10009, 1111.110009, 1111.111001, 1111.111100, 1111.111110, 1111.111111, 1111.111111, 1111.111111, 1111.111111, 1111.111111, 1111.111111, 1111.111111, 1111.111111, 1111.111111, 1111.111111, 1111.111111, 1111.111111, 1111.111111, 1111.111111, 1111.111111, 1111.111111, 1111.111111, 1111.111111, 1111.111111, 1111.111111, 1111.111111, 1111.111111, 1111.111111, 1111.111111, 1111.111111, 1111.111111, 1111.111111, 1111.111111, 1111.111111, 1111.111111, 1111.111111, 1111.111111, 1111.111111, 1111.111111, 1111.111111, 1111.111111, 1111.111111, 1111.111111, 1111.111111,

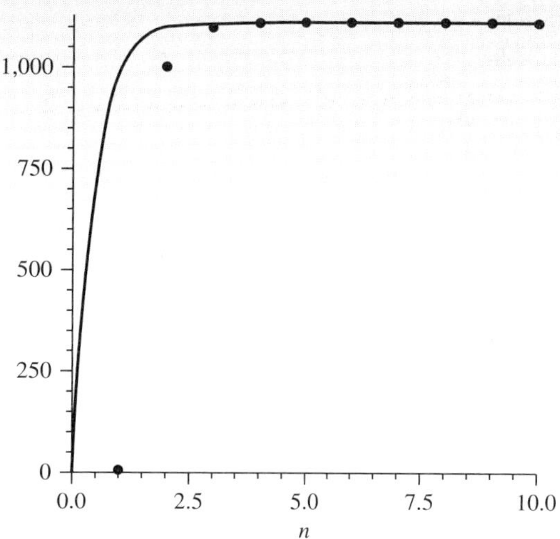

**10.**
$$-\frac{2414}{13}\left(\frac{9987}{10000}\right)^n + \frac{3350}{13}$$

72, 72.2414, 72.48248618, 72.72325895, 72.96371871, 73.20386588, 73.44370085, 73.68322404, 73.92243585, 74.16133668, 74.39992694, 74.63820703, 74.87617736, 75.11383833, 75.35119034, 75.58823379, 75.82496909, 76.06139663, 76.29751681, 76.53333004, 76.76883671, 77.00403722, 77.23893197, 77.47352136, 77.70780578, 77.94178563, 78.17546131, 78.40883321, 78.64190173, 78.87466726, 79.10713019, 79.33929092, 79.57114984, 79.80270735, 80.03396383, 80.26491968, 80.49557528, 80.72593103, 80.95598732, 81.18574454, 81.41520307, 81.64436331, 81.87322564, 82.10179045, 82.33005812, 82.55802904, 82.78570360, 83.01308219, 83.24016518, 83.46695297

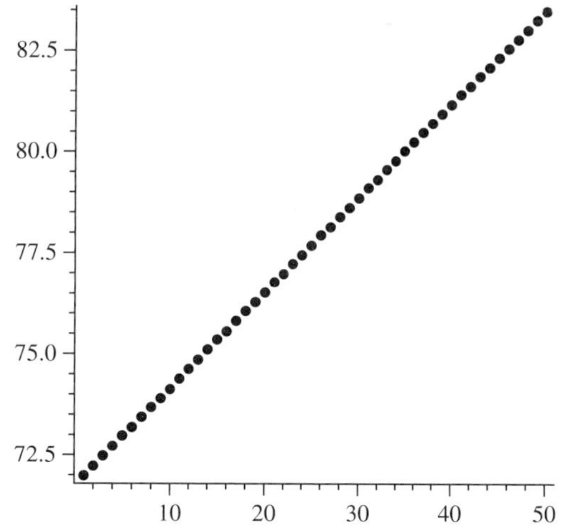

## Section 3.2

1. **a.** 0, unstable  **b.** 0, stable
   **c.** 0, stable  **d.** No equilibrium value
   **e.** 84, stable  **f.** 500, stable
   **g.** 500, stable  **h.** 55.55, stable

2. **a.** Not stable  **b.** Stable
   **c.** Stable  **d.** Not stable

## Section 3.3

1. 0.2, 0.32, 0.4352, 0.49160192, 0.4998589446,
   0.4999999602, 0.5000000000, 0.5000000000,
   0.5000000000, 0.5000000000, 0.5000000000,
   0.5000000000, 0.5000000000, 0.5000000000,
   0.5000000000, 0.5000000000, 0.5000000000,
   0.5000000000, 0.5000000000, 0.5000000000

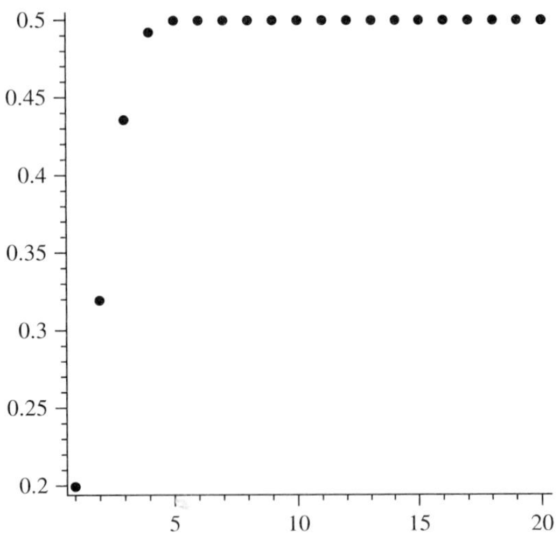

Stable at 0.5

2. 0.2, 0.48, 0.7488, 0.56429568, 0.7375981965,
   0.5806412910, 0.7304909466, 0.5906217705,
   0.7253630841, 0.5976344409, 0.7214025480,
   0.6029427351, 0.7182083799, 0.6071553087,
   0.7155532194, 0.6106104288, 0.7132959990,
   0.6135144504, 0.7113434088, 0.6160018908

Two-Cycle

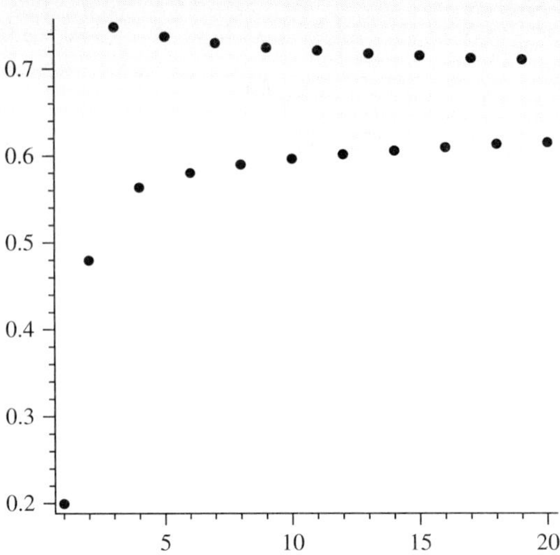

3. 0.2, 0.576, 0.8792064, 0.3823290223, 0.8501527475,
   0.4586149923, 0.8938342119, 0.3416206088,
   0.8096974866, 0.5547148805, 0.8892226148,
   0.3546207220, 0.8239135158, 0.5222881234,
   0.8982116623, 0.3291388992, 0.7949033432,
   0.5869152655, 0.8728046519, 0.3996600895

Four-cycle

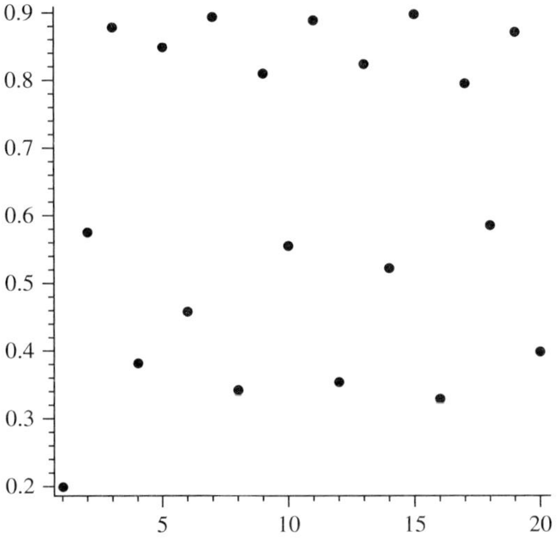

4. 0.2, 0.592, 0.8936832, 0.3515500907, 0.8434617107,
   0.4885259972, 0.9245128850, 0.2582185987,
   0.7087044898, 0.7638370129, 0.6674431134,
   0.8212623741, 0.5431248018, 0.9181189306,
   0.2781532715, 0.7429009078, 0.7066968513,
   0.7669227232, 0.6613833614, 0.8286350194

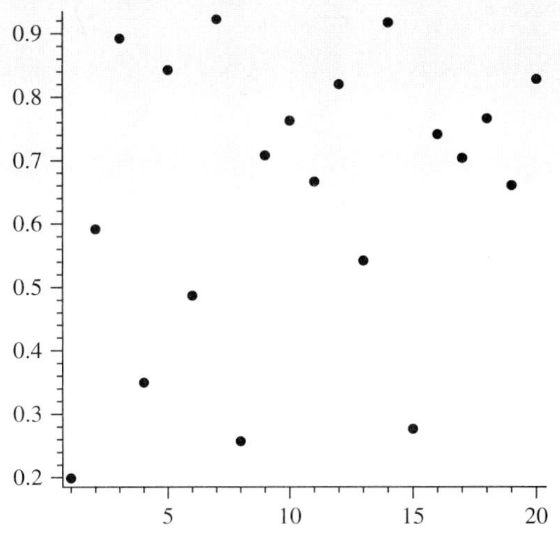

Chaos

**5.** Equilibrium values are 0,5 and stable at 5.

**6.** Equilibrium values are 2,0 and stable at 0.

**7.** Real equilibrium value at 0 and unstable

**8.** Equilibrium values at –1,2, and –1 is stable.

**9.** About 15 days.

### Section 3.4

**1.** The system goes to about 107 and 142, respectively.

**2.** (0,0) and (1500,2000). Only (0,0) is achievable if we start at (0,0).

# Chapter 4

### Section 4.1

**1.** **a.** $y = 0.7x - 0.1$
   **b.** $y = 0.1583248212x^2$

**2.** **a.** $y = 3.703309539x$
   **b.** $y = 4.06933x - 25.407127$
   **c.** $y = 0.04487x^2$

**3.** **a.** $y = 10.4328x - 175.70395$
   **b.** $y = \dfrac{454692}{10133225}\,x^2$
   **c.** $y = \dfrac{37343125}{9062813948}\,x^3$
   **d.** $y = -0.01361706\,x^3 + 1,2910969\,x^2 - 28.7102007\,x + 199.6338129$

**4.** $\text{period} = 0.000000000544 \times \text{distance}^{\left(\frac{3}{2}\right)}$

### Section 4.3

**1.** $0.2916 \le c_{max} \le 0.8416$

**2.** **a.** $43.07 \le c_{max} \le 60.82$
   **b.** $13.914 \le c_{max} \le 19.669$

# Chapter 5

### Section 5.1

**1.** If we plot $W$ on the "**y**" axis and *Flu* on the "*x*" axis, we would obtain a straight line that passes through the origin with a slope equal to $k,\ k > 0$. Thus, $W = k(Flu),\ k > 0$.

**2.** 70.5 miles

**3.** a) yes b) yes, c) yes d) yes e) yes f) yes

### Section 5.2

**1.** **a.** F

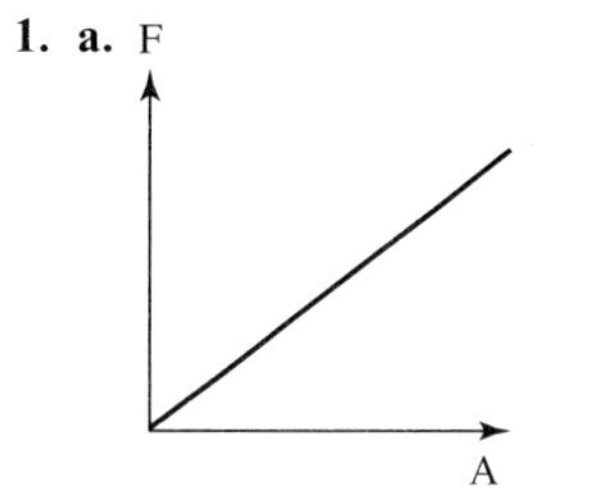


   **b.** I

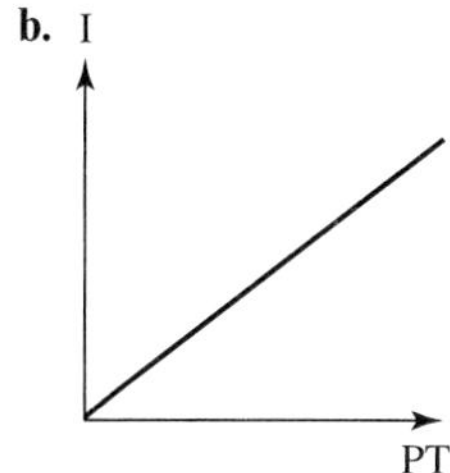


   **c.** D

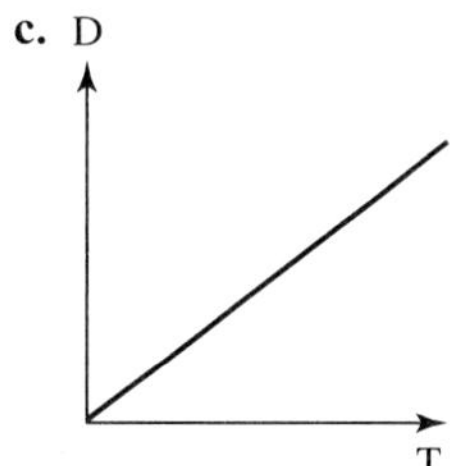


   **d.** F

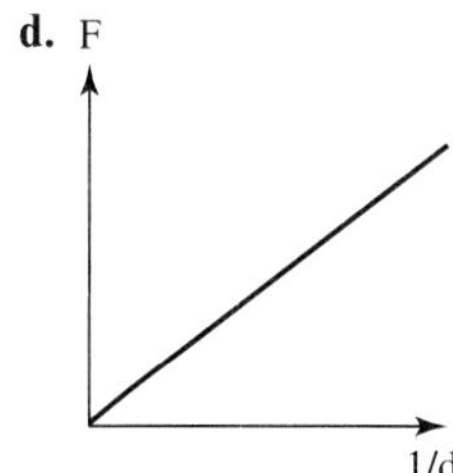


**2.** No

**3.** 209.1 million

**4.** 137.5

**5.** 118.8 ft.

# Chapter 6

## Section 6.1

**1. a.** Scatterplot is concave up and increasing.

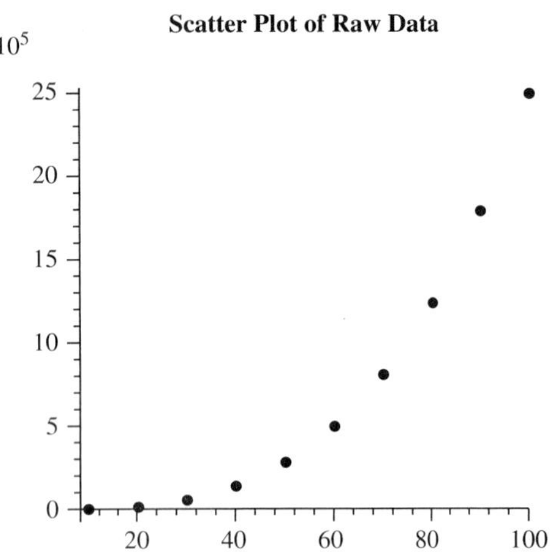

**Scatter Plot of Raw Data**

**b.** $y \to y, x \to x^2 \to x^3$. Stop linear relationship found.
$y = 2.457853223 \, x^3$

**c.** The model is $y = 2.457853223 \, x^3$.

**d.** Residual plot shows a pattern.

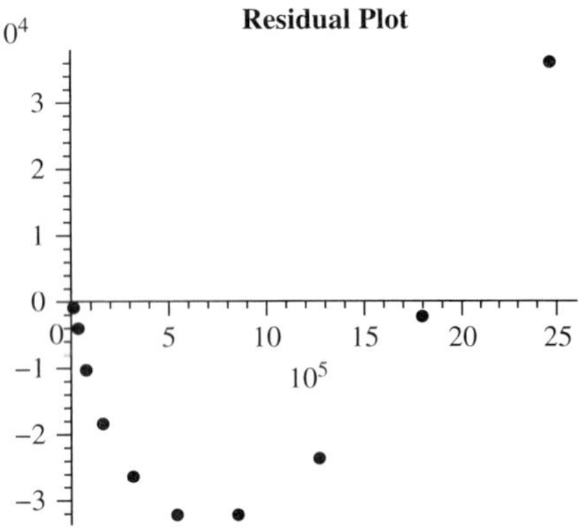

**Residual Plot**

**e.** Using the natural logs, we obtain the model after it is transformed back into the real space,

$$y = 1.250297116 \, x^{3.149942645}$$

The plot of the data with the model:

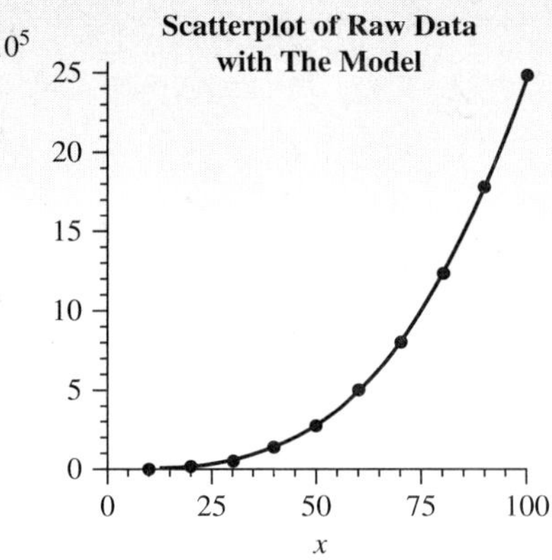

**Scatterplot of Raw Data with The Model**

> *errors:=seq(Ydata[i]−model1[i],i=1..10);*

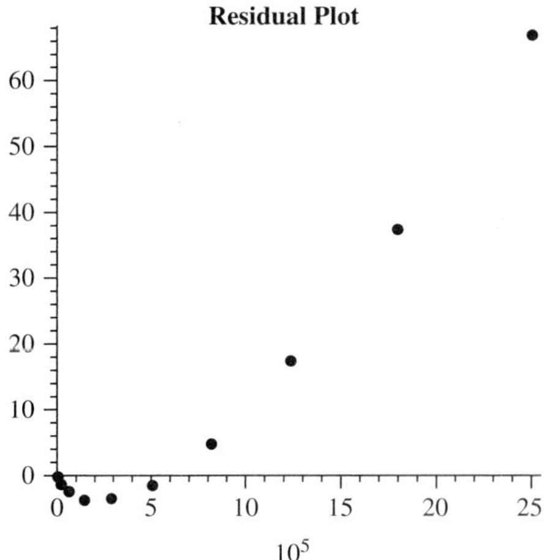

**Residual Plot**

A pattern is seen in the residuals.

**2.** The scatterplot is concave up and increasing. We transform $x$ using the ladder of powers and stop at $x^3$. The model via least squares is $y = 0.008419645855 \, x^3$. Residual are fanning out.

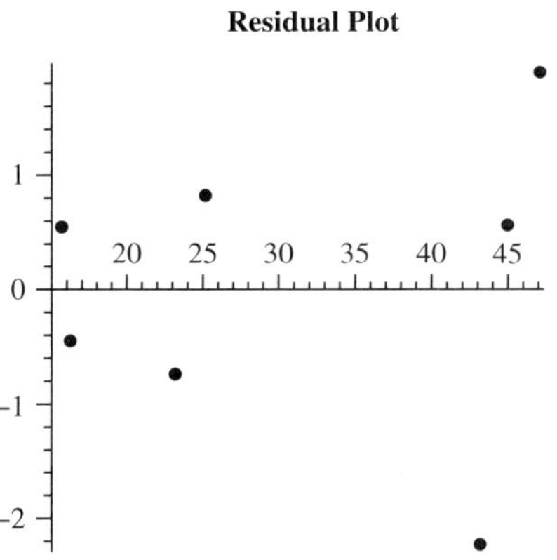

**Residual Plot**

We use the ln–ln transform to approximate the model as

$$model := 0.008714651137\ x^{2.98683873}$$

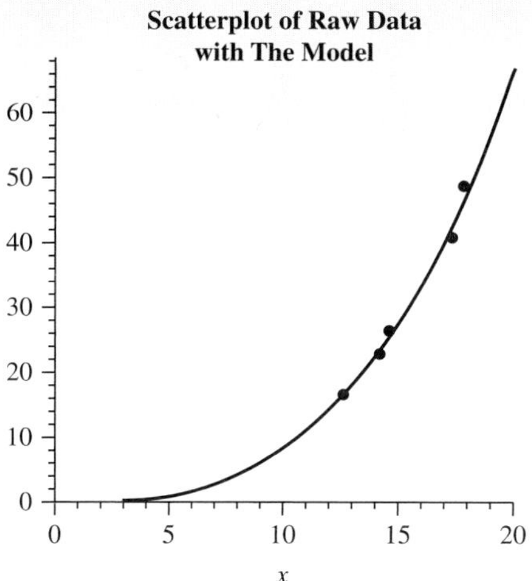

Plot of data with model.

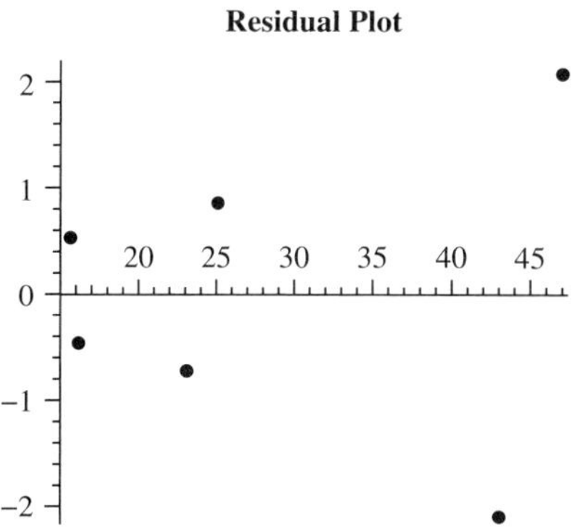

Residual plot is still fanning out.

**3.** Scatterplot is increasing and concave down.

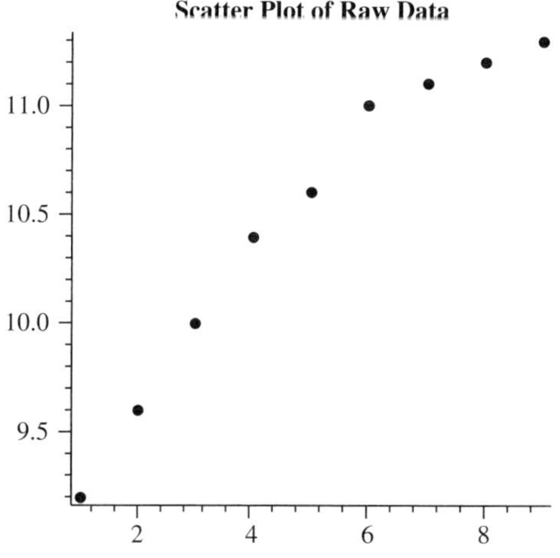

Ladder suggests the following: $x \to x,\ y \to y^2$ (stop).

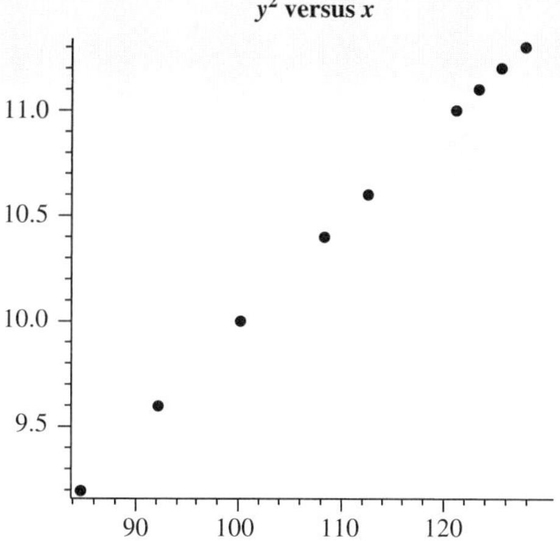

The model $y^2 = kx$ is equivalent to $y = 4.588664208\ \sqrt{x}$. There is a linear trend in the residuals.

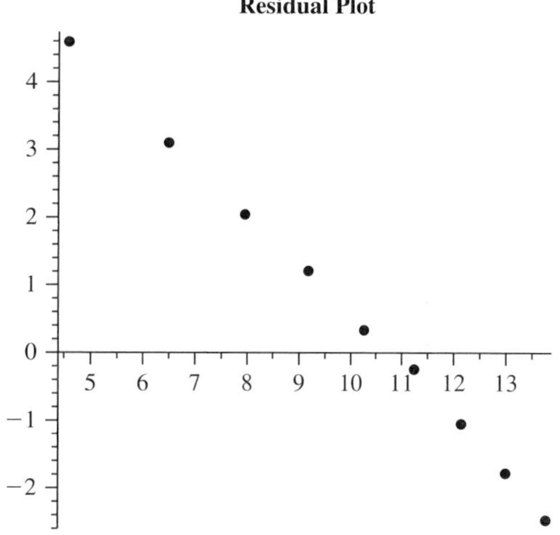

We use the ln–ln transform to get the model.

$$model := 9.067632068\ x^{0.1007174918}$$

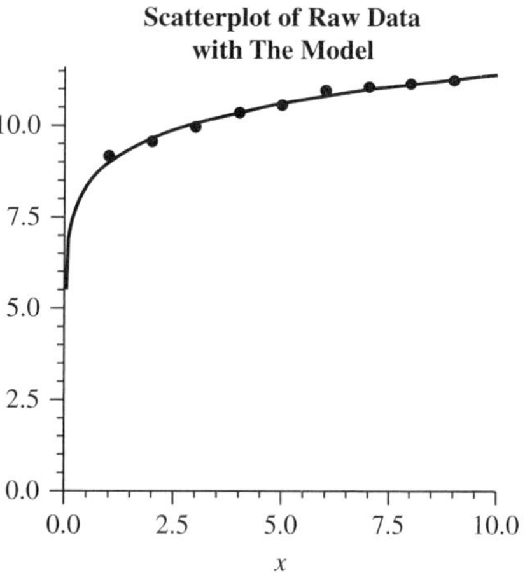

Plot of the data and the model.

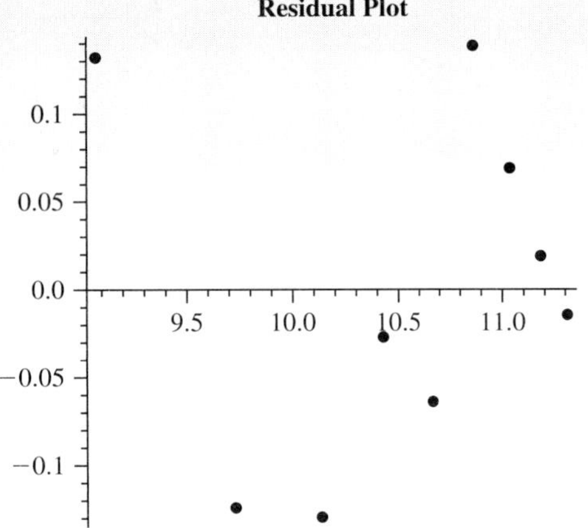

Residual appear random.

4. Scatterplot is concave down and increasing. We transform $y$ to $y^2$.

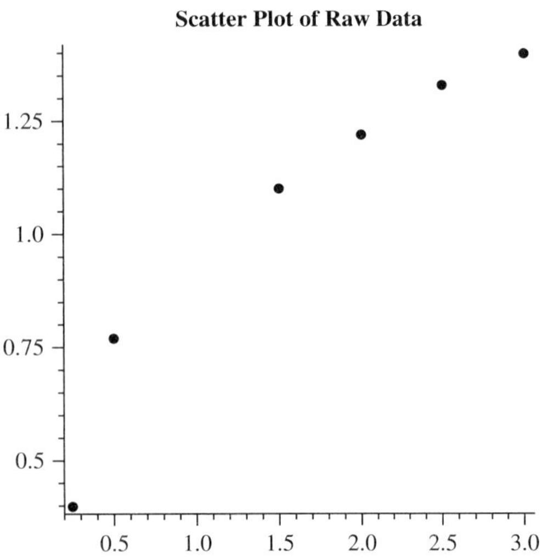

The model is

$$y = 0.8558787590\,\sqrt{x}$$

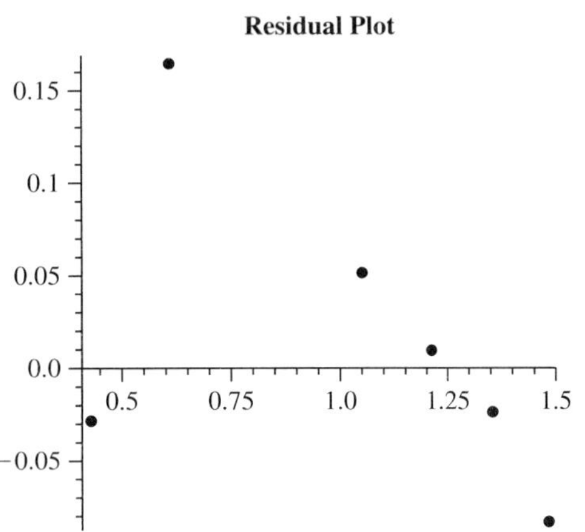

We use the ln–ln transform to obtain the model

$$model := 0.8830063922\,x^{0.4687226109}$$

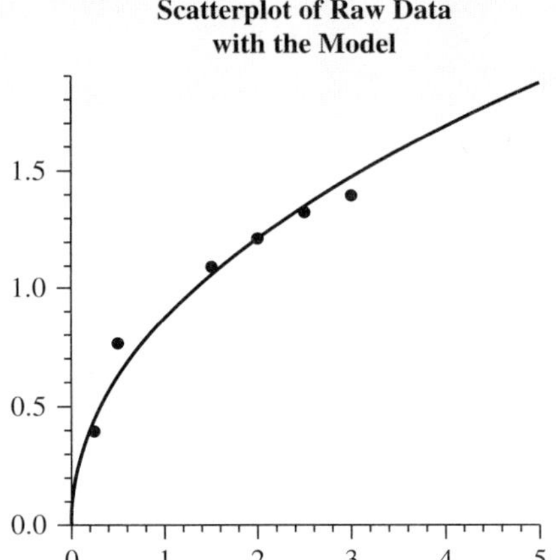

Plot of data and model.

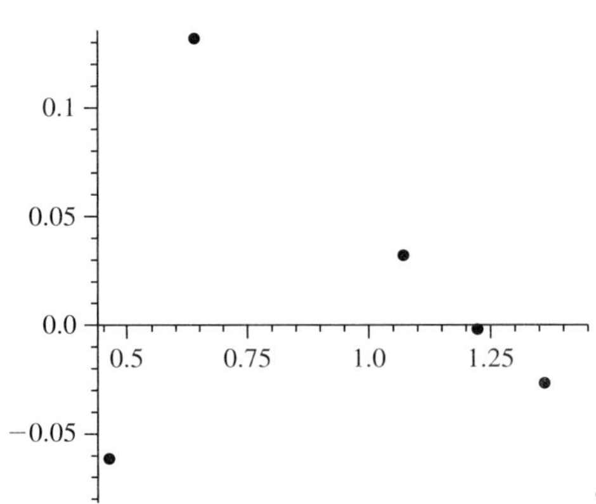

Residuals look as before.

## Section 6.2

We provide both the high order equation and a plot.

1. $eq\_fit := y = 1.333333333\,x^3 - 9.\,x^2 + 26.66666667\,x - 12.$

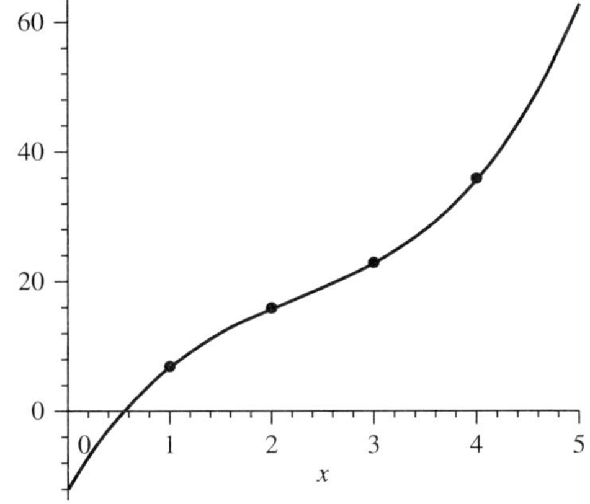

**3.** $eq\_fit := y = 0.02104955773\ x^5 - 0.8957571554\ x^4$
$+ 5.496536893\ x^3 + 246.8287942\ x^2$
$- 4340.568375\ x + 20417.86733$

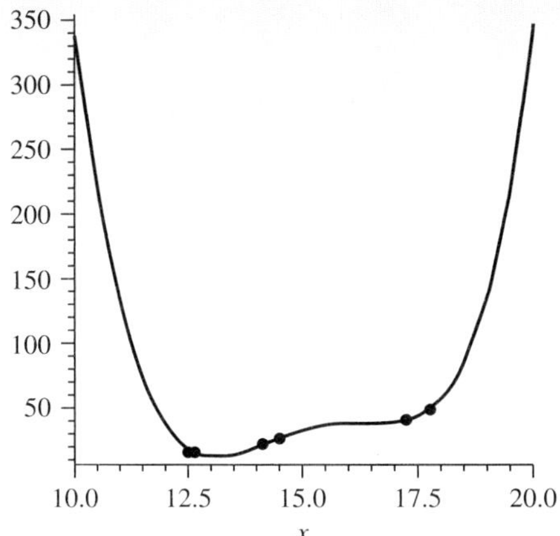

**5.** $eq\_fit := y = 0.06748572355\ x^5 - 0.6650852547\ x^4$
$+ 2.487164217\ x^3 - 4.382280119\ x^2$
$+ 3.826328786\ x - 0.3190240951$

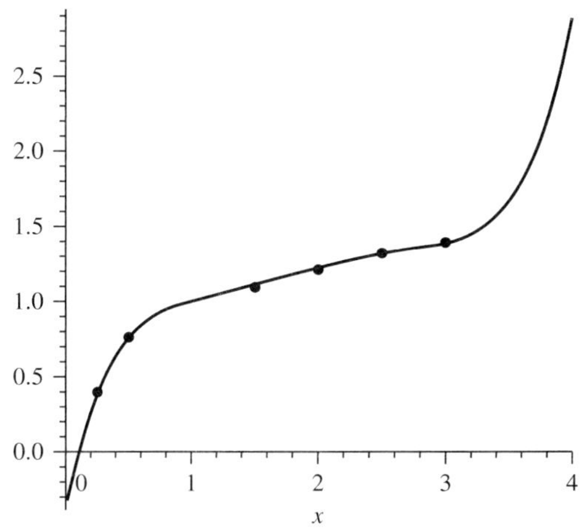

## Section 6.3

We provide a table, comments, and the selected least-squares-fit equation.

**1.** $ddproc(Xdata, Ydata);$

$$\begin{bmatrix} 1. & 7. & 0. & 0. \\ 2. & 16. & 9. & 0. \\ 3. & 23. & 7. & -1. \\ 4. & 36. & 13. & 3. \end{bmatrix}$$

The fourth column representing the second derivative shows a sign change. We might conclude that a third-order polynomial is best.

$eqfit := y = 1.333333333\ x^3 - 9.\ x^2 + 26.66666667\ x - 12.$

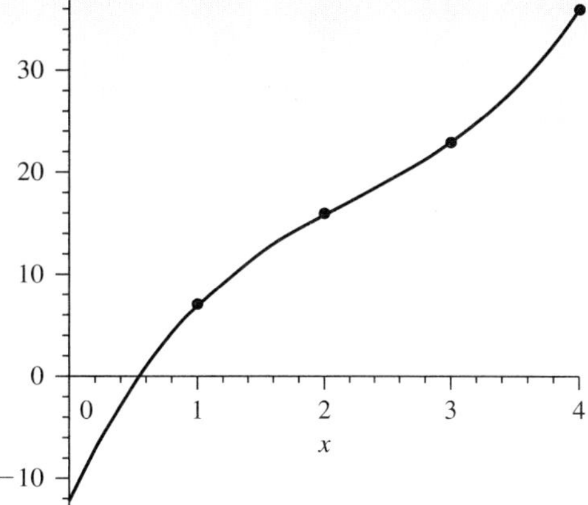

**3.** $ddproc(Xdata, Ydata);$

$$\begin{bmatrix} 12.5 & 17. & 0. & 0. & 0. & 0. \\ 12.62 & 16.5 & -4.000 & 0. & 0. & 0. \\ 14.12 & 23. & 4.333 & 5.128 & 0. & 0. \\ 14.5 & 26.5 & 9.333 & 2.667 & -1.231 & 0. \\ 17.25 & 41. & 5.273 & -1.299 & -.8575 & 0.07858 \\ 17.75 & 49. & 16.00 & 3.301 & 1.269 & 0.4149 \end{bmatrix}$$

The negative in the third column reflects an error in data (most likely). We would not expect a change in concavity, so try a second-order polynomial.

$eqfit := y = 0.4726495295\ x^2 - 8.467587110\ x + 48.67975254$

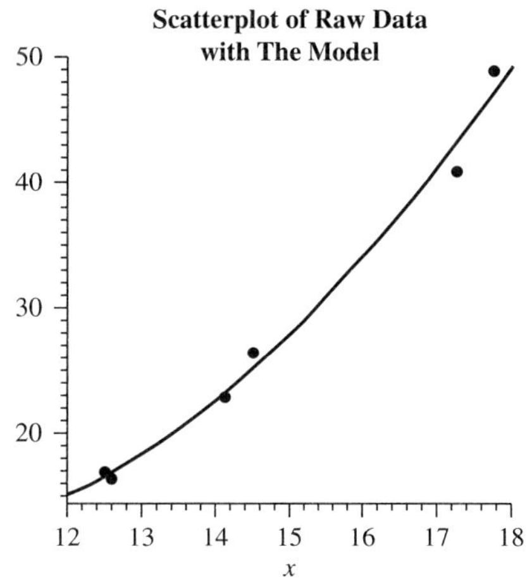

**5.** *ddproc( Xdata, Ydata );*

$$\begin{bmatrix} 0.25 & 0.4 & 0. & 0. & 0. & 0. \\ 0.5 & 0.77 & 1.480 & 0. & 0. & 0. \\ 1.5 & 1.1 & 0.3300 & -.9200 & 0. & 0. \\ 2. & 1.22 & 0.2400 & -0.06000 & 0.4914 & 0. \\ 2.5 & 1.33 & 0.2200 & -0.02000 & 0.02000 & -.2095 \\ 3. & 1.4 & 0.1400 & -0.08000 & -0.04000 & -0.02400 \end{bmatrix}$$

Negatives indicate that errors will propagate in further columns. Try quadratic unless a change of concavity is suspected and then go to cubic.

$$eqfit := y = -0.1103565082\,x^2 + 0.6794805289\,x + 0.3337026963$$

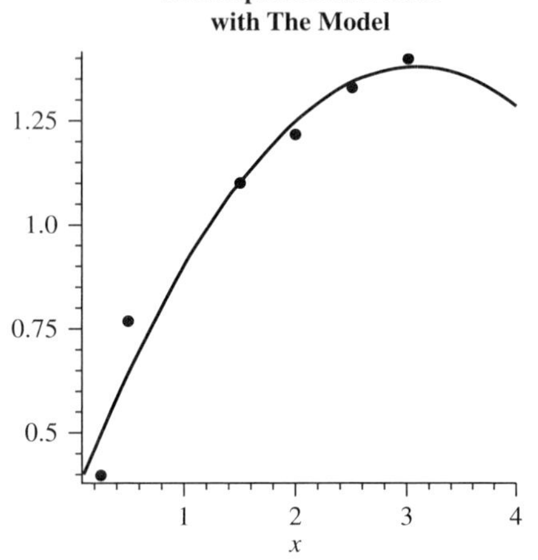

**Scatterplot of Raw Data with The Model**

**Section 6.4**

**1.**

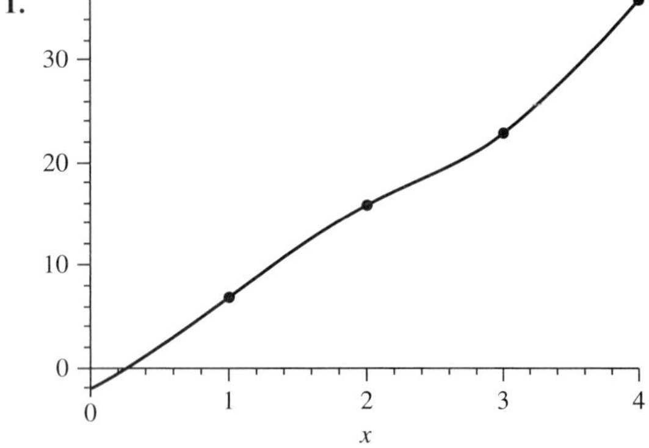

> *spline( xobs, yobs, x, cubic );*

$$\begin{cases} -2 + \dfrac{107}{15}x + \dfrac{14}{5}x^2 - \dfrac{14}{15}x^3 & x < 2 \\[2mm] -\dfrac{154}{5} + \dfrac{151}{3}x - \dfrac{94}{5}x^2 + \dfrac{8}{3}x^3 & x < 3 \\[2mm] 88 - \dfrac{1027}{15}x + \dfrac{104}{5}x^2 - \dfrac{26}{15}x^3 & otherwise \end{cases}$$

**3.**

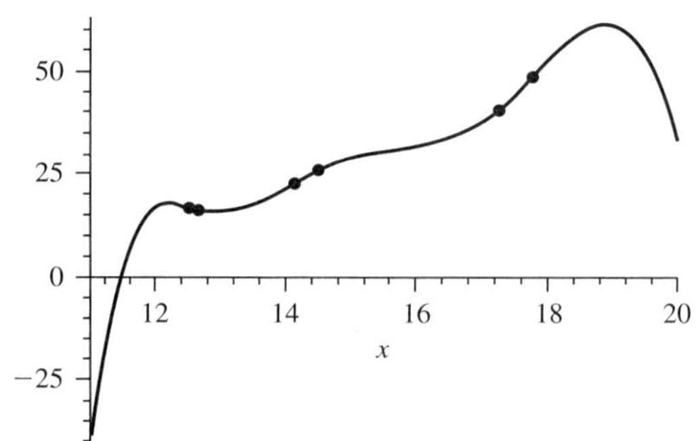

> *s:=unapply(%,x);*

$s := x \rightarrow piecewise$
$(x < 12.625,\ 70.58091120 - 4.28647289600000026\,x$
$+ 18.3342653200000001\,(x - 12.5)^3,\ x < 14.125,$
$59.76655936 - 3.42705420699999985\,x$
$+ 6.87534949621465952\,(x - 12.625)^2$
$-1.13450520100000008\,(x - 12.625)^3,\ x < 14.5,$
$-111.7678139 + 9.54108417199999970\,x$
$+ 1.77007609153490408\,(x - 14.125)^2$
$-6.19754221700000050\,(x - 14.125)^3,\ x < 17.25,$
$-93.18377020 + 8.25405311899999994\,x$
$-5.20215890020767713\,(x - 14.5)^2$
$+ 1.49746924000000004\,(x - 14.5)^3,$
$-193.8762167 + 13.6160125599999998\,x$
$+ 7.15196233162632478\,(x - 17.25)^2$
$-4.76797488700000028\,(x - 17.25)^3)$

**5.**

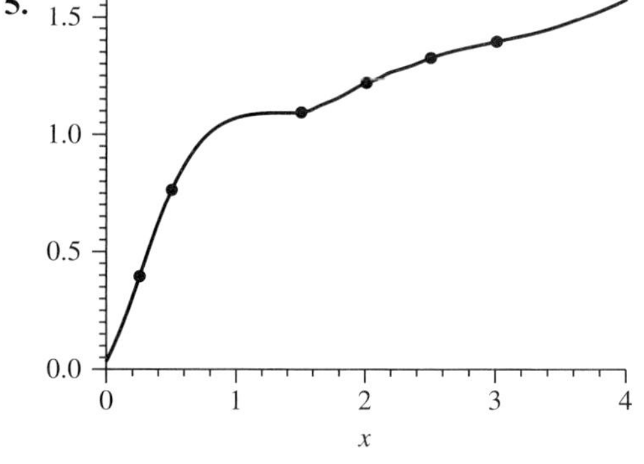

> *s :=unapply(%,x);*

$s := x \rightarrow piecewise$
$(x < 0.5,\ -0.0025000000 + 1.61000000000000010x$
$-2.08000000000000008\,(x - 0.25)^3,\ x < 1.5,$

$$0.1600000000 + 1.2199999999999997\,x$$
$$-1.56000000000000006\,(x - 0.5)^2$$
$$+ 0.669999999999999928\,(x - 0.5)^3,$$
$$x < 2, 0.9350000000 + 0.110000000000000000\,x$$
$$+ 0.450000000000000066\,(x - 1.5)^2$$
$$-0.380000000000000002\,(x - 1.5)^3, x < 2.5,$$
$$0.6700000000 + 0.275000000000000022\,x$$
$$-0.119999999999999994\,(x - 2)^2$$
$$+ 0.0200000000000000004\,(x - 2)^3,$$
$$0.9050000000 + 0.170000000000000012\,x$$
$$-0.0899999999999999968\,(x - 2.5)^2$$
$$+ 0.0599999999999999978\,(x - 2.5)^3)$$

$$
\begin{cases}
-0.0025000000 & x < 0.5 \\
\quad + 1.61000000000000010\,x \\
\quad - \\
\quad 2.08000000000000008 \\
\quad (x - 0.25)^3 \\[4pt]
0.1600000000 & x < 1.5 \\
\quad + 1.2199999999999997\,x \\
\quad - \\
\quad 1.56000000000000006\,(x - 0.5)^2 \\
\quad + \\
\quad 0.669999999999999928 \\
\quad (x - 0.5)^3 \\[4pt]
0.9350000000 & x < 2 \\
\quad + 0.110000000000000000\,x \\
\quad + \\
\quad 0.450000000000000066 \\
\quad (x - 1.5)^2 - \\
\quad 0.380000000000000002 \\
\quad (x - 1.5)^3 \\[4pt]
0.6700000000 & x < 2.5 \\
\quad + 0.275000000000000022\,x \\
\quad - \\
\quad 0.119999999999999994\,(x - 2)^2 \\
\quad + \\
\quad 0.0200000000000000004\,(x - 2)^3 \\[4pt]
0.9050000000 & otherwise \\
\quad + 0.170000000000000012\,x \\
\quad - \\
\quad 0.0899999999999999968 \\
\quad (x - 2.5)^2 + \\
\quad 0.0599999999999999978 \\
\quad (x - 2.5)^3
\end{cases}
$$

# Chapter 7

## Section 7.1

**1.** Let $A$ = number of drink A produced; let $B$ = number of drink B produced.

$$\text{Max } Z = 15A + 18B$$
Subject to
$$3A + 2.5B \le 40$$
$$2A + 2.5B \le 35$$
Nonnegativity

**2.** Let $x$ = number of steamships produced, $y$ = number of sailboats produced, $w$ = number of submarines produced.

$$\text{Max } Z = 6x + 3y + 2w$$
Subject to
$$x + 3y + w \le 45$$
$$2x + 3y + 3w \le 50$$
$$4x + 2y + w \le 60$$
Nonnegativity

**3.** Let $S$ = number of small units, $M$ = number of medium units, $L$ = number of large units produced.

$$\text{Minimize } Z = 9S + 12M + 15L$$
Subject to
$$2S + 2M + L \ge 10$$
$$2S + 3M + L \ge 12$$
$$S + M + 5L \ge 14$$
Nonnegativity

**4.** In thousands, let $x_1$ = number of ad units for daytime TV, $x_2$ = number of ad units for prime-time TV, $x_3$ = number of ad units for radio, $x_4$ = number of ad units for magazines.

$$\text{Max } Z = 400x_1 + 900x_2 + 500x_3 + 200x_4$$
Subject to
$$40x_1 + 75x_2 + 30x_3 + 15x_4 \le 800$$
$$30x_1 + 40x_2 + 20x_3 + 10x_4 \ge 2000$$
$$40x_1 + 75x_2 \le 500$$
$$x_1 \ge 3$$
$$x_2 \ge 2$$
$$5 \le x_3 \le 10$$
$$5 \le x_4 \le 10$$

**5.** $\text{Max } Z = 0.08Xwa + 0.08Xwb + 0.05Xpa + 0.05Xpb$

Subject to
$$Xwa + Xpa \le 5{,}000$$
$$Xwb + Xpb \le 10{,}000$$
$$0.2Xwa - 0.8Xwb \ge 0$$
$$0.9Xpa - 0.1Xpb \ge 0$$
Nonnegativity

## Section 7.2

**1.** Max: $Z = 21$, $x = 6$, $y = 3$; min: $Z = 6$, $x = 3$, $y = 0$

**2.** Max $Z = 144$, $x = 24$, $y = 0$; min: $x = y = Z = 0$

**3.** Max is unbounded. Min: $x = y = 3$, $Z = 33$

**4.** Max and min are unbounded.

**5.** Max $Z = 400$, $x = 80$, $y = 0$. Min: $x = y = z = 0$

## Sections 7.4 and 7.5

1. Max: $Z = 21$, $x = 6$, $y = 3$. Min: $Z = 6$, $x = 0$, $y = 2$.

2. Max $Z = 144$, $x = 24$, $y = 0$. Min: $Z = 0$, $x = 0$, $y = 0$

3. Min: $x = y = 3$, $Z = 33$. Max: unbounded

4. Max and min are both is unbounded.

5. Max $Z = 120$, $x = 0$, $y = 40$. Min: $Z = 0$, $x = 0$, $y = 0$

## Section 7.6

1. $E1 = 1$, $E4 = 0.92$, $E5 = 0.776$, $E2 = 0.7414$, $E3 = 0.389$

## Section 7.7

1. For the maximization problem, cost coefficients $x$ currently at 2, $1.2 < x <$ infinity; $y$ currently at 3, $-1.667 < y < 5$. RHS coefficients $b1$: infinity $< b1 < 21$; $b2$: $-2 < b2 < 40.5$; $b3$: $-3 < b3 <$ infinity; $b4$: $10 < b4 < 86.5$.

2. For the maximization problem, cost coefficients $x$ currently at 6, $4 < x <$ infinity; $y$ currently at 4, *infinity* $< y < 6$. RHS coefficients $b1$: $-24 < b1 <$ infinity; $b2$: $0 < b2 < 40$; $b3$: $48 < b3 <$ infinity.

# Chapter 8

## Section 8.1

1. Max at $f(0) = 10$.

2. Max at $f(0) = 10$

3. Min $f(4) = -28.5$

4. Critical points are 0,0,1. We have a min at $f(1) = -1$. There is an inflection point at $x = 0$. Min at $f\left(\frac{3}{4}\right) = -\frac{18}{16}$

## Section 8.2

1. 2,000 by road

2. $f' = 0$ at $x = 0$; max is at $x = 1$, $f(1) = 1$

3. Extreme points at 0 and 10.6667. The point $x = 0$ yields at relative maximum, and $x = 10.667$ yields a relative minimum

4. Maximum at $x = 1$.

5. The function is maximized at 0 and $2/3r_o$. This confirms the conjecture.

## Section 8.4

1. $f:=x->-x^2-2*x;$
$$x \rightarrow -x^2 - 2x$$

**a.** DICHOTOMOUS(f, $-2$, 1, .2, .01);

The interval [a,b] is [$-2.00$, 1.00] and user specified tolerance level is 0.20000.
The first 2 experimental endpoints are x1 $= -0.510$ and x2 $= -0.490$.

<br>

| Iteration | x(1) | x(2) | f(x1) | f(x2) | Interval |
|---|---|---|---|---|---|
| 1 | $-0.5100$ | $-0.4900$ | 0.7599 | 0.7399 | [$-2.0000$, 1.0000] |
| 2 | $-1.2550$ | $-1.2350$ | 0.9350 | 0.9448 | [$-2.0000$, $-0.4900$] |
| 3 | $-0.8825$ | $-0.8625$ | 0.9862 | 0.9811 | [$-1.2550$, $-0.4900$] |
| 4 | $-1.0688$ | $-1.0488$ | 0.9953 | 0.9976 | [$-1.2550$, $-0.8625$] |
| 5 | $-0.9756$ | $-0.9556$ | 0.9994 | 0.9980 | [$-1.0688$, $-0.8625$] |

The midpoint of the final interval is $-0.965625$ and f(midpoint) = 0.999.
The maximum of the function is 0.995 and the x value $=$ $-1.068750$

**b.** FIBSearch(f, $-2$, 1, .6);

The interval [a, b] is [$-2.00$, 1.00] and user specified tolerance level is 0.60000.
The first 2 experimental endpoints are x1$= -0.800$ and x2 $= -0.200$.

| Iteration | x(1) | x(2) | f(x1) | f(x2) | Interval |
|---|---|---|---|---|---|
| 2 | $-1.4000$ | $-0.8000$ | 0.9600 | 0.3600 | [$-2.0000$, $-0.2000$] |
| 3 | $-0.8000$ | $-0.8000$ | 0.8400 | 0.9600 | [$-1.4000$, $-0.2000$] |

The midpoint of the final interval is $-0.800000$ and f(midpoint) = 0.960.
The maximum of the function is 0.990 and the x value $=$ $-1.100000$

**c.** GOLD(f, $-2$, 1, .6);

The interval [a,b] is [$-2.00$, 1.00] and user specified tolerance level is 0.60000.
The first 2 experimental endpoints are x1$= -0.854$ and x2 $= -0.146$.

| Iteration | x(1) | x(2) | f(x1) | f(x2) | Interval |
|---|---|---|---|---|---|
| 2 | $-1.2918$ | $-0.8540$ | 0.9787 | 0.2707 | [$-2.0000$, $-0.146$] |
| 3 | $-0.8540$ | $-0.5837$ | 0.9149 | 0.9787 | [$-1.2918$, $-0.146$] |
| 4 | $-1.0213$ | $-0.8540$ | 0.9787 | 0.8267 | [$-1.2918$, $-0.5837$] |
| 5 | $-1.1245$ | $-1.0213$ | 0.9995 | 0.9787 | [$-1.2918$, $-0.8540$] |

The midpoint of the final interval is $-1.072886$ and f(midpoint) = 0.995.
The maximum of the function is 0.9995 and the x value $=$ $-1.1245$

**d.** NEWTON's

*Newton(f,$-$.5,10);*
$$-.5$$
$$-1.000000000$$
$$-1.000000000$$

2. $f:=x->-x^2-3*x;$
$$x \rightarrow -x^2 - 3x$$

**a.** DICHOTOMOUS(f, $-3$, 1, .2, .01);

The interval [a,b] is [−3.00, 1.00] and user specified tolerance level is 0.20000.
The first 2 experimental endpoints are x1 = −1.010 and x2 = −0.990.

| Iteration | x(1) | x(2) | f(x1) | f(x2) | Interval |
|---|---|---|---|---|---|
| 1 | −1.0100 | −0.9900 | 2.0099 | 1.9899 | [−3.0000, 1.0000] |
| 2 | −2.0050 | −1.9850 | 1.9950 | 2.0148 | [−3.0000, −0.9900] |
| 3 | −1.5075 | −1.4875 | 2.2499 | 2.2498 | [−2.0050, −0.9900] |
| 4 | −1.7562 | −1.7362 | 2.1843 | 2.1942 | [−2.0050, −1.4875] |
| 5 | −1.6319 | −1.6119 | 2.2326 | 2.2375 | [−1.7562, −1.4875] |
| 6 | −1.5697 | −1.5497 | 2.2451 | 2.2475 | [−1.6319, −1.4875] |

The midpoint of the final interval is −1.559688 and f(midpoint) = 2.246.
The maximum of the function is 2.250 and the x value = −1.487500

**b.** *FIBSearch(f,−3,1,.6);*

The interval [a,b] is [−3.00, 1.00] and user specified tolerance level is 0.60000.
The first 2 experimental endpoints are x1 = −1.500 and x2 = −0.500.

| Iteration | x(1) | x(2) | f(x1) | f(x2) | Interval |
|---|---|---|---|---|---|
| 2 | −2.0000 | −1.5000 | 2.2500 | 1.2500 | [−3.0000, −0.5000] |
| 3 | −1.5000 | −1.0000 | 2.0000 | 2.2500 | [−2.0000, −0.5000] |
| 4 | −1.5000 | −1.5000 | 2.2500 | 2.0000 | [−2.0000, −1.0000] |

The midpoint of the final interval is −1.500000 and f(midpoint) = 2.250.
The maximum of the function is 2.250 and the x value = −1.500000

**c.** GOLD(f, −3, 1, .6);

The interval [a,b] is [−3.00, 1.00] and user specified tolerance level is 0.60000.
The first 2 experimental endpoints are x1 = −1.472 and x2 = −0.528.

| Iteration | x(1) | x(2) | f(x1) | f(x2) | Interval |
|---|---|---|---|---|---|
| 2 | −2.0557 | −1.4720 | 2.2492 | 1.3052 | [−3.0000, −0.5280] |
| 3 | −1.4720 | −1.1116 | 1.9412 | 2.2492 | [−2.0557, −0.5280] |
| 4 | −1.6950 | −1.4720 | 2.2492 | 2.0991 | [−2.0557, −1.1116] |
| 5 | −1.4720 | −1.3345 | 2.2120 | 2.2492 | [−1.6950, −1.1116] |

The midpoint of the final interval is −1.403312 and f(midpoint) = 2.241.
The maximum of the function is 2.243 and the x value = −1.583638

**d.** Newton(f, 1, 10);

$$1$$
$$-1.500000000$$
$$-1.500000000$$

**3.** *f:=x−>−x^2−2*x;*

$$f := x \rightarrow -x^2 - 2x$$

**a.** DICHOTOMOUS(f, −2, 1, .2, .01);

The interval [a,b] is [−2.00, 1.00] and user specified tolerance level is 0.20000.
The first 2 experimental endpoints are x1 = −0.510 and x2 = −0.490.

| Iteration | x(1) | x(2) | f(x1) | f(x2) | Interval |
|---|---|---|---|---|---|
| 1 | −0.5100 | −0.4900 | 0.7599 | 0.7399 | [−2.0000, 1.0000] |
| 2 | −1.2550 | −1.2350 | 0.9350 | 0.9448 | [−2.0000, −0.4900] |
| 3 | −0.8825 | −0.8625 | 0.9862 | 0.9811 | [−1.2550, −0.4900] |
| 4 | −1.0688 | −1.0488 | 0.9953 | 0.9976 | [−1.2550, −0.8625] |
| 5 | −0.9756 | −0.9556 | 0.9994 | 0.9980 | [−1.0688, −0.8625] |

The midpoint of the final interval is −0.965625 and f(midpoint) = 0.999.
The maximum of the function is 0.995 and the x value = −1.068750

**b.** FIBSearch(f, −2, 1, .6);

The interval [a,b] is [−2.00, 1.00] and user specified tolerance level is 0.60000.
The first 2 experimental endpoints are x1 = −0.800 and x2 = −0.200.

| Iteration | x(1) | x(2) | f(x1) | f(x2) | Interval |
|---|---|---|---|---|---|
| 2 | −1.4000 | −0.8000 | 0.9600 | 0.3600 | [−2.0000, −0.2000] |
| 3 | −0.8000 | −0.8000 | 0.8400 | 0.9600 | [−1.4000, −0.2000] |

The midpoint of the final interval is −0.800000 and f(midpoint) = 0.960.
The maximum of the function is 0.990 and the x value = −1.100000

**c.** GOLD(f, −2, 1, .6);

The interval [a,b] is [−2.00, 1.00] and user specified tolerance level is 0.60000.
The first 2 experimental endpoints are x1 = −0.854 and x2 = −0.146.

| Iteration | x(1) | x(2) | f(x1) | f(x2) | Interval |
|---|---|---|---|---|---|
| 2 | −1.2918 | −0.8540 | 0.9787 | 0.2707 | [−2.0000, −0.1460] |
| 3 | −0.8540 | −0.5837 | 0.9149 | 0.9787 | [−1.2918, −0.1460] |
| 4 | −1.0213 | −0.8540 | 0.9787 | 0.8267 | [−1.2918, −0.5837] |
| 5 | −1.1245 | −1.0213 | 0.9995 | 0.9787 | [−1.2918, −0.8540] |

The midpoint of the final interval is −1.072886 and f(midpoint) = 0.995.
The maximum of the function is 0.915 and the x value = −1.291772

**d.** *Newton(f,–3,20);*

$$-3$$
$$-1.$$
$$-1.$$

**4.** Since the programs were written to find a maximum, we change sign of $f(x)$ so we can use the programs as written.
*f:=x->x–exp(x);*

$$x \to x - e^x$$

**a.** DICHOTOMOUS(f, $-1$, 3, .2, .01);

The interval [a,b] is [–1.00, 3.00] and user specified tolerance level is 0.20000.
The first 2 experimental endpoints are x1 = 0.990 and x2 = 1.010.

| Iteration | x(1) | x(2) | f(x1) | f(x2) | Interval |
|---|---|---|---|---|---|
| 1 | 0.9900 | 1.0100 | –1.7012 | –1.7356 | [–1.0000, 3.0000] |
| 2 | –0.0050 | 0.0150 | –1.0000 | –1.0001 | [–1.0000, 1.0100] |
| 3 | –0.5025 | –0.4825 | –1.1075 | –1.0997 | [–1.0000, 0.0150] |
| 4 | –0.2538 | –0.2338 | –1.0296 | –1.0253 | [–0.5025, 0.0150] |
| 5 | –0.1294 | –0.1094 | –1.0080 | –1.0058 | [–0.2538, 0.0150] |
| 6 | –0.0672 | –0.0472 | –1.0022 | –1.0011 | [–0.1294, 0.0150] |

The midpoint of the final interval is –0.057188 and f(midpoint) = –1.002.
The maximum of the function is –1.000 and the x value = 0.015000

**b.** FIBSearch(f, –1, 3, .1);

The interval [a,b] is [–1.00, 3.00] and user specified tolerance level is 0.10000.
The first 2 experimental endpoints are x1 = 0.527 and x2 = 1.473.

| Iteration | x(1) | x(2) | f(x1) | f(x2) | Interval |
|---|---|---|---|---|---|
| 2 | –0.0545 | 0.5273 | –1.1670 | –2.8884 | [–1.0000, 1.4727] |
| 3 | –0.4182 | –0.0545 | –1.0015 | –1.1670 | [–1.0000, 0.5273] |
| 4 | –0.0545 | 0.1636 | –1.0764 | –1.0015 | [–0.4182, 0.5273] |
| 5 | –0.2000 | –0.0545 | –1.0015 | –1.0141 | [–0.4182, 0.1636] |
| 6 | –0.0545 | 0.0182 | –1.0187 | –1.0015 | [–0.2000, 0.1636] |
| 7 | 0.0182 | 0.0909 | –1.0015 | –1.0002 | [–0.0545, 0.1636] |
| 8 | 0.0182 | 0.0182 | –1.0002 | –1.0043 | [–0.0545, 0.0909] |

The midpoint of the final interval is 0.018182 and f(midpoint) = –1.000.
The maximum of the function is –1.001 and the x value = –0.054545

**c.** GOLD(f,–1, 3, .1);

The interval [a,b] is [–1.00, 3.00] and user specified tolerance level is 0.10000.

The first 2 experimental endpoints are x1 = 0.528 and x2 = 1.472.

| Iteration | x(1) | x(2) | f(x1) | f(x2) | Interval |
|---|---|---|---|---|---|
| 2 | –0.0557 | 0.5280 | –1.1675 | –2.8859 | [–1.0000, 1.4720] |
| 3 | –0.4163 | –0.0557 | –1.0015 | –1.1675 | [–1.0000, 0.5280] |
| 4 | –0.0557 | 0.1673 | –1.0758 | –1.0015 | [–0.4163, 0.5280] |
| 5 | –0.1934 | –0.0557 | –1.0015 | –1.0148 | [–0.4163, 0.1673] |
| 6 | –0.0557 | 0.0295 | –1.0175 | –1.0015 | [–0.1934, 0.1673] |
| 7 | 0.0295 | 0.0821 | –1.0015 | –1.0004 | [–0.0557, 0.1673] |
| 8 | –0.0031 | 0.0295 | –1.0004 | –1.0035 | [–0.0557, 0.0821] |
| 9 | –0.0231 | –0.0031 | –1.0000 | –1.0004 | [–0.0557, 0.0295] |

The midpoint of the final interval is –0.013095 and f(midpoint) = –1.000.
The maximum of the function is –1.002 and the x value = –0.055696

**d.** *Newton (f, –1, 20);*

$$-1$$
$$0.718281828$$
$$0.2058711269$$
$$0.0198090911$$
$$0.00019491102$$
$$1.90103 \; 10^{-8}$$
$$1.030036 \; 10^{-11}$$
$$1.030036 \; 10^{-11}$$

The value of $x$ is essentially 0.

# Chapter 9

## Section 9.3

**1. a.** PD, convex    **b.** Indefinite, neither
**c.** ND, concave    **d.** Indefinite, neither
**e.** Indefinite, neither    **f.** Indefinite, neither
**g.** Indefinite neither    **h.** Indefinite, neither

**2. a.** Saddle at (0,0)
**b.** Global min at (0,0)
**c.** Global max at (0,0)
**d.** *f:=3·x+5·y–4·x²+y²–5·x·y;*

$$f := 3\,x + 5\,y - 4\,x^2 + y^2 - 5\,x\,y$$

> *Hessian(f,[x,y]);*

$$\begin{bmatrix} -8 & -5 \\ -5 & 2 \end{bmatrix}$$

The Hessian is indefinite.

> *dfx:=diff(f,x);*

$$dfx := 3 - 8\,x - 5\,y$$

> *dfy:=diff(f,y);*

$$dfy := 5 + 2\,y - 5\,x$$

The answer is neither a max nor a min, since the Hessian is indefinite.

> *solve({dfx=0,dfy=0},{x,y});*

$$\left\{x = \frac{31}{41}, y = -\frac{25}{41}\right\}$$

**e.** The critical point $(0.756, -0.609)$ is a saddle point because the Hessian of the function evaluated at the critical point is indefinite.

**f.** $f := a \cdot x^2 + b \cdot x \cdot y + c \cdot y^2;$

$$f := a\,x^2 + b\,x\,y + c\,y^2$$

> *Hessian(f,[x,y]);*

$$\begin{bmatrix} 2a & b \\ b & 2c \end{bmatrix}$$

Convex : $a > 0, c > 0\ 4ac > b^2$
Concave: $a < 0, c < 0, 4ac < b^2$

**3. a.** $f := x^2 + 3 \cdot x \cdot y - y^2;$

$$f := x^2 + 3\,x\,y - y^2$$

*dfx:=diff(f,x);*

$$dfx := 2\,x + 3\,y$$

*dfy:=diff( f,y);*

$$dfy := 3\,x - 2\,y$$

*solve({dfx=0,dfy=0},{x,y});*

$$\{x = 0, y = 0\}$$

**b.** $(0, 0)$ is a min.

**c.** $f := -x^2 - x \cdot y - 2 \cdot y^2;$

$$f := -x^2 - x\,y - 2\,y^2$$

*dfx:=diff(f,x);*

$$dfx := -2\,x - y$$

*dfy:=diff(f,y);*

$$dfy := -x - 4\,y$$

*solve({dfx=0,dfy=0},{x,y});*

$$\{x = 0, y = 0\}$$

**d.** $(0.756, -0.609)$ is a saddle point.

**e.** $f := 2 \cdot x + 3 \cdot y + 3 \cdot z - x \cdot y + x \cdot z - y \cdot z - x^2 - 3 \cdot y^2 - z^2;$

$$f := 2\,x + 3\,y + 3\,z - x\,y + x\,z - y\,z - x^2 - 3\,y^2 - z^2$$

*A:=Hessian(f,[x,y,z]);*

$$A := \begin{bmatrix} -2 & -1 & 1 \\ -1 & -6 & -1 \\ 1 & -1 & -2 \end{bmatrix}$$

*IsDefinite(A,'query'='negative_definite');*

*true*

*dfx:=diff(f,x);*

$$dfx := 2 - y + z - 2\,x$$

*dfy:=diff(f,y);*

$$dfy := 3 - x - z - 6\,y$$

*dfz:=diff(f, z);*

$$dfz := 3 + x - y - 2\,z$$

*solve({dfx=0,dfy=0,dfz=0},{x,y,z});*

$$\left\{x = 0, y = 0, z = \frac{3}{2}\right\}$$

**4.** $f := exp(x-y) + x^2 + y^2;$

$$f := e^{x-y} + x^2 + y^2$$

*Hessian(f,[x,y]);*

$$\begin{bmatrix} e^{x-y} + 2 & -e^{x-y} \\ -e^{x-y} & e^{x-y} + 2 \end{bmatrix}$$

*dfx:=diff(f,x);*

$$dfx := e^{x-y} + 2\,x$$

*dfy:=diff(f,y);*

$$dfy := -e^{x-y} + 2\,y$$

*fsolve({dfx=0,dfy=0},{x,y});*

$$\{x = -.2835716452, y = 0.2835716452\}$$

*subs({x=-.2835716452,y=0.2835716452},Hessian(f,[x,y]);*

$$\begin{bmatrix} e^{-.5671432904} + 2 & -e^{-.5671432904} \\ -e^{-.5671432904} & e^{-.5671432904} + 2 \end{bmatrix}$$

*evalm(%);*

$$A := \begin{bmatrix} 2.567143290 & -.5671432904 \\ -.5671432904 & 2.567143290 \end{bmatrix}$$

Point is a minimum.

### Section 9.4

**1. a.** $f := 2 \cdot x1 \cdot x2 - 2 \cdot x1^2 - x2^2;$

$$f := 2\,x1\,x2 - 2\,x1^2 - x2^2$$

> *(kt,MP,z1,z2,z3):=STEEPEST(50,.05,1.0,1.0,f):*

Initial Condition: ( 1.0000, 1.0000)

| Iter | Gradient Vector | G magnitude | G x[k] | Step Length |
|---|---|---|---|---|
| 1 | $(-2.0000, 0.0000)$ | 2.0000 | $(1.0000, 1.0000)$ | 0.2500 |
| 2 | $(0.0000, -1.0000)$ | 1.0000 | $(0.5000, 1.0000)$ | 0.5000 |
| 3 | $(-1.0000, 0.0000)$ | 1.0000 | $(0.5000, 0.5000)$ | 0.2500 |
| 4 | $(0.0000, -0.5000)$ | 0.5000 | $(0.2500, 0.5000)$ | 0.5000 |

| Iter | Gradient Vector | G magnitude | G x[k] | Step Length |
|------|------|------|------|------|
| 5 | $(-0.5000, 0.0000)$ | 0.5000 | $(0.2500, 0.2500)$ | 0.2500 |
| 6 | $(0.0000, -0.2500)$ | 0.2500 | $(0.1250, 0.2500)$ | 0.5000 |
| 7 | $(-0.2500, 0.0000)$ | 0.2500 | $(0.1250, 0.1250)$ | 0.2500 |
| 8 | $(0.0000, -0.1250)$ | 0.1250 | $(0.0625, 0.1250)$ | 0.5000 |
| 9 | $(-0.1250, 0.0000)$ | 0.1250 | $(0.0625, 0.0625)$ | 0.2500 |
| 10 | $(0.0000, -0.0625)$ | 0.0625 | $(0.0312, 0.0625)$ | 0.5000 |
| 11 | $(-0.0625, 0.0000)$ | 0.0625 | $(0.0312, 0.0312)$ | 0.2500 |
| 12 | $(0.0000, -0.0312)$ | 0.0312 | $(0.0156, 0.0312)$ | |

Approximate Solution: $(0.0156, 0.0312)$
Maximum Functional Value: $-0.0005$
Number gradient evaluations: 13
Number function evaluations: 12

**b.** Newtons(f,10,.5,55,55);

Hessian: $[\,-4.000\ 2.000\,]$
$[\,2.000\ -2.000\,]$
eigenvalues: $-5.236\ -0.764$
pos def: false
new x = 0.000 new y = 0.000
Hessian: $[\,-4.000\ 2.000\,]$
$[\,2.000\ -2.000\,]$
eigenvalues: $-5.236\ -0.764$
pos def: false
new x = 0.000 new y = 0.000
final new x = 0.000 final new y = 0.000
final fvalue is 0.000

**2. a.** $f:=3*x1*x2-4*x1^2-2*x2^2;$

$$f := 3\ x1\ x2 - 4\ x1^2 - 2\ x2^2$$

$$>(kt,MP,z1,z2,z3):=STEEPEST(50,.05,1.0,1.0,f):$$

Initial Condition: $(1.0000, 1.0000)$

| Iter | Gradient Vector | G magnitude | G x[k] | Step Length |
|------|------|------|------|------|
| 1 | $(-5.0000, -1.0000)$ | 5.0990 | $(1.0000, 1.0000)$ | 0.1494 |
| 2 | $(0.5287, -2.6437)$ | 2.6960 | $(0.2529, 0.8506)$ | 0.1884 |
| 3 | $(-1.7625, -0.3525)$ | 1.7974 | $(0.3525, 0.3525)$ | 0.1494 |
| 4 | $(0.1864, -0.9319)$ | 0.9503 | $(0.0891, 0.2998)$ | 0.1884 |
| 5 | $(-0.6212, -0.1242)$ | 0.6336 | $(0.1242, 0.1242)$ | 0.1494 |
| 6 | $(0.0657, -0.3285)$ | 0.3350 | $(0.0314, 0.1057)$ | 0.1884 |
| 7 | $(-0.2190, -0.0438)$ | 0.2233 | $(0.0438, 0.0438)$ | 0.1494 |
| 8 | $(0.0232, -0.1158)$ | 0.1181 | $(0.0111, 0.0373)$ | 0.1884 |
| 9 | $(-0.0772, -0.0154)$ | 0.0787 | $(0.0154, 0.0154)$ | 0.1494 |
| 10 | $(0.0082, -0.0408)$ | 0.0416 | $(0.0039, 0.0131)$ | |

Approximate Solution: $(0.0039, 0.0131)$
Maximum Functional Value: $-0.0003$
Number gradient evaluations: 11
Number function evaluations: 10

**b.** $f := 3\ x1\ x2 - 4\ x1^2$

$$-2\ x2^2$$

*Newtons(f,100,.5,2,2);*

Hessian: $[\,-8.000\ 3.000\,]$
$[\,3.000\ 0.000\,]$
eigenvalues: $-9.000\ 1.000$
pos def: false
new x = 0.000 new y = 0.000
Hessian: $[\,-8.000\ 3.000\,]$
$[\,3.000\ 0.000\,]$
eigenvalues: $-9.000\ 1.000$
pos def: false
new x = 0.000 new y = 0.000
final new x = 0.000 final new y = 0.000
final fvalue is 0.000

**3. a.** $f:=-x1^3+3*x1+84*x2-6*x2^2;$

$$f := -x1^3 + 3\ x1 + 84\ x2 - 6\ x2^2$$

*Newtons(f,100,.01,1,1);#4: init (0.5,1)*

Hessian: $[\,-6.000\ 0.000\,]$
$[\,0.000\ -12.000\,]$
eigenvalues: $-12.000\ -6.000$
pos def: false
new x = 1.000 new y = 7.000
Hessian: $[\,-6.000\ 0.000\,]$
$[\,0.000\ -12.000\,]$
eigenvalues: $-12.000\ -6.000$
pos def: false
new x = 1.000 new y = 7.000
final new x = 1.000 final new y = 7.000
final fvalue is 296.000

# Chapter 10

**Section 10.1**

**1. a.** $\left[x = \dfrac{4}{5}, y = \dfrac{8}{5}, \lambda_1 = \dfrac{8}{5}, x^2 + y^2 = \dfrac{16}{5}\right]$
$[x = 0.8000000000, y = 1.600000000,$
$\lambda_1 = 1.600000000, x^2 + y^2 = 3.200000000]$

**b.** $\left[x = \dfrac{12}{5}, y = \dfrac{4}{5}, \lambda_1 = -\dfrac{6}{5}, (x-3)^2 + (y-2)^2 = \dfrac{9}{5}\right]$
$[x = 2.400000000, y = 0.8000000000, \lambda_1 = -1.200000000,$
$(x - 3.)^2 + (y - 2.)^2 = 1.800000000]$

**c.** $[x = -.7071067810, y = 0.7071067810, \lambda_1 = -1.,$
$x^2 + y^2 + 4.x\,y = -.9999999994], [x = 0.7071067810,$
$y = 0.7071067810, \lambda_1 = 3., x^2 + y^2 + 4.x\,y = 2.999999998]$

**d.** [11.6799999999999998, [$x$ = 1.59999999999999986,
$y$ = 1.20000000000000016]]

$$\left[x = \frac{8}{5}, y = \frac{6}{5}, \lambda_1 = \frac{9}{5}, \lambda_2 = \frac{56}{25}, x^2 + y^2 + 4xy = \frac{292}{25}\right],$$

$[x = 0, y = 2, \lambda_1 = -3, \lambda_2 = 8, x^2 + y^2 + 4\,x\,y = 4]$
$[x = 1.600000000, y = 1.200000000, \lambda_1 = 1.800000000,$
$\lambda_2 = 2.240000000, x^2 + y^2 + 4.x\,y = 11.68000000],$
$[x = 0., y = 2., \lambda_1 = -3., \lambda_2 = 8., x^2 + y^2 + 4.x\,y = 4.]$

**3.** $[x = 0., y = \mathrm{RootOf}(-1 + 2\_Z^2 - z^2, label = \_L3),$
$\lambda_1 = 0.5000000000,$
$x^2 + y^2 + z^2 = \mathrm{RootOf}(-1 + 2\_Z^2 - z^2, label = \_L3)^2 + z^2],$
$x = \mathrm{RootOf}(\_Z^2 - 1 - z^2, label=\_L4), y = 0., \lambda_1 = 1.,$
$x^2 + y^2 + z^2 = \mathrm{RootOf}(\_Z^2 - 1 - z^2, label = \_L4)^2 + z^2]$

**5.** [14.2189500386216583, [$x$ = 2.22540333075842822,
$y$ = 3.32379000772471445]]
$[x = 2.225403331, y = 3.323790007, \lambda_1 = 1.447943664,$
$\lambda_2 = -.431754164, -2.\,x^2 - 1.y^2 + x\,y + 8.x$
$+ 3.\,y = 14.21895004]$

**6.** $x = 6, y = 4, w = 3, Z = 72, \lambda = 6.$

## Section 10.2

**1.** $x = 5.447368421, y = 4.368421053, \lambda_1 = -0.236, \lambda_2 = 0,$
$Z = 30.64473684$

**2.** $x = 3.973684211, y = 2.684210526, \lambda_1 = -0.868, \lambda_2 = 0,$
$Z = 26.22368421$

**3.** $x = -3, y = -2, \lambda_1 = 0, \lambda_2 = 0, Z = -6$

**4.** $x = -0.5, y = 2.75, Z = 6.375, \lambda_1 = 2.50, \lambda_2 = 6.25$

**5.** $x = 2.139387691, y = 3.069693846, \lambda_1 = 2.4936,$
$\lambda_2 = -0.1835, Z = -6.4136$

**6.** (0,0)

**7.** $[x = 2.285714286, y = 3.428571429, \lambda_1 = 3.428571429,$
$-1.\,(x - 4.)^2 + x\,y - 1.\,(y - 4.)^2 = 4.571428571]$

# Chapter 11

## Section 11.1

**1.** $x = 21, y = 35$

**2.** $x = 40, y = -9$

**3.** $x = 8, y = -16$

**4.** $6x + y = 50$, infinite solutions

**5.** $x = 2.2353, y = 1.35294, z = 0.19607$

**6.** No solution

**7.** $x = 4.846, y = 2.538, z = 1.846$

**8.** $x = 2, y = 3, z = 13$

**9.** $x = -5, y = 6, z = 9, t = 6$

## Section 11.2

**7.** > $ReducedRowEchelonForm(\langle A \rangle);$

$$\begin{bmatrix}
1 & 0 & 0 & 0 & 0 & 0 & 0 & 0 & \frac{3}{4} \\
0 & 1 & 0 & 0 & 0 & 0 & 0 & 0 & -\frac{9}{4} \\
0 & 0 & 1 & 0 & 0 & 0 & 0 & 0 & \frac{39}{2} \\
0 & 0 & 0 & 1 & 0 & 0 & 0 & 0 & -9 \\
0 & 0 & 0 & 0 & 1 & 0 & 0 & 0 & -\frac{3}{4} \\
0 & 0 & 0 & 0 & 0 & 1 & 0 & 0 & \frac{27}{4} \\
0 & 0 & 0 & 0 & 0 & 0 & 1 & 0 & \frac{3}{2} \\
0 & 0 & 0 & 0 & 0 & 0 & 0 & 1 & 3
\end{bmatrix}$$

> $spline([1,2,3],[9,27,48],x,cubic);$

$$\begin{cases} -9 + \dfrac{39}{2}x - \dfrac{9}{4}x^2 + \dfrac{3}{4}x^3 & x < 2 \\[2ex] 3 + \dfrac{3}{2}x + \dfrac{27}{4}x^2 - \dfrac{3}{4}x^3 & otherwise \end{cases}$$

## Section 11.3

**1.** $Cu + 2AgNO_3 \rightarrow 2Ag + Cu(NO_3)_2$
Copper (Cu), silver (Ag), nitrogen (N) and oxygen (O) are
the elements.

**2.** $Zn + 2HCl \rightarrow ZnCl_2 + H_2$
Zinc (Zn), hydrogen (H), and chlorine (Cl) are the elements.

**3.** $3Ca(OH)_2 + 2H_3PO_4 \rightarrow 6H_2O + Ca_3(PO_4)_2$
Calcium (Ca), hydrogen (H), oxygen (O), and phosphorus
(P) are the elements.

# Chapter 12

## Section 12.2

**1. a.**

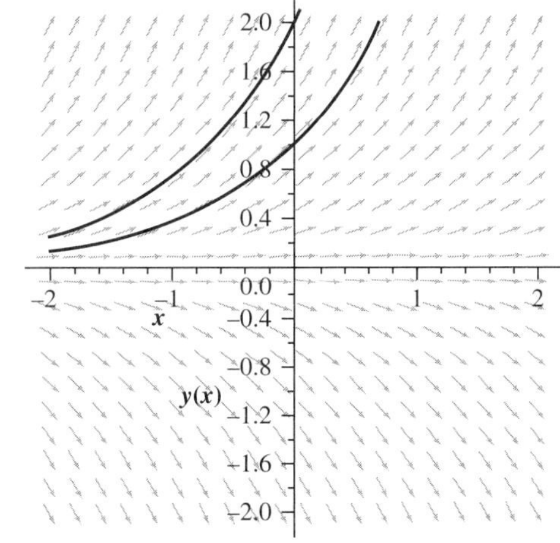

**b.**

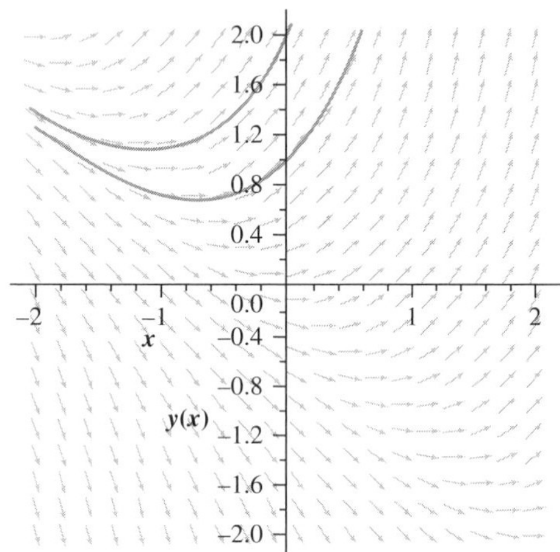

**c.** $DEplot(diff(y(x),x)=x+y(x),y(x),x=-2..2,y=-2..2,$
$initsarrows=LARGE,color=black,stepsize=0.05);$

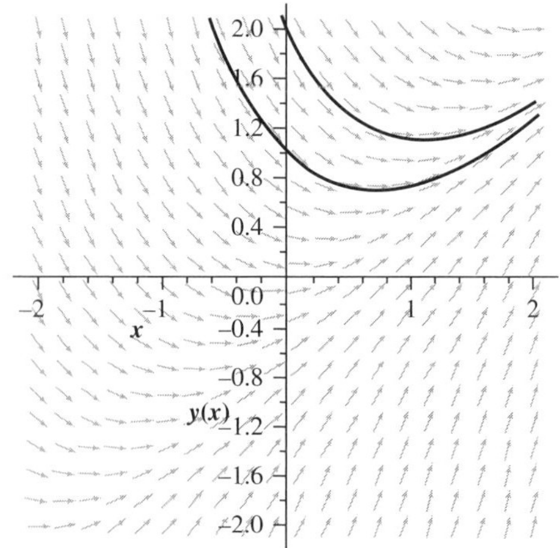

**d.** $> DEplot(diff(y(x),x)=x-y(x),y(x),x=-2..2,y=-2..2,$
$inits,arrows=LARGE,color=black,stepsize=0.05);$

**e.** $> DEplot(diff(y(x),x)=x{\cdot}y(x),y(x),x=-2..2,y=-2..2,$
$inits,arrows=LARGE,color=black,stepsize=0.05);$

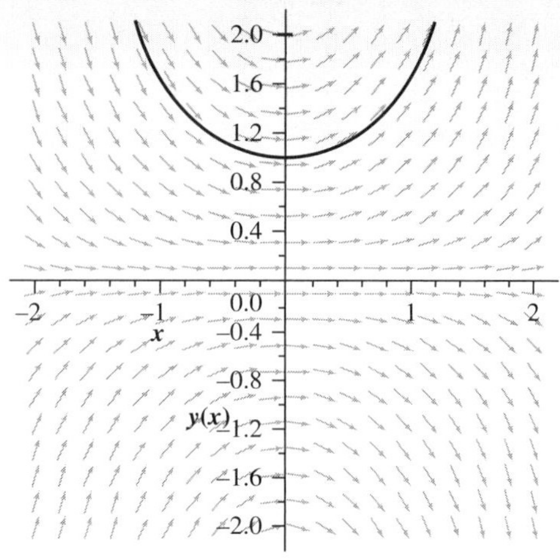

**f.** $> DEplot\left(diff(y(x),x)=\dfrac{1}{y(x)},y(x),x=-2..2,y=-2..2,\right.$
$[[y(0)=2],[y(0)=2]],arrows=LARGE,color=black,$
$stepsize=0.05);$

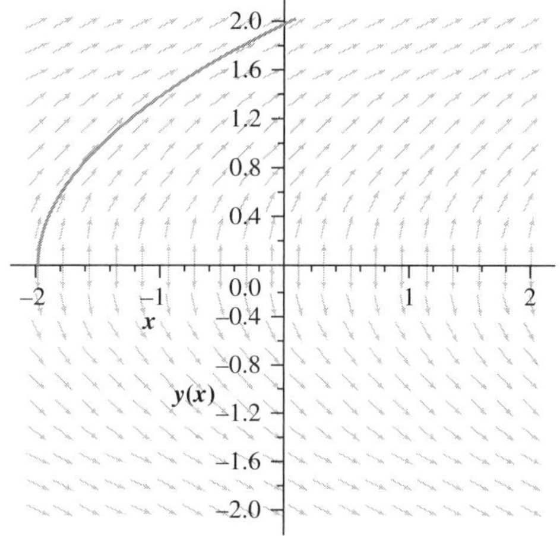

**2.** >*DEplot(diff(y(x),x)=(y(x)+2)·(y(x)−3),y(x),x=−4..4,*
*y=−4..4,[[y(0)=2],[y(3)=3]],arrows=LARGE,color=*
*black,stepsize=0.05);*

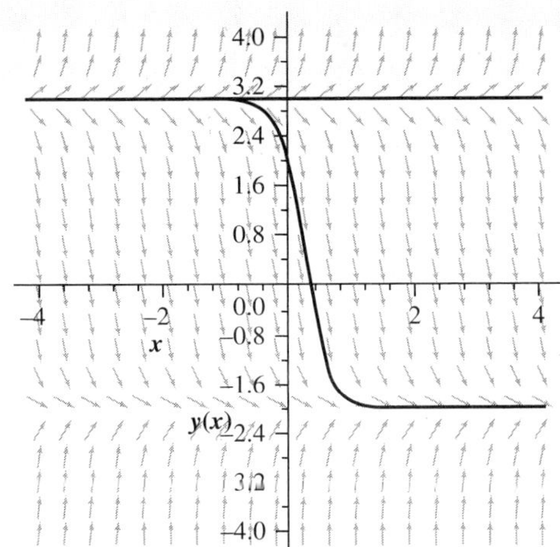

**b.** *DEplot(diff(y(x),x)=(y(x)²−4),y(x),x=−4..4,y=−4..4,*
*[[(0)=3],[y(1)=−1]],arrows=LARGE,color=black,*
*stepsize=0.05);*

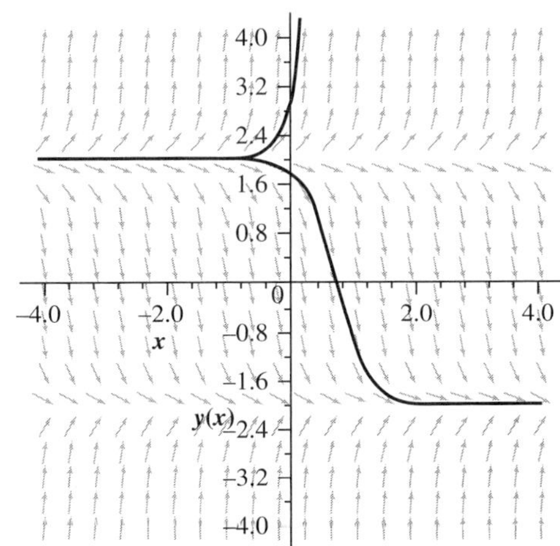

**c.** *DEplot(diff(y(x),x)=(y(x)³−4),y(x),x=−4..4,y=−4..4,*
*[[(0)=3],[y(1)=−1]],arrows=LARGE,color=black,*
*stepsize=0.05);*

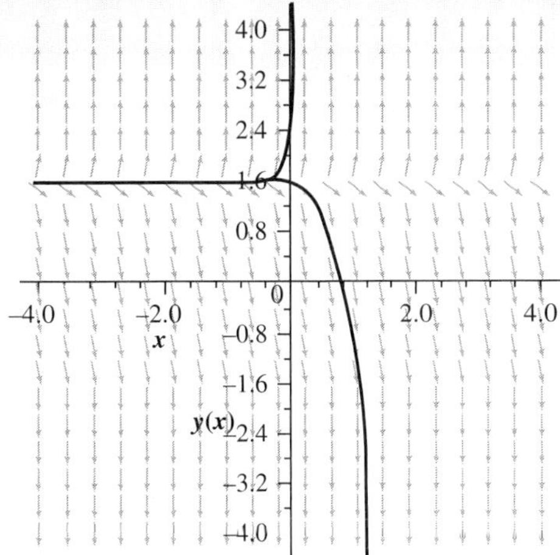

> *DEplot(diff(y(x),x)=(x−2·y(x)),y(x),x=−4..4,*
*y=−4..4,y(0)=3],[y(1)=−1]],arrows=LARGE,color=*
*black,stepsize=0.05);*

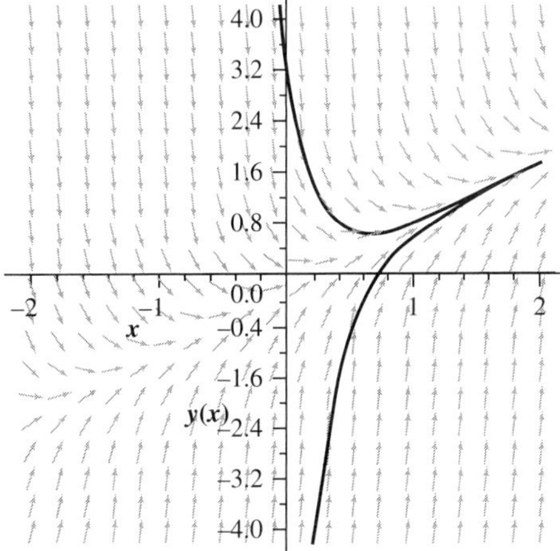

## Section 12.4

**1.** > *dsolve ({ode, ics});*

$$y(t) = -1 - t + 2\,e^t$$

> *subs(t=3,−1−t+2·exp(t));*

$$-4 + 2\,e^3$$

> *ans1:=evalf(%);*

$$ans1 := 36.17107384$$

> *Digits:=20:*
*ans2:= dsolve({ode,ics},numeric,method=classical*
*[foreuler],output=array ( [0.01,0.1,1,1.1,*
*2,3]),stepsize=0.1);*

$$ans2 := \begin{bmatrix} [\,t\ y(t)\,] \\ 0.01 & 1.01 \\ 0.1 & 1.1 \\ 1. & 3.1874849202 \\ 1.1 & 3.60623341222 \\ 2. & 10.454999898651200184 \\ 3. & 30.898804537772814636 \end{bmatrix}$$

> $err1 := \dfrac{100\cdot(ans1-30.898804537772814636)}{ans1};$

$$err1 := 14.575926956298473454$$

> Digits:=20:
ans2:= dsolve({ode, ics},numeric,method=classical
[foreuler],output=array([0.01,0.1,1,1.1,2,3]),
stepsize=0.05);

$$ans2 := \begin{bmatrix} [\,t\ y(t)\,] \\ 0.01 & 1.01 \\ 0.1 & 1.1050 \\ 1. & 3.3065954102888402680 \\ 1.1 & 3.7505214398434463955 \\ 2. & 11.079977424249292491 \\ 3. & 33.358371788245891342 \end{bmatrix}$$

> $err2 := \dfrac{100\cdot(ans1-33.35837178824)}{ans1};$

$$err2 := 7.7761088047353986388$$

> Digits:=20:
ans2:=dsolve({ode, ics}, numeric,method=classical
[foreuler],output=array([0.01,0.1,1,1.1,2,3]),
stepsize=0.01);

$$ans2 := \begin{bmatrix} [\,t\ y(t)\,] \\ 0.01 & 1.01 \\ 0.1 & 1.1092442508224090200 \\ 1. & 3.4096276588430521869 \\ 1.1 & 3.8755944021944506392 \\ 2. & 11.632035703659880910 \\ 3. & 35.576932523848776648 \end{bmatrix}$$

> $err3 := \dfrac{100\cdot(ans1-35.7569325)}{ans1};$

$$err3 := 1.1449517593374854758$$

**2.** > Digits:=20:
ans2:=dsolve({ode,ics},numeric,method=classical[rk4],
output=array([0.01,0.1,.5,1,1.1,2,3]),
stepsize = 0.25);

$$ans2 := \begin{bmatrix} [\,t\ y(t)\,] \\ 0.01 & 1.0201003341666666667 \\ 0.1 & 1.2103416666666666667 \\ 0.5 & 2.2973989380730523004 \\ 1. & 4.4364198784026464997 \\ 1.1 & 5.0081726873641515010 \\ 2. & 13.777330547145722877 \\ 3. & 39.167886768114832214 \end{bmatrix}$$

> Digits:=20:
ans2:= dsolve({ode,ics},numeric, method=classical
[foreuler],output=array([0.01,0.1,1,1.1,2,3]),
stepsize=0.1);

$$ans2 := \begin{bmatrix} [\,t\ y(t)\,] \\ 0.01 & 1.02 \\ 0.1 & 1.2 \\ 1. & 4.1874849202 \\ 1.1 & 4.70623341222 \\ 2. & 12.454999898651200184 \\ 3. & 33.898804537772814636 \end{bmatrix}$$

> $err1 := \dfrac{100\cdot(ans1-33.898804)}{ans1};$

$$err1 := 13.459599976892644792$$

> Digits:=20:
ans2 := dsolve({ode, ics},numeric,method = classical
[foreuler],output=array([0.01,0.1,1,1.1,2,3]),
stepsize = 0.05);

$$ans2 := \begin{bmatrix} [\,t\ y(t)\,] \\ 0.01 & 1.02 \\ 0.1 & 1.2050 \\ 1. & 4.3065954102888402680 \\ 1.1 & 4.8505214398434463955 \\ 2. & 13.079977424249292491 \\ 3. & 36.358371788245891342 \end{bmatrix}$$

> $err2 := \dfrac{100\cdot(ans1-36.358371788)}{ans1};$

$$err2 := 7.1805589742227775441$$

> Digits:=20:
ans2:= dsolve({ode,ics},numeric,method=classical
[foreuler],output=array([0.01,0.1,1,1.1,2,3]),
>stepsize=0.01);

$$ans2 := \begin{bmatrix} & [\, t\ y(t)\,] \\ 0.01 & 1.02 \\ 0.1 & 1.2092442508224090200 \\ 1. & 4.4096276588430521869 \\ 1.1 & 4.9755944021944506392 \\ 2. & 13.632035703659880909 \\ 3. & 38.576932523848776647 \end{bmatrix}$$

$$> err3 := \frac{100 \cdot (ans1 - 38.5769325)}{ans1};$$

$$err3 := 1.5167859546191988921$$

**3. a.** *Digits:=20:*
   *ans2:= dsolve({ode,ics},numeric,method=classical*
       *[foreuler],output=array([0.01,0.1,1,1.1,2,3]),*
       *stepsize=0.05);*

$$ans2 := \begin{bmatrix} & [\, t\ v(t)\,] \\ 0.01 & 0.32 \\ 0.1 & 3.07200 \\ 1. & 16.226133416744068927 \\ 1.1 & 16.805799323932179940 \\ 2. & 19.287896550589210233 \\ 3. & 19.865630829422367002 \end{bmatrix}$$

$$> err2 := \frac{100 \cdot (ans1 - 19.8656308)}{ans1};$$

$$err2 := -3.5490411365732322777$$

Terminal velocity is 19.999.

**4.** *Digits:=20 :*
   *ans2:= dsolve({ode,ics},numeric,method=classical*
       *[heunform],output=array([0.01,0.1,.5,1,1.1,2,3]),*
       *stepsize=0.25);*

$$ans2 := \begin{bmatrix} & [\, t\ y(t)\,] \\ 0.01 & 1.0201000000000000000 \\ 0.1 & 1.2100000000000000000 \\ 0.5 & 2.2832031250000000000 \\ 1. & 4.3897113800048828125 \\ 1.1 & 4.9556310749053955078 \\ 2. & 13.524494379877069150 \\ 3. & 38.141416324020201490 \end{bmatrix}$$

$$y(0.5) = 2.283203125.$$

**5.** *Digits:=20 :*
   *ans2:= dsolve({ode, ics}, numeric,method=classical*
       *[foreuler],output=array([0.01,0.1,1,1.1,2,3,4,5,6,7]),*
       *stepsize=0.1);*

$$ans2 := \begin{bmatrix} & [\, t\ v(t)\,] \\ 0.01 & 0.32 \\ 0.1 & 3.2 \\ 1. & 16.501975424680381645 \\ 1.1 & 17.061659356731520582 \\ 2. & 19.388191203523000183 \\ 3. & 19.892993889726336795 \\ 4. & 19.981284499827668251 \\ 5. & 19.996726636022889250 \\ 6. & 19.999427484618205031 \\ 7. & 19.999899866356236636 \end{bmatrix}$$

**7.** *Digits:=20:*
   *ans2:= dsolve({ode,ics},numeric,method=classical[rk4],*
       *output=array([0.01,0.1,.5,1,1.1,2,3]),*
       *stepsize=0.25);*

$$ans2 := \begin{bmatrix} & [\, t\ y(t)\,] \\ 0.01 & 1.0201003341666666667 \\ 0.1 & 1.2103416666666666667 \\ 0.5 & 2.2973989380730523004 \\ 1. & 4.4364198784026464997 \\ 1.1 & 5.0081726873641515010 \\ 2. & 13.777330547145722877 \\ 3. & 39.167886768114832214 \end{bmatrix}$$

Estimate for $y(0.5)$ is 2.2973989280730523004.

**8.** *Digits:=20:*
   *ans2:= dsolve({ode,ics},numeric,method=classical[rk4],*
       *output=array([0.01,0.1,.5,1,1.1,2,3,5,10,15,20,25,30]),*
       *stepsize=0.25);*

$$ans2 := \begin{bmatrix} & [\, t\ v(t)\,] \\ 0.01 & 0.31745359872000000000 \\ 0.1 & 2.9571072000000000000 \\ 0.5 & 11.011276800000000000 \\ 1. & 15.960142761689088000 \\ 1.1 & 16.557457308008153686 \\ 2. & 19.183977674703346559 \\ 3. & 19.835169315126350648 \\ 5. & 19.993274724062458237 \\ 10. & 19.999997738533178196 \\ 15. & 19.999999999239550580 \\ 20. & 19.999999999999744288 \\ 25. & 19.99999999999999914 \\ 30. & 19.999999999999999999 \end{bmatrix}$$

**9.** > *Digits:=20:*

> *ans2:= dsolve({ode9, ics9},numeric,method=classical[rk4],output=array([0.01,0.1,.2,.3,.4,.5,.6,.7,.8,.9,1,1.1,2,3]),stepsize=1);*

$$ans2 := \begin{bmatrix} & [\,t\ p(t)\,] \\ 0.01 & 2.1823939866402952357 \\ 0.1 & 3.7417919856283322187 \\ 0.2 & 4.4534652419440640000 \\ 0.3 & 2.2152217966943749062 \\ 0.4 & 3.0027811446456320000 \\ 0.5 & -1736.1070184707641602 \\ 0.6 & -68809.898553809108992 \\ 0.7 & -1.1160658422227225101\ 10^6 \\ 0.8 & -1.1057720976714369532\ 10^7 \\ 0.9 & -7.8747607659843454887\ 10^7 \\ 1. & -4.4053535500000000000\ 10^8 \\ 1.1 & -1.1748975740378927409\ 10^{126} \\ 2. & -1.1748974460230148156\ 10^{141} \\ 3. & -7.6967428740439053007\ 10^{2259} \end{bmatrix}$$

> *Digits:=20:*

*ans2:= dsolve({ode9, ics9}, numeric,method=classical[rk4], output=array([0.01,0.1,.2,.3,.4,.5,.6,.7,.8,.9,1,1.1,2,3]),stepsize=.5);*

$$ans2 := \begin{bmatrix} & [\,t\ p(t)\,] \\ 0.01 & 2.1823939866402952357 \\ 0.1 & 3.7417919856283322187 \\ 0.2 & 4.4534652419440640000 \\ 0.3 & 2.2152217966943749062 \\ 0.4 & 3.0027811446456320000 \\ 0.5 & -1736.1070184707641602 \\ 0.6 & -4.1958534911767819853\ 10^{39} \\ 0.7 & -1.3540293071321619192\ 10^{44} \\ 0.8 & -5.8990229171177889635\ 10^{46} \\ 0.9 & -4.4030353329038449720\ 10^{48} \\ 1. & -1.2495016071440306552\ 10^{50} \\ 1.1 & -2.0610889145664881088\ 10^{789} \\ 2. & -1.0695629927060573950\ 10^{12795} \\ 3. & -5.5132899474042789918\ 10^{3275497} \end{bmatrix}$$

> *Digits:=20:*

> *ans2:= dsolve({ode9,ics9},numeric,method=classical[rk4],output=array{[0.01,0.1,.2,.3,.4,.5,.6,.7,.8,.9,1,1.1,2,3]),stepsize=.1);*

$$ans2 := \begin{bmatrix} & [\,t\ p(t)\,] \\ 0.01 & 2.1823939866402952357 \\ 0.1 & 3.7417919856283322187 \\ 0.2 & 4.6274220225660867275 \\ 0.3 & 4.8964328567794413161 \\ 0.4 & 4.9715463270025305207 \\ 0.5 & 4.9922092710096557588 \\ 0.6 & 4.9978689339588796589 \\ 0.7 & 4.9994172274781347692 \\ 0.8 & 4.9998406437119887933 \\ 0.9 & 4.9999564256839505317 \\ 1. & 4.9999880851232018708 \\ 1.1 & 4.9999967420240247175 \\ 2. & 4.9999999999721599095 \\ 3. & 4.9999999999999999348 \end{bmatrix}$$

Note that the smaller the step-size the better the estimates.

# Chapter 13

### Section 13.2

**1.** Rest point is *(0,0)*. It is unstable.

> *DEplot([diff(x(t),t)=x(t),diff(y(t),t)=y(t)], [x(t),y(t)],t=−2..2,x=−1..2,y=−1..2,arrows= LARGE,color=BLACK);*

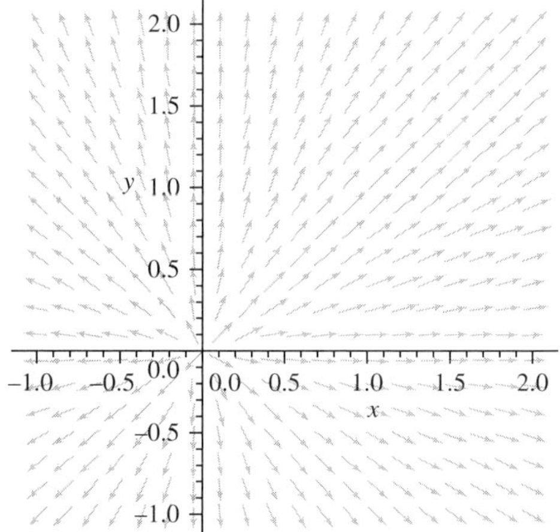

**2.** Rest point is *(0,0)*. It is unstable.

> *DEplot( [diff(x(t),t)=−x(t),diff(y(t),t)=2*y(t)],[x(t),y(t)],t=−2..2,x=−1..2,y=−1..2, arrows=LARGE,color=BLACK);*

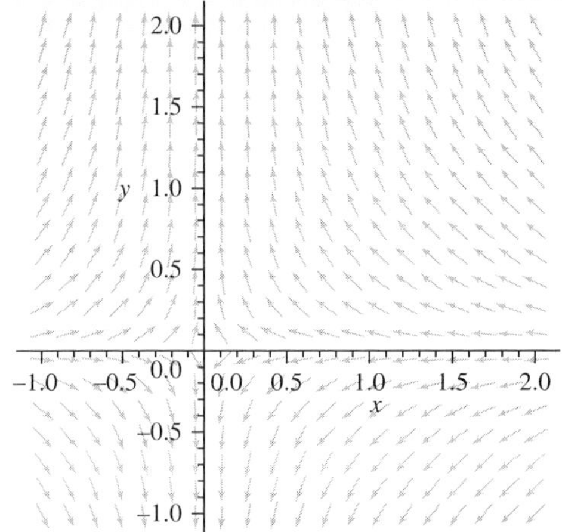

**3.** Rest point is *(0,0)*. It is stable.

> *DEplot( [diff(x(t),t)=y(t),diff(y(t),t)=−2*x(t)],[x(t),y(t)],t=−2..2,x=−1..2,y=−1..2, arrows=LARGE,color=BLACK);*

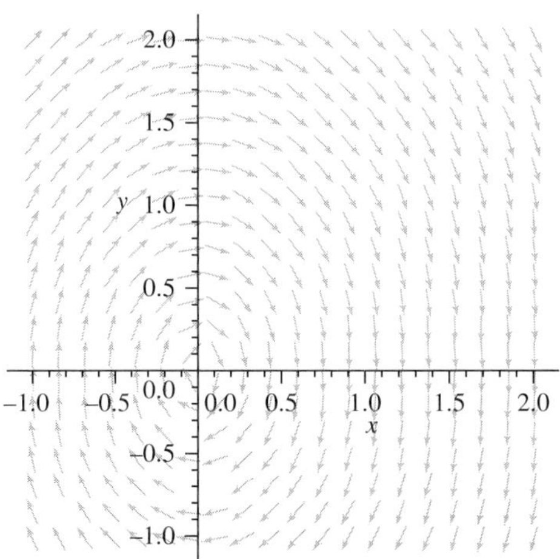

**4.** Rest point is *(1,0)*. It is asymptotically stable.

> *DEplot( [diff(x(t),t)=−x(t)+1,diff(y(t),t)=−2*y(t)], [x(t),y(t)],t=−2..2,x=−1..2,y=−1..2, arrows=LARGE,color=BLACK);*

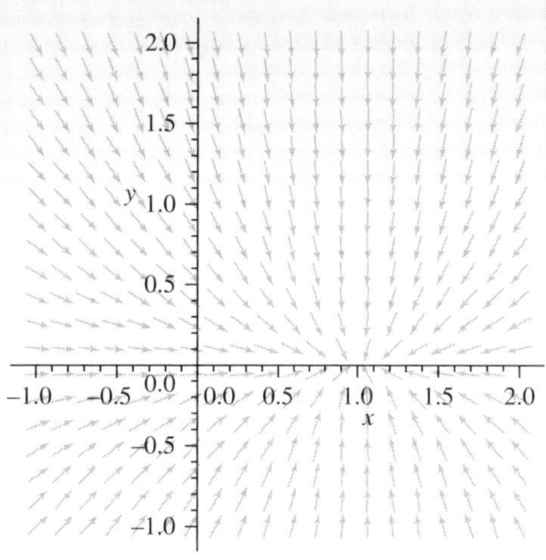

## Section 13.5

**1. a.** > *IVP3:={eqn1,x(0)=200,eqn2,y(0)=500};*

$$IVP3 := \left\{ \frac{d}{dt} x(t) = -0.5\, x(t) + y(t), \right.$$

$$\frac{d}{dt} y(t) = 0.25\, x(t) - 0.5\, y(t),\ x(0) = 200,$$

$$\left. y(0) = 500 \right\}$$

> *nsol3:=dsolve(IVP3,{y(t),x(t)},numeric,method=classical [foreuler],output=array([0,.1,.2,.3,.4,.5,.6,.7,.8,.9,1,1.1,1.2, 1.3,1.4,1.5,1.6,1.7,1.8,1.9,2.0]),stepsize=.10);*

$$nsol3 := [[\ t\ x(t)\ y(t)\ ]],[$$

| 0.   | 200.               | 500.               |
|------|--------------------|--------------------|
| 0.1  | 240.               | 480.               |
| 0.2  | 276.               | 462.               |
| 0.3  | 308.399999999999978 | 445.800000000000012 |
| 0.4  | 337.559999999999946 | 431.220000000000027 |
| 0.5  | 363.803999999999974 | 418.098000000000014 |
| 0.6  | 387.423600000000022 | 406.288200000000018 |
| 0.7  | 408.681240000000002 | 395.659379999999998 |
| 0.8  | 427.813116000000036 | 386.093441999999982 |
| 0.9  | 445.031804399999998 | 377.484097799999972 |
| 1.   | 460.528623960000004 | 369.735688019999940 |
| 1.1  | 474.475761564000038 | 362.762119218000009 |
| 1.2  | 487.028185407600006 | 356.485907296200026 |
| 1.3  | 498.325366866840000 | 350.837316566580000 |
| 1.4  | 508.492830180155976 | 345.753584909921984 |
| 1.5  | 517.643547162140408 | 341.178226418929796 |
| 1.6  | 525.879192445926378 | 337.060403777036810 |
| 1.7  | 533.291273201333752 | 333.354363399333124 |
| 1.8  | 539.962145881200398 | 330.018927059399800 |
| 1.9  | 545.965931293080302 | 327.017034353459848 |
| 2.0  | 551.369338163772227 | 324.315330918113886 |

Graphs:

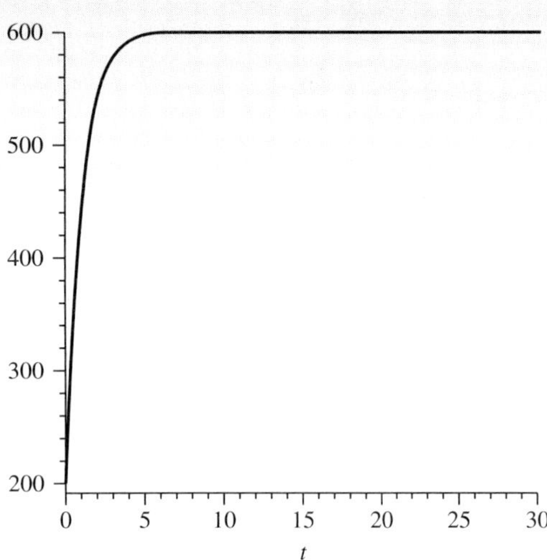

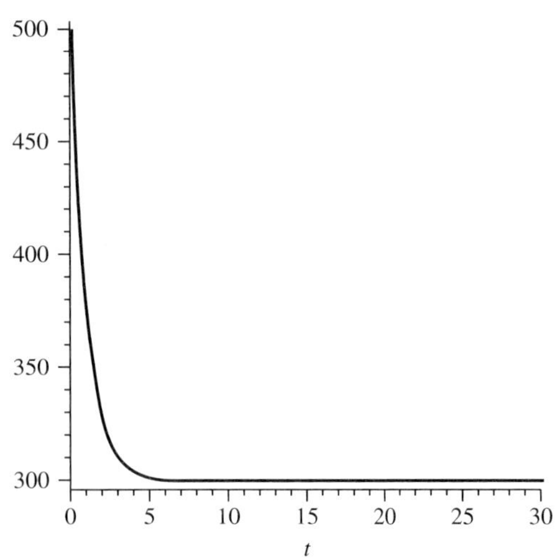

Steady-state solutions at $x = 600$, $y = 300$.

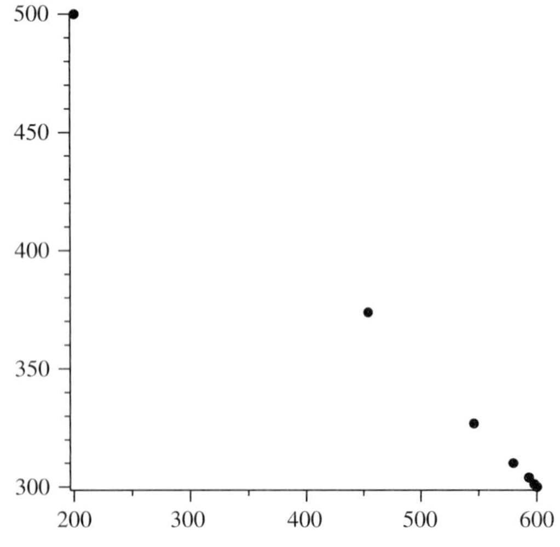

**2. a.**

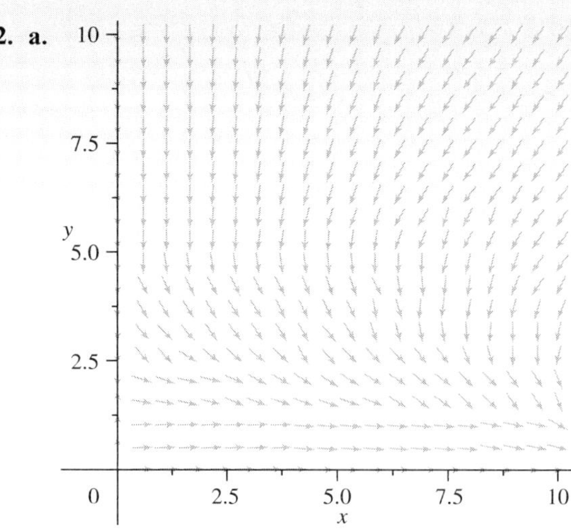

**b.** $x = 0$, $y = 0$, unstable; and $x = 4.5$, $y = 5.25$, unstable

**c.**

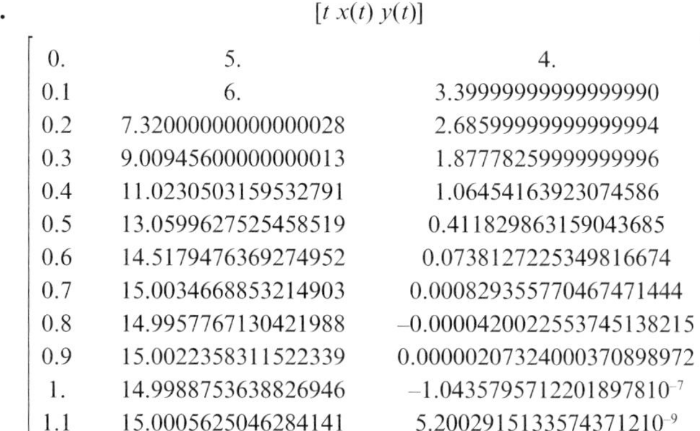

$$[t\ x(t)\ y(t)]$$

| $t$ | $x(t)$ | $y(t)$ |
|---|---|---|
| 0. | 5. | 4. |
| 0.1 | 6. | 3.39999999999999990 |
| 0.2 | 7.32000000000000028 | 2.68599999999999994 |
| 0.3 | 9.00945600000000013 | 1.87778259999999996 |
| 0.4 | 11.0230503159532791 | 1.06454163923074586 |
| 0.5 | 13.0599627525458519 | 0.411829863159043685 |
| 0.6 | 14.5179476369274952 | 0.0738127225349816674 |
| 0.7 | 15.0034668853214903 | 0.000829355770467471444 |
| 0.8 | 14.9957767130421988 | $-0.0000420022553745138215$ |
| 0.9 | 15.0022358311522339 | 0.0000020732400037089 8972 |
| 1. | 14.9988753638826946 | $-1.0435795712201897810^{-7}$ |
| 1.1 | 15.0005625046284141 | $5.2002915133574371210^{-9}$ |

Supports, $x$ winning out and $y$ going to 0.

## Section 13.6

**2.** Equilibrium is hares $= 1{,}000$ or 0 and foxes $= 12{,}000$ or 0; $(0,0)$ or $(1000,12000)$.

**3.** Equilibrium is $I = 0$, $S = 300$.

**4.** $(0,0)$

**5.** a and b on the graph

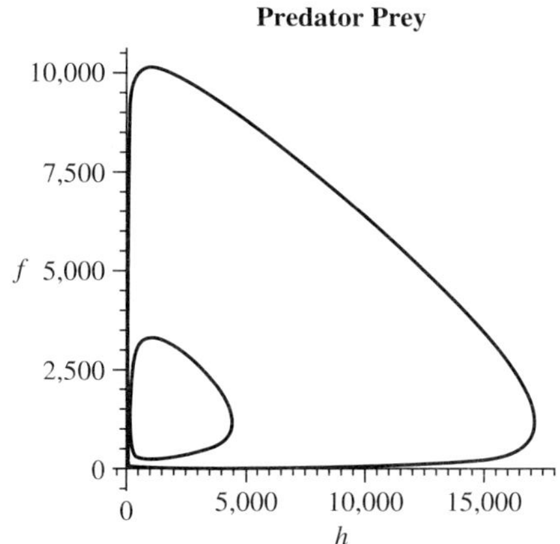

**Predator Prey**

# Chapter 14

## Section 14.1

**1. a.** 0.058399    **b.** 0.078126    **c.** $\mu = 7.5,\ \sigma^2 = 1.875$

**2. a.** 0.354294    **b.** 0.114265    **c.** 0.6,.54

**3. a.**

| $X$ | 3 | 5 | 6 | 8 | 10 |
|---|---|---|---|---|---|
| $P(X = x)$ | 0.17 | 0.34 | 0.28 | 0.12 | 0.09 |
| $P(X \le x)$ | 0.17 | 0.51 | 0.79 | 0.91 | 1 |

    **b.** 0.79, 0.49    **c.** 0.49    **d.** 5.75, 3.7275

**4.** 0.014729

## Section 14.2

**1.** $P(X \ge 3) = 1 - P(X \le 2) = 1 - 0 = 1$

**4.** 0.949335057

**5.** 0.994

## Section 14.3

**2.** $> overbook(5,6,.6,100,200);$

    346.0032000

  $> overbook(5,6,.8,100,200);$

    401.3568000

**3.** $> overbook(5,6.,.6,100,250);$

    343.6704000

  $> overbook(5,6,.8,100,250);$

    388.2496000

**4.** $> overbook(5,6,.6,100,150);$

    348.3360000

  $> overbook(5,6,.8,100,150);$

    414.4640000

## Section 14.4

**1.** Long-term behavior is stable with $D(n)$ at 8.54% and $B(n)$ at 91.46%, where $D(n)$ is dining and $B(n)$ is the deli bar.

**2.** Adding the pizza parlor makes the new values $D(n) = 7.4\%,\ B(n) = 18.53,\ P(n) = 74.07\%.$

# Chapter 15

## Section 15.1

**1. a.** $R = 1 - F(t) = 1 - e^{\frac{-y}{3}}$    **b.** 0.32967

**2. a.** 0.97044

    **b.** Memoryless, so 0.977044.

**3.** Minimum reliability is 0.9.

**4. a.** 0.90022        **b.** 0.8995

## Section 15.2

**1.** 0.9772

**2.** 0.954499876

**3.**

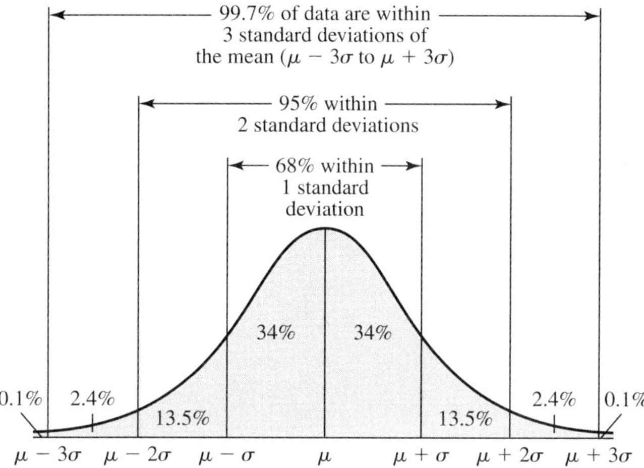


**4.** 0.158655

## Section 15.4

**1.** The regression equation is

  $y = -45.55 + 1.71143\,x$
  R-Sq = 96.1%
  Analysis of Variance
  Source DF SS MS F P
  Regression 1 398030 398030 294.74 0.000
  Residual Error 12 16205 1350
  Total 13 414236
  Residuals appear random.

**2.** The regression equation is

  $y = 186.61 - 0.17717\,x$
  R$-$Sq = 67.8%
  Analysis of Variance
  Source DF SS MS F P
  Regression 1 9222.8 9222.8 23.14 0.001
  Residual Error 11 4384.9 398.6
  Total 12 13607.7
  Residuals appear random.

# Chapter 16

### Section 16.1

**1.** Possible random numbers from Maple are

> *random:=**proc**(n)**option** remember **for** i **from** 1 **to** n **do***

$$x(i):=evalf\left(\frac{rand\ ()}{999999999999}\right)\ \textbf{end do end;}$$

> *seq(x(i),i=1..20);*

0.2240171515, 0.2008401063, 0.8685719066,
0.5704134665, 0.9920881460, 0.04437752746,
0.4780291372, 0.01294305199, 0.6408831565,
0.2487102377, 0.9700117830, 0.04705144220,
0.5745759906, 0.1717448313, 0.6488303816,
0.4114303688, 0.09704265386, 0.7719464925,
0.2477991384, 0.8950382012

**2.** Random numbers between –10 and 10 are

*for i **from** 1 **to** 20 **do** y(i):=evalf (−10+(20)·x(i))**end do;***

$$y(1) := -5.519656970$$
$$y(2) := -5.983197874$$
$$y(3) := 7.37143813$$
$$y(4) := -1.40826933$$
$$y(5) := 9.84176292$$
$$y(6) := -9.112449451$$
$$y(7) := -.439417256$$
$$y(8) := -9.741138960$$
$$y(9) := 2.81766313$$
$$y(10) := -5.025795246$$
$$y(11) := 9.40023566$$
$$y(12) := -9.058971156$$
$$y(13) := 1.49151981$$
$$y(14) := -6.565103374$$
$$y(15) := 2.97660763$$
$$y(16) := -1.771392624$$
$$y(17) := -8.059146923$$
$$y(18) := 5.43892985$$
$$y(19) := -5.044017232$$
$$y(20) := 7.90076402$$

**3.** Random exponential variables are

> *for i **from** 1 **to** 20 **do** expX(i):=−0.5·ln(x(i));end do;*

$$expX(1) := 0.7480163305$$
$$expX(2) := 0.8026230895$$
$$expX(3) := 0.07045245135$$
$$expX(4) := 0.2806969006$$
$$expX(5) := 0.003971659395$$
$$expX(6): = 1.557511038$$
$$expX(7) := 0.3690417960$$
$$expX(8) := 2.173598080$$
$$expX(9) := 0.2224540609$$
$$expX(10) := 0.6957333820$$
$$expX(11) := 0.01522353007$$
$$expX(12) := 1.528256880$$
$$expX(13) := 0.2770614590$$
$$expX(14) := 0.8808727215$$
$$expX(15) := 0.2162919750$$
$$expX(16) := 0.4440577432$$
$$expX(17) := 1.166302334$$
$$expX(18) := 0.1294200208$$
$$expX(19) := 0.6975683935$$
$$expX(20) := 0.05544443935$$

**4.** Normal 0,1 variables are

> *seq(X1(i),i=1..20);*

1.521070484, −.5768899232, −1.042158196, −.1882229151,
0.8014649537, 1.578949306, −.9882787283, −.9383631568,
0.7018566620, 1.956211964, −.9212502179, 0.1331107801,
−1.379775780, 0.03929782162, −1.090981430,
0.1086640465, 1.269057050, 0.02160515710,
−.6052083964, −1.145969309

> *seq(X2(i),i=1..20);*

−.4092496291, 0.09498239192, 0.4725129005,
−.8563409978, −.7432717739, 1.365049016,
−.4002408685, −2.159175315, 0.8682968698,
−1.378212065, 0.4190439735, 2.668672851, 1.175383344,
−0.09933300662, −.4249319179,
0.7566029424, −.4004936206, 0.1509887558,
−0.02918414136, −1.073491864

**5.** Normal 5, 0.5, found by using the formula $z = \dfrac{x - \mu}{\sigma}$.

*seq(.5*X1(i)+5,i=1..20);*

5.760535242, 4.711555038, 4.478920902, 4.905888542,
5.400732477, 5.789474653, 4.505860636, 4.530818422,
5.350928331, 5.978105982, 4.539374891, 5.066555390,
4.310112110, 5.019648911, 4.454509285, 5.054332023,
5.634528525, 5.010802579, 4.697395802, 4.427015346

### Section 16.2

**1.** The approximate area with $n = 2,000$ is 3.12116. The
exact area is 3.14.

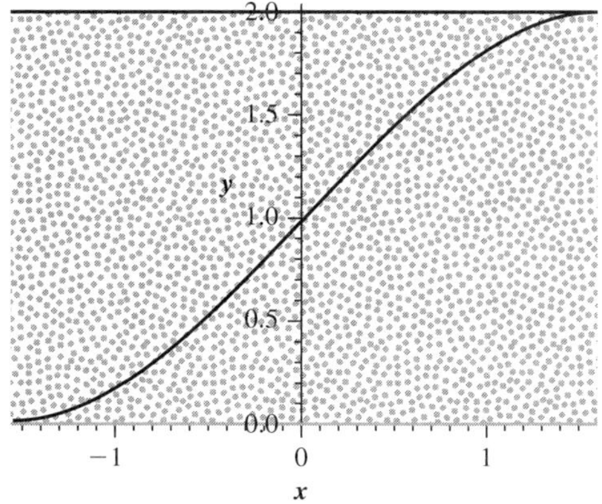

**2.** Area is approximately with $n = 2,000$ trials is 0.9855 sq units. The exact area is 0.9890. sq units.

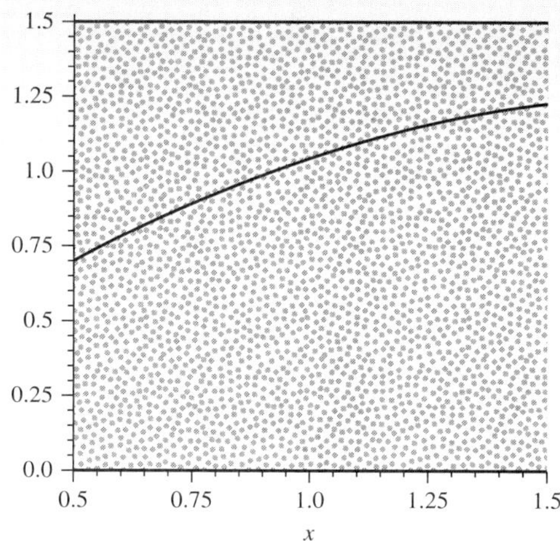

**3.** Approximate area with $n = 2,000$ is 0.7885.

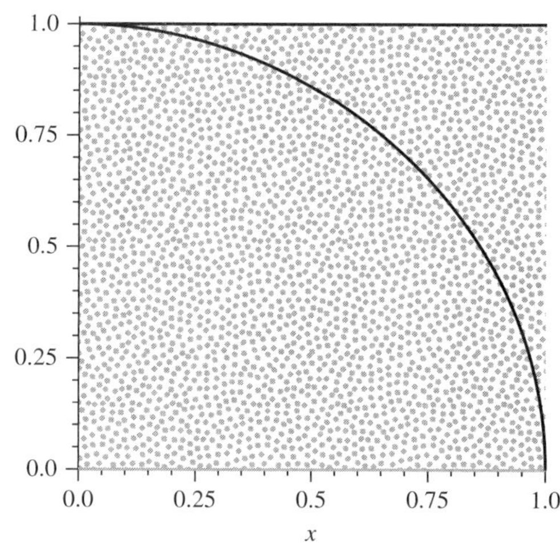

The real area is 0.7853981634.

**5.** We use $x$ and $y$ between [0,1]. Approximate volume is 0.734 cubic units. The real volume is 0.66666667 or $\frac{2}{3}$.

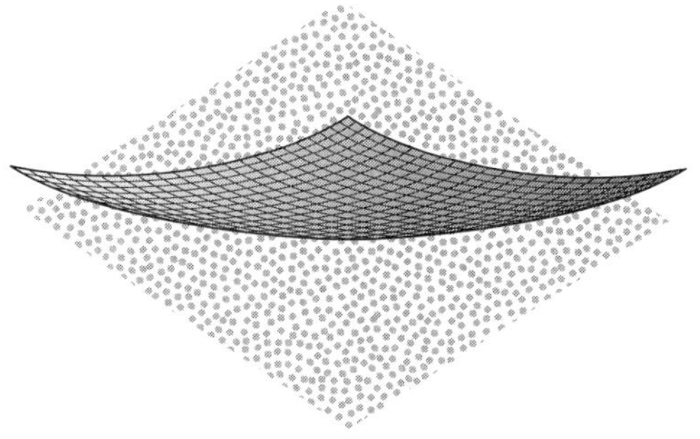

**Section 16.3**

**1.** HHTHTTTHTHHTHTHHHHTT, $P(\text{H})=0.55$, $P(\text{T})=0.45$

**2.** Rolls of a die: 3,5,6,2,3,1,2,5,4,3,2,1,6,6,5,4,2,3,1,4

**Section 16.4**

**1.** $n = 6, p = 0.9871120530$

# Chapter 17

**Section 17.1**

**1.** 40

**2.** 65

**3.** 350

**4.** 4

**5.** Rose plays 4/9 a, 5/9 b; Colin plays 5/9 a, 4/9 b; $V = 0.22222$

**6.** Rose plays 3/4 a, 1/4 b; Colin plays 1/2 a, 1/2 b; $V = 0.25$

**7.** $V = 4$; Rose plays c, Colin plays c.

**8.** Rose and Colin both play ½ a, 0 b, ½ **c**; $V = 0.7$

**9.** $V = 5$

**10.** Rose plays 0a, 0.5715b, 0.4286c; Colin plays .2857a, .7143b; $V = 3.1428517$.

**11.** $V = 1.1428$ when Rose plays 0.5714a, 0.4286 b, and Colin plays 0 a, 0.2857 b, and 0.71142 c.

**12.** $V = 5$: Rose plays b, and Colin plays b.

**Section 17.2**

**1.** Rose: $V = 2.4615$ when Rose plays 0.3846 a and 0.6154 b. Colin: $V = 3$ when Colin plays a and Rose plays a.

**2.** Rose: $V = 1.88889$ when Rose plays 1/9 a and 8/9 b. Colin plays a and Rose a to get (3,5); $V = 5$.

**3.** Rose: $V = 1.667$, when Rose plays 2/3 a and 1/3 b. Colin $V = 2$, Colin a and Rose a.

**Section 17.3**

**1.** Nash arbitration point (2.058,3.8557) using security level (1.5,2.46538).

**2.** Nash arbitration point is (3,5) using security level (2,1.8889).

**3.** Nash arbitration point is (1.916,2.1906) using security level (1,1.667).